Microcontrollers
and
Embedded System Design

Microcontrollers
and
Embedded System Design

Nilesh Bhaskarrao Bahadure (Ph.D.)

Professor
School of Electronics Engineering
Sanjay Ghodawat University
Kolhapur, Maharashtra

Saurabh Chandrakar

Senior Engineer
BHEL Corporate R&D
Hyderabad, Telangana

TECHSAR™

NEW DELHI

Published and Printed by

Techsar Pvt. Ltd.
4435-36/7, Ansari Road, Daryaganj
New Delhi–110 002 (India)
E-mail: tech_info@techsarworld.com
Website: www.techsarworld.com

ISBN: 978-93-85909-70-2

Reprint 2025

Dedicated to

My Parents Smt. Kamal B. Bahadure & Shri Bhaskarrao M. Bahadure,
to my grandparents Late Smt. Parvathi M. Bahadure & Late Shri Motiramji S. Bahadure,
to my wife Shilpa N. Bahadure,
to my beautiful daughters Nishita and Mrunmayi
and to all my beloved students.

Nilesh B. Bahadure

Dedicated to

My Parents Dr Surendra Kumar Chandrakar and Smt. Bhuneshwari Chandrakar, my wife Priyanka Chandrakar, and brother Shri Pranav Chandrakar.

Saurabh Chandrakar

Preface

This book explores the beauty of Microcontroller 8051. Intel's Microprocessor 8051 and families is the most widely used architecture in modern microcomputer systems. Most of the manufacturers use the 8051 Microcontroller to make handheld devices. Intel's 8051 family of microcontrollers is also used in a wide variety of other electronic equipment.

This book is meant for undergraduate and graduate level courses in Microcontrollers and Embedded System Design. The Intel's 8051 Microcontroller is also one of the most preferred Microcontroller ICs used by graduate, undergraduate students, engineers and hobbyist for making electronics based projects. Problems on interfacing and programming at the end of each subsection will enable the students to face the exams confidently.

All the necessary care has been taken to maintain balanced coverage of both the theory and programs with a wide variety of applications, interfacing on the different devices. It also gives comprehensive study of advanced microcontrollers like 8096, PIC and ARM.

Any suggestions from the reviewers to improve the utility of the book will be gladly accepted, please feel to mail me at nbahadure@gmail.com/nilesh_bahadure@rediffmail.com I am open to accept and provide solutions to any queries within the scope of this book.

Authors

Acknowledgements

My heartfelt thanks to Shri Vijay Kumar Gupta, Chairman of Beekay Industries Bhilai and BIT Trust, for his encouragement and support. I would like to thank my mentors Dr. Arun Kumar Ray, Dean School of Electronics Engineering KIIT University and Dr. Har Pal Thethi, Associate Director, Lovely Professional University Punjab. I thank Prof. R. M. Potdar and Prof. D.P. Mishra for their advice, assistance, prompting, and encouragement throughout the preparation of the book.

I would also like to thank Dr. Samridhi Kulkarni and my colleagues in Bhilai Institute of Technology Raipur, Prof. Md. Khwaja Mohiddin, Prof. Sankalp Verma, Prof. Shravan Kumar Singh, and Prof. Vinay Kumar for providing valuable suggestions and lots of encouragement throughout the project.

Thanks to our family members, our parents, wife, children and our well wishers for their kind support, without this book was not possible, it just a dream. I also thank my students, who always been with me, for relating problems and to finding solutions too.

I also thank Mr. Amit Tamrakar, LAN Administrator, BIT Raipur for his help and suggestions.

Last but not the least I am grateful to Prof. (Dr.) Anupam Shukla, Professor, ABV - International Institute of Information Technology, Gwalior (Madhya Pradesh) and Dr. C. G. Dethe, Director UGC Staff College Nagpur.

Most importantly, I would like to give Lord Ganesha the glory for all of the efforts I have put into the preparation of the book. If not for God's awesome creation of the universes, I would not have the zeal for Microprocessor that I have.

Nilesh B. Bahadure

First of all I do take the opportunity to greet and thank my mentor Prof. Nilesh B. Bahadure for motivating me on the topics related to Microcontrollers and Embedded System Design. I appreciate that he believed in me, always stood behind me and keep pushing to achieve more. I express my deepest warm feelings to my parents Dr Surendra Kumar Chandrakar and Smt. Bhuneshwari Chandrakar, my brother Shri Pranav Chandrakar and all my friends for always being an inspiration for me.

Saurabh Chandrakar

Contents

Section C: ARM Microcontroller

Basics of 8051 Microcontroller

1

Chapter

In this chapter readers will get insights of the microcontroller based systems. Microcontrollers are found almost everywhere. A microcontroller is a fast driven electronic programmable chip. After completion of the chapter, the reader would be able to understand basics of microcontroller, difference between a microprocessor (CPU) and a microcontroller, CISC, RISC, Harvard architecture, Princeton architecture, memory and types of memories.

1.1 INTRODUCTION

Microcontrollers, as the name suggests, are small controllers. They are like single-chip computers that are often embedded into other systems to function as processing/controlling units. For example, the remote control you use probably has microcontrollers inside that do decoding and other controlling functions. They are also used in automobiles, washing machines, microwave ovens, toys, etc., where automation is needed.

1.1.1 Key Features of Microcontrollers

- **High Integration of Functionality**

  Microcontrollers are also called single-chip computers because they have on-chip memory and I/O circuitry and other circuitries that enable them to function as small standalone computers without any other supporting circuitry.

- **Field Programmability, Flexibility**

  Microcontrollers often use EEPROM or EPROM as their storage device to allow field programmability so they are flexible to use. Once the program is tested to be correct, then large quantities of microcontrollers can be programmed to be used in embedded systems.

- **Easy to Use**

  Assembly language is often used in microcontrollers and since they usually follow the RISC architecture, the instruction set is small. The development package of microcontrollers often includes an assembler, a simulator, a programmer to "burn" the chip, and a demonstration board. Some packages include a high level language compiler such as a C compiler and more sophisticated libraries.

1.2 THE 8051 MICROCONTROLLER

The 8051 is an 8-bit microcontroller originally designed in the 1980s by Intel that has gained great popularity since its introduction. Its standard form includes several standard on-chip peripherals, including timers, counters, and UARTs, plus 4 KB of on-chip program memory and 128 bytes (note: bytes, not KB) of data memory, making single-chip implementation possible. Its hundreds of derivatives, manufactured by several different companies (like Philips) include even more on-chip peripherals, such as analog-digital converters, pulse-width modulators, I2C bus interfaces, etc. Costing only a few dollars per IC, the 8051 is estimated to be used in a large percentage of all embedded system products.

The 8051 memory architecture includes 128 bytes of data memory that are accessible directly by its instructions. A 32-byte segment of this 128-byte memory block is bit addressable by a subset of the 8051 instructions, namely, the bit-instructions. External memory of up to 64 KB is accessible by a special "MOVX" instruction. Up to 4 KB of program instructions can be stored in the internal memory of the 8051, or the 8051 can be configured to use up to 64 KB of external program memory. The majority of the 8051 instructions are executed within 12 clock cycles.

1.2.1 Comparison Between Microprocessor and Microcontroller

In a microprocessor based system, microprocessor (central processing unit) serves as the main processor and additional devices like RAM, ROM, timers, counters, serial ports are to be interfaced externally. A microprocessor based system is functional only when the above said external devices and other peripherals are interfaced with the microprocessor.

Whereas in a microcontroller based system RAM, ROM, timers, counters, serial ports are available on the same chip, so microcontroller based systems are cost effective and ideally suited for small applications.

By comparing, it is found that a microprocessor based system is more versatile because the user or programmer decides his own resources, i.e., size of RAM/ROM, number of timers/counters, number of serial ports and interrupts. Whereas in a microcontroller based system these resources are fixed and are available on chip.

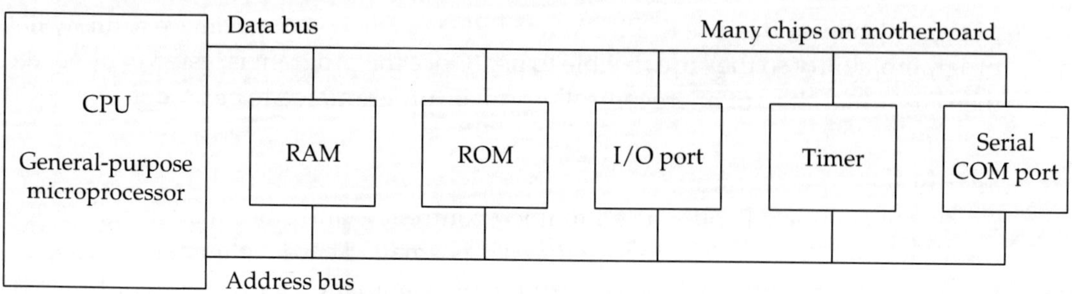

Fig. 1.1 *Microprocessor based system*

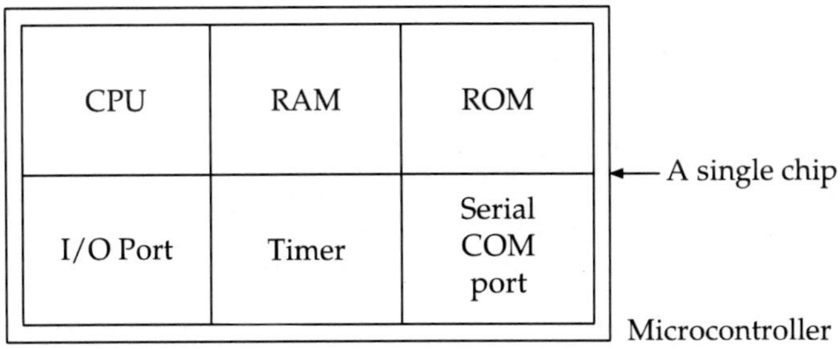

Fig. 1.2 *Microcontroller based system*

S. No.	Microprocessor	Microcontroller
1.	An electronic component made from transistors, capacitors, diodes and other small elements on a single semiconductor integrated circuit.	An on-chip computer for optimizing control of the electronic devices.
2.	Here CPU serves as the main processor and the additional features like RAM, ROM, timers, counters, serial ports are to be interfaced externally. It is functional only when external RAM, ROM and other important peripherals are interfaced.	Here, in microcontroller based system, RAM, ROM, timers, counters, serial ports are available on-chip.
3.	It is not cost effective since the extra features are added externally which adds to the cost.	It is cost effective and ideally suited for small applications.
4.	It is more flexible and versatile as the users decide their own resource, i.e., size of the RAM, size of ROM, number of timers/counters, serial ports and interrupt sources.	Here, all these resources are fixed and available on-chip.
5.	It is a subset of microcontroller.	It incorporates features of microprocessor. Thus, it is a superset of microprocessor.
6.	It works as a general-purpose programmable device in the Intel-Core family or Pentium family processors.	Microcontrollers from different families find their applications in small embedded applications such as in robotics, camcorders, portable devices, etc.
7.	It doesn't handle real-time tasks at all times.	It handles real-time tasks.

8.	Clock speed of μP is higher as compared to μC. Today's μP operates above 1 GHz as it perform complex tasks.	It operates from a few MHz to 50 MHz.
9.	μP consists of more number of instructions.	μC consists of less number of instructions
10.	It has less function pins on the IC.	Many function pins on the IC.
11.	μP has program and data to be stored in the same memory.	μC has separate memory to store data and program.
12.	μP has no Boolean operation which performs directly.	μC has Boolean operation which performs directly.
13.	μP takes many instructions to read and write data from external memory.	μC takes few instructions to read and write data from external memory.
14.	μP has more Opcodes.	μC has less Opcodes.
15.	μP operates on nibbles, bytes, etc. These instruction sets are "processive intensive" which implies powerful addressing modes with instructions which leads to large data volumes.	μC has instructions which leads to the control of inputs and outputs that operates on a single bit.
16.	It is based on von-Neumann architecture.	It is based on Harvard architecture.

Advantages of Microcontroller over Microprocessor

1. Size is reduced
2. Complexity is reduced
3. Faster access operation to its devices
4. User-friendly
5. Cheaper than μP
6. Its program is difficult to copy because of copy protection functions and an emulator is required to copy it.
7. Power consumed is low
8. Easily portable
9. Very much specific oriented
10. Dedicated execution of programs without much intervention
11. All required components are on-chip such as serial ports, interrupts, timers/counters
12. Highly reliable
13. Common software/hardware design environment is available from the manufacturer,
14. Design time is decreased.

Disadvantages of Microcontroller over Microprocessor

1. Work oriented operating system is required
2. Modularity is limited
3. Less adaptive
4. Understanding of functionality is quite difficult.

Conclusion

Microprocessor = CPU

Microcontroller = CPU + Peripherals + Memory

Peripherals = Ports + Clock + Timers + UARTS + ADC converters + LCD drivers + DAC + other stuff.

Memory = EEPROM + SRAM + EPROM + FLASH.

1.2.2 Applications of Microcontrollers

Microcontrollers are found in different types of devices from basic electronic toys to highly advanced car and aircraft systems. A few of them are listed below:

1. Ignition system for vehicles (two-wheelers)
2. Appliances
3. Cell phones
4. Buzzer sounds in fridges
5. Electronic toys
6. Washing machine
7. Remote controllers
8. TV sets
9. Sewing machine
10. Camcorders
11. Digital camera
12. Microwave oven
13. Paging system
14. AC (four wheelers) as an embedded system
15. Electronic geysers
16. Electronic watches
17. Portable devices
18. Home computers
19. Fax
20. Play video console
21. Colour/laser printers
22. Cars (airbags, ignition system, parking system, etc.)

1.2.3 Manufacturers of Microcontrollers

Table 1.1 *Manufactures of microcontrollers*

S No.	Manufacturer	ICs name
1.	Intel	8051, 8031, 8048, 8052, 8751, 8752 etc.
2.	Motorola	68HC04, 68HC05, 68HC11 etc.
3.	ARM Processor	ARM 7, ARM7TDMI, Cortex – M3 etc.
4.	PIC	PIC 16C84 etc.
5.	Hitachi	H8 , Super H
6.	Atmel	AVR (8-bit), AVR 32(32-bit)
7.	ST Microelectronics	STM 8S(8-bit), STM 10(16-bit) , STM 32(32-bit)

Others include Philips, ADM and Zialog's etc.

1.3 RISC AND CISC CONTROLLERS

There are some differences that an electronics engineer must know regarding the design aspect of microcontrollers. RISC stands for *reduced instruction set computing* and CISC stands for *complex instruction set computing*. As the name implies RISC performs the same operations as CISC can perform, but on less instructions. In other words, we can say that with only a few or less number of instructions compared to CISC, RISC can perform all the operations initiated by CISC computing approach.

The CISC Approach

- The primary goal of CISC architecture is to complete a task in as few lines of assembly as possible. This is achieved by the processor hardware that is capable of understanding and executing a series of operations.
- For this particular task, a CISC processor would come prepared with a specific instruction (we'll call it "MULT"). When executed, this instruction loads the two values into separate registers, multiplies the operands in the execution unit, and then stores the product in the appropriate register. Thus, the entire task of multiplying two numbers can be completed with one instruction.

MULT 2:3, 5:2

MULT is what is known as a "complex instruction". It operates directly on the computer's memory banks and does not require the programmer to explicitly call any loading or storing functions. It closely resembles a command in a higher level language. For instance, if we let "a" represent the value of 2:3 and "b" represent the value of 5:2, then this command is identical to the C statement "a = a * b."

Advantages

- One of the primary advantages of this system is that the compiler has to do very little work to translate a high-level language statement into assembly.
- Because the length of the code is relatively short, very little RAM is required to store instructions.
- The emphasis is put on building complex instructions directly into the hardware.

The RISC Approach

- RISC processors only use simple instructions that can be executed within one clock cycle.
- Thus, the "MULT" command described above could be divided into three separate commands: "LOAD", which moves data from the memory bank to a register, "PROD". which finds the product of two operands located within the registers, and "STORE", which moves data from a register to the memory banks. In order to perform the exact series of steps described in the CISC approach, a programmer would need to code four lines of assembly:

  ```
  LOAD A, 2:3
  LOAD B, 5:2
  PROD A, B
  STORE 2:3, A
  ```

- At first, this may seem like a much less efficient way of completing the operation. Because there are more lines of code, more RAM is needed to store the assembly level instructions.
- The compiler must also perform more work to convert a high-level language statement into code of this form.

Advantages

- Because each instruction requires only one clock cycle to execute, the entire program will execute in approximately the same amount of time as the multi-cycle "MULT" command.
- These RISC "reduced instructions" require less number of transistors of hardware space than the complex instructions, leaving more room for general purpose registers. Because all of the instructions execute in a uniform amount of time (i.e., one clock), pipelining is possible.
- Separating the "LOAD" and "STORE" instructions actually reduces the amount of work that the computer must perform. After a CISC-style "MULT" command is executed, the processor automatically erases the registers.
- If one of the operands needs to be used for another computation, the processor must reload the data from the memory bank into a register.
- In RISC, the operand will remain in the register until another value is loaded in its place.

Conclusion

CISC	RISC
Emphasis on hardware	Emphasis on software
Includes multi-clock complex instructions	Single-clock, reduced instruction only
Memory-to-memory: "LOAD" and "STORE" incorporated in instructions	Register to register: "LOAD" and "STORE" are independent instructions
Small code sizes, more cycles per second	Less cycles per second, large code sizes
Transistors used for storing complex instructions	Spends more transistors on memory registers

1.4 HARVARD ARCHITECTURE AND NEUMANN ARCHITECTURE

In the late 1940s, the United States Government asked Harvard and Princeton universities to come up with a computer architecture to be used in computing distances of naval artillery shell for defense applications. Princeton suggested computer architecture with a single memory interface. It is also known as von Neumann architecture after the name of the chief scientist of the project, i.e., John von Neumann (1903-1957; born in Budapest, Hungary).

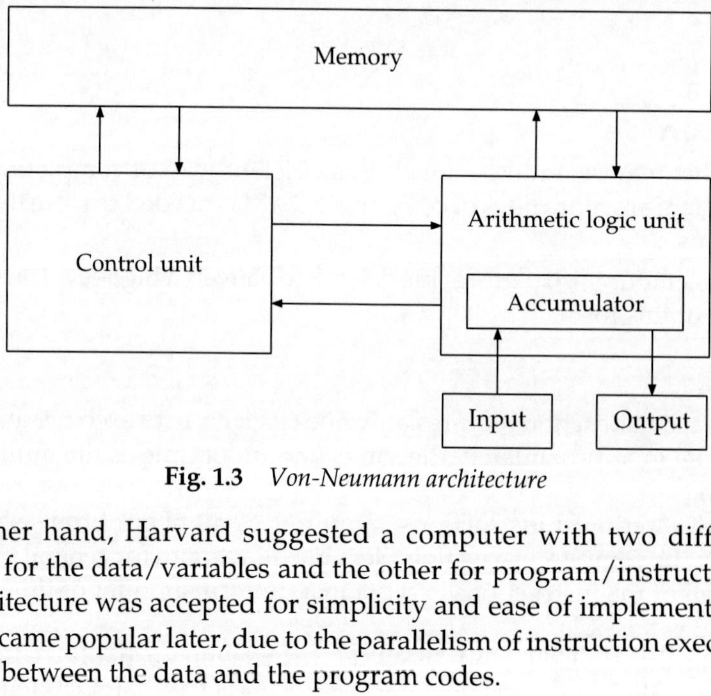

Fig. 1.3 *Von-Neumann architecture*

On the other hand, Harvard suggested a computer with two different memory interfaces, one for the data/variables and the other for program/instructions. Although Princeton architecture was accepted for simplicity and ease of implementation, Harvard architecture became popular later, due to the parallelism of instruction execution and easy differentiation between the data and the program codes.

Fig. 1.4 *Harvard architecture*

S. No.	Von-Neumann Architecture	Harvard Architecture
1.	Here, the data and the program are stored in the same memory.	The data and the program are stored in separate memories.
2.	Used in traditional x86 architecture processors found in PCs and servers, and embedded systems with only control functions.	Used in advanced RISC machine processors found in mobile communication systems, audio, speech, image processing systems and modern embedded systems.
3.	No barrel shifter	Presence of a barrel shifter (helps in shifting and rotating operations of the data)
4.	Optimization of the program in lesser size is possible.	Here, the optimization of the program happens in big sizes.
5.	No exclusive multiplier is present.	It has an exclusive multiplier as MAC (Multiply Accumulate).
6.	Execution of the code is done serially.	Execution of the code is done in parallel.
7.	Takes more clock cycles.	Takes less clock cycles.
8.	It is much more flexible and allows for many concepts unavailable to Harvard architecture such as self-programming, word processing.	It is less flexible as compared to the von-Neumann architecture.
9.	If any computer is using this architecture without cache, the CPU can either read/write data to/from the memory or it can read an instruction. The simultaneous operation of both of these operations cannot be done as the data and instructions use the same system bus.	If any computer is using this architecture, both reading of an instruction and accessing of the data memory at the same time without cache is possible.

1.5 OPERATING SYSTEMS FOR EMBEDDED SYSTEMS

The most common operating system that we use include Windows from Microsoft, Macintosh from Apple and a wide variety of Linux operating systems. But, when we come across the word RTOS, we often get confused with this word. These are generally the operating systems that are used for more specialized applications where the response required is very much close to the time, i.e., perfect synchronization with time. Different time sharing architectures are used by the operating systems where a particular task is assigned a small period of time before the task is switched on to the other one. The switching process often appears as real time to users as it is too fast.

Comparison between the OS and the RTOS

S. No.	Operating System (OS)	Real Time Operating System (RTOS)
1.	Focus area is computing throughput. (No. of processes completing their execution per unit time)	Focus area is very fast response time.
2.	Used for system/applications which are not time critical.	Used for the time critical applications, i.e., need for the "deterministic" timing behaviour. It means consumption of known and expected amounts of time.
3.	Coding is simple as compared to RTOS.	Coding of an RTOS is stricter.
4.	Allows multitasking using time sharing design.	Uses either time sharing design or even driven design.
5.	Task scheduling is not always priority based	Here, task scheduling is always priority based.
6.	More weight and big in size as compared to RTOS.	Less weight and small in size as compared to OS.
7.	It has unbound dispatch latency, i.e., more number of threads to schedule, the latencies will get added up.	It has bound latency, i.e., a thread will be executed within a specified time limit.
8.	Kernel of OS is not pre-emptible.	Kernel of RTOS is pre-emptible.

1.6. THE MICROCONTROLLER SYSTEM 51 FAMILY

The microcontroller system 8051 is an 8-bit microcontroller. The family members that belong to the 8051 microcontrollers are also 8-bit microcontrollers. These microcontrollers are fabricated by using HMOS / NMOS technology. The CHMOS version of the microcontroller is also available. This version is architecturally and functionally similar to its counterpart fabricated using HMOS technology. The only difference is that CHMOS version has additional features like power down and idle mode. In idle mode the system is ON without consuming power. They also have better speed, low power consumption, high noise immunity compared to HMOS technology.

Microcontroller 8051 core is available in many variations. Table 1.2 shows some of the microcontroller 8051 family members with the number of pins, size of data RAM, size of program, ROM, number of timers/counters, number of serial ports and number of interrupt sources available. The numbering convention system used for the microcontroller 8051 family is shown in Figure 1.5.

Note: All the ATMEL's microcontrollers are fabricated using CMOS technology.

The table below shows the comparison of different microcontrollers with their on-chip capacity of RAM, ROM, serial ports, timers/counters and interrupts

Table: 1.2 *Comparison between of microcontroller 8051 and its family*

Device	Pins in DIP	ON-Chip Data Memory (RAM) in Bytes	ON-Chip Program Memory (ROM)	Timers / Counters	I/O Port Lines	Serial Ports	Interrupt Sources
8031	40	128	0 KB	2	32	1	6
8032	40	256	0 KB	3	32	1	6
8051	40	128	4 KB ROM	2	32	1	6
8052	40	256	8 KB ROM	3	32	1	8
8054	40	256	16 KB ROM	3	32	1	8
8751	40	128	4 KB EPROM	2	32	1	6
8752	40	256	8 KB EPROM	3	32	1	8
8754	40	256	16 KB EPROM	3	32	1	8
AT89C51	40	128	4 KB Flash ROM	2	32	1	6
AT89C52	40	256	8 KB Flash ROM	3	32	1	8
AT89C1051	20	64	1 KB Flash ROM	1	15	--	6
AT89C2051	20	128	2 KB Flash ROM	2	15	1	6
AT89C4051	20	128	4 KB Flash ROM	2	15	1	6
AT89C51RD2	40	256	8 KB Flash ROM	3	32	1	8

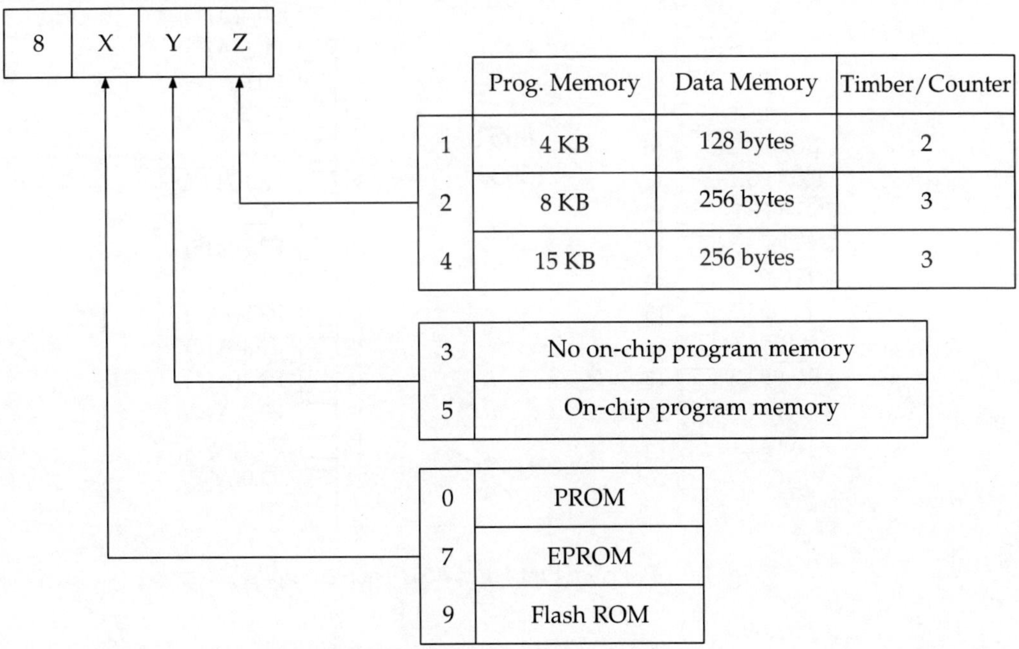

Fig. 1.5 *Numbering convention system used for the microcontroller 8051 family*

1.7 FEATURES OF INTEL'S 8051 MICROCONTROLLER

1. It is a 40-pin IC fabricated using HMOS technology.
2. It is also available in CHMOS versions.
3. It is an 8-bit microcontroller and belongs to the family of 8-bit microcontrollers.
4. It consists of 128-bytes on-chip data memory (RAM) and 4 KB on-chip program memory (ROM).
5. It also consists of two 16-bit timers, 32 I/O lines and 6 interrupt sources.
6. The 32 I/O lines are organized into 48-bit parallel ports P0, P1, P2 and P3.
7. Using 16 address lines it is also able to handle 64 KB data memory (RAM) or 64 KB program memory (ROM) externally (off-chip).
8. It also consists of standard UART (Universal Asynchronous Receiver Transmitter).
9. Operates at 12 MHz frequency (11.0592 MHz).

1.8 PIN CONFIGURATION OF 8051 MICROCONTROLLER

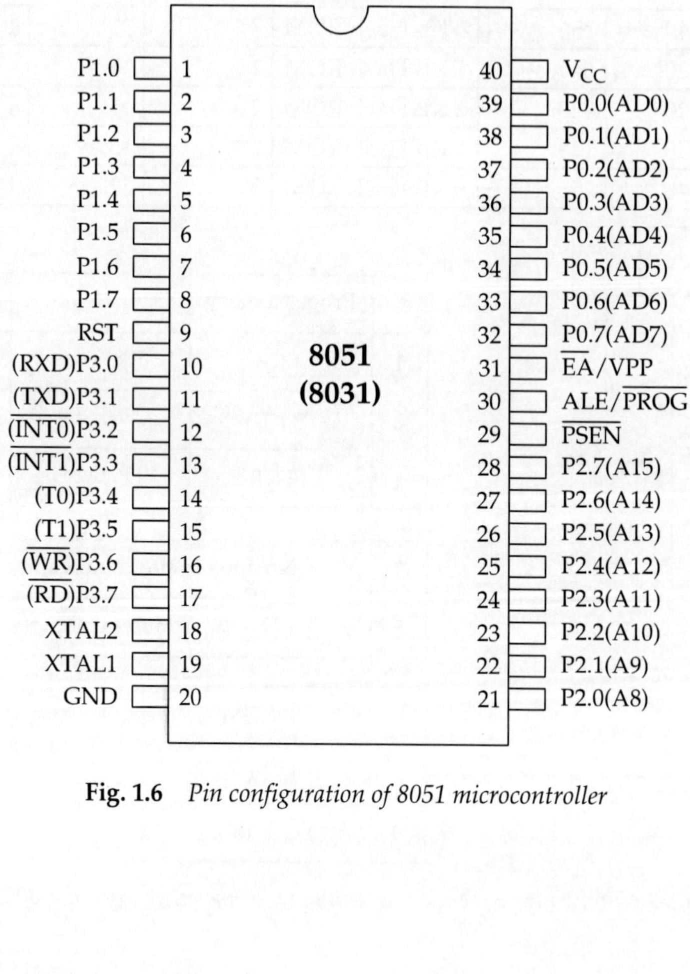

Fig. 1.6 *Pin configuration of 8051 microcontroller*

μC 8051 is a 40-pin IC fabricated using HMOS technology. CMOS version is also available. The pin configuration of 8051 consists of $\overline{RD}, \overline{WR}, \overline{EA}, \overline{PSEN}$, ALE, RST and 32 I/O pins configured as 4 parallel ports. The description of the pin configuration is given below:

1. **VCC**
   - Type: Input
   - Pin No. 40
   - This pin is connected to the +5V single power supply. +5V is the power requirement of 8051 microcontroller.
2. **VSS/Gnd**
   - Type: Input
   - Pin No. 20
   - This pin is connected to the ground.
3. **XTAL2/XTAL1**
   - Type: Input
   - Pin No. 18 and 19
   - XTAL2 and XTAL1 are the crystal oscillator pins and are connected to the external oscillator. With two capacitors of 30pF each, one side of the capacitor is connected to the ground and the other side to the crystal oscillator pins respectively.

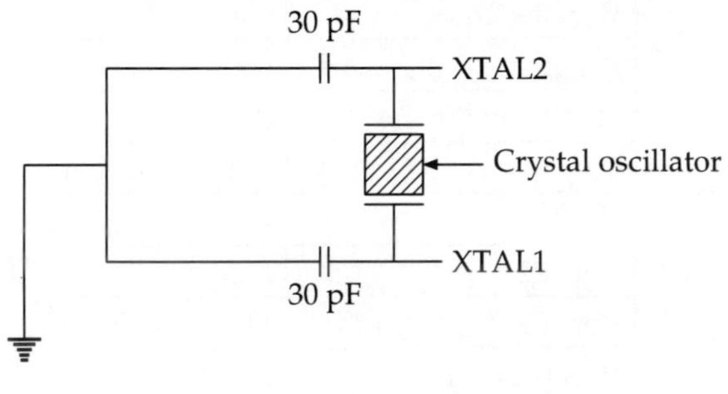

Fig. 1.7 *Generation of clock frequency*

4. **RST (Reset)**
   - Type: Input
   - Pin No. 9
   - RST is an active high level pin, normally held low.
   - When a high level signal is applied to RST pin, then it terminates all current activities and resets the microcontroller, i.e., all the registers are set to their default values so that old values or executed values in the register are lost. This activity is referred to as "power on reset".
   - The reset values of the different register is shown in Table 1.3.

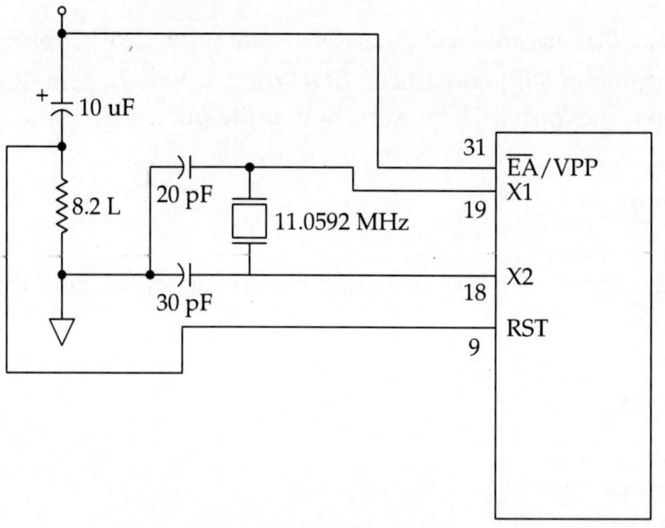

Fig. 1.8 *Reset operation of 8051 microprocessor*

Table 1.3 *Reset Values of 8051 registers*

Registers of 8051	Reset Value
Register A	00h
Register B	00h
Program status word	00h
Stack pointer	07h
Port P0	FFh
Port P1	FFh
Port P2	FFh
Port P3	FFh
Program counter	0000h
DPTR	0000h

Conclusion

- By observing the reset value we know that PC sets to 0000. So that Microcontroller 8051 begins its execution from the memory location 0000 after reset.
- To ensure "power on reset", reset signal must be active for at least two machine cycles. In other words, we say that the high pulse applied on reset must be held high or stay high for at least 2 M/C before it goes low, i.e., 2.17 μs for the crystal frequency.

5. **External access $\overline{EA}$**
   - Type: Input
   - Pin No. 31
   - Stands for the external access
   - In the 8051 µC, following types of memory are supported:
     - 64 kB – Data memory,
     - 64 kB – Program memory,
     - 128-byte – On-chip RAM with 128 byte on-chip SFR and 4 KB internal program Memory.
   - In 8051 µC, $\overline{EA}$ signal is used for external access and is related with program memory and CROM (Code ROM).
   - If $\overline{EA}$ is set to 1 then 1st 4 KB memory, i.e., from 0000H to 0FFFH is accessed from the internal program memory. The remaining 60 KB, i.e., from 1000H to FFFFH is accessed, from external program memory.
   - If $\overline{EA}$ is set to 0, then it indicates external access which means that all 64 KB program memory is accessed from the external memory connected with 8051 µC.
   - In the µC 8031/32, $\overline{EA}$ is always connected to the ground to indicate access of instruction codes from external program memory because 8031/32 is a ROM less.
   - $\overline{EA}$ is never left unconnected.

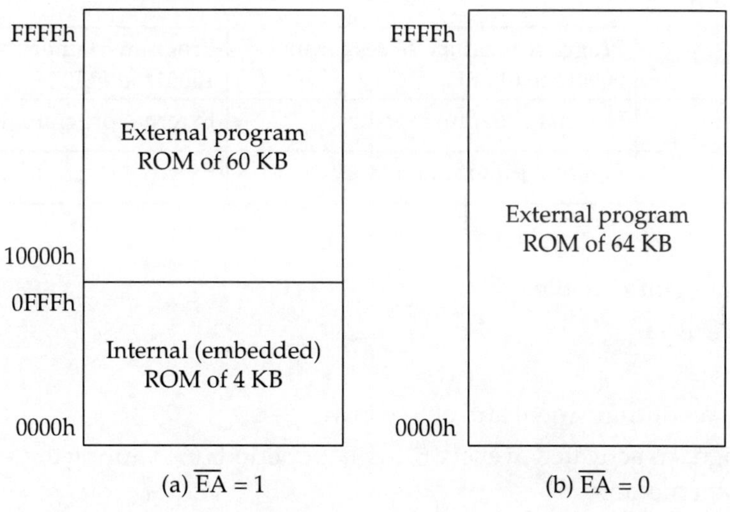

Fig. 1.9 *Use of EA pin to access program memory*

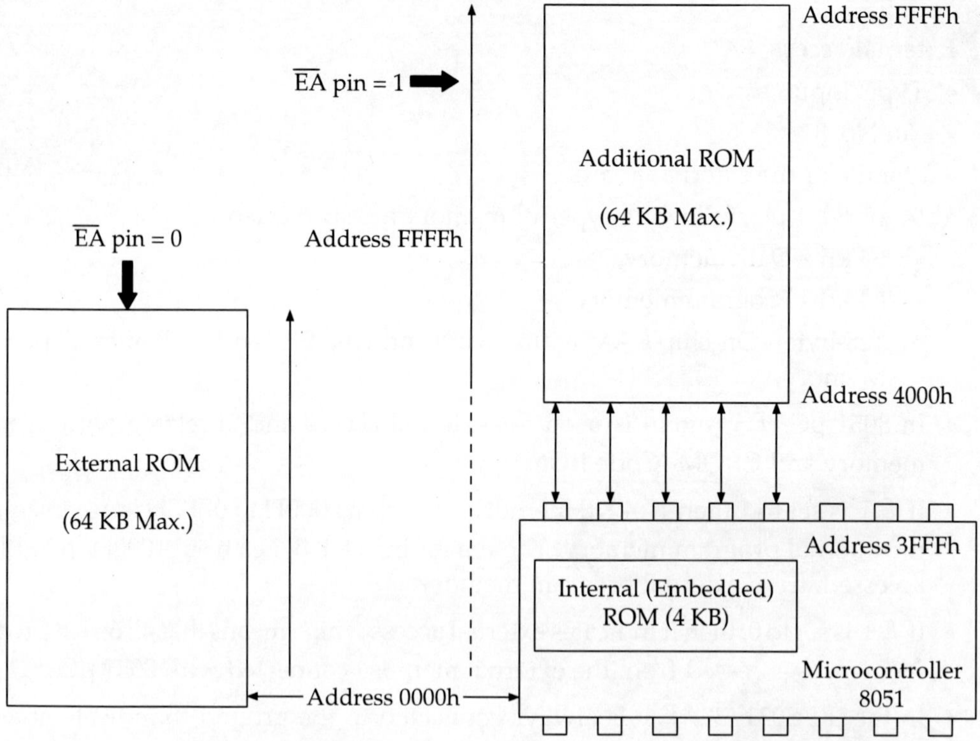

Fig. 1.10 *Use of EA to access program memory*

Conclusion

Signal on $\overline{EA}$	**Program Memory Access from 0000H to 0FFFH**	**Program Memory Access from 1000H to FFFFH**
$\overline{EA} = 1$	Internal program memory	External program memory
$\overline{EA} = 0$	External program memory	External program memory

6. ***$\overline{PSEN}$***
   - Program Store Enable
   - Type: Output
   - Pin No. 29
   - This is an output signal and active low.
   - This signal is activated in every 6 oscillator periods (6T) during the access of external program memory.
   - This signal is a read strobe to access the external program memory; which means it is connected to the $\overline{RD}$.

- In 8051/8031 μC, $\overline{PSEN}$ is connected to the $\overline{RD}$ of the memory to indicate external memory is a program memory.

7. **ALE**
   - Pin No. 30
   - Port 0 in the 8051 microcontroller has dual configuration.
   - It is used as an I/O port and also serves as lower order address data line; it means that address and data are multiplexes to save the pins.
   - This signal is used for de-multiplexing lower order address and data lines using external latch IC 74LS373.
   - If ALE = 1, then A0-A7 is accessed.
   - If ALE = 0, then data line D0-D7 is accessed.
8. **Parallel Ports**
   - Out of the 40 pins, 32 are set aside for ports P0, P1, P2 and P3.
   - Each port shares 8 pins.
   - So, these are 8-bit ports.
   - Upon reset or by default, these pins are configured as an I/P port because the value FFH is returned on port pins. If they are configured as an output port, they must be programmed by writing 00H on them.
   - Practically using them as an input port, they must be programmed by writing FFH on them and by default on reset they are configured as an output port.

A. **PORT P0**

- Port P0 consists of 8 pins, i.e., P0.0 to P0.7 making it an 8-bit port.
- P0 is configured as an I/O port.
- To access port 0 pins as an I/O port external 10 KΩ "pull up resistor" must be connected with them because P0 pins are an open drain.
- Open drain is a terminology used for "MOS" technology and is similar to open collector used in the TTL technology.
- Figure 1.11 shows Port P0 pins with 10 KΩ pull up resistors
- The basic structure of individual pins of P0 is shown in Figure 1.12.

Dual Role of Port P0

- As seen in the pin configuration with P0, address and data lines are also used which indicate the dual role of Port P0.
- If P0 is not used as an I/O port, then it may provide lower order address line A7-A0 or data lines D7-D0.
- The status of ALE decides whether it is an address or data.
- If ALE =1, then the address lines are accessed else data lines.

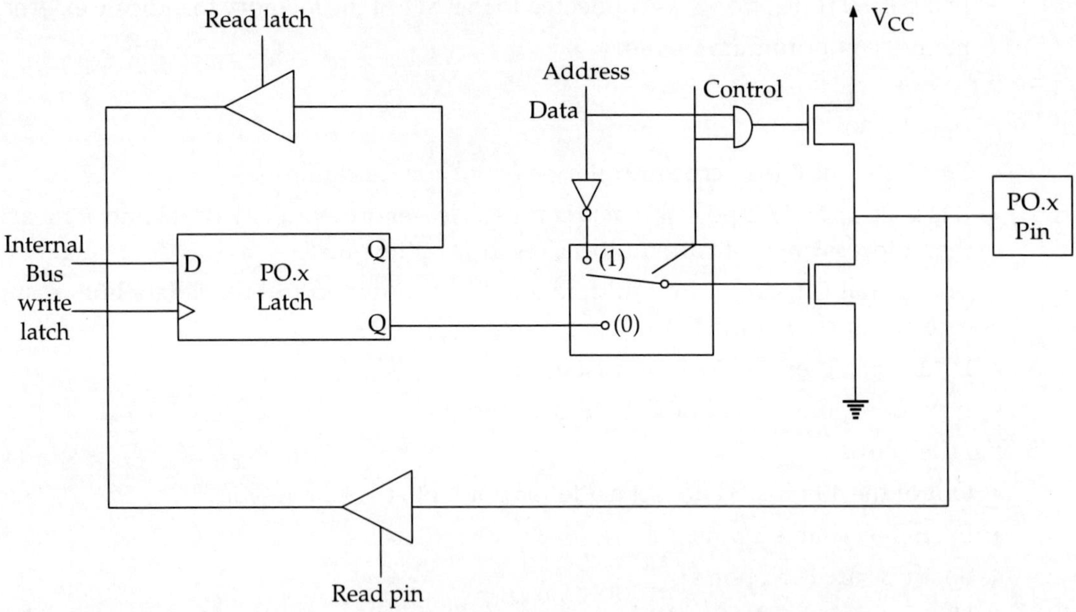

Fig. 1.11 *Port P0 structure*

B. **Port P1**

- Port P1 consists of 8 pins (P1.0 to P1.7) making them an 8-bit port. Port P1 is configured as an I/O port.
- Like Port P0, Port P1 doesn't need any pull up resistor, since they already have this resistor internally.
- The basic structure of individual pins of Port P1 is shown in Figure 1.13.

C. **Port P2**

- Like Port P1, Port P2 also doesn't need any pull up resistor, since they have them internally.
- Port P2 consists of 8 pins (P2.0 to P2.7) making them as an 8-bit port. Port P2 is configured as an I/O port.
- But there is a dual role of Port P2. As seen from the pin configuration, Port P2 pins also serve as a higher order address A8-A15.
- 8051 microcontroller is capable to address up to 64 KB external memory, to access 64 KB memory requires 16 address lines. Port P0 will provide lower order address line A7-A0 of 16 address lines A15-A0.
- Now, this is the job of Port P2 to provide remaining 8 address lines i.e., higher order address lines A15-A8 and this indicates the dual role of Port P2.
- The basic structure of individual pins of Port P2 is shown in Figure 1.14.

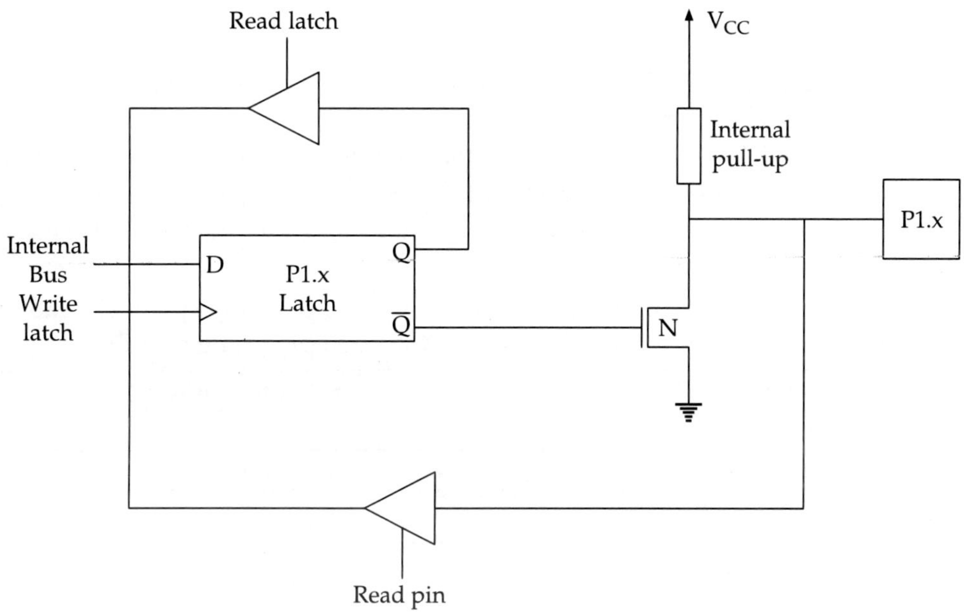

Fig. 1.12 *Port P1 structure*

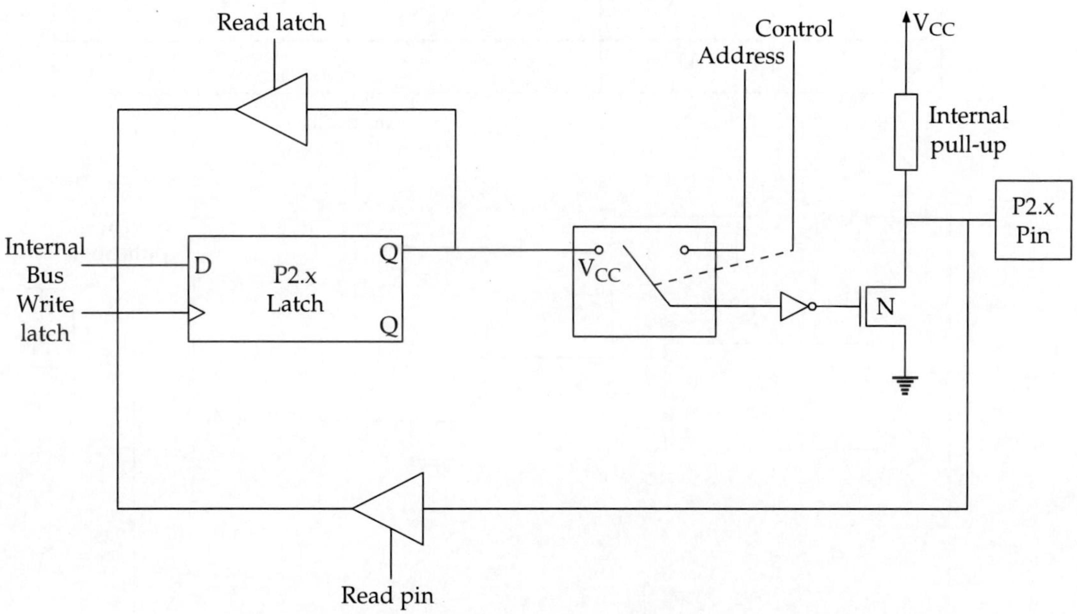

Fig. 1.13 *Port P2 structure*

D. **Port P3**

- Like Ports P1 and P2, P3 also doesn't need any pull up resistor since it has pull up resistor already internally.
- Port P3 consists of 8 pins, (P3.0 to P3.7) making them an 8-bit port. Port P3 is configured as an I/O port, but Port P3 was employed because Port P3 is used to provide some important and special functions for 8051 which includes interrupts, serial communications, timers, etc.
- There is a dual role of Port P3 for which we use the above port. The table given below displays the role and functionality of each pin of the port P3.

Port P3 Pins	Pin Number	Dual Configuration
P3.0	10	RXD
P3.1	11	TXD
P3.2	12	$\overline{INTO}$
P3.3	13	$\overline{INT1}$
P3.4	14	T0
P3.5	15	T1
P3.6	16	$\overline{WR}$
P3.7	17	$\overline{RD}$

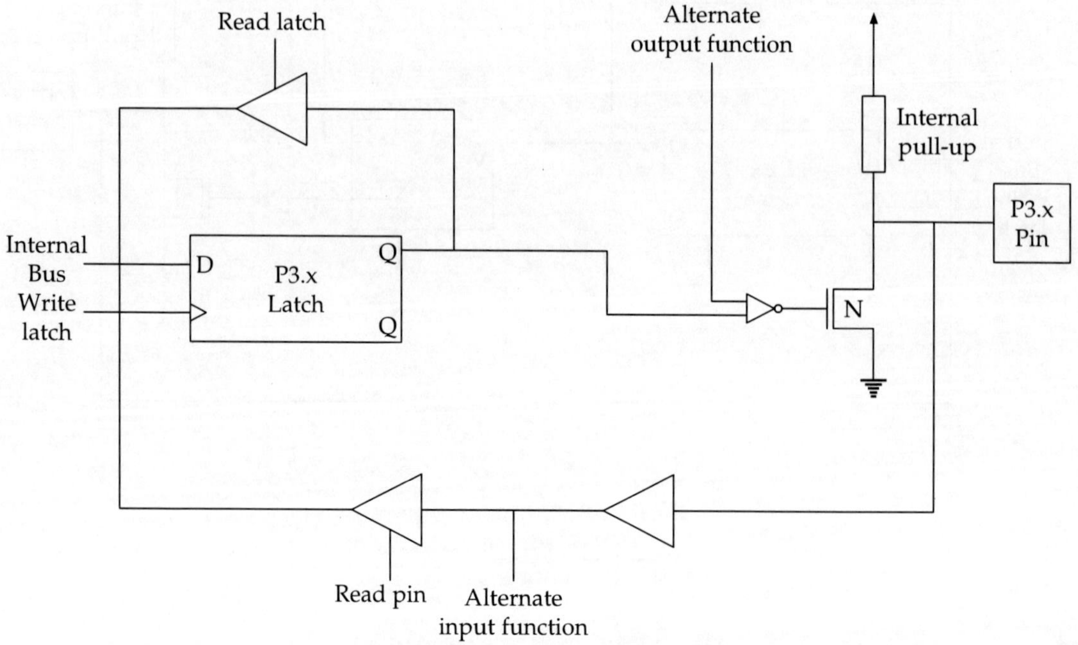

Fig. 1.14 *Port P3 structure*

- Port P3 pins (P3.0, P3.1) are also marked as RXD and TXD which indicates serial communication.
- Port pins P3.2 and P3.3 marked as $\overline{INT0}$ and $\overline{INT1}$ indicates external interrupt.
- Port P3 pins P3.4 and P3.5 are set aside for timers T0 and T1 and finally 3 pins P3.6 and P3.7 are $\overline{WR}$ and $\overline{RD}$ which are used to access external memory connected with 8051.
- The basic structure of individual pins of Port P0 is shown in Figure 1.15.

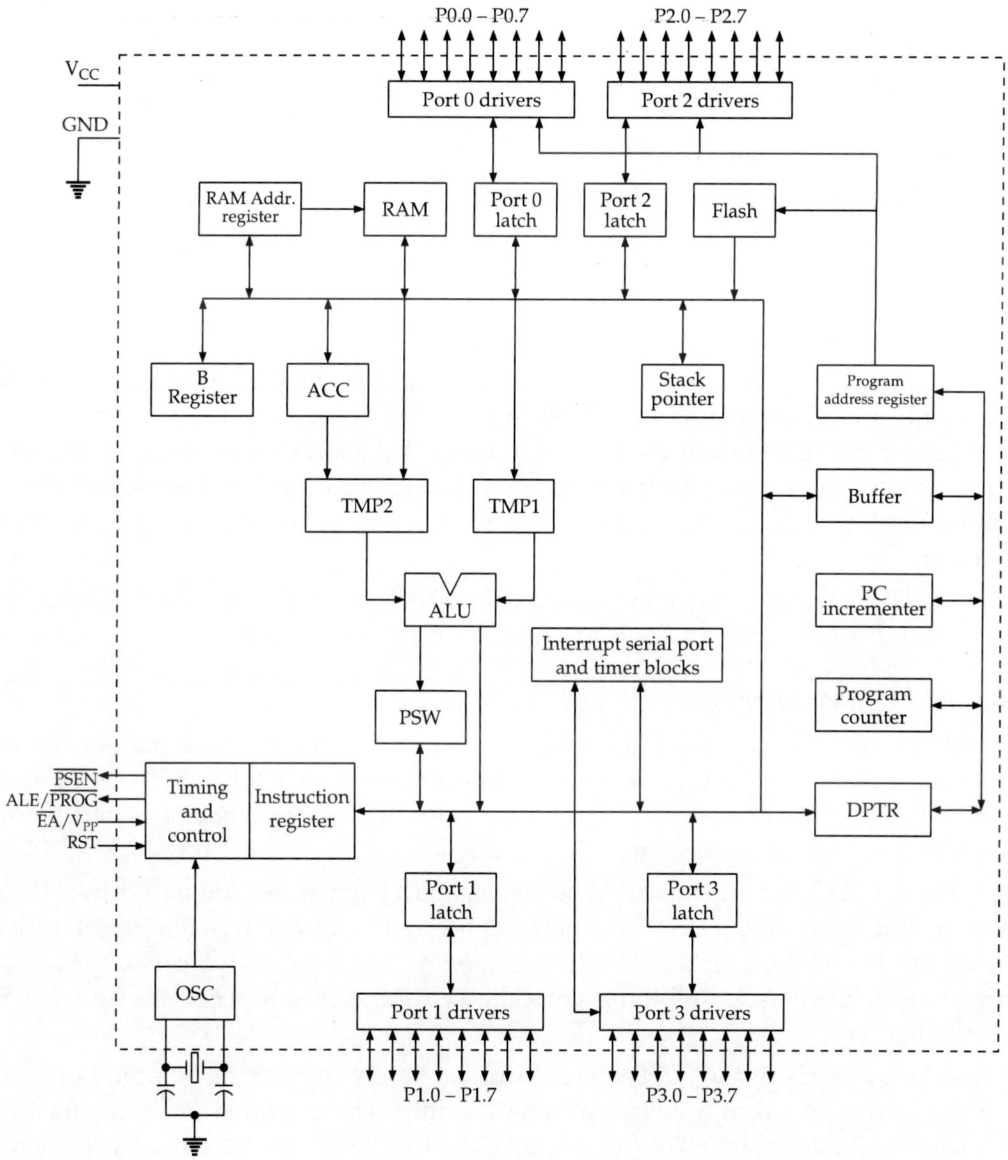

Fig. 1.15 *Block diagram of 8051*

1.9 ARCHITECTURE OR BLOCK DIAGRAM OF 8051 MICROCONTROLLER

The functional block diagram of an 8-bit microcontroller is shown in Figure 1.16. The architecture of 8051 is broadly classified into the blocks as given below:

- Arithmetic and logic unit (ALU)
- Timing and control unit
- RAM/ROM
- Registers
- Drivers and latches for the ports P0, P1, P2 and P3

Arithmetic and Logic Unit (ALU)

All the arithmetic and logical operations on the 8-bit operand is performed using ALU unit. After execution of any arithmetic and logical operation, the result is automatically transferred to the accumulator register with few exceptions.

Boolean Processor

In the 8051 microcontroller, inbuilt Boolean processor is available. In other words, Boolean processor is integrated on the chip. Boolean processor consists of its own instruction set, registers and bit addressable memory. The carry flag serves as an accumulator for the Boolean processor. Some of the instructions of Boolean processor include set bit, clear bit, complement bit, etc. Jump instruction or branching instruction is also supported by this processor.

Note: In 8086 microprocessor onwards, a microprocessor can also perform bit operations. But due to absence of Boolean processor it takes more time.

Data and Program Memory

There are two types of memory available in the 8051 microcontroller. Data memory (internal or on-chip RAM) and internal or on-chip program memory (ROM/EPROM). Program memory is used to hold instruction/program codes or simply opcodes and data memory is used to hold data of a program.

There are 128 bytes on-chip RAM available, apart from an additional 128 byte RAM is also available; addresses from 80h to FFh are referred as special function register (SFRs). Microcontroller 8051 consists of 16 address lines, so it is capable to access off-chip 64 KB data/program memory. A 16-bit register, called DPTR, is used to access the contents from off-chip memory.

The first models of the 8051 microcontroller family did not have internal program memory. It was added as an external separate chip. These models are recognizable by their label beginning with 80X1 (for example 8031 or 8032). All later models have a few Kbyte ROM embedded. Even though such an amount of memory is sufficient for writing most of the programs, there are situations when it is necessary to use additional memory as well. A typical example is so-called lookup tables. The lookup tables are generally used to store predefined data, for example, the code of seven-segment display, temperature reading, etc. They are used in cases when equations describing some processes are too complicated or when there is no time for solving them. In such cases, all necessary estimates and approximates are executed in advance and the final results are put in the tables, i.e., in the program memory referred as lookup table contents.

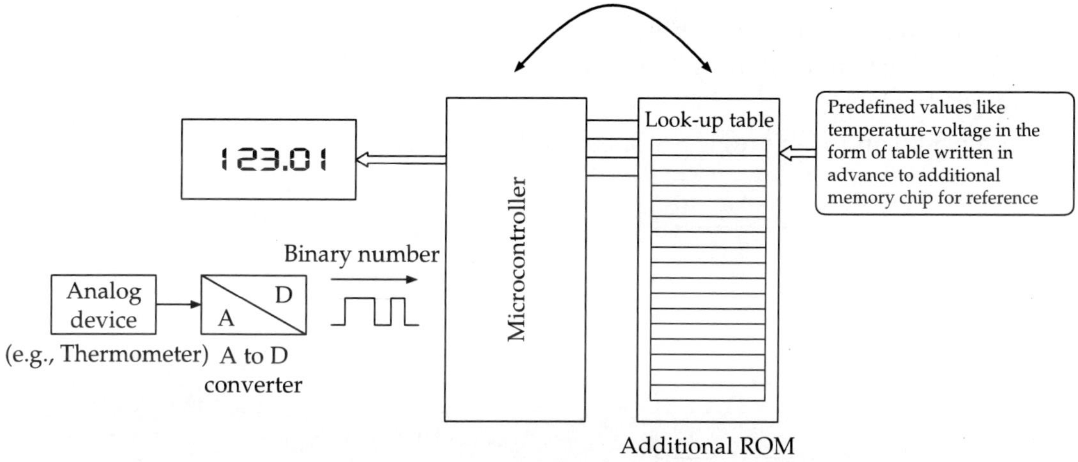

Fig. 1.16 *Use of look-up table for holding predefined values of temperature – voltage*

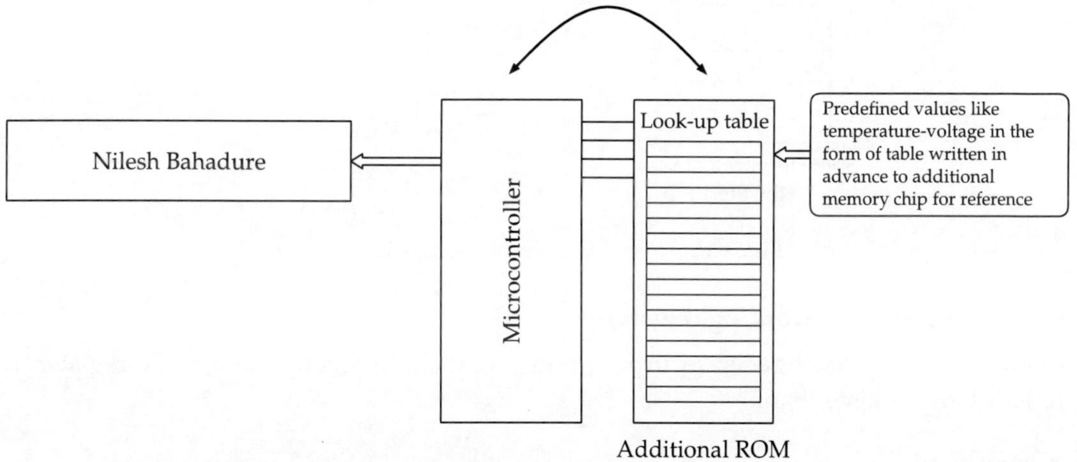

Fig. 1.17 *Use of look-up table to display predefined message*

For detail information on memory and its types, refer section 1.14.

Timing and control unit

All the operations in the 8051 microcontroller are always synchronous with the timing and control unit. The timing and control unit also has signals like ALE, /PSEN, /RD, /WR, RST, and /EA.

Register structure of 8051 Microcontroller

In the CPU registers were used to load the information temporarily. In the 8051 microcontroller, large number of CPU registers are available which are broadly classified as:

General-purpose registers or working registers, stack pointer registers, program counter registers and DPTR (data pointer register).

Along with this, a separate special function register is also available.

The most widely used general-purpose registers are A (accumulator), B, R0, R1, R2, R3, R4, R5, R6, R7, PC and DPTR. All these registers are 8-bit wide except PC and DPTR which are 16-bit registers.

Some 8-bit and 16-bit registers of 8051 microcontroller are shown in Figure 1.19.

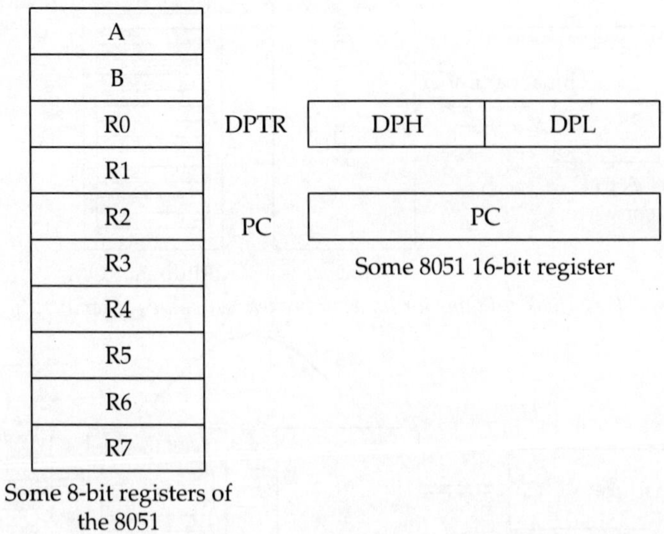

Fig. 1.18 *Register model of 8051 microcontroller*

General-purpose or working registers

The most widely used general-purpose registers include accumulator, B, SP register, R0 to R7 of four register bank.

Accumulator

It is an 8-bit register. It is one of the important registers in the 8051 microcontroller. In any arithmetic and logical operations, one of the operand is always loaded in the accumulator and after execution of the instruction the result is also loaded in the accumulator.

It is the only register in 8051 which has a direct path to ALU.

Accessing it is always faster compared to accessing main memory.

B – Register

The register B is an 8-bit register. This register is available as a general-purpose register, if it is not employed or used for the multiplication and divide operations.

While multiplication, register B is used to hold one of the operands and after multiplication higher byte of the result is stored in it. Similarly, during divide operation, register B holds divisor and after executing division operation, the result of operation, i.e., remainder is stored in register B and the quotient is stored in the accumulator.

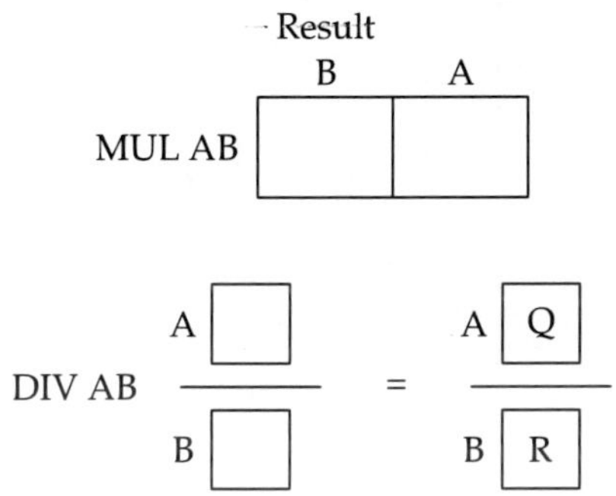

Stack Pointer (SP)

The stack pointer is an 8-bit register. The content of stack pointer registers is incremented by 1 during Push and Call operation and is decremented by 1 during POP and RET operations.

Upon Power Up or after Reset, stack pointer is initialized at 07H so that after Push or Call Operation the contents are transferred in the memory location 08 and above (after register bank 0).

The stack pointer is initialized anywhere within 128 bytes on-chip RAM but in good programming practices, it is always initialized in the memory location from 30 to 7FH so that there is no interference of stack memory with the register banks and bit addressable RAM.

Program Counter (PC) and Data Pointer Register (DPTR)

These are the only 16-bit registers available in the 8051 μC. PC register is used to access the instruction opcodes from the code memory. The PC is 16-bit so that it is capable to access 64 KB of instruction opcodes. The content of the PC indicates the instruction to be fetched from the code memory location.

To access the content from the off-chip memory, 16-bit register DPTR is used. Because DPTR is a 16-bit register, maximum size of external or off-chip memory is 6 KB Data Memory or program memory.

To access off-chip program memory, use the instruction operand "MOVC" and to access the data contents from off-chip memory, use the instruction (MOVX). The DPTR has 2 bytes: Lower byte – DPL and higher byte – DPH.

For the DPTR, there is an instruction "*INC DPTR*" to increment the contents of DPTR but there is no decrement instruction for DPTR. It should be noted that there is no on-chip address for the program counter and there is no direct address available to DPTR but individual bytes of DPTR is accessed by the on-chip address (DPL-82 and DPH-83).

1.10 REGISTERS OF 8051 MICROCONTROLLER

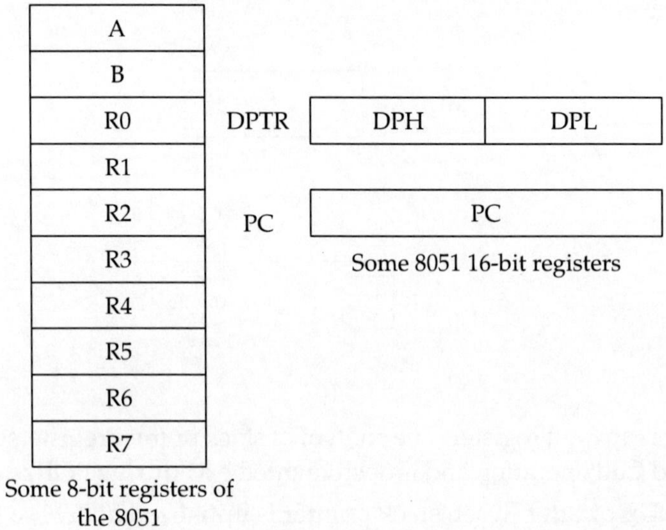

Fig. 1.19 *Registers of 8051 microcontroller*

1.10.1 Flag Register of 8051 Microcontroller

The flag register in the 8051 μC is referred to as program status word (PSW). PSW is a bit addressable register and 0D0H is the on-chip address of PSW. In the flag register of 8051, 4 flag bits are called conditional or status flag bits; namely, carry, auxiliary carry, overflow and parity. Depending on certain operations like arithmetic or logical, these flag bits are affected and hence they are called conditional or status flags. Apart from the status flag, there are other 2 bits called RS1 and RS0 used for register bank selection.

CY	AC	F0	RS1	RS0	OV	–	P

Carry flag	PSW.7	CY
Auxiliary carry flag	PSW.6	AC
Available to the user for general purpose	PSW.5	–
Register bank selector bit 1	PSW.4	RS1
Register bank selector bit 0	PSW.3	RSO
Overflow flag bit	PSW.2	OV
User defined flag bit	PSW.1	–
Parity flag bit	PSW.0	P

RS1	RS0	Register bank	Address
0	0	0	00h – 07h
0	1	1	0h –0Fh
1	0	2	10h – 17h
1	1	3	18h – 1Fh

Fig. 1.20 *Flag register of 8051*

By default or after reset, RS1 and RS0 sets to 00 and hence register bank 0 is selected. RS1 and RS0 is the D4 and D3 bit of PSW. Since PSW is a Bit addressable, these bits are also accessed using PSW.4 and PSW.3. The remaining 2 bits, i.e., D5th bit and D1st bit are the user's flag bits. So the role of this flag bit is fixed by the user.

Bits	**SFR Symbol**	**Function**
D7	PSW.7	Carry flag
D6	PSW.6	Auxiliary carry flag
D5	PSW.5	Users can defined its role, i.e., user's flag bit
D4	PSW.4	Register bank select bit 1
D3	PSW.3	Register bank select bit 0
D2	PSW.2	Overflow flag
D1	PSW.1	User's can define its role.
D0	PSW.0	Parity flag bit

1. **Carry Flag (CY)**

If after addition or subtraction, the bit is generated out of D7th bit then it is called CY bit. In other words, after addition or subtraction of the two 8-bit numbers if the 9th bit is generated or required as a borrow then this bit is automatically copied in the CY flag bit.

It is also possible to set and reset CY flag bit using the instruction SETB C (Set Carry) and CLR C (Clear Carry).

2. **Auxiliary Carry (AC)**

After addition or subtraction of the two 8-bit numbers if the carry is generated out of D3 bit, than it is called auxiliary carry. This flag bit is used in the BCD arithmetic operation.

3. **Parity**

Parity indicates count of one's in accumulator. If the number of one's in the accumulator is odd, than parity is 1 otherwise 0, i.e., for even number of one's the parity flag bit is zero.

4. **Overflow**

After Sign Arithmetic operation, the carry will be generated out of D7 bit then overflow flag bit is set indicating the result is out of the range. For the unsigned number if the result is out of the range, then the carry flag is set.

Similarly, for the sign number if the result is out of the range, then the overflow flag bit is set.

In general, if there is a carry bit from either of the D6th or D7th bit, then the overflow flag bit is set but if there is a carry from both of these bits then this particular flag bit is reset. However, if there is no carry from either of D6th or D7th and from both of these bits then the OV flag bit is reset only.

5. **RS1 and RS0**

These 2 bits stand for register bank select bit 1 and 0.

RS1	RS0	Register Bank	Memory Locations
0	0	RB0	00-07
0	1	RB1	08-0F
1	0	RB2	10-17
1	1	RB3	18-1F

Example 1.1

Show the status of different flag bits after performing the following arithmetic operation and also state the register bank which is selected.

1. **7F + 05**

Solution:

$$\begin{array}{r} \text{7F } 0111 \longrightarrow 1111 \\ \text{05 } 0000 \longrightarrow 0101 \\ \hline \text{84H } 1000 \longrightarrow 0100 \end{array}$$

The status of different flag bits

- **Carry:** Since there is no carry from the D7th bit, the carry flag bit is reset.
- **Auxiliary Carry:** Since there is a carry generating from the D3rd bit, the AC flag bit is set.
- **Parity:** Since the number of one's is even, the parity flag bit is reset.
- **Overflow:** Since there is a carry from the D6 to D7 , the overflow flag bit is set.
- **RS1 and RS0:** Since the D4 and D3 bits are 00, the register bank '0' is selected.

2. $(-127)_d + (2)_d$

Solution:

$$
\begin{array}{rcl}
(-127)_d & \longrightarrow & 1000\ 0001 \\
(2)_d\ 0000 & \longrightarrow & 0010 \\
\hline
83\text{ H }1000 & \longrightarrow & 0011
\end{array}
$$

The status of different flag bits

- **Carry:** Since there is no carry from the D7th bit, the carry flag bit is reset.
- **Auxiliary Carry:** Since there is no carry generating from the D3rd bit, the AC flag bit is reset.
- **Parity:** Since the number of one's is odd, the parity flag bit is set.
- **Overflow:** Since there is no carry generated either from the D6 or D7, the overflow flag bit is reset.
- **RS1 and RS0:** Since the D4 and D3 bits are 00, the register bank '0' is selected.

3. $(-125)_d + (-3)_d$

Solution:

$$
\begin{array}{rcl}
(-125)_d\ 1000 & \longrightarrow & 0011 \\
(-3)_d\ 1111 & \longrightarrow & 1101 \\
\hline
80\text{H }1000 & \longrightarrow & 0000
\end{array}
$$

The status of different flag bits

- **Carry:** Since there is a generation of carry from the D7th bit, the carry flag bit is set.
- **Auxiliary Carry:** Since there is a carry generating from the D3rd bit, the AC flag bit is set.
- **Parity:** Since the number of one's is odd, the parity flag bit is set.
- **Overflow:** Since there is a carry generating both from the D6 and D7, the overflow flag bit is reset.
- **RS1 and RS0:** Since the D4 and D3 bits are 00, the register bank '0' is selected.

4. $(-127)_d + (3)_d$

Solution:

$(-127)_d$ 1000 ⟶ 0001
$(3)_d$ 0000 ⟶ 0011

84H 1000 ⟶ 0100

The status of different flag bits

- **Carry:** Since there is no carry from the D7th bit, the carry flag bit is reset.
- **Auxiliary Carry:** Since there is no carry generating from the D3rd bit, the AC flag bit is reset.
- **Parity:** Since the number of one's is even, the parity flag bit is reset.
- **Overflow:** Since there is no carry either from the D6 or D7, the overflow flag bit is reset.
- **RS1 and RS0:** Since the D4 and D3 bits are 00, the register bank '0' is selected.

5. $(-125)_d + (-5)_d$

Solution:

$(-125)_d$ 1000 ⟶ 0011
$(-5)_d$ 1111 ⟶ 1011

7EH 0111 ⟶ 1110

The status of different flag bits

- **Carry:** Since there is a carry generating from the D7th bit, the carry flag bit is set.
- **Auxiliary Carry:** Since there is no carry generating from the D3rd bit, the AC flag bit is reset.
- **Parity:** Since the number of one's is even, the parity flag bit is reset.
- **Overflow:** Since there is a carry generating from the D7th bit, the overflow flag bit is set.
- **RS1 and RS0:** Since the D4 and D3 bits are 11, the register bank '3' is selected.

6. $(-2)_d + (-5)_d$

Solution:

$(-2)_d$ 1111 ⟶ 1110
$(-5)_d$ 1111 ⟶ 1011

F9H 1111 ⟶ 1001

The status of different flag bits

- **Carry:** Since there is a carry generating from the D7th bit, the carry flag bit is set.

- **Auxiliary Carry:** Since there is a carry generating from the D3rd bit, the AC flag bit is set.
- **Parity:** Since the number of one's is even, the parity flag bit is reset.
- **Overflow:** Since there is a carry generating both from the D6th and the D7th bit, the overflow flag bit is reset.
- **RS1 and RS0:** Since the D4 and D3 bits are 11, the register bank '3' is selected.

1.11 8051 MICROCONTROLLER'S INTERNAL RAM AND SPECIAL FUNCTION REGISTERS (SFRs)

There is an additional 128 bytes on-chip RAM available if the 8051 µC addresses from 80H to 0FFh. This additional on-chip RAM location is used for special functions and hence, they are referred as special function register. These SFRs are used to control certain operations and also to show the status of different operations performing in the 8051 µC.

For example, TMOD is an SFR used to set timer 0 or timer 1 and not for any other operation and hence, this dedicated operation called special function operation and the register used for the purpose is called SFR.

The SFRs available in the 8051 µC with their name and on-chip addresses is listed below.

Table 1.4 *SFRs and their addresses*

SFR Symbol	Name of the Register	Address(H)
A*	Accumulator	0E0
B*	Register - B	0F0
PSW*	Program Status Word	0D0H
SP	Stack pointer	81
DPTR	Data pointer	
DPL	Data pointer lower byte	82
DPH	Data pointer higher byte	83
P0*	Port 0	80
P1*	Port 1	90
P2*	Port 2	A0
P3*	Port 3	B0
IE*	Interrupt enable	0A8
IP*	Interrupt priority	0B8
PCON	Power control	87
TCON*	Timer control	88
TMOD	Timer mode	89

TL0	Timer zero lower byte	8A
TH0	Timer zero higher byte	8C
TL1	Timer one lower byte	8B
TH1	Timer one higher byte	8D
SCON*	Serial control	98
SBUF	Serial buffer	99

* Indicates bit addressable

Table 1.5 *SFR maps and their addresses for the microcontroller 8051 and 8051*

F8H								
F0H	B*							
E8H								
E0H	ACC*							
D8H								
D0H	PSW*							
C8H	(T2CON)*		(RCAP2L)	(RCAP2H)	(TL2)	(TH2)		
C0H								
B8H	IP*							
B0H	P3*							
A8H	IE*							
A0H	P2*							
98H	SCON*	SBUF						
90H	P1*							
88H	TCON*	TMOD	TL0	TL1	TH0	TH1		
80H	P0*	SP	DPL	DPH				PCON

* indicates bit addressable

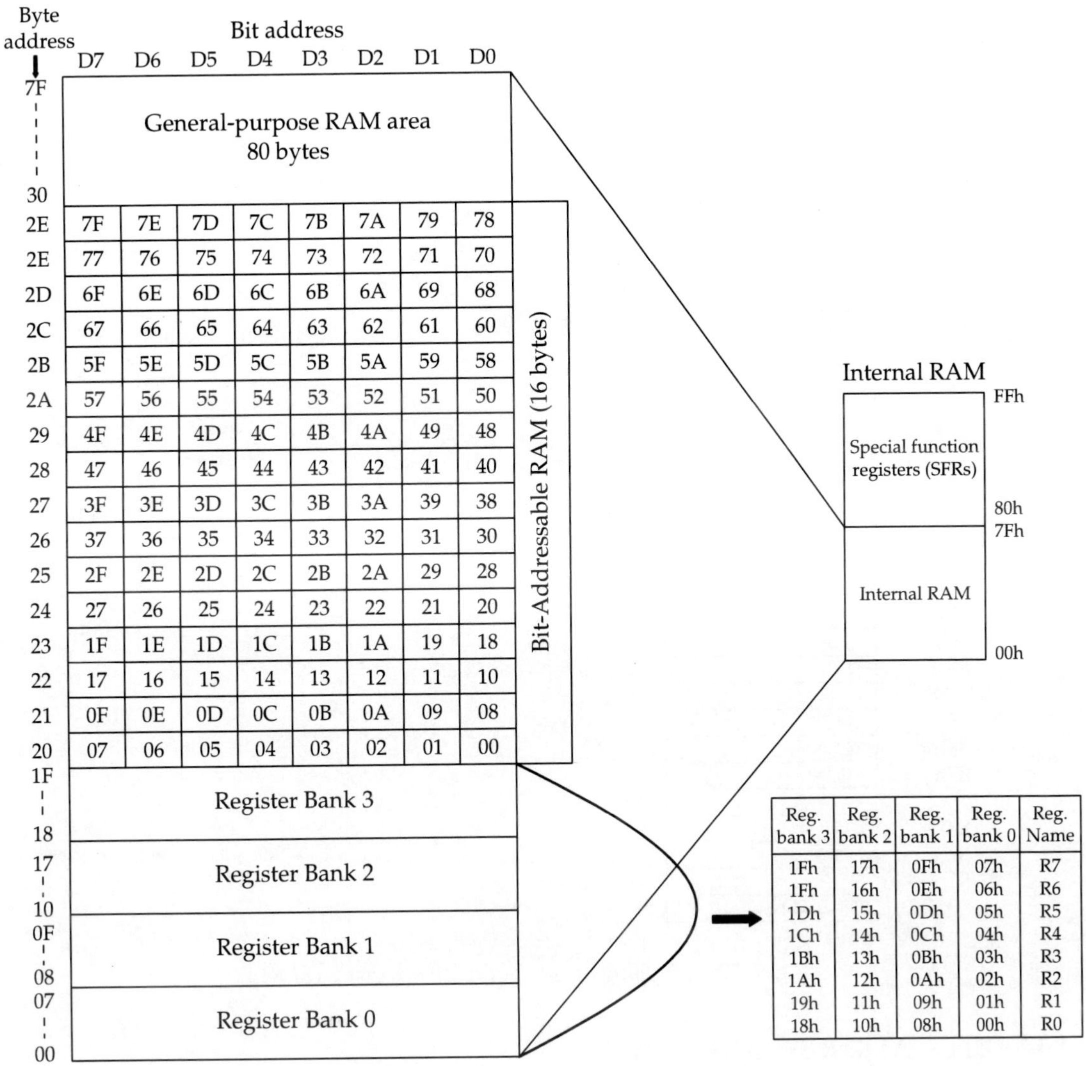

Fig. 1.21 *Mapping of internal RAM*

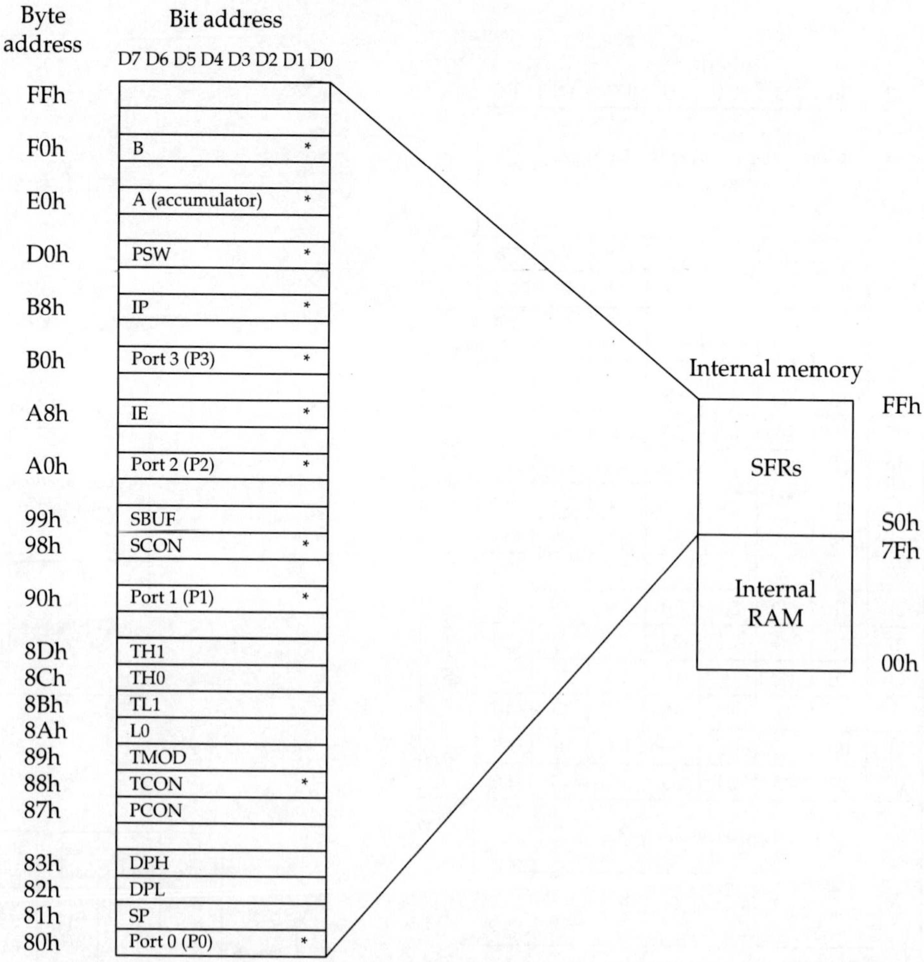

Fig. 1.22 *Mapping of special function register (SFR)*

1.12 BIT – ADDRESSABLE RAM

The 8051 contains 210 bit-addressable locations of which 128 are at locations 20H to 2FH while the rest are in the SFRs. Each of the 128 bits from 20H to 2FH has a unique number (address) attached to it, as shown in the table above. The 8051 instruction set allows you to set or reset any single bit in this section of RAM. With the general-purpose RAM from 30H to 7FH and the register banks from 00H to 1FH, you may only read or write a full byte (8 bits) at these locations. However, with bit-addressable RAM (20H to 2FH) you can read or write any single bit in this region by using the unique address for that bit. We will later see that this is a very powerful feature.

The 8051, being a communications-oriented microcontroller, gives the user the ability to access a number of *bit variables*. These variables may be either 1 or 0. There are 128 bit variables available to the user, numbered 00h through 7Fh. The user may use these variables with commands such as SETB and CLR.

For example, to set bit number 24 (hex) to 1 you would execute the instruction:

SETB 24h

It is important to note that bit memory is really an important part of internal RAM. In fact, the 128 bit variables occupy the 16 bytes of internal RAM from 20h through 2Fh. Thus, if you write the value FFh to internal RAM address 20h you have effectively set bits 00h through 07h. That is to say that ***MOV 20h, #0FFh*** is equivalent to:

SETB 00h

SETB 01h

SETB 02h

SETB 03h

SETB 04h

SETB 05h

SETB 06h

SETB 07h

As illustrated above, bit memory is not really a new type of memory. It's really just a subset of internal RAM. But since the 8051 provides special instructions to access these 16 bytes of memory on a bit-by-bit basis, it is useful to think of it as a separate type of memory. However, always keep in mind that it is just a subset of internal RAM-and that operations performed on internal RAM can change the values of the bit variables.

Bit variables 00h through 7Fh are for user-defined functions in their programs. However, bit variables 80h and above are actually used to access certain SFRs on a bit-by-bit basis. For example, if output lines P0.0 through P0.7 are all clear (0) and you want to turn on the P0.0 output line you may either execute:

MOV P0, #01h

or you may execute:

SETB 80h

Both these instructions accomplish the same thing. However, using the SETB command will turn on the P0.0 line without affecting the status of any of the other P0 output lines. The MOV command effectively turns off all the other output lines which, in some cases, may not be acceptable.

Programming Tip: If your program does not use bit variables, you may use internal RAM locations 20h through 2Fh for your own use. But if you plan to use bit variables, be very careful about using addresses from 20h through 2Fh as you may end up overwriting the value of your bits!

Bit position →	D7	D6	D5	D4	D3	D2	D1	D0
2E	7F	7E	7D	7C	7B	7A	79	78
2E	77	76	75	74	73	72	71	70
2D	6F	6E	6D	6C	6B	6A	69	68
2C	67	66	65	64	63	62	61	60
2B	5F	5E	5D	5C	5B	5A	59	58
2A	57	56	55	54	53	52	51	50
29	4F	4E	4D	4C	4B	4A	49	48
28	47	46	45	44	43	42	41	40
27	3F	3E	3D	3C	3B	3A	39	38
26	37	36	35	34	33	32	31	30
25	2F	2E	2D	2C	2B	2A	29	28
24	27	26	25	24	23	22	21	20
23	1F	1E	1D	1C	1B	1A	19	18
22	17	16	15	14	13	12	11	10
21	0F	0E	0D	0C	0B	0A	09	08
20	07	06	05	04	03	02	01	00

Fig. 1.23 *Bit-addressable RAM*

1.13 REGISTER BANKS AND STACK OF 8051 MICROCONTROLLER
OR
MEMORY ORGANIZATION OF ON-CHIP 128 BYTE RAM OF 8051 MICROCONTROLLER

In the 8051 microcontroller, 8 registers of R0-R7 are available in the group of 4 Register Bank. These registers R0-R7 are also referred to as Scratch-Pad Registers. All these registers are 8-bit wide and they are located on the first 32 bytes on-chip RAM location. So accessing these registers by their name R0-R7 means effectively we are using on-chip RAM location from 00-1FH.

By default or after reset, register bank 0 is selected. In order to select any other register bank, we need to program and change corresponding bit in the program status word.

For example; suppose register bank 0 is selected where writing a byte in the R4 register means that the byte will be transferred in the 04H memory location. Also suppose that register bank 1 is selected and we are writing a byte in the R4 register means that the byte will be transferred in the 0CH memory location. Obviously, it is advantageous to access memory location by their register names. So effectively, a programmer can set 32 registers.

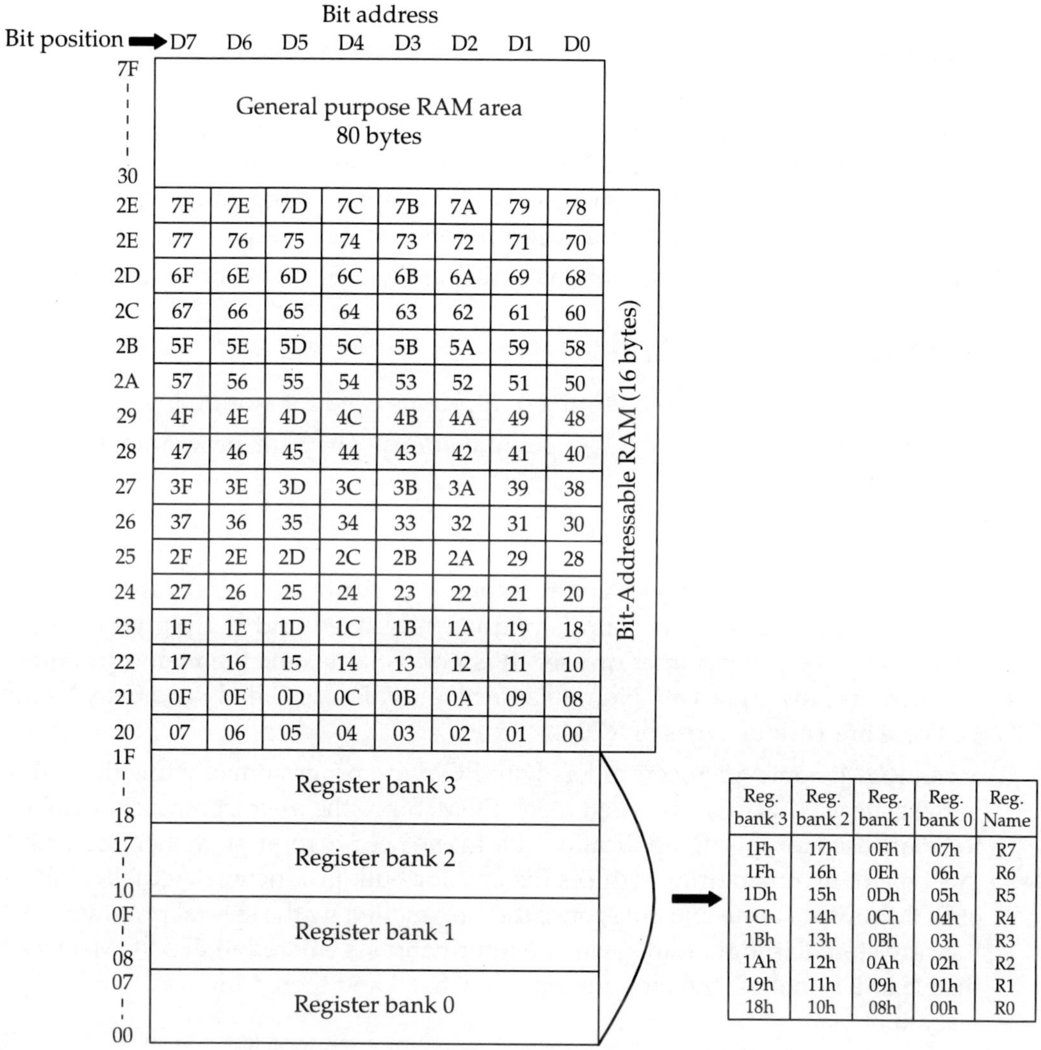

Fig. 1.24 *Internal (on-chip) RAM of 128 bytes*

Figure 1.24 shows the classification of 128-byte on-chip RAM and it is organized or grouped as: First 32 bytes of the 128 byte on-chip RAM is dedicated to the register bank. So 4 register banks are available and each consists of 8 registers from R0-R7.

Next 16 bytes, i.e., from memory location from 20 to 2FH are set aside for bit addressable RAM.

Last 80 bytes, i.e., from 30 to 7FH are used for reading/ writing general storage data. They are also called scratch-pad RAM.

Example 1.2

How many instructions are supported or possible in the 8051 microcontroller and its family?

Solution: The size of the instruction register in 8051 microcontroller is 8-bit, so the size of the opcode in the 8051 microcontroller is 8-bit, therefore, unique possible combination of the codes in 8051 microcontrollers are $2^8 = 256$ and hence, only 256 unique combinations of the opcode are possible in the 8051 microcontroller, therefore, maximum 256 instructions are possible to be executed by 8051 microcontrollers and its family.

Actual number of basic instructions available in the 8051 microcontrollers are 91.

1.14 MEMORY AND ITS TYPES

Memory is combination of cells consisting of flip-flops used for storing information.

In a microcontroller, there are two types of memory. They are, program memory and data memory respectively.

Program Memory

Program memory is also known as 'control store' and 'firmware'. Firmware refers to the situation when the software program is permanently fixed and not permitted for any modifications by the programmer or user. It is non-volatile, i.e., the memory content is not lost when the power goes off. Non-volatile memory is also called Read Only Memory (ROM). There are various types of ROM.

1. **Mask ROM:** Some microcontrollers with ROM are programmed while they are still in the factory. This ROM is called Mask ROM. Since the microcontrollers with Mask ROM are used for specific application, there is no need to reprogram them. Sometimes, this type of manufacturing reduces the cost for bulk production. Actually, this type of ROM consists of fusible links, once they are melted by the special programmer kit, they are stored as logic 1, this way the information is entered in this ROM. Once the fusible link is melted, then it is permanently fixed and there is no way to reconfigure it again.
2. **PROM (Programmable Memory):** User programmable (one-time programmable) memory. If the information burned into PROM is wrong, it needs to be discarded since internal fuses are blown permanently. To enter the information or data into this memory, special equipment is needed called ROM burner or ROM programmer.
3. **Reprogrammable Program Memory (or) Erasable PROM (EPROM):** Microcontrollers with EPROM were introduced in late 1970s. These devices are electrically programmable but are erased with UV radiation. The construction of a EPROM memory cell is somewhat like a MOSFET but with a control and float semiconductor as shown in Figure 1.26.

   All EPROM chips have a window that is used to shine ultraviolet (UV) radiation to erase its contents and hence, also referred to as UV-EPROM.

   This type of memory is approximately 200,000 times more programmable.

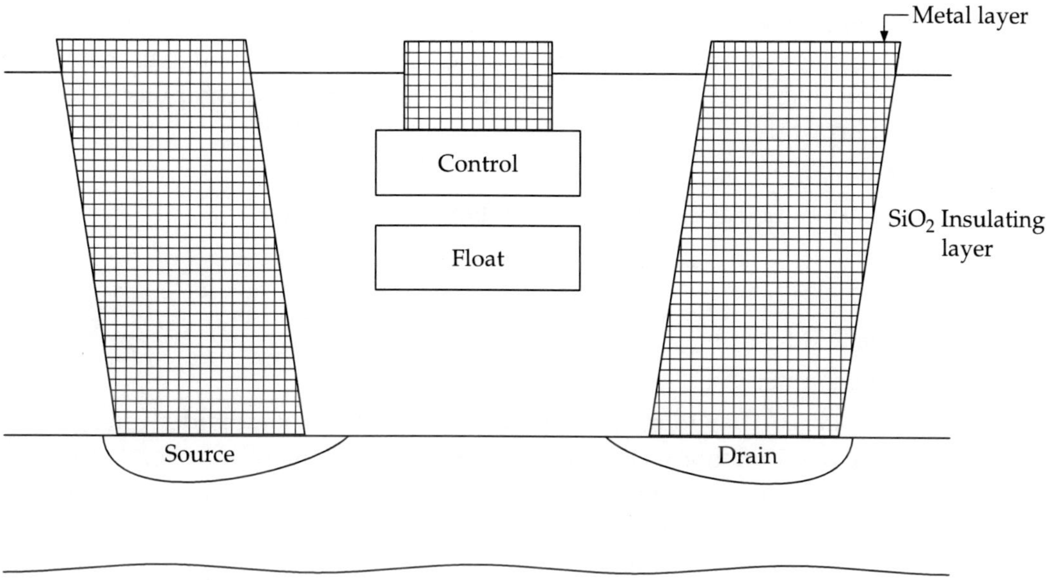

Fig. 1.25 *Structure of an EPROM memory*

In the unprogrammed state, the 'float' does not have any charge and the MOSFET is in the OFF state. To program the cell, the 'control' above the 'float' is raised to a high enough potential such that a charge leaks to the float through SiO_2 insulating layer. Hence, a channel is formed between 'source' and 'drain' in the silicon substrate and the MOSFET becomes 'ON'. The charge in the 'float' remains for a long time (typically over 30 years). The charge can be removed by exposing the float to UV radiation. For UV erasable version, the packaging is done in a ceramic enclosure with a glass window. The UV erasable version of EPROM memory is shown in Figure 1.27.

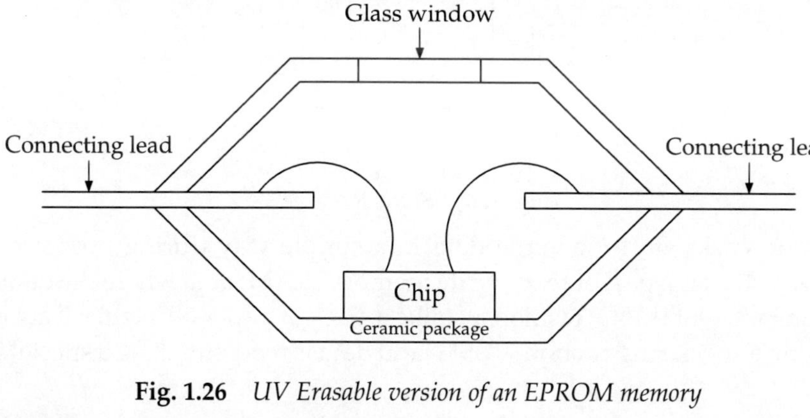

Fig. 1.26 *UV Erasable version of an EPROM memory*

4. **EEPROM (Electrically Erasable ROM):** Method of erasure is electrical. Moreover, one can select which byte to be erased. Cost per bit is much higher than for UV-EPROM. This type of memory is approximately 500,000 times more programmable.

5. **Flash Memory EPROM:** First, the process of erasure of the entire contents takes less than a second, or one might say in a flash, hence its name, flash memory. When flash memory's contents are erased, the entire device is erased. Even though flash memories are writeable, like EPROMs they find their widest use in microcomputer systems for storage of firmware.

Data Memory

Data memory can be classified into the following categories:

- Bits
- Registers
- Variable RAM
- Program counter stack

Microcontroller can have ability to perform manipulation of individual bits in certain registers (bit manipulation). This is a unique feature of a microcontroller, not available in a microprocessor. Eight bits make a byte. Memory bytes are known as file registers. Registers are some special RAM locations that can be accessed by the processor very easily. Figure 1.28 shows the static RAM memory cell using the MOSFET transistor and its equivalent is shown in Figure 1.29. This consists of two CMOS inverters connected back to front, so as to form a latch.

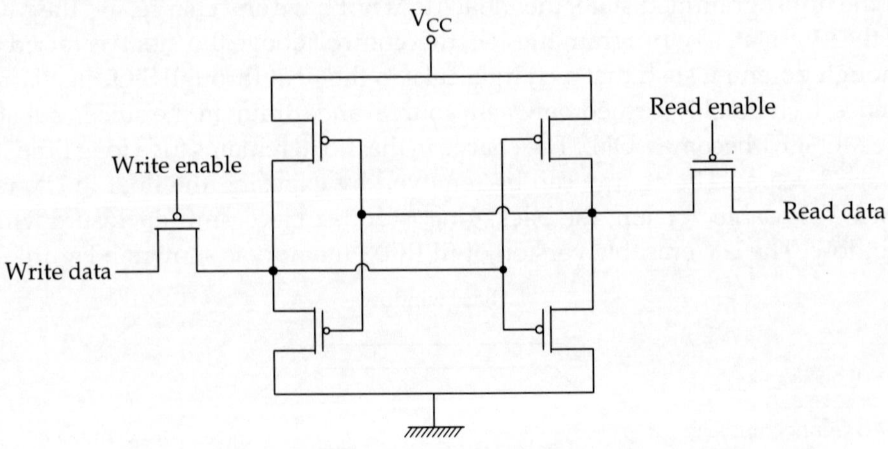

Fig. 1.27 *Static RAM cell*

Processor stacks store/save the data in a simple way during program execution. Processor stack is a part of read/write memory (RAM) area where the data is saved in a Last in First Out (LIFO) fashion just like a stack of paper on a table. Data is stored by executing a special instruction 'PUSH' and data is read out using a special instruction 'POP'.

I/O Registers: In addition to the data memory, some special-purpose registers are required that are used in input/output and control operations. These registers are called

I/O registers. These are important for microcontroller peripheral interface and control applications.

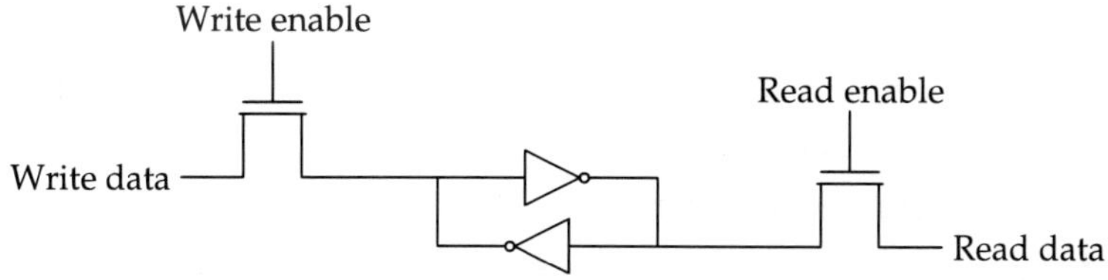

Fig. 1.28 *Static RAM cell equivalent*

RAM (Random Access Memory) is infinite times programmable and erasable. RAM is called volatile memory since cutting off the power to the IC will mean the loss of data. Also referred to as R/WM (Read and Write Memory). The basic type of RAM is given below:

1. **SRAM (Static RAM):** Storage cells are made of flip-flops and therefore they do not require refreshing to keep their data. A cell handling one bit requires 6 or 4 transistors each, which is too many. SRAMs are widely used for cache memory and battery-backed memory systems.
2. **DRAM (Dynamic RAM):** Uses MOS capacitors to store a bit. It requires constant refreshing due to leakage (every 2 ms–4 ms).

   Advantages

   (a) High density (capacity)

   (b) Cheaper cost per bit

   (c) Lower power consumption.

   Disadvantages

   (a) While it is being refreshed, data cannot be accessed

   (b) Larger access times

   (c) Too many pins due to large capacity

REVIEW QUESTIONS

1. Explain the differences between Harvard and von-Neumann architectures.
2. Draw and explain the classification of internal RAM.
3. Draw and explain architecture of microcontroller 8051.
4. Draw and explain program status word with suitable example.
5. What is a special function register? Explain in details.
6. Explain what is a Boolean processor with suitable example.
7. Explain the significance of an accumulator and register B.
8. Explain the register banks in details; also explain how the microcontroller 8051 can access the register bank?

9. Explain how to reset a microcontroller. Show the reset value of registers of the 8051 microcontroller.
10. Explain the difference between data memory and program memory used in the microcontroller 8051.
11. Explain in details, how microcontroller 8051 will access external ROM and internal ROM. Show with proper illustrations.
12. Explain the significance of EA# and PSEN# signals.
13. Explain bit-addressable RAM in details.
14. Draw and explain on-chip area of RAM of microcontroller 8051.
15. Explain the differences between microprocessor and microcontrollers.
16. Explain the importance of ports P0, P1, P2 and P3.
17. Explain the dual functionality of port P3.
18. What do you mean by Harvard and Princeton architectures? Explain in details.
19. What is microcontroller? Explain different families of microcontroller 8051.
20. Explain the following registers of 8051 with their importance:
    (a) Accumulator
    (b) Program counter
    (c) Flag register
    (d) Registers R0 to R7
    (e) Stack pointer

Instruction Set and Programming of 8051

2

Chapter

Microcontroller consists of all varieties of instructions, that are basically used for the data transfer operations, arithmetic operations, logical operations, Boolean operations, branching operations, etc. This chapter provides the basic understanding of each and every instruction. These instructions of any processor are backbone of the processor, and getting complete understanding of each and every instruction ensures the mastery in the processor system.

This chapter also includes basic examples that illustrate the working of instructions more clearly, so readers become more competent over how to use instructions and understand the world of 8051 microcontroller.

2.1 ADDRESSING MODES OF THE 8051 MICROCONTROLLER

To perform any operation in the microcontroller, we have to give corresponding instruction to the microcontroller. Along with the instruction we have to provide three pieces of information:

Operation to be performed (opcode)
Address of source of data
Address of destination of result

The method by which address of source of data and address of destination of result is given along with the instruction are called addressing modes.

The 8051 microcontroller provides five distinct addressing modes:

1. Immediate addressing mode (IAM)
2. Register addressing mode (RAM)
3. Direct addressing mode (DAM)
4. Indirect or register indirect addressing mode (RIAM)
5. Indexed or base register plus index register addressing modes

The 8051 architecture supports several distinct physical address spaces.

1. On-chip program memory (ROM)
2. On-chip data memory (RAM)
3. External program memory (Ext. ROM)
4. External data memory (Ext. RAM)
5. On-chip special function registers (SFRs)

Most of the instructions can operate with variables from these physical address spaces.

2.1.1 Immediate Addressing Mode

In this addressing mode, the source operand is constant. In this mode, the operand is given directly along with the instruction. The immediate data must be preceded by the pound sign #. This addressing mode can be used to load data into any register, including 16-bit register DPTR.

Examples:

MOV A, #45H
MOV B, #50H
MOV R0, #78H
MOV DPTR, #5678H
MOV P1, #55H

2.1.2 Register Addressing Mode

If 8-bit data is required to execute the instruction and this 8-bit data is present in any register and the name of this register is given along with the instruction, then it is called register addressing mode.

Examples:

MOV A, B
MOV A, R0
MOV R2, A
MOV R7, A
MOV R0, R2 - Invalid instruction, register-to-register movement within the register bank is not permitted.

Programming Tips: Register-to-register movement or within register bank and between register banks is not permitted. For this type of data transfer, use direct addressing mode.

2.1.3 Direct Addressing Mode

If 8-bit data is required to execute the instruction and this 8-bit data is present in on-chip RAM or in on-chip SFRs. The address of on-chip RAM or address of on-chip ASFRs address is given directly along with the instruction, then, such instructions are called direct addressing modes. MSB bit of the given 8-bit address is distinguished between on-chip RAM address and on-chip SFR address. In the on-chip RAM address range is from 00h to 7Fh (MSB = 0) and in the on-chip SFR it is from 80h to FFh (MSB = 1).

The allocation of 128 bytes of on-chip RAM is given below:

(*i*) RAM locations 00h to 1Fh are set aside for register banks and stack.

(*ii*) RAM locations 20h to 2Fh are set aside for bit addressable RAM to save single bit information.

(*iii*) RAM locations 30h to 7Fh are set aside for storing byte information.

Although it is possible we can define all 128 bytes of on-chip RAM, but in practice we use 30h to 7Fh memory locations to avoid any interference to register bank and bit addressable RAM memory.

Examples:

MOV A, 30H

MOV B, 56H

MOV 0E0H, #45H

MOV 98H, #56H

MOV R4, R5 --- INVALID

MOV 04H, 05H

MOV 0E0H, 0F0H

Use of direct addressing mode in stack

Pushing the value of the register on to the stack is permitted only by using direct addressing mode.

For example, to push the contents of the accumulator register into the stack, the instruction to perform this is

PUSH 0E0h

Similarly, for retrieving the contents back to the register using POP is written as:

POP 0E0h instead of POP A

Example 2.1

Assume SP = 07h, show the contents of memory location for the following instructions:

PUSH 0E0h

PUSH 02h

PUSH 07h

Assume the contents of accumulator = 35h, register R2 = 56h and register R7 = 55h.

Solution:

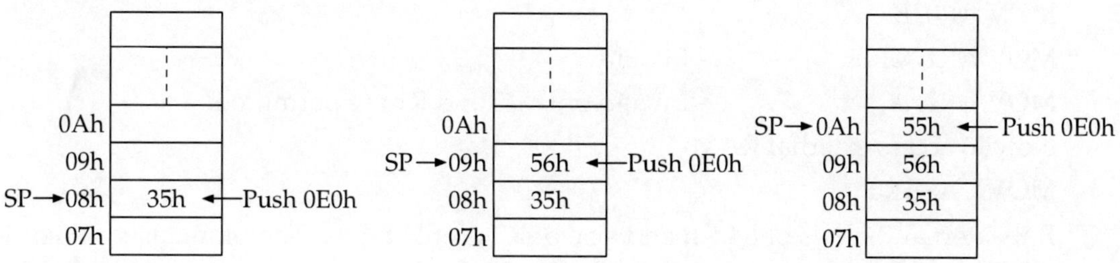

Example 2.2

Show that how to save the contents of an accumulator, R2 and R7 in the stack memory and then retrieve the contents back into the registers B, R5 and R6.

Solution:

```
PUSH 0E0H
PUSH 02H
PUSH 07H
POP 06H
POP 05H
POP 0F0H
```

Instruction	Comment
PUSH 0E0H	Saving the contents of accumulator in the stack memory, i.e., at 09H.
PUSH 02H	Saving the contents of register R2 in the stack memory, i.e., at 0AH.
PUSH 07H	Saving the contents of register R7 in the stack memory, i.e., at 0BH.
POP 06H	Retrieving the content of register R7 into the register R6. The stack pointer points to memory location 0AH.
POP 05H	Retrieving the content of register R5 into register R2. The stack pointer points to memory location 09H.
POP 0F0H	Retrieving the content of the accumulator into register B. The stack pointer points to memory location 08H.

2.1.4 Register Indirect Addressing Mode (RIAM)

If 8-bit data is required to execute the instruction and this 8-bit data is present in on-chip RAM or SFR and the address of this on-chip RAM location or SFR is present in the register R0 or R1 and the name of this register is given along with the instruction preceded by @ sign, then it is called register indirect addressing mode.

Only registers R0 and R1 are used in RIAM and names of the registers R0 and R1 are always given along with @ sign.

Examples:

```
MOV A, @R0
MOV @R0, B
MOV @R1, @R0          -- Invalid
MOV @R2, A            -- Invalid only R0 and R1 are permitted.
```

Now to access external RAM

```
MOVX A, @R0
```

But external RAM is of 64 KB and size of R0 is only 8-bits. Therefore, using R0 or R1 register 8-bit addresses are given and only first 256 bytes or memory location is accessible.

DPTR register is used to access full 64 KB of external RAM memory using
MOVX A, @DPTR

Example 2.3

Transfer contents of memory location 25h to the accumulator using DAM and RIAM.

Solution:

Using Direct Addressing Mode (DAM)

MOV A, 25H

Using Register Indirect Addressing Mode (RIAM)

MOV R0, #25H

MOV A, @R0

Example 2.4

Give instruction to transfer the contents of memory location 50h to the memory location 60h using DAM and RIAM.

Solution:

Using DAM

MOV 60H, 50H

Using RIAM

MOV R0, #50H

MOV 60H, @R0

2.1.5 Indexed Addressing Mode

Index addressing mode is used for defining lookup table values. The values defined by lookup table are always present in ROM location (external ROM) and this content is accessed using instruction MOVC.

The instruction format to access lookup table contents from the external ROM memory is

MOVC A, @A+DPTR

C stands for code memory. In this instruction 8-bit value of accumulator is added with 16-bit DPTR and the contents of this 16-bit address are transferred to the accumulator. Such instructions are called index addressing mode.

2.2 ADVANTAGES OF REGISTER INDIRECT ADDRESSING MODE OVER DIRECT ADDRESSING MODE

Accessing data bytes from RAM is a dynamic process rather than static process used in direct addressing mode.

For example, if data 55h is to be copied in 5 memory locations from 40h to 44h, it is easily proved that register indirect addressing mode is more efficient compared to DAM.

1st method is to copy the data 55h in 5 memory locations, 40h to 44h.

Using direct addressing modes

8085 program	8051 program
MVI A, 55H	MOV A, #55H
STA 0040H	MOV 40H, A
STA 0041H	MOV 41H, A
STA 0042H	MOV 42H, A
STA 0043H	MOV 43H, A
STA 0044H	MOV 44H, A

2nd method using RIAM without loop

8085 program	8051 program
MVI A, 55H	MOV A, #55H
LXI H, 0040H	MOV R0, #40H
MOV M, A	MOV @R0, A
INX H	INC R0
MOV M, A	MOV @R0, A
INX H	INC R0
MOV M, A	MOV @R0, A
INX H	INC R0
MOV M, A	MOV @R0, A
INX H	INC R0
MOV M, A	MOV @R0, A

3rd method using RIAM with loop

8085 program	8051 program
MVI A, 55H	MOV A, #55H
LXI H, 0040H	MOV R0, #40H
MVI C, 05H	MOV R2, #05H
L1: MOV M, A	L1: MOV @R0, A
INX H	INC R0
DCR C	DJNZ R2, L1
JNZ L1	

In the 2nd method, two instructions are repeated many times and in the method 3 using a counter the same program is designed very efficiently, but method 3 is possible because of register indirect addressing mode. In direct addressing mode, looping is not possible.

Limitation of register indirect addressing mode

Here only two registers are permitted, i.e., R0 and R1 to access the memory location, both these registers are 8-bit wide so using 8-bit register only on-chip RAM or SFR can be accessed.

To access external RAM or ROM, 16-bit address is given in DPTR and it is accessed through index addressing mode.

2.3 SYMBOLS/NOMENCLATURE USED FOR DATA OR MEMORY LOCATION IN 8051 MICROCONTROLLER

# (Pound sign)	For immediate value or data
@Ri	i = 0 or 1 memory location using RIAM
Rn	n = 0 to 7, registers R0 to R7 from register bank
Direct	On-chip internal RAM and SFR
Bit	Bit addressable address of RAM
Rel	Relative memory location (8-bit address used with branching instructions)
Addr/addr16	16-bit address
Addr11/sadd	11-bit address/absolute address

2.4 INSTRUCTION SET AND PROGRAMMING OF 8051 MICROCONTROLLER

To perform any operation in the microcontroller, we have to give corresponding instruction to the microcontroller. The instructions of microcontroller 8051 are classified as:

- Data transfer group
- Arithmetic group
- Logical group
- Boolean instruction group
- Branching instruction group

2.4.1 Data Transfer Group

In data transfer group, data is transferred to destination from source. Old contents are not lost but the old content of destination is lost. The size of the source and destination must be same. The opcodes used in the data transfer instructions are:

MOV, MOVX, MOVC, PUSH, POP, XCH AND XCHD

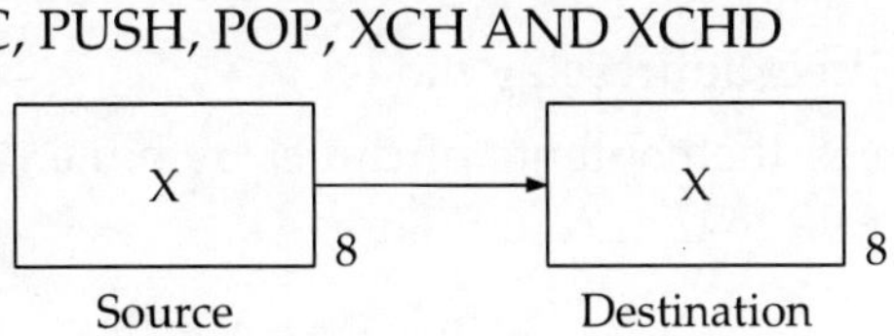

1. **MOV A, #DATA**

   Copy 8-bit immediate data to the accumulator

   Bytes = 2

   Cycles = 1

   Addressing mode: Immediate addressing mode

   This instruction copies 8-bit data given along with the instruction to the accumulator.

   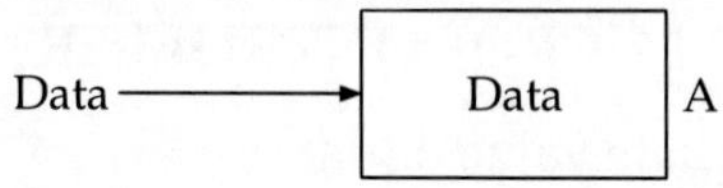

   Examples:

   MOV A, #45H

   MOV A, #12 (MOV A, #0CH)

2. **MOV A, Rn**

   Transfer or copy the contents of register Rn to the accumulator

   Bytes = 1

   Cycles = 1

   This instruction copies or transfers the contents of the register Rn, i.e., R0 to R7 to the accumulator. After execution old values of the Rn remain intact but the old value of the accumulator is lost.

   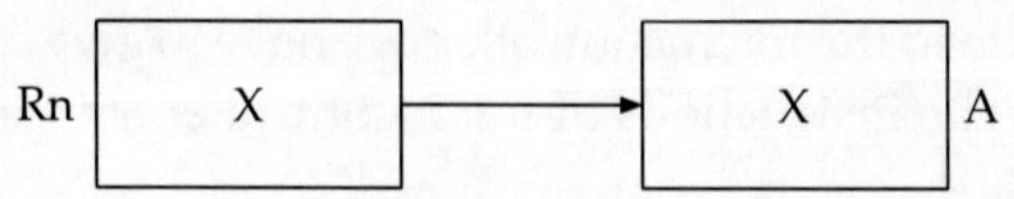

   Examples:

   MOV A, R0

   MOV A, R2

   MOV A, R7

3. **MOV A, DIRECT**

   Copy direct byte to the accumulator

   Bytes = 02

   Cycles = 1

   Addressing mode: Direct addressing mode

   This instruction copies the contents of direct byte, i.e., contents of internal RAM memory location or SFR to the accumulator.

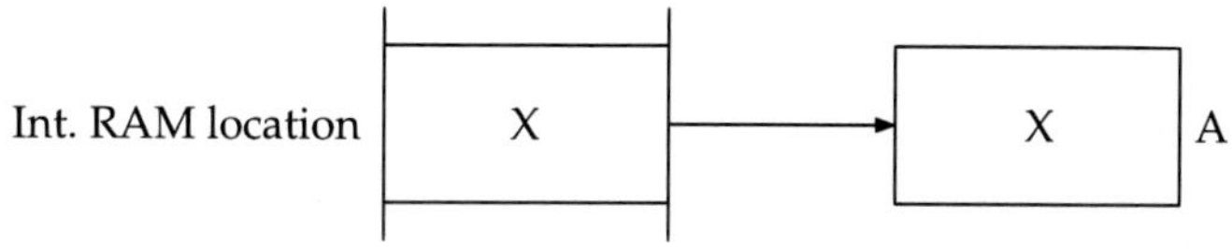

Examples:

MOV A, 45H

MOV A, 78H

MOV A, 82H

MOV A, 0F0H

4. **MOV A, @Ri**

   Copy indirect byte to the accumulator

   Bytes = 1

   Cycles = 1

   Addressing mode: Indirect addressing mode

   This instruction copies the contents of internal RAM location or SFR pointed by register R0 or R1 to the accumulator.

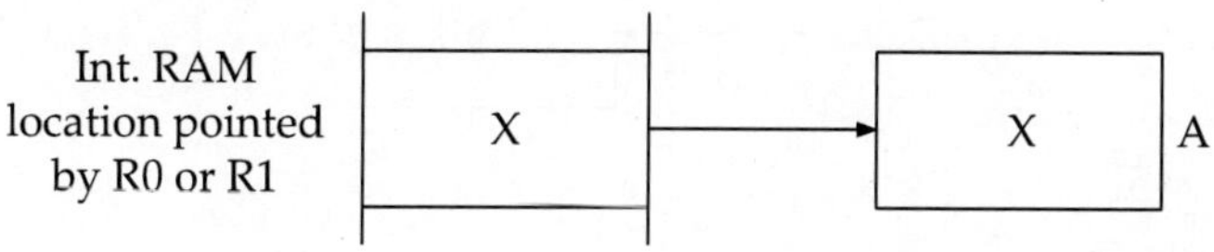

Examples:

MOV A, @R0

MOV A, @R1

5. **MOV Rn, A**

   Copy accumulator to the register Rn

   Bytes = 1

   Cycles = 1

   Addressing mode: Register addressing mode

   This instruction copies the contents of the accumulator to the register Rn (n = 0 to 7)

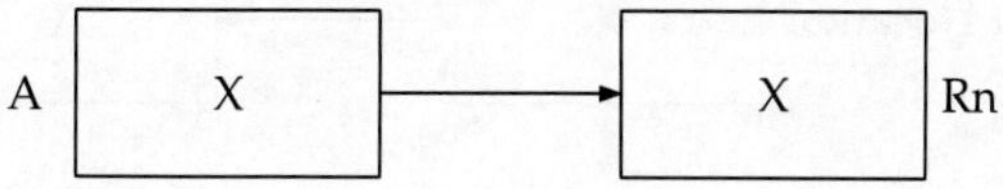

Examples:

MOV R2, A

MOV R7, A

Example 2.5

Copy data 23h to the accumulator and then copy the contents of the accumulator to the register R2 of register bank 2

Solution:

```
MOV A, #23H
SETB PSW.4
MOV R2, A
```

6. **MOV Rn, #DATA**

   Copy 8-bit immediate data to the register

   Bytes = 2

   Cycles = 1

   Addressing mode: Immediate addressing mode

   This instruction copies 8-bit immediate data given along with the instruction to the register Rn (n = 0 to 7).

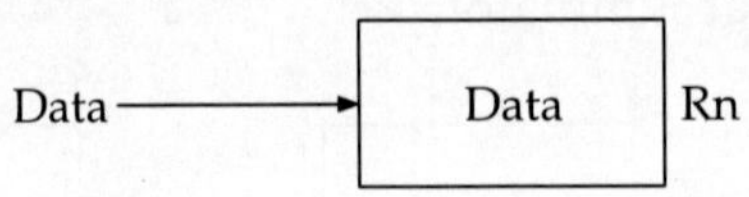


```
MOV R2, #45H
MOV R7, #89H
```

7. **MOV Rn, DIRECT**

   Copy direct byte to the register

   Bytes = 2

   Cycles = 2

   Addressing mode: Direct addressing mode

   This instruction copies the contents of internal RAM location or SFR to the register Rn (n = 0 to 7). The address of internal RAM location or SFR is given directly along with the instruction.

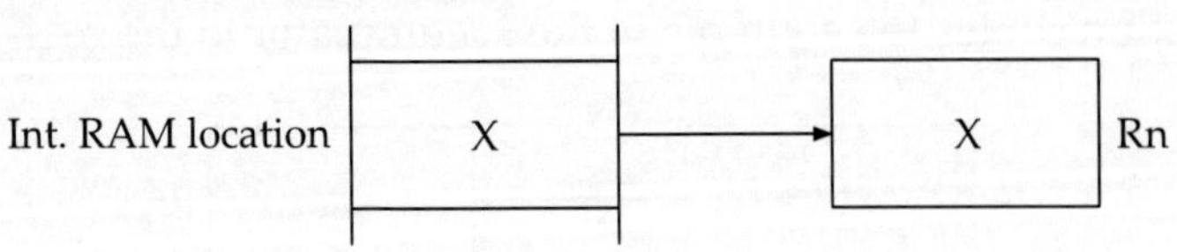

Examples:

```
MOV R2, 56H
MOV R5, 0E0H
```

8. **MOV DIRECT, A**

Copy the accumulator to direct byte

Bytes = 2

Cycles = 1

Addressing mode: Direct addressing mode

This instruction copies the contents of the accumulator to the internal RAM location or SFR whose 8-bit address is directly given along with the instruction.

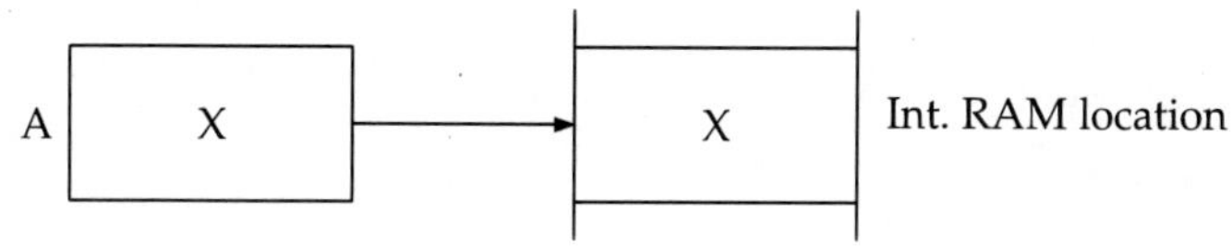

Examples:

MOV 67H, A

MOV 0F0H, A

9. **MOV DIRECT, Rn**

Copy registers Rn to the direct byte

Bytes = 2

Cycles = 2

Addressing mode: Direct addressing mode

This instruction copies the contents of register Rn (n = 0 to 7) to the internal RAM location or SFR, whose 8-bit address is given directly along with the instruction.

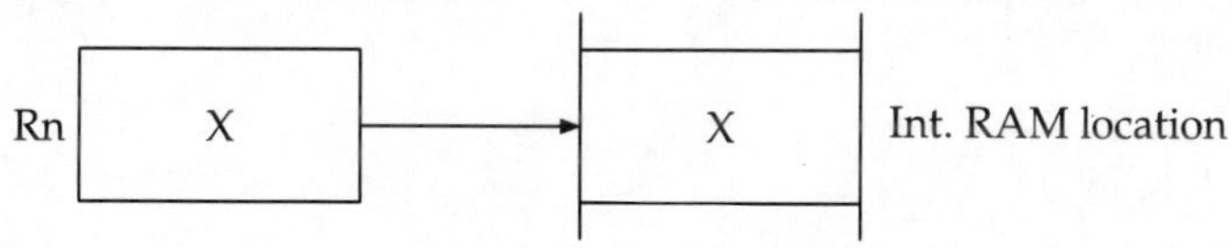

Examples:

MOV 0E0H, R2

MOV 0F0H, R5

10. **MOV DIRECT, #DATA**

Copy 8-bit immediate data to the direct byte

Bytes = 3

Cycles = 2

Addressing mode: Immediate and direct addressing modes

This instruction copies the 8-bit immediate data given along with the instruction to the internal RAM location or SFR. The 8-bit address of this internal RAM location or SFR is given directly along with the instruction.

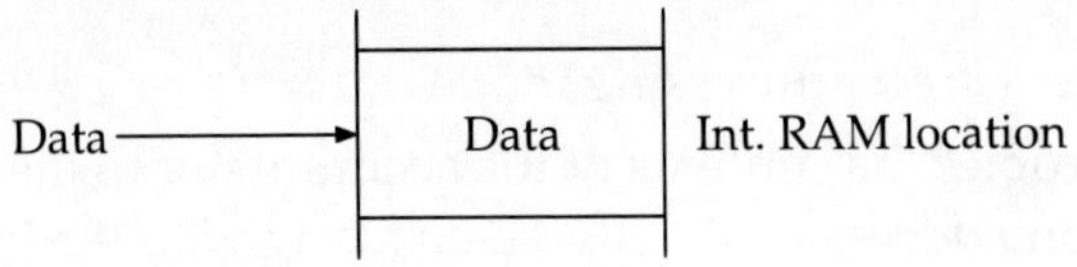

Examples:

MOV 45H, #45H

MOV 0E0H, #67H

11. **MOV DIRECT, DIRECT**

Copy direct byte to direct byte

Bytes = 3

Cycles = 2

Addressing mode: Direct addressing mode

This instruction copies the contents of internal RAM location or SFR to the internal RAM location or SFR. The source and destination 8-bit address of internal RAM location or SFR is given directly along with the instruction.

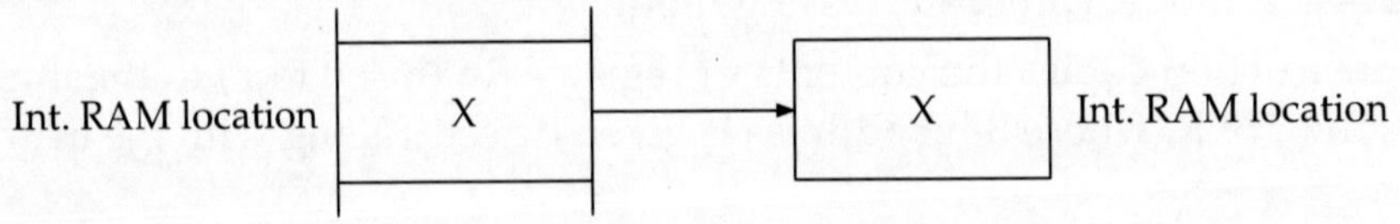

Examples:

MOV 45H, 67H

MOV 02, 03

MOV 0E0H, 0F0H

MOV 1FH, 30H

12. **MOV DIRECT, @Ri**

Copy indirect byte to direct byte

Bytes = 2

Cycles = 2

Addressing mode: Direct and indirect addressing modes

This instruction copies the contents of internal RAM location or SFR pointed or specified by register R0 or R1 to the internal RAM location or SFR, whose 8-bit address is given directly along with the instruction.

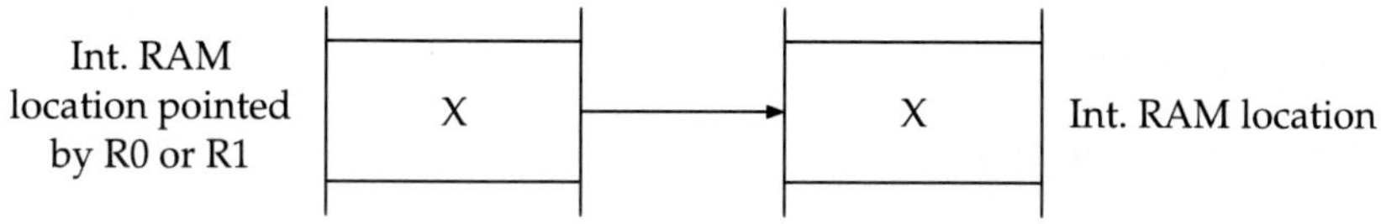

Examples:

MOV 50H, @R0

MOV 0D0H, @R1

13. **MOV @Ri, A**

Copy the accumulator to the indirect byte

Bytes = 1

Cycles = 1

Addressing mode: Indirect addressing mode

This instruction copies the contents of the accumulator to the internal RAM location or SFR pointed register R0 or R1.

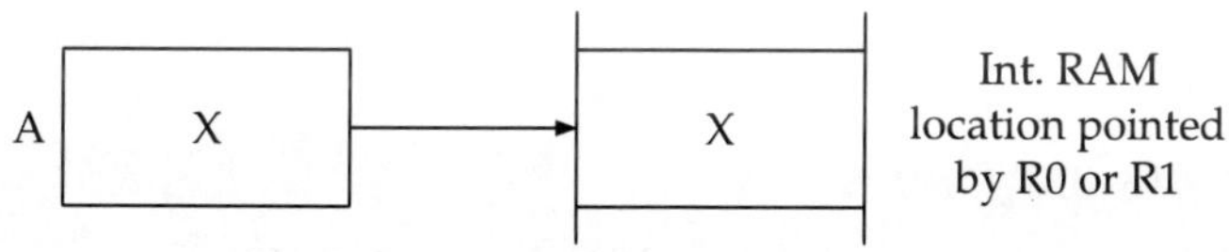

Examples:

MOV @R1, A

MOV @R0, A

14. **MOV @Ri, #DATA**

Copy 8-bit immediate data to the indirect byte

Bytes = 2

Cycles = 1

Addressing modes: Immediate and indirect addressing modes

This instruction copies 8-bit immediate data given along with the instruction to the internal RAM location or SFR pointed by R0 or R1.

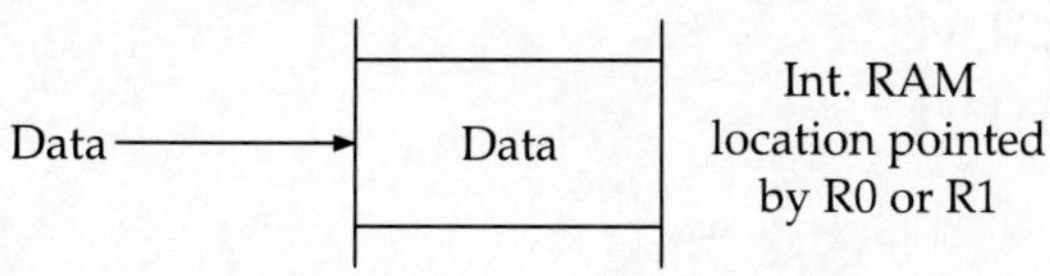

Examples:

MOV @R0, #56H

MOV @R1, #0A0H

15. **MOV @Ri, DIRECT**

    Copy direct byte to indirect byte

    Bytes = 02

    Cycles = 02

    Addressing mode: Direct and indirect addressing modes

    This instruction copies the contents of internal RAM location or SFR given directly along with the instruction to the internal RAM location or SFR pointed by register R0 or R1.

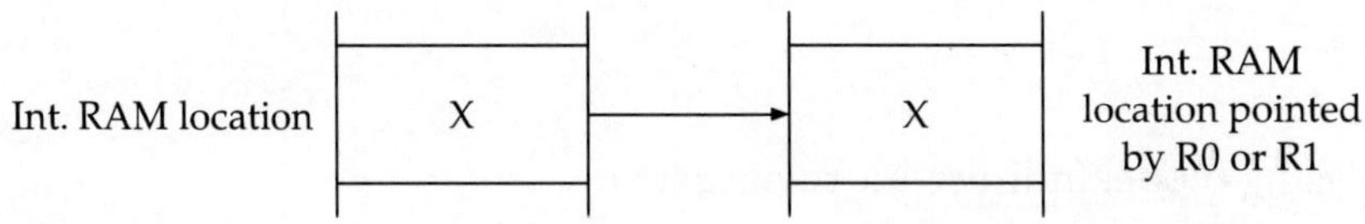

Examples:

    MOV @R0, 45H
    MOV @R1, 0E0H

16. **MOV DPTR, #16-BIT DATA**

    Copy 16-bit immediate data to the 16-bit register DPTR

    Bytes = 3

    Cycles = 2

    Addressing mode: Immediate addressing mode

    This instruction copies 16-bit immediate value given along with the instruction to the DPTR. This is the only instruction in the 8051 microcontroller available to copy 16-bit number to the 16-bit register DPTR.

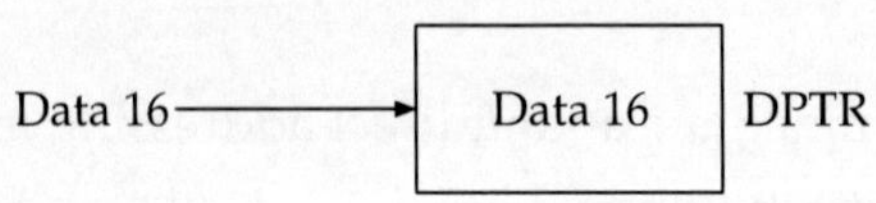

Examples:

    MOV DPTR, #2345H
    MOV DPTR, #6543H

Example 2.6

Copy 16-bit data 6543H to the DPTR using three methods.

Solution:

Method – I

    MOV DPTR, #6543H

Method – II

MOV DPH, #65H

MOV DPL, #43H

Method – III

MOV 82H, #43H

MOV 83H, #65H

17. **MOVX, A @Ri**

Copy contents of external RAM to the accumulator

Bytes = 1

Cycles = 2

Addressing mode: Register indirect addressing mode

This instruction copies the contents of external RAM location pointed by R0 or R1 (0000H to 00FFH) to the accumulator.

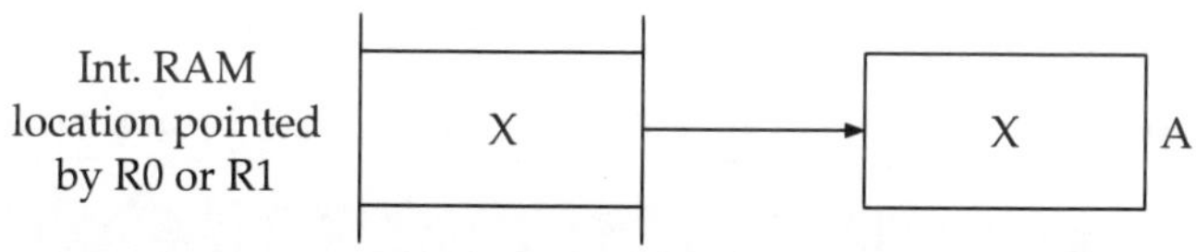

The examples include: (MOVX A, @R0), (MOVX A, @R1).

Example 2.7

Assume external RAM location F2H consists of data byte 07H. Copy this content to the accumulator and register R2.

Solution:

Instructions	Comments
MOV R0, #0F2H	Loading the data byte F2H to the register R0.
MOVX A, @R0	Copying the data externally and indirectly pointed by R0 to the accumulator.
MOV R2, A	Copying the data from the accumulator to the register R0.

18. **MOVX @Ri, A**

Copy accumulator to the external RAM location

Bytes = 1

Cycles = 2

Addressing mode: Register Indirect addressing mode

This instruction copies the contents of the accumulator to the external RAM location (0000H to 00FFH) pointed by R0 or R1.

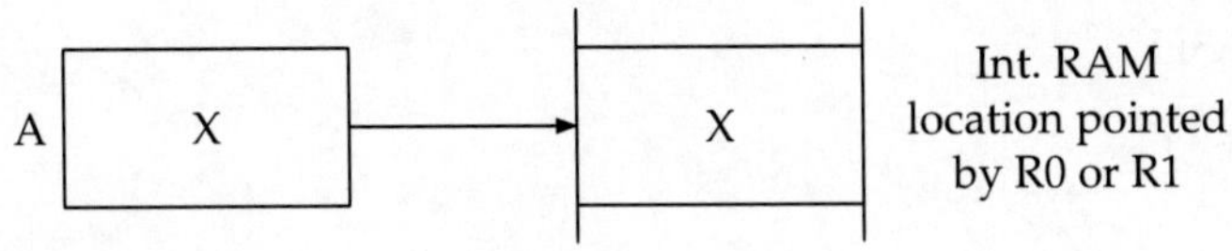

Examples:

MOVX @R0, A
MOVX @R1, A

Example 2.8

Copy data byte 78H to the accumulator and then transfer to the external RAM location 80H.

Solution:

Instructions	Comments
MOV A, #78H	Loading the data 78H immediately to the accumulator.
MOV R0, #80H	Loading the data 80H immediately to register R0.
MOVX @R0, A	Copying the data byte 78H to the external RAM location 80H pointed by R0.

19. **MOVX A, @DPTR**

    Copy contents of external RAM location (16-bit address) to the accumulator
    Bytes = 1
    Cycles = 2
    Addressing modes: Register indirect addressing mode
    This instruction copies the contents of external RAM location pointed by DPTR (0000H to 00FFH) to the accumulator.
    Using this instruction, all 64 KB of external data memory is to be accessed.

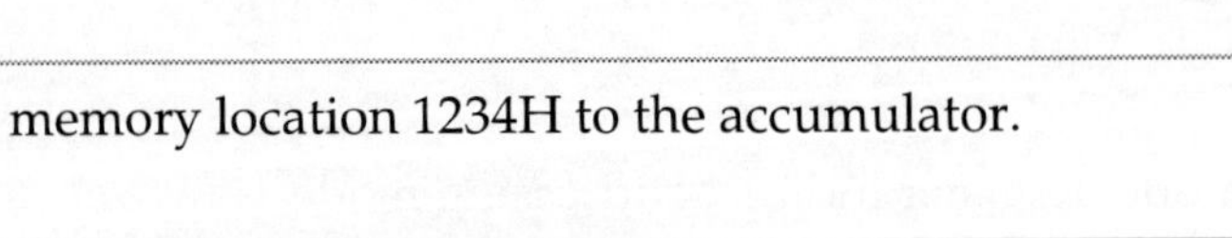

The examples include only (MOVX A, @DPTR).

Example 2.9

Copy contents of memory location 1234H to the accumulator.

Solution:

Instructions	Comment
MOV DPTR, #1234H	The memory location 1234H is pointed by the DPTR.
MOVX A, @DPTR	Copying the data inside the memory location pointed by DPTR externally to the accumulator.

20. **MOVX @DPTR, A**

Copy the accumulator to the external RAM location (16-bit address) pointed by DPTR

Bytes = 1

Cycles = 2

Addressing modes: Register indirect addressing mode

This instruction copies the contents of the accumulator to the external RAM location pointed by DPTR (0000H to 00FFH).

Using this instruction, all 64 KB of external data memory (RAM) is to be accessed.

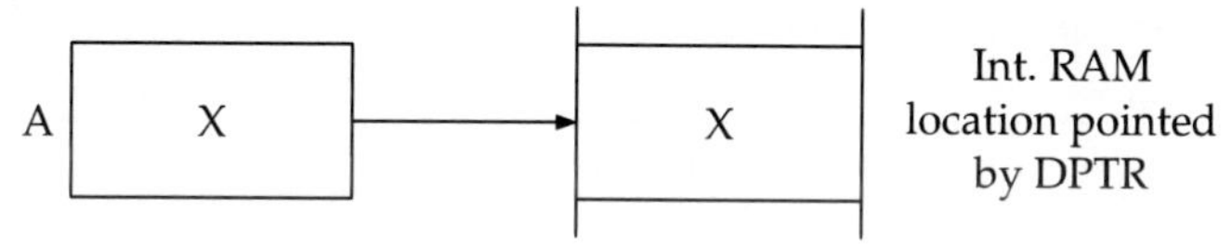

The examples include only (MOVX @DPTR, A).

Example 2.10

Copy data byte 55H to the accumulator and then transfer the contents of the accumulator to the memory location 8000H.

Solution:

Instructions	Comments
MOV A, #55H	Loading the data 55H immediately to the accumulator.
MOV DPTR, #8000H	The memory location 8000H is pointed by DPTR.
MOVX @DPTR, A	Copying the accumulator content externally to the memory location pointed by DPTR.

21. **MOVC A, @A + DPTR**

Copy external ROM location contents to the accumulator

Bytes = 1

Cycles = 2

Addressing modes: Register indirect addressing mode

This instruction copies the contents of external ROM location calculated by adding the accumulator, i.e., relative displacement with 16-bit value in DPTR. This resultant address is also called effective address.

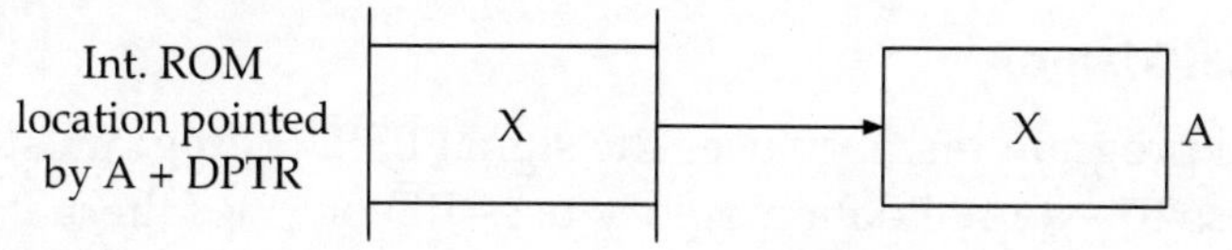

The examples include only (MOVC A, @ A + DPTR).

Example 2.11

Copy the contents of external ROM 2005 to the Register R2.

Solution:

Method – I

Instructions	Comments
CLR A	Clearing the accumulator value to 00H.
MOV DPTR, #2005H	The memory location 2005H is pointed by DPTR.
MOVC A, @ A+DPTR	Copying external ROM contents to the accumulator.
MOV R2,A	Copying the value stored inside the accumulator to register R2.

Method – II

Instructions	Comments
MOV A, #05H	Loading the accumulator value with 05H.
MOV DPTR, #2000H	The memory location 2000H is pointed by DPTR.
MOVC A, @ A+DPTR	Copying external ROM contents to the accumulator.
MOV R2, A	Copying the value stored inside the accumulator to the register R2.

22. **MOVC A, @A + PC**

    Copy internal ROM contents to the accumulator

    Bytes = 1

    Cycles = 2

    Addressing modes: Register indirect addressing mode

    This instruction copies the contents of internal PROM location calculated by adding the accumulator, i.e. relative displacement with 16-bit value in PC. This resultant address is also called effective address indexed by ACC + PC to the accumulator.

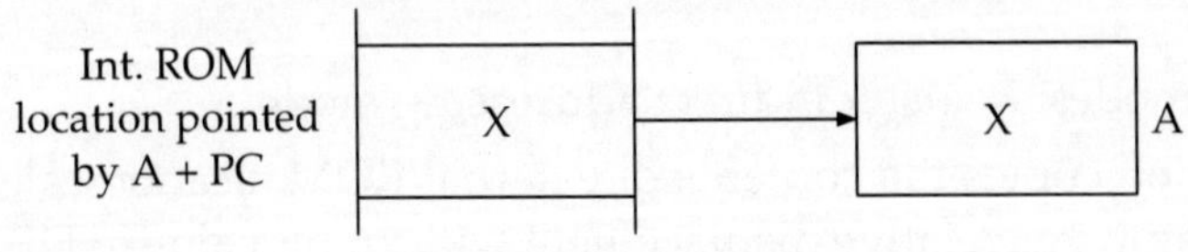

The examples include only (MOVC A, @ A + PC).

Accessing External Memory

Access to external program memory uses the signal $\overline{\text{PSEN}}$ (Program store enable) as the read strobe. Access to external data memory uses $\overline{\text{RD}}$ or $\overline{\text{WR}}$ (alternate function of P3.7 and P3.6).

For external program memory, 16-bit address is always used. For example:

MOVC A, @ A+DPTR

MOVC A, @ A+PC

Access to external data memory can be either 8-bit address or 16-bit address:

8-bit address

MOVX A, @Ri where Ri is either R0 or R1

MOVX @Ri, A

16 bit address

MOVX A, @DPTR

MOVX @DPTR, A

The external memory access in 8051 can be shown by a schematic diagram (Figure 2.1).

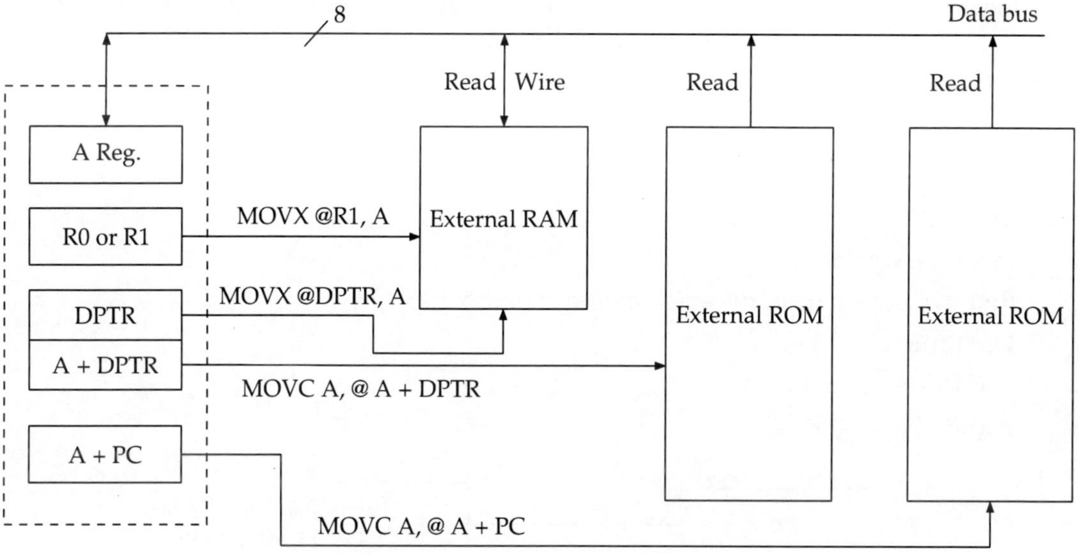

Fig. 2.1 *Schematic diagram of external memory access*

23. **PUSH DIRECT**

Bytes: 2

Machine Cycles: 2

Copy the contents of the direct byte to the stack memory pointed by stack pointer.

Operation:

(SP +1) ⟶ (SP)

(Direct) ⟶ (SP)

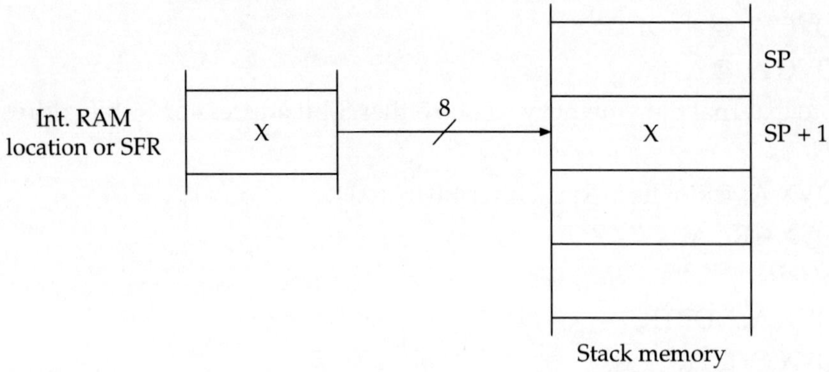

In this instruction, SP is first incremented by 1 and then contents of internal RAM location or SFRs is copied in the stack memory pointed by stack pointer register. Pushing the data on the stack supports only 1 addressing mode, i.e., direct addressing mode.

Some of the examples include PUSH 0E0H, PUSH 0F0H, etc.

24. **POP DIRECT**

Bytes: 2

Machine Cycles: 2

Retrieve or pop from the stack memory to the direct byte.

Operation:

(SP) ⟶ (Direct)

(SP-1) ⟶ (SP)

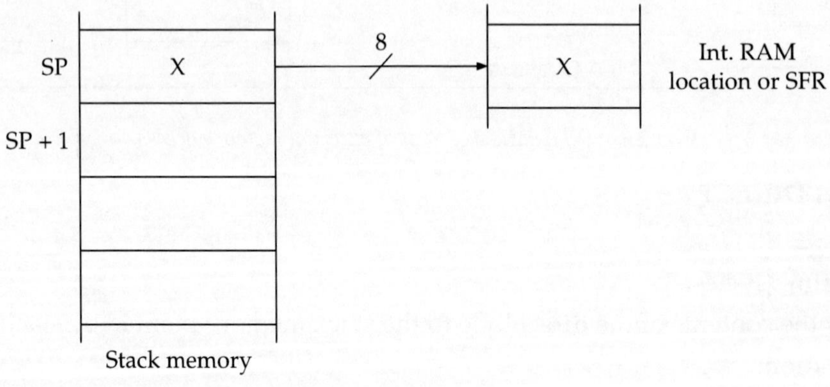

This instruction copies the contents of internal RAM location pointed by stack pointer to the given direct byte and then the value in the stack pointer is decremented by one. Popping the data from the stack supports only 1 addressing mode, i.e., direct addressing mode.

Some of the examples include POP 0E0H, POP 0F0H, etc.

Example 2.12

Assume that the stack pointer points to the memory location 08H. Save the contents of registers R2, R5 and the accumulator to the stack memory. Show the contents of the memory and stack pointer register.

Assume R2 = 07H, R5 = 08H and accumulator = 55H.

Solution:

Instructions	Comments
PUSH 02H	Pushing the data of register R2 into the stack memory. The stack pointer is incremented by one and moves to the memory location 09H.
PUSH 05H	Pushing the data of register R5 into the stack memory. The stack pointer is incremented by one and moves to the memory location 0AH.
PUSH 0E0H	Pushing the data of the accumulator into the stack memory. The stack pointer is incremented by one and moves to memory location 0BH.

Example 2.13

Assume the registers of register bank 2 that has the data bytes stored as follows:

R2 = 55H,

R7 = 65H,

ACC = 90H

Save this content in the stack memory. Show the contents in the memory and stack pointer.

Solution:

Instructions	Comments
SETB PSW.4	Selecting register bank 2
PUSH 02H	Pushing register R2 contents into the stack pointer. The stack pointer is incremented by one and moved to memory location 11H.
PUSH 07H	Pushing register R7 contents into the stack pointer. The stack pointer is incremented by one and moved to memory location 12H.
PUSH 0E0H	Pushing the accumulator contents into the stack pointer. The stack pointer is incremented by one and moved to memory location 13H.

Conclusion: Thus, for pushing and popping the contents to and fro from the stack memory are the PUSH and POP instructions and they both belong to direct addressing mode.

25. **XCH A, Rn**

    Bytes: 1

    Machine Cycles: 1

    Exchange the contents of register Rn (where n varies from 0 to 7) with the accumulator.

    Operation:

    This instruction exchanges the contents of the accumulator with register Rn.

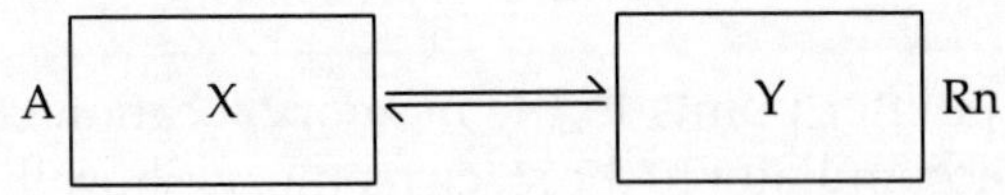

Example: Assume (A) = 55H

And R2 = 86H

Then XCH A, R2

Will result in (A) = 86H and R2 = 55H.

26. **XCH A, DIRECT**

Bytes: 2

Machine Cycles: 1

Exchange the contents of the register accumulator with direct byte (internal RAM + SFRs).

Operation:

This instruction exchanges the contents of the accumulator with the contents of internal RAM or SFRs.

The internal RAM location or SFR address is given directly along with the instruction.

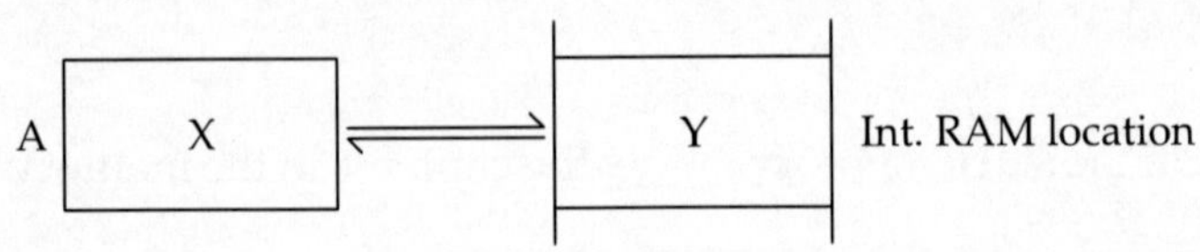

Example:

Assume (A) = 55H and 35H memory location has data byte 86H and XCH A, 35H is executed. So after the execution the result will be

(A) = 86H and 35H will have the data byte 55H.

27. **XCH A, @Ri**

Bytes: 1

Machine Cycles: 1

Exchange the contents of the accumulator with the contents of internal RAM pointed by R0 or R1.

Operation:

This instruction exchanges the contents of the accumulator with the contents of internal RAM location pointed by R0 or R1.

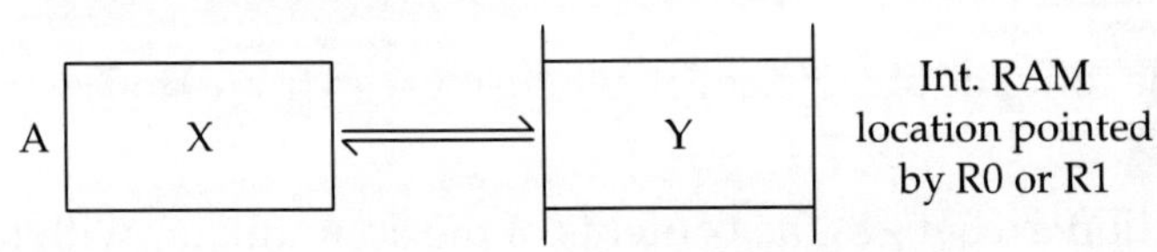

Example 2.14

Initialize R0 at 7FH and assume the contents of memory 7FH are 07H and the accumulator A is having the data byte 12H. Exchange these two contents.

Solution:

Instructions	Comments
MOV R0, #7FH	Loading the data byte 7FH to the register R0.
XCH A, @R0	Exchanging the accumulator and register R0 contents. After the execution, the accumulator content is 07H and register R0 content is 12H.

28. **XCHD A, @Ri**

Bytes: 1

Machine Cycles: 1

Exchange the lower digit/lower nibble of accumulator with lower digit/nibble of memory pointed by R0 or R1.

Operation:

This instruction exchanges the lower digit or lower nibble of the value in the accumulator with lower digit/nibble of the value from memory location pointed by R0 or R1.

Higher digit/nibble of the accumulator and memory location is unchanged.

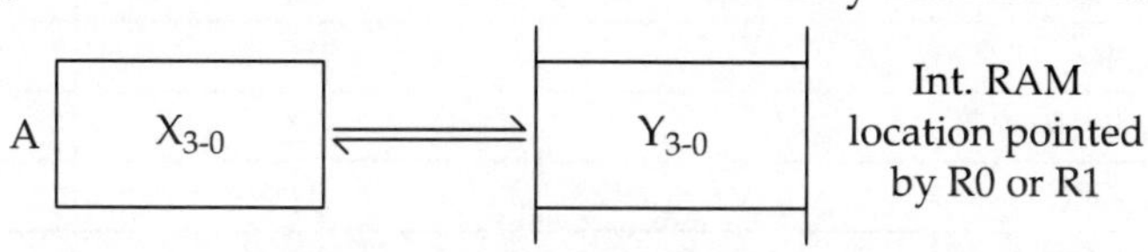

Assume (A) =56H and (35H) is 87H. Then following instructions are executed:

```
MOV R0, #35H
XCHD A, @R0
```

Then after the execution, the contents inside register R0 and the accumulator are (A) = 57H and (35H) is 86H.

Programmer Tips: Here generally no flags are affected. But when the data byte is transferred or copied to the PSW register, then only flags are changed. **For example:** MOV PSW, #23H or MOV 0D0H, #23H

Table 2.1 *Summary of Data Transfer Group*

Sr. No.	Instruction	No. of Bytes	No. of Cycles
01	MOV A, Rn	01	01
02	MOV A, #DATA	02	01
03	MOV A, DIRECT	02	01
04	MOV A, @Ri	01	01
05	MOV Rn, A	01	01

06	MOV Rn, #DATA	02	01
07	MOV Rn, DIRECT	02	02
08	MOV DIRECT, A	02	01
09	MOV DIRECT, Rn	02	02
10	MOV DIRECT, #DATA	03	02
11	MOV DIRECT, DIRECT	03	02
12	MOV DIRECT, @Ri	02	02
13	MOV @Ri, A	01	01
14	MOV @Ri, #DATA	02	01
15	MOV @Ri, DIRECT	02	02
16	MOV DPTR, #16 – BIT DATA	03	02
17	MOVX A, @Ri	01	02
18	MOVX @Ri, A	01	02
19	MOVX A, @DPTR	01	02
20	MOVX @DPTR, A	01	02
21	MOVC A, @A+DPTR	01	02
22	MOVC A, @A + PC	01	02
23	PUSH DIRECT	02	02
24	POP DIRECT	02	02
25	XCH A, Rn	01	01
26	XCH A, DIRECT	02	01
27	XCH A, @Ri	01	01
28	XCHD A, @Ri	01	01

2.4.2 Arithmetic Group

The operations performed in the arithmetic Instruction include addition, addition with carry, subtraction with borrow, BCD addition, increment, decrement, multiplication and division.

Addition Instructions

Format:

ADD A, Source

Where sources can be

Rn

#data

Direct

@Ri

1. **ADD A, Rn**

   Bytes: 1

   Machine Cycles: 1

   Add the accumulator with register Rn. Put the result in the accumulator.

   Operation:

   This instruction adds the 8-bit contents of the accumulator with the 8-bit contents of register Rn (where n varies from 0 to 7) and puts the result in the accumulator.

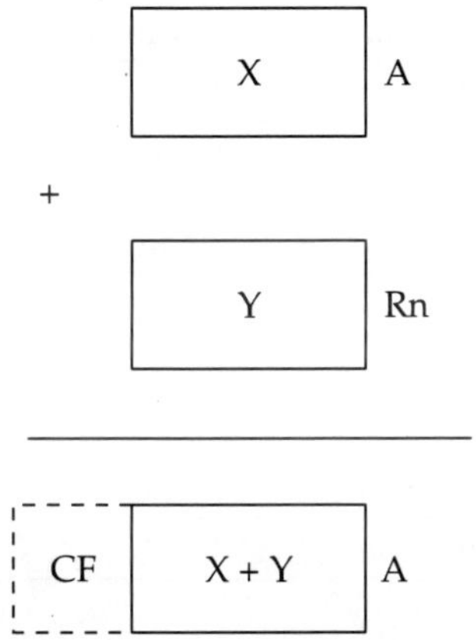

   Example:

   ADD A, R2

Example 2.15

Copy data 35H and 47H in the accumulator and register R2 respectively and then add the 2 contents.

Solution:

Instructions	Comments
MOV A, #35H	Loading the data 35H immediately to the accumulator.
MOV R2, #47H	Loading the data 47H immediately to the register R2
ADD A, R2	Adding the data 35H and 47H
L1: SJMP L1	A short jump to the label L1

Programming Tips: There is no HLT (Halt) instruction in the microcontroller 8051, so L1: SJMP L1 is generally used to write at the end of the program indicating the end of the program.

Program 2.1

Write an assembly language program to add 2 8-bit values 45H and 12H present in the memory locations 30H and 31H respectively. Put the result in the memory location 32H.

Solution:

Method – I

Program	Comments
MOV R1, #30h	Loading immediate data byte 30H to register R1.
MOV A, @R1	Copying the data indirectly to the accumulator pointed by register R1.
MOV R0, #31H	Loading immediate data byte 31H to register R0.
ADD A, @R0	Adding the value of accumulator with the data byte present in the memory location pointed indirectly by register R0.
MOV 32H, A	Copying the accumulator data to memory location 32H.
L1: SJMP L1	A short jump to the label L1.

Method – II

Program	Comments
MOV A, 30H	Copying the data inside the memory location 30H to the accumulator.
MOV R0, 31H	Copying the data inside the memory location 31H to register R0.
ADD A, R0	Adding the accumulator content with register R0.
MOV 32H,A	Storing the result of addition to memory location 32H.
L1: SJMP L1	A short jump to the label L1.

2. **ADD A, #DATA**

   Bytes: 2

   Machine Cycles: 1

   Add 8-bit immediate value with the accumulator. Put the result in the accumulator.

   Operation:

   This instruction adds 8-bit immediate value given along with the instruction with the accumulator and puts the result in the accumulator.

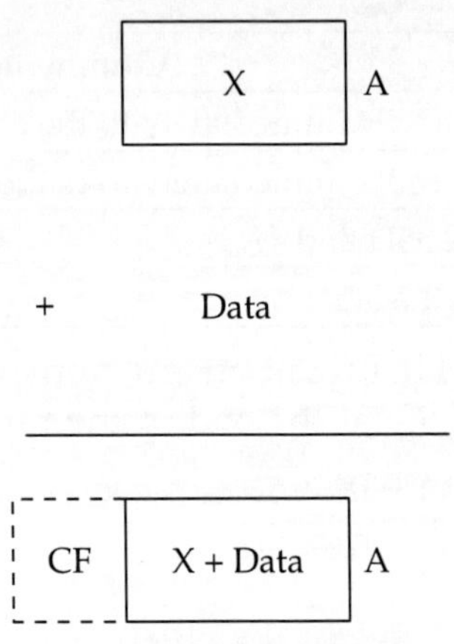

Some of the examples include: (ADD A, #55H), (ADD A, #0F2H), etc.

Program 2.2

Write an assembly language program to add two 8-bit numbers 45H and 12H. Put the result in register R2.

Solution:

Program	Comments
MOV A, #45H	Loading the data 45H immediately into the accumulator.
ADD A, #12H	Adding the data 12H with the data byte stored inside the accumulator. The result is stored in the accumulator.
MOV R2, A	Copy the result into register R2.
L1: SJMP L1	A short jump to the label L1.

3. **ADD A, DIRECT**

Bytes: 2

Machine Cycles: 1

Add direct byte with the accumulator. Put the result in the accumulator.

Operation:

This instruction adds the contents of internal RAM location or SFR with the contents of the accumulator and puts the result in the same accumulator.

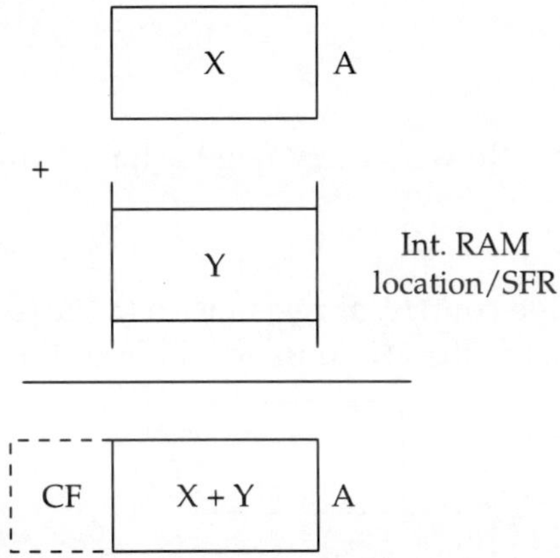

Some of the examples include: (ADD A, 35H), (ADD A, 56H), (ADD A, 0F0H), etc.

4. **ADD A, @Ri**

Bytes: 1

Machine Cycles: 1

Add contents of internal RAM pointed by R0 or R1 with the accumulator. Put the result in the accumulator.

Operation:

This instruction adds the contents of internal RAM location or SFR pointed by R0 or R1 with the contents of the accumulator. Put the result in the same accumulator.

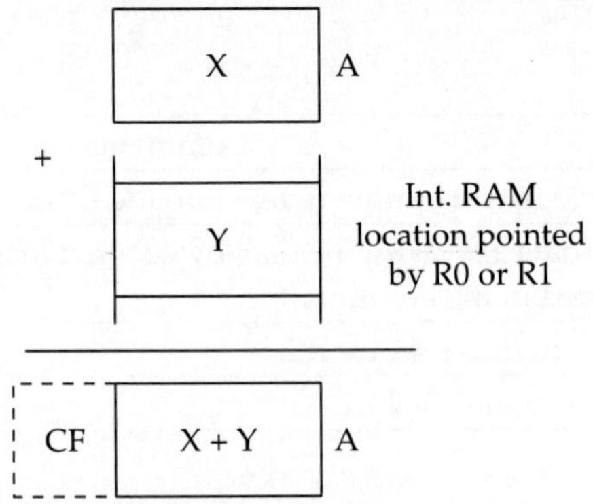

The examples include: (ADD A, @ R0), (ADD A, @R1).

Addition with carry

Format:

ADDC A, Source

Sources can be: Rn, #data, Direct, @Ri

5. **ADDC A, Rn**

Bytes: 1

Machine Cycles: 1

Add contents of register Rn with carry flag to the accumulator. Put the result in the accumulator.

Operation:

This instruction adds the content of register Rn (n varies from 0 to 7) with the carry flag to the 8-bit contents of the accumulator and put the result in the accumulator.

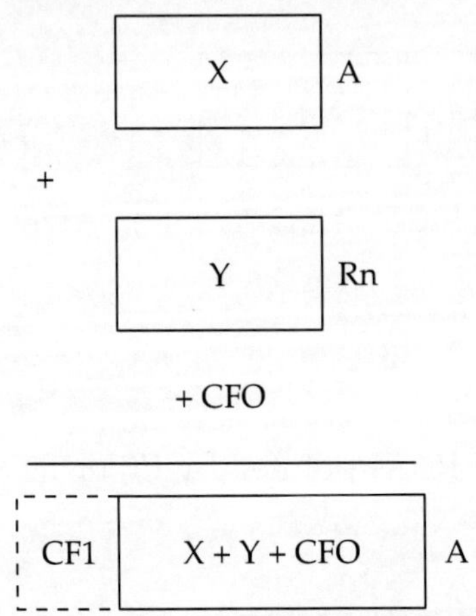

(CF0 indicates previous carry flag bit)

Some of the examples include: (ADDC A, R5), (ADDC A, R7), (ADDC A, R0), etc.

6. **ADDC A, #DATA**

   Bytes: 2

   Machine Cycles: 1

   Add 8-bit immediate value with the carry flag to the accumulator. Put the result in the accumulator.

   Operation:

   This instruction adds the 8-bit immediate data with the carry flag to the 8-bit contents of the accumulator and the result is stored in the accumulator.

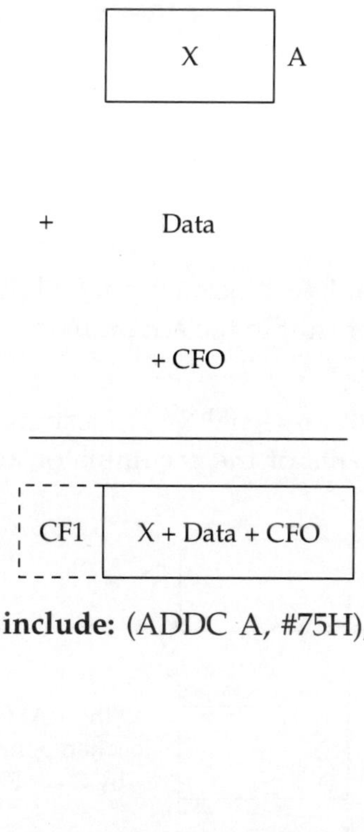

   Some of the examples include: (ADDC A, #75H), (ADDC A, #0F0H), (ADDC A, #0E0H) etc.

7. **ADDC A, DIRECT**

   Bytes: 2

   Machine Cycles: 1

   Add 8 direct byte with carry flag to the accumulator. Put the result in the accumulator.

   Operation:

   This instruction adds the 8-bit internal RAM or SFR with the carry flag to the 8-bit contents of the accumulator and the result is stored in the accumulator.

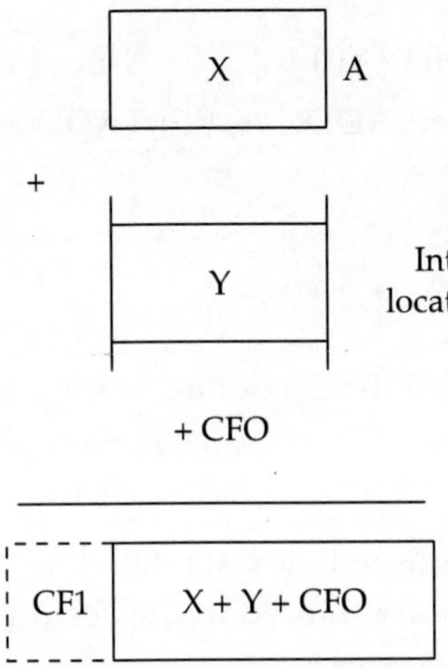

Some of the examples include: (ADDC A, 50H), (ADDC A, 0F0H), (ADDC A, 0E0H), etc.

8. **ADDC A, @Ri**

Bytes: 1

Machine Cycles: 1

Add the contents of internal RAM location pointed by R0 or R1 with the carry flag to the accumulator. Put the result in the accumulator.

Operation:

This instruction adds the 8-bit internal RAM location pointed by R0 or R1 with the carry flag to the 8-bit contents of the accumulator and the result is stored in the accumulator.

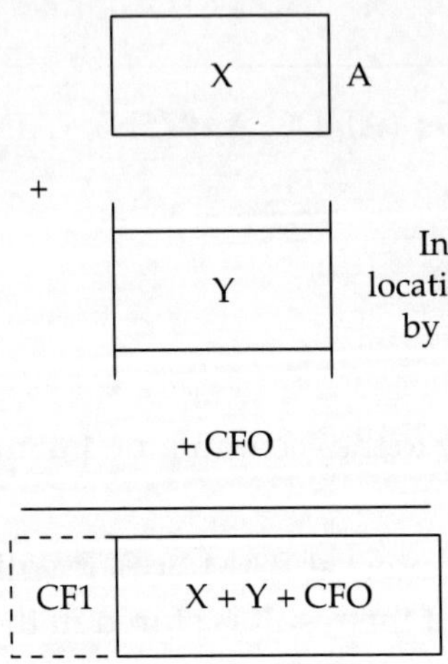

The examples include: (ADDC A, @R0), (ADDC A, @R1).

Program 2.3

Write an assembly language program to add two 16-bit numbers 5678H and 9ABCH respectively. Put the result in register R0, accumulator and register R6. R6 consists of lower byte, accumulator higher byte. Carry is generated to register R0.

Solution:

Program	Comments
CLR C	Clear the carry flag. (Optional)
MOV A, #78H	Loading the data 78H immediately to the accumulator.
ADD A, #0BCH	Adding the accumulator content to the immediate data byte BCH.
MOV R6, A	The stored result is moved to register R6.
MOV A, #56H	Loading the data 56H immediately to the accumulator.
ADDC A, #9AH	Adding the accumulator content to the immediate data byte 9AH with the carry flag.
MOV R2, A	The stored result is moved to register R2.
CLR A	Clearing the accumulator, i.e., memory location 0E0H has the data byte content 00H.
ADDC A, 0E0H	Adding the accumulator content with itself and the carry flag if generated after the addition of the higher nibbles.
MOV R0, A	Copy the stored result into register R0.
MOV A, R2	The higher data byte is retrieved back into the accumulator.
L1: SJMP L1	A short jump to label L1.

Program 2.4

Write an assembly language program to add contents of memory location 40H to 45H. Put the result in accumulator and register R6. Accumulator lower byte and carry are generated to register R6 (the higher byte).

Solution:

Note: *The instruction ADDC A, A is also valid instead of ADDC A, 0E0H.*

Program	Comments
CLR C	Clear the carry flag.
MOV R6, #00H	Loading the data 00H into register R6.
MOV R0, #40H	Loading the data 40H into register R0.
MOV R2, #05H	Loading the counter value 05H into register R2.
MOV A, @R0	Copying the data of memory location pointed by register R0 indirectly into the accumulator.
L2: INC R0	Incrementing the register content by one.
ADD A, @R0	Adding the accumulator content with the memory location pointed by register R0 generated after addition of one.

JNC L1	Jump to label L1 if there is no generation of the carry.
INC R6	Increment register R6 content by one.
L1: DJNZ R2, L2	Decrement by one and jump to the label L2 if the zero flag is not set. If the zero flag is set, move to the next instruction.
J1: SJMP J1	A short jump to the label J1.

Program 2.5

Repeat the previous problem and write an assembly language problem if the higher byte is stored in the accumulator and the lower byte in register R0.

Solution:

Instructions	Comments
CLR C	Clear the carry flag.
MOV R6, #00H	Loading the data 00H into register R6.
MOV R0, #40H	Loading the data 40H into register R0.
MOV R2, #05H	Loading the counter value 05H into register R2.
MOV A, @R0	Copying the data of memory location pointed by register R0 indirectly into the accumulator.
L2: INC R0	Incrementing the register content by one.
ADD A, @R0	Adding the accumulator content with the memory location pointed by register R0 generated after addition of one.
JNC L1	Jump to label L1 if there is no generation of the carry.
INC R6	Increment register R6 content by one.
L1: DJNZ R2, L2	Decrement by one and jump to the label L2 if the zero flag is not set. If the zero flag is set, move to the next instruction.
XCH A, R6	Exchanging the contents of the accumulator with register R6 content.
J1: SJMP J1	A short jump to the label J1.

Conclusion: From the above problem, we can see that the higher byte and lower byte storage is swapped with the help of "XCH" instruction.

Program 2.6

Write an assembly language problem to add two 16-bit numbers present in the memory location 31-30H and in the registers R3-R2 and puts the result in memory location 72-71-70H.

Solution:

Instructions	Comments
CLR C	Clear the carry flag.
MOV A, 30H	Copying the data stored in the memory location 30H into the accumulator.

ADD A, R2	Adding the accumulator content with register R2 data byte content.
MOV 70H, A	Copying the accumulator data byte into memory location 70H after addition (lower byte).
MOV A, 31H	Copying the data stored in memory location 31H into the accumulator
ADDC A, R3	Adding the accumulator content with register R3 data byte content and the carry flag.
MOV 71H, A	Copying the data stored in memory location 71H into the accumulator after addition (higher byte).
CLR A	Clearing the accumulator
ADC A, A	Adding the accumulator to itself with the addition of carry flag.
MOV 72H, A	Copy the accumulator content into the memory location 72H indicating the generation of the carry flag.
J1: SJMP J1	A short jump to the label J1.

9. **DAA**

   Bytes: 1

   Machine Cycles: 1

   Full Form: Decimal Adjust Accumulator

   This instruction adjusts the contents of the accumulator to the binary coded decimal. After addition of two 8-bit numbers, the result is always in the hexadecimal format. To convert the result of the BCD number addition to the BCD equivalent, use instruction DAA.

   Logic Used in DAA:

   1. After addition, if the 4 LSBs of result is greater than 9 (1001) or auxiliary carry flag is set, then 6 (0110) are added to the 4 LSBs of the result.
   2. After addition of 4 MSBs of result is greater than 9 (1001) or carry flag is set, then 6 (0110) is added to the 4 MSBs of the result.

   Example:

   A is 56H and (R2) is 67H. (A+ R2) is BDH and since both higher and lower nibbles are greater than 9, the value 6 is added to both the nibbles indicating the result to be generated as $(127)_{10}$ in BCD after using the instruction DAA on addition of the accumulator and register R2.

Program 2.7

Write an assembly language problem to 5 BCD numbers present in the memory location 40 to 44H. Put the result in accumulator and register R5. The accumulator must store lower byte and register R5 must store higher byte.

Solution:

Instructions	Comments
CLR C	Clear the carry flag.
MOV R0, #40H	Loading the data byte 40H into register R0.
MOV R2, #05H	Loading the data byte 05H into register R2.
CLR A	Clearing the accumulator to the value 00H.
MOV R5, A	Copying the data stored in the accumulator into register R5.
BACK: ADDC A, @R0	Adding the accumulator content indirectly with the data byte stored in the memory location pointed by register R0.
DAA	Converting the result from the hexadecimal byte into the binary coded decimal.
JNC L1	Jump to label L1 if the carry flag is reset.
INC R5	Incrementing register R5 by one.
L1: INC R0	Incrementing register R0 by one.
DJNZ R2, BACK	Decrement by one and jump to the back if the zero flag is not set. If it is set, move to the next instruction.
L2: SJMP L2	A short jump to the label L2.

Program 2.8

Write an assembly language problem to 5 BCD numbers present in the memory location 40 to 44H. Put the result in accumulator and register R5. The accumulator must store higher byte and register R5 must store lower byte.

Solution:

Instructions	Comments
CLR C	Clear the carry flag.
MOV R0, #40H	Loading the data byte 40H into register R0.
MOV R2, #05H	Loading the data byte 05H into register R2.
CLR A	Clearing the accumulator to the value 00H.
MOV R5, A	Copying the data stored in the accumulator into register R5.
BACK: ADDC A, @R0	Adding the accumulator content indirectly with the data byte stored in the memory location pointed by register R0.
DAA	Converting the result from the hexadecimal byte into the binary coded decimal.
JNC L1	Jump to label L1 if the carry flag is reset.

INC R5	Incrementing register R5 by one.
L1: INC R0	Incrementing register R0 by one.
DJNZ R2, BACK	Decrement by one and jump to the back if the zero flag is not set. If it is set, move to the next instruction.
XCH A, R5	Exchanging the contents of register R5 with that of the accumulator.
L2: SJMP L2	A short jump to the label L2.

Subtraction instructions

Format:

SUBB A, Source

Where source can be Rn, #Data, Direct, @Ri.

In a microprocessor, there are two separate instructions for the subtraction. One is without borrow and the second is with borrow whereas in a microcontroller, there is only 1 instruction, i.e., subtraction with borrow.

To make subtract without borrow, out of subtract with borrow before subtraction clear the carry flag (CLR C).

10. **SUBB A, Rn**

Bytes: 1

Machine Cycles: 1

Subtract register Rn with borrow from the accumulator. Put the result in the accumulator.

Operation:

This instruction subtracts the contents of register Rn (where n varies from 0 to 7) with borrow from the accumulator and puts the result in the same accumulator.

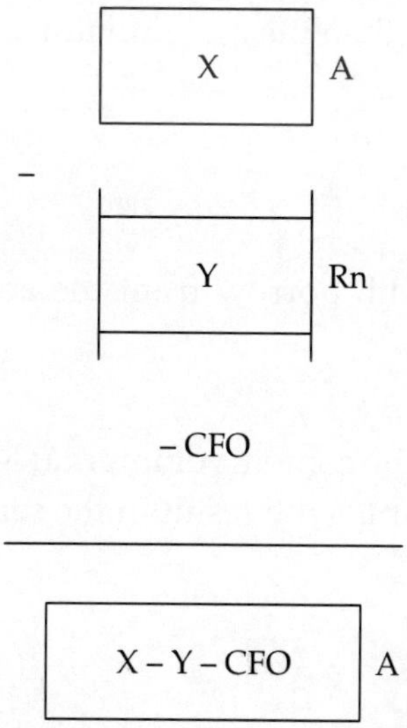

For example: Assume (A) = 56H , (R2) = 05H, carry = 00H and the instruction SUBB A, R2 is executed, then the accumulator will be 51H.

11. **SUBB A, #DATA**

Bytes: 2

Machine Cycles: 1

Subtract 8-bit immediate value with borrow from the accumulator. Put the result in the accumulator.

Operation:

This instruction subtracts the contents of 8-bit data with borrow flag from the accumulator and puts the result in the same accumulator.

X	A

− Data

− CFO

X − Data − CFO	A

For example: Assume (A) = 56H , (8-bit data) = 15H, carry = 01H and the instruction SUBB A, #15H is executed, then the accumulator will be 40H.

12. **SUBB A, DIRECT**

Bytes: 2

Machine Cycles: 1

Subtract the direct byte with borrow from the accumulator. Put the result in the accumulator.

Operation:

This instruction subtracts the contents of internal RAM location or SFR with borrow from the accumulator and puts the result in the same accumulator.

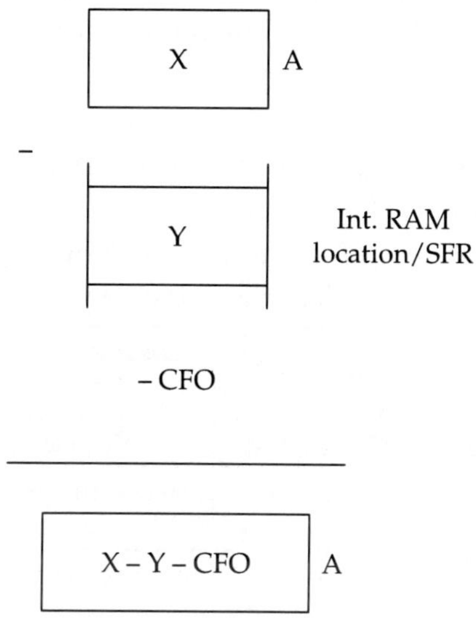

For example: Assume (A) = 56H, (30H) = 45H, carry = 01H and the instruction SUBB A, 30H is executed then the accumulator will be 10H.

13. **SUBB A, @Ri**

Bytes: 1

Machine Cycles: 1

Subtract indirect byte pointed by Ri with borrow flag from the accumulator. Put the result in the accumulator.

Operation:

This instruction subtracts the contents of internal RAM location pointed by R0 or R1 with borrow flag from the accumulator and puts the result in the same accumulator.

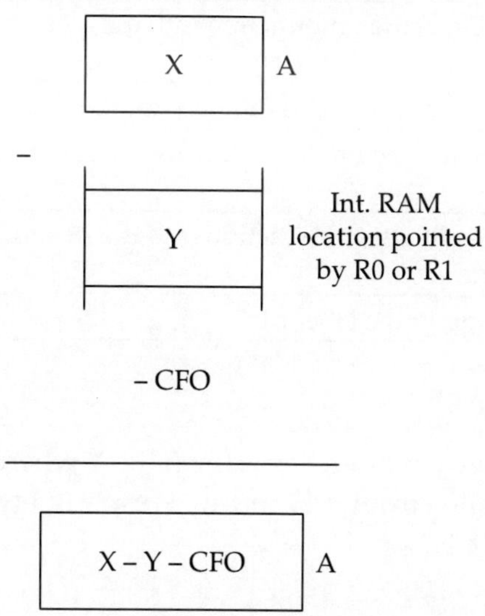

For example: Assume (A) = 56H, (R0) = 05H, (05H) = 36H, carry = 00H and the instruction SUBB A, @R0 is executed then the accumulator will be 20H.

Program 2.9

Write an assembly language program to perform 8-bit subtraction 23H – 45H. Put the result in the memory location 30H.

Solution: The very first instruction to note on performing any subtraction assembly language program is CLR C, i.e., to clear the carry flag.

Instructions	Comments
CLR C	Clear the carry flag.
MOV A, #23H	Loading the data 23H immediately to the accumulator.
SUBB A, #45H	Subtracting the accumulator data byte with the data byte 45H and with borrow.
MOV 30H, A	Store the result after subtraction into the memory location 30H from the accumulator.
L1: SJMP L1	A short jump to the label L1.

Program 2.10

Write an assembly language program to perform 16-bit subtraction 5678H – 1234H. Put the result in the memory location 31H and 30H.

Solution:

Instructions	Comments
CLR C	Clear the carry flag.
MOV A, #78H	Loading the data 23H immediately to the accumulator.
SUBB A, #34H	Subtracting the accumulator data byte with the data byte 45H and with borrow.
MOV 30H, A	Store the result after subtraction into the memory location 30H from the accumulator.
MOV A, #56H	Loading the data 56H immediately to the accumulator.
SUBB A, #12H	Subtracting the accumulator data byte with the data byte 12H and with borrow.
MOV 31H, A	Store the result after subtraction into the memory location 31H from the accumulator.
L1: SJMP L1	A short jump to the label L1.

Program 2.11

Write an assembly language program to perform X – Y where X and Y are 24-bit numbers present in the memory location from 30H and 40H respectively. Put the result in the external memory location from 2000H.

Solution:

Instructions	Comments
CLR C	Clear the carry flag.
MOV DPTR, #2000H	The memory location 2000H is pointed by the DPTR.
MOV A, 30H	Copying the data inside the memory location 30H to the accumulator.
SUBB A, 40H	Subtracting the accumulator data byte with the data byte present in memory location 40H and with borrow.
MOVX @DPTR, A	Store the result after subtraction externally into the memory location pointed by the data pointer, i.e., at 2000H.
INC DPTR	Increment the data pointer memory location by one. The DPTR points to memory location 2001H.
MOV A, 31H	Copying the data inside memory location 31H to the accumulator.
SUBB A, 41H	Subtracting the accumulator data byte with the data byte present in memory location 41H and with borrow.
MOVX @DPTR, A	Store the result after subtraction externally into the memory location pointed by the data pointer, i.e., at 2001H.
INC DPTR	Increment the data pointer memory location by one. The DPTR points to memory location 2002H.
MOV A, 32H	Copy the data byte present in memory location 32H into the accumulator.
SUBB A, 42H	Subtracting the accumulator data byte with the data byte present in memory location 42H and with borrow.
MOVX @DPTR, A	Store the result after subtraction externally into the memory location pointed by the data pointer, i.e., at 2002H.
L1: SJMP L1	A short jump to the label L1.

Program 2.12

Write an assembly language program to perform X –Y where X and Y are 32-bit numbers present in the external memory location from 80H and B0H. Put the result in the external memory location from 4000H.

Solution:

Instructions	Comments
CLR C	Clear the carry flag
MOV DPTR, #4000H	Memory location 4000H is pointed by the DPTR.
MOV R0, #80H	Loading the data 80H to register R0.
MOV R1, #0B0H	Loading the data B0H into register R1.
MOV R2, #04H	Loading the counter value with 04H.
L1: MOVX A, @R1	Copy the data externally from the memory location pointed by register R1 into the accumulator .

MOV R3, A	Copying the data from the accumulator into register R3.
MOVX A, @R0	Copy the data externally from the memory location pointed by register R0 into the accumulator.
SUBB A, R3	Subtracting the content inside the accumulator with the register R3 and the borrow.
MOVX @DPTR, A	Storing the result after subtraction from the accumulator into the memory location pointed by the data pointer 4000H.
INC R0	Incrementing register R0 content by one.
INC R1	Incrementing register R1 content by one.
INC DPTR	Incrementing the data pointer memory location by one.
DJNZ R2, L1	Decrement register R2 content by one and check whether zero flag is set or not. If not, then jump to the label L1. If zero flag is set, i.e., the data byte is 00H then move to the next instruction.
L2: SJMP L2	A short jump to the label L2.

Increment instructions

Format:

INC Source
Sources can be: Accumulator, register Rn, direct, @Ri

14. **INC A**

Bytes: 1
Machine Cycles: 1
Increment the content of the accumulator by one. Put the result in the same accumulator.

Operation:
This instruction increments the contents of the accumulator by one and puts the result in the same accumulator.

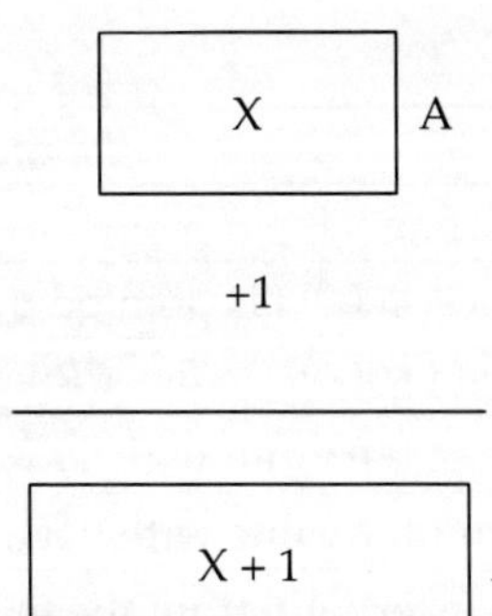

For example: Assume A = 0F0H. INC A gives 0F1H.

15. **INC Rn**

Bytes: 1

Machine Cycles: 1

Increment the content of register by one. Put the result in the same register.

Operation:

This instruction increments the contents of register Rn by one and puts the result in the same register.

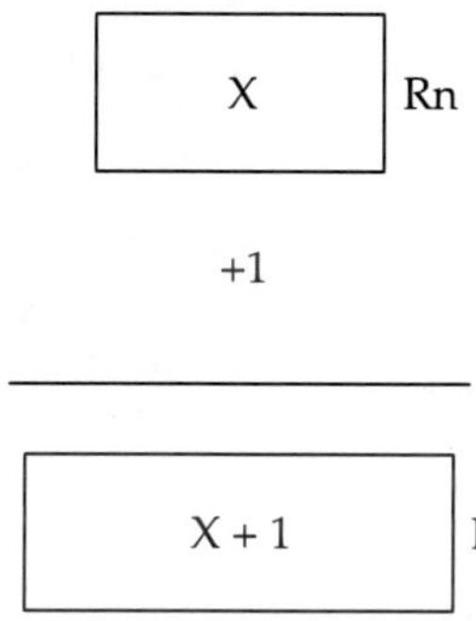

For example: Assume R6 = 010H. INC R6 gives 011H.

16. **INC DIRECT**

Bytes: 2

Machine Cycles: 1

Increment the contents of internal RAM or SFRs by one. Put the result in the same memory.

Operation:

This instruction increments the contents of internal RAM or SFRs by one which is given directly along with the instruction and puts the result in the same memory location.

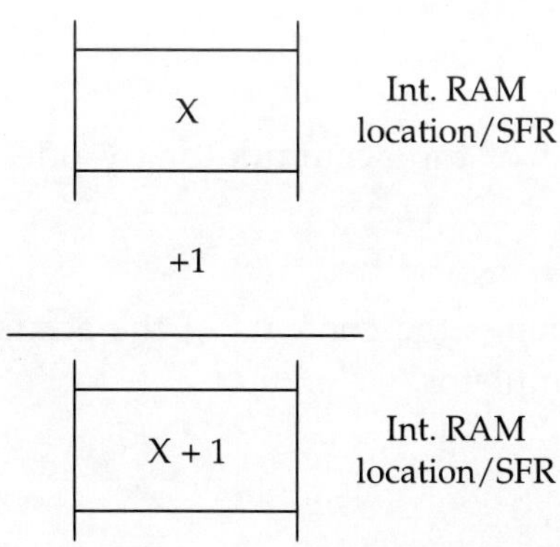

For example: Assume (35H) = 50H. INC 35H gives 51H.

17. **INC @Ri**

Bytes: 1

Machine Cycles: 1

Increment the contents of internal RAM or SFR pointed by R0 or R1 by one. Put the result in the same memory.

Operation:

This instruction increments the contents of internal RAM pointed by R0 or R1 by one which is given directly along with the instruction and puts the result in the same memory location.

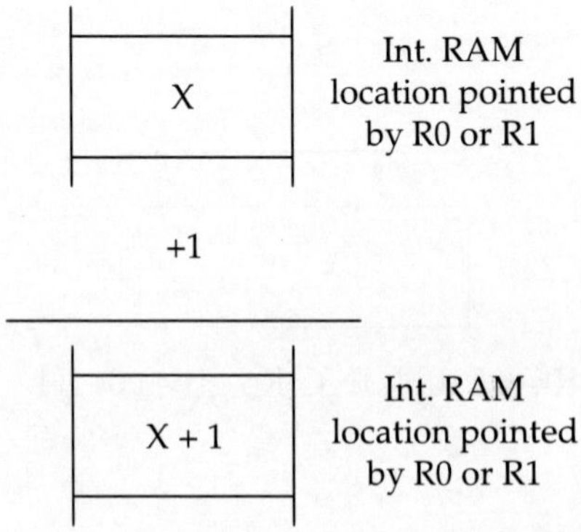

For example: Assume (R0) = 35H, (35H) = 50H. INC @R0 gives 51H.

Decrement instructions

Format:

DEC Source

Sources can be: Accumulator, register Rn, direct, @Ri

18. **DEC A**

Bytes: 1

Machine Cycles: 1

Decrement the content of the accumulator by one. Put the result in the same accumulator.

Operation:

This instruction decrements the contents of the accumulator by one and puts the result in the same accumulator.

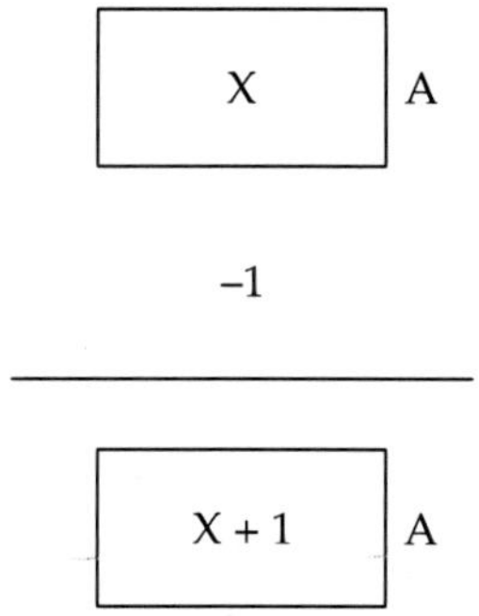

For example: Assume A = 0F0H. DEC A gives 0EFH.

19. **DEC Rn**

    Bytes: 1

    Machine Cycles: 1

    Decrement the content of the register by one. Put the result in the same register.

    Operation:

    This instruction decrements the contents of register Rn by one and puts the result in the same register.

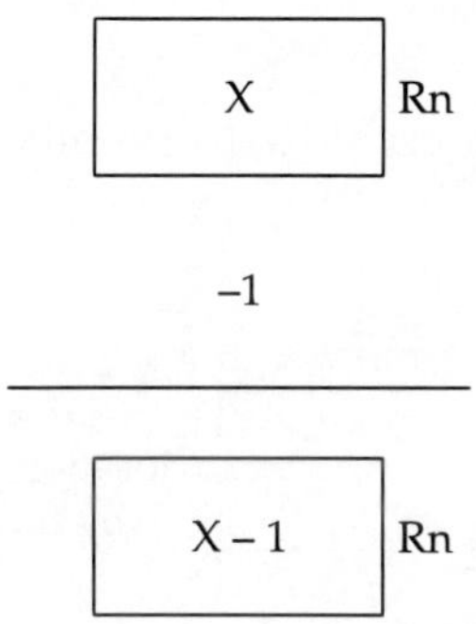

For example: Assume R6 = 010H. DEC R6 gives 0FH.

20. **DEC DIRECT**

    Bytes: 2

    Machine Cycles: 1

    Decrement the contents of internal RAM or SFRs by one. Put the result in the same memory.

Operation:

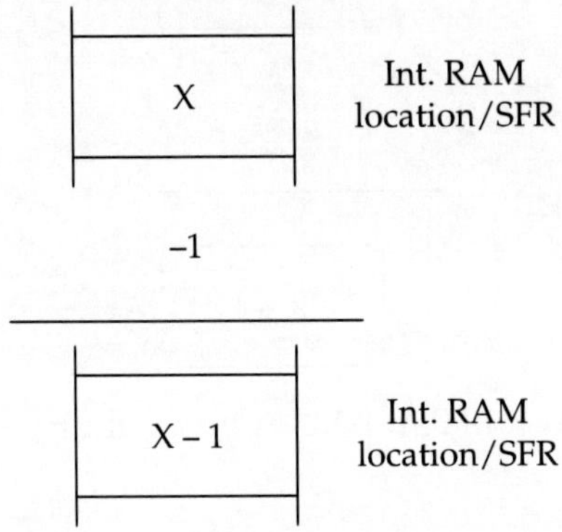

This instruction decrements the contents of internal RAM or SFRs by one which is given directly along with the instruction and puts the result in the same memory location.

For example: Assume (35H) = 50H. DEC 35H gives 4FH.

21. **DEC @Ri**

Bytes: 1

Machine Cycles: 1

Decrement the contents of internal RAM pointed by R0 or R1 by one. Put the result in the same memory.

Operation:

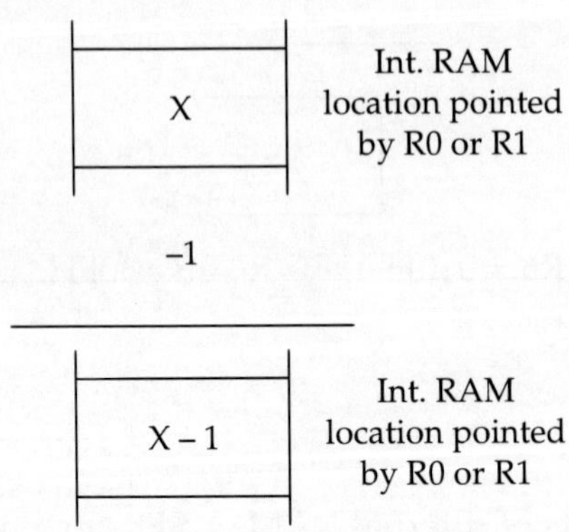

This instruction decrements the contents of internal RAM pointed by R0 or R1 by one which is given directly along with the instruction and puts the result in the same memory location.

For example: Assume (R0) = 35H, (35H) = 50H. DEC @R0 gives 4FH.

22. **INC DPTR**

Bytes: 1

Machine Cycles: 2

Increment the 16-bit DPTR register by one.

Operation:

X_{16}	DPTR

+1

$X_{16} + 1$	DPTR

This instruction increments the contents of 16-bit register DPTR by one and puts the result in the same register.

This is the only 16-bit register available to be incremented and there is a no decrement operation for DPTR.

Example 2.16

Assume DPTR = 10FEH. Give the result in the DPTR after executing

INC DPTR

INC DPTR

INC DPTR

Solution:

Instructions	**Results**
INC DPTR	10FFH
INC DPTR	1100H
INC DPTR	1101H

Example 2.17

Assume DPL = 5DH and DPH = 5CH. Give the result in the DPTR after executing

INC DPTR

INC DPTR

INC DPTR

Solution:

Instructions	Results
INC DPTR	5C5EH
INC DPTR	5C5FH
INC DPTR	5C60H

Example 2.18

Assume that DPL = FEH and DPH = 13H. Give the result in the DPTR after executing

INC 82H
INC 82H
INC 83H
INC DPTR

Solution:

Instructions	Results
INC 82H	13FFH
INC 82H	1300H
INC 83H	1400H
INC DPTR	1401H

23. **MUL AB**

Bytes: 1
Machine Cycles: 4
Multiply the contents of the accumulator with register B.
Operation:

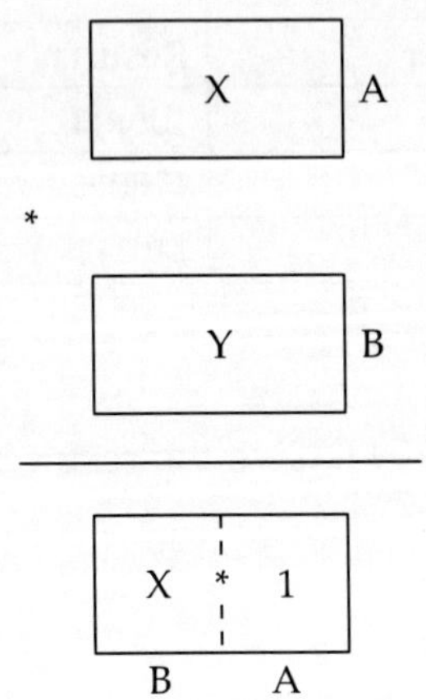

This instruction multiplies 8-bit unsigned number of the accumulator with register B. The product is 16-bit. Higher byte of the product is left in register B and lower byte of the product is left in the accumulator.

If the result is 8-bit, i.e., up to FFH, then overflow flag is '0', otherwise OV = 1.

Carry flag is always cleared.

No other flags are affected.

For example: Assume the accumulator content = 80 and register B = 20. After this instruction MUL AB is executed then the result is 1600 in decimal or 0640H in hexadecimal.

The higher 8-byte is stored in register B, i.e., 06H and the lower byte is stored in the accumulator, i.e., 40H.

Since getting a value in register B, hence the overflow flag is set and carry flag is as usual reset.

Program 2.13

Write an assembly language program to multiply 45H and 67H. Put the result in the memory 70H and 71H.

Solution:

Instructions	Comments
MOV A, #45H	Loading the data 45H immediately into the accumulator.
MOV B, #67H	Loading the data 67H immediately into register B.
MUL AB	Multiplying the accumulator content with register B content.
MOV 71H, B	Copying the higher data byte into register B.
MOV 70H, A	Copying the lower data byte into the accumulator.
L1: SJMP L1	A short jump to the label L1.

Program 2.14

Write an assembly language program to multiply two 8-bit numbers residing at memory locations 30H and 31H. Put the result in register R7-R6 of register bank 1.

Solution:

Instructions	Comments
SETB PSW.3	Selecting the register bank 1.
MOV A, 30H	Copying the data byte present in memory location 30H into the accumulator.
MOV B, 31H	Copying the data byte present in the memory location 31H into register B.

MUL AB	Multiplying the accumulator content with the register B content.
MOV R6, 0F0H	Copying the higher data byte into register B.
MOV R7, 0E0H	Copying the lower data byte into the accumulator.
L1: SJMP L1	A short jump to the label L1.

24. **DIV AB**

Bytes: 1

Machine Cycles: 4

Dividing the contents of the accumulator with register B. The quotient will remain in the accumulator and the remainder in register B.

Operation:

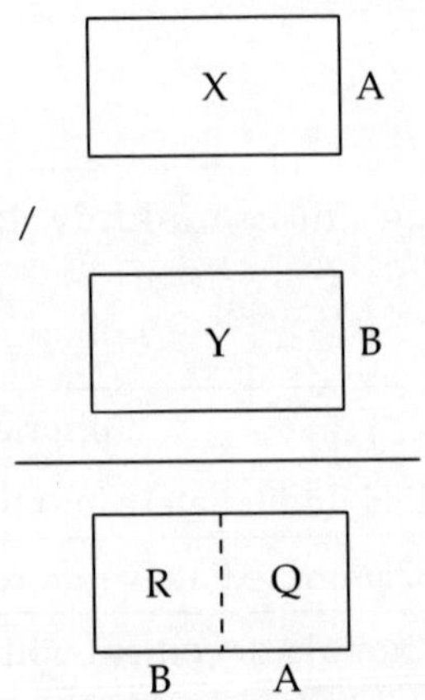

This instruction divides 8-bit unsigned number in the accumulator by 8-bit unsigned number in register B.

If register B is 00H then A/B is infinity then overflow flag is set otherwise overflow flag is reset.

Carry flag is always cleared.

No other flags are affected.

For example: Assume the accumulator content = 85H and register B = 17H. After this instruction DIV AB is executed then the quotient will be 05H which will be stored in the accumulator and the remainder will be 00H which will be stored in register B.

Program 2.15

Write an assembly language program to divide data byte 97H by 23H. Put the result in the memory locations 40H and 41H. The memory location 40H should contain the quotient part and 41H to be that of the remainder part.

Solution:

Instructions	Comments
MOV A, #98H	Loading the data 98H immediately into the accumulator.
MOV 0F0H, #23H	Loading the data 23H immediately into register B.
DIV AB	Dividing the accumulator content with register B content.
MOV 30H, B	Copying the remainder into register B.
MOV 31H, A	Copying the quotient into the accumulator.
L1: SJMP L1	A short jump to the label L1.

Program 2.16

Write an assembly language program to perform X/Y where X and Y are 8- bit numbers present in the external memory locations 80H and 81H. Put the result, i.e., the quotient part at the external memory 82H and the remainder part in the 83H.

Solution:

Instructions	Comments
MOV R0, #81H	Loading the data 81H immediately into register R0.
MOVX A, @R0	Copying the data byte present in memory location 81 H pointed by register R0 into the accumulator externally.
MOV B, A	Copying the accumulator content into register B.
DEC R0	Decrement register R0 content by one. So the new value of register R0 is 80H.
MOVX A, @R0	Copying the data byte present in memory location 80H pointed by register R0 into the accumulator externally.
DIV A,B	Dividing the accumulator content by register B content.
INC R0	Increment register R0 content by one. So the new value of register R0 is 81H.
INC R0	Increment register R0 content by one. So the new value of the register R0 is 82H.
MOVX @R0, A	Copying the accumulator content into the data byte present in the memory location 82H pointed by register R0 externally.
INC R0	Increment the register R0 content by one. So the new value of the register R0 is 83H.
MOV A,B	Copying the register B content into the accumulator.
MOVX @R0, A	Copying the accumulator content into the data byte present in memory location 83H pointed by register R0 externally.
L1: SJMP L1	A short jump to the label L1

Note: In microcontroller 8051, only two instructions have 4 cycles, MUL AB and DIV AB, all other instructions either had one machine cycle or two machine cycles.

Table 2.2: *Summary of Arithmetic Instructions*

S. No.	Instruction	No. of Bytes	No. of Cycles	Flags affected
01	ADD A, Rn	01	01	CY, AC, P and OV
02	ADD A, #DATA	02	01	CY, AC, P and OV
03	ADD A, DIRECT	02	01	CY, AC, P and OV
04	ADD A, @Ri	01	01	CY, AC, P and OV
05	ADDC A, Rn	01	01	CY, AC, P and OV
06	ADDC A, #DATA	02	01	CY, AC, P and OV
07	ADDC A, DIRECT	02	01	CY, AC, P and OV
08	ADDC A, @Ri	01	01	CY, AC, P and OV
09	DA A	01	01	Only CY is affected
10	SUBB A, Rn	01	01	CY, AC, P and OV
11	SUBB A, #DATA	02	01	CY, AC, P and OV
12	SUBB A, DIRECT	02	01	CY, AC, P and OV
13	SUBB A, @Ri	01	01	CY, AC, P and OV
14	INC A	01	01	No flags are affected.
15	INC Rn	02	01	No flags are affected.
16	INC DIRECT	02	01	No flags are affected.
17	INC @Ri	03	01	No flags are affected.
18	DEC A	01	01	No flags are affected.
19	DEC Rn	01	01	No flags are affected.
20	DEC DIRECT	01	01	No flags are affected.
21	DEC @Ri	01	01	No flags are affected.
22	INC DPTR	01	02	No flags are affected.
23	MUL AB	01	04	Over flow flag is affected and the carry flag is reset.
24	DIV AB	01	04	Over flow flag is affected and the carry flag is reset.

2.4.3 Logical Group

The different logical operation preferred by microcontroller 8051 includes bitwise ANDing, ORing, EXoring, Clearing Accumulator, Complementing Accumulator, and Rotation of Accumulator towards Left/Right With/Without Carry.

Format:

ANL A, Source

Where source can be: Rn, #Data, Direct, @Ri.

1. **ANL A, Rn**

   Bytes: 1

   Machine Cycles: 1

   ANDing the contents of register Rn with the accumulator. Put the result in the accumulator.

   Operation:

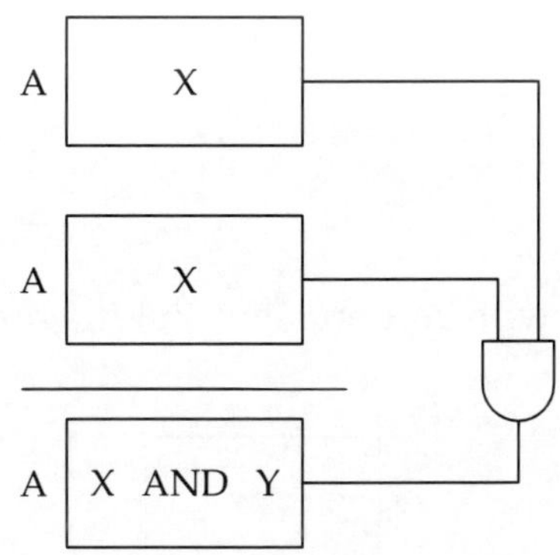

   This instruction performs bitwise ANDing the contents of register Rn (n varies from 0 to 7) with the accumulator and puts the result in the same accumulator, i.e., destination register.

   For example: Assume (R0) = 35H, ACC = 50H. ANL A, R0 gives 10H.

2. **ANL A, #DATA**

   Bytes: 2

   Machine Cycles: 1

   ANDing the 8-bit immediate number with the accumulator. Put the result in the accumulator.

Operation:

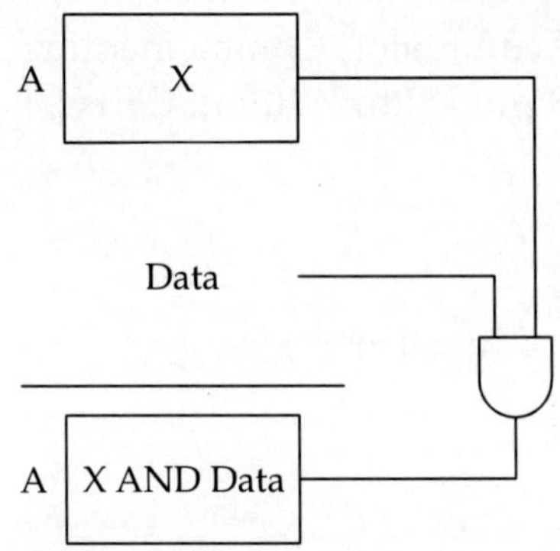

This instruction performs bitwise ANDing of 8-bit immediate number given along with the instruction directly with the accumulator and puts the result in the same accumulator, i.e., in the destination register.

For example: Suppose the accumulator content is 9AH and the instruction ANL A, #56H, then the accumulator content is 12H.

3. **ANL A, DIRECT**

   Bytes: 2

   Machine Cycles: 1

   ANDing the direct byte with accumulator. Put the result in the accumulator.

   Operation:

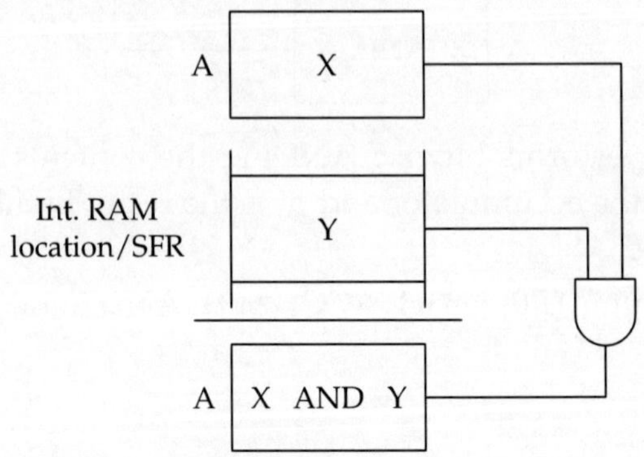

This instruction performs bitwise ANDing of the contents of internal RAM location or SFRs given along with the instruction with the accumulator contents and puts the result in the same accumulator, i.e., in the destination register.

Example: Suppose the accumulator content is 9AH and the instruction ANL A, 56H is executed and the memory location 56H contains the data byte 07H, then the accumulator content is 06H.

4. **ANL A, @Ri**

   Bytes: 1

   Machine Cycles: 1

   ANDing the indirect byte with the accumulator. Put the result in the accumulator.

   Operation:

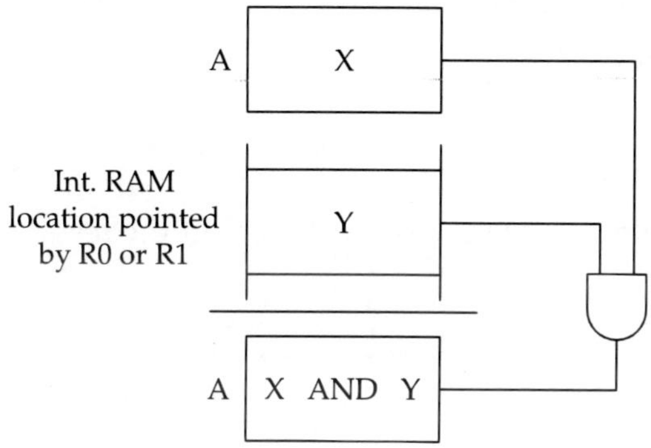

This instruction performs bitwise ANDing the contents of internal RAM location pointed by R0 or R1 with the accumulator contents and puts the result in the same the accumulator, i.e., in the destination register.

Example: Suppose the accumulator content is C5H and the instruction ANL A, @R0 is executed and the memory location pointed by R0 contains the data byte 75H, then the accumulator content is 45H.

5. **ANL DIRECT, A**

   Bytes: 2

   Machine Cycles: 1

   ANDing the accumulator to the direct byte. Put the result in the direct byte.

   Operation:

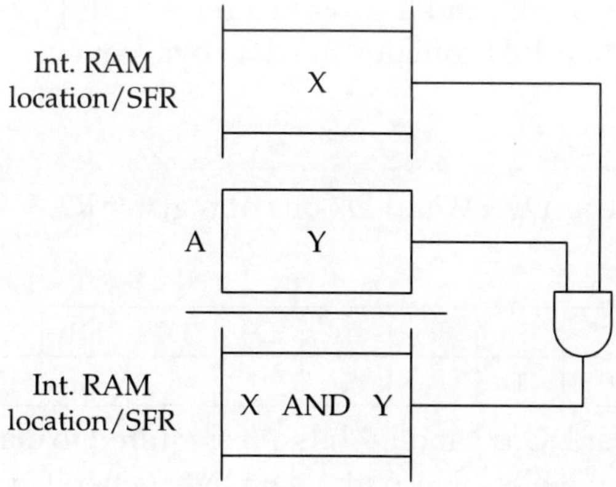

This instruction performs bitwise ANDing the contents of internal RAM location or special function register with the accumulator contents and puts the result in the same internal RAM location or the special function register.

Example: Suppose the accumulator content is 42H and the memory location 10H contains the data byte 56H and the instruction ANL 10H, A is executed then the internal RAM location 10H contains the data byte content 42H.

6. **ANL DIRECT, #DATA**

Bytes: 3

Machine Cycles: 2

ANDing immediate value to the direct byte. Put the result in the direct byte.

Operation:

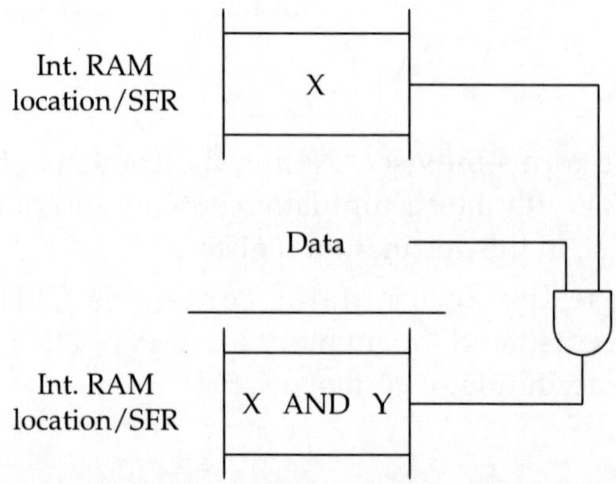

This instruction performs bitwise ANDing the contents of internal RAM location or SFR with 8-bit immediate value given along with the instruction and puts the result in the same internal RAM location or SFR, i.e., the destination register.

Example: Suppose the 8-bit immediate value is 42H and the memory location 10H contains the data byte 56H and the instruction ANL 10H, #42H is executed then the internal RAM location 10H contains the data byte content 42H.

Example 2.19

Give an instruction to clear D0, D3 and D7 bits of register R2.

Solution:

D7	D6	D5	D4	D3	D2	D1	D0
0	1	1	1	0	1	1	0

Since we need to clear DO,D3 and D7 bits, it is required to mask, i.e., disable these bits and unmask the rest of the bits. Hence, the data byte generated is 76H. So, first the data byte in R2 will move into the accumulator and then it will be AND operated with 76H. Hence, the required data bits of the above data byte will definitely be masked in the R2 on transferring.

Instructions	Comment
MOV A, R2	Copy the data byte of R2 in the accumulator.
ANL A, #76H	AND operate the accumulator content with the data byte 76H. Put the result in the accumulator.
MOV R2, A	Copy the accumulator data byte into R2.
L1: SJMP L1	A short jump to the label L1.

Example 2.20

Give an instruction to mask D0, D1 and D4 bits of the port P0.

Solution:

D7	D6	D5	D4	D3	D2	D1	D0
1	1	1	0	1	1	0	0

Since we need to clear DO, D1 and D4 bits, it is required to mask, i.e., disable these bits and unmask the rest of the bits. Hence, the data byte generated is ECH. So, first the data byte in the port P0 will move into the accumulator and then it will be AND operated with ECH. Hence, the required data bits of the above data byte will definitely be masked in the port P0 on transferring.

Instructions	Comment
MOV A, 80H	Copy the data byte of port P0 in the accumulator.
ANL A, #0ECH	AND operate the accumulator content with the data byte ECH. Put the result in the accumulator.
MOV 80H, A	Copy the accumulator data byte into the port P0.
L1: SJMP L1	A short jump to the label L1.

Example 2.21

Give an instruction to reset D6, D4 and D0 bit of the PSW.

Solution:

D7	D6	D5	D4	D3	D2	D1	D0
1	0	1	0	1	1	1	0

Since we need to clear D0, D4 and D6 bits, it is required to mask, i.e., disable these bits and unmask the rest of the bits. Hence, the data byte generated is AEH. So first the data byte in the PSW will move into the accumulator and then it will be AND operated with AEH. Hence, the required data bits of the above data byte will definitely be masked in the PSW on transferring.

Instructions	Comment
MOV A, 0D0H	Copy the data byte of PSW in the accumulator.
ANL A, #0AEH	AND operate the accumulator content with the data byte AEH. Put the result in the accumulator.
MOV 0D0H, A	Copy the accumulator data byte back into the PSW.
L1: SJMP L1	A short jump to the label L1.

Example 2.22

Give an instruction to mask D3 and D5 bits of register B and SFR SCON.

Solution: SCON is a bit addressable register and the SFR address of the SCON register is 98H. Hence, each bit of the SCON register can be masked or unmasked by using bit addressable instructions such as SETB SCON.3, i.e., D3 bit of the SCON register is set.

D7	D6	D5	D4	D3	D2	D1	D0
1	1	0	1	0	1	1	1

Since we need to clear D3 and D5 bits, it is required to mask, i.e., disable these bits and unmask the rest of the bits. Hence, the data byte generated is D7H. So first the data byte in R5 will move into the accumulator and then it will be AND operated with D7H. Similarly, the data byte in the SCON register will move into the accumulator and then it will be AND operated with the same byte. Hence, the required data bits of the above data byte will definitely be masked in register B and the SFR SCON on transferring.

Instructions	Comment
MOV A, 0F0H	Copy the data byte of register B in the accumulator.
ANL A, #0D7H	AND operate the accumulator content with the data byte D7H. Put the result in the accumulator.
MOV 0F0H, A	Copy the accumulator data byte into register B.
MOV A, 98H	Copy the data byte of SCON register in the accumulator.
ANL A, #0D7H	AND operate the accumulator content with the data byte D7H. Put the result in the accumulator.
MOV 98H, A	Copy the accumulator data byte into the SCON register.
L1: SJMP L1	A short jump to the label L1.

ORing instructions

Format:

ORL A, Source

Sources can be: Register Rn, #Data, Direct and @Ri

7. **ORL A, Rn**

Bytes: 1

Machine Cycles: 1

ORing the contents of register Rn with the accumulator. Put the result in the accumulator.

Operation:

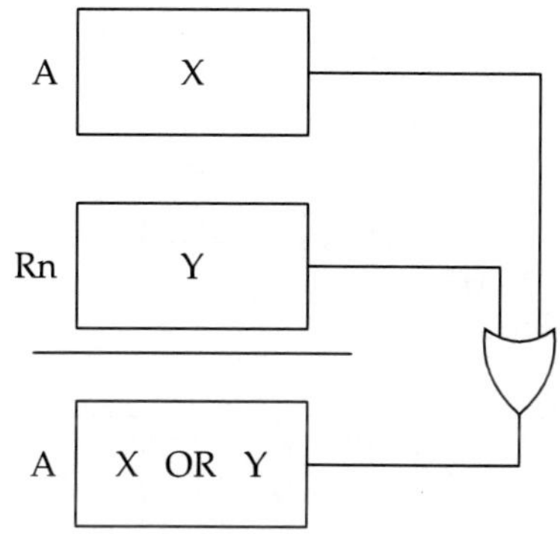

This instruction performs bitwise ORing the contents of register Rn (n varies from 0 to 7) with the accumulator and puts the result in the same accumulator, i.e., destination register.

Example: Assume (R0) = 35H, ACC = 50H. ORL A, R0 gives 75H.

8. **ORL A, #DATA**

Bytes: 2

Machine Cycles: 1

ORing the contents of 8-bit immediate data with the accumulator. Put the result in the accumulator.

Operation:

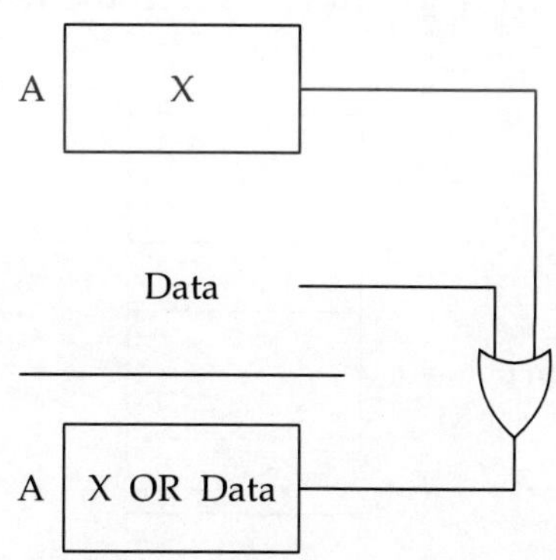

This instruction performs bitwise ORing of 8-bit immediate number given along with the instruction directly with the accumulator and put the result in the same accumulator, i.e., the destination register.

Example: Assume (8-bit immediate data) = 99H, ACC = 50H. ORL A, #99H gives D9H.

9. **ORL A, DIRECT**

Bytes: 2

Machine Cycles: 1

ORing the direct byte with accumulator. Put the result in the accumulator.

Operation:

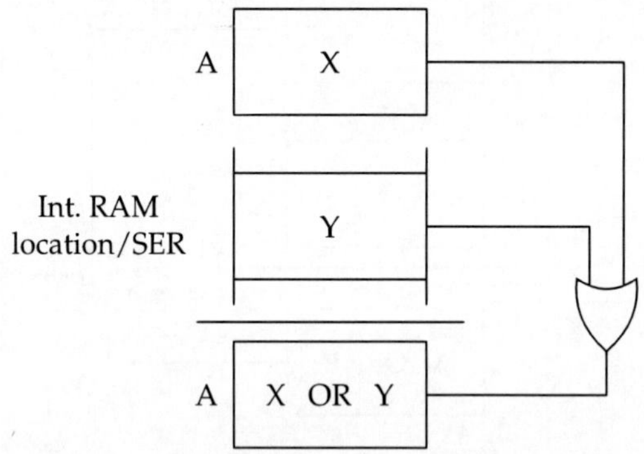

This instruction performs bitwise ORing of the contents of internal RAM location or SFRs given along with the instruction with the accumulator contents and puts the result in the same accumulator, i.e., the destination register.

Example: Suppose the accumulator content is 9AH and the instruction ORL A, 56H is executed and the memory location 56H contains the data byte 07H, then the accumulator content is 9FH.

10. **ORL A, @Ri**

Bytes: 1

Machine Cycles: 1

ORing the indirect byte with accumulator. Put the result in the accumulator.

Operation:

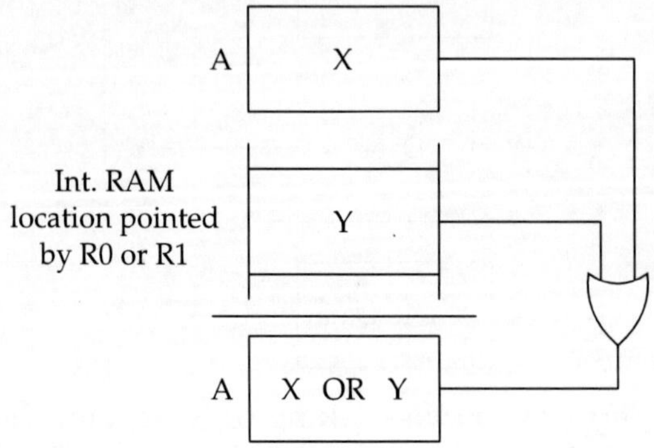

This instruction performs bitwise ORing of the contents of internal RAM location pointed by R0 or R1 with the accumulator contents and puts the result in the same accumulator, i.e., the destination register.

Example: Suppose the accumulator content is C5H and the instruction ORL A, @R0 is executed and the memory location pointed by R0 contains the data byte 75H, then the accumulator content is F5H.

11. **ORL DIRECT, A**

Bytes: 2

Machine Cycles: 1

ORing the accumulator to the direct byte. Put the result in the direct byte.

Operation:

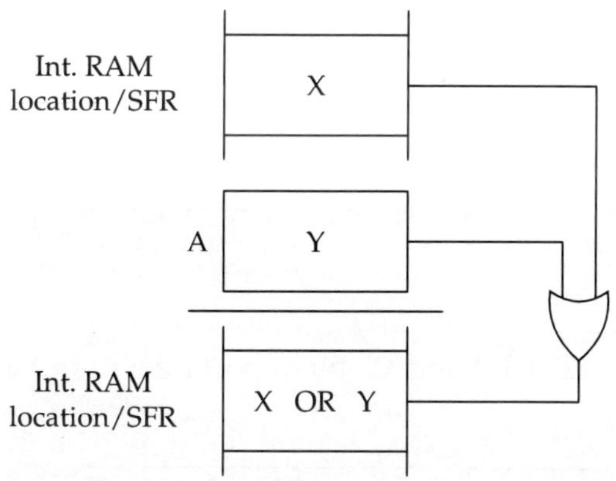

This instruction performs bitwise ORing the contents of internal RAM location or SFR with the accumulator contents and puts the result in the same internal RAM location or SFR.

Example: Suppose the accumulator content is 42H and the memory location 10H contains the data byte 56H and the instruction ORL 10H, A is executed then the internal RAM location 10H contains the data byte content 56H.

12. **ORL DIRECT, #DATA**

Bytes: 3

Machine Cycles: 2

ORing immediate value to the direct byte. Put the result in the direct byte.

Operation:

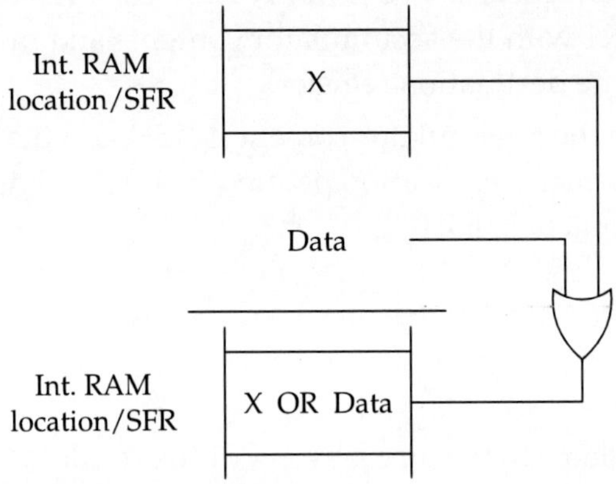

This instruction performs bitwise ORing of the contents of internal RAM location or SFR with 8-bit immediate value given along with the instruction and puts the result in the same internal RAM location or SFR, i.e., the destination register.

Example: Suppose the 8-bit immediate value is 42H and the memory location 10H contains the data byte 5CH and the instruction ORL 10H, #42H is executed then the internal RAM location 10H contains the data byte content 5EH.

Example 2.23

Give an instruction to set D0, D2 and D5 bit of port P0 without affecting other bits.

Solution:

D7	D6	D5	D4	D3	D2	D1	D0
0	0	1	0	0	1	0	1

Since we need to set D0, D2 and D5 bits, it is required to unmask, i.e., enable these bits and mask the rest of the bits. Hence, the data byte generated is 25H. So first the data byte in the Port P0 will move into the accumulator and then it will be OR operated with 25H. Hence, the required data bits of the above data byte will definitely be set in the port P0 on transferring.

Instructions	Comment
MOV A, 80H	Copy the data byte of port P0 in the accumulator.
ANL A, #25H	OR operate the accumulator content with the data byte 25H. Put the result in the accumulator.
MOV 80H, A	Copy the accumulator data byte back into the port P0.
L1: SJMP L1	A short jump to the label L1.

Example 2.24

Give an instruction to set 1st, 3rd and 5th bit of register B without affecting other bits.

Solution:

D7	D6	D5	D4	D3	D2	D1	D0
0	0	0	1	0	1	0	1

Since we need to set 1st , 3rd and 5th bits, i.e., D0, D2 and D4 bits, it is required to unmask, i.e., enable these bits and mask the rest of the bits. Hence, the data byte generated is 15H. So first the data byte in register B will move into the accumulator and then it will be OR operated with 15H. Hence, the required data bits of the above data byte will definitely be set in register B on transferring.

Instructions	Comment
MOV A, 0F0H	Copy the data byte of register B in the accumulator.
ANL A, #15H	OR operate the accumulator content with the data byte 15H. Put the result in the accumulator.
MOV 0F0H, A	Copy the accumulator data byte back into register B.
L1: SJMP L1	A short jump to the label L1.

Example 2.25

Give an instruction to unmask D1, D6 and D7 bit of memory location 10H without affecting operation.

Solution:

D7	D6	D5	D4	D3	D2	D1	D0
1	1	0	0	0	0	1	0

Since we need to set D0, D2 and D4 bits, it is required to unmask, i.e., enable these bits and mask the rest of the bits. Hence, the data byte generated is C2H. So first the data byte in the memory location 10H will move into the accumulator and then it will be OR operated with C2H. Hence, the required data bits of the above data byte will definitely be set in the memory location 10H on transferring.

Instructions	Comment
MOV A, 0F0H	Copy the data byte of register B in the accumulator.
ANL A, #15H	OR operate the accumulator content with the data byte 15H. Put the result in the accumulator.
MOV 0F0H, A	Copy the accumulator data byte back into register B.
L1: SJMP L1	A short jump to the label L1.

EX-ORing instructions

Format:

XRL A, Source

Sources can be: Register Rn, #Data, Direct and @Ri

13. **XRL A, Rn**

Bytes: 1

Machine Cycles: 1

XORing the contents of register Rn with the accumulator. Put the result in the Accumulator.

Operation:

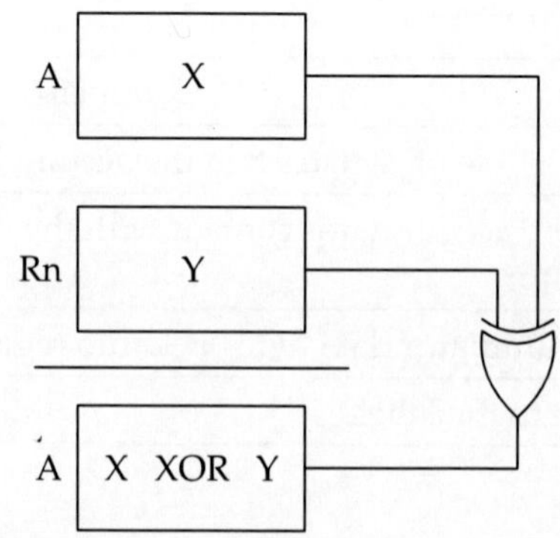

This instruction performs bitwise XORing of the contents of register Rn (n varies from 0 to 7) with the accumulator and puts the result in the same accumulator, i.e., destination register.

For example: Assume (R0) = 35H, ACC = 50H. XRL A, R0 gives 65H.

14. **XRL A, #DATA**

Bytes: 2

Machine Cycles: 1

Ex-ORing the contents of 8-bit immediate data with the accumulator. Put the result in the accumulator.

Operation:

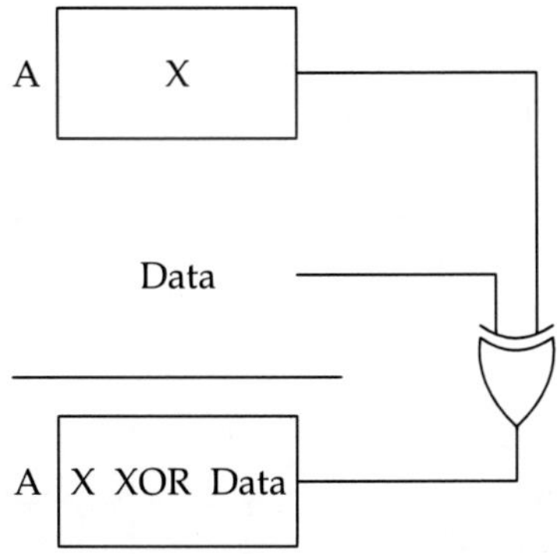

This instruction performs bitwise Ex-ORing of 8-bit immediate number given along with the instruction directly with the accumulator and puts the result in the same accumulator, i.e., the destination register.

For example: Assume (8-bit immediate data) = 99H, ACC = 50H. XRL A, #99H gives C9H.

15. **XRL A, DIRECT**

Bytes: 2

Machine Cycles: 1

Ex-ORing the direct byte with the accumulator. Put the result in the accumulator.

Operation:

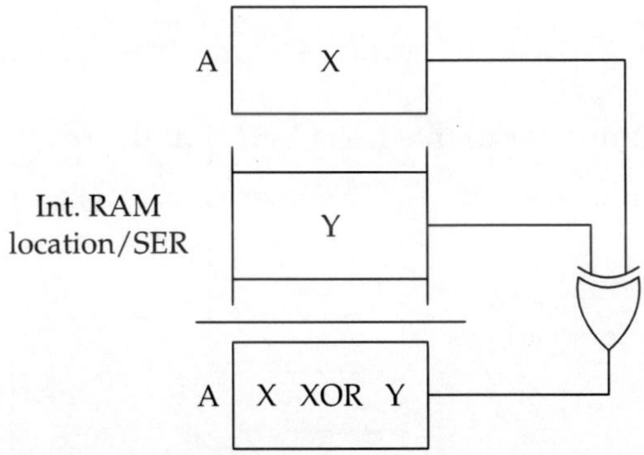

This instruction performs bitwise Ex-ORing of the contents of internal RAM location or SFRs given along with the instruction with the accumulator contents and puts the result in the same accumulator, i.e., destination register.

For example: Suppose the accumulator content is 9AH and the instruction XRL A, 56H is executed and the memory location 56H contains the data byte 07H, then the accumulator content is 9DH.

16. **XRL A, @Ri**

Bytes: 1

Machine Cycles: 1

Ex-ORing the indirect byte with the accumulator. Put the result in the accumulator.

Operation:

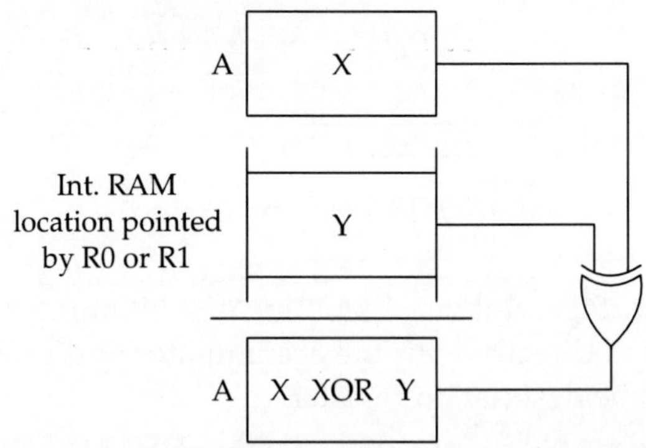

This instruction performs bitwise Ex-ORing of the contents of internal RAM location pointed by R0 or R1 with the accumulator contents and puts the result in the same accumulator, i.e., destination register.

Example: Suppose the accumulator content is C5H and the instruction XRL A, @R0 is executed and the memory location pointed by R0 contains the data byte 75H, then the accumulator content is B0H.

17. **XRL DIRECT, A**

Bytes: 2

Machine Cycles: 1

Ex-ORing the accumulator to the direct byte. Put the result in the direct byte.

Operation:

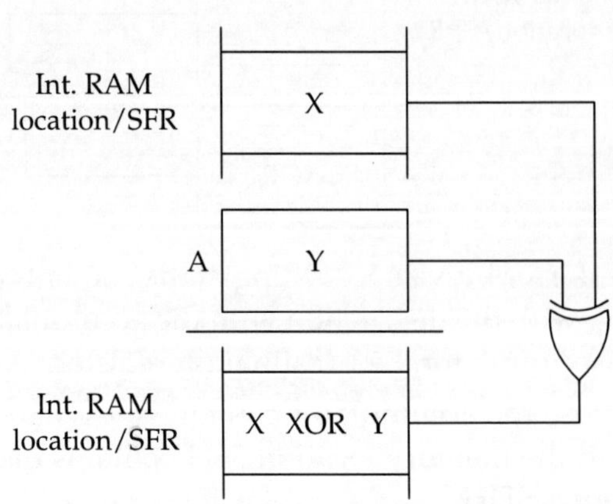

This instruction performs bitwise Ex-ORing of the contents of internal RAM location or SFR with the accumulator contents and puts the result in the same internal RAM location or SFR.

Example: Suppose the accumulator content is 42H and the memory location 10H contains the data byte 56H and the instruction XRL 10H, A is executed then the internal RAM location 10H contains the data byte content 14H.

18. **XRL DIRECT, #DATA**

Bytes: 3

Machine Cycles: 2

Ex-ORing immediate value of the direct byte. Put the result in the direct byte.

Operation:

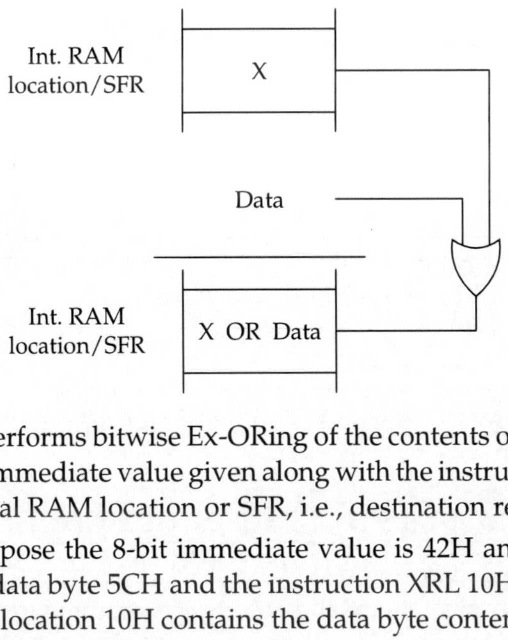

This instruction performs bitwise Ex-ORing of the contents of internal RAM location or SFR with 8-bit immediate value given along with the instruction and puts the result in the same internal RAM location or SFR, i.e., destination register.

For example: Suppose the 8-bit immediate value is 42H and the memory location 10H contains the data byte 5CH and the instruction XRL 10H, #42H is executed then the internal RAM location 10H contains the data byte content 1EH.

Programmer Tips: If Ex-ORing operation is performed with the same byte itself, then the result is 00H. It means then, Ex-ORing operation is also used to check the numbers in the 2 operands are equal or not.

Example: Suppose by using XRL A, R5 and the contents of the accumulator and Register R5 is same then XOR operation gives result 00H in the accumulator indicating both the numbers are same and if the result is non-zero in the accumulator, then it indicates both the numbers are different. This is because Ex-ORing of the same bits will reset the particular bit and the different bits will set the bit.

It is also possible by checking the numbers whether they are equal or not we make other decisions using JZ instruction or JNZ instruction.

Example 2.26

Give an instruction to complement D0 and D5th bit of memory location 35H without affecting other bits.

Solution:

D7	D6	D5	D4	D3	D2	D1	D0
0	0	1	0	0	0	0	1

Since we need to complement DO, D5 bits, it is required to unmask, i.e., enable these bits and mask the rest of the bits. Hence, the data byte generated is 21H. So first the data byte in the memory location 35H will move into the accumulator and then it will be Ex-OR operated with 21H. Hence, the required data bits of the above data byte will definitely be complemented in the memory location 35H on transferring.

Instructions	Comment
MOV A, 35H	Copy the data byte of memory location 35H in the accumulator.
XRL A, #21H	Ex-OR operate the accumulator content with the data byte 21H. Put the result in the accumulator.
MOV 35H, A	Copy the accumulator data byte back into the memory location 35H.
L1: SJMP L1	A short jump to the label L1.

It is required to unmask these bits because on Ex-OR operation with the 'set' bit will complement the bit, i.e., 1 becomes 0 and vice versa.

Example 2.27

Give an instruction to invert D6th and D7th bit of register B without affecting other bits.

Solution:

D7	D6	D5	D4	D3	D2	D1	D0
1	1	0	0	0	0	0	0

Since we need to complement D6 and D7 bits, it is required to unmask, i.e., enable these bits and mask the rest of the bits. Hence, the data byte generated is C0H. So first the data byte in register B will move into the accumulator and then it will be Ex-OR operated with C0H. Hence, the required data bits of the above data byte will definitely be complemented in register B on transferring.

Instructions	Comment
MOV A, 0F0H	Copy the data byte of register B in the accumulator.
XRL A, #C0H	Ex-OR operate the accumulator content with the data byte C0H. Put the result in the accumulator.
MOV 0F0H, A	Copy the accumulator data byte back into register B.
L1: SJMP L1	A short jump to the label L1.

Example 2.28

Give an instruction to invert D5 and D7 bit of the contents of external RAM location 2000H without affecting other bits.

Solution:

D7	D6	D5	D4	D3	D2	D1	D0
1	0	1	0	0	0	0	0

Since we need to complement D5 and D7 bits, it is required to unmask, i.e., enable these bits and mask the rest of the bits. Hence, the data byte generated is A0H. So first the data byte in the external memory location 2000H pointed by DPTR register will move into the accumulator and then it will be Ex-OR operated with A0H. Hence, the required data bits of the above data byte will definitely be complemented in the external RAM location 2000H on transferring.

Instructions	Comment
MOV DPTR, #2000H	Memory location 2000H is pointed by the data pointer register.
MOVX A, @DPTR	Copy the data byte of external memory location 2000H in the accumulator.
XRL A, #0A0H	Ex-OR operate the accumulator content with the data byte A0H. Put the result in the accumulator.
MOVX @DPTR, A	Copy the accumulator data byte back into the external memory location 2000H.
L1: SJMP L1	A short jump to the label L1.

Conclusion on ANDing, ORing & XORing Operations:

AND Operation	ANDing operation is used to clear/mask/reset the particular bits.
OR Operation	ORing operation is used to set particular bits.
EX-OR Operation	EX-ORing operation is used to complement or invert particular bit.

Boolean Operation for ANDing:

A*0 = 0	To reset or clear the bits. This operation is used to perform ANDing with zero.
A*1 = A	To keep bit unchanged. This operation is used to perform ANDing with one.

Boolean Operation For ORing:

A+0 = A	To keep bit unchanged. This operation is used to perform ORing with zero.
A+1 = 1	To set a bit. This operation is used to perform ORing with one.

Boolean Operation For Ex-ORing:

A XOR 0 = A	To keep bit unchanged. This operation is used to perform Ex-ORing with zero.
A XOR 1 = Á	To invert a bit. This operation is used to perform EX-ORing with one.

19. **RL A**

Bytes: 1

Machine Cycles: 1

Rotate the contents of the accumulator towards left. Put the result in the accumulator.

Operation:

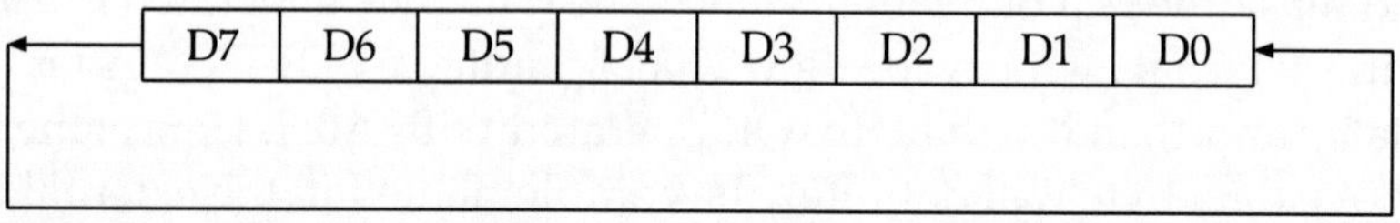

This instruction rotates the content of the accumulator towards left without carry, so that D7 bit is entered in the D0 position.

Generally, rotating the accumulator data byte left will multiply the data byte by 2.

Here, no flags are affected accept the carry flag.

Example: Assume (A) = 35H and RL A instruction is executed, then the accumulator content will be 6AH after the execution of the above instruction.

20. **RLC A**

Bytes: 1

Machine Cycles: 1

Rotate the contents of the accumulator towards left with carry. Put the result in the Accumulator.

Operation:

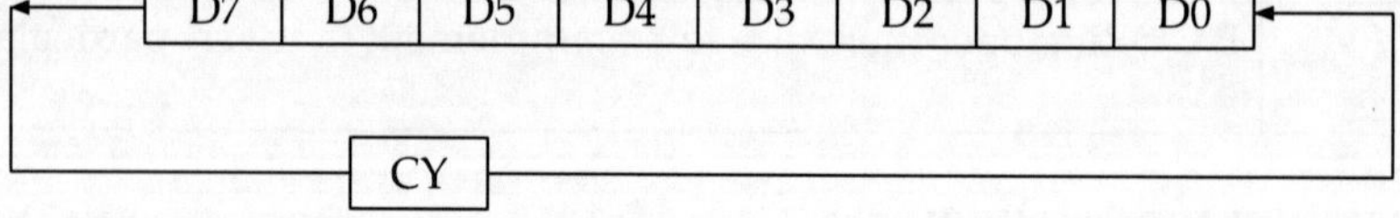

This instruction rotates the content of the accumulator towards left with carry, so that original value of D7 bit is entered in the carry flag and the carry flag value is stored in the D0 position.

Here, only carry flag is affected.

For example: Assume (A) = 35H , carry flag value is set and RLC A instruction is executed, then the accumulator content will be 6BH and carry flag is reset after the execution of the above instruction.

21. **RR A**

Bytes: 1

Machine Cycles: 1

Rotate the contents of the accumulator towards right. Put the result in the accumulator.

Operation:

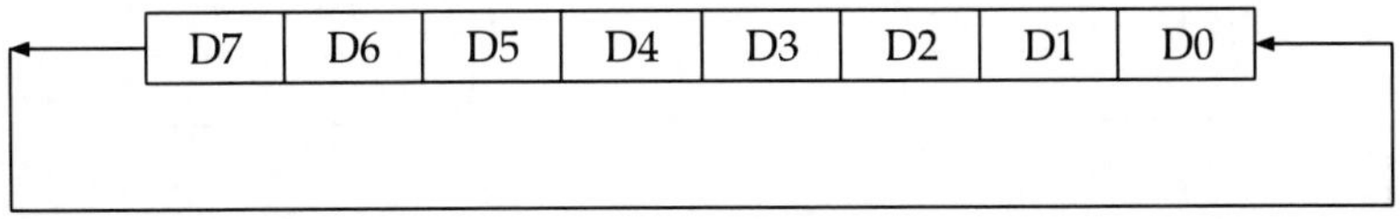

This instruction rotates the content of the accumulator towards right without carry, so that D0 bit is entered in the D7 position.

Generally, rotating the accumulator data byte right will divide the data byte by 2.

Here, except carry flag no other flags are affected.

Example: Assume (A) = 35H and RR A instruction is executed, then the accumulator content will be 9AH after the execution of the above instruction.

22. **RRC A**

Bytes: 1

Machine Cycles: 1

Rotate the contents of the accumulator towards right with carry. Put the result in the Accumulator.

Operation:

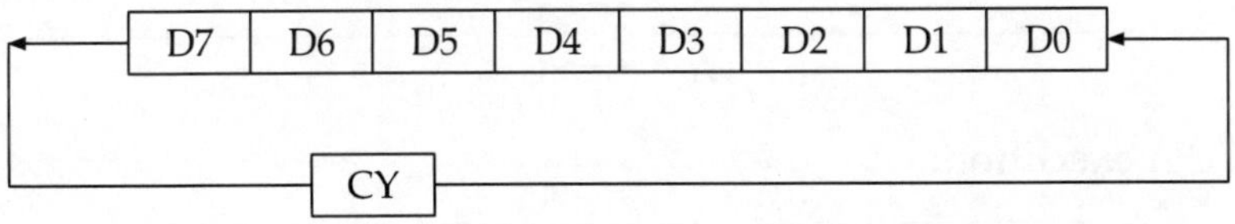

This instruction rotates the content of accumulator towards right with carry, so that original value of D0 bit is entered in the carry flag and the carry flag value is stored in the D7 position.

Here, only carry flag is affected.

Example: Assume (A) = 35H, carry flag value is reset and RLC A instruction is executed, then the accumulator content will be 1AH and carry flag is set after the execution of the above instruction.

Example 2.29

Assume the accumulator data byte to be 56H and the carry flag to be set. Then the following instructions are executed.

RLC A
RLC A
RLC A

Solution: From the above problem, the accumulator data byte is 56H and the carry flag is set.

Before 1st RLC A execution:

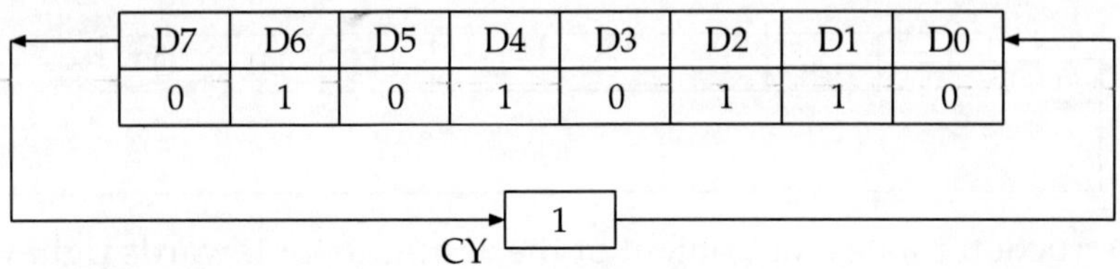

After 1st RLC A execution:

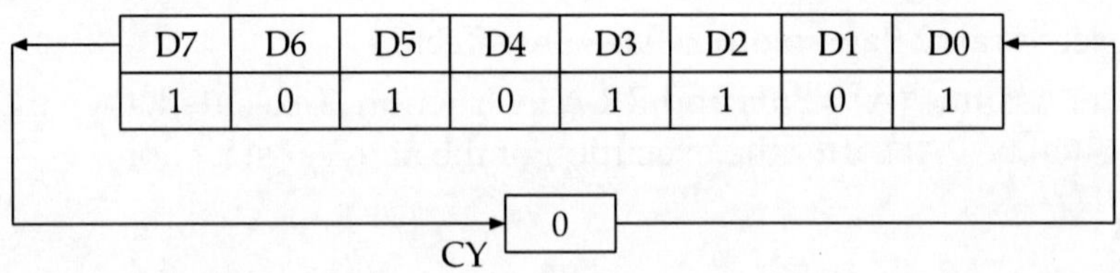

After 2nd RLC A execution:

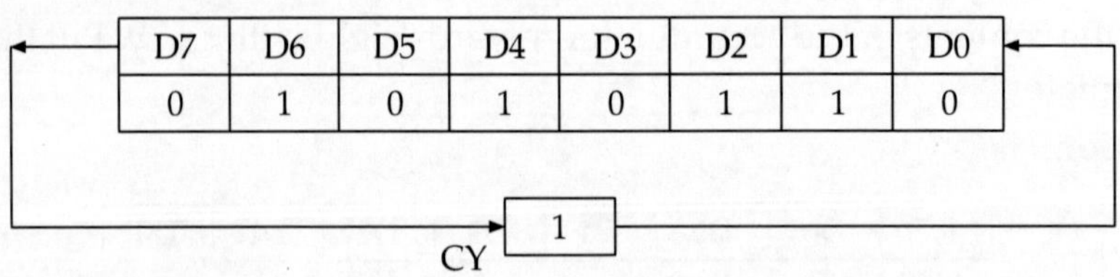

After 3rd RLC A execution:

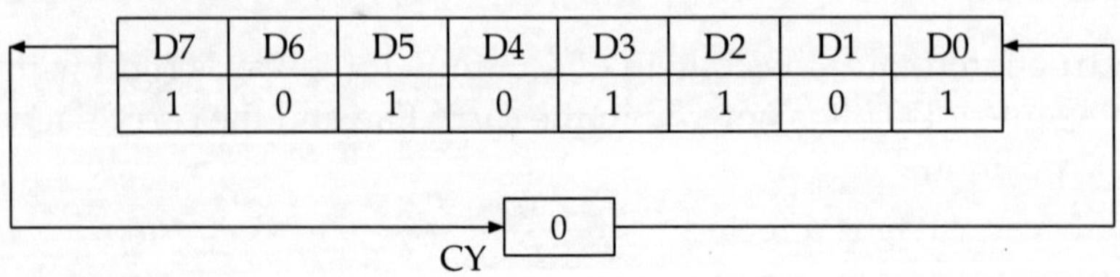

Conclusion:

RLC A Execution Number	Data Byte	Carry Flag
First	ADH	0
Second	5AH	1
Third	B5H	0

Example 2.30

Assume the accumulator data byte to be 56H and the carry flag to be set. Then the following instructions are executed.

RL A

RL A

RL A

Solution: From the above problem, the accumulator data byte is 56H and the carry flag is set.

Before 1st RL A execution:

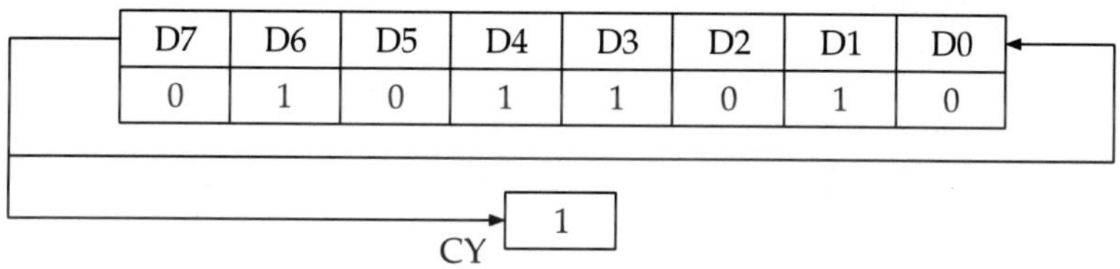

After 1st RL A execution:

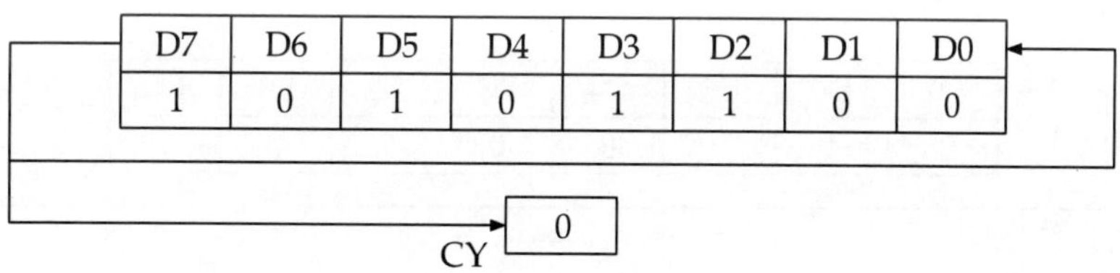

After 2nd RL A execution:

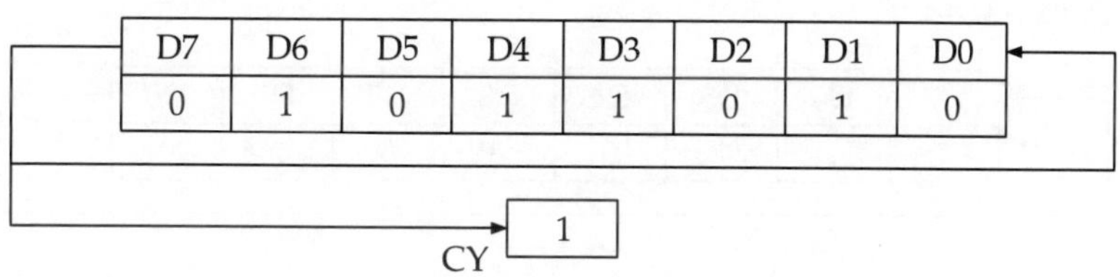

After 3rd RL A execution:

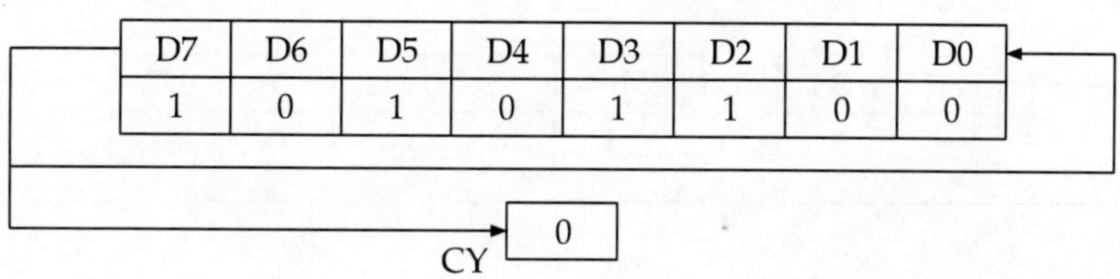

Conclusion:

RL A Execution Number	Data Byte	Carry Flag
First	ACH	0
Second	59H	1
Third	B2H	0

Example 2.31

Assume the accumulator data byte to be CFH and the carry flag to be set. Then the following instructions are executed.

RR A

RR A

RR A

Solution: From the above problem, the accumulator data byte is C5H and the carry flag is set.

Before 1st RR A execution:

D7	D6	D5	D4	D3	D2	D1	D0
1	1	0	0	0	1	0	1

CY 1

After 1st RR A execution:

D7	D6	D5	D4	D3	D2	D1	D0
1	1	1	0	0	0	1	0

CY 1

After 2nd RR A execution:

D7	D6	D5	D4	D3	D2	D1	D0
0	1	1	1	0	0	0	1

CY 0

After 3rd RR A execution:

D7	D6	D5	D4	D3	D2	D1	D0
1	0	1	1	1	0	0	0

CY 1

Conclusion:

RR A Execution Number	Data Byte	Carry Flag
First	E2H	1
Second	71H	0
Third	B8H	1

Example 2.32

Assume the accumulator data byte to be CFH and the carry flag to be set. Then the following instructions are executed.

```
RRC A
RRC A
RRC A
```

Solution:

From the above problem, the accumulator data byte is C5H and the carry flag is set.

Before 1st RRC A execution:

D7	D6	D5	D4	D3	D2	D1	D0
1	1	0	0	0	1	0	1

CY 1

After 1st RRC A execution:

D7	D6	D5	D4	D3	D2	D1	D0
1	1	1	0	0	0	1	0

CY 1

After 2nd RRC A execution:

D7	D6	D5	D4	D3	D2	D1	D0
1	1	1	1	0	0	0	1

CY 0

After 3rd RRC A execution:

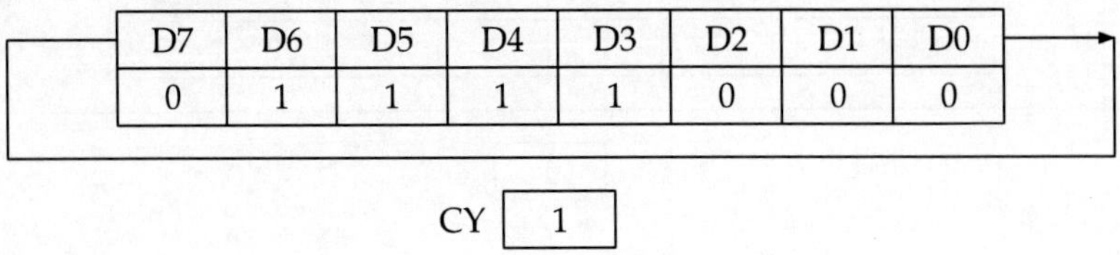

Conclusion:

RRC A Execution Number	Data Byte	Carry Flag
First	E2H	1
Second	F1H	0
Third	78H	1

23. **CLR A**

Bytes: 1

Machine Cycles: 1

Operation:

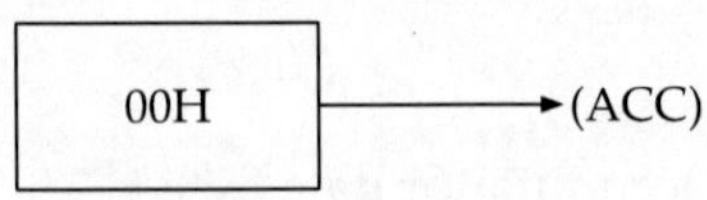

This instruction clears the contents of the accumulator, i.e., the accumulator is copied with 00H.

No flags are affected.

Example: Suppose the accumulator content is 56H and CLR A instruction is executed then the accumulator data byte is 00H.

24. **CPL A**

Bytes:1

Machine Cycles: 1

Operation:

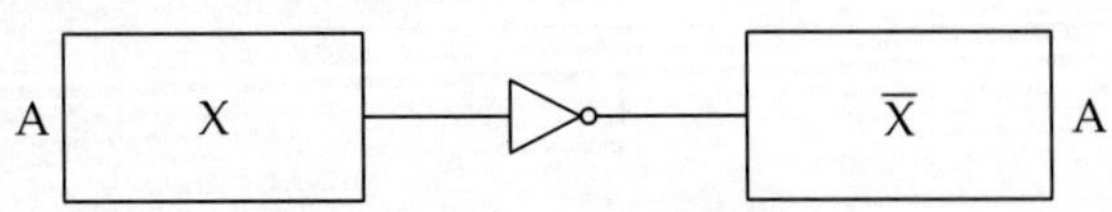

This instruction bitwise complements or inverts the contents of the accumulator and puts the result in the accumulator.

No flags are affected.

For example: Suppose the accumulator content is 95H and CPL A instruction is executed then the accumulator data byte is 6AH.

Example 2.33

Complement the contents of external memory location 3000H.

Solution:

Instruction	Comments
MOV DPTR, #3000H	Memory location 3000H is pointed by the DPTR.
MOVX A, @DPTR	Copy the data byte content of external memory location into the accumulator.
CPL A	Complement the value of the accumulator.
MOVX @DPTR, A	Move the data byte of the accumulator after inversion back into the same external memory location pointed by the data pointer register.
$: SJMP $	A short jump to the $ sign.

25. **SWAP A**

    Bytes: 1

    Machine Cycles: 1

    Interchange the contents of lower and higher nibble within the accumulator.

    Operation:

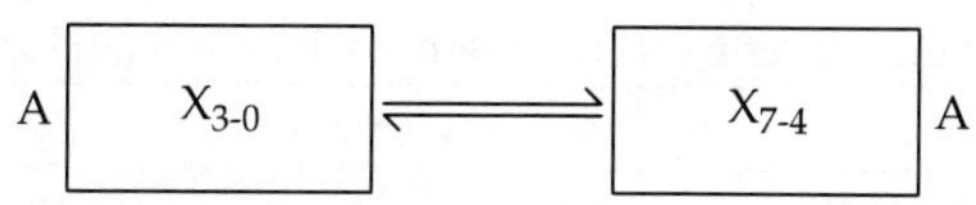

This instruction interchanges lower nibble and higher nibble within the accumulator and puts the result in the accumulator.

Here, no flags are affected.

For example: Suppose the accumulator content is 9AH and the instruction SWAP A is executed, then the accumulator data byte after the execution of the above instruction is A9H. Hence, the lower and higher nibbles are interchanged.

This instruction is equivalent to 4 times rotation either towards left or towards right without carry.

Program 2.17

Write an assembly language program to convert packed BCD number present in the memory location 64H into 2 ASCII equivalent and put the result in R7-R6.

Solution:

Method – I

Instructions	Comments
MOV A, 64H	Copy the data byte from the memory location 64H into the accumulator.
MOV B, A	Copy the same data byte into the register B.

ANL A, 0FH	AND operate the accumulator with the data byte 0FH. This means masking the higher nibble.
ADD A, #30H	Adding the accumulator data byte with the immediate value 30H.
MOV R6, A	Copy the accumulator data byte into R6.
MOV A, B	Retrieve the original data byte stored in register B into the accumulator.
ANL A, F0H	AND operate the accumulator data byte with the data byte value F0H. This means masking the lower nibble.
RL A	Rotating the accumulator left by one without the carry.
RL A	Rotating the accumulator left by one without the carry.
RL A	Rotating the accumulator left by one without the carry.
RL A	Rotating the accumulator left by one without the carry.
ADD A, #30H	Adding the immediate value 30H with the accumulator.
MOV R7, A	Copy the accumulator value into R7.
$:SJMP $	A short jump to the label $.

Method – II

Instructions	Comments
MOV A, 64H	Copy the data byte from the memory location 64H into the accumulator.
MOV B, A	Copy the same data byte into register B.
ANL A, 0FH	AND operate the accumulator with the data byte 0FH. This means masking the higher nibble.
ADD A, #30H	Adding the accumulator data byte with the immediate value 30H.
MOV R6, A	Copy the accumulator data byte into R6.
MOV A, B	Retrieve the original data byte stored in register B into the accumulator.
SWAP A	Exchanging the higher and lower nibbles.
ANL A, 0FH	AND operate the accumulator data byte with the data byte value 0FH. This means masking the higher nibble.
ADD A, #30H	Adding the immediate value 30H with the accumulator.
MOV R7, A	Copy the accumulator value into R7.
$:SJMP $	A short jump to the label $.

Programming Tips: Method 1 takes more time to execute the program than method 2, since method 1 has more number of instructions and takes more machine cycles to execute the above program. As per the observation, in method 2 we are using SWAP A instruction instead of 4 times rotation in method 1. So, here we are saving 3 machine cycles in method 2. A good programmer is not who solves the program quickly and takes more time to execute, but is the person who saves time by saving the machine cycle execution time.

Table 2.3 *Summary of Logical Instruction Group*

Sr. No.	Instruction	No. of Bytes	No. of Cycles	Flags affected
01	ANL A, Rn	01	01	All flags with CY = 0 & AC =1
02	ANL A, #DATA	02	01	All flags with CY = 0 & AC =1
03	ANL A, DIRECT	02	01	All flags with CY = 0 & AC =1
04	ANL A, @Ri	01	01	All flags with CY = 0 & AC =1
05	ANL DIRECT, A	02	01	All flags with CY = 0 & AC =1
06	ANL DIRECT, #DATA	03	02	All flags with CY = 0 & AC =1
07	ORL A, Rn	01	01	All flags with CY = 0 & AC =0
08	ORL A, #DATA	02	01	All flags with CY = 0 & AC =0
09	ORL A, DIRECT	02	01	All flags with CY = 0 & AC =0
10	ORL A, @Ri	01	01	All flags with CY = 0 & AC =0
11	ORL DIRECT, A	02	01	All flags with CY = 0 & AC =0
12	ORL DIRECT, #DATA	03	02	All flags with CY = 0 & AC =0
13	XRL A, Rn	01	01	All flags with CY = 0 & AC =0
14	XRL A, #DATA	02	01	All flags with CY = 0 & AC =0
15	XRL A, DIRECT	02	01	All flags with CY = 0 & AC =0
16	XRL A, @Ri	01	01	All flags with CY = 0 & AC =0
17	XRL DIRECT, A	02	01	All flags with CY = 0 & AC =0
18	XRL DIRECT, #DATA	03	02	All flags with CY = 0 & AC =0
19	CLR A	01	01	No flags are affected
20	CPL A	01	01	No flags are affected
21	RL A	01	01	Only carry flag is affected
22	RLC A	01	01	Only carry flag is affected
23	RR A	01	01	Only carry flag is affected
24	RRC A	01	01	Only carry flag is affected
25	SWAP A	01	01	Only carry flag is affected

2.4.4 Boolean Instruction Group

In the microcontroller 8051, 128 byte internal RAM is available addresses from 00H to 7FH. 16 memory location of this on-chip 128 byte internal RAM from 20H to 2FH is a bit addressable (16 * 8 = 128) addresses from 00H to 7FH.

Bit position →	D7	D6	D5	D4	D3	D2	D1	D0
2F	7F	7E	7D	7C	7B	7A	79	78
2E	77	76	75	74	73	72	71	70
2D	6F	6E	6D	6C	6B	6A	69	68
2C	67	66	65	64	63	62	61	60
2B	5F	5E	5D	5C	5B	5A	59	58
2A	57	56	55	54	53	52	51	50
29	4F	4E	4D	4C	4B	4A	49	48
28	47	46	45	44	43	42	41	40
27	3F	3E	3D	3C	3B	3A	39	38
26	37	36	35	34	33	32	31	30
25	2F	2E	2D	2C	2B	2A	29	28
24	27	26	25	24	23	22	21	20
23	1F	1E	1D	1C	1B	1A	19	18
22	17	16	15	14	13	12	11	10
21	0F	0E	0D	0C	0B	0A	09	08
20	07	06	05	04	03	02	01	00

Fig. 2.2 *Bit-addressable RAM*

Memory locations 20H to 2FH are also byte addressable. Bit addressable RAM and their allocated memory locations are shown in Figure 2.2. From the above figure, D0 bit of memory location 20H has the address 00H. Bit D1 of memory location 20H has 01H and bit D7 of memory location 20H has 07H.

Similarly, 08H memory location belongs to D0 bit of memory location 21H and 0FH memory location belongs to bit D7 of memory location 21H and so on up to 78H memory location belongs to D0 bit of 25H memory location belongs to D0 bit of 2F memory location and memory location 7FH belongs to bit D7 of memory location 2FH. There are two ways to access the information in the microcontroller 8051. First is for the byte operation, 128 byte internal RAM addresses from 00H to 7FH accesses using direct and indirect addressing mode. Bit addressable RAM from 20H to 2FH again having the bit addressable location 00H to 7FH is accessed using only one addressing mode, direct addressing mode and there is a specified bit manipulation instruction available to access bit addressable RAM.

For example:

SETB bit: Where bit can be any memory location lying between 00H to 7FH.SETB 67H. This instruction means setting the bit of the memory location 67H. So the above memory location will have the data bit '1' which is set.

CLR bit: This instruction means resetting the bit of the memory location lying between 00H to 7FH. So the above memory location will have the data bit '0' which is reset.

1. **CLR C**

   Bytes: 1

   Machine Cycles: 1

   Clears carry flag.

   Operation:

   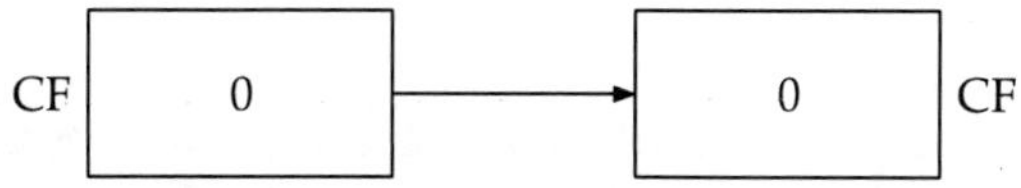

   This instruction clears the content of carry flag bit. So, replace carry flag with zero. Here, always after the execution of this instruction carry flag will be 00H.

2. **CLR BIT**

   Bytes: 2

   Machine Cycle: 1

   Clears indicated bit (bit addressable RAM or SFR).

   Operation:

   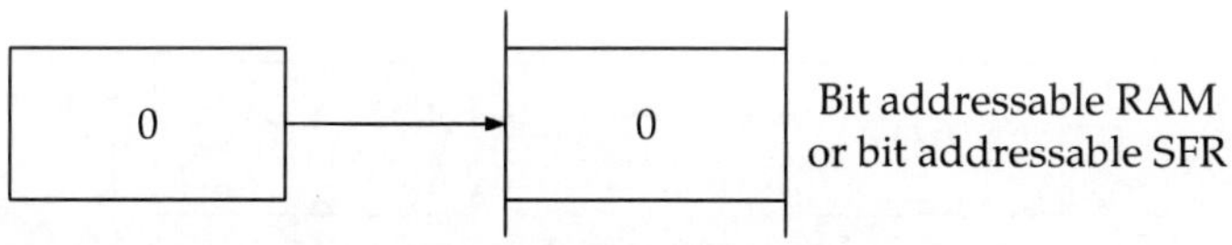

   This instruction clears contents of indicated bit addressable RAM/SFR and puts the result in the same bit addressable RAM or bit addressable SFR.

3. **SETB C**

   Bytes: 1

   Machine Cycle: 1

   Set carry flag.

   Operation:

   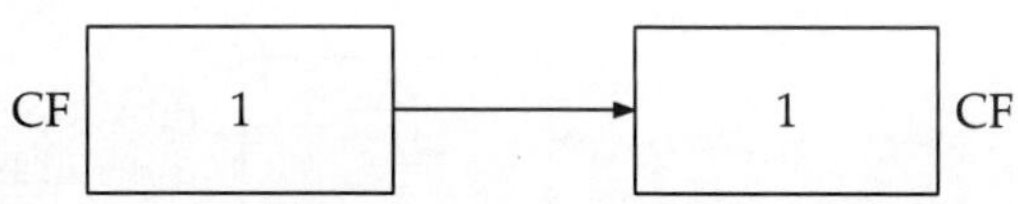

   Set the contents of the carry flag. Put the result in the carry flag.

   This instruction will set the content of the carry flag bit. So, replace the content of the carry flag bit with '1'.

4. **SETB BIT**

   Bytes: 2

   Machine Cycle: 1

Set the indicated bit, i.e., set the contents of bit addressable RAM or SFR.

Operation:

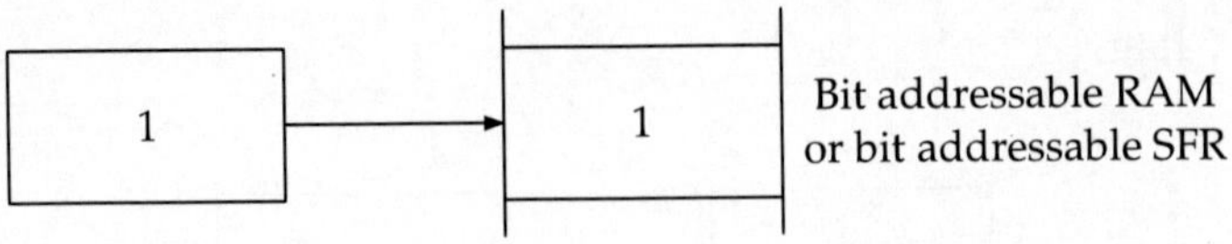

This instruction will set the contents of indicated bit addressable RAM/ SFR and puts the result in the same bit addressable RAM or bit addressable SFR.

Example:

SETB 30H will set the bit to '1' of memory location 30H.

5. **CPL C**

Bytes: 1

Machine Cycle: 1

Complement the carry flag bit.

Operation:

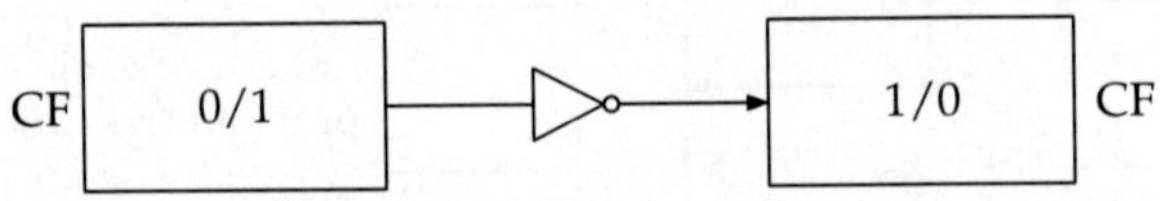

This instruction will complement or invert the contents of the carry flag and put the result in the same carry flag bit.

6. **CPL BIT**

Bytes: 2

Machine Cycle: 1

Complement the indicated bit, i.e., contents of bit addressable RAM or bit addressable SFR.

Operation:

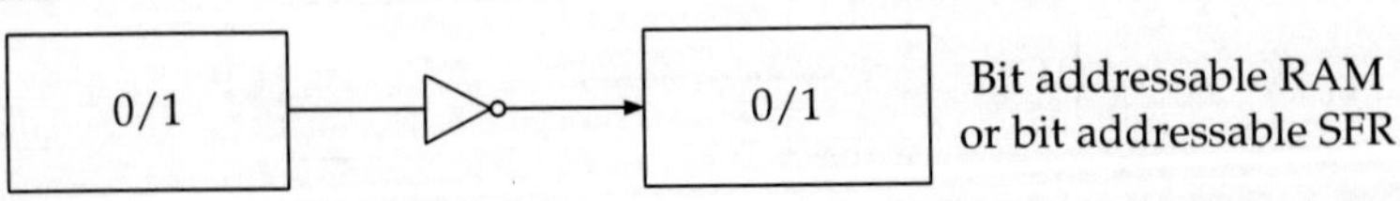

This instruction complements the indicated bit, i.e., the contents of bit addressable RAM or bit addressable SFR and puts the result in the same bit addressable RAM or bit addressable SFR.

Example 2.34

Assume the accumulator data byte to be 56H and CPL ACC.0 instruction is executed. Indicate the data byte present in the accumulator after the execution of the above instruction.

Solution: The accumulator data byte is 56H and CPL ACC.0 means the D0th bit of the accumulator is complemented, i.e., the bit '0' changes to 1. Hence, the accumulator data byte present after the execution of the above instruction is 57H.

Example 2.35

Assume the port content is 8AH and the instruction SETB P0.0 is executed. Indicate the data byte stored in the port P0 after the execution of the above instruction.

Solution: Since the instruction P0.0 is executed, hence the D0th bit of the port is set and hence the content stored inside the port P0 is 8BH.

Example 2.36

Give the result of the accumulator and register B after execution of the program given below.

```
ORG 0000H
MOV B, #56H
SETB B.0
CPL B.1
MOV A, B
CLR ACC.0
$: SJMP $
END
```

Solution: The first instruction ORG 0000H indicates that the program will start from the memory location 0000H.

The next instruction will immediately load the data byte 56H into register B.

SETB B.0 will set the D0th bit of register B irrespective of the bit content whether 0 or 1. So the D0th bit is set and hence the data byte present in register B is 57H.

CPL B.1 will complement the bit in the D1st position. Here, bit '1' becomes '0' and '0' becomes '1'. Since the D1st bit is 1, this particular bit is changed to 0. So, the data byte stored in register B is 55H.

MOV A, B will move the data byte 55H in the accumulator.

CLR ACC.0 will clear, i.e., reset the D0th bit of the accumulator. Hence, the data bit stored in the accumulator is now 54H.

$: SJMP $ will now jump to the $ label position.

END will now end the above program.

Thus, after the execution of the above program, the data byte stored in the accumulator will be 54H and the data byte stored in register B will be 55H.

Example 2.37

Give the data byte stored in the accumulator after the execution of the program given below.

```
ORG 0000H
SETB PSW.4
SETB PSW.3
MOV A, #56H
ANL A, #0FH
MOV R0, #76H
MOV 19H, @R0
CPL A
SETB ACC.2
XRL A, 19H
SWAP A
ANL A, 18H
$: SJMP $
END
```

Solution: The first instruction ORG 0000H will start the memory location from 0000H.

The next instruction PSW.4 will set the D4th bit of the accumulator.

PSW.3 will also set the D3rd bit of the accumulator. Since both D4 and D3 bits of the accumulator are set, hence register bank 3 is selected.

MOV A, #56H will load the data 56H immediately into the accumulator.

ANL A, #0FH will mask the higher byte and hence the data byte loaded into the accumulator will be 06H.

MOV R0, #76H. This will load the data byte 76H into the memory location 18H. This is because register bank 3 is selected. Hence, the above data will be loaded into the memory location 18H.

MOV 19H, @R0 instruction will copy the data byte 76H into the memory location 19H.

CPL A will complement the content of the accumulator. Hence, the accumulator data byte will be F9H.

SETB ACC.2 will set the data bit present in the accumulator. Hence, the data byte loaded into the accumulator will be FDH.

XRL A, 19H will XOR the data byte present in the accumulator, i.e., FDH with the data byte present in the memory location 19H, i.e., 76H. The final result stored in the accumulator will be 8BH.

SWAP A will exchange the lower and higher nibbles. So, the data byte is now present in the accumulator after the execution of the above instruction B8H.

ANL A, 18H will AND operate the accumulator data byte, i.e., B8H with the data byte present in the memory location 18H, i.e., 76H.

$ SJMP $ will move the control into the label $.

END will terminate the execution of the above program.

7. **MOV C, BIT**

   Bytes: 2

   Machine Cycles: 1

   Copy contents of bit addressable RAM or bit addressable SFR to carry flag.

   Operation:

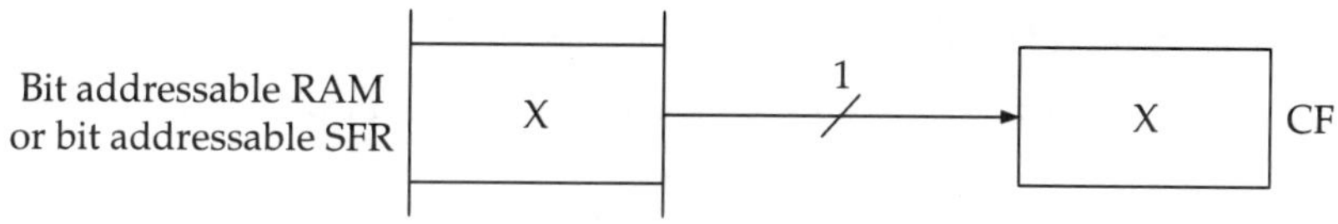

   This instruction copies the one bit value of the indicated bit addressable RAM or indicated bit addressable SFR to the carry flag.

8. **MOV BIT, C**

   Bytes: 2

   Machine Cycles: 2

   Copy contents of carry flag to the bit addressable RAM or bit addressable SFR.

   Operation:

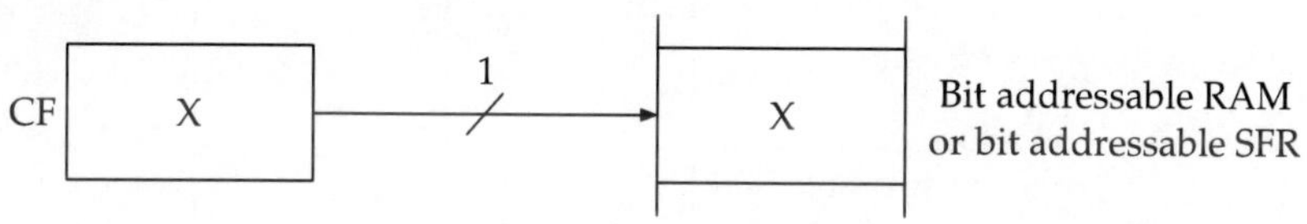

   This instruction copies the one bit value of the carry flag to the indicated bit addressable RAM or indicated bit addressable SFR.

9. **ANL C, BIT**

   Bytes: 2

   Machine Cycles: 2

   Perform ANDing 1 bit of information from bit addressable RAM with the carry flag. Put the result in the carry flag.

Operation:

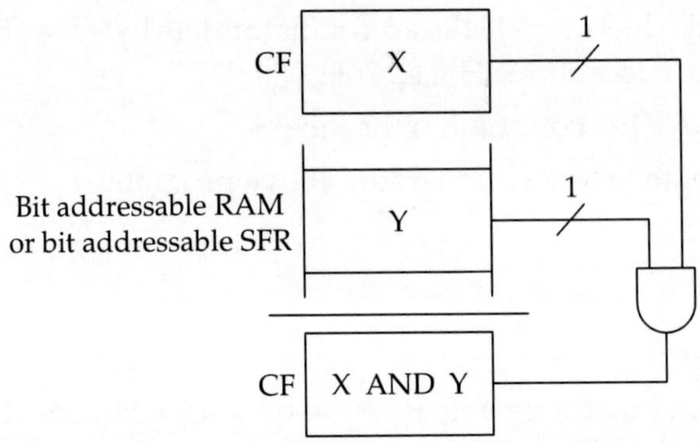

This instruction performs ANDing of the contents of the carry bit with the contents of bit addressable RAM or SFR and puts the result in the carry flag.

Assume accumulator content is 56H, carry flag is set and the instruction ANL C, ACC.6 is executed then the carry flag will store the value '1', i.e., carry flag is set.

10. **ANL C, /BIT**

Bytes: 2

Machine Cycles: 2

Perform ANDing 1 bit of information with inverted bit of bit addressable RAM. Put the result in the carry flag.

Operation:

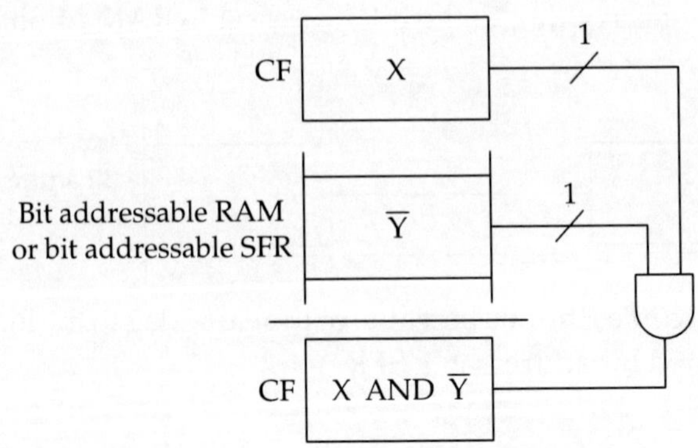

This instruction performs 1 bit ANDing on the contents of the carry flag bit with the contents of inverted value of bit addressable RAM or SFR and puts the result in the carry flag.

Assume the accumulator content is 56H , carry flag is set and the instruction ANL C, /ACC.5 is executed then the carry flag will store the value '1', i.e., carry flag is set. This is because after inversion of the bit 0 in ACC.5 , the bit is now changed to 1.

11. **ORL C, BIT**

Bytes: 2

Machine Cycles: 2

Perform ORing 1 bit of information from bit addressable RAM with the carry flag. Put the result in the carry flag.

Operation:

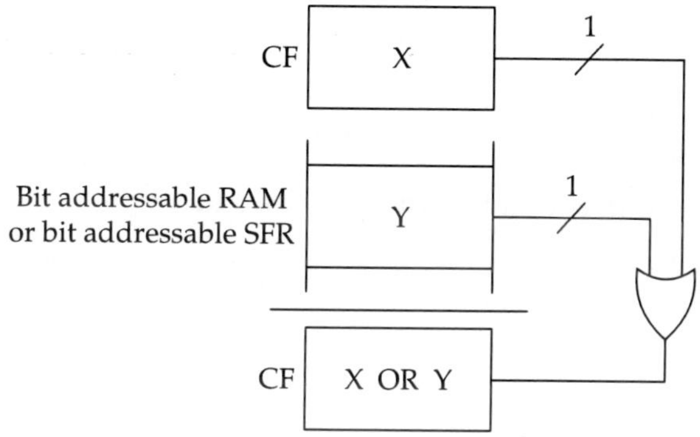

This instruction performs ORing of the contents of the carry bit with the contents of bit addressable RAM or SFR and puts the result in the carry flag.

Assume port P0 has the data byte 85H, carry flag is reset and the instruction ORL C, P0.7 is executed, then the carry flag is set, since P0.7 bit is 1 and we know that if either bit of the OR operation is 1, then the resultant bit will also be 1.

12. **ORL C, /BIT**

Bytes: 2

Machine Cycles: 2

Perform ORing 1 bit of information from bit addressable RAM with the carry flag. Put the result in the carry flag.

Operation:

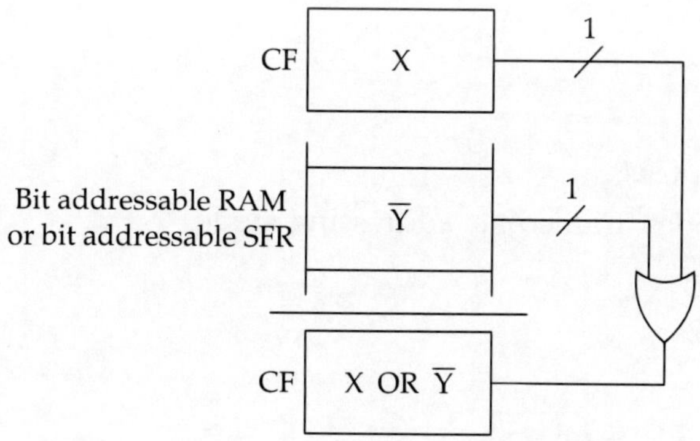

This instruction performs ORing of the contents of the carry bit with the contents of bit addressable RAM or SFR and puts the result in the carry flag.

Assume port P0 has the data byte 85H, carry flag is reset and the instruction ORL C,/P0.7 is executed, then the carry flag is reset, since P0.7 bit is 1 and the inverted bit will make it to zero and carry flag is reset. Hence, finally the carry flag is reset.

13. **JC REL**

Bytes: 2

Cycles: 2

Jump if carry is set

Addressing modes: Immediate addressing mode

Operation:

(PC) ← (PC) + 2

IF (CF) = 1

THEN

(PC) ← (PC) + rel

If the carry flag is set, JC branches to the address indicated; otherwise, it proceeds with the next instruction. The branch destination is computed by adding the signed relative-displacement in the second instruction byte to the PC, after incrementing the PC twice. No flags are affected.

Example: The carry flag is cleared. The following instruction sequence,

JC LABEL1

CPL C

JC LABEL2

sets the carry and causes program execution to continue at the instruction identified by the label LABEL2.

14. **JNC REL**

Bytes: 2

Cycles: 2

Jump if carry not set

Addressing modes: Immediate addressing mode

Operation:

(PC) → (PC) + 2

IF (C) = 0

THEN (PC) → (PC) + rel

If the carry flag is a 0, JNC branches to the address indicated; otherwise, it proceeds with the next instruction. The branch destination is computed by adding the signal relative-displacement in the second instruction byte to the PC, after incrementing the PC twice to point to the next instruction. The carry flag is not modified.

Example: The carry flag is set. The following instruction sequence,

JNC LABEL1

CPL C

JNC LABEL2

clears the carry and causes program execution to continue at the instruction identified by the label, i.e., LABEL2.

15. **JB BIT, REL**

Bytes: 3

Cycles: 2

Jump if bit set

Addressing modes: Direct addressing mode

Operation:

(PC) + 3 → (PC)

If (bit) =1

Then

(PC) + rel → (PC)

If the indicated bit is a one, JB jumps to the address indicated; otherwise, it proceeds with the next instruction.

The branch destination is computed by adding the signed relative-displacement in the third instruction byte to the PC, after incrementing the PC to the first byte of the next instruction. The bit tested is not modified. No flags are affected.

Example: The data present at input port 1 is 11001010B. The accumulator holds 56 (01010110B). The following instruction sequence,

JB P1.2, LABEL1

JB ACC. 2, LABEL2

causes program execution to branch to the instruction at label, i.e., LABEL2.

16. **JNB BIT, REL**

Bytes: 3

Cycles: 2

Jump if bit not set

Addressing modes: Direct addressing mode

Operation:

(PC) + 3 → (PC)

If (bit) =0

Then

(PC) + rel → (PC)

If the indicated bit is a 0, JNB branches to the indicated address; otherwise, it proceeds with the next instruction. The branch destination is computed by adding the signed relative-displacement in the third instruction byte to the PC, after incrementing the PC to the first byte of the next instruction. *The bit tested is not modified.* No flags are affected.

Example: The data present at input port 1 is 11001010B. The accumulator holds 56H (01010110B). The following instruction sequence,

JNB P1.3, LABEL1

JNB ACC.3, LABEL2

causes program execution to continue at the instruction at label, i.e., LABEL2.

17. **JBC BIT, REL**

Bytes: 3

Cycles: 2

Jump if bit is set and clear bit

Addressing modes: Direct addressing mode

Operation:

(PC) + 3 → (PC)

If (bit) =1

Then

0 → (bit)

(PC) + rel → (PC)

If the indicated bit is one, JBC branches to the address indicated; otherwise, it proceeds with the next instruction. *The bit will not be cleared if it is already a zero.* The branch destination is computed by adding the signed relative-displacement in the third instruction byte to the PC, after incrementing the PC to the first byte of the next instruction. No flags are affected.

Note: When this instruction is used to test an output pin, the value used as the original data will be read from the output data latch, *not* the input pin.

Example: The accumulator holds 56H (01010110B). The following instruction sequence,

JBC ACC.3, LABEL1

JBC ACC.2, LABEL2

causes program execution to continue at the instruction identified by the label, i.e., LABEL2, with the accumulator modified to 52H (01010010B).

Table 2.4 *Summary of Boolean Instruction Group*

Sr. No.	Instruction	No. of Bytes	No. of Cycles
01	CLR C	01	01
02	CLR BIT	02	01
03	SETB C	01	01
04	SETB BIT	02	01
05	CPL C	01	01
06	CPL BIT	02	01
07	MOV C, BIT	02	01
08	MOV BIT, C	02	02
09	ANL C, BIT	02	02
10	ANL C, /BIT	02	02
11	ORL C, BIT	02	02
12	ORL C, /BIT	02	02
13	JC REL	02	02
14	JNC REL	02	02
15	JB BIT, REL	03	02
16	JNB BIT, REL	03	02
17	JBC BIT, REL	03	02

2.4.5 Branching Instruction Group

1. **ACALL ADDR11**

   Bytes = 2

   Cycles = 2

   Addressing modes: Immediate and indirect addressing mode

   ACALL unconditionally calls a subroutine located at the indicated address. The instruction increments the PC twice to obtain the address of the following instruction, then pushes the 16-bit result onto the stack (low-order byte first) and increments the stack pointer twice. The destination address is obtained by successively concatenating the five high-order bits of the incremented PC, opcode bits 7 through 5, and the second byte of the instruction. The subroutine called must therefore start within the same 2 K block of the program memory as the first byte of the instruction following ACALL. No flags are affected.

   Operation:

   PC + 2 → PC

(SP) + 1 → (SP)

PC_{7-0} → (SP)

(SP) + 1 → (SP)

PC_{15-8} → (SP)

Page address → PC_{10-0}

Encoding:

A10	A9	A8	1	0	0	0	1	A7	A6	A5	A4	A3	A2	A1	A0

Note: As the given address is copied in the program counter, i.e., the given number is copied directly to the PC, so immediate addressing mode and the address of the next instruction following to the ACALL instruction is saved in the stack memory, so indirect addressing mode.

Example: Initially, SP equals 07H. The label SUBRTN is at the program memory location 0345 H. After executing the following instruction,

ACALL SUBRTN

at location 0123H, SP contains 09H, internal RAM locations 08H and 09H will contain 25H and 01H, respectively, and the PC contains 0345H.

2. **LCALL ADDR16**

Bytes = 3

Cycles = 2

Addressing modes: Direct and immediate addressing modes

Long call

LCALL calls a subroutine located at the indicated address. The instruction adds three to the program counter to generate the address of the next instruction and then pushes the 16-bit result onto the stack (low byte first), incrementing the stack pointer by two. The high-order and low-order bytes of the PC are then loaded, respectively, with the second and third bytes of the LCALL instruction. Program execution continues with the instruction at this address. The subroutine may therefore begin anywhere in the full 64 KB program memory address space. No flags are affected.

Operation:

(PC) ← (PC) + 3

(SP) ← (SP) + 1

((SP)) ← (PC7-0)

(SP) ← (SP) + 1

((SP)) ← (PC15-8)

(PC) ← addr15-0

Example: Initially, the stack pointer equals 07H. The label SUBRTN is assigned to program memory location 1234H. After executing the instruction,

LCALL SUBRTN

at location 0123H, the stack pointer will contain 09H, internal RAM locations 08H and 09H will contain 26H and 01H, and the PC will contain 1234H.

3. RET

Bytes = 1

Cycles = 2

Addressing mode: Indirect addressing mode

Return from subroutine

RET pops the high- and low-order bytes of the PC successively from the stack, decrementing the stack pointer by two. Program execution continues at the resulting address, generally the instruction immediately following the ACALL or LCALL. No flags are affected.

Operation:

$((SP)) \rightarrow (PC_{15-8})$

$(SP) - 1 \rightarrow (SP)$

$((SP)) \rightarrow (PC_{7-0})$

$(SP) - 1 \rightarrow (SP)$

Example: The stack pointer originally contains the value 0BH. Internal RAM locations 0AH and 0BH contain the values 23H and 01H, respectively. The following instruction,

RET

leaves the stack pointer equal to the value 09H. Program execution continues at location 0123H.

4. RETI

Bytes = 1

Cycles = 2

Addressing modes: Indirect addressing mode

Return from interrupt

RETI pops the high- and low-order bytes of the PC successively from the stack and restores the interrupt logic to accept additional interrupts at the same priority level as the one just processed. The stack pointer is left decremented by two. No other registers are affected; the PSW is *not* automatically restored to its pre-interrupt status. Program execution continues at the resulting address, which is generally the instruction immediately after the point at which the interrupt request was detected. If a lower- or same-level interrupt was pending when the RETI instruction is executed, that one instruction is executed before the pending interrupt is processed.

Operation:

$((SP)) \rightarrow (PC_{15-8})$

$(SP) - 1 \rightarrow (SP)$

$((SP)) \rightarrow (PC_{7-0})$

$(SP) - 1 \rightarrow (SP)$

Example: The stack pointer originally contains the value 0BH. An interrupt was detected during the instruction ending at location 0122H. Internal RAM locations 0AH and 0BH contain the values 23H and 01H, respectively. The following instruction, RETI leaves the stack pointer equal to 09H and returns program execution to location 0123H.

5. **AJMP ADDR11**

Bytes = 2

Cycles = 2

Addressing modes: Immediate addressing mode

AJMP transfers program execution to the indicated address, which is formed at run-time by concatenating the high-order five bits of the PC (after incrementing the PC twice), opcode bits 7 through 5, and the second byte of the instruction. The destination must therefore be within the same 2 K block of program memory as the first byte of the instruction following AJMP.

Operation:

$PC + 2 \rightarrow PC$

Page address $\rightarrow PC_{10-0}$

Encoding:

A10	A9	A8	0	0	0	0	1	A7	A6	A5	A4	A3	A2	A1	A0

Example: The label JMPADR is at program memory location 0123H. The following instruction,

AJMP JMPADR

is at location 0345H and loads the PC with 0123H.

6. **LJMP ADDR16**

Bytes = 3

Cycles = 2

Addressing modes: Immediate addressing mode

Long Jump

LJMP causes an unconditional branch to the indicated address, by loading the high-order and low-order bytes of the PC (respectively) with the second and third instruction bytes. The destination may therefore be anywhere in the full 64K program memory address space. No flags are affected.

Operation:

(PC) + 3 → (PC)

Example: The label JMPADR is assigned to the instruction at program memory location 1234H. The instruction,

LJMP JMPADR

at location 0123H will load the program counter with 1234H.

7. **SJMP REL**

Bytes = 2

Cycles = 2

Addressing modes: Immediate addressing mode

Short Jump

Program control branches unconditionally to the address indicated. The branch destination is computed by adding the signed displacement in the second instruction byte to the PC, after incrementing the PC twice. Therefore, the range of destinations allowed is from 128 bytes preceding these instructions 127 bytes following it.

Operation:

(PC) + 2 → (PC)

(PC) + rel → (PC)

Example: The label RELADR is assigned to an instruction at program memory location 0123H. The following instruction

SJMP RELADR

assembles into location 0100H. After the instruction is executed, the PC contains the value 0123H.

Note: Under the above conditions the instruction following SJMP is at 102H. Therefore, the displacement byte of the instruction is the relative offset (0123H-0102H) = 21H. Put another way, an SJMP with a displacement of 0FEH is a one-instruction infinite loop.

8. **JMP @A+DPTR**

Bytes = 1

Cycles = 2

Addressing modes: Index addressing mode

Jump indirect

JMP @A+DPTR adds the eight-bit unsigned contents of the accumulator with the 16-bit data pointer and loads the resulting sum to the program counter. This is the address for subsequent instruction fetches. Sixteen-bit addition is performed (modulo 216): a carry-out from the low-order eight bits propagates through the higher-order bits. Neither the accumulator nor the data pointer is altered. No flags are affected.

Operation:

+ (DPTR) → (PC)

Example: An even number from 0 to 6 is in the accumulator. The following sequence of instructions branches to one of four AJMP instructions in a jump table starting at JMP_TBL.

```
            MOV DPTR, # JMP_TBL
            JMP @A + DPTR
JMP_TBL:    AJMP LABEL0
            AJMP LABEL1
            AJMP LABEL2
            AJMP LABEL3
```

If the accumulator equals 04H when starting this sequence, execution jumps to label LABEL2. Because AJMP is a 2-byte instruction, the jump instructions start at every other address.

9. **JZ REL**

Bytes = 2

Cycles = 2

Addressing modes: Immediate addressing mode

Jump if Accumulator Zero

If all bits of the accumulator are 0, JZ branches to the address indicated; otherwise, it proceeds with the next instruction. The branch destination is computed by adding the signed relative-displacement in the second instruction byte to the PC, after incrementing the PC twice. The accumulator is not modified. No flags are affected.

Operation:

(PC) + 2 → (PC)

If (A) = 00h

Then

(PC) + rel → (PC)

Example: The accumulator originally contains 01H. The following instruction sequence,

```
JZ LABEL1
DEC A
JZ LABEL2
```

changes the accumulator to 00H and causes program execution to continue at the instruction identified by the label, i.e., LABEL2.

10. JNZ REL

Bytes = 2

Cycles = 2

Addressing modes: Immediate addressing mode

Jump if the accumulator Not Zero

If any bit of the accumulator is a one, JNZ branches to the indicated address; otherwise, it proceeds with the next instruction. The branch destination is computed by adding the signed relative-displacement in the second instruction byte to the PC, after incrementing the PC twice. The accumulator is not modified. No flags are affected.

Operation:

(PC) + 2 → (PC)

If (A) ≠ 00h

Then

(PC) + rel → (PC)

Example: The accumulator originally holds 00H. The following instruction sequence

```
JNZ LABEL1
INC A
JNZ LABEL2
```

sets the accumulator to 01H and continues at label, i.e., LABEL2.

11. CJNE A, DIRECT, REL

Bytes = 3

Cycles = 2

Addressing modes: Immediate addressing mode

Compare the accumulator with direct byte and Jump if Not Equal.

CJNE compares the magnitudes of the first two operands and branches if their values are not equal. This instruction compares the contents of the accumulator with the contents of internal RAM location or SFR and branches to the destination address if their values are not same. Destination is computed by adding the signed relative-displacement in the last instruction byte to the PC, after incrementing the PC to the start of the next instruction. The carry flag is set if the unsigned integer value of the accumulator is less than the unsigned integer value of direct byte; otherwise, the carry is cleared. Neither operand is affected.

Operation:

(PC) + 3 → (PC)

If (A) <> (DIRECT)

Then

(PC) + Relative Offset → (PC)

If (A) < (DIRECT)

Then

1 → CY

Else

0 → CY

Example: The accumulator contains 34H. Register 7 contains 56H. The first instruction in the sequence,

```
          CJNE R7, # 60H, NOT_EQ      ; R7 = 60H.
NOT_EQ:   JC REQ_LOW                  ; IF R7 < 60H.
                                      ; R7 > 60H.
```

sets the carry flag and branches to the instruction at label NOT_EQ. By testing the carry flag, this instruction determines whether R7 is greater or less than 60H.

If the data presented to port 1 is also 34H, then the following instruction,

WAIT: CJNE A, P1, WAIT

clears the carry flag and continues with the next instruction in sequence, since the accumulator does equal the data read from P1. (If some other value was input on P1, the program loops at this point until the P1 data changes to 34H.)

12. CJNE A, #DATA, REL

Bytes = 3

Cycles = 2

Addressing modes: Immediate addressing mode

Compare the accumulator with immediate data and Jump if Not Equal.

CJNE compares the magnitudes of the first two operands and branches if their values are not equal. This instruction compares the contents of the accumulator with the immediate data given along with the instruction and branches to the destination address if their values are not same. Destination is computed by adding the signed relative-displacement in the last instruction byte to the PC, after incrementing the PC to the start of the next instruction. The carry flag is set if the unsigned integer value of the accumulator is less than the immediate data; otherwise, the carry is cleared. Neither operand is affected.

Operation:

(PC) + 3 → (PC)

If (A) <> data

Then

(PC) + Relative Offset → (PC)

If (A) < data

Then

1 → CY

Else

0 → CY

Example: The accumulator contains 34H. Register 7 contains 56H. The first instruction in the sequence,

```
            CJNE A, # 60H, NOT_EQ
NOT_EQ:     JC REQ_LOW
```

sets the carry flag and branches to the instruction at label NOT_EQ. By testing the carry flag, this instruction determines whether the accumulator is greater or less than 60H.

13. CJNE Rn, #DATA, REL

Bytes = 3

Cycles = 2

Addressing modes: Immediate addressing mode

Compare register Rn with immediate data and Jump if Not Equal.

CJNE compares the magnitudes of the first two operands and branches if their values are not equal. This instruction compares the contents of the register Rn (n = 0 to 7) with the immediate data given along with the instruction and branches to the destination address if their values are not same. Destination is computed by adding the signed relative-displacement in the last instruction byte to the PC, after incrementing the PC to the start of the next instruction. The carry flag is set if the unsigned integer value of register Rn is less than the immediate data; otherwise, the carry is cleared. Neither operand is affected.

Operation:

(PC) + 3 → (PC)

If (Rn) <> data

Then

(PC) + Relative Offset → (PC)

If (Rn) < data

Then

1 → CY

Else

0 → CY

Example: Register 7 contains 56H. The first instruction in the sequence,

```
            CJNE R7, # 60H, NOT_EQ
NOT_EQ:     JC REQ_LOW
```

sets the carry flag and branches to the instruction at label NOT_EQ. By testing the carry flag, this instruction determines whether register R7 is greater or less than 60H.

14. CJNE @Ri, #DATA, REL

Bytes = 3

Cycles = 2

Addressing modes: Immediate addressing mode

Compare contents of internal RAM or SFR and Jump if Not Equal.

CJNE compares the magnitudes of the first two operands and branches if their values are not equal. This instruction compares the contents of the internal RAM location or SFR pointed by register R0 or R1 with the immediate data given along with the instruction and branches to the destination address if their values are not same. Destination is computed by adding the signed relative-displacement in the last instruction byte to the PC, after incrementing the PC to the start of the next instruction. The carry flag is set if the unsigned integer value of memory location is less than the immediate data; otherwise, the carry is cleared. Neither operand is affected.

Operation:

(PC) + 3 → (PC)

If ((Ri)) <> data

Then

(PC) + Relative Offset → (PC)

If ((Ri)) < data

Then

1 → CY

Else

0 → CY

Example: Memory location 34H contains 56H. The first instruction in the sequence,

```
            MOV R0, #34H
            CJNE @R0, #55H, NOT_EQ
NOT_EQ:     JC REQ_LOW
```

clears the carry flag and branches to the instruction at label NOT_EQ. By testing the carry flag, this instruction determines whether the contents of memory are greater or less than 55H.

15. DJNZ Rn, REL

Bytes = 2

Cycles = 2

Addressing modes: Register and immediate addressing modes

DJNZ decrements the contents of register Rn (n = 0 to 7) 1, and branches to the address indicated by the second operand if the resulting value is not zero. An original value of 00H underflows to 0FFH. No flags are affected.

The branch destination is computed by adding the signed relative-displacement value in the last instruction byte to the PC, after incrementing the PC to the first byte of the following instruction.

Operation:

(PC) + 2 → PC

(Rn) – 1 → (Rn)

If (Rn) > 0 or (Rn) < 0

Then

(PC) + Relative offset (rel) → PC

Note: When this instruction is used to modify an output port, the value used as the original port data will be read from the output data latch, *not* the input pins.

Example: The following instruction sequence,

```
            MOV R2, # 08H
TOGGLE:     CPL P1.7
            DJNZ R2, TOGGLE
```

toggles P1.7 eight times, causing four output pulses to appear at bit 7 of output port 1. Each pulse lasts three machine cycles; two for DJNZ and one to alter the pin.

16. DJNZ DIRECT, REL

Bytes = 2

Cycles = 2

Addressing modes: Direct and Immediate addressing modes

DJNZ decrements the contents of internal RAM location or SFR and branches to the address indicated by the second operand if the resulting value is not zero. An original value of 00H underflows to 0FFH. No flags are affected.

The branch destination is computed by adding the signed relative-displacement value in the last instruction byte to the PC, after incrementing the PC to the first byte of the following instruction.

Operation:

(PC) + 2 → PC

(DIRECT) – 1 → (DIRECT)

If (DIRECT) > 0 or (DIRECT) < 0

Then

(PC) + Relative offset (rel) → PC

Example: Internal RAM locations 40H, 50H, and 60H contain the values 01H, 70H, and 15H, respectively. The following

DJNZ 40H, LABEL_1
DJNZ 50H, LABEL_2
DJNZ 60H, LABEL_3

cause a jump to the instruction at label, i.e., LABEL_2 with the values 00H, 6FH, and 15H in the three RAM locations. The first jump was *not* taken because the result was zero. This instruction provides a simple way to execute a program loop a given number of times or for adding a moderate time delay (from 2 to 512 machine cycles) with a single instruction.

17. NOP

Bytes: 1

Cycles: 1

Addressing mode: Implicit addressing mode

No Operation

Execution continues at the following instruction. Other than the PC, no registers or flags are affected.

Encoding: 0 0 0 0 0 0 0 0 = 00h

Operation: NOP

(PC) + 1 → (PC)

Example: A low-going output pulse on bit 7 of port 2 must last exactly 5 cycles. A simple SETB/CLR sequence generates a one-cycle pulse, so four additional cycles must be inserted. This may be done (assuming no interrupts are enabled) with the following instruction sequence,

CLR P2.7
NOP
NOP
NOP
NOP
SETB P2.7

Table 2.5 *Summary of Branching Instruction Group*

Sr. No.	Instruction	No. of Bytes	No. of Cycles
01	ACALL ADDR11	02	02
02	LCALL ADDR16	03	02
03	RET	01	02

04	RETI	01	02
05	AJMP ADDR11	02	02
06	LJMP ADDR16	03	02
07	SJMP REL	02	02
08	JMP @A+DPTR	01	02
09	JZ REL	02	02
10	JNZ REL	02	02
11	CJNE A, DIRECT, REL	03	02
12	CJNE A, #DATA, REL	03	02
13	CJNE Rn, #DATA, REL	03	02
14	CJNE @Ri, #DATA, REL	03	02
15	DJNZ Rn, REL	02	02
16	DJNZ DIRECT, REL	03	02
17	NOP	01	01

2.5 ASSEMBLER DIRECTIVES/ PSEUDO INSTRUCTION/ DUMMY INSTRUCTIONS

Before introducing to you about the assembler directives, it is important to know about the word assembler.

Assembler:

It is a software program used to convert assembly language program into machine language equivalent or binary code. Here, the source code is converted into object code with the help of the assembler. We used to call assembler as Asmblr.

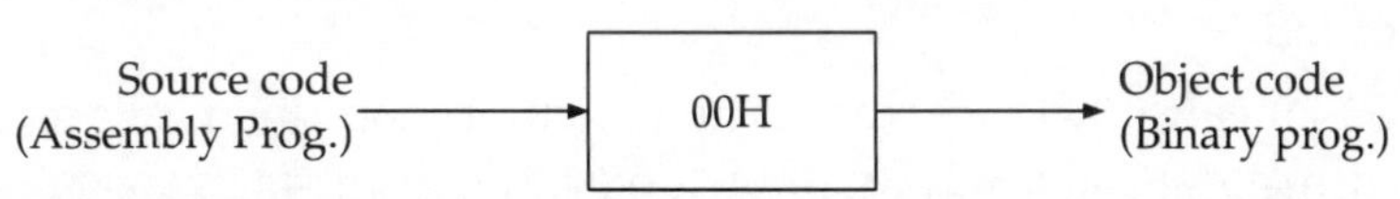

Assembler Directive:

It is the instruction used to inform the assembler about conversion of assembly program into machine language program. It is also called pseudo or dummy instruction because these instructions are not executed by microcontroller. Rather, they are executed by the assembler. In other words, we say that assembler directives is used for assembling and they are not consuming any memory location because they are not executed by microcontroller.

Different types of assembler directives are as follows:

1. **ORG: (ORIGIN)**

   ORG assembler directive is used to initialize starting memory location of the program. There is no memory space allocated for the ORG assembler directive.

   In the program, ORG is always initialized at the beginning.

   In some of the assemblers, .ORG is used instead of ORG assembler directive. ORG assembler directive is always followed by a number which may be represented in decimal or hexadecimal. If the given number is in the decimal, then writing 'D' at the end of the number is optional. But when given number is in the hexadecimal, then put 'H' at the end of the number. Whenever the number is given in any format, assembler is always converted into the hexadecimal.

   FORMAT OF ORG:

   ORG 100, i.e., ORG 0064H, ORG 100H, ORG $

2. **DB: (Define Byte):**

   This assembler directive is used to allocate 1 memory location to each byte. So one memory location is reserved. Along with DB assembler directive, 8-bit number is given either in decimal, binary or in hexadecimal. If the number is given in the decimal format, then it is optional to end the number with 'D'. But when the number is given in the binary format, then put 'B' at the end of the number. If the number is given in hexadecimal, then put 'H' at the end of the number. Regardless of this number system, the assembler always converts the given number into the hexadecimal. Apart from this number system, it is also possible to allocate or give ASCII numbers or characters with this DB assembler directive. To give the ASCII value, it always includes either in single or double quotation marks. Quotation mark indicates the given string represented in ASCII. Its simple application is used to display program in LCD screen. So we are using DB, to save more number of bytes.

   Format of DB:

   DB 12($(12)_d$ means (0CH). So the starting memory location is allocated.)

   DB 1001 0101B (binary is represented as 95H in hexadecimal.)

   DB 45H(given number in hexadecimal.)

   DB 'BHEL' / DB "BHEL", both are possible. Since the single inverted comma is possible in simulator and the double inverted comma is possible.

   For example: ORG 0500H

   0500H ⟶ DATA1: DB 45H, 67H, 01H.

   Given the data starting memory location is 0500H.

0500H	45H
0501H	67H
0502H	01H

Note: There is no version for .DB.

3. **EQU: (EQUATE)**

   This assembler directive is used to assign constant value to any value. For example, Label EQU 1000H, then after executing this, the value of label becomes 1000H.

   Format:

   Label EQU value or expression which may be register, accumulator, jump.

   Use of Equate:

   Consider the example of counter value. Assume counter value is 25, then using EQU it may be defined as

   ```
   COUNT EQU 25
   :       :   :
   MOV     R2, #Count
   ```

   In the above example, the value of counter, i.e., register R2 is 25 and suppose that we need this counter value in different places in the program, then the task is simplified by defining counter value using EQU only once in the program by a programmer and the assembler is responsible for searching all its occurrences in the program and replacing the constant value with 25. So there is no need to search all its occurrences manually.

   In other words, we say that programmer will define constant value only once using EQU and the assembler is responsible to search its occurrences and replace the value in many different places.

4. **END**

   END means the end of the assembly process. So, no instruction either belonging to the assembler or microcontroller is executed after END.

   END gives only once in a program and that to be in the last indicating end of assembly process. Some assembler will support .END instead of END assembler directive.

2.6 DELAY CALCULATION FOR 8051 MICROCONTROLLER

Machine cycles for 8051 microcontroller

The central processing unit (CPU) takes certain amount of time or number of clock cycles to execute an instruction. In 8051 microcontroller, these clock cycles are referred to as machine cycles. In 8051 microcontroller, the length of the machine cycle depends on the frequency of the crystal oscillator connected to the 8051 microcontroller. In the 8051 microcontroller, one machine cycle lasts 12 oscillator periods, therefore to calculate machine cycle for the 8051 we have to take 1/12th of the crystal frequency. In other words, internal operating frequency of the 8051 microcontroller is 1/12th of the external frequency connected to the 8051 microcontroller system.

Example 2.38

Find period of the machine cycle for the frequency

(*a*) 11.0592 MHz

(*b*) 16 MHz

Solution: (*a*)

$$\text{Internal frequency} = \frac{11.0592 \text{ MHz}}{12} = 921.6 \text{ kHz}$$

Therefore,

$$\text{machine cycle} = 1/f = \frac{1}{921.6 \text{ kHz}} = 1.085\ \mu\text{s}$$

(*b*)

$$\text{Internal frequency} = \frac{16 \text{ MHz}}{12} = 1.333 \text{ MHz}$$

Therefore,

$$\text{machine cycle} = 1/f = \frac{1}{1.333 \text{ MHz}} = 0.75\ \mu\text{s}$$

Example 2.39

For an 8051 system of 11.0592 MHz, find how long it takes to execute each of the following instructions:

MOV R4, #56H
DJNZ R2, AGAIN
DJNZ 30H, REPEAT
LJMP 2000H
DIV AB

Solution: Crystal frequency = external frequency = 11.0592 MHz

$$\text{Internal frequency} = \frac{11.0592 \text{ MHz}}{12} = 921.6 \text{ kHz}$$

Therefore,

$$\text{Machine cycle} = 1/f = \frac{1}{921.6 \text{ kHz}} = 1.085\ \mu\text{s}$$

Instruction	Machine cycles	Time to execute
MOV R4, #56h	1	1 × 1.085 μ sec = 1.085 μ sec
DJNZ R2, AGAIN	2	2 × 1.085 μ sec = 2.17 μ sec
DJNZ 30H, REPEAT	2	2 × 1.085 μ sec = 2.17 μ sec
LJMP 2000H	2	2 × 1.085 μ sec = 2.17 μ sec
DIV AB	4	4 × 1.085 μ sec = 4.34 μ sec

Example 2.40

Find the size of the delay in the following program; assume crystal frequency as 11.0592 MHz.

```
DELAY:  MOV R2, #200
HERE:   DJNZ R2, HERE
        RET
```

Solution: Crystal frequency = external frequency = 11.0592 MHz

$$\text{Internal frequency} = \frac{11.0592 \text{ MHz}}{12} = 921.6 \text{ kHz}$$

Therefore,

$$\text{Machine cycle} = 1/f = \frac{1}{921.6 \text{ kHz}} = 1.085 \ \mu\text{s}$$

Instruction	Machine cycle
MOV R2, #200	1
DJNZ R2, HERE	2
RET	2

Therefore, time delay = [1 × 1.085 μs + 2 × 200 × 1.085 μs + 2 × 1.085 ms]

Here the inner loop HERE: DJNZ R2, HERE is executed 200 times, and hence the inner time is calculated as 2 * 200.

= [1 + 400 + 2] × 1.085 μ sec

= 403 × 1.085 μs = 436.255 μs

Example 2.41

Find the maximum possible delay for the previous program.

Solution:

```
DELAY:  MOV R2, #255
HERE:   DJNZ R2, HERE
        RET
```

The size of the register R2 is 8-bit, so maximum value to be loaded in the register R2 is 255 decimal.

Therefore,

Time delay = [1 × 1.085 μs + 2 × 255 × 1.085 μs + 2 × 1.085 μs]

Here the inner loop HERE: DJNZ R2, HERE is executed 200 times, and hence the inner time is calculated as 2 × 255.

= [1 + 510 + 2] × 1.085 μs

= 513 × 1.085 μs = 556.605 μs

So, the maximum delay for the above program is 556.605 μs (microsecond) approximately.

Example 2.42

For an 8051 system of 11.0592 MHz, find time delay for the following program:

```
        MOV R2, #255
REPEAT: NOP
        NOP
        NOP
        NOP
        NOP
        DJNZ R2, REPEAT
        RET
```

Solution:

INSTRUCTION	MACHINE CYCLE
MOV R2, #255	1
REPEAT: NOP	1
NOP	1
NOP	1
NOP	1
NOP	1
NOP	1
DJNZ R2, REPEAT	2
RET	2

Crystal frequency = external frequency = 11.0592 MHz

$$\text{Internal frequency} = \frac{11.0592 \text{ MHz}}{12} = 921.6 \text{ kHz}$$

Therefore, $\text{machine cycle} = 1/f = \frac{1}{921.6 \text{ kHz}} = 1.085 \ \mu\text{s}$

Therefore,

$$\text{Time delay} = [1 \times 1.085 \ \mu\text{s} + (1 + 1 + 1 + 1 + 1 + 2) \times 255 \times 1.085 \ \mu\text{s} + 2 \times 1.85 \ \mu \text{ sec}]$$

$$= [1 + 7 \times 255 + 2] \times 1.085 \ \mu\text{s}$$

Here the inner loop consists 5 NOP instructions and DJNZ instruction so total machine cycles within the inner loop is 5+2 = 7 and it is executed 255 times, and hence the inner time is calculated as 7 × 255.

$$= 1788 \times 1.085 \ \mu\text{s} = 1939.98 \ \mu\text{s}$$

Example 2.43

For a machine cycle of 1.085 μs (microsecond), find the time delay in the following subroutine program:

```
        MOV R2, #250
REPEAT: MOV R3, #200
HERE:   NOP
        NOP
        DJNZ R3, HERE
        DJNZ R2, REPEAT
        RET
```

Solution:

Time delay = [1 + {(1 + 1 + 2) × 200 + 1 + 2} × 250 + 2] × 1.085 μs]

= 217817.005 μs

REVIEW QUESTIONS

1. What is assembler? Explain assembler directives in details.
2. Explain the following instructions of microcontroller 8051 with suitable examples.
   (a) SUBB A, R2
   (b) MUL AB
   (c) INC DPTR
   (d) DEC DPL
   (e) DA A
3. What are addressing modes? Explain addressing modes with suitable examples.
4. Write an assembly language program to perform X + Y, where X and Y are 16-bit numbers present in the internal RAM location from 30h. Store the result at the external RAM location 2001 h – 2000 h.
5. Explain the importance of register indirect addressing mode over direct addressing mode.
6. Explain bit addressable RAM; also explain any five instructions related to Boolean operation.
7. Write three instructions for clearing an accumulator.
8. Suppose that there is no instruction available for clearing the carry flag, then how can a carry flag be cleared. Show the instructions to perform this.
9. Bit addressing capability enhances the features of microcontroller. Justify this statement.

10. Differentiate between absolute call and long call.
11. Explain pseudo instructions with suitable examples.
12. Show the instruction to transfer the contents of internal RAM to register B using indirect addressing and direct addressing mode.
13. Explain different instructions to transfer the contents of external and internal ROM to the accumulator.
14. Show the instruction to transfer the contents of external RAM location 5000 h to the registers B and R7 of register bank 2.
15. Show how to transfer the contents of register R2 of register bank 1 to the register R3 of register bank 2 using any two methods.

Assembly Language Programming of 8051 Microcontroller

3 Chapter

This chapter provides the readers the basic foundation of writing an assembly language program using the different instructions of microcontroller 8051. For the benefit of readers, a very basic programming to logical one is included in this chapter and also for better understanding flowcharts are drawn.

Programming is always a step-by-step approach of writing instructions, and requires logical thinking. This chapter deals with the same approach, so readers may not only enjoy the art of programming, but also get insight on how to write efficient programs for one of the most popular microcontrollers.

3.1 ASSEMBLY LANGUAGE PROGRAMMING

Program 3.1

Write an ALP (assembly language program) to move a block of data bytes present in the internal memory location with starting address 40H and ending address 50H to the destination memory with starting address 60H without overlapping.

Solution: Assembly language program:

	Mnemonic	Comments
	MOV R1,#40H	; Source starting address
	MOV R2,#50H	; Source ending address
	MOV R0,#60H	; Destination starting address
	CLR C	; Clearing the carry flag
	MOV A,R2	; Copying the data 50H to the accumulator
	SUBB A,R1	; Subtracting the accumulator data byte with register R1 content
	MOV R2,A	; Copying the result after subtraction into register R2 indicating the counter.
L2:	*MOV A,@R1*	; Copying the first data byte of source into the accumulator
	MOV @R0,A	; Transferring the content of source into the destination register.

```
        INC R1         ; Incrementing the register R1 by one.
        INC R0         ; Incrementing the register R0 by one.
        DJNZ R2,L2     ; Decrementing the register R2 content by one
                         and jump to L2 if (R2) is not equal to zero. If
                         (R2) is zero then go to the next instruction.
LAST:   SJMP LAST      ; End of the Program.
```

Flowchart:

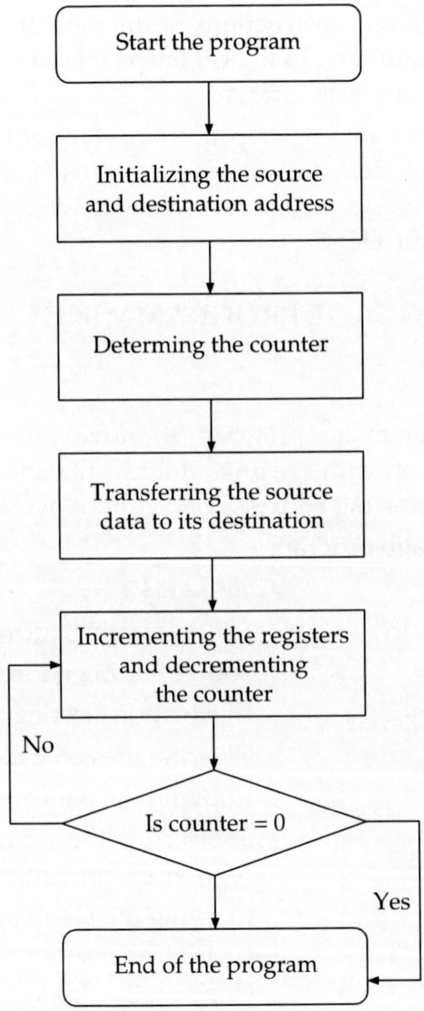

Approach: In the above program, first the registers are initialized with the respective memory location addresses along with the calculation of the counter value. Then, finally the source address data is transferred to the destination address till the counter is reset to zero indicating the last byte of the source data is transferred to its destination. Finally, the source data is transferred to the destination without overlapping, i.e., no memory location address is overlapped.

Program 3.2

Write an ALP to move a block of data bytes present in the internal memory location with starting address 40H and ending address 50H to the destination memory with the starting address 45H with overlapping.

Solution: Assembly language program:

	Mnemonic	Comments
	MOV R1,#40H	; Source starting address
	MOV R2,#50H	; Source ending address
	MOV R0,#55H	; Destination starting address
	CLR C	; Clearing the carry flag
	MOV A,R2	; Copying the data 50H to the accumulator
	SUBB A,R1	; Subtracting the accumulator data byte with register R1 content
	MOV R2,A	; Copying the result after subtraction into register R2 indicating the counter.
	MOV A,R1	; Copying the data byte of register R1 to the accumulator
	ADD A,R2	; Adding the accumulator data byte with register R2
	MOV R1,A	; Copying the accumulator data byte to register R1
	MOV A,R0	; Copying the data byte of register R0 to the accumulator
	ADD A,R2	; Adding the accumulator data byte with register R2
	MOV R0,A	; Transferring the accumulator data byte to register R0
	INC R2	; Incrementing the register R2 by one.
L2:	*MOV A,@R1*	; Copying the first data byte of source into the accumulator
	MOV @R0,A	; Transferring the content of source into the destination register.
	DEC R1	; Decrementing the register R1 by one.

```
        DEC R0          ; Decrementing the register R0 by one.
        DJNZ R2,L2      ; Decrementing the register R2 content by one
                          and jump to L2 if (R2) is not equal to zero. If
                          (R2) is zero then go to the next instruction.
LAST:   SJMP LAST       ; End of the Program.
```

Flowchart:

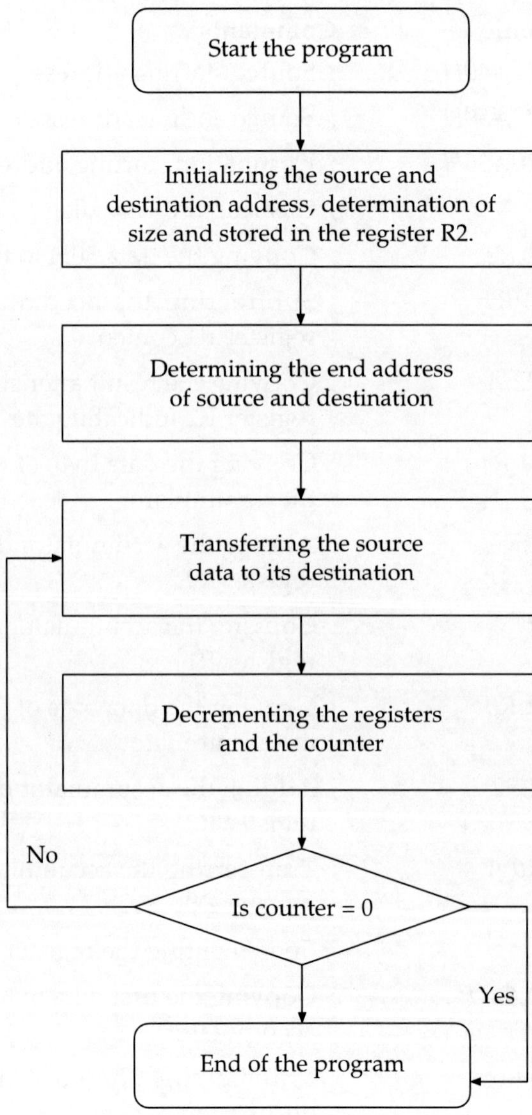

Approach:

In the above program, first the registers are initialized with the respective memory location addresses along with the calculation of the counter value. The most important crux of the above program is to transfer the data byte from the ending source address to that of the ending destination address. Then, finally the source address data is transferred to the destination address till the counter is reset to zero indicating the last byte of the source data is transferred to its destination. Finally, the source data is transferred to the destination with overlapping, i.e., the memory location addresses are overlapped on transferring.

Program 3.3

Write an ALP to move a block of data bytes present in external memory with starting address 3000h to the destination memory with starting address 4000h and size of array is 0Ah.

Solution: Assembly language program:

	Mnemonics	**Comment**
	ORG 0000H	; Initializing the starting memory location of the program
	COUNT EQU 0AH	; The assembler directive EQU is used to assign value 0Ah to the counter.
	MOV R0, COUNT	; Copying the counter value into the register R2.
	MOV 82H,#00H	; Loading the data byte 00h into the memory location 82H.
L2:	*MOV 83H,#30H*	; Initializing the external memory location with the DPTR register.
	MOVX A, @DPTR	; Copying the data byte present in the external memory location into the accumulator.
	MOV 83H,#40H	; Initializing the new external memory location with the DPTR register.
	MOVX @DPTR,A	; Transferring the accumulator data byte externally into the DPTR memory location.
	INC DPTR	; Incrementing the value of DPTR by one.
	DJNZ R0,L2	; Decrementing the register R0 content by one and jump to L2 if (R0) is not equal to zero. If (R0) is zero then go to the next instruction.
	LAST: *SJMP LAST*	; End of the Program.

Flowchart:

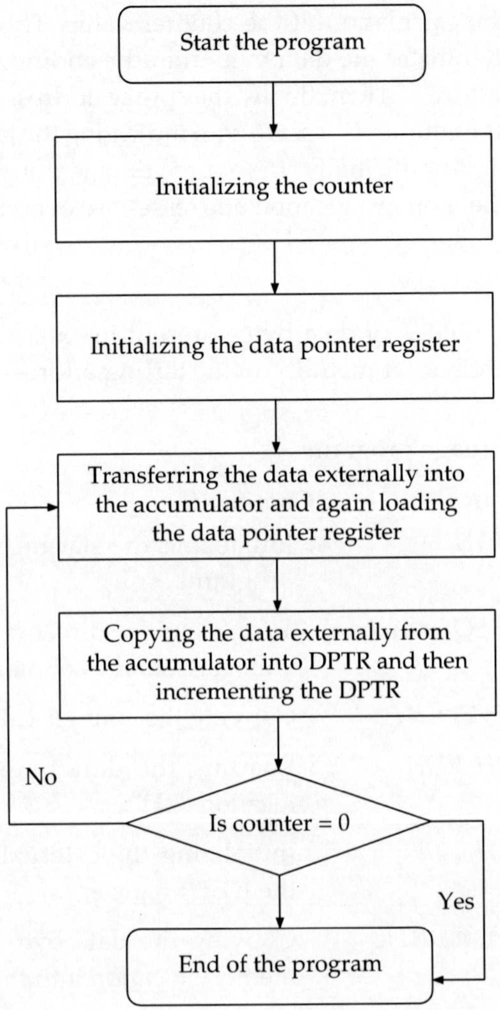

Approach:

- First the counter is initialized to indicate the value of the data bytes to be transferred.
- Then the external memory location is assigned to the data pointer register.
- The data from the external memory location is transferred to the accumulator.
- Then the data is transferred to the destination memory location externally after some logic.
- The process is repeated until all the data bytes are transferred into the required destination register.

Program 3.4

Write an ALP to exchange a block of data bytes present in the external memory. Starting address of the first is 3000h and starting address of other block is 4000h and size of array is 0Ah.

Solution: Assembly language program:

	Mnemonics	**Comment**
	ORG 0000H	; Initializing the starting memory location of the program
	COUNT EQU 0AH	; The assembler directive EQU is used to assign value 0Ah to the counter.
	MOV R0, COUNT	; Copying the counter value into register R2.
	MOV 82H,#00H	; Loading the data byte 00h into the memory location 82H.
L2:	*MOV 83H,#30H*	; Initializing the external memory location with the DPTR register.
	MOVX A,@DPTR	; Copying the data byte present in the external memory location into the accumulator.
	MOV R1,A	; Transferring the accumulator data byte into register R1.
	MOV 83H,#40H	; Initializing the new external memory location with the DPTR register.
	MOVX A,@DPTR	; Copying the data byte of the new external memory location into the accumulator.
	XCH A, R1	; Exchanging the contents of the accumulator and the register R1.
	MOVX @DPTR,A	; Copying the accumulator data byte externally into the new external memory location.
	MOV 83H,#30H	; Initializing the new external memory location with the DPTR register.
	MOV A,R1	; Copying the data byte of register R1 into the accumulator.
	MOVX @DPTR,A	; Transferring the accumulator data byte externally into the DPTR memory location.

```
        INC DPTR        ; Incrementing the value of DPTR by one.
        DJNZ R0,L2      ; Decrementing the register R0 content by
                          one and jump to L2 if (R0) is not equal to
                          zero. If (R0) is zero then go to the next
                          instruction.
LAST:   SJMP LAST       ; End of the program.
```

Flowchart:

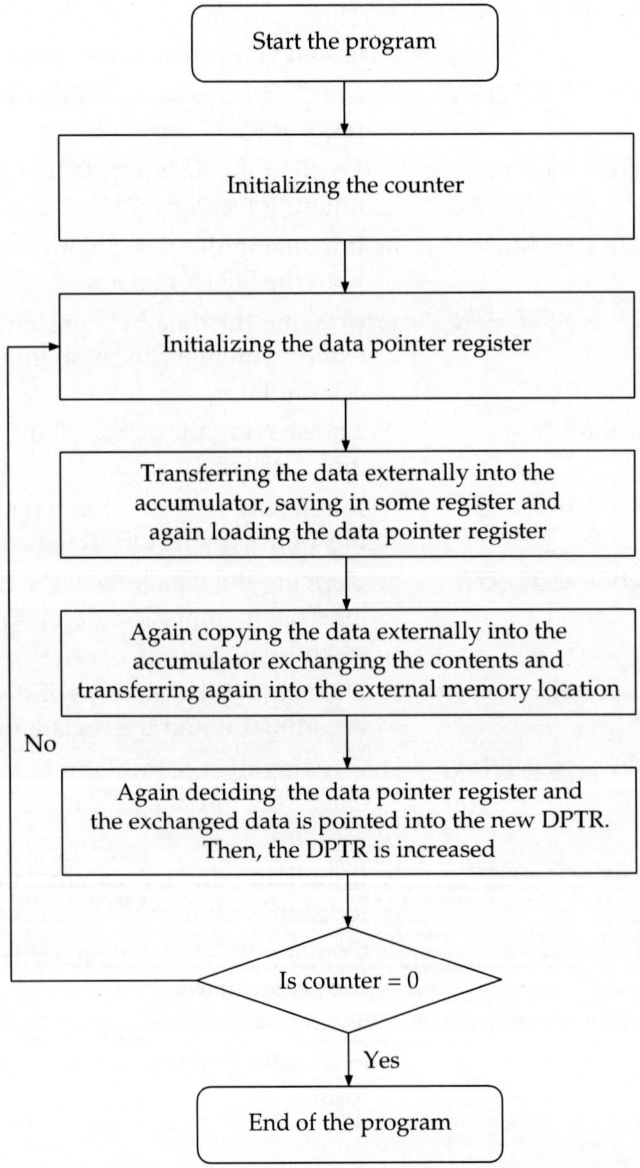

Approach:

- First the external memory location is allocated and then the data is transferred externally into the accumulator.
- Then it is copied into some register and again the new external memory location is allocated.
- Finally, the data byte transferred from external memory location again will be exchanged with the data byte saved in the register and then will transfer into the required data pointer register.
- Finally, the above process will continue by manipulation of the memory location addresses into the required destined memory locations until all the data bytes are exchanged.

Program 3.5

Write an ALP to sort a given array present in the external memory with a starting address 3000h and size of the array is 0Ah using bubble sort technique in ascending order.
Solution: Before discussing the above program, it is important to know what actually the bubble sorting technique is about. It's a simple sorting algorithm that works on repeated sorting of the list that needs to be sorted on comparison with each pair of adjacent items and if they are found in wrong order, they need to be sorted. The above definition will be more clear by looking into the above example.

Let us take the array of numbers "9 3 7 6 4", and sort the array from lowest number to greatest number using bubble sort. In each step, elements written in bold are being compared. Five passes will be required.

First Pass:

(**9 3** 7 6 4) → (**3 9** 7 6 4), Here, algorithm compares the first two elements, and swaps since 9 > 3.
(3 **9 7** 6 4) → (3 **7 9** 6 4), Swap since 9 > 7
(3 7 **9 6** 4) → (3 7 **6 9** 4), Swap since 9 > 6
(3 7 6 **9 4**) → (3 7 6 **4 9**), Swap since 9 > 4.

Second Pass:

(**3 7** 6 9 4) → (**3 7** 6 9 4), Since the array is already sorted, so no need to interchange.
(3 **7 6** 9 4) → (3 **6 7** 9 4), Swap since 7 > 6.
(3 6 **7 9** 4) → (3 6 **7 9** 4), Since the array is already sorted, so no need to interchange.
(3 6 7 **9 4**) → (3 6 7 **4 9**), Swap since 9 > 4.

Now, the array is already sorted, but our algorithm does not know if it is complete. The algorithm needs one whole pass without any swap to know it is sorted.

Third Pass:

(**3 6** 7 4 9) → (**3 6** 7 4 9), Since the array is already sorted, so no need to interchange.
(3 **6 7** 4 9) → (3 **6 7** 4 9), Since the array is already sorted, so no need to interchange.

(3 6 **7 4** 9) → (3 6 **4 7** 9), Swap since 7 > 4.

(3 6 4 **7 9**) → (3 6 4 **7 9**), Since the array is already sorted, so no need to interchange.

Fourth Pass:

(**3 6** 4 7 9) → (**3 6** 4 7 9), Since the array is already sorted, so no need to interchange.

(3 **6 4** 7 9) → (3 **4 6** 7 9), Swap since 6 > 4.

(3 4 **6 7** 9) → (3 4 **6 7** 9), Since the array is already sorted, so no need to interchange.

(3 4 6 **7 9**) → (3 4 6 **7 9**), Since the array is already sorted, so no need to interchange.

Fifth Pass:

(**3 4** 6 7 9) → (**3 4** 6 7 9), Since the array is already sorted, so no need to interchange.

(3 **4 6** 7 9) → (3 **4 6** 7 9), Since the array is already sorted, so no need to interchange.

(3 4 **6 7** 9) → (3 4 **6 7** 9), Since the array is already sorted, so no need to interchange.

(3 4 6 **7 9**) → (3 4 6 **7 9**), Since the array is already sorted, so no need to interchange.

So, here in the fifth pass, we can see that the array is already sorted. Hence, we can conclusively say that the array is finally sorted. So, I think you are now familiar with the bubble sort technique. Now, we will get into the above program.

Assembly Language Program:

	Mnemonics	**Comment**
	ORG 0000H	; Initializing the starting memory location of the program.
	COUNT EQU 09H	; The assembler directive EQU is used to assign value 09h to the count.
	MOV R1, #COUNT	; Loading the counter value into the register R1.
OUTER:	*MOV R0,#COUNT*	; Loading the counter value into the register R0.
	MOV DPTR, #3000H	; Allocating the data pointer with memory location 3000H.
INNER:	*CLR C*	; Clearing the carry flag
	MOVX A, @DPTR	; Copying the data byte externally into the accumulator.
	MOV R2,A	; Storing the data byte from the accumulator into register R2.
	INC DPTR	; Incrementing the data pointer register.
	MOVX A, @DPTR	; Copying the data byte externally into the accumulator.

```
        MOV R3,A          ; Storing the data byte from
                            the accumulator into register R3.
        SUBB A,R2         ; Subtracting the value of register
                            and borrow with the accumulator.
        JNC L1            ; Jump to label L1 if carry flag is reset
                            else go to the next instruction.
        XCH A,R2          ; Exchanging the contents of
                            register with that of the
                            accumulator.
        MOVX @DPTR,A      ; Copying the data byte externally
                            from the accumulator into data
                            pointer register.
        DEC 82H           ; Decrementing the data pointer
                            register by one.
        MOV A,R3          ; Copying the data byte from
                            register R3 into the accumulator.
        MOVX @DPTR,A      ; Copying the data byte externally
                            from the accumulator into data
                            pointer register.
L1:     INC DPTR          ; Incrementing the data pointer
                            register.
        DEC R0            ; Decrementing register R0 by
                            one.
        MOV A,R0          ; Copying register R0 content
                            into the accumulator.
        CJNE A,#00H,INNER ; Comparing the value of the
                            accumulator and jump to
                            INNER label if (A) is equal to 00H,
                            then go to the next instruction else
                            jump to the label INNER.
        DEC R1            ; Decrementing the value of
                            register R1 by one.
        MOV A,R1          ; Copying register R1 content
                            into the accumulator.
        CJNE A,#00H,OUTER ; Comparing the value of the
                            accumulator and jump to
                            OUTER label if (A) is equal to 00H,
                            then go to the next instruction else
                            jump to the label OUTER.
LAST:   SJMP LAST         ; End of the program.
```

Flowchart:

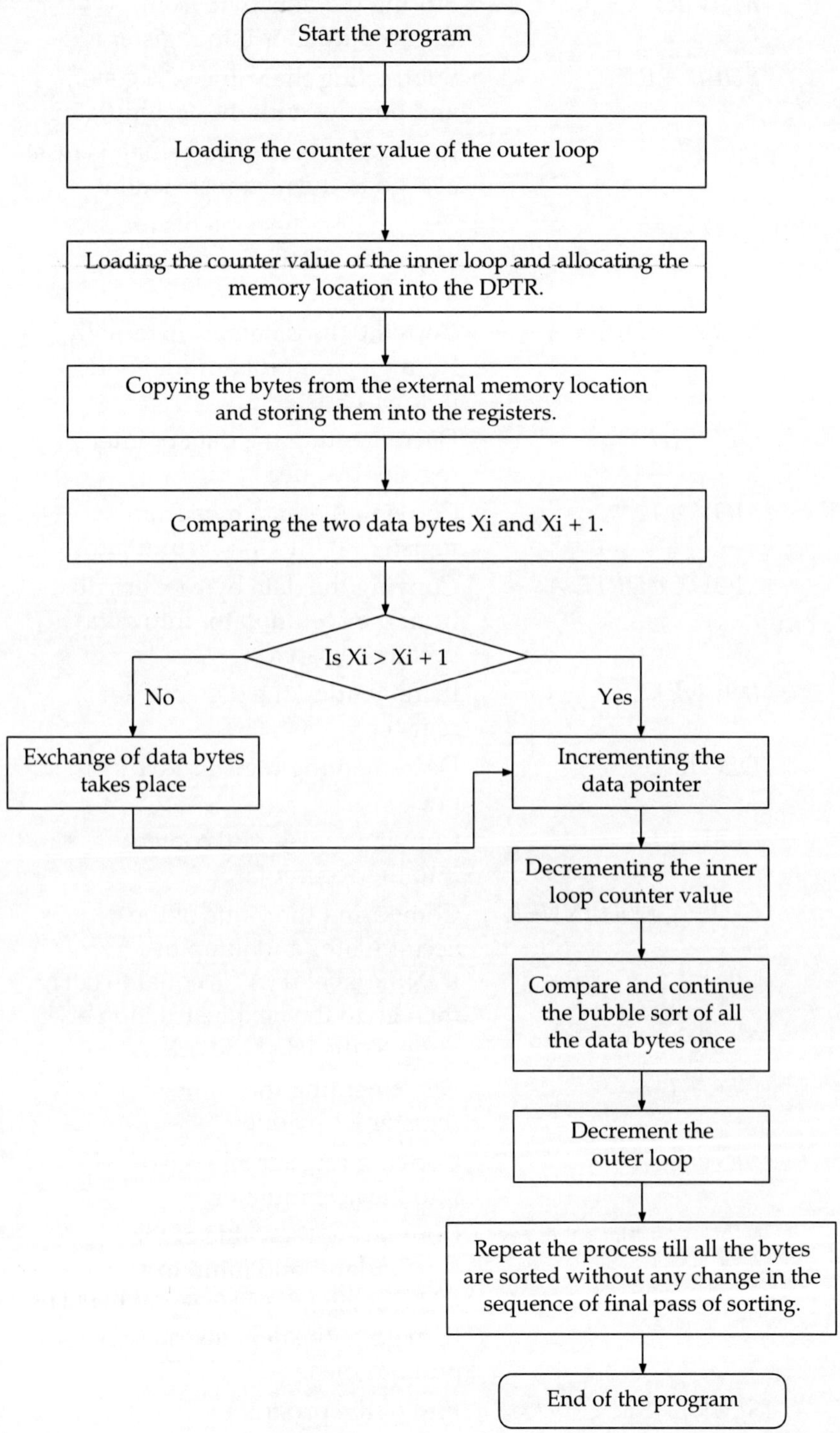
Start the program
Loading the counter value of the outer loop
Loading the counter value of the inner loop and allocating the memory location into the DPTR.
Copying the bytes from the external memory location and storing them into the registers.
Comparing the two data bytes Xi and Xi + 1.
Is Xi > Xi + 1
No
Yes
Exchange of data bytes takes place
Incrementing the data pointer
Decrementing the inner loop counter value
Compare and continue the bubble sort of all the data bytes once
Decrement the outer loop
Repeat the process till all the bytes are sorted without any change in the sequence of final pass of sorting.
End of the program

Approach:

- First loading of the inner loops is done for the inner passes of all the data bytes once by comparing 2 data bytes at a time.
- Then the outer loop counter is initialized for checking that all the data bytes are sorted without any change in the sequence of final pass of sorting.
- Then 2 data bytes are stored in the registers externally from the memory location pointed by DPTR register indicating the data bytes that need to be sorted.
- Then the 2 data bytes are then compared for checking of higher values among them.
- If the first data byte is larger than the other data byte(Xi>Xi+1), then the exchange of location of the data bytes will follow indicating that the data bye stored in the ith position is greater then the $i+1^{th}$ position.
- If the first byte is smaller than the second byte then the data pointer register is incremented.
- Then the inner count value is decreased indicating that 2 data bytes are compared while checking of the rest of the data bytes of the first pass.
- Finally, the first pass of the data bytes is done.
- The process is repeated until all the data bytes are bubble sorted, i.e., in the last pass none of the positions of the data bytes are exchanged indicating that all the data bytes of the external memory location are sorted.

Program 3.6

Write an ALP to sort a given array present in the internal memory with a starting address 80h and size of the array is 0Ah using bubble sort technique.

Solution: Assembly language program:

Mnemonics	**Comment**
ORG 0000H	; Initializing the starting memory location of the program.
COUNT EQU 09H	; The assembler directive EQU is used to assign value 09h to the count.
MOV R1,#80H	; Assigning the memory location 80H indicating the data bytes starting to be sorted.
MOV DPTR, #3000H	; Allocating the data pointer with memory location 3000H.

```
        MOV A, #COUNT   ; Loading the counter value 09H
                          into the accumulator.
        MOVX @DPTR,A    ; Copying the data externally into
                          DPTR from the accumulator for
                          counter value.
        MOV R0, #COUNT  ; Loading the counter value into
                          register R0.
INNER:  CLR C           ; Clearing the carry flag
        MOV A, R1       ; Copying the data byte from
                          register R1 into the
                          accumulator.
        MOV R2,A        ; Storing the data byte from the
                          accumulator into register R2.
        INC R1          ; Incrementing register R1 by
                          one.
        MOV A, R1       ; Storing the data byte from
                          register R1 into the
                          accumulator
        MOV R3,A        ; Storing the data byte from the
                          accumulator into register R3.
        SUBB A,R2       ; Subtracting the value of register
                          R2 and borrow with the
                          accumulator value.
        JNC L1          ; Jump to label L1 if carry flag is reset,
                          else go to the next instruction.
        XCH A,R2        ; Exchanging the contents of
                          register R2 with that of the
                          accumulator.
```

```
        MOV R1,A            ; Copying the data byte from the
                              accumulator into register R1.
        DEC R1              ; Decrementing register R1
                              value by one.
        MOV A,R3            ; Copying the data byte from
                              register R3 into the accumulator.
        MOV R1,A            ; Copying the data byte from the
                              accumulator into register R1.
L1:     INC R1              ; Incrementing register R1 by
                              one.
        DEC R0              ; Decrementing register R0 by
                              one.
        MOV A,R0            ; (R0) (A)
        CJNE A,#00H,INNER   ; Comparing the value of the
                              accumulator and move to the next
                              instruction if (A) is equal to 00H
                              else jump to the label INNER.
        MOVX A,@DPTR        ; Copying the DPTR data byte
                              externally into the
                              accumulator.
        DEC A               ; (A) - 1              (A)
        MOVX @DPTR,A        ; Copying the data externally into
                              DPTR from the accumulator for
                              counter value.
        CJNE A,#00H,INNER   ; Comparing the value of the
                              accumulator and move to the next
                              instruction if (A) is equal to
                              00H, else jump to the label
                              INNER.
LAST:   SJMP LAST           ; End of the program.
```

Flowchart:

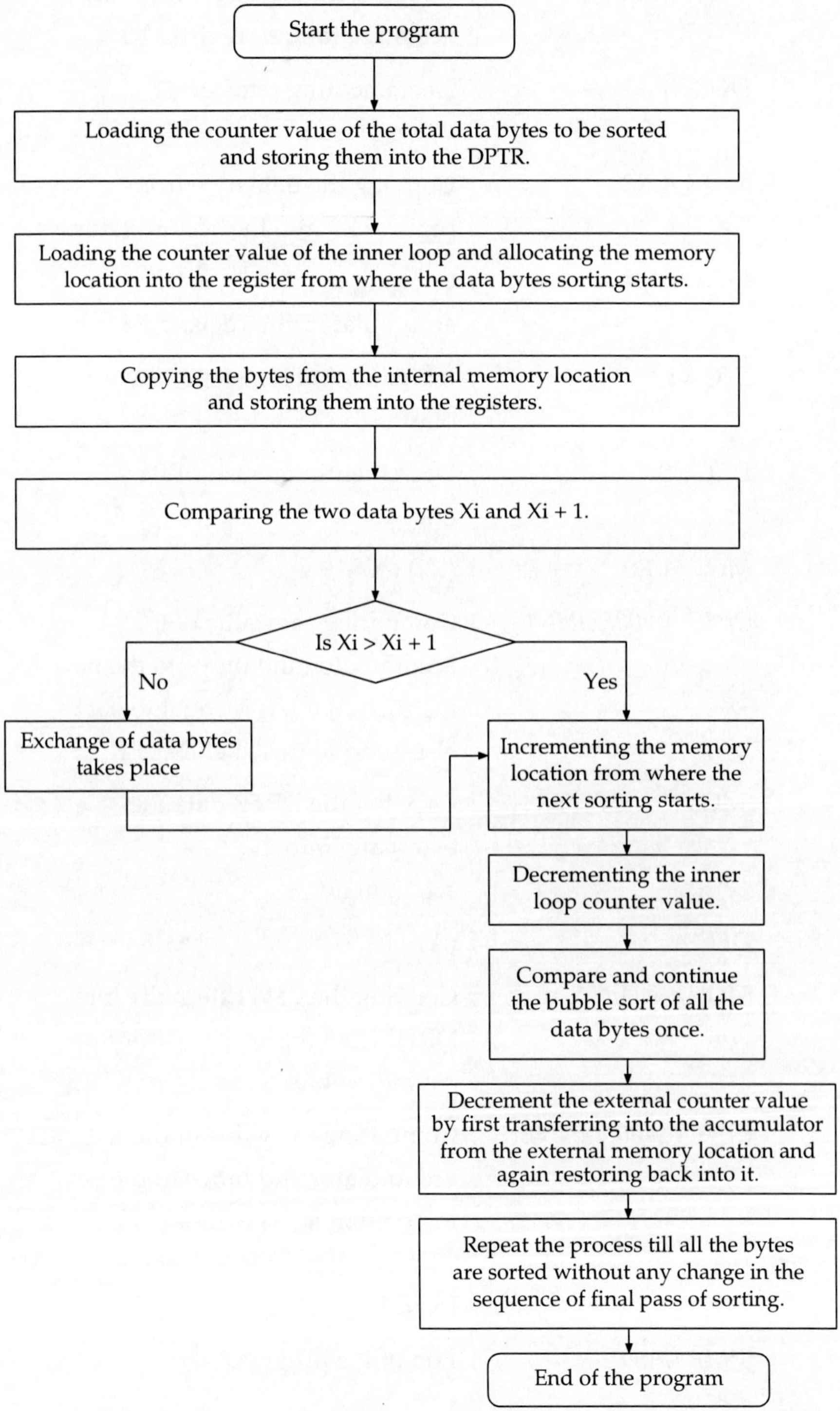
Start the program
Loading the counter value of the total data bytes to be sorted and storing them into the DPTR.
Loading the counter value of the inner loop and allocating the memory location into the register from where the data bytes sorting starts.
Copying the bytes from the internal memory location and storing them into the registers.
Comparing the two data bytes Xi and Xi + 1.
Is Xi > Xi + 1
No
Yes
Exchange of data bytes takes place
Incrementing the memory location from where the next sorting starts.
Decrementing the inner loop counter value.
Compare and continue the bubble sort of all the data bytes once.
Decrement the external counter value by first transferring into the accumulator from the external memory location and again restoring back into it.
Repeat the process till all the bytes are sorted without any change in the sequence of final pass of sorting.
End of the program

Approach:

- First the memory location address is stored in one of the registers.
- One of the important points to understand here is the usage of data pointer register. The DPTR memory location has been allocated and the count value of the required data bytes to be bubble sorted has been stored.
- Then the 2 data bytes are stored in the registers internally from the memory location pointed by DPTR register indicating the data bytes that need to be sorted.
- The 2 data bytes are then compared for checking of higher values among them.
- If the first data byte is larger than the other data byte (Xi>Xi+1), then the exchange of location of the data bytes will follow indicating that the data byte stored in the ith position is greater than the $i+1^{th}$ position.
- If the first byte is smaller than the second byte then the register memory location has been incremented.
- Then the inner count value is decreased indicating that the 2 data bytes are compared and checking of the rest of the data bytes of the first pass.
- The process repeats until all the data bytes of the first pass are sorted.
- Finally, the first pass of the data bytes are done.
- Then the counter value stored in the external memory location is transferred in the accumulator which is decremented by one and again stored back externally into the DPTR memory location.
- The process is repeated until all the data bytes are bubble sorted, i.e., in the last pass none of the positions of the data bytes are exchanged indicating that all the data bytes of the internal memory location are sorted.

Program 3.7

Write an ALP to add 'x' bytes stored in external RAM. The starting address is 3000H and the number of bytes is 12 or 0CH.

Solution: Assembly language program:

Mnemonic	**Comments**
ORG 0000H	; Initializing the starting memory
COUNT EQU 0CH	; The assembler directive EQU is used to assign value 09h to the count.

```
        MOV R1,#COUNT    ; Copying the count value in register
                           R1
        CLR A            ; Resetting the accumulator content.
        MOV R0,A         ; Resetting register R0 for carrying of
                           SUM
        MOV R3,A         ; For the carry flag if generated.
        MOV DPTR,#3000H  ; Assigning the external memory location
                           3000H into the data pointer register.
L1:     MOVX A,@DPTR     ; Copying the data byte externally into the
                           accumulator.
        ADD A,R0         ; Adding register R0 data byte with the
                           accumulator
        MOV R0,A         ; Storing a copy into register R0
        JNC L2           ; Copying the result after subtraction into
                           register R2 indicating the counterchecking
                           if the carry is generated.
        INC R3           ; If CY = 1, then R3 data byte is increased
                           by one.
L2:     INC DPTR         ; Incrementing the data pointer by one.
        DJNZ R1,L2       ; Decrementing register R1 content by
                           one and jump to L2 if (R1) is not equal to zero.
                           If (R1) is zero then go to the next
                           instruction.
LAST:   SJMP LAST        ; End of the program.
```

Flowchart:

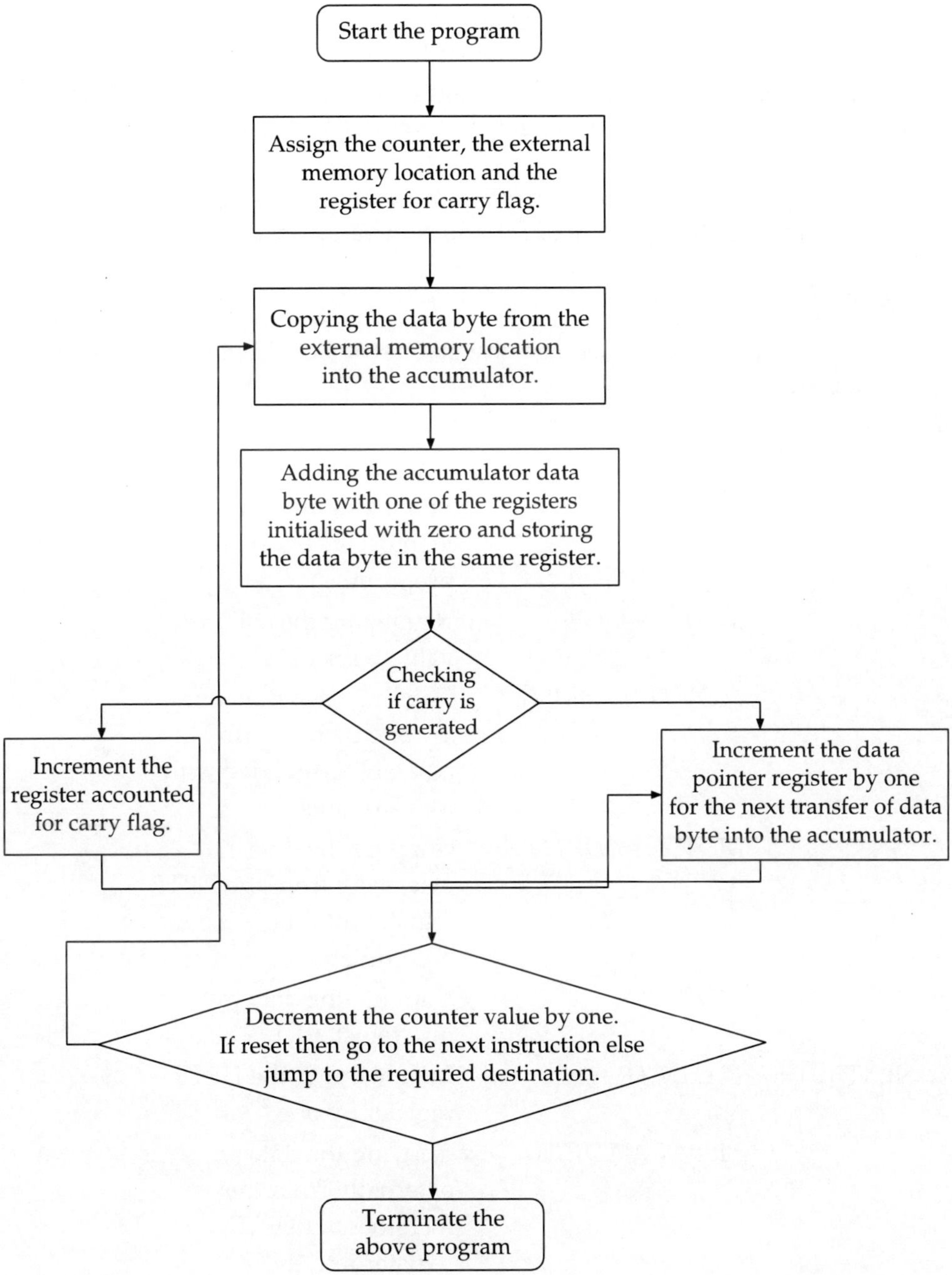
Start the program
Assign the counter, the external memory location and the register for carry flag.
Copying the data byte from the external memory location into the accumulator.
Adding the accumulator data byte with one of the registers initialised with zero and storing the data byte in the same register.
Checking if carry is generated
Increment the register accounted for carry flag.
Increment the data pointer register by one for the next transfer of data byte into the accumulator.
Decrement the counter value by one. If reset then go to the next instruction else jump to the required destination.
Terminate the above program

Approach:

- First the counter value and the external memory location is initialized.
- Then, the data from the external memory location will be copied into the accumulator and added with the register initialized with the data byte 00h.
- Then, it is checked whether the carry flag is being generated or not. If yes, then the register allocated for the carry will be incremented or else will be moved into the next instruction.
- The data pointer is then incremented by one and the counter value is checked whether the necessary data bytes required for addition had been done or not.

Program 3.8

Write an ALP to find the smallest element in a given string of 12 bytes and store the smallest element at location FEDCH.

Solution: Assembly language program:

	Mnemonics	**Comment**
	ORG FE00H	; Initializing the starting memory location of the program.
	MOV R3,#0BH	; Initializing the counter value with 12.
	MOV DPTR,#MSG	; Copying each data one by one by using db assembler directive with MSG as label.
	MOVX A,@DPTR	; Copying the data externally from the DPTR register into the accumulator.
	MOV R1,A	; Copying the data byte into register R1.
NEXTBYTE:	*INC DPTR*	; Incrementing the DPTR register by one.
	MOVX A,@DPTR	; Copying the data externally from the incremented DPTR register into the accumulator.
	CLR C	; Clearing the carry flag.

```
          MOV R2,A           ; Copying the accumulator
                               content into register
                               R2.
          SUBB A,R1          ; Subtracting the value of
                               register R1 content from the
                               accumulator data byte.
          JNC SKIP           ; Jump to SKIP label if carry flag is
                               reset else move on to
                               the next instruction.
          MOV A,R2           ; Copying register R2
                               data byte into the
                               accumulator.
          MOV R1,A           ; Copying the accumulator
                               data byte into register
                               R1.
SKIP:     DJNZ R3,NEXTBYTE   ; Decrementing
                               register R3 content by
                               one and jump to NEXT
                               BYTE if (R3) is not equal
                               to zero. If (R3) is zero
                               then go to the next
                               instruction.
          MOV DPL,#DCH       ; Loading the data byte
                               DCH into the DPL
                               register. The new
                               memory location will be
                               FEDCH.
          MOV A,R1           ; Copying the smallest
                               element from register
                               R1 into the accumulator.
          MOVX @DPTR,A       ; Transferring the data
                               byte externally from the
                               accumulator into the
                               DPTR register.
ENDPROG   SJMP ENDPROG       ; End of the program.
MSG:      db 04,05,06,07,08,02,0A,0B,0C,0D,0E,0F
                             ; Collection of string of
                               data bytes.
```

Flowchart:

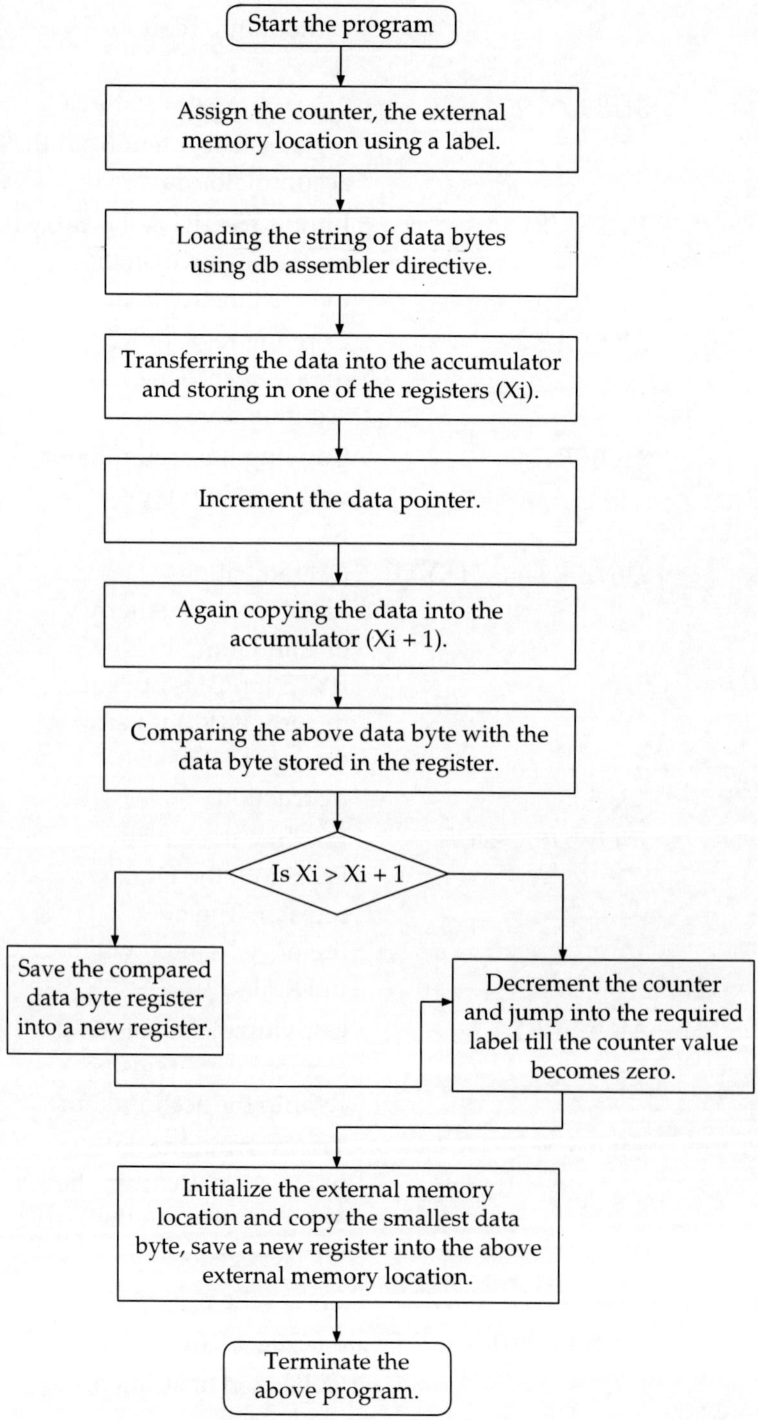
Start the program
Assign the counter, the external memory location using a label.
Loading the string of data bytes using db assembler directive.
Transferring the data into the accumulator and storing in one of the registers (Xi).
Increment the data pointer.
Again copying the data into the accumulator (Xi + 1).
Comparing the above data byte with the data byte stored in the register.
Is Xi > Xi + 1
Save the compared data byte register into a new register.
Decrement the counter and jump into the required label till the counter value becomes zero.
Initialize the external memory location and copy the smallest data byte, save a new register into the above external memory location.
Terminate the above program.

Approach:

- First the counter value is loaded into one of the registers.
- A string of data bytes is stored in a message "MSG" label from where one by one the data bytes will be transferred externally into the internal register for checking of the smallest elements.
- The first element to be compared will be transferred externally into the accumulator and stored in one of the registers.
- After incrementation of the DPTR register by one, the next data byte is again transferred into the accumulator where the present data byte is compared with that of the data byte stored in one of the registers. If no carry is generated, the counter is decreased by one else there will be exchange of the data bytes in the registers.
- The smallest data byte will be saved in one of the registers after comparison with collection of string of data bytes, and the smallest data byte will be transferred into the desired external memory location after assigning that particular memory location with the DPTR register.

Program 3.9

Write an ALP to find the largest element in a given string of 12 bytes and store the largest element at location FEDCH.

Solution: Assembly Language Program:

	Mnemonics	**Comment**
	ORG FE00H	; Initializing the starting memory location of the program.
	MOV R3,#0BH	; Initializing the counter value with 12.
	MOV DPTR,#MSG	; Copying each data one by one using db assembler directive with MSG as label.
	MOVX A,@DPTR	; Copying the data externally from the DPTR register into the accumulator.
	MOV R1,A	; Copying the data byte into register R1.
NEXTBYTE:	*INC DPTR*	; Incrementing the DPTR register by one.
	MOVX A,@DPTR	; Copying the data externally from the

```
                                      incremented DPTR
                                      register into the
                                      accumulator.
          CLR C                     ; Clearing the carry flag.
          MOV R2,A                  ; Copying the accumulator
                                      content into register R2.
          SUBB A,R1                 ; Subtracting the value of
                                      register R1 content from the
                                      accumulator data byte.
          JC SKIP                   ; Jump to SKIP label if carry flag is
                                      set else move on to the next
                                      instruction.
          MOV A,R2                  ; Copying the register R2
                                      data byte into the
                                      accumulator.
          MOV R1,A                  ; Copying the accumulator
                                      data byte into register
                                      R1.
SKIP:     DJNZ R3,NEXTBYTE          ; Decrementing the
                                      register R3 content by
                                      one and jump to NEXT
                                      BYTE if (R3) is not equal
                                      to zero. If (R3) is zero
                                      then go to the next
                                      instruction.
          MOV DPL,#0DCH             ; Loading the data byte
                                      DCH into the DPL
                                      register. The new
                                      memory location will
                                      be FEDCH.
          MOV A,R1                  ; Copying the smallest
                                      element from register
                                      R1 into the accumulator.
          MOVX @DPTR,A              ; Transferring the data
                                      byte externally from the
                                      accumulator into the
                                      DPTR register.
ENDPROG:  SJMP ENDPROG              ; End of the program.
MSG:      db 04,05,06,07,08,02,0A,0B,0C,0D,0E,0F
                                    ; Collection of string of
                                      data bytes.
```

Flowchart:

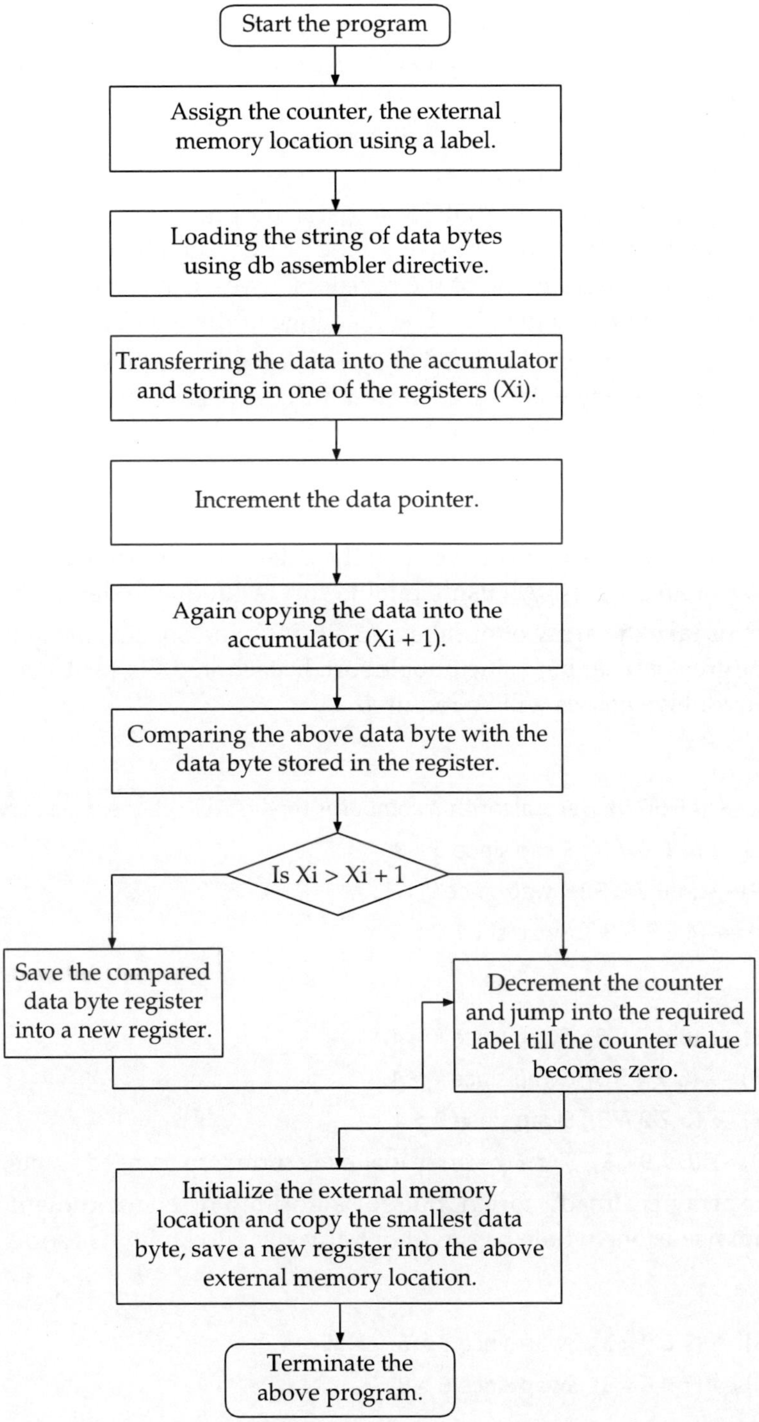
Start the program
Assign the counter, the external memory location using a label.
Loading the string of data bytes using db assembler directive.
Transferring the data into the accumulator and storing in one of the registers (Xi).
Increment the data pointer.
Again copying the data into the accumulator (Xi + 1).
Comparing the above data byte with the data byte stored in the register.
Is Xi > Xi + 1
Save the compared data byte register into a new register.
Decrement the counter and jump into the required label till the counter value becomes zero.
Initialize the external memory location and copy the smallest data byte, save a new register into the above external memory location.
Terminate the above program.

Approach:

- First the counter value is loaded into one of the registers.
- A string of data bytes are stored in a message "MSG" label from where one by one the data bytes will be transferred externally into the internal register for checking of the smallest elements.
- The first element to be compared will be transfered externally into the accumulator and stored in one of the registers.
- After incrementation of the DPTR register by one, the next data byte is again transferred into the accumulator where the present data byte is compared with that of the data byte stored in one of the registers. If yes, carry is generated, the counter is decreased by one else there will be exchange of the data bytes in the registers.
- The largest data byte will be saved in one of the registers on or not of exchange, and the smallest data byte will be transferred into the desired external memory location after assigning that particular memory location with the DPTR register.

Program 3.10

Write an ALP to sort a given array present in the external memory with a starting address 3000h and size of an array is 0Ah using bubble sort technique in descending order.

Solution: Let us take the array of numbers "3 4 6 7 9", and sort the array from the lowest number to the greatest number using bubble sort. In each step, elements written in bold are being compared. Five passes will be required.

First Pass

(**3 4** 6 7 9) → (**4 3** 6 7 9), Here, algorithm compares the first two elements, and swaps since 4 > 3.

(4 **3 6** 7 9) → (4 **6 3** 7 9), Swap since 6 > 3.

(4 6 **3 7** 9) → (4 6 **7 3** 9), Swap since 7 > 3.

(4 67 **3 9**) → (4 6 7 **9 3**), Swap since 9 > 3.

Second Pass:

(**4 6** 7 9 3) → (**6 4** 7 9 3), Swap since 6 > 4.

(6 **4 7** 9 3) → (6 **7 4** 9 3), Swap since 7 > 4.

(6 7 **4 9** 3) → (6 7 **9 4** 3), Swap since 9 > 4.

(6 7 9 **4 3**) → (6 7 9 **4 3**), Since the array is already sorted, so no need to interchange.

Now, the array is already sorted, but our algorithm does not know if it is complete. The algorithm needs one whole pass without any swap to know it is sorted.

Third Pass:

(**6 7** 9 4 3) → (**7 6** 9 4 3), Swap since 7 > 6.

(7 **6 9** 4 3) → (7 **9 6** 4 3), Swap since 6 > 9.

(7 9 **6 4** 3) → (7 9 **6 4** 3), Since the array is already sorted, so no need to interchange.

(7 9 6 **4 3**) → (7 9 6 **4 3**), Since the array is already sorted, so no need to interchange.

Fourth Pass:

(**7 9** 6 4 3) → (**9 7** 6 4 3), Swap since 9 > 7.

(9 **7 6** 4 3) → (9 **7 6** 4 3), Since the array is already sorted, so no need to interchange.

(9 7 **6 4** 3) → (9 7 **6 4** 3), Since the array is already sorted, so no need to interchange.

(9 7 6 **4 3**) → (9 7 6 **4 3**), Since the array is already sorted, so no need to interchange.

Fifth Pass:

(**9 7** 6 4 3) → (**9 7** 6 4 3), Since the array is already sorted, so no need to interchange.

(9 **7 6** 4 3) → (9 **7 6** 4 3), Since the array is already sorted, so no need to interchange.

(9 7 **6 4** 3) → (9 7 **6 4** 3), Since the array is already sorted, so no need to interchange.

(9 7 6 **4 3**) → (9 7 6 **4 3**), Since the array is already sorted, so no need to interchange.

So, here in the fifth pass, we can see that the array is already sorted. Hence, we can conclusively say that the array is finally sorted. So, I think you are now familiar with the bubble sort technique. Now, we will get into the above program.

Assembly Language Program:

	Mnemonics	**Comment**
	ORG 0000H	; Initializing the starting memory location of the program.
	COUNT EQU 09H	; The assembler directive EQU is used to assign value 09h to the count.
	MOV R1, #COUNT	; Loading the counter value into register R1.
OUTER:	*MOV R0,#COUNT*	; Loading the counter value into register R0.
	MOV DPTR, #3000H	; Allocating the data pointer with memory location 3000H.
INNER:	*CLR C*	; Clearing the carry flag
	MOVX A, @DPTR	; Copying the data byte externally into the accumulator.
	MOV R2,A	; Storing the data byte from the accumulator into register R2.
	INC DPTR	; Incrementing the data pointer register.
	MOVX A, @DPTR	; Copying the data byte externally into the accumulator.

```
        MOV R3,A             ; Storing the data byte from the
                               accumulator into register R3.
        SUBB A,R2            ; Subtracting the value of register
                               and borrow with the accumulator.
        JC L1                ; Jump to label L1 if carry flag is set
                               else go to the next instruction.
        XCH A,R2             ; Exchanging the contents of
                               register with that of the
                               accumulator.
        MOVX @DPTR,A         ; Copying the data byte externally
                               from the accumulator into the data
                               pointer register.
        DEC 82H              ; Decrementing the data pointer
                               register by one.
        MOV A,R3             ; Copying the data byte from
                               register R3 into the accumulator.
        MOVX @DPTR,A         ; Copying the data byte externally
                               from the accumulator into the data
                               pointer register.
L1:     INC DPTR             ; Incrementing the data pointer
                               register.
        DEC R0               ; Decrementing register R0 by
                               one.
        MOV A,R0             ; Copying register R0 content
                               into the accumulator.
        CJNE A,#00H,INNER    ; Comparing the value of the
                               accumulator and jump to INNER
                               label if (A) is equal to 00H, then go
                               to the next instruction else jump to
                               the label INNER.
        DEC R1               ; Decrementing the value of
                               register R1 by one.
        MOV A,R1             ; Copying the register R1 content
                               into the accumulator.
        CJNE A,#00H,OUTER    ; Comparing the value of the
                               accumulator and jump to OUTER
                               label if (A) is equal to 00H, then go
                               to the next instruction else jump to
                               the label OUTER.
LAST:   SJMP LAST            ; End of the program.
```

Flowchart:

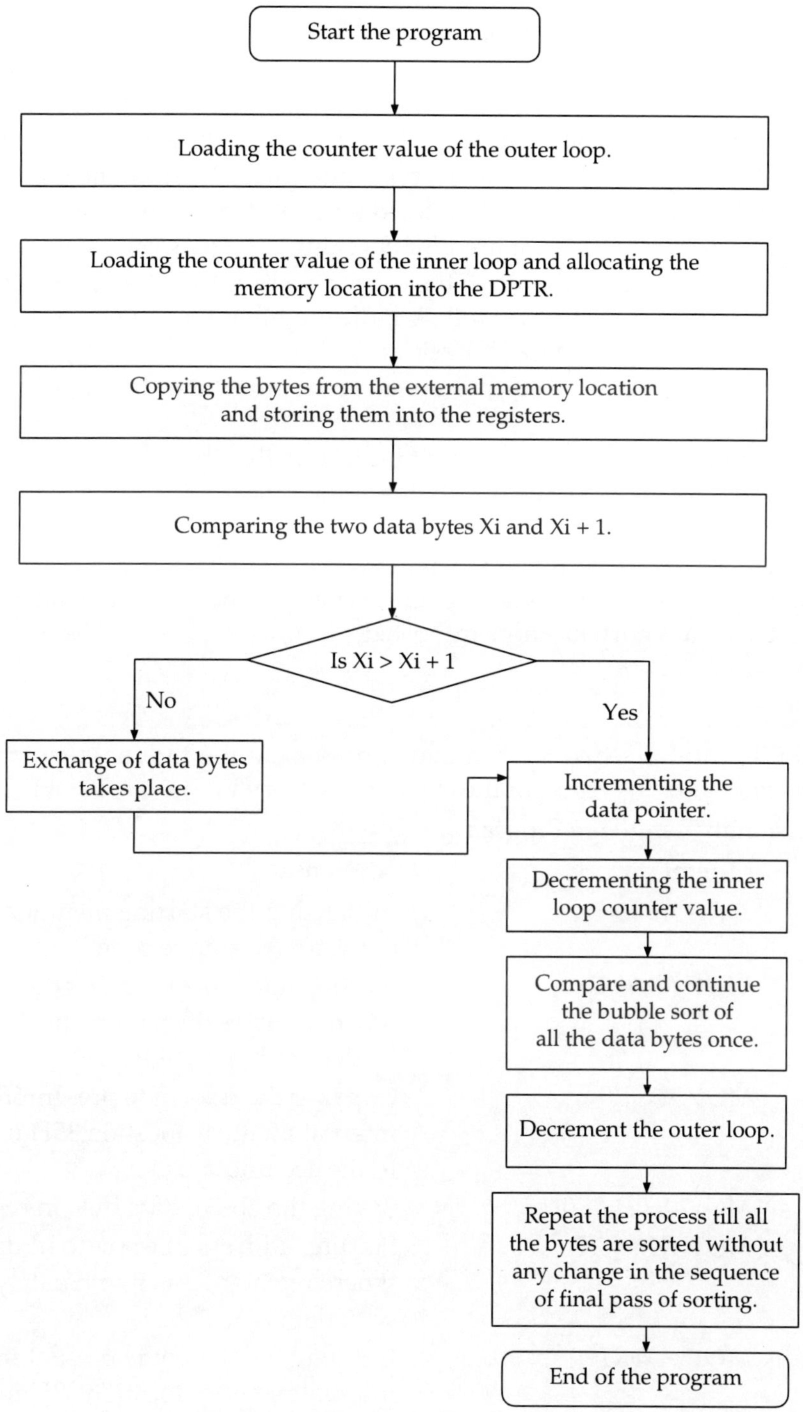
Start the program
Loading the counter value of the outer loop.
Loading the counter value of the inner loop and allocating the memory location into the DPTR.
Copying the bytes from the external memory location and storing them into the registers.
Comparing the two data bytes Xi and Xi + 1.
Is Xi > Xi + 1
No
Yes
Exchange of data bytes takes place.
Incrementing the data pointer.
Decrementing the inner loop counter value.
Compare and continue the bubble sort of all the data bytes once.
Decrement the outer loop.
Repeat the process till all the bytes are sorted without any change in the sequence of final pass of sorting.
End of the program

Approach:

- First loading of the inner loops is done for the inner passes of all the data bytes once by comparing 2 data bytes at a time.
- Then the outer loop counter is initialized for checking that all the data bytes are sorted without any change in the sequence of final pass of sorting.
- Then the 2 data bytes are stored in the registers externally from the memory location pointed by DPTR register indicating the data bytes that need to be sorted.
- The 2 data bytes are then compared for checking of higher values among them.
- If the first data byte is smaller than the other data byte (Xi<Xi+1), then the exchange of location of the data bytes will follow indicating that the data byte stored in the ith position is greater than the $i + 1^{th}$ position.
- If the first byte is larger than the second byte, then the data pointer register is incremented.
- Then the inner count value is decreased indicating that the 2 data bytes are compared and checking of the rest of the data bytes of the first pass.
- Finally, the first pass of the data bytes is done.
- The process is repeated until all the data bytes are bubble sorted, i.e., in the last pass none of the positions of the data bytes are exchanged indicating that all the data bytes of the external memory location are sorted.

Program 3.11

Write an ALP to multiply two 8-bit numbers present in the internal memory locations 35H and 38H and store the result in the external memory location ABCDH and AB78H.

Solution: Assembly Language Program:

Mnemonic	**Comments**
ORG AB00H	; Initializing the starting memory location of the above source code.
MOV R0,#35H	; Starting address of the first data byte where register R0 is immediately loaded with data byte 35H.
MOV A,@R0	; Copying the data byte present in the internal memory location 35H indirectly to the accumulator.
MOV 0F0H,0E0H	; Saving the above data byte in register B.
MOV R1,#38H	; Starting address of the second data byte where register R1 is immediately loaded with data byte 38H.
MOV A,@R1	; Copying the data byte present in the internal memory location 38H indirectly to the accumulator.

```
        MUL AB                  ; Multiplying the 2 data bytes.
        MOV DPTR,#0ABCDH        ; Initializing the external memory
                                  location with ABCDH.
        MOVX@DPTR,A             ; Transferring the data byte into the
                                  external memory location.
        MOV DPL,#78H            ; Initializing the external memory location
                                  with AB78H.
        MOV A,B                 ; Copying the data byte from register A to
                                  B.
        MOVX @DPTR,A            ; Transferring the data byte into the
                                  external memory location.
LAST:   SJMP LAST               ; End of the program.
```

Flowchart:

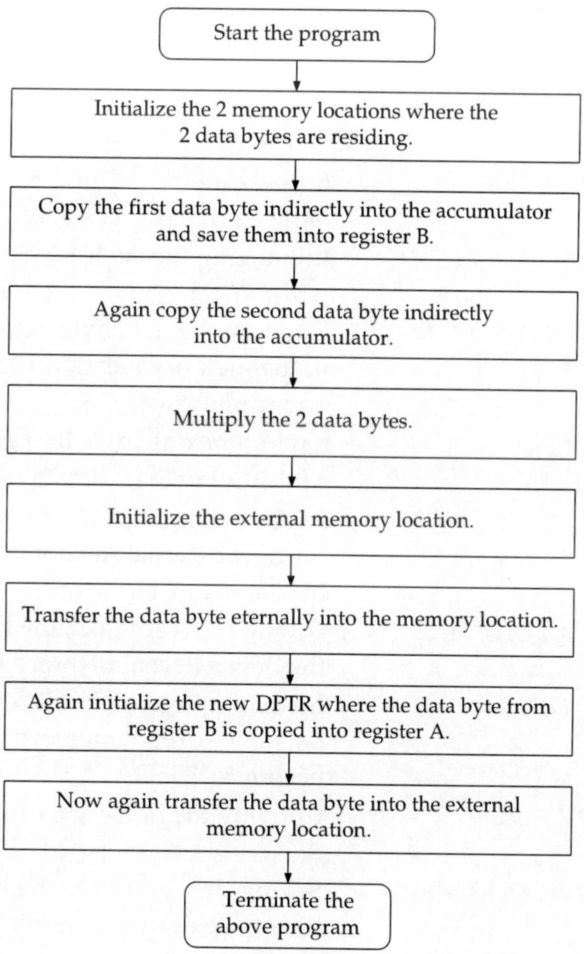

Approach:

- First initialize the 2 memory locations where the data byte needs to be multiplied.
- Transfer the first data byte indirectly into the accumulator and save them into the register B.
- Copy the second data byte indirectly into the accumulator and multiply with the data byte saved in register B.
- Multiply the above 2 data bytes.
- Initialize the external memory location and transfer the data byte externally into the above memory location.
- Again copy the second data byte and transfer the second data byte into the required external memory location.

Program 3.12

Write an ALP to square the data byte present in the external memory location 1234H

Solution: Assembly Language Program:

	Mnemonic	Comments
	ORG AB00H	; Initializing the starting memory location of the above source code.
	MOV DPTR,#1234H	; Initializing the external memory location with 1234H.
	MOVX A,@DPTR	; Copying the data byte externally present in the memory location 1234H to the accumulator.
	MOV B,A	; Saving the data byte in register B.
	MUL AB	; Multiplying the 2 data bytes.
	INC DPTR	; Incrementing the DPTR register by one. So, new external memory location becomes 1235H.
	MOVX @DPTR,A	; Saving the result, i.e., the lower byte in the new external memory location.
	INC DPTR	; Incrementing the DPTR register by one. So, new external memory location becomes 1236H.
	MOV A,B	; Copying the data byte from register B to A.
	MOVX @DPTR,A	; Saving the result, i.e., the higher byte in the new external memory location.
LAST:	*SJMP LAST*	; End of the program.

Flowchart:

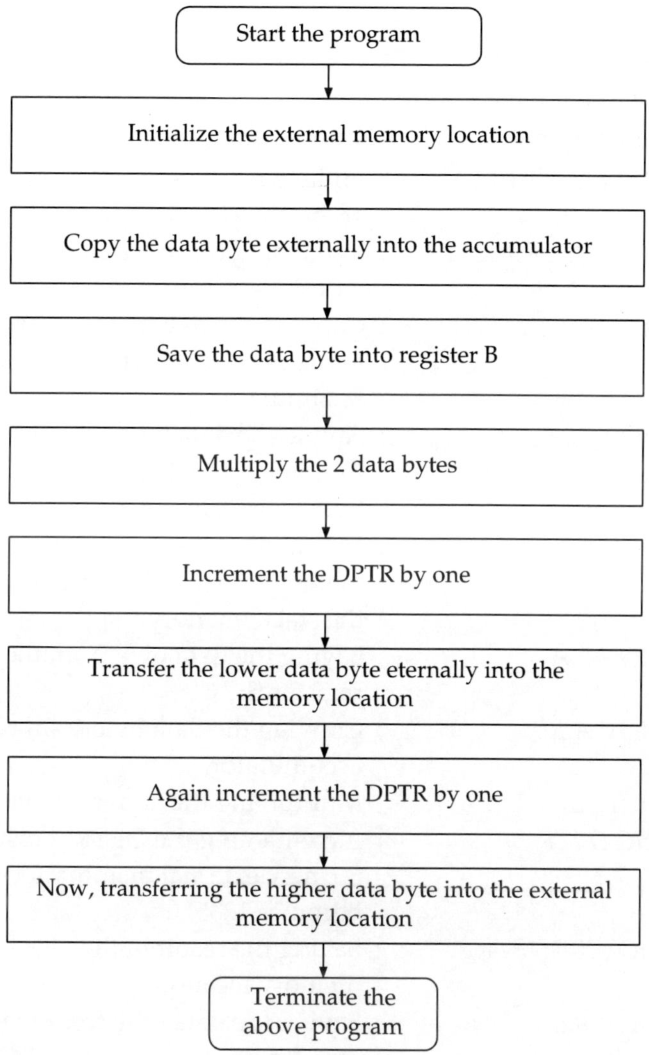

Approach:

- First, transfer the data byte from the external memory location into the accumulator.
- Save the data byte in register B.
- Multiply the above two data bytes.
- Increment the data pointer by one.
- Save the lower data byte into the new external memory location.
- Increment the data pointer by one.
- Save the higher data byte into the new external memory location.

Program 3.13

Write an ALP to cube the data byte present in the external memory location 1234H.

Solution: Assembly Language Program:

Mnemonic	**Comments**
ORG AB00H	; Initializing the starting memory location of the above source code.
MOV DPTR,#1234H	; Initializing the external memory location with 1234H.
MOVX A,@DPTR	; Copying the data byte externally present in the memory location 1234H to the accumulator.
MOV B,A	; Saving the data byte in register B.
MOV R0,A	; Copying the data byte from the accumulator to register R0.
MUL AB	; Multiplying the 2 data bytes.
PUSH 0F0H	; Pushing the data byte from register B into the stack memory.
MOV B,A	; Saving the data byte from the accumulator to register B.
MOV A,R0	; Copying the data byte from register R0 to the accumulator.
MUL AB	; Multiplying the above 2 data bytes
INC DPTR	; Incrementing the DPTR register by one. So, new external memory location becomes 1235H.
MOVX @DPTR,A	; Saving the result in the new external memory location.
MOV R2,B	; Saving the data byte from register B to register R2.
POP 0F0H	; Popping out the data byte from the stack memory.
MOV A,R0	; Restoring the data byte from register R0 to the accumulator.
MUL AB	; Multiplying the above 2 data bytes
ADD A,R2	; Adding the accumulator data byte to register R2.
INC DPTR	; Incrementing the DPTR register by one. So, new external memory location becomes 1236H.

```
          MOVX @DPTR,A        ; Saving the result in the new external
                                memory location.
          MOV A,B             ; Copying the data byte from register B to
                                A.
          INC DPTR            ; Incrementing the data pointer register by
                                one.
          MOVX @DPTR,A        ; Saving the result in the new external
                                memory location.
ENDPROG: SJMP ENDPROG         ; End of the program.
```

Flowchart:

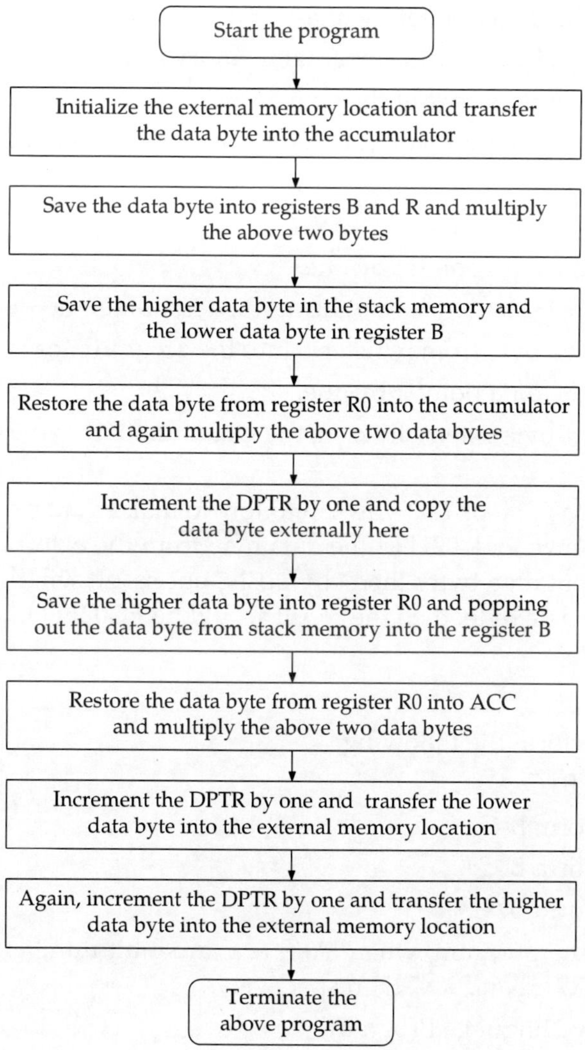

Approach:

- First initialize the external memory location and then transfer data byte externally into the accumulator.
- Then save the above data byte in registers B and R0.
- Multiply the above data bytes.
- Save the higher data byte in the stack memory.
- Copy the lower data byte in register B.
- Restoring the data byte from the register R0 into the accumulator.
- Again multiplying the above two data bytes.
- Incrementing the data pointer by one.
- Copying the data byte into the external memory location.
- Saving the higher data byte of the above 2 multiplication result into register R2.
- Popping out the data byte stored in the stack memory into register B.
- Restoring the data byte from register R0 into the accumulator.
- Multiplying the above 2 data bytes.
- Incrementing the data pointer by one.
- Copying the lower data byte into the new external memory location.
- Copying the data byte from register B into the accumulator.
- Incrementing the data pointer by one.
- Moving the data byte externally from the accumulator into the new external memory location.

These are the basic steps that you can follow to make a cube of any data byte. In the above example we took 03H as the data byte from the external memory location. If we take different data bytes then obviously, our result will be different. Say if FFH is taken as the data byte, then the result would be FD02FFH.

Program 3.14

Write an ALP to perform the following:

If A=0 perform B+C;

else if A=1 perform B-C;

else if A=2 perform B*C;

else if A=3 perform B/C,

else terminate the program, where A, B & C are 8-bit numbers residing in memory locations XX70H,XX71H and XX72H respectively.

Solution: Assembly Language Program:

	Mnemonic	**Comments**
	ORG 1200H	; Initializing the starting memory location of the above source code.
	MOV R0,#XX70H	; Initializing the internal memory location with XX70H.
	MOV A,@R0	; Copying the data byte present in the memory location XX70H to the accumulator.
	MOV R1,A	; Saving the data byte in register R1.
	INC R0	; Incrementing the memory location by one.
	MOV A,@R0	; Copying the data byte present in the memory location XX71H to the accumulator.
	MOV B,A	; Save the data byte in register B.
	INC R0	; Incrementing the memory location by one.
	CJNE A,#00H,sub	; Compare the accumulator data byte with 00H. If equal, then jump to the next instruction else jump to the sub label.
	ADD A,B	; Adding the accumulator data byte with register B.
	MOV B,#00H	; Refreshing the data byte of register B with 00H.
	JNC leave	; Jump if no carry to leave label else move to the next instruction.
	MOV B,#01H	; Load the data byte of register B with 01H.
Leave:	*SJMP end*	; Short jump to the label end.
Sub:	*CJNE A,#01H,mul*	; Compare the accumulator data byte with 01H. If equal, then jump to the next instruction else jump to the mul label.
	CLR C	; Clearing the carry flag.
	SUBB A,B	; Subtracting the accumulator data byte with register B data byte.

```
          MOV B,#00H         ; Refreshing the data byte of register B
                               with 00H.
          JNC Leave_a        ; Jump if no carry to leave_a label else
                               move to the next instruction.
          MOV B,#0FFH        ; Loading the data byte of register B with
                               FFH.
Leave_a:  SJMP end           ; Short jump to the label end.
Mul:      CJNE A,#02H,div    ; Compare the accumulator data byte with
                               02H. If equal, then jump to the next
                               instruction else jump to the div label.
          MUL AB             ; Multiply the data byte of register B with
                               accumulator.
          SJMP end           ; Short jump to the label end.
Div:      CJNE A,#03H,other  ; Compare the accumulator data byte with
                               03H. If equal, then jump to the next
                               instruction else jump to the other label.
          DIV AB             ; Divide the data byte of register B with
                               accumulator.
          SJMP end           ; Short jump to the label end.
Other:    SJMP ENDPROG       ; Short jump to the label ENDPROG label.
end:      INC R0             ; Increment the memory location by one.
                               The new memory location becomes
                               XX73H.
          MOV @R0,A          ; Copy the accumulator content to
                               memory location pointed by register R0
                               indirectly.
          INC R0             ; Increment the memory location by one.
                               The new memory location becomes
                               XX74H.
          MOV A,B            ; Copy the data byte from register B to the
                               accumulator.
          MOV @R0,A          ; Copy the accumulator content to
                               memory location pointed by register R0
                               indirectly.
ENDPROG:  SJMP ENDPROG       ; Short jump to label end program
                               indicating the end of the program.
```

Flowchart:

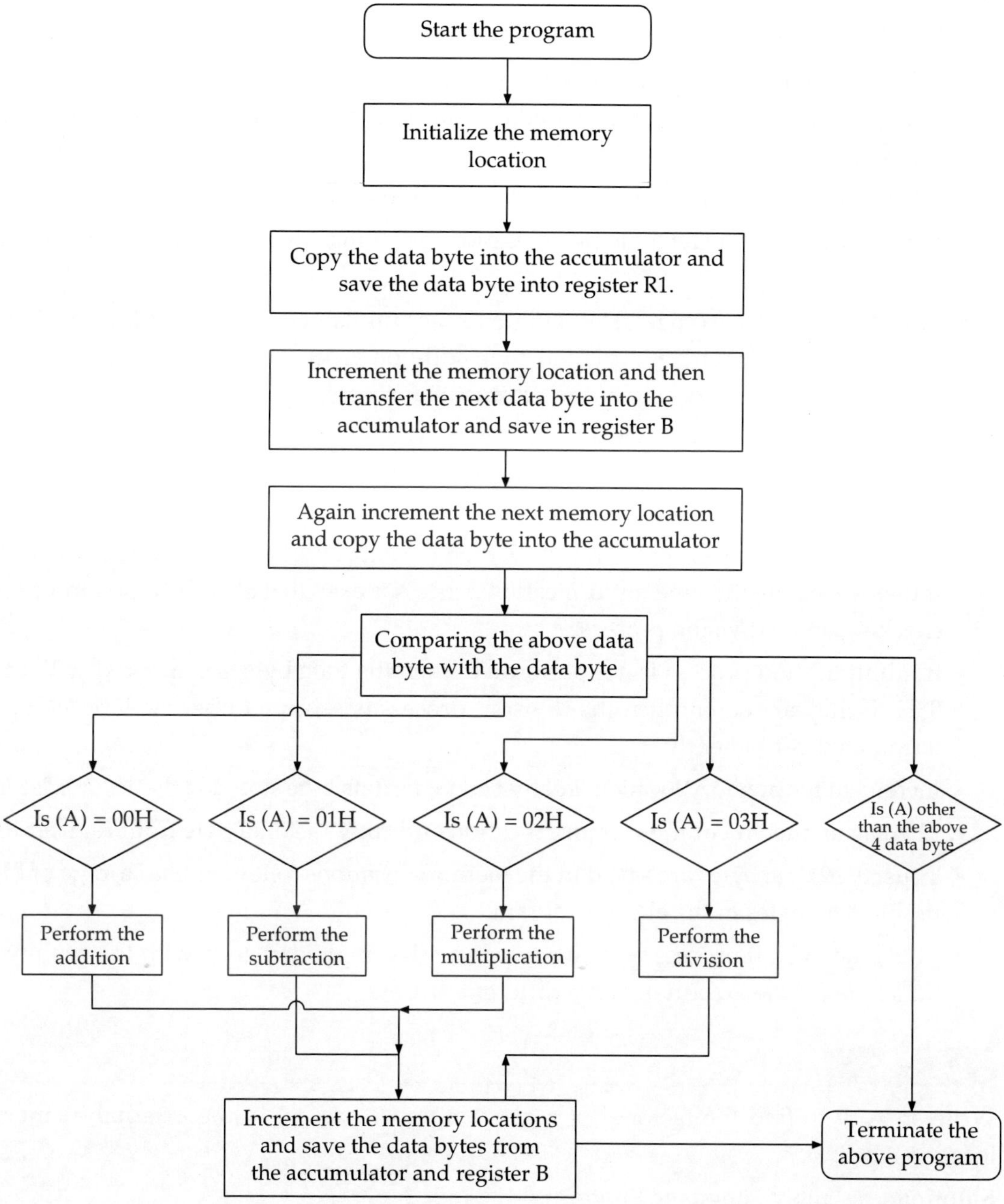
Start the program
Initialize the memory location
Copy the data byte into the accumulator and save the data byte into register R1.
Increment the memory location and then transfer the next data byte into the accumulator and save in register B
Again increment the next memory location and copy the data byte into the accumulator
Comparing the above data byte with the data byte
Is (A) = 00H
Is (A) = 01H
Is (A) = 02H
Is (A) = 03H
Is (A) other than the above 4 data byte
Perform the addition
Perform the subtraction
Perform the multiplication
Perform the division
Increment the memory locations and save the data bytes from the accumulator and register B
Terminate the above program

Approach:

- Initialize the memory location from where the data byte needs to be transferred.
- Then, the data byte to be compared will be saved in one of the registers, say R1, for different process of execution whether addition, subtraction, multiplication and division.
- Increment the memory location and save the next data byte into register B.
- Again, increment the next memory location and transfer the next data byte into the accumulator.
- Copy the data byte from register R1 into the accumulator and compare the data byte with 00H. If the data bytes are equal, then addition process is performed. If not, then the loop jumps for checking of the subtraction process.
- If addition process is not performed, then the data byte is compared with the 01H. If they are equal, then the subtraction process is executed else, the loop jumps for checking of the multiplication process.
- If subtraction process is not performed, then the data byte is compared with the 02H. If they are equal, then the multiplication process is executed else, the loop jumps for checking of the division process.
- If multiplication process is not performed, then the data byte is compared with the 03H. If they are equal, then the division process is executed else, the loop will be terminated permanently.
- Increment the memory location and first copy the data byte stored in the accumulator.
- Then again increment the memory location and copy the data byte from register B.
- Thus, the 2 data bytes are saved in the memory locations following the storage of the usable data bytes in the above program.
- Thus, the overall process will be comparing the single data byte with the required data byte for the execution of the different process.

Program 3.15

Write an ALP to find the factorial of a given number present in the external memory location ABCDH.

Solution: Assembly Language Program:

Mnemonic	**Comments**
ORG AB00H	; Initializing the starting memory location of the above source code.

```
           MOV DPTR,#0ABCDH    ; Initializing the external memory
                                 location with ABCDH.
           MOVX A,@DPTR        ; Copying the data byte externally
                                 present in the memory location
                                 ABCDH to the accumulator.
           MOV R2,A            ; Copy the data byte from the accumulator
                                 to register R2.
           MOV R6,#01H         ; Copy the data byte 01H to register R6.
           LCALL FACTORIAL     ; Give a long jump call to the FACTORIAL
                                 subroutine
           MOV R6,A            ; Copy the accumulator data byte into
                                 register R6.
FACTORIAL: MOV A,R6            ; Copy register R6 data byte into
                                 register R6.
           CJNE R2,#00H,GO     ; Compare the register R2 data byte
                                 with 00H and if equal then move to the
                                 next instruction else jump to label GO.
           SJMP UP             ; Short jump to label UP.
GO:        MOV B,R2            ; Copy the data byte of register R2 to
                                 register B.
           MUL AB              ; Multiply the data bytes stored in
                                 registers A and B.
           DJNZ R2,GO          ; Decrement and jump to label GO if
                                 register R2 data byte is not equal to
                                 00H.
UP:        RET                 ; Return from the subroutine.
ENDPROG:   SJMP ENDPROG        ; Short jump to label end program
                                 indicating the end of the program.
           END                 ; End of the above program.
```

Flowchart:

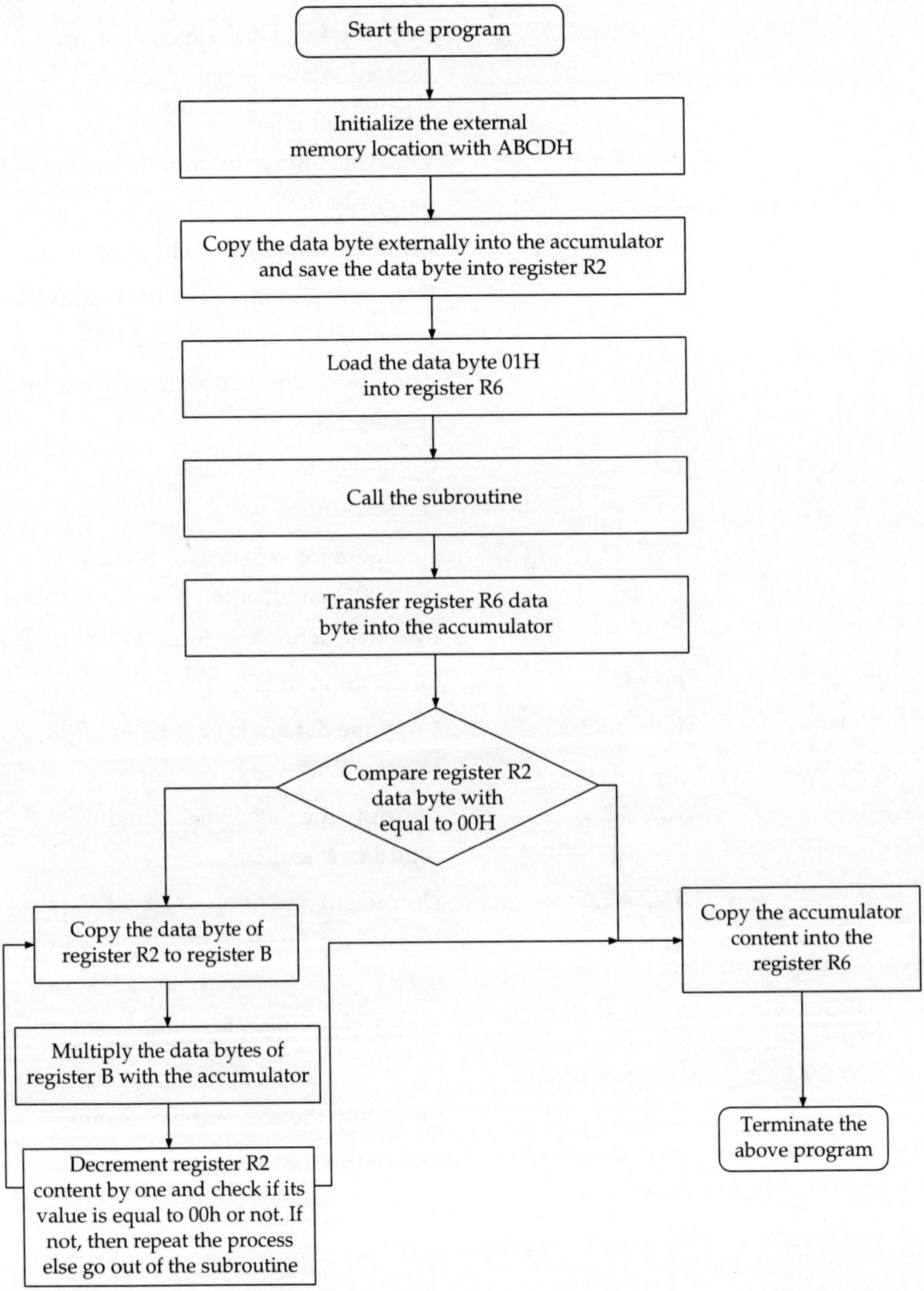
Start the program
Initialize the external memory location with ABCDH
Copy the data byte externally into the accumulator and save the data byte into register R2
Load the data byte 01H into register R6
Call the subroutine
Transfer register R6 data byte into the accumulator
Compare register R2 data byte with equal to 00H
Copy the data byte of register R2 to register B
Multiply the data bytes of register B with the accumulator
Decrement register R2 content by one and check if its value is equal to 00h or not. If not, then repeat the process else go out of the subroutine
Copy the accumulator content into the register R6
Terminate the above program

Approach:

- Initialize the external memory location with ABCDH.
- Copy the data byte externally from the above memory location into the accumulator.
- Save the above data byte in register R2.
- Load the data byte 01H into register R6.
- Make a call to the factorial subroutine.
- Transfer register R6 data byte, i.e., 01H into the accumulator.
- Compare the register R2 data byte with 00H. If not equal then go the GO label.
- Copy the register R2 data byte into register B.
- Multiply the data bytes present in the accumulator and register B.
- Decrement the register R2 data byte by one and repeat the process of multiplication till register R2 data byte is reduced to zero.
- Then loop out from the subroutine and save the final result of the factorial data byte into register R2.

Program 3.16

Write an ALP to check port 0. The data byte is present in the external memory location 6789H.

Solution: Assembly Language Program:

Mnemonic	Comments
ORG 7000H	; Initializing the starting memory location of the above source code.
MOV DPTR,#6789H	; IO mapping addresses of say digital input switches connected to port 0.
MOVX A,@DPTR	; Reading the port 0 switches and copying the data byte into the accumulator.
INC DPTR	; Incrementing the data pointer by one.
RL A	; Rotating the accumulator data byte by one towards left without carry.
MOVX @DPTR,A	; Sending the status switches to the output (say LED for display)
ENDPROG: *SJMP ENDPROG*	; Short jump to label end program indicating the end of the program.
END	; End of the assembly program

Flowchart:

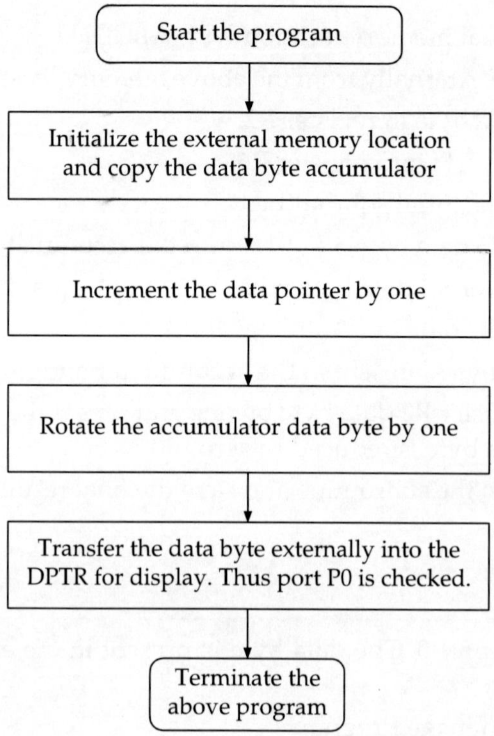

Approach:

- First initialize the external memory location and transfer the data byte into the accumulator externally.
- Increment the data pointer by one.
- Rotate the accumulator content by one towards left without carry.
- Transfer the data byte externally into the DPTR memory location say LED for display.
- Thus, by rotating the accumulator content by one, the data byte present in port P0 is checked.

Program 3.17

Write an ALP to find the LCM of the data bytes present in the external memory locations 5000H and 5001H respectively. Store the result in the memory location 5102H.

Solution: Assembly Language Program:

Mnemonic	**Comments**
ORG AB00H	; Initializing the starting memory location of the above source code.

```
          MOV DPTR,#5000H     ; Initializing the external memory
                                location with 5000H.
          MOVX A,@DPTR        ; Copying the data byte externally
                                present in the memory location 5000H
                                to the accumulator.
          MOV R0,A            ; Copy the data byte from accumulator
                                to register R0.
          INC DPTR            ; Incrementing the data pointer by one.
          MOVX A,@DPTR        ; Copying the data byte externally
                                present in the memory location 5001H
                                to the accumulator.
          MOV R1,A            ; Copy the accumulator data byte into
                                register R1.
          MOV A,R0            ; Copy register R0 data byte into the
                                accumulator.
          MOV R2,A            ; Copy the accumulator data byte into
                                register R2.
UP:       MOV B,R1            ; Copy register R1 data byte into the
                                register B.
          MOV A,B             ; Copy the data byte of register B to
                                the accumulator.
          JZ GO               ; Jump to label GO if zero flag is set.
          MOV A,R2            ; Copy register R2 data byte into the
                                accumulator.
          ADD A,R0            ; Adding the data byte of register R0
                                with the accumulator data byte.
          MOV R2,A            ; Copy the accumulator content into
                                register R2.
          SJMP UP             ; Short jump to label UP.
GO:       MOV A,R2            ; Copying the data byte of register R2
                                into the accumulator.
          INC DPH             ; Increment the DPH register by one.
          INC DPL             ; Increment the DPL register by one.
          MOVX @DPTR,A        ; Copying the data byte of the accumulator
                                into the DPTR memory location.
ENDPROG:  SJMP ENDPROG        ; Short jump to label end program
                                indicating the end of the program.
          END                 ; End of the assembly program.
```

Flowchart:

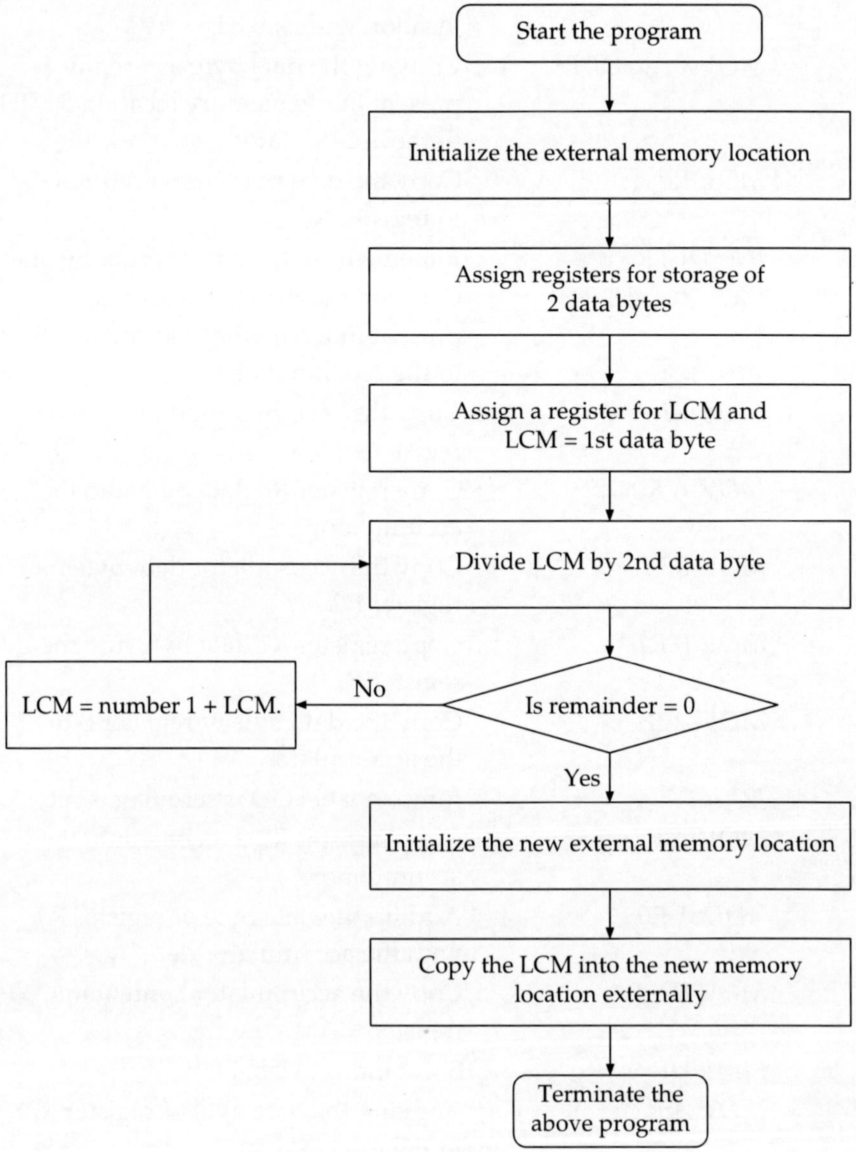

Approach:

- First, initialize the external memory location.
- Then, assign the registers for the 2 data bytes which will be copied externally.
- Then divide number 1 with number 2.

- Check the remainder. If not found equal, keep on checking the loop till the remainder becomes zero.
- During such process increment the LCM with number 1 and save as LCM only after assigning a register for LCM.
- Finally, initialize the new external memory location where the final LCM needs to be transferred.
- Then, transfer the data byte externally to the required address indicating the LCM value.

Program 3.18

Write an ALP to find the HCF of the data bytes present in the external memory locations 5000H and 5001H respectively. Store the result in the memory location 5102H.

Solution: Assembly Language Program:

Mnemonic	**Comments**
ORG AB00H	; Initializing the starting memory location of the above source code.
MOV DPTR,#5000H	; Initializing the external memory location with 5000H.
MOVX A,@DPTR	; Copying the data byte externally present in the memory location 5000H to the accumulator.
MOV R0,A	; Copy the data byte from the accumulator to register R0.
INC DPTR	; Incrementing the data pointer by one.
MOVX A,@DPTR	; Copying the data byte externally present in the memory location 5001H to the accumulator.
MOV R1,A	; Copy the accumulator data byte into register R1.
MOV A,R0	; Copy the data byte from register R0 to the accumulator.
MOV R2,A	; Copy the accumulator data byte into register R2.

```
UP:       MOV A,R1         ; Copy the data byte from register R1 to the
                           accumulator.
          MOV R3,A         ; Copy the accumulator data byte into
                           register R3.
          MOV B,R1         ; Copy the register R1 data byte into
                           register B.
          MOV A,R0         ; Copy the data byte of register R0 to the
                           accumulator.
          DIV AB           ; Dividing register B data byte to the
                           accumulator data byte.
          MOV A,B          ; Copy the register B data byte into the
                           accumulator.
          JZ GO            ; Jump to label GO if zero flag is set.
          MOV R1,A         ; Copy the accumulator content into
                           register R1.
          MOV A,R3         ; Copy the register R3 data byte into the
                           accumulator.
          MOV R0,A         ; Copy the accumulator content into
                           register R0.
          SJMP UP          ; Short jump to label UP.
GO:       MOV A,R3         ; Copying the data byte of register R3
                           into the accumulator.
          INC DPH          ; Increment the DPH register by one.
          INC DPL          ; Increment the DPL register by one.
          MOVX @DPTR,A     ; Copying the data byte of the accumulator
                           into the DPTR memory location.
ENDPROG: SJMP ENDPROG      ; Short jump to label end program
                           indicating the end of the program.
          END              ; End of the assembly program.
```

Flowchart:

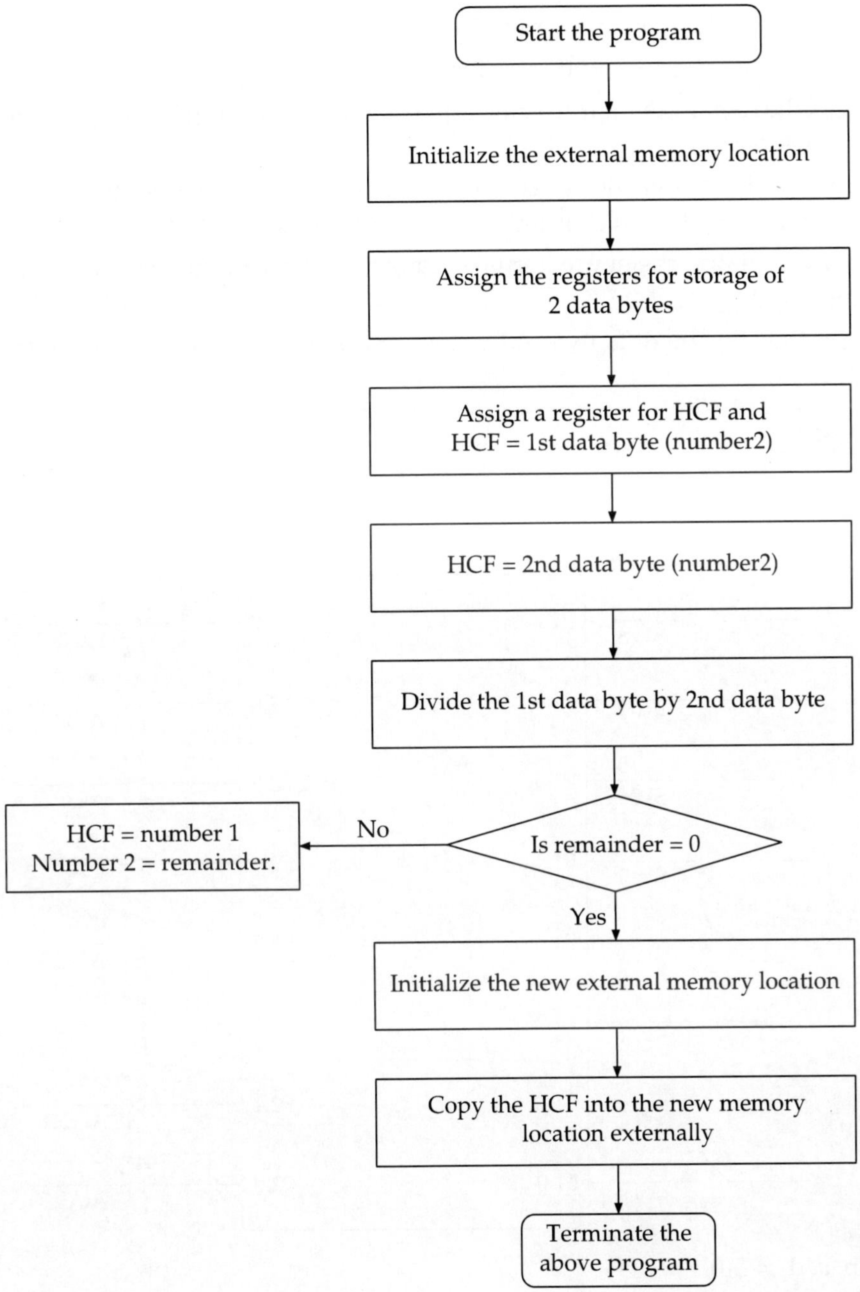
Start the program
Initialize the external memory location
Assign the registers for storage of 2 data bytes
Assign a register for HCF and HCF = 1st data byte (number2)
HCF = 2nd data byte (number2)
Divide the 1st data byte by 2nd data byte
Is remainder = 0
No
HCF = number 1
Number 2 = remainder.
Yes
Initialize the new external memory location
Copy the HCF into the new memory location externally
Terminate the above program

Approach:

- First, initialize the external memory location.
- Then, assign the registers for the 2 data bytes which will be copied externally.
- Then divide number 1 with number 2.
- Check the remainder. If not found equal, keep on checking the loop till the remainder becomes zero.
- During such process, the first data byte, i.e., number 1 will be assigned as HCF whereas the second data byte, i.e., number 2 will be assigned as remainder.
- Finally, initialize the new external memory location where the final HCF needs to be transferred.
- Then, transfer the data byte externally to the required address indicating the HCF value.

Program 3.19

Consider port P1 as an input port with port address FF22H and port P2 as an output port with port address FF23H. Realize the figure and write an ALP code in 8051 for the function:

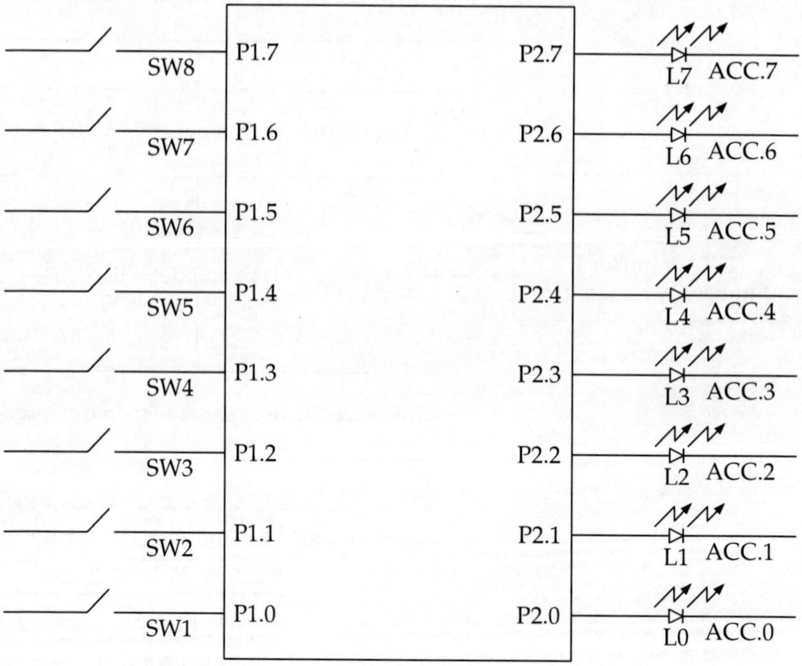

f(a, b, c, d) = Σ(0, 3, 7, 11)

Solution: Port P1 is connected to the switches such that port P0 has address FF22H. The connection of the pins are such that the following switches are connected to port P1 bits.

Switches	**Port Number**
SW1	P1.0
SW2	P1.1
SW3	P1.2
SW4	P1.3
SW5	P1.4
SW6	P1.5
SW7	P1.6
SW8	P1.7

The input variables are assigned as switches SW1 as d, SW2 as c, SW3 as b, SW4 as a. The digital output with address FF23H is connected with 8 LEDs L0 to L7 and any data written from the accumulator to this address will cause enable/disable of the LEDs. The mapping of the accumulator bits and LEDs are as follows:

After writing the data from the accumulator to LEDs, the status of LEDs is as follows:

LEDs	**Port Number**
L0	P2.0
L1	P2.1
L2	P2.2
L3	P2.3
L4	P2.4
L5	P2.5
L6	P2.6
L7	P2.7

Therefore, for the given Boolean expression, assigning LED L1 as output f, we have to create the status of expression in 2nd bit of the accumulator, i.e., ACC.1.

Truth Table

	Inputs			**Output**
a	b	c	d	f
0	0	0	0	1
0	0	0	1	0
0	0	1	0	0
0	0	1	1	1
0	1	0	0	0

0	1	0	1	0
0	1	1	0	0
0	1	1	1	1
1	0	0	0	0
1	0	0	1	0
1	0	1	0	0
1	0	1	1	1
1	1	0	0	0
1	1	0	1	0
1	1	1	0	0
1	1	1	1	0

Assembly Language Program:

Mnemonic	**Comments**
ORG AB00H	; Initializing the starting memory location of the above source code.
MOV DPTR,#0FF23H	; Pointing the address of digital output LED L1 as FF23H.
CLR A	; Clearing the content of the accumulator to 00H.
MOVX @DPTR,A	; Copy the data byte from the accumulator externally into the memory location pointed by data pointer register.
MOV P1,#0FFH	; Initializing port P1 as an input port.
MOV C,90H	; Copying the data bit from 90H memory location into carry flag. The carry flag bit is d.
CPL C	; Complementing the carry flag. The bit becomes d′.
ANL C,/91H	; ANDing the carry flag bit with the inverted bit of memory location 91H. The bit becomes c′d′.
ANL C,/92H	; ANDing the carry flag bit with the inverted bit of memory location 92H. The bit becomes b′c′d′.

```
ANL C,/93H      ; ANDing the carry flag bit with the
                  inverted bit of memory location 93H.
                  The bit becomes a'b'c'd'.
MOV 00,C        ; Copying the resulted carry flag data
                  bit into the memory location 00H.
MOV C,90H       ; Copying the data bit from 90H memory
                  location into carry flag. The carry flag
                  bit is d.
ANL C,91H       ; ANDing the carry flag bit with the bit
                  of memory location 91H. The bit
                  becomes cd.
ANL C,/92H      ; ANDing the carry flag bit with the
                  inverted bit of memory location 92H.
                  The bit becomes b'cd.
ANL C,/93H      ; ANDing the carry flag bit with inverted
                  bit of memory location 93H. The bit
                  becomes a'b'cd.
MOV 01H,C       ; Copying the resulted carry flag data
                  bit into the memory location 01H.
MOV C,90H       ; Copying the data bit from 90H memory
                  location into carry flag. The carry flag
                  bit is d.
ANL C,91H       ; ANDing the carry flag bit with the bit
                  of memory location 91H. The bit
                  becomes cd.
ANL C,92H       ; ANDing the carry flag bit with the bit
                  of memory location 92H. The bit
                  becomes bcd.
ANL C,/93H      ; ANDing the carry flag bit with the
                  inverted bit of memory location 93H.
                  The bit becomes a'bcd.
MOV 02H,C       ; Copying the resulted carry flag data
                  bit into the memory location 02H.
MOV C,90H       ; Copying the data bit from 90H memory
                  location into carry flag. The carry flag
                  bit is d.
```

```
ANL C,91H       ; ANDing the carry flag bit with the bit
                  of memory location 91H. The bit
                  becomes cd.
ANL C,/92H      ; ANDing the carry flag bit with the
                  inverted bit of memory location 92H.
                  The bit becomes b'cd.
ANL C,93H       ; ANDing the carry flag bit with the bit
                  of memory location 93H. The bit
                  becomes ab'cd.
ORL C,02H       ; ORing the carry flag with the data bit
                  present in the memory location 02H.
                  The output bit is ab'cd + a'bcd.
ORL C,01H       ; ORing the carry flag with the data bit
                  present in the memory location 01H.
                  The output bit is ab'cd + a'bcd + a'b'cd.
ORL C,00H       ; ORing the carry flag with the data bit
                  present in the memory location 00H.
                  The output bit is ab'cd + a'bcd + a'b'cd
                  + a'b'c'd'.
CLR A           ; Clearing the accumulator.
RLC A           ; Rotating the accumulator towards left
                  by one position including carry flag.
RLC A           ; Rotating the accumulator towards left
                  by one position including carry flag.
                  Finally, the LED will glow/disable
                  based on the output of the input
                  status.
MOVX @DPTR,A    ; Copying the data byte externally into
                  the memory location pointed by the
                  data pointer register.
ENDPROG:        SJMP ENDPROG
                Short jump to label end program
                indicating the end of the program.
END             ; End of the above program.
```

Flowchart:

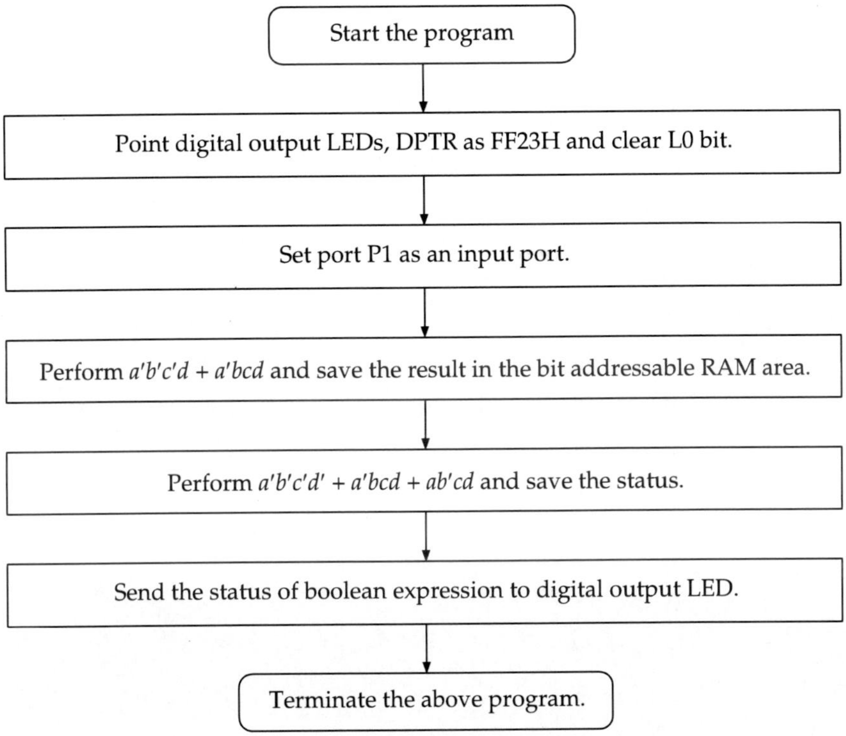

Approach:

- First assign the address of the digital output.
- Set port P1 as an input port.
- Then start reading of the ports connected to the respective switches, i.e., ports P1.0, P1.1, P1.2 and P1.3.
- Perform the required Boolean expression and save the status in the carry flag.
- Clear the accumulator and rotate the accumulator twice so as to store the status of the resulted expression into the external register thus terminating the program.

Part (b): Write an ALP code in 8051 for the above function

$f(a, b, c, d) = \Pi\, 0, 1, 2, 3, 5, 6, 7, 9, 10, 11, 13, 14$

Solution: The above expression is similar as

$f(a, b, c, d) = \Pi\, (4, 8, 12, 15)$

Considering the same ports as input and output ports with the same addresses.

Truth Table

Inputs			Output	
a	b	c	d	f
0	0	0	0	0
0	0	0	1	0
0	0	1	0	0
0	0	1	1	0
0	1	0	0	1
0	1	0	1	0
0	1	1	0	0
0	1	1	1	0
1	0	0	0	1
1	0	0	1	0
1	0	1	0	0
1	0	1	1	0
1	1	0	0	1
1	1	0	1	0
1	1	1	0	0
1	1	1	1	1

Assembly Language Program:

Mnemonic	Comments
ORG AB00H	; Initializing the starting memory location of the above source code.
MOV DPTR,#0FF23H	; Pointing the address of digital output LED L1 as FF23H
CLR A	; Clearing the content of the accumulator to 00H.
MOVX @DPTR,A	; Copy the data byte from the accumulator externally into the memory location pointed by data pointer register.
MOV P1,#0FFH	; Initializing port P1 as an input port.
MOV C,90H	; Copying the data bit from 90H memory location into carry flag. The carry flag bit is d.

CPL C	; Complementing the carry flag. The bit becomes d′.
ANL C,/91H	; ANDing the carry flag bit with the inverted bit of memory location 91H. The bit becomes c′d′.
ANL C,92H	; ANDing the carry flag bit with the inverted bit of memory location 92H. The bit becomes bc′d′.
ANL C,/93H	; ANDing the carry flag bit with the inverted bit of memory location 93H. The bit becomes a′bc′d′.
MOV 00,C	; Copying the resulted carry flag data bit into the memory location 00H.
MOV C,90H	; Copying the data bit from 90H memory location into carry flag. The carry flag bit is d.
CPL C	; Complementing the carry flag. The bit becomes d′.
ANL C,/91H	; ANDing the carry flag bit with the bit of memory location 91H. The bit becomes c′d′.
ANL C,/92H	; ANDing the carry flag bit with the inverted bit of memory location 92H. The bit becomes b′c′d′.
ANL C,93H	; ANDing the carry flag bit with inverted bit of memory location 93H. The bit becomes ab′c′d′.
MOV 01H,C	; Copying the resulted carry flag data bit into the memory location 01H.
MOV C,90H	; Copying the data bit from 90H memory location into carry flag. The carry flag bit is d.
CPL C	; Complementing the carry flag. The bit becomes d′.

```
ANL C,/91H     ; ANDing the carry flag bit with the bit
                 of memory location 91H. The bit
                 becomes c'd'.
ANL C,92H      ; ANDing the carry flag bit with the bit
                 of memory location 92H. The bit
                 becomes bc'd'.
ANL C,93H      ; ANDing the carry flag bit with the
                 inverted bit of memory location 93H.
                 The bit becomes abc'd'.
MOV 02H,C      ; Copying the resulted carry flag data
                 bit into the memory location 02H.
MOV C,90H      ; Copying the data bit from 90H memory
                 location into carry flag. The carry flag
                 bit is d.
ANL C,91H      ; ANDing the carry flag bit with the bit
                 of memory location 91H. The bit
                 becomes cd.
ANL C,92H      ; ANDing the carry flag bit with the
                 inverted bit of memory location 92H.
                 The bit becomes bcd.
ANL C,93H      ; ANDing the carry flag bit with the bit
                 of memory location 93H. The bit
                 becomes abcd.
ORL C,02H      ; ORing the carry flag with the data bit
                 present in the memory location 02H.
                 The output bit is abcd + abc'd'.
ORL C,01H      ; ORing the carry flag with the data bit
                 present in the memory location 01H.
                 The output bit is abcd + abc'd'+ ab'c'd'.
ORL C,00H      ; ORing the carry flag with the data bit
                 present in the memory location 00H.
                 The output bit is abcd + abc'd'+
                 b'c'd'+a'bc'd'.
CLR A          ; Clearing the accumulator.
RLC A          ; Rotating the accumulator towards left
                 by one position including carry flag.
```

```
          RLC A                 ; Rotating the accumulator towards left
                                  by one position including carry flag.
                                  Finally, the LED will glow/disable based
                                  on the output of the input status.
          MOVX @DPTR,A          ; Copying the data byte externally into
                                  the memory location pointed by the
                                  data pointer register.
ENDPROG:  SJMP ENDPROG          ; Short jump to label end program
                                  indicating the end of the program.
          END                   ; End of the assembly program.
```

Flowchart:

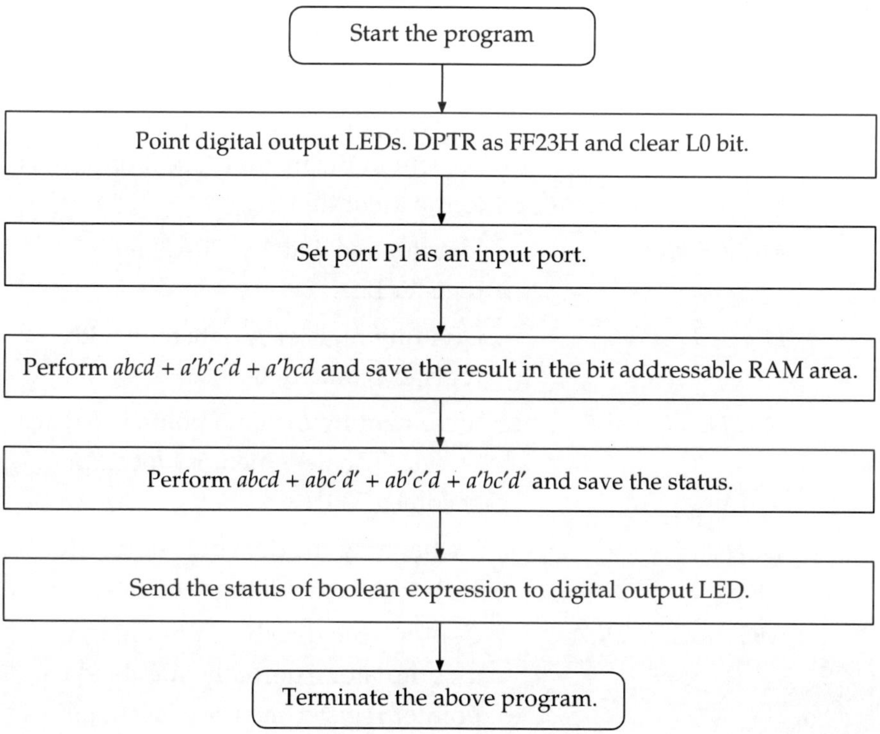

Approach:

- First assign the address of the digital output.
- Set port P1 as an input port.
- Then start reading of the ports connected to the respective switches, i.e., ports P1.0, P1.1, P1.2 and P1.3.

- Perform the required Boolean expression and save the status in the carry flag.
- Clear the accumulator and rotate it twice so as to store the status of the resulted expression into the external register thus terminating the program.

Program 3.20

Write an assembly language program to convert a hexadecimal number into its equivalent decimal number. The hexadecimal number is present in the external memory location F300H. Store the result from the memory location F300H onwards.

Solution: Assembly Language Program:

Mnemonic	Comments
ORG F000H	; Initializing the starting memory location of the above source code.
MOV DPTR,#0F300H	; Initializing the external memory location with F300H.
MOVX A,@DPTR	; Copying the data byte externally present in the memory location F300H to the accumulator.
MOV B,#0AH	; Loading the data byte 0AH into register B.
DIV AB	; Dividing register A data byte with register B data byte.
INC DPTR	; Incrementing the data pointer register by one. The new memory location becomes F301H.
XCH A,B	; Exchanging the data byte of register B with the accumulator data byte.
MOVX @DPTR,A	; Copying the data byte present in the accumulator externally into the data pointer register memory location.
XCH A,B	; Exchanging the data byte of register B with the accumulator data byte.
MOV B,#0AH	; Loading the data byte 0AH into register B.
DIV AB	; Dividing register A data byte with register B data byte.

```
            INC DPTR          ; Incrementing the data pointer register
                                by one. The new memory location
                                becomes F302H.
            XCH A,B           ; Exchanging the data byte of register B
                                with the accumulator data byte.
            MOVX @DPTR,A      ; Copying the data byte present in the
                                accumulator externally into the data
                                pointer register memory location.
            MOV R2,A          ; Saving the data byte into register
                                R2.
            XCH A,B           ; Exchanging the data byte of register B
                                with the accumulator data byte.
            SWAP A            ; Interchanging the contents of lower
                                and higher nibble within the ACC.
            INC DPTR          ; Incrementing the data pointer register
                                by one. The new memory location
                                becomes F302H.
            MOVX @DPTR,A      ; Copying the data byte present in the
                                accumulator externally into the data
                                pointer register memory location.
            ADD A,R2          ; Adding the data byte value of register
                                R2 with that of the accumulator.
            INC DPTR          ; Incrementing the data pointer register
                                by one. The new memory location
                                becomes F303H.
            MOVX @DPTR,A      ; Copying the data byte present in the
                                accumulator externally into the data
                                pointer register memory location.
                                The hexadecimal byte is
                                converted into the decimal one.
ENDPROG:    SJMP ENDPROG      ; Short jump to label end program
                                indicating the end of the program.
            END               ; End of the assembly program.
```

Flowchart:

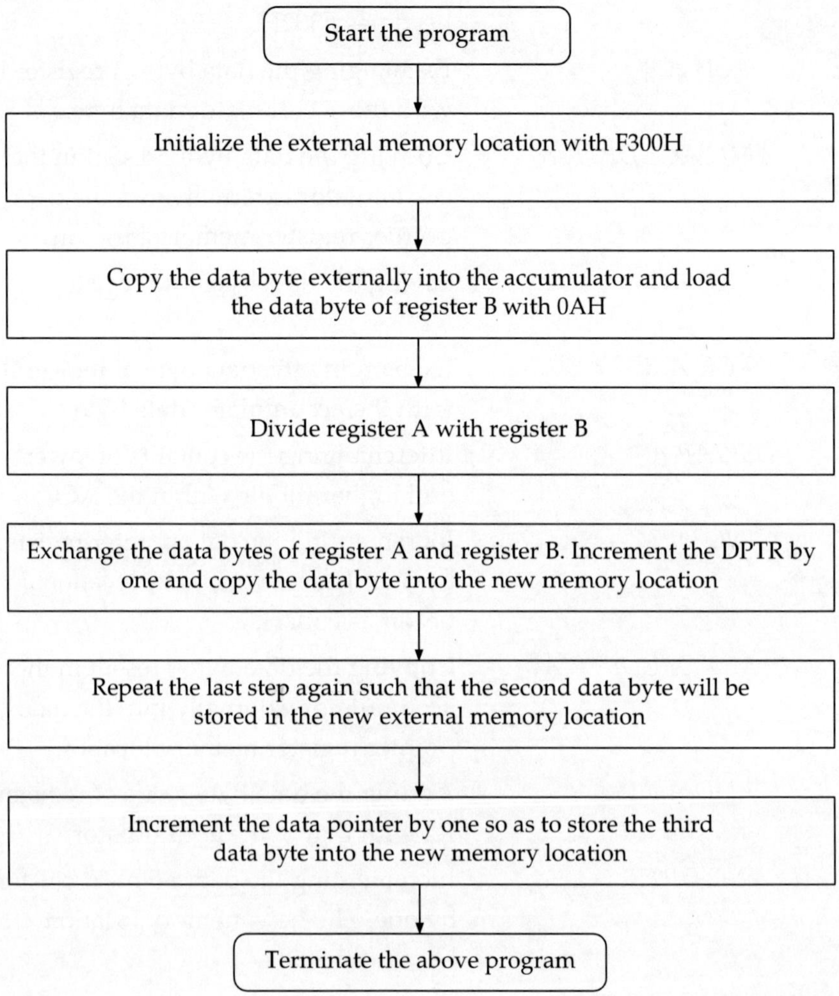

Approach:

- First, initialize the external memory location and copy the data byte whose equivalent decimal is to be done.
- Then, assign register B data byte with 0AH and divide it with register A data byte.
- Increment the data pointer by one, exchange the data bytes of register A and register B and copy the accumulator data byte into the DPTR memory location.
- Again, exchange the data bytes so as to restore the original data bytes present in the register B and the accumulator.
- Then, assign register B data byte with 0AH and divide it with register A data byte.

- Increment the data pointer by one, exchange the data bytes of register A and register B and copy the accumulator data byte into the DPTR memory location.
- Again, exchange the data bytes so as to restore the original data bytes present in the register B and the accumulator.
- Then, increment the data pointer by one and transfer the data byte externally into the DPTR memory location.

Program 3.21

Write an assembly language program to convert a decimal number into its equivalent hexadecimal number. The decimal number is present in the external memory location F300H.

Solution: Assembly Language Program:

Mnemonic	Comments
ORG F000H	; Initializing the starting memory location of the above source code.
MOV DPTR,#0F300H	; Initializing the external memory location with F300H.
MOVX A,@DPTR	; Copying the data byte externally present in the memory location F300H to the accumulator.
ANL A,#0F0H	; ANDing the accumulator data byte with 0F0H.
SWAP A	; Exchanging the data byte of register B with the accumulator data byte and vice versa.
MOV B,#0AH	; Loading the register B data byte with 0AH.
MUL AB	; Multiplying the accumulator data byte with register B data byte.
MOV R1,A	; Copying the data byte present in the accumulator externally into the data pointer register memory location.
MOVX A,@DPTR	; Again, copying the data byte externally present in the memory location F300H to the accumulator.
ANL A,#0FH	; ANDing the accumulator data byte with 0FH.

```
          ADD A,R1          ; Adding the accumulator data byte with
                              register R1 data byte.
          INC DPTR          ; Incrementing the data pointer register
                              by one. The new memory location
                              becomes F301H.
          MOVX @DPTR,A      ; Copying the resultant hexadecimal
                              data byte present in the accumulator
                              externally into the data pointer
                              register memory location.
ENDPROG: SJMP ENDPROG       ; Short jump to label end program
                              indicating the end of the program.
          END               ; End of the assembly program.
```

Flowchart:

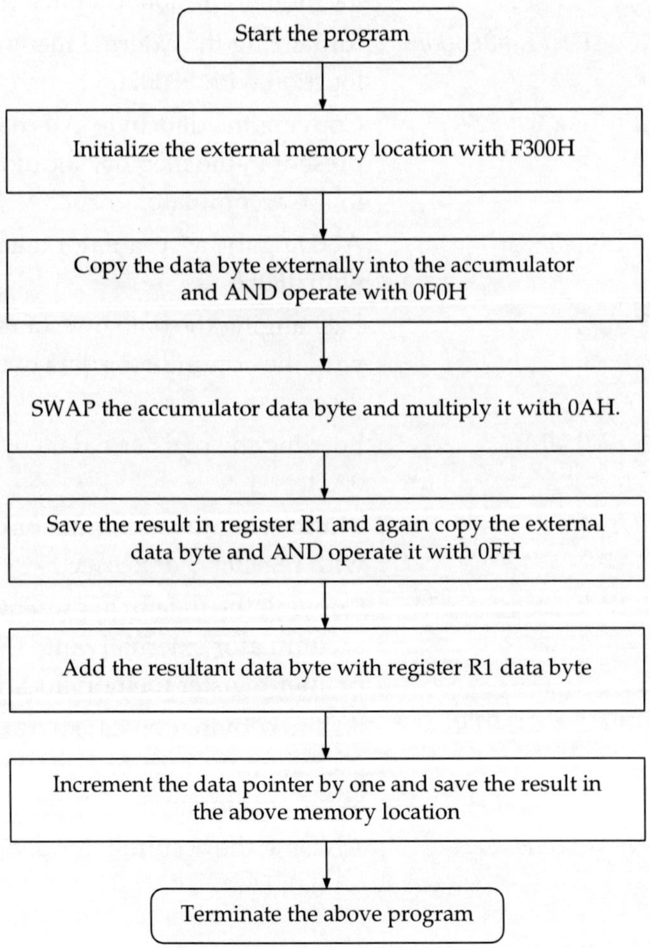

Approach:

- First, initialize the external memory location and copy the data byte whose equivalent hexadecimal conversion is to be done.
- Then, assign register B data byte with 0F0H and AND operate it with register A data byte.
- Swap the data byte of register A and multiply it with register B data byte containing 0AH.
- Save the data byte of the accumulator into register R1 after the multiplication.
- Again, copy the original data byte externally and AND operate now with 0FH.
- Add the accumulator data byte with register R1 data byte.
- Increment the DPTR register and save the result in the above memory location.

Program 3.22

Write an assembly language program to convert a packed BCD number present in the accumulator into 2 ASCII equivalents and put the result in the registers R7-R6. The BCD number is present in the external memory location F300H.

Solution: Assembly Language Program:

Mnemonic	**Comments**
ORG F000H	; Initializing the starting memory location of the above source code.
MOV DPTR,#0F300H	; Initializing the external memory location with F300H.
MOVX A,@DPTR	; Copying the data byte externally present in the memory location F300H to the accumulator.
MOV B,A	; Copying the data byte from the accumulator into register B.
SWAP A	; Exchanging the data byte of register B with the accumulator data byte and vice versa.
ANL A,#0F0H	; ANDing the accumulator data byte with 0F0H.
ORL A,#30H	; ORing the accumulator data byte with 30H.
MOV R7,A	; Storing the packed higher nibble BCD data byte from the accumulator into register R7.

```
          MOV A,B            ; Copying the data byte from
                               register B to the accumulator.
          ANL A,#0FH         ; ANDing the accumulator data byte
                               with 0FH
          ORL A,#30H         ; ORing the accumulator data byte with
                               30H.
          MOV R6,A           ; Storing the packed lower nibble BCD
                               data byte from the accumulator into
                               register R6.
ENDPROG:  SJMP ENDPROG       ; Short jump to label end program
                               indicating the end of the program.
          END                ; End of the above program.
```

Flowchart:

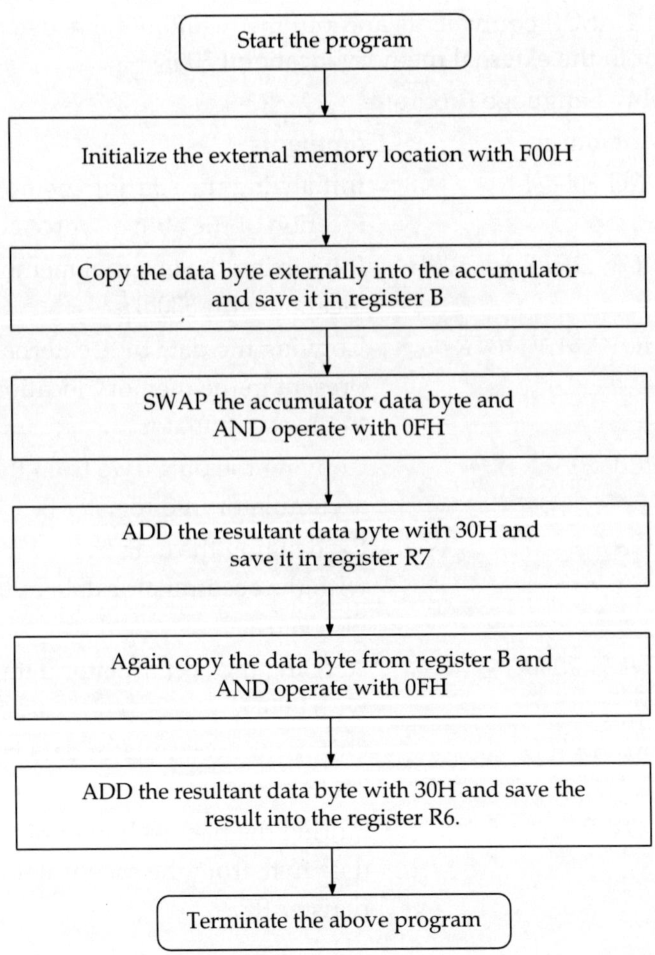

Approach:

- First, initialize the external memory location and copy the data byte from the above memory location into the accumulator and save it into register B.
- Exchange the data bytes of the accumulator and AND operate with 0FH.
- ADD the resultant data byte with 30H and save it in register R7. The higher nibble BCD has been converted into ASCII.
- Again restore the original data byte and AND operate with 0FH.
- ADD the resultant data byte with 30H and save it in register R6. The lower nibble BCD has been converted into ASCII.

Program 3.23

Explain the code written below and state what the program is doing.

```
        ORG 0000H
        MOV DPTR,#5678H
        MOVX A,@DPTR
        ANL A,#0F0H
        CJNE A,#30H,JUMP
        MOV A,DPL
        SWAP A
        INC DPTR
        MOVX @DPTR,A
        SJMP $
JUMP:   MOV A,DPL
        SWAP A
        MOV DPL,A
        MOV A,DPH
        RL A
        RL A
        RL A
        RL A
        XCH A,DPL
        MOV DPH,A
        MOVX A,@DPTR
        XRL A,#00H
        MOVX @DPTR,A
        ORG 5678H
```

```
DB 32H
SJMP $
END
```

Solution: In the above program, the DPTR memory location contains the data byte 32H since memory location 5678H has the data byte 32H. Now, explaining the above sequence of codes:

- MOV DPTR,#5678H: Initializing the data pointer register with memory location 5678H.
- MOVX A,@DPTR: Copying the data byte 32H externally from the DPTR register into the accumulator.
- ANL A,#0F0H: AND operating the accumulator data byte 32H with the data byte 0F0H. The resultant data byte in the accumulator will be 30H.
- CJNE A,#30H,JUMP: Comparing the accumulator data byte 30H with the data byte 30H. Since the accumulator data byte value is equal to 30h, the control will flow to the next instruction.
- MOV A,DPL: Copying the data byte 78H into the accumulator.
- SWAP A: Exchanging the data byte position of the accumulator. The resultant data byte will be 87H.
- INC DPTR: Incrementing the data pointer register by one. Now, the data pointer register is 5679H.
- MOVX @DPTR,A: Copying the data byte 87H into the memory location pointed by the data pointer register.
- SJMP $: Short jump to the label $.
- END: Termination of the above program.

Conclusion: Thus, after the execution of the above program, the data byte which was transferred externally into the accumulator got swapped. So, the above code was to swap the nibble positions of the 8-bit data byte.

Program 3.24

Explain the code written below and state what the program is doing.

```
ORG 0000H
MOV DPTR,#5678H
MOVX A,@DPTR
ANL A,#0F0H
CJNE A,#30H,JUMP
MOV A,DPL
SWAP A
```

```
        INC DPTR
        MOVX @DPTR,A
        SJMP $
JUMP:   MOV A,DPL
        SWAP A
        MOV DPL,A
        MOV A,DPH
        RL A
        RL A
        RL A
        RL A
        XCH A,DPL
        MOV DPH,A
        MOVX A,@DPTR
        XRL A,#00H
        MOVX @DPTR,A
        ORG 5678H
        DB 42H
        SJMP $
        END
```

Solution: In the above program, the DPTR memory location contains the data byte 42H since memory location 5678H has the data byte 32H. Now, explaining the above sequence of codes:

- MOV DPTR,#5678H: Initializing the data pointer register with memory location 5678H.
- MOVX A,@DPTR: Copying the data byte 42H externally from the DPTR register into the accumulator.
- ANL A,#0F0H: AND operating the accumulator data byte 42H with the data byte 0F0H. The resultant data byte in the accumulator will be 30H.
- CJNE A,#30H,JUMP: Comparing the accumulator data byte 40H with the data byte 30H. Since the accumulator data byte value is equal to 40H, the control will move to JUMP label and the execution will start from the above memory location.
- MOV A,DPL: Copying the data byte 78H into the accumulator.
- SWAP A: Exchanging the data byte position of the accumulator. The resultant data byte will be 87H.
- MOV DPL,A: Copying the data byte 87H into the DPL register.

- MOV A,DPH: Copying the data byte 56H into the accumulator.
- RL A: Rotating the accumulator left by one without carry. The final data byte becomes ACH.
- RL A: Rotating the accumulator left by one without carry. The final data byte becomes 59H.
- RL A: Rotating the accumulator left by one without carry. The final data byte becomes B2H.
- RL A: Rotating the accumulator left by one without carry. The final data byte becomes 65H.
- XCH A,DPL: Exchanging the data bytes of accumulator and the data pointer register. The accumulator data byte becomes 87H and the DPL register has the data byte 65H.
- MOV DPH, A: Copying the data byte of the accumulator into the DPH Register. Thus, the DPH register has the data byte 87H. So, finally the memory location of the above DPTR register is 8765H.
- MOVX A,@DPTR: Copying the 8-bit data byte of the accumulator into register A.
- XRL A,#00H: XOR operating the accumulator data byte with 00H. So, finally the data byte stored in the accumulator is 00H.
- MOVX @DPTR, A: Copying the data byte 00H into the memory location pointed by the data pointer register.
- SJMP $: Short jump to the label $.
- END: Termination of the above program.

Conclusion: Thus, after the execution of the above program, the 16-bit data byte of the DPTR register got swapped. So, the above code was to swap the nibble positions of the 16-bit data byte, i.e., the memory location of the DPTR register.

Program 3.25

Explain the code written below and state what the program is doing.

```
ORG 0000H
MOV DPTR,#5678H
MOVX A,@DPTR
ANL A,#0F0H
MOV R0,#00H
MOV 01H,00H
JB ACC.7,NEG
INC R0
MOV A,#0FFH
INC DPTR
```

```
        MOVX @DPTR,A
        SJMP $
NEG:    INC R1
        MOV A,#00H
        INC DPTR
        MOVX @DPTR,A
        ORL A,#30H
        RL A
        RL A
        RL A
        RL A
        XCH A,DPL
        MOV DPH,A
        ORG 5678H
        DB 42H
        SJMP $
        END
```

Solution: In the above program, the DPTR memory location contains the data byte 42H since the memory location 5678H has the data byte 42H. Now, explaining the above sequence of codes:

- MOV DPTR,#5678H: Initializing the data pointer register with memory location 5678H.
- MOVX A,@DPTR: Copying the data byte 42H externally from the DPTR register into the accumulator.
- ANL A,#0F0H: AND operating the accumulator data byte 42H with the data byte 0F0H. The resultant data byte in the accumulator will be 40H.
- MOV R0,#00H: Loading the data byte of register R0 with 00H.
- MOV 01H,00H: Copying the data byte 00H into register R1.
- JB ACC.7,NEG: Jump to NEG if ACC.7 is set. Since, ACC.7 is 0, the control will move to the next memory location.
- INC R0: Incrementing register R0 data byte by one. Hence, the data byte present in register R0 will be 01H.
- MOV A, #0FFH: Loading the data byte FFH into the accumulator.
- INC DPTR: Incrementing the data pointer register by one. The new data pointer register memory location will be 5679H.
- MOVX @DPTR,A: Copying the data byte 0FFH into the memory location pointed by the data pointer register.

- SJMP $: Short jump to the label $.
- END: Termination of the above program.

Conclusion: In the above program, the data byte 42H is being checked whether it is positive or negative. By inspection, since the data byte is less than 07Fh, the above data byte will be positive. As the above data byte is positive, the data byte 0FFH is being stored in the accumulator and the present data byte is being transferred to the new DPTR memory location.

Program 3.26

Explain the code written below and state what the program is doing.

```
ORG 0000H
MOV DPTR,#5678H
MOVX A,@DPTR
ANL A,#0F0H
MOV R0,#00H
MOV 01H,00H
JB ACC.7,NEG
INC R0
MOV A,#0FFH
INC DPTR
MOVX @DPTR,A
SJMP $
NEG: INC R1
MOV A,#00H
INC DPTR
MOVX @DPTR,A
ORL A,#30H
RL A
RL A
RL A
RL A
XCH A,DPL
MOV DPH,A
ORG 5678H
DB A2H
SJMP $
END
```

Solution: In the above program, the DPTR memory location contains the data byte A2H since the memory location 5678H has the data byte A2H. Now, explaining the above sequence of codes:

- MOV DPTR, #5678H: Initializing the data pointer register with memory location 5678H.
- MOVX A, @DPTR: Copying the data byte A2H externally from the DPTR register into the accumulator.
- ANL A, #0F0H: AND operating the accumulator data byte A2H with the data byte 0F0H. The resultant data byte in the accumulator will be A0H.
- MOV R0, #00H: Loading the data byte of register R0 with 00H.
- MOV 01H, 00H: Copying the data byte 00H into register R1.
- JB ACC.7, NEG: Jump to NEG if ACC.7 is set. Since, ACC.7 is 1, the control will move to the NEG label.
- INC R1: Incrementing the register R1 data byte by one. Hence, the data byte present in register R1 will be 01H.
- MOV A, #00H: Loading the data byte 00H into the accumulator.
- INC DPTR: Incrementing the data pointer register by one. The new data pointer register memory location will be 5679H.
- MOVX @DPTR, A: Copying the data byte 00H into the memory location pointed by the data pointer register.
- ORL A,#30H: ORing the accumulator data byte with 30H. Hence, the data byte present in the accumulator will be 30H.
- RL A: Rotating the accumulator left by one without carry. The final data byte becomes 60H.
- RL A: Rotating the accumulator left by one without carry. The final data byte becomes C0H.
- RL A: Rotating the accumulator left by one without carry. The final data byte becomes 81H.
- RL A: Rotating the accumulator left by one without carry. The final data byte becomes 03H.
- XCH A, DPL: Exchanging the DPL register with the accumulator data byte. So, the data byte present in the accumulator will be 79H and in DPL register will be 03H. So, the new DPTR memory location becomes 5603H.
- MOV DPH,A: Copying the data byte 79H into the DPH register. Hence, the new DPTR register will be 7903H.
- SJMP $: Short jump to the label $.
- END: Termination of the above program.

Conclusion: In the above program, the data byte A2H is checked whether it is positive or negative. By inspection, since the data byte is more than 07Fh, the above data byte will be negative. As the above data byte is negative, so the control jumps to the NEG label and we can see the new DPTR memory location we got. Hence, this program will check that the above data byte is negative. Thus, in order to monitor the data byte being positive or negative, just check the D7th bit whether it is set or not.

Program 3.27

Explain the code written below and state what the program is doing.

```
        ORG 0000H
        MOV DPTR,#1234H
        MOVX A,@DPTR
        ANL A,#0FH
        MOV R0,#00H
        MOV 01H,00H
        JB ACC.0,ODD
        INC R0
        MOV A,#0FFH
        INC DPTR
        MOVX @DPTR,A
        SJMP $
ODD:    INC R1
        MOV A,#00H
        INC DPTR
        MOVX @DPTR,A
        ORL A,#30H
        RL A
        RL A
        XCH A,DPL
        MOV DPH,A
        ORG 1234H
        DB A1H
        SJMP $
        END
```

Solution: In the above program, the DPTR memory location contains the data byte A1H since the memory location 1234H has the data byte A1H. Now, explaining the above sequence of codes:

- MOV DPTR, #1234H: Initializing the data pointer register with memory location 1234H.
- MOVX A, @DPTR: Copying the data byte A1H externally from the DPTR register into the accumulator.
- ANL A, #0FH: AND operating the accumulator data byte A1H with the data byte 0FH. The resultant data byte in the accumulator will be 01H.
- MOV R0, #00H: Loading the data byte of register R0 with 00H.
- MOV 01H, 00H: Copying the data byte 00H into the register R1.
- JB ACC.0, ODD: Jump to ODD if ACC.0 is set. Since, ACC.0 is 1, the control will move to the ODD label.
- INC R1: Incrementing register R1 data byte by one. Hence, the data byte present in register R1 will be 01H.
- MOV A, #00H: Loading the data byte 00H into the accumulator.
- INC DPTR: Incrementing the data pointer register by one. The new data pointer register memory location will be 5679H.
- MOVX @DPTR, A: Copying the data byte 00H into the memory location pointed by the data pointer register.
- ORL A,#30H: ORing the accumulator data byte with 30H. Hence, the data byte present in the accumulator will be 30H.
- RL A: Rotating the accumulator left by one without carry. The final data byte becomes 60H.
- RL A: Rotating the accumulator left by one without carry. The final data byte becomes C0H.
- XCH A, DPL: Exchanging the DPL register with the accumulator data byte. So, the data byte present in the accumulator will be 34H and in DPL register will be C0H. So, the new DPTR memory location becomes 12C0H.
- MOV DPH,A: Copying the data byte 34H into the DPH register. Hence, the new DPTR register will be 34C0H.
- SJMP $: Short jump to the label $.
- END: Termination of the above program.

Conclusion: In the above program, the data byte A1H is being checked whether it is positive or negative. After AND operating with 0FH, we can conclude that the above data byte is ODD. As the above data byte is ODD the new DPTR memory location is calculated which is found to be 34C0H.

Program 3.28

Explain the code written below and state what the program is doing.

```
        ORG 0000H
        MOV DPTR,#1234H
        MOVX A,@DPTR
        ANL A,#0FH
        MOV R0,#00H
        MOV 01H,00H
        JB ACC.0,ODD
        INC R0
        MOV A,#0FFH
        INC DPTR
        MOVX @DPTR,A
        SJMP $
ODD:    INC R1
        MOV A,#00H
        INC DPTR
        MOVX @DPTR,A
        ORL A,#30H
        RL A
        RL A
        XCH A,DPL
        MOV DPH,A
        ORG 1234H
        DB 22H
        SJMP $
        END
```

Solution: In the above program, the DPTR memory location contains the data byte 22H since the memory location 1234H has the data byte 22H. Now, explaining the above sequence of codes:

- MOV DPTR, #1234H: Initializing the data pointer register with memory location 1234H.
- MOVX A, @DPTR: Copying the data byte 22H externally from the DPTR register into the accumulator.
- ANL A, #0FH: AND operating the accumulator data byte 22H with the data byte 0FH. The resultant data byte in the accumulator will be 02H.
- MOV R0, #00H: Loading the data byte of register R0 with 00H.
- MOV 01H, 00H: Copying the data byte 00H into the register R1.

- JB ACC.0, ODD: Jump to ODD if ACC.0 is set. Since, ACC.0 is 0, the control will move to the next memory location.
- INC R0: Incrementing register R0 data byte by one. Hence, the data byte present in register R0 will be 01H.
- MOV A, #0FFH: Loading the data byte FFH into the accumulator.
- INC DPTR: Incrementing the data pointer register by one. The new data pointer register memory location will be 1235H.
- MOVX @DPTR, A: Copying the data byte 0FFH into the memory location pointed by the data pointer register.
- SJMP $: Short jump to the label $.
- END: Termination of the above program.

Conclusion: In the above program, the data byte 22H is checked whether it is positive or negative. After AND operating with 0FH, we can conclude that the above data byte is EVEN. As the above data byte is EVEN the new DPTR memory location is calculated which is found to be 1235H. Hence, to check whether any particular data byte is EVEN or ODD, just monitor the D0th bit of the accumulator.

Program 3.29

After the execution of the program given below, estimate the register R0 data byte and R1 data byte.

```
        ORG 3000H
        MOV R0,#00H
        MOV R1,#00H
        MOV R3,#08H
        MOV DPTR,#A000H
        MOVX A,@DPTR
LOOP:   CLRC
        RLC A
        JC GO
        INC R0
        SJMP NEW
GO:     INC R1
NEW:    DJNZ R3,LOOP
        ORG A000H
        DB 12H
        SJMP $
        END
```

Solution: In the above program, the DPTR memory location contains the data byte 12H since the memory location A000H has the data byte 12H. Now, explaining the above sequence of codes:

- MOV R0, #00H: Loading the data byte 00H in register R0.
- MOV R1, #00H: Loading the data byte 00H in register R1.
- MOV R3, #08H: Loading the data byte 08H in register R3.
- MOV DPTR, #A000H: Initializing the external memory location with A000H.
- MOVX A,@DPTR: Loading the data byte 12H from the external memory location into the accumulator.

LOOP1:

- CLR C: Clear the carry flag.
- RLC A: Rotating the accumulator left by one without carry.

  Before 1st RLC A execution: 12H in the accumulator

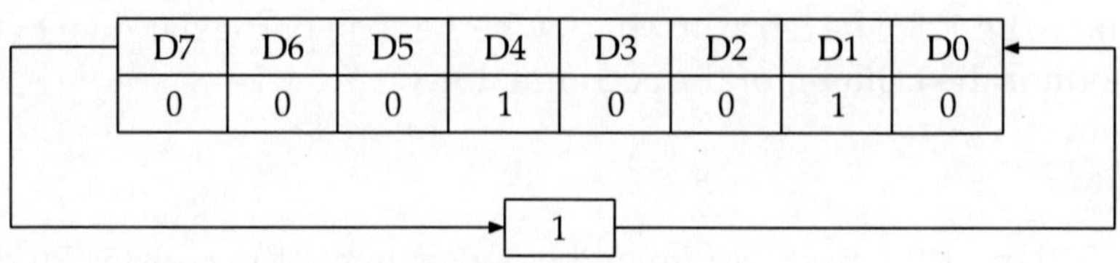

  After 1st RLC A execution: 24H in the accumulator

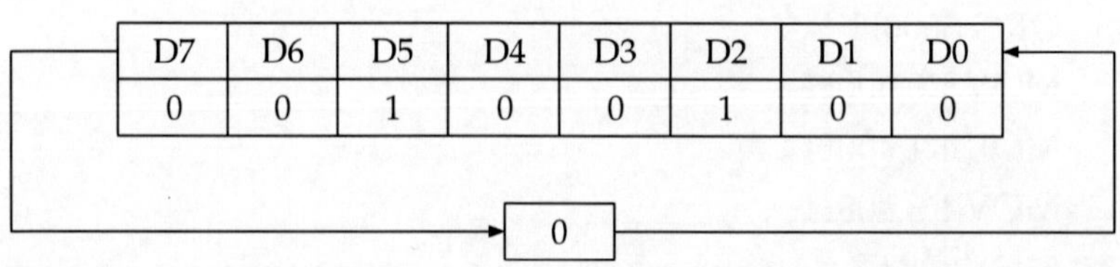

- JC GO: Jump to GO if carry flag is set. But carry flag is reset, hence the control will move onto the next memory location.
- INC R0: Incrementing register R0 by one. The data byte present in register R0 is 01H.
- SJMP NEW: Short jump to NEW label.
- NEW: DJNZ R3, LOOP: Decrementing register R3 by one and checking if zero flag is set or not. In the above case, the data byte present in register R3 is 07H and zero flag is reset. Hence, the control will move onto the LOOP label.

LOOP2:

- CLR C: Clear the carry flag.
- RLC A: Rotating the accumulator left by one without carry.

Before 2nd RLC A execution: 24H in the accumulator

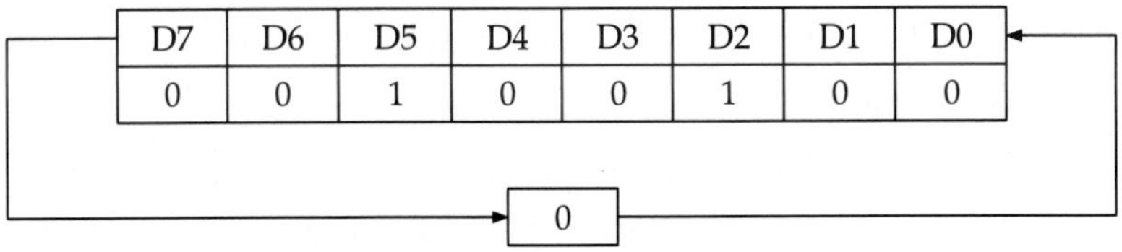

After 2nd RLC A execution: 48H in the accumulator

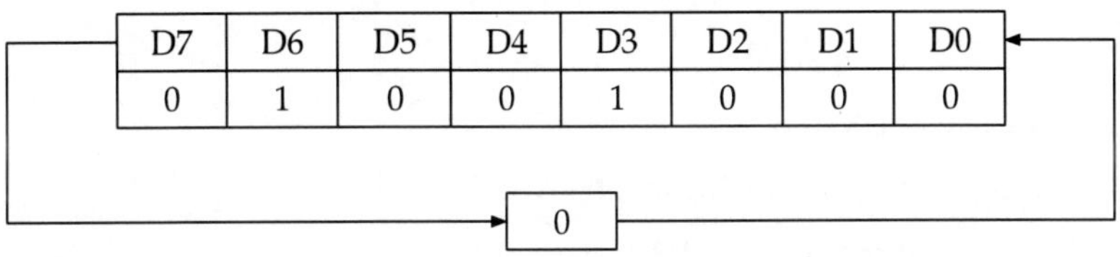

- JC GO: Jump to GO if carry flag is set. But carry flag is reset, hence the control will move onto the next memory location.
- INC R0: Incrementing register R0 by one. The data byte present in register R0 is 02H.
- SJMP NEW: Short jump to NEW label.
- NEW: DJNZ R3, LOOP: Decrementing register R3 by one and checking if zero flag is set or not. In the above case, the data byte present in register R3 is 06H and zero flag is reset. Hence, the control will move onto the LOOP label.

LOOP3:

- CLR C: Clear the carry flag.
- RLC A: Rotating the accumulator left by one without carry.

Before 3rd RLC A execution: 48H in the accumulator

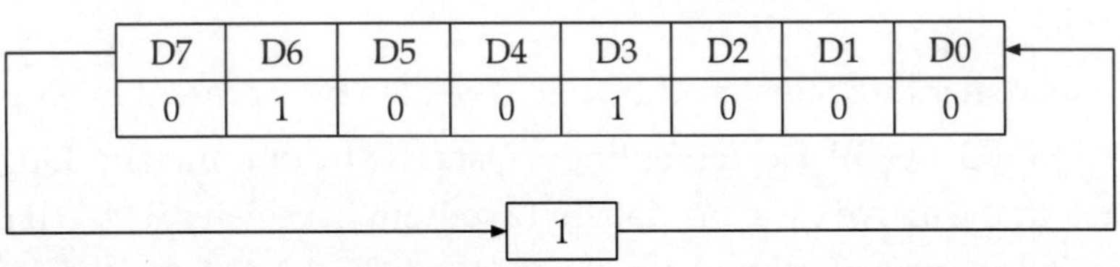

After 3rd RLC A execution: 90H in the accumulator

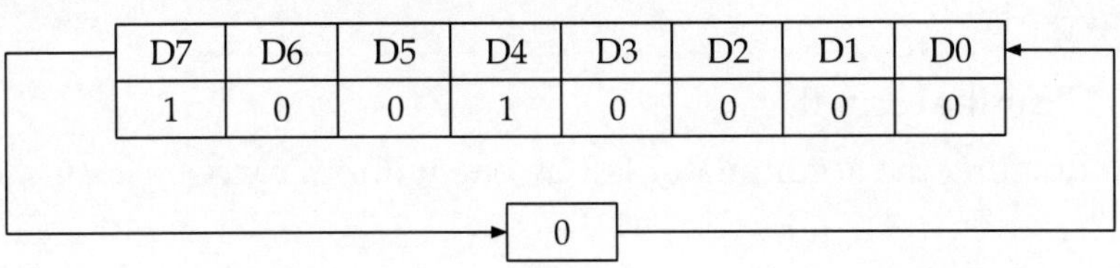

- JC GO: Jump to GO if carry flag is set. But carry flag is reset, hence the control will move onto the next memory location.
- INC R0: Incrementing register R0 by one. The data byte present in register R0 is 03H.
- SJMP NEW: Short jump to NEW label.
- NEW: DJNZ R3, LOOP: Decrementing register R3 by one and checking if zero flag is set or not. In the above case, the data byte present in register R3 is 05H and zero flag is reset. Hence, the control will move onto the LOOP label.

LOOP4:

- CLR C: Clear the carry flag.
- RLC A: Rotating the accumulator left by one without carry.

Before 4th RLC A execution: 90H in the accumulator

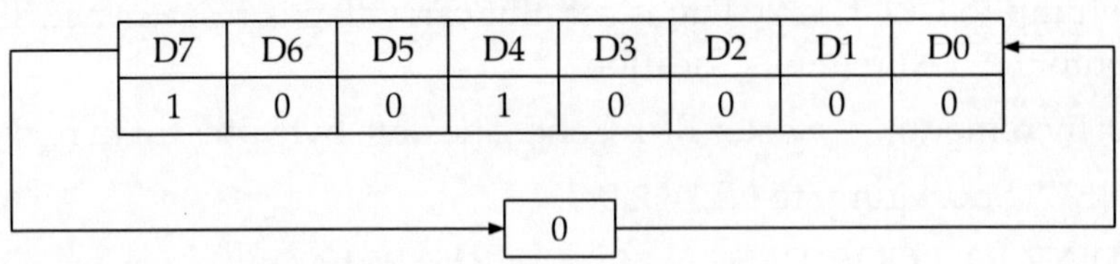

After 4th RLC A execution: 20H in the accumulator

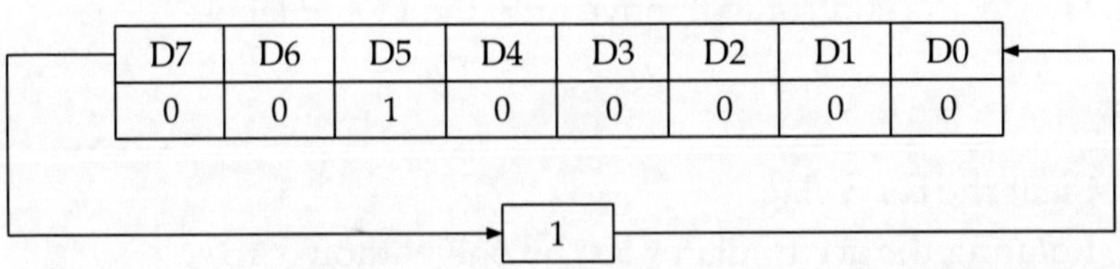

- JC GO: Jump to GO if carry flag is set. But carry flag is set, hence the control will jump to label GO.
- INC R1: Incrementing register R1 by one. The data byte present in register R1 is 01H.
- NEW: DJNZ R3, LOOP: Decrementing register R3 by one and checking if zero flag is set or not. In the above case, the data byte present in register R3 is 04H and zero flag is reset. Hence, the control will move onto the LOOP, label.

LOOP5:

- CLR C: Clear the carry flag.
- RLC A: Rotating the accumulator left by one without carry.

Before 5th RLC A execution: 20H in the accumulator

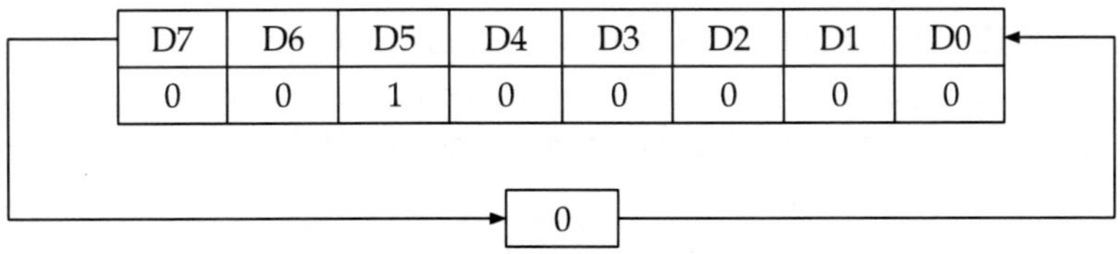

After 5th RLC A execution: 40H in the accumulator

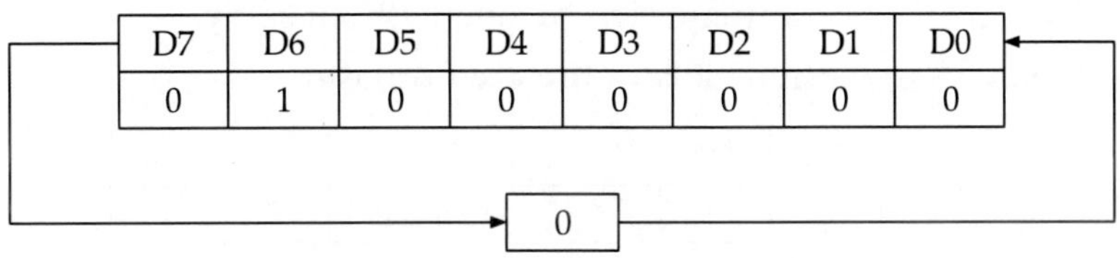

- JC GO: Jump to GO if carry flag is set. But carry flag is reset, hence the control will move onto the next memory location.
- INC R0: Incrementing register R0 by one. The data byte present in register R0 is 04H.
- SJMP NEW: Short jump to NEW label.
- NEW: DJNZ R3, LOOP: Decrementing register R3 by one and checking if zero flag is set or not. In the above case, the data byte present in register R3 is 03H and zero flag is reset. Hence, the control will move onto the LOOP label.

LOOP6:

- CLR C: Clear the carry flag.
- RLC A: Rotating the accumulator left by one without carry.

Before 6th RLC A execution: 40H in the accumulator

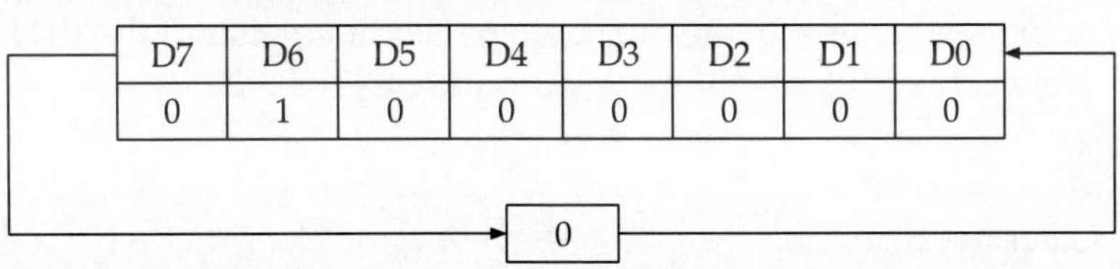

After 6th RLC A execution: 80H in the accumulator

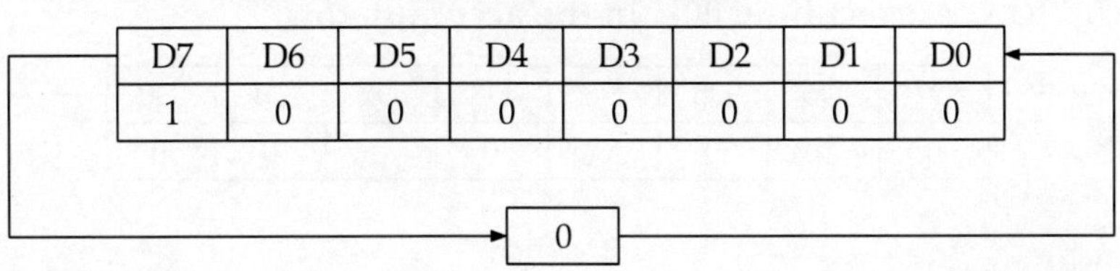

- JC GO: Jump to GO if carry flag is set. But carry flag is reset, hence the control will move onto the next memory location.

- INC R0: Incrementing register R0 by one. The data byte present in register R0 is 05H.
- SJMP NEW: Short jump to NEW label.
- NEW: DJNZ R3, LOOP: Decrementing register R3 by one and checking if zero flag is set or not. In the above case, the data byte present in register R3 is 02H and zero flag is reset. Hence, the control will move onto the LOOP label.

LOOP7:

- CLR C: Clear the carry flag.
- RLC A: Rotating the accumulator left by one without carry.

Before 7th RLC A execution: 80H in the accumulator

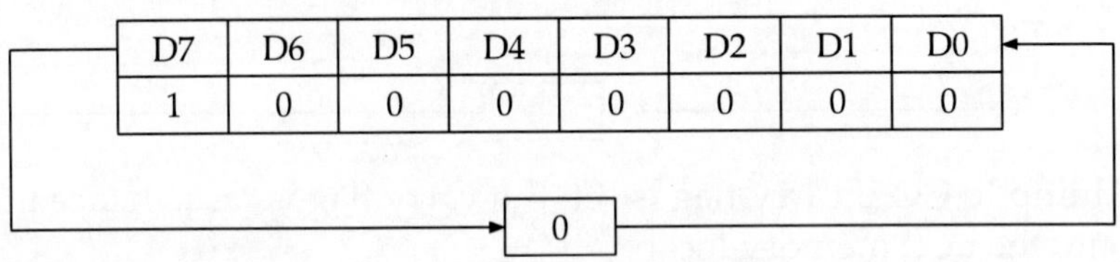

After 7th RLC A execution: 00H in the accumulator

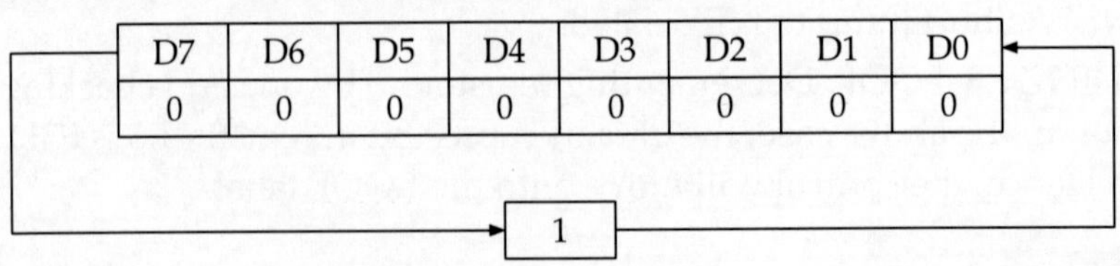

- JC GO: Jump to GO if carry flag is set. But carry flag is set, hence the control will jump onto the label GO.
- INC R1: Incrementing register R1 by one. The data byte present in register R1 is 02H.
- NEW: DJNZ R3, LOOP: Decrementing register R3 by one and checking if zero flag is set or not. In the above case, the data byte present in register R3 is 01H and zero flag is reset. Hence, the control will move onto the LOOP label.

LOOP8:

- CLR C: Clear the carry flag.
- RLC A: Rotating the accumulator left by one without carry.

Before 8th RLC A execution: 00H in the accumulator

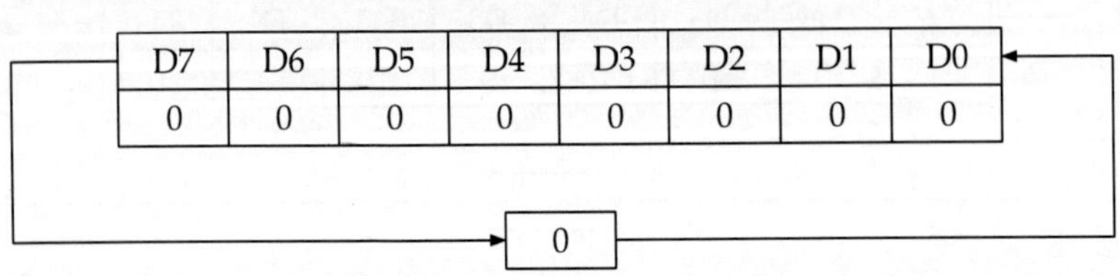

After 8th RLC A execution: 00H in the accumulator

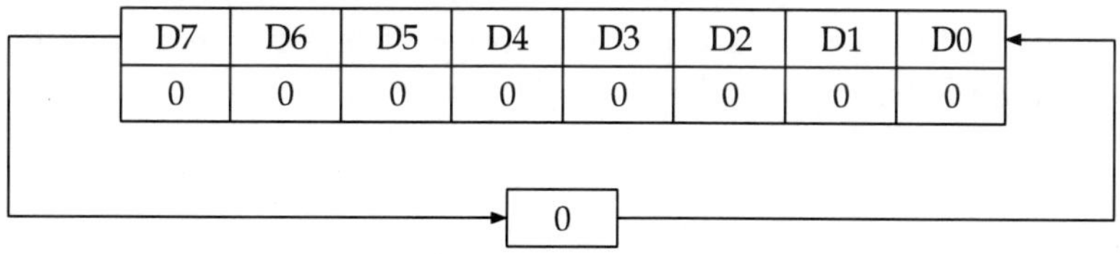

- JC GO: Jump to GO if carry flag is set. But carry flag is reset, hence the control will move onto the next memory location.
- INC R0: Incrementing register R0 by one. The data byte present in register R0 is 06H.
- SJMP NEW: Short jump to NEW label.
- NEW: DJNZ R3, LOOP: Decrementing register R3 by one and checking if zero flag is set or not. In the above case, the data byte present in register R3 is 00H and zero flag is set. Hence, the control will move onto the next memory location.
- $:SJMP $: Short jump to the label $.

  END: Termination of the above program.

Conclusion: Hence, after the execution of the above program, the data bytes present in the registers R0 and R1 are 06H and 02H respectively. Thus, the above program will count the number of times 1's or 0's appear in the carry flag.

Program 3.30

After the execution of the below program, determine the data byte present in the accumulator.

```
ORG 0000H
MOV A,#0FH
CPL A
ADD A,#01H
SJMP $
END
```

Solution: The above program will find the two's complement of 8-bit number. The data byte present in the accumulator after the execution of the above program is determined as follows:

- ORG 0000H: Initializing the program from the memory location 0000H.
- MOV A,#0FH: Loading the accumulator with the data byte 0FH.
- CPL A: Complementing the value of the accumulator. The data byte present in the accumulator is F0H.
- ADD A,#01H: Adding the data byte 01H with the accumulator data byte F0H. The resultant data byte present in the accumulator is F1H.

- $:SJMP $: Short jump to $.
- END: Terminate the execution of the above program.

Conclusion: Thus, after the execution of the above program, the data byte present in the accumulator will be F1H.

Program 3.31

Predict the DPTR memory location after the execution of the program given below:

```
ORG 3000H
MOV DPTR,#2345H
MOV A,DPL
CPL A
ADD A,#01H
MOV DPL,A
MOV A,DPH
CPL A
ADDC A,#00H
MOV DPH,A
MOVX A,@DPTR
XRL A,#00H
ORL A,#60H
MOVX @DPTR,A
SJMP $
END
```

Solution: The DPTR memory location after the execution of the above program is predicted as follows:

- ORG 3000H: Initialize the program with memory location 3000H.
- MOV DPTR,#2345H: Initialize the DPTR with external memory location 2345H.
- MOV A,DPL: Copy the data byte 45H into the accumulator.
- CPL A: Complementing the accumulator data byte. The resultant data byte stored in the accumulator is BAH.
- ADD A, #01H: Adding the accumulator content with data byte 01H. The final data byte present in the accumulator will be BBH.
- MOV DPL,A: Copy the data byte BBH into the DPL register.
- MOV A,DPH: Copy the data byte 23H into the accumulator.
- CPL A: Complementing the value of accumulator data byte. The final data byte stored in the accumulator will be DCH.

- ADDC A,#00H: Adding the data byte 00H to DCH with carry. The final data byte after addition will be DCH.
- MOV DPH,A: Copy the data byte DCH into the DPH register. Thus, the final external memory location will be DCBBH.
- MOVX A,@DPTR: Copying the data byte present in the external memory location into the accumulator.
- XRL A,#00H: XORing the data byte of accumulator with 00H. The final data byte after XORing will be 00H.
- ORL A,#30H: ORing the accumulator data byte with 30H. The final data byte after addition will be 30H.
- MOVX @DPTR,A: Copying the data byte 30H externally into the memory location pointed by the data pointer register.
- $:SJMP $: Short jump to $.
- END: Terminate the execution of the above program.

Conclusion: Thus, the data byte present in the accumulator after the execution of the above program will be 30H and the above data byte will be copied externally into the memory location DCBBH pointed by the DPTR.

Program 3.32

Estimate the behaviour of the program after the execution of the program when

NUM1 = 24H and NUM2 = 24H.
NUM1 = 32H and NUM2 = 30H.
NUM1 = 30H and NUM2 = 32H.

```
ORG 3000H
MOV DPTR,#4000H
MOVX A,@DPTR
MOV R0,A
INC DPTR
MOVX A,@DPTR
CLR C
SUBB A,R0
JZ SAME
JNC LARGE
SETB 78H
SJMP $
LARGE:          SETB 7FH
```

```
SJMP $
SAME:         CLR 7FH
CLR 78H
SJMP $
ORG 4000H
DB NUM1, NUM2
END
```

Solution: The above program will demonstrate the comparison of 2 numbers whether they are equal, greater or less than the other one. The data bytes present in the memory location 4000H and 4001H are 24H and 24H respectively.

(*a*) The first case is when the 2 numbers are equal, i.e., NUM1 = 24 , NUM2 = 24.

- ORG 3000H: Initialize the program from the memory location.
- MOV DPTR,#4000H: Initialize the data pointer register having external memory location 4000H.
- MOVX A,@DPTR: Copying the data byte 24H from the external memory location into the accumulator.
- MOV R0,A: Saving the data byte 24H into register R0.
- INC DPTR: Incrementing the data pointer register by one. The new memory location becomes 4001H.
- MOVX A, @DPTR: Copying the data byte 24H from the external memory location into the accumulator.
- CLR C: Clearing the carry flag.
- SUBB A, R0: Subtracting the value of the accumulator data byte, i.e., 24H with register R0 data byte 24H. The final data byte becomes 00H and the zero flag is set.
- JZ SAME: Jump to label SAME if zero flag is set. Here, zero flag is set. Hence, the control will move onto the label SAME.
- CLR 7FH: Clearing the bit address 7FH, i.e., the bit residing in the above memory location is '0'.
- CLR 78H: Clearing the bit address 78H, i.e., the bit residing in the above memory location is '0'.
- S: SJMP $: Short jump to the label $.
- End: Terminate the execution of the above program.

Conclusion: From the execution of the above program, we can conclude that the bit addresses 7FH and 78H become reset, since the data bytes residing in the 2 memory locations are SAME.

(*b*) The second case is when the 2 numbers are not equal and NUM1 is greater than NUM2, i.e., NUM1 = 32H , NUM2 = 30H.

- ORG 3000H: Initialize the program from the memory location.
- MOV DPTR,#4000H: Initialize the data pointer register having external memory location 4000H.
- MOVX A,@DPTR: Copying the data byte 32H from the external memory location into the accumulator.
- MOV R0,A: Saving the data byte 32H into register R0.
- INC DPTR: Incrementing the data pointer register by one. The new memory location becomes 4001H.
- MOVX A, @DPTR: Copying the data byte 30H from the external memory location into the accumulator.
- CLR C: Clearing the carry flag.
- SUBB A, R0: Subtracting the value of the accumulator data byte, i.e., 24H with register R0 data byte 24H. The final data byte becomes 02H and both zero and carry flags are reset.
- JZ SAME: Jump to label SAME if zero flag is set. Here, zero flag is reset, hence the control will move onto the next memory location.
- JNC LARGE: Jump to label LARGE when the carry flag is reset. Here, carry flag is reset. Hence, the control will jump to LARGE.
- SETB 7FH: Setting the bit address 7FH, i.e., the bit residing in the above memory location is '1'.
- S: SJMP $: Short jump to the label $.
- End: Terminate the execution of the above program.

Conclusion: From the execution of the above program, we can conclude that the bit address 7FH becomes set, since the data bytes residing in the two memory locations are not equal and NUM1 is greater than NUM2.

(*c*) The third case is when the 2 numbers are not equal and NUM1 is smaller than NUM2, i.e., NUM1 = 30H , NUM2 = 32H.

- ORG 3000H: Initialize the program from the memory location.
- MOV DPTR, #4000H: Initialize the data pointer register having external memory location 4000H.
- MOVX A, @DPTR: Copying the data byte 30H from the external memory location into the accumulator.
- MOV R0, A: Saving the data byte 30H into register R0.
- INC DPTR: Incrementing the data pointer register by one. The new memory location becomes 4001H.
- MOVX A, @DPTR: Copying the data byte 32H from the external memory location into the accumulator.

- CLR C: Clearing the carry flag.
- SUBB A, R0: Subtracting the value of the accumulator data byte, i.e., 24H with register R0 data byte 24H. The final data byte becomes FEH. So, zero flag is reset and the carry flag is set.
- JZ SAME: Jump to label SAME if zero flag is set. Here, zero flag is reset. Hence, the control will move onto the next memory location.
- JNC LARGE: Jump to label LARGE when the carry flag is reset. Here, carry flag is set. Hence, the control will move onto the next memory location.
- SETB 78H: Setting the bit address 78H, i.e., the bit residing in the above memory location is '1'.
- S: SJMP $: Short jump to the label $.
- End: Terminate the execution of the above program.

Conclusion: From the execution of the above program, we can conclude that the bit address 78H becomes set, since the data bytes residing in the 2 memory locations are not equal and NUM1 is smaller than NUM2.

Program 3.33

Determine the behaviour of the program when the data byte present in the memory location 70H is NUM1 has different data bytes for different cases as

(*a*) 00H.
(*b*) 01H.
(*c*) 02H.
(*d*) 03H.
(*e*) 04H.
(*f*) 08H.

The code written is shown below:

```
        ORG 3000H
        MOV R0,70H
        CJNE R0,#00H,J1
        MOV C,08H
        ANL C,10H
        SJMP FINAL
J1:     CJNE R0,#01H,J2
        MOV C,0FH
        ANL C,17H
        SJMP FINAL
```

```
J2:     CJNE R0,#02H,J3
        CPL 0FH
        MOV C,0FH
        SJMP FINAL
J3:     CJNE R0,#03H,J4
        MOV C,18H
        ORL C,1FH
        SJMP FINAL
J4:     CJNE R0,#04H,J5
        CLR C
        ORL C,20H
J5:     CJNE R0,#08H,J6
        SETB C
        ANL C,7FH
J6:     SETB C
FINAL:  MOV 00H,C
        SJMP $
        ORG 70H
        DB NUM1
        END
```

Solution: The above program will check the particular data byte in the memory location 70H and depending on the data byte value of the above memory location whereas bit addressable manipulations will be done. So, depending on the data byte value different operations are being performed:

(a) When data byte present in the memory location 70H is 00H.

- ORG 3000H: Initialize the program from the above memory location.
- MOV R0,70H: Copy the data byte 00H from memory location 70H into register R0.
- CJNE R0,#00H,J1: Comparing the data byte 00H in register R0 with the immediate data byte 00H. Since both the data bytes are same, the control will move onto the next memory location.
- MOV C, 08H: Copy the data bit present in memory location 08H into the carry flag.
- ANL C, 10H: AND operate the data bit present in the carry flag with 10H.
- SJMP FINAL: Short jump to the label FINAL.

- FINAL: MOV 00H,C: Copy the resultant carry flag bit into the bit memory location 00H.
- $:SJMP $: Short jump to the label $.
- END: Terminate the execution of the above program.

Conclusion: After the execution of the above program, we conclude that the LSB of data bytes present in the memory locations 21H and 22H are AND operated. Hence, finally the final carry flag bit is copied into the bit memory location 00H.

(*b*) When data byte present in the memory location 70H is 01H.

- ORG 3000H: Initialize the program from the above memory location.
- MOV R0, 70H: Copy the data byte 01H from the memory location 70H into register R0.
- CJNE R0, #00H, J1: Comparing the data byte 01H in register R0 with the data byte immediate data byte 00H. Since both the data bytes are not same, the control will jump onto the label J1.
- CJNE R0, #01H, J2: Comparing the data byte 01H in register R0 with the data byte immediate data byte 00H. Since both the data bytes are same, the control will move onto the next memory location.
- MOV C, 0FH: Copy the data bit present in the memory location 0FH into the carry flag.
- ANL C, 17H: AND operate the data bit present in the carry flag with bit memory location 17H.
- SJMP FINAL: Short jump to the label FINAL.
- FINAL: MOV 00H, C: Copy the resultant carry flag bit into the bit memory location 00H.
- $: SJMP $: Short jump to the label $.
- END: Terminate the execution of the above program.

Conclusion: After the execution of the above program, we conclude that the MSB of data bytes present in the memory location 21H and 22H are AND operated. Hence, finally the final carry flag bit is copied into the bit memory location 00H.

(*c*) When data byte present in the memory location 70H is 02H.

- ORG 3000H: Initialize the program from the above memory location.
- MOV R0, 70H: Copy the data byte 02H from the memory location 70H into the register R0.
- CJNE R0, #00H, J1: Comparing the data byte 02H in register R0 with the data byte immediate data byte 00H. Since both the data bytes are not same, the control will jump onto the label J1.

- CJNE R0, #01H, J2: Comparing the data byte 02H in register R0 with the data byte immediate data byte 01H. Since both the data bytes are not same, the control will jump onto the label J2.
- CJNE R0, #02H, J3: Comparing the data byte 02H in register R0 with the data byte immediate data byte 02H. Since both the data bytes are same, the control moves onto the next memory location.
- CPL 0FH: Complement the data bit present in the bit memory location 0FH.
- MOV C, 0FH: Copying the data bit present in the memory location 0FH into the carry flag.
- SJMP FINAL: Short jump to the label FINAL.
- FINAL: MOV 00H, C: Copy the resultant carry flag bit into the bit memory location 00H.
- $: SJMP $: Short jump to the label $.
- END: Terminate the execution of the above program.

Conclusion: After the execution of the above program, we conclude that the MSB of data byte present in the memory location 21H is complemented. Hence, finally the complemented carry flag bit is copied into the bit memory location 00H.

(*d*) When data byte present in the memory location 70H is 03H.

- ORG 3000H: Initialize the program from the above memory location.
- MOV R0, 70H: Copy the data byte 03H from the memory location 70H into register R0.
- CJNE R0, #00H, J1: Comparing the data byte 03H in register R0 with the data byte immediate data byte 00H. Since both the data bytes are not same, the control will jump onto the label J1.
- CJNE R0, #01H, J2: Comparing the data byte 03H in register R0 with the data byte immediate data byte 01H. Since both the data bytes are not same, the control will jump onto the label J2.
- CJNE R0, #02H, J3: Comparing the data byte 03H in register R0 with the data byte immediate data byte 02H. Since both the data bytes are not same, the control will jump onto the label J3.
- CJNE R0, #03H, J4: Comparing the data byte 03H in register R0 with the data byte immediate data byte 03H. Since both the data bytes are same, the control will move onto the next memory location.
- MOV C, 18H: Copying the data bit present in the bit addressable memory location 18H into the carry flag.
- ORL C, 1FH: ORing the data bit present in the bit addressable memory location 1FH with the carry flag bit.
- SJMP FINAL: Short jump to the label FINAL.

- FINAL: MOV 00H, C: Copy the resultant carry flag bit into the bit memory location 00H.
- $: SJMP $: Short jump to the label $.
- END: Terminate the execution of the above program.

Conclusion: After the execution of the above program, we conclude that the LSB of data byte present in the memory location 23H is ORed with the MSB of the data byte present in the same memory location. Hence, finally the carry flag bit is copied into the bit memory location 00H.

(*e*) When data byte present in the memory location 70H is 04H.

- ORG 3000H: Initialize the program from the above memory location.
- MOV R0, 70H: Copy the data byte 04H from the memory location 70H into register R0.
- CJNE R0, #00H, J1: Comparing the data byte 04H in register R0 with the data byte immediate data byte 00H. Since both the data bytes are not same, the control will jump onto the label J1.
- CJNE R0, #01H, J2: Comparing the data byte 04H in register R0 with the data byte immediate data byte 01H. Since both the data bytes are not same, the control will jump onto the label J2.
- CJNE R0, #02H, J3: Comparing the data byte 04H in register R0 with the data byte immediate data byte 02H. Since both the data bytes are not same, the control will jump onto the label J3.
- CJNE R0, #03H, J4: Comparing the data byte 04H in register R0 with the data byte immediate data byte 03H. Since both the data bytes are not same, the control will jump onto the label J4.
- CJNE R0, #04H, J5: Comparing the data byte 04H in register R0 with the data byte immediate data byte 04H. Since both the data bytes are same, the control will move onto the next memory location.
- CLR C: Clearing the carry flag.
- ORL C, 20H: ORing the data bit present in the bit addressable memory location 20H with the carry flag bit.
- SJMP FINAL: Short jump to the label FINAL.
- FINAL: MOV 00H, C: Copy the resultant carry flag bit into the bit memory location 00H.
- $: SJMP $: Short jump to the label $.
- END: Terminate the execution of the above program.

Conclusion: After the execution of the above program, we conclude that after clearing of the carry flag bit, the LSB of data byte present in the memory location 24H is ORed with carry flag. Hence, finally the carry flag bit is copied into the bit memory location 00H.

(*f*) When data byte present in the memory location 70H is 08H.

- ORG 3000H: Initialize the program from the above memory location.
- MOV R0, 70H: Copy the data byte 08H from the memory location 70H into register R0.
- CJNE R0, #00H, J1: Comparing the data byte 08H in register R0 with the data byte immediate data byte 00H. Since both the data bytes are not same, the control will jump onto the label J1.
- CJNE R0, #01H, J2: Comparing the data byte 08H in register R0 with the data byte immediate data byte 01H. Since both the data bytes are not same, the control will jump onto the label J2.
- CJNE R0, #02H, J3: Comparing the data byte 08H in register R0 with the data byte immediate data byte 02H. Since both the data bytes are not same, the control will jump onto the label J3.
- CJNE R0, #03H, J4: Comparing the data byte 08H in register R0 with the data byte immediate data byte 03H. Since both the data bytes are not same, the control will jump onto the label J4.
- CJNE R0, #04H, J5: Comparing the data byte 08H in register R0 with the data byte immediate data byte 04H. Since both the data bytes are not same, the control will jump onto label J5.
- CJNE R0, #08H, J6: Comparing the data byte 08H in register R0 with the data byte immediate data byte 08H. Since both the data bytes are same, the control will move onto the next memory location.
- SETB C: Setting the carry flag.
- ANL C, 7FH: ANDing the data bit present in the bit addressable memory location 7FH with the carry flag bit.
- SJMP FINAL: Short jump to the label FINAL.
- FINAL: MOV 00H, C: Copy the resultant carry flag bit into the bit memory location 00H.
- $: SJMP $: Short jump to the label $.
- END: Terminate the execution of the above program.

Conclusion: After the execution of the above program, we conclude that after setting of the carry flag bit, the MSB of data byte present in the memory location 2FH is AND operated with carry flag. Hence, finally the carry flag bit is copied into the bit memory location 00H.

Program 3.34

Write an assembly language program to count positive and negative bytes from the 10 bytes present in the internal RAM memory location from 30h. Put the result of positive count and negative count in the registers R6 and R7 respectively.

Solution: Bit D_7 is used for the identification of the number, whether the number is positive or negative.

If bit $D_7 = 0$, number is positive

If bit $D_7 = 1$, number is negative

Assembly Language Program		Comments
	MOV R0, #30H	; Initialize memory location
	MOV R2, #0AH	; Initialize counter for 10 bytes
	MOV R6, #00H	; Count for positive bytes
	MOV R7, #00H	; Count for negative bytes
REPEAT:	MOV A, @R0	; Get byte from memory location
	INC R0	; Pointed to the next memory location
	RL A	; Rotate the contents of the accumulator towards left
	JC NEG	; If carry, the number is negative, otherwise positive
	INC R6	; Increment positive count
	SJMP NEXT	; Decrement counter and repeat
NEG:	INC R7	; Increment negative count
NEXT:	DJNZ R2, REPEAT	; Decrement counter and repeat if not zero
LAST:	SJMP LAST	; End of the program

Program 3.35

Write an assembly language program to count even and odd bytes from the 10 bytes present in the internal RAM memory location from 30h. Put the result of even count and odd count in the registers R6 and R7 respectively.

Solution: Bit D_0 is used for the identification of the number, whether the number is even or odd.

If bit $D_0 = 0$, number is even

If bit $D_1 = 1$, number is odd

Assembly Language Program		Comments
	MOV R0, #30H	; Initialize memory location
	MOV R2, #0AH	; Initialize counter for 10 bytes
	MOV R6, #00H	; Count for even bytes
	MOV R7, #00H	; Count for odd bytes
REPEAT:	MOV A, @R0	; Get byte from memory location
	INC R0	; Pointed to the next memory location
	RR A	; Rotate the contents of the accumulator towards right

```
        JC ODD              ; If carry, the number is odd, otherwise even
        INC R6              ; Increment even count
        SJMP NEXT           ; Decrement counter and repeat
NEG:    INC R4              ; Increment odd count
NEXT:   DJNZ R2, REPEAT     ; Decrement counter and repeat if not zero
LAST:   SJMP LAST           ; End of the program
```

Program 3.36

Write an assembly language program to find and count number divisible by 2 from the 10 bytes present in the external RAM location from 2000h, store the count value in register R7.

Solution: To find the numbers divisible by 2 checks D_0 bit, if D_0 bit is zero then the number is divisible by 2, otherwise the number is not divisible by 2.

Assembly Language Program **Comments**

```
        MOV DPTR, #2000H    ; Initialize memory location
        MOV R2, #0AH        ; Initialize counter for 10 bytes
        MOV R7, #00H        ; Count for number divisible by 2
REPEAT: MOVX A, @DPTR       ; Get byte from external memory location
        INC DPTR            ; Pointed to the next memory location
        RR A                ; Rotate the contents of the accumulator towards
                              right
        JC NEXT             ; If carry, the number is not divisible by 2
        INC R7              ; Increment counter for number divisible by 2
NEXT:   DJNZ R2, REPEAT     ; Decrement counter and repeat if not zero
LAST:   SJMP LAST           ; End of the program
```

Program 3.37

Write an assembly language program to find and count number divisible by 4 from the 10 bytes present in the external RAM location from 2000h, store the count value in register R7.

Solution: To find the numbers divisible by 2 checks D_1 and D_0 bits, if both D_1 and D_0 bits is zero then the number is divisible by 4, otherwise the number is not divisible by 4.

Assembly Language Program **Comments**

```
        MOV DPTR, #2000H    ; Initialize memory location
        MOV R2, #0AH        ; Initialize counter for 10 bytes
        MOV R7, #00H        ; Count for number divisible by 4
REPEAT: MOVX A, @DPTR       ; Get byte from external memory location
        MOV R3, #02H        ; Initialize counter for two times rotation
```

```
         INC DPTR          ; Pointed to the next memory location
AGAIN:   RR A              ; Rotate the contents of the accumulator towards
                             right
         JC NEXT           ; If carry, the number is not divisible by 4
         DJNZ R3, AGAIN    ; Rotate again to check D1 bit 0
         INC R7            ; Increment counter for number divisible by 4
NEXT:    DJNZ R2, REPEAT   ; Decrement counter and repeat if not zero
LAST:    SJMP LAST         ; End of the program
```

Program 3.38

Write an assembly language program to find the largest byte from the 10 bytes present in the internal RAM location 50h, put the largest byte in register R6.

Solution: To find the largest byte, assume initially the accumulator is zero and then compare with the contents of the memory location. If the carry is generated then the contents of memory are greater than the contents of the accumulator, so replace the contents of the accumulator with the contents of memory location (largest byte).

```
Assembly Language Program   Comments
         MOV R0, #50H      ; Initialize memory location
         MOV R2, #0AH      ; Initialize counter for 10 bytes
         CLR A             ; Clear accumulator
REPEAT:  CLR C             ; Clear carry flag
         SUBB A, @R0       ; Subtract contents of memory from the accumulator
         JNC SMALL         ; If no carry, the number in the memory is small
         MOV A, @R0        ; If carry, the number in the memory is greater than
                             the accumulator, so transfer to the accumulator
SMALL:   INC R0            ; Go for next memory location
         DJNZ R2, REPEAT   ; Decrement counter and repeat if not zero
         MOV R6, A         ; Store the largest byte to the register R6
LAST:    SJMP LAST         ; End of the program
```

Program 3.39

Write an assembly language program to search for the byte 0Dh from the 10 bytes present in the memory location from 40h (internal RAM), and if 0Dh is present then set carry flag otherwise reset carry flag.

Solution: To search for the byte, put the byte in the accumulator and subtract the contents of the accumulator from the contents of the memory. If the zero flag is set after subtraction then the contents of the accumulator and the contents of the memory location are same, i.e., a byte is present in the memory location.

It is important to note that after subtraction, the contents of the accumulator get changed, so it is necessary to load the accumulator again with the byte before subtraction.

```
          Assembly Language Program    Comments
          MOV R0, #40H                 ; Initialize starting memory location
          MOV R2, #10                  ; Initialize counter for 10 bytes
REPEAT:   MOV A, #0DH                  ; Load 0Dh to the accumulator for search
          CLR C                        ; Clear carry flag
          SUBB A, @R0                  ; Subtract with borrow the contents of memory
                                         location from the accumulator
          JNZ NEXT                     ; If ZF = 0, go for next round
          SETB C                       ; If ZF =1, set carry flag and exit
          SJMP LAST                    ; If ZF =1, exit from the loop
NEXT:     CLR C                        ; Clears carry flag
          INC R0                       ; Increment memory location for the next round
          DJNZ R2, REPEAT              ; Decrement counter and repeat if not zero
LAST:     SJMP LAST                    ; End of the program
```

Program 3.40

Write an assembly language program cam to search for the byte 0Dh and count from the 10 bytes present in the internal RAM location 40h, and put the count value in register R7.

Solution: To search for the byte, put the byte in the accumulator and subtract the contents of the accumulator from the contents of the memory. If the zero flag is set after subtraction then the contents of the accumulator and the contents of the memory location is same, i.e., a byte is present in the memory location.

It is important to note that after subtraction, the contents of accumulator gets changed, so it is necessary to load the accumulator again with the byte before subtraction.

Method – I

```
          Assembly Language Program    Comments
          MOV R0, #40H                 ; Initialize starting memory location
          MOV R2, #10                  ; Initialize counter for 10 bytes
          MOV R7, #00H                 ; Initialize counter for count of 0Dh
REPEAT:   MOV A, #0DH                  ; Load 0Dh to the accumulator for search
          CLR C                        ; Clear carry flag
          SUBB A, @R0                  ; Subtract with borrow the contents of memory
                                         location from accumulator
          JNZ NEXT                     ; If ZF = 0, go for next round
```

```
        INC R7             ; If ZF =1, 0Dh is present, so increment count in
                             register R7
NEXT:   INC R0             ; Increment memory location for the next round
        DJNZ R2, REPEAT    ; Decrement counter and repeat if not zero
LAST:   SJMP LAST          ; End of the program
```

Method – II

Assembly Language Program **Comments**

```
        MOV R0, #40H       ; Initialize starting memory location
        MOV R2, #10        ; Initialize counter for 10 bytes
        MOV R7, #00H       ; Initialize counter for count of 0Dh
        CJNE @R0, #0DH, NO ; Compare contents of internal RAM with data
                           ; Byte 0Dh
        INC R7             ; If ZF =1 (number equal), 0Dh is present, so
                           ; Increment count in register R7
NO:     INC R0             ; Increment memory location for the next round
        DJNZ R2, REPEAT    ; Decrement counter and repeat if not zero
LAST:   SJMP LAST          ; End of the program
```

Program 3.41

Write an assembly language program to solve the equation $2P + 4Q \longrightarrow R7$, where P and Q are 8-bit unsigned numbers present in the internal RAM location 30h and 31h respectively. Put the result in the register R7. Use add and shift method, ignore carry flag.

Solution: Rotate left instruction is sued to multiply the number by 2^n, where n is the number of times rotation.

For example, if the number is X and it is rotated towards left by 2 times then the number becomes

$$X * 2^2 = 4X$$

In this case use rotates with carry, because it facilitates 9-bit operation.

The given equation is

$$2P + 4Q \longrightarrow R7$$

The rearrangement of the equation in terms of 2^n is given as

$$2^1P + 2^2Q \longrightarrow R7$$

Assembly Language Program **Comments**

```
        MOV A, 30H         ; Get the contents of memory location 30h (P)
        CLR C              ; Clears carry flag
        RLC A              ; Rotate towards left (2P)
```

```
        MOV R0, A        ; Store contents of the accumulator (2P) to register
                           R0
        MOV A, 31H       ; Get the contents of memory location 31h (Q)
        CLR C            ; Clears carry flag
        RLC A            ; Rotate towards left (2Q)
        RLC A            ; Rotate again towards left (4Q)
        ADD A, R0        ; Add 2P with 4Q
        MOV R7, A        ; Store the result in register R7
LAST:   SJMP LAST        ; End of the program
```

Program 3.42

Write an assembly language program to solve the equation P + Q/2 + R/4 → R7. Where P, Q and R are 8-bit unsigned numbers present in the internal RAM location 30h, 31h and 32h respectively, put the result in the register R7. Use add and shift method, ignore carry.

Solution: Rotate towards right instruction is used to divide the number by 2^n, where n is the number of times rotation.

For example, if the number X is rotated towards right by 2 times then the number X becomes

$$X/2^2 = X/4$$

The given equation is

$$P + Q/2 + R/4 \longrightarrow R7$$

The rearrangement of the equation in terms of 2^n is given as

$$P + Q/2^1 + R/2^2 \longrightarrow R7$$

Assembly Language Program **Comments**

```
        MOV A, 31H       ; Get the contents of memory location 31h (Q)
        CLR C            ; Clears carry flag
        RRC A            ; Rotate accumulator towards left (Q/2)
        MOV R0, A        ; Store value Q/2 in register R0
        MOV A, 32H       ; Get the contents of memory location 32h (R)
        CLR C            ; Clears carry flag
        RRC A            ; Rotate the accumulator towards right (R/2)
        RRC A            ; Rotate the contents of the accumulator towards
                           right again (R/4)
        ADD A, R0        ; Add the contents of R0 (Q/2) with the contents
                           of the accumulator (R/4)
        ADD A, 30H       ; Add the contents of the accumulator (Q/2 + R/4)
                           with the contents of memory location 30h (P)
```

```
          MOV R7, A          ; Store the result in register R7
LAST:     SJMP LAST          ; End of the program
```

Program 3.43

Show the contents of internal RAM locations after execution of the following program.

```
SETB PSW. 3
MOV R2, #45H
MOV R3, #55H
MOV R6, #0DH
MOV R7, #0FFH
```

Solution: The 8-bit format of the PSW register is shown below

D7	D6	D5	D4	D3	D2	D1	D0
CY	AC	F0	RS1	RS0	OV	–	P
0	0	0	1	0	0	0	0

According to PSW format when SETB PSW.3 is executed then RS0 = 1, so RS1 RS0 = 0 1, so register bank is selected.

RS1	RS0	Register Bank	Address
0	0	0	00h – 07h
0	1	1	08h – 0Fh
1	0	2	10h – 17h
1	1	3	18h – 1Fh

From the register bank 1, memory location 08h to 0Fh is selected.

Address of Internal RAM location	Contents of memory location
08h	--
09h	--
0Ah	45h
0Bh	55h
0Ch	--
0Dh	--
0Eh	0Dh
0Fh	FFh

Program 3.44

Write a sub program to add 5 to the accumulator 50 times. Save the result in register R6. Ignore carry if any.

Solution:

Label	Assembly Language Program	Comments
	MOV SP, #07H	; Initialize stack pointer (optional if SP is initialized from memory location 07h, because by default SP is initialized at 07h)
	CLR A	; Clears the accumulator
	MOV R2, #50	; Initialized counter for 50 times addition of data byte 05h
	ACALL SUBP	; Jump on the subroutine
	MOV R6, A	; Save the result in register R7
LAST:	*SJMP LAST*	; End of the program
SUBP:	*ADD A, #05H*	; Add 05h to the accumulator
	DJNZ R2, SUBP	; Repeat the process for 50 times
	RET	; Return to the main program

Program 3.45

Write a program to complement the contents of port P0 600 times.

Solution: In 8051 microcontroller registers R0 to R7 are used for counter and each of these registers are 8-bit wide, so hold only 8-bit value of the counter, so counter can be executed only up to 256 times.

Therefore, two registers are used to implement counter if the counter value exceeds 256 times.

Label	Assembly Language Program	Comments
	MOV A, #55H	; Initialize the accumulator with alternate 1's and 0's
	MOV R2, #10	; Initialize counter 1
AGAIN2:	*MOV R3, #60*	; Initialize counter 2
AGAIN1:	*MOV P0, A*	; Transfer the accumulator to port P0
	CPL A	; Complement the accumulator
	DJNZ R3, AGAIN1	; Decrement counter 2, and jump to AGAIN1 label if not zero
	DJNZ R2, AGAIN2	; Decrement counter 1, and jump to AGAIN2
		; Label if not zero
LAST:	*SJMP LAST*	

Program 3.46

Write a program to clear 50 external RAM locations from 0100h.

Solution:

Assembly Language Program		Comments
	CLR A	; Clears the accumulator (A = 00h)
	MOV R2, #50	; Initialize counter
	MOV DPTR, #0100H	; Initialize external RAM location
AGAIN:	*MOVX @DPTR, A*	; Copy the accumulator to the external RAM
	INC DPTR	; Pointed to the next memory location
	DJNZ R2, AGAIN	; Decrement counter and repeat the process if not zero
LAST:	*SJMP LAST*	; End of the program

Program 3.47

Write a program to copy data bytes 00h, 01h, 02h, …. FEh, FFh in the external RAM location from 1000h.

Solution:

Assembly Language Program		Comments
	CLR A	; Clears the accumulator (A = 00h)
	MOV R2, #00	; Initialize counter (for 256 times)
	MOV DPTR, #1000H	; Initialize external RAM location
AGAIN:	*MOVX @DPTR, A*	; Copy the accumulator to the external RAM
	INC A	; Increment accumulator by one
	INC DPTR	; Pointed to the next memory location
	DJNZ R2, AGAIN	; Decrement counter and repeat the process if not zero
LAST:	*SJMP LAST*	; End of the program

Program 3.48

Write a program to copy data bytes 00, 01, 02, …. 98, 99 in the external RAM location from 1000h.

Solution:

Assembly Language Program		Comments
	CLR A	; Clears the accumulator (A = 00h)
	MOV R2, #100	; Initialize counter (for 100 (64h) values)

```
            MOV DPTR, #1000H  ; Initialize external RAM location
AGAIN:      MOVX @DPTR, A     ; Copy accumulator to the external RAM
            ADD A, #01H       ; Add data byte 01h to the accumulator
            DA A              ; Adjust the contents of the accumulator into BCD
            INC DPTR          ; Pointed to the next memory location
            DJNZ R2, AGAIN    ; Decrement counter and repeat the process if
                                not zero
LAST:       SJMP LAST         ; End of the program
```

Program 3.49

Write a program to copy data bytes 00h, 11h, 22h, 33h,….EEh, FFh in the external RAM location from 1000h.

Solution:

Assembly Language Program		Comments
	CLR A	; Clears the accumulator (A = 00h)
	MOV R2, #16	; Initialize counter (for 16 values)
	MOV DPTR, #1000H	; Initialize external RAM location
AGAIN:	*MOVX @DPTR, A*	; Copy the accumulator to the external RAM
	ADD A, #11H	; Add the accumulator with data byte 11h
	INC DPTR	; Pointed to the next memory location
	DJNZ R2, AGAIN	; Decrement counter and repeat the process if not zero
LAST:	*SJMP LAST*	; End of the program

Program 3.50

Write a program to copy data bytes 00h, 11h, 22h, 33h,….EEh, FFh, 00h, 11h, 22h, 33h,…. EEh, FFh, 00h, 11h, 22h, 33h,…..EEh, FFh in the external RAM memory location from 1000h.

Solution:

Method – I

Assembly Language Program		Comments
	MOV R2, #16	; Initialize counter 1 (for 16 values)
	MOV R3, #03	; Initialize counter 2 for three times repetition
	MOV DPTR, #1000H	; Initialize external RAM location
RESET:	*CLR A*	; Clears the accumulator (A = 00h)
AGAIN:	*MOVX @DPTR, A*	; Copy the accumulator to the external RAM

```
        ADD A, #11H          ; Add the accumulator with data byte 11h
        JNC L1               ; If no carry, pointed to the next memory
        DJNZ R3, RESET       ; If carry, decrement counter 2 and resets the
                               accumulator again and continue
L1:     INC DPTR             ; Pointed to the next memory location
        DJNZ R2, AGAIN       ; Decrement counter and repeat the process if
                               not zero
LAST:   SJMP LAST            ; End of the program
```

Method – II

Assembly Language Program **Comments**

```
        MOV R2, #48          ; Initialize counter (for 48 values, 3 times 00h to
                               FFh)
        MOV DPTR, #1000H     ; Initialize external RAM location
RESET:  CLR A                ; Clears the accumulator (A = 00h)
AGAIN:  MOVX @DPTR, A        ; Copy the accumulator to the external RAM
        ADD A, #11H          ; Add the accumulator with data byte 11h
        JC RESET             ; If no carry, pointed to the next memory
L1:     INC DPTR             ; Pointed to the next memory location
        DJNZ R2, AGAIN       ; Decrement counter and repeat the process if
                               not zero
LAST:   SJMP LAST            ; End of the program
```

Program 3.51

Write a program to copy 10 bytes from the internal RAM location starting at 30h to the internal RAM location starting at 40h.

Solution:

Assembly Language Program **Comments**

```
        MOV R2, #10          ; Initialize counter
        MOV R0, #30H         ; Initialize address of source memory location
        MOV R1, #40H         ; Initialize address of destination memory
                               location
AGAIN:  MOV A, @R0           ; Read contents from source memory location
        MOV @R1, A           ; Store contents of accumulator to destination
                               memory location
        INC R0               ; Increment source memory location
        INC R1               ; Increment destination memory location
```

```
        DJNZ R2, AGAIN       ; Decrement counter and repeats if not zero
LAST:   SJMP LAST            ; End of the program
```

Program 3.52

Write an assembly language program to find 2's complement of 8-bit number present in the external ROM location 2000h, put the result in register R7.

Solution:

Assembly Language Program **Comments**

```
        CLR A                ; Clears accumulator
        MOV DPTR, #2000H     ; Initialize address of external ROM location
        MOVC A, @A+DPTR      ; Read contents of external ROM location
        CPL A                ; Find 1's complement
        ADD A, #01H          ; Find 2's complement
        MOV R7, A            ; Store the result in register R7
LAST:   SJMP LAST            ; End of the program
```

Program 3.53

Write an assembly language program to find 2's complement of 16-bit number present in the register R7-R6, assume MSB in register R7 and LSB in register R6, put the result in the same register.

Solution:

Assembly Language Program **Comments**

```
        MOV A, R6            ; Get LSB byte in accumulator
        CPL A                ; Find 1's complement
        MOV R6, A            ; Store 1's complement in register R6
        MOV A, R7            ; Get MSB byte in accumulator
        CPL A                ; Find 1's complement
        MOV R7, A            ; Store 1's complement in register R7
        MOV A, R6            ; Get 1's complement of LSB byte in accumulator
        ADD A, #01H          ; Find 2's complement by adding 1
        MOV R6, A            ; Store 2's complement of LSB byte in register R6
        MOV A, R7            ; Get 1's complement of MSB byte in the accumulator
        ADD C A, #00H        ; Find 2's complement by adding 00h with carry
                               from the LSB byte, if any
```

```
        MOV R7, A          ; store 2's complement of MSB byte in register
                           R7
LAST:   SJMP LAST          ; end of the program
```

Program 3.54

Write an assembly language program to find number of 1's in a given byte. Assume that the byte is 95h.

Solution:

```
Assembly Language Program   Comments
        MOV R2, #08        ; Initialize counter for 8 times rotation
        MOV R1, #00H       ; Keep track of 1's
        MOV A, #95H        ; Load data byte 95h to the accumulator
AGAIN:  RLC A              ; Rotate the accumulator towards left with carry
        JNC NEXT           ; If no carry (bit is 0), go for the next bit
        INC R1             ; If carry (bit is 1), increment register R1
NEXT:   DJNZ R2, AGAIN     ; Decrement counter and repeat if not zero
LAST:   SJMP LAST          ; End of the program
```

Program 3.55

Write an assembly language program to count number of 1's and 0's in a byte. The byte is present in the external RAM location 2000h. Put the result of count of 1's and 0's in the registers R7 and R6 respectively.

Solution:

```
Assembly Language Program   Comments
        MOV R2, #08        ; Initialize counter for 8 times rotation
        MOV DPTR, #2000H   ; Initialize address of ext. RAM
        MOV R7, #00H       ; Keep track of 1's in register R7
        MOV R6, #00H       ; Keep track of 0's in register R6
        MOVX A, @DPTR      ; Load data byte from ext. RAM to accumulator
AGAIN:  RLC A              ; Rotate the accumulator towards left with carry
        JNC NEXT           ; If no carry (bit is 0), go for the next bit
        INC R7             ; If carry (bit is 1), increment register R7
NEXT:   DJNZ R2, AGAIN     ; Decrement counter and repeat if not zero
        MOV A, #08H        ; Load accumulator with 08h (8 bits)
        CLR C              ; Clears carry flag
        SUBB A, R7         ; Subtract count of number of 1's from 08h
```

```
        MOV R6, A           ; Store result of accumulator (number of 0's) in
                            ; Register R6
LAST:   SJMP LAST           ; End of the program
```

Program 3.56

Write an assembly language program to read 50 values from the port P0 and store them at the external RAM memory location from 2000h.

Solution:

```
Assembly Language Program   Comments
        MOV R2, #50         ; Initialize counter for 50 values
        MOV DPTR, #2000H    ; Initialize address of ext. RAM memory location
        MOV P0, #0FFH       ; Initialize port P0 as an input port
BACK:   MOV A, P0           ; Read values from port P0 to the accumulator
        MOVX @DPTR, A       ; Stores the contents of accumulator to ext. RAM
        INC DPTR            ; Pointed to the next memory location
        DJNZ R2, BACK       ; Decrement counter and repeat if not zero
LAST:   SJMP LAST           ; End of the program
```

Program 3.57

Write an assembly language program to toggle all the bits of port P2 continuously. Assume a suitable delay between each toggle.

Solution:

Method – I

```
Assembly Language Program   Comments
        ORG 0000h           ; Initialize starting memory location for the
                              program
BACK:   MOV A, #55H         ; Load accumulator with 55h (alternate 1's and
                              0's)
        MOV P2, A           ; Transfer contents of accumulator to port P2
                              (toggle port P2 bits)
        LCALL DELAY         ; Get delay
        MOV A, #0AAH        ; Load accumulator with AAh (alternate 1's and
                              0's)
        MOV P2, A           ; Transfer contents of accumulator on port P2
                              (toggle port P2 bits)
        LCALL DELAY         ; Get delay
```

```
         SJMP BACK          ; Repeat
         ORG 0300H
DELAY:   MOV R2, #0FFH
AGAIN:   DJNZ R2, AGAIN
         RET
         END
```

Method – II

Assembly Language Program		Comments
	ORG 0000H	; Initialize starting address for the program
	MOV A, #55H	; Initialize accumulator with value 55h (alternate 1's and 0's)
BACK:	MOV P2, A	; Transfer contents of accumulator on port P2
	LCALL DELAY	; Get delay
	CPL A	; Complement contents of accumulator
	SJMP BACK	; Repeat
	ORG 0300H	
DELAY:	MOV R2, #0FFH	
AGAIN:	DJNZ R2, AGAIN	
	RET	
	END	

Program 3.58

Write an assembly language program to toggle port bit P2.0, assume suitable delay between each toggle.

Solution:

Assembly Language Program		Comments
	ORG 0000H	; Initialize starting address of the program
	CLR P2.0	; Clear port P2.0 bit
BACK:	CPL P2.0	; Complement port P2.0 bit (toggle)
	LCALL DELAY	; Get delay
	SJMP BACK	; Repeat
	ORG 0300H	
DELAY:	MOV R2, #0FFH	
AGAIN:	DJNZ R2, AGAIN	
	RET	
	END	

Program 3.59

Write an assembly language program to find the sum of the series 1 – 2 + 3 – 4 + 5 – 6 + …. up to N number of times, where N is the count present in internal RAM memory location 45h. Store the sum in the internal RAM memory location 46h.

Solution:

Assembly Language Program		Comments
	CLR A	; Clears the accumulator
	MOV R0, A	; Clears register R0 (copy 00h to register R0)
ADDD:	*INC R0*	; Increments register R0 by one
	ADD A, R0	; Add the accumulator with register R0 (00h + 01 = 01)
	DJNZ 45H, SUB	; Decrement counter and perform subtraction if not zero
	AJMP STORE	; If counter zero, exit from the program and save the result
SUB:	*INC R0*	; Increments register R0 by one
	CLR C	; Clears carry flag
	SUBB A, R0	; Subtracts register R0 with borrow from accumulator
	DJNZ 45H, ADDD	; Decrements counter and performs addition if not zero
STORE:	*MOV 46H, A*	; If counter zero, exit from the program and save the result
LAST:	*SJMP LAST*	; End of the program

Program 3.60

Write an ALP to add 2 m*n matrix stored in the external memory location 7000H and 8000H onwards in row wise order and store the result from the external memory location 9000I I onwards whcrc m and n are the two values stored in 5000H and 6000H respectively.

Solution: Let X and Y be the 2 matrices of the dimensions m*n and the matrix addition is performed as shown below:

Let

$$X = \begin{pmatrix} x_{11} & x_{12} & x_{13} & \cdots & x_{1n} \\ x_{21} & x_{22} & x_{23} & \cdots & x_{2n} \\ x_{31} & x_{32} & x_{33} & \cdots & x_{3n} \\ \vdots & \vdots & \vdots & \vdots & \vdots \\ x_{m_1} & x_{m_2} & x_{m_3} & \cdots & x_{m_n} \end{pmatrix}$$

And

$$Y = \begin{pmatrix} y_{11} & y_{12} & y_{13} & \cdots & y_{1n} \\ y_{21} & y_{22} & y_{23} & \cdots & y_{2n} \\ y_{31} & y_{32} & y_{33} & \cdots & y_{3n} \\ \vdots & \vdots & \vdots & \vdots & \vdots \\ y_{m_1} & y_{m_2} & y_{m_3} & \cdots & y_{m_n} \end{pmatrix}$$

Then X + Y is equal to

$$X + Y = \begin{pmatrix} x_{11}+y_{11} & x_{12}+y_{12} & x_{13}+y_{13} & \cdots & x_{1n}+y_{1n} \\ x_{21}+y_{21} & x_{22}+y_{22} & x_{23}+y_{23} & \cdots & x_{2n}+y_{2n} \\ x_{31}+y_{31} & x_{32}+y_{32} & x_{33}+y_{33} & \cdots & x_{3n}+y_{3n} \\ \vdots & \vdots & \vdots & \vdots & \vdots \\ x_{m_1}+y_{m_1} & x_{m_2}+y_{m_2} & x_{m_3}+y_{m_3} & \cdots & x_{m_n}+y_{m_n} \end{pmatrix}$$

In µC 8051, the matrix elements must be entered in the data memory either row wise or column wise. This will be much clear after looking in the above example

$$\text{Let } X = \begin{bmatrix} 1 & 2 & 3 \\ 4 & 5 & 6 \\ 7 & 8 & 9 \end{bmatrix} \text{ and } Y = \begin{bmatrix} 10 & 11 & 13 \\ 12 & 14 & 15 \\ 16 & 17 & 18 \end{bmatrix}$$

For Matrix X:

Memory Location	**Data Byte**
7000H	01H
7001H	02H
7002H	03H
7003H	04H
7004H	05H
7005H	06H
7006H	07H
7007H	08H
7008H	09H

For Matrix Y:

Memory Location	Data Byte
8000H	10H
8001H	11H
8002H	13H
8003H	12H
8004H	14H
8005H	15H
8006H	16H
8007H	17H
8008H	18H

So, the matrix addition will be performed if each location is accessed, the contents are added and the result is saved into the destination.

Assembly Language Program:

Instructions	**Comments**
ORG 0000H	; Initialize the memory location from where the program starts.
MOV DPTR,#5000H	; The external memory location is initialized by the register with memory location 5000H.
MOVX A,@DPTR	; Copying the data byte from the external memory location pointed by the DPTR memory location 5000H into the accumulator.
MOV B,A	; Saving the data byte present in the accumulator into register B.
MOV DPH,#60H	; The new external memory location is 6000H.
MOVX A,@DPTR	; Copying the data byte from the external
	; Memory location pointed by the DPTR memory location 6000H into the accumulator.

```
        MUL AB          ; Multiplying the data bytes stored in register B
                          and the accumulator. This is generally for the
                          count value of performing the number of times
                          for the addition.
        MOV R0,A        ; Saving the data byte present in the accumulator
                          into register R0.
L1:     MOV DPH,#70H    ; The new external memory location is 7000H.
        MOVX A,@DPTR    ; Copying the data byte from the external
                          memory location pointed by the DPTR memory
                          location 7000H into the accumulator.
        MOV R1,A        ; Saving the data byte present in the accumulator
                          into register R1.
        MOV DPH, #80H   ; The new external memory location is 8000H.
        MOVX A,@DPTR    ; Copying the data byte from the external
                          memory location pointed by the DPTR memory
                          location 8000H into the accumulator.
        ADD A, R1       ; Adding the data byte present in the accumulator
                          with the data byte of register R1.
        MOV DPH, #90H   ; The new external memory location is 9000H.
        MOVX @DPTR,A    ; Copying the data byte of the accumulator
                          externally into the memory location pointed by
                          the DPTR register.
        INC DPTR        ; Incrementing the DPTR by one.
        DJNZ R0,L1      ; Decrement register R0 by one and jump to
                          label L1 if the zero flag is reset.
$:      SJMP $          ; Short jump to the label $.
```

Flowchart:

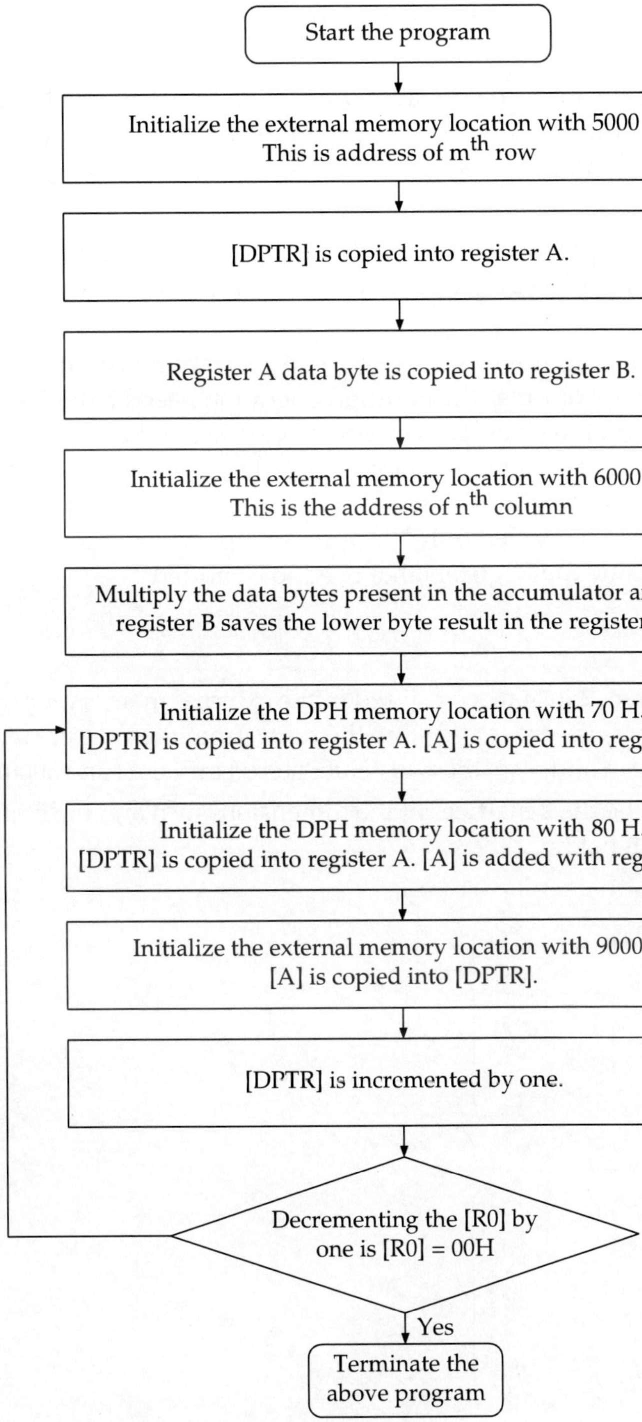
Start the program
Initialize the external memory location with 5000 H.
This is address of mth row
[DPTR] is copied into register A.
Register A data byte is copied into register B.
Initialize the external memory location with 6000 H.
This is the address of nth column
Multiply the data bytes present in the accumulator and the register B saves the lower byte result in the register R0.
Initialize the DPH memory location with 70 H.
[DPTR] is copied into register A. [A] is copied into register R1.
Initialize the DPH memory location with 80 H.
[DPTR] is copied into register A. [A] is added with register R1.
Initialize the external memory location with 9000 H.
[A] is copied into [DPTR].
[DPTR] is incremented by one.
Decrementing the [R0] by one is [R0] = 00H
Yes
Terminate the above program

Approach:

- In the above program, it is first necessary to calculate the counter value since this will ensure the number of times the loop will run which is nothing but the product of m*n matrix.
- Then the data byte from the 2 external memory locations are brought inside the registers of the controller indicating the mth row and the nth column.
- Then the mth row and the nth column data bytes are added and stored externally into the memory location required.
- The counter value is decreased by one after each addition operation. Thus, at the end the mth row and the nth column data bytes will be added up indicating the completion of addition operation.
- In the above program, it was very important to grab the count value first so as to indicate the number of times the addition operation needs to be performed. Hence, the data bytes from the external memory location are copied into the register B and the accumulator. Then, only the above 2 data bytes are getting multiplied for the counter by using ***MUL AB*** instruction. We can also multiply by using successive addition approach. After that only the addition operation is performed by copying the data bytes inside the Registers that need to be added.

Program 3.61

Write an ALP to subtract 2 m*n matrix stored in the external memory location 7000H and 8000H onwards in row wise order and store the result from the external memory location 9000H onwards where m and n are the two values stored in 5000H and 6000H respectively.

Solution: Let X and Y be the 2 matrices of the dimensions m*n and the matrix subtraction is performed as shown below:

Let

$$X = \begin{pmatrix} x_{11} & x_{12} & x_{13} & \cdots & x_{1n} \\ x_{21} & x_{22} & x_{23} & \cdots & x_{2n} \\ x_{31} & x_{32} & x_{33} & \cdots & x_{3n} \\ \vdots & \vdots & \vdots & \vdots & \vdots \\ x_{m_1} & x_{m_2} & x_{m_3} & \cdots & x_{m_n} \end{pmatrix}$$

And

$$Y = \begin{pmatrix} y_{11} & y_{12} & y_{13} & \cdots & y_{1n} \\ y_{21} & y_{22} & y_{23} & \cdots & y_{2n} \\ y_{31} & y_{32} & y_{33} & \cdots & y_{3n} \\ \vdots & \vdots & \vdots & \vdots & \vdots \\ y_{m_1} & y_{m_2} & y_{m_3} & \cdots & y_{m_n} \end{pmatrix}$$

Then X + Y is equal to

$$X + Y = \begin{pmatrix} x_{11} - y_{11} & x_{12} - y_{12} & x_{13} - y_{13} & \cdots & x_{1n} - y_{1n} \\ x_{21} - y_{21} & x_{22} - y_{22} & x_{23} - y_{23} & \cdots & x_{2n} - y_{2n} \\ x_{31} - y_{31} & x_{32} - y_{32} & x_{33} - y_{33} & \cdots & x_{3n} - y_{3n} \\ \vdots & \vdots & \vdots & \vdots & \vdots \\ x_{m_1} - y_{m_1} & x_{m_2} - y_{m_2} & x_{m_3} - y_{m_3} & \cdots & x_{m_n} - y_{m_n} \end{pmatrix}$$

In µC 8051, the matrix elements must be entered in the data memory either row wise or column wise. This will be much clear after looking in the above example

$$\text{Let } X = \begin{bmatrix} 21 & 22 & 23 \\ 24 & 25 & 26 \\ 27 & 28 & 29 \end{bmatrix} \quad \text{and } Y = \begin{bmatrix} 30 & 31 & 33 \\ 32 & 34 & 35 \\ 36 & 37 & 38 \end{bmatrix}$$

For Matrix X:

Memory Location	Data Byte
7000H	21H
7001H	22H
7002H	23H
7003H	24H
7004H	25H
7005H	26H
7006H	27H
7007H	28H
7008H	29H

For Matrix Y:

Memory Location	Data Byte
8000H	30H
8001H	31H
8002H	32H
8003H	33H
8004H	34H
8005H	35H
8006H	36H
8007H	37H
8008H	38H

So, the matrix subtraction will be performed if each location is accessed, the contents are added and the result is saved into the destination.

Assembly Language Program:

Label	Instructions	Comments
	ORG 0000H	; Initialize the memory location from where the program starts.
	MOV DPTR, #5000H	; The external memory location is initialized by the register with memory location 5000H.
	MOVX A,@DPTR	; Copying the data byte from the external memory location pointed by the DPTR memory location 5000H into the accumulator.
	MOV B, A	; Saving the data byte present in the accumulator into register B.
	MOV DPH, #60H	; The new external memory location is 6000H.
	MOVX A,@DPTR	; Copying the data byte from the external memory location pointed by the DPTR memory location 6000H into the accumulator.
	MUL AB	; Multiplying the data bytes stored in register B and the accumulator. This is generally for the count value of performing the number of times for the subtraction.
	MOV R0, A	; Saving the data byte present in the accumulator into register R0.
L1:	*MOV DPH, #70H*	; The new external memory location is 7000H.
	MOVX A,@DPTR	; Copying the data byte from the external memory location pointed by the DPTR memory location 7000H into the accumulator.
	MOV R1, A	; Saving the data byte present in accumulator into register R1.
	MOV DPH, #80H	; The new external memory location is 8000H.
	MOVX A,@DPTR	; Copying the data byte from the external memory location pointed by the DPTR memory location 8000H into the accumulator.
	SUBB A, R1	; Adding the data byte present in the accumulator with the data byte of register R1.
	MOV DPH, #90H	; The new external memory location is 9000H.
	MOVX @DPTR, A	; Copying the data byte of the accumulator externally into the memory location pointed by the DPTR register.
	INC DPTR	; Incrementing the DPTR by one.

	DJNZ R0, L1	; Decrement register R0 by one and jump to label L1 if the zero flag is reset.
$:	*SJMP $*	; Short jump to the label $.

Flowchart:

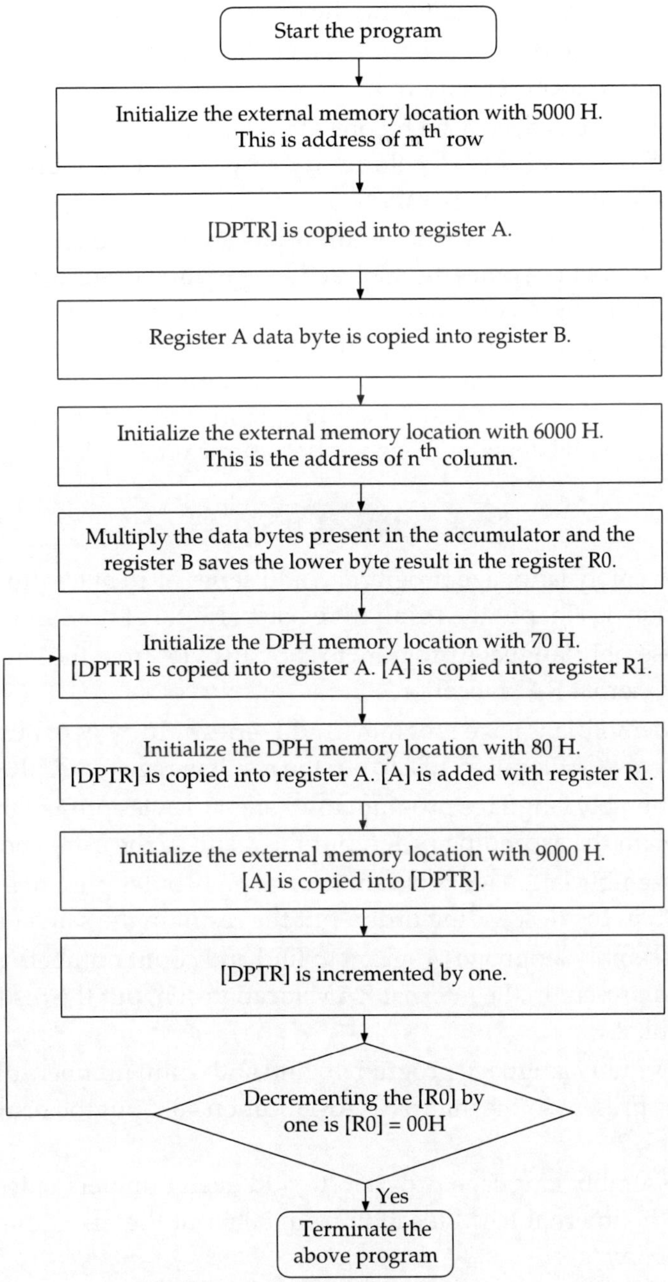

Approach:

- In the above program, it is first necessary to calculate the counter value since this will ensure the number of times the loop will run which is nothing but the product of m*n matrix.
- Then the data byte from the 2 external memory locations are brought inside the registers of the controller indicating the mth row and the nth column.
- Then the mth row and the nth column data bytes are subtracted and stored externally into the memory location required.
- The counter value is decreased by one after each subtraction operation. Thus, at the end, the mth row and the nth column data bytes will be subtracted indicating the completion of subtraction operation.
- In the above program, it was very important to grab the count value first so as to indicate the number of times the subtraction operation needs to be performed. Hence, the data bytes from the external memory location are copied into register B and the accumulator. Then, only the above 2 data bytes are getting multiplied for the counter by using ***MUL AB*** instruction. We can also multiply by using successive subtraction approach. After that only the subtraction operation is performed by copying the data bytes inside the registers that needs to be subtracted.

REVIEW QUESTIONS

1. Write an assembly language program to add series of 10 bytes present in the external RAM location 3000h, put the result in registers R7-R6 of register bank 0.
2. Write an assembly language program to copy 10 bytes from the internal RAM location 30h to the external RAM location 40h.
3. Write an assembly language program to add series of 10 bytes of BCD numbers present in the external ROM location 1000h, put the result in registers R7-R6 of register bank 1.
4. Write an assembly language program to arrange 10 bytes present in the internal RAM location 30h in the ascending order, put the result in the same memory location.
5. Write an assembly language program to arrange 10 bytes present in the internal RAM location 30h in the descending order, put the result in the same memory location.
6. Write an assembly language program to find and count numbers divisible by 8 from the 10 bytes present in the internal RAM location 40h, put the result in register R7 of register bank 2.
7. Write an assembly language program to find and count numbers divisible by 32 from the 10 bytes present in the internal RAM location 40h, put the result in register R7 of register bank 2.
8. Write an assembly language program to add even numbers only from the 10 bytes present in the internal RAM location from 50h, put the result in register R7, ignore carry flag if any.

9. Write an assembly language program to find smallest bytes from the 10 bytes present in the external ROM location 3000h, put the result in register R6.
10. Write an assembly language program to solve the equation $P + 2Q + 8R \rightarrow 40h$. Where P, Q and R are 8-bit unsigned numbers present in registers R2, R3 and R4 of register bank 0 respectively, put the result in register R7, ignore carry flag, if any. Use add and shift method.
11. Write an assembly language program to solve the equation $P/2 + Q/4 + R/8 \rightarrow 40h$. Where P, Q and R are 8-bit unsigned numbers present in registers R2, R3 and R4 of register bank 0 respectively, put the result in register R7, ignore carry flag, if any. Use add and shift method.
12. Write an assembly language program to solve the equation $P + 2Q + R/8 \rightarrow 40h$. Where P, Q and R are 8-bit unsigned numbers present in registers R2, R3 and R4 of register bank 0 respectively, put the result in register R7, ignore carry flag, if any. Use add and shift method.
13. Write an assembly language program to find the average of two 8-bit numbers present in the external ROM locations 3000h and 3001h respectively, put the result in the external ROM location 3002h.
14. Write an assembly language program to copy 10 bytes from the internal RAM 30 in the reverse order from the internal RAM 40h.
15. Write an assembly language program to find two's complement of 24-bit number present in the registers R2-R3-R4, put the result in the same register.
16. Write an assembly language program to convert 8-bit hexadecimal number present in the register R5, put the result in register R6-R5.
17. Write an assembly language program to count numbers divisible by 3 from the 10 bytes present in the external ROM location 3000h, put the result in register R4.
18. Write an assembly language program to count numbers divisible by 5 from the 10 bytes present in the external ROM location 3000h, put the result in register R4.
19. Write an assembly language program to find the square of 0 to 9 and put the result in the internal RAM location from 60h.
20. Write an assembly language program to display ASCII code of 0 to 9 on the output port P2.
21. Write an assembly language program to convert decimal numbers 0 to 9 into equivalent ASCII, and put the result in the internal RAM location from 60h.
22. Write an assembly language program to read 30 bytes from external RAM location 0200h and store them in the reverse order from internal RAM location 30h.
23. Write an assembly language program to add the series as $1 + 3 + 5 + 7 + \ldots$ up to N number of times, where N is the count present in the external RAM location 50h. Put the result in the external RAM location 51h.

Chapter 4

Timers of 8051 Microcontroller

Sometimes it is required to interface slower devices or peripherals with the microcontroller in order to communicate properly with them, delay must be introduced during the data transferring. We have already seen that delay is generated using assembly programming but it is not so accurate, in some cases delay plays an important role and it must be highly accurate as possible. Time delay generated by using the internal timing facility of the microcontroller is very close to requirement and highly accurate. This chapter deals with time delay generation, counting of pulses from the external device and, of course, generation of pulses. At the end of this chapter, readers will get a valuable insight about how to generate delay, time measurement, counting of pulses, etc.

4.1 INTRODUCTION OF TIMING IN 8051 MICROCONTROLLER

Microcontrollers (and many other electrical systems) use crystals to synchronize operations. The 8051 uses it to synchronize its operation. Effectively, the 8051 operates using what are called "machine cycles". A single machine cycle is the minimum amount of time in which a single 8051 instruction can be executed. Although many instructions take multiple cycles.

A cycle is, in reality, 12 pulses of the crystal. That is to say, if an instruction takes one machine cycle to execute, it will take 12 pulses of the crystal to execute. Since we know the crystal is pulsing 11,059,000 times per second and that one machine cycle is 12 pulses, we can calculate how many instruction cycles the 8051 can execute per second:

11,059,000 / 12 = 921,583

This means that the 8051 can execute 921,583 single-cycle instructions per second.

Since a large number of 8051 instructions are single-cycle instructions it is often considered that the 8051 can execute roughly 1 million instructions per second, although in reality it is less and, depending on the instructions being used, an estimate of about 600,000 to 700,000 instructions per second is more realistic.

For example, if you are using exclusively 2-cycle instructions you would find that the 8051 would execute 460,791 instructions per second. The 8051 also has two really slow instructions that require full 4 cycles to execute. If you were to execute nothing but those instructions you would find performance to be about 230,395 instructions per second. It is again important to emphasize that not all instructions execute in the same amount of time. The fastest instructions require one machine cycle (12 crystal pulses), many others require two machine cycles (24 crystal pulses), and the two very slow math operations require four machine cycles (48 crystal pulses).

Programming Tips:

Many 8051 derivative chips change instruction timing. For example, many optimized versions of the 8051 execute instructions in 4 oscillator cycles instead of 12; such a chip would be effectively 3 times faster than the 8051 when used with the same 11.059 MHz crystal.

Since all the instructions require different amounts of time to execute, a very obvious question comes to mind: How can one keep track of time in a time-critical application if we have no reference to time in the outside world?

Luckily, the 8051 includes timers which allow us to time events with high precision.

4.2 TIMERS OF 8051 MICROCONTROLLERS

The 8051 comes equipped with two timers, both of which may be controlled, set, read, and configured individually. The 8051 timers have three general functions:

1. keeping time and/or calculating the amount of time between events,
2. counting the events themselves (act as a counter), and
3. generating baud rates for the serial port.

The first two uses will be discussed in this chapter while the use of timers for baud rate generation will be discussed in the chapter relating to serial ports.

How does a timer count?

The answer to this question is simple: A timer always counts up. It doesn't matter whether the timer is being used as a timer, a counter, or a baud rate generator. A timer is always incremented by the microcontroller.

Programming Tips:

Some derivative chips actually allow the program to configure whether the timers count up or down. However, since this option only exists on some derivatives it is beyond the scope of this tutorial which is aimed at the standard 8051. It is only mentioned here in the event that you absolutely need a timer to count backwards, you will know that you may be able to find an 8051-compatible microcontroller that does it.

4.3 USE OF TIMERS TO MEASURE TIME

Obviously, one of the primary uses of timers is to measure time. We will discuss this use of timers first and will subsequently discuss the use of timers to count events. When a timer is used to measure time it is also called an "interval timer" since it measures the time of the interval between two events.

How long does a timer take to count?

First, it's worth mentioning that when a timer is in interval timer mode (as opposed to event counter mode) and correctly configured, it will increment by 1 every machine cycle. As you will recall from the previous chapter, a single machine cycle consists of 12 crystal pulses. Thus, a running timer will be incremented:

11,059,000 / 12 = 921,583

921,583 times per second.

Unlike instructions, some of which require 1 machine cycle, others 2, and others 4, the timers are consistent. They are incremented once per machine cycle. Thus, if a timer has counted from 0 to 50,000 you may calculate:

50,000 / 921,583 = .0542

.0542 second has passed.

In plain English, about half of a tenth of a second, or one twentieth of a second. Obviously it's not very useful to know .0542 second has passed. If you want to execute an event once per second you would have to wait for the timer to count from 0 to 50,000 18.45 times. How can you wait "half of a time"? You can't. So we come to another important calculation.

Let's say we want to know how many times the timer will be incremented in .05 second. We can do simple multiplication: .05 * 921,583 = 46,079.15. This tells us that it will take .05 second (1/20th of a second) to count from 0 to 46,079. Actually, it will take it .049999837 second—so we're off by .000000163 second—however, that's close enough. Consider that if you were building a watch based on the 8051 and made the above assumption, your watch would only gain about one second every 2 months. Again, I think that's accurate enough for most application—I wish my watch only gained one second every two months! Obviously, this is a little more useful. If you know it takes 1/20th of a second to count from 0 to 46,079 and you want to execute some event every second you simply wait for the timer to count from 0 to 46,079 twenty times; then you execute your event, reset the timer, and wait for the timer to count up another 20 times. In this manner you will effectively execute your event once per second, accurate to within thousandths of a second. Thus, we now have a system with which to measure time. All we need to review is how to control the timers and initialize them to provide us with the information we need.

4.4 TIMERS SPECIAL FUNCTION REGISTERS (SFRs)/ TIMERS 0 AND 1

As mentioned before, the 8051 has two timers which each function essentially the same way. One timer is timer0 and the other is timer1. The two timers share two SFRs (TMOD and TCON) which control the timers, and each timer also has two SFRs dedicated solely to itself (TH0/TL0 and TH1/TL1).

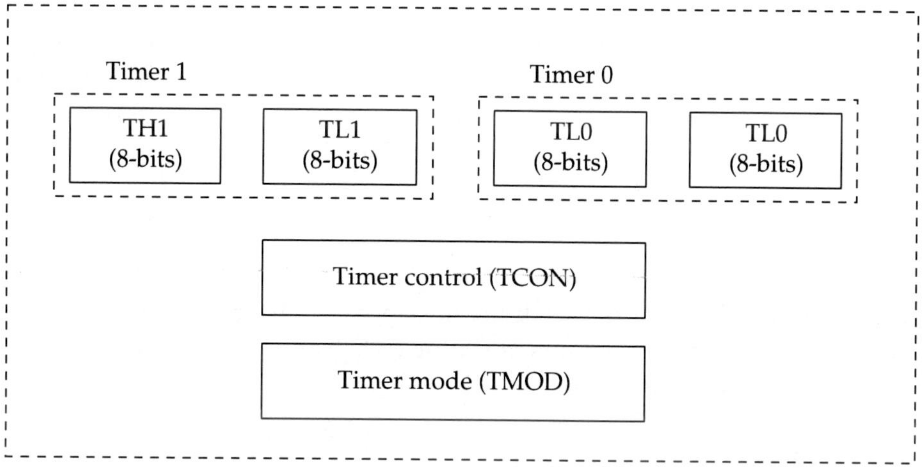

Fig. 4.1 *Timer 0 and 1 for the 8051 family*

We've given SFRs names to make it easier to refer to them, but in reality an SFR has a numeric address. It is often useful to know the numeric address that corresponds to an SFR's name. The SFRs relating to timers are shown in Table 4.1.

Table 4.1 *SFR and their addresses*

SFR name	Description	SFR address
TL0	Timer 0 lower byte	8Ah
TH0	Timer 0 higher byte	8Ch
TL1	Timer 1 lower byte	8Bh
TH1	Timer 1 higher byte	8Dh
TCON	Timer control register	88h
TMOD	Timer mode register	89h

When you enter the name of an SFR into an assembler, it internally converts it to a number. For example, the command:

MOV TH0, #25h

Moves the value 25h into the TH0 SFR. However, since TH0 is the same as SFR address 8Ch this command is equivalent to:

MOV 8Ch, #25h

Now, back to the timers. Timer0 has two SFRs dedicated exclusively to itself: TH0 and TL0.

Without making things too complicated to start off with, you may just think of this as the high and low byte of the timer. That is to say, when timer 0 has a value of 0, both TH0 and TL0 will contain 0. When Timer0 has the value 1000, TH0 will hold the high byte of the value (3 decimal) and TL0 will contain the low byte of the value (232 decimal).

Reviewing low/high byte notation, recall that you must multiply the high byte by 256 and add the low byte to calculate the final value. That is to say:

TH0 * 256 + TL0 = 1000

3 * 256 + 232 = 1000

Timer 1 works the exact same way, but its SFRs are TH1 and TL1. Since there are only two bytes devoted to the value of each timer it is apparent that the maximum value a timer may have is 65,535. If a timer contains the value 65,535 and is subsequently incremented, it will reset—or *overflow*—back to 0.

4.5 TMOD SFR

TMOD (timer mode). The TMOD SFR is used to control the mode of operation of both the timers. Each bit of the SFR gives the microcontroller specific information concerning how to run a timer. The high four bits (bits 4 through 7) relate to Timer1 whereas the low four bits (bits 0 through 3) perform the exact same functions, but for timer 0. The format of TMOD register and working of individual bits of TMOD is shown below:

TMOD

GATE	$C/\overline{T}$	M1	M0	GATE	$C/\overline{T}$	M1	M0

Timer 1 Timer 0

First half (from MSB) of the TMOD register is for controlling and managing TIMER 1 and second half (from LSB) is for TIMER0.

GATE When TRx (in TCON register) is set and GATE = 1, TIMER / COUNTERx will run only while INTx pin is high (hardware control).

When GATE = 0, TIMER / COUNTERx will run only while TRx = 1 (software control).

C/T Timer or counter selector. Cleared for timer operation (input from internal system clock).

Set for counter operation (input from Tx input pin)

M1 Mode selector bit 1

M0 Mode selector bit 0

M1	M0	Mode	Operating Mode
0	0	Mode 0	13-bit timer mode
0	1	Mode 1	16-bit timer mode
1	0	Mode 2	8-bit auto reload mode
1	1	Mode 3	Split timer mode

GATE: Gating control when set

Start and stop timer 1/0 by means of a signal provided to the pin INT1/INT0 (P3.3 and P3.2).

GATE is used to start and stop the timer. Some timers do this by software, some by hardware and some have both software as well as hardware controls. The timers in the 8051 microcontroller have both.

The hardware way of starting and stopping the timers is achieved by applying an external signal on the pin INT1, i.e., to start timer 1 set INT1 and to stop timer reset INT1. Similarly, start and stop timer 0 by setting and resetting INT0, INT1 and INT0 is accepting signals for start and stop the timers only when GATE = 1.

The start and stop of the timers are controlled by software using TR (timer run) bits TR0 and TR1. These bits belong to TCON register. To start the timers use instruction SETB TR1 and SETB TR0 for timer 1 and timer 0 respectively. Similarly, for stopping the timers use instruction CLR TR1 and CLR TR0 for timer 1 and timer 0 respectively. These instructions start and stop the timers as long as the GATE = 0.

C/$\overline{T}$ (Counter/Timer):

This bit in the TMOD register is used to decide whether the timer is used as a delay generator or an event counter.

If C/$\overline{T}$ = 0 (Timer)

Timer count pulses from internal oscillator, i.e., it is used as a timer for time delay generation. The clock source for the timer delay is the crystal frequency of the 8051. This means that crystal frequency attached to the 8051 also decides the speed at which the 8051 timer ticks.

If C/$\overline{T}$ = 1 (Counter)

Timer counts pulses applied to the pins T1 (P3.5) for timer 1 and T0 (P3.4) for timer 0.

When C/#T = 1, the timer is used as a counter and gets its pulses from outside the 8051. These outside pins are T0 (timer 0 input) and T1 (timer 1 input).

When C/#T =1 and clock pulse are applied to the T0 input then counter counts up for each clock pulse coming from that pin for timer 0. Similarly if the clock pulses are applied to T1 input then counter counts up for each clock pulse coming from that pin for timer 1.

M1	M0	Mode	Operating Mode
0	0	Mode 0	13-bit timer mode. 8-bit timer counter THx with TLx as 5-bit prescaler [D15 – D8 (THx) – D4 – D0 (TLx)]
0	1	Mode 1	16-bit timer mode. 8-bit timer counter THx with 8-bit timer counter TLx are cascaded; there is no prescaler

1	0	Mode 2	8-bit auto reload mode. 8-bit auto reload timer / counter; THx holds a value that is to be reloaded into TLx each time it overflows.
1	1	Mode 3	Split timer mode.

4.5.1 13-Bit Time Mode (Mode 0)

Timer mode "0" is a 13-bit timer. This is a relic that was kept around in the 8051 to maintain compatibility with its predecessor, the 8048. Generally, the 13-bit timer mode is not used in new development. When the timer is in 13-bit mode, TLx will count from 0 to 31. When TLx is incremented from 31, it will "reset" to 0 and increment THx. Thus, effectively, only 13 bits of the two timer bytes are used: bits 0-4 of TLx and bits 0-7 of THx. This also means, in essence, the timer can only contain 8192 values. If you set a 13-bit timer to 0, it will overflow back to zero 8192 machine cycles later.

4.5.2 16-Bit up Counter Time Mode (Mode 1)

Timer mode "1" is a 16-bit timer. This is a commonly used mode. It functions just like 13-bit mode except that all 16 bits are used.

TLx is incremented from 0 to 255. When TLx is subsequently incremented from 255, it resets to 0 and causes THx to be incremented by 1. Since this is a full 16-bit timer, the timer may contain up to 65536 distinct values. If you set a 16-bit timer to 0, it will overflow back to 0 after 65,536 machine cycles.

The following are the characteristics and operation of mode 1.

1. It is a 16-bit timer mode, therefore it allows values of 0000h to FFFFh to be loaded into the timer registers TLx and THx.
2. After THx and TLx are loaded with a 16-bit initial count value, the timer must be started, this is done by SETB TRO for timer 0 and SETB TR1 for timer 1 using software control similarly, by setting INT1 and INT0 pins for timer 1 and timer 0 in case of hardware control of the timers. 8051 Microcontroller supported both types of control, i.e., software control as well as hardware control to start the timer.
3. After the timer is started, it starts to count up. It counts up until it reaches the limit of FFFFh (8051 timer always counts up no matter for what purpose timer is being used), when it rolls over from FFFFh to 0000h, it sets high a flag bit called TFx (timer flag bit). This timer flag bit can be monitored using the instruction JNB TFx, Label. When this timer flag bit is raised, one option would be to stop the timer with the instruction CLR TR0 or CLR TR1 for timer 0 and timer 1 respectively under software control and by resetting INT0 or INT1 for timer 0 and timer 1 respectively under hardware control of the timers.
4. After the timer reaches the limit and rolls over in order to repeat the process, the registers THx and TLx must be reloaded with the original value and timer flag bit

TFx must be reset to 0. It is necessary to reset the timer flag bit, otherwise the next round will not be started. To check whether maximum limit is reached or not, only option is to keep monitoring on the TFx flag and hence once the maximum limit is reached, it must be reset again for the next round.

Operation of the timer 0/1 in the 16-bit timer mode is shown in Figure 4.2.

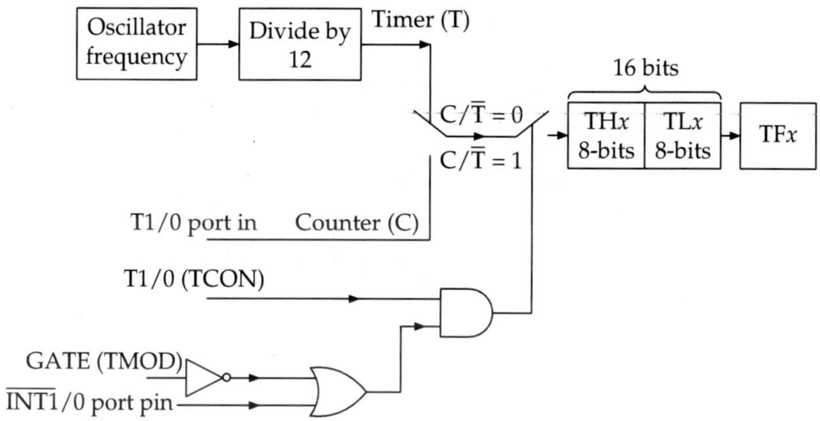

Fig. 4.2 *Circuit for timer 0/1 operating as a 16-bit up counter in mode 1*

4.5.3 8-Bit Timer Mode/Auto-Reload Mode (Mode 2)

The instruction MOV TMOD, #02h / MOV 89h, #02h would put timer 0 into auto-reload mode, i.e., mode 2 defining 8-bit timer using TL0 as the working register and TH0 as the automatic reload register. The circuit arrangement is shown in Figure 4.3

In mode 2, the working register is only 8-bit wide and so the base number is 8-bit wide. When the TLx register rolls over it is automatically reloaded with the contents of the THx register, whose contents remain the same, so the base number goes into the THx register.

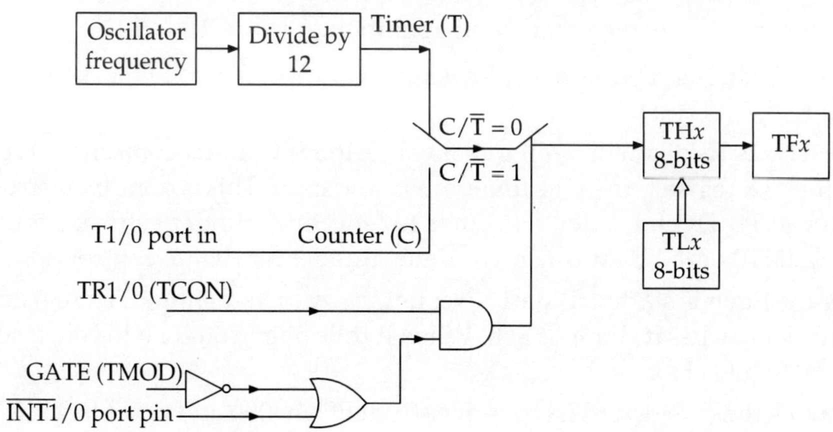

Fig. 4.3 *Circuit for timer 1/0 to operate as an 8-bit auto-reload mode*

Timer mode "2" is an 8-bit auto-reload mode. When a timer is in mode 2, THx holds the "reload value" and TLx is the timer itself. Thus, TLx starts counting up. When TLx reaches 255 and is subsequently incremented, instead of resetting to 0 (as in the case of modes 0 and 1), it will be reset to the value stored in THx. For example, let's say TH0 holds the value FDh and TL0 holds the value FEh. If we were to watch the values of TH0 and TL0 for a few machine cycles this is what we had seen:

Machine cycle	TH0 value	TL0 value
1	FDh	FEh
2	FDh	FFh
3	FDh	FDh
4	FDh	FEh
5	FDh	FFh
6	FDh	FDh
7	FDh	FEh

As you can see, the value of TH0 never changed. In fact, when you use mode 2 you almost always set THx to a known value and TLx is the SFR that is constantly incremented. What's the benefit of auto-reload mode? Perhaps you want the timer to always have a value from 200 to 255. If you use mode 0 or 1, you would have to check in code to see if the timer had overflown and, if so, reset the timer to 200. This takes precious instructions of execution time to check the value and/or to reload it. When you use mode 2 the microcontroller takes care of this for you. Once you have configured a timer in mode 2 you don't have to worry about checking to see if the timer has overflowed nor do you have to worry about resetting the value—the microcontroller hardware will do it all for you. The auto-reload mode is commonly used for establishing a baud rate which we will talk more about in the "Serial Communications" chapter.

The following are the characteristics and operation of mode 2.

1. It is an 8-bit timer, therefore it allows only values of 00h to FFh to be loaded into the timer register THx.
2. After THx is loaded with 8-bit initial value, the 8051 microcontroller gives a copy of it to the TLx register, then the timer must be started. This is done by instruction SETB TR0 or SETB TR1 for timer 0 or timer 1 respectively under software control and by setting INT0 or INT1 for timer 0 or timer 1 under hardware control.
3. After the timer is started, it starts to count up by incrementing TLx register. It counts up until it reaches its limit of FFh. When it rolls over from FFh to 00h, it sets high the timer flag bit (TFx).
4. When the timer register TLx rolls over from FFh to 00h and timer flag bit TFx is set to 1, TLx is reloaded automatically with the original value kept by the THx register. To repeat the process, we must simply clear TFx and let it go without any need by the

programmer to reload the original value. This makes mode 2 an auto-reload mode, in contrast with mode 1 in which the programmer has to reload the THx and TLx. This reloading leaves THx unchanged, still holding a copy of the original value.

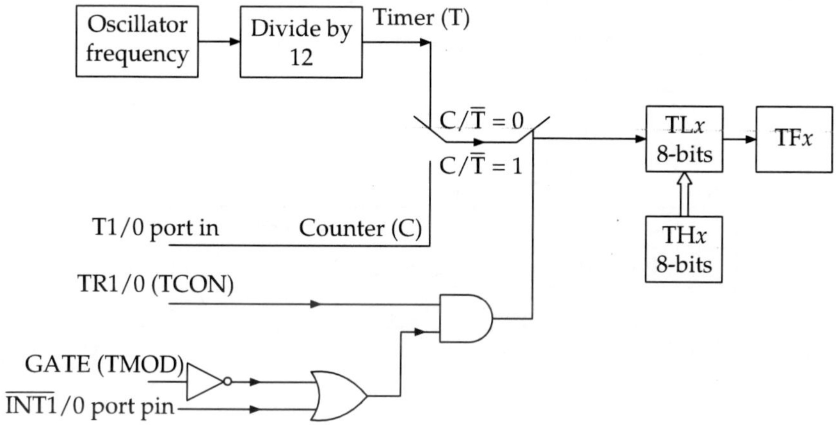

4.5.4 Split Timer Mode (Mode 3)

Timer mode "3" is a split-timer mode. When timer 0 is placed in mode 3, it essentially becomes two separate 8-bit timers. That is to say, timer 0 is TL0 and timer 1 is TH0. Both timers count from 0 to 255 and overflow back to 0. All the bits that are related to timer 1 will now be tied to TH0. While timer 0 is in split mode, the real timer 1 (i.e., TH1 and TL1) can be put into modes 0, 1 or 2 normally, however, you may not start or stop the real timer 1 since the bits that do that are now linked to TH0. The real timer 1, in this case, will be incremented every machine cycle no matter what. The only real use I can see of using split timer mode is if you need to have two separate timers and, additionally, a baud rate generator. In such case you can use the real timer 1 as a baud rate generator and use TH0/TL0 as two separate timers.

4.6. TCON (TIMER CONTROL) SPECIAL FUNCTION REGISTER (SFR)

As you may notice, we have only defined 4 of the 8 bits. That's because the other 4 bits of the SFR don't have anything to do with timers—they have to do with interrupts and they will be discussed in the chapter that addresses interrupts. A new piece of information in this chart is the column "bit address". This is because this SFR is "bit-addressable". What does this mean? It means if you want to set the bit TF1, which is the highest bit of TCON, you could execute the command:

```
MOV TCON, #80h
```

or, since the SFR is bit-addressable, you could just execute the command:

```
SETB TF1
```

This has the benefit of setting the high bit of TCON without changing the value of any of the other bits of the SFR. Usually, when you start or stop a timer you don't want to modify other values in TCON, so you take advantage of the fact that the SFR is bit-addressable.

The 8-bit format of the TCON register is shown below:

D7	D6	D5	D4	D3	D2	D1	D0
TF1	TR1	TF0	TR0	IE1	IT1	IE0	IT0

Bit no.	Name	Meaning
D7	TF1	This bit is automatically set when timer 1 overflows
D5	TF0	This bit is automatically set when timer 0 overflows
D6	TR1	This bit turns the timer 1 ON = 1 Timer 1 is turned ON = 0 Timer 1 is turned OFF
D4	TR0	This bit turns the timer 0 ON = 1 Timer 0 is turned ON = 0 Timer 0 is turned OFF

4.6.1 Rollover

The timer clock increments in timer clock cycles from the base number up to the maximum value of the 16-bit register, which is FFFFh. One more increment would cause the register to rollover to 0000h and set the timer flag (TF) to 1. The TF is a bit in the TCON register SFR.

TCON							
TF1	TR1	TF0	TR0	IE1	IT1	IE0	IT0

The four least significant bits (LSB) are concerned with the trigger profiles of the interrupt signals INT0 and INT1. This will be examined later in the interrupt section. On power – up the default values of all the TCON registers are zero and so the timer flags TF1 and TF0 are zero; the control action of timer 0 is the same using TR0. Timer 1 is turned on by making TR1 = 1 and it is turned off by making TR1 = 0

As soon as TR0 = 1, the timer 0 register TH0 – TL0 starts incrementing upwards from their base number. Upon rollover the TF0 flag sets to 1 and this indicates that the delay has been completed. A possible assembly language routine for the 0.5 ms delay could be:

```
DELAY:   MOV TH0, #0FEH    ; Load timer 0 higher byte with FEh
         MOV TL0, #33H     ; Load timer 0 lower byte with 33h
         SETB TR0          ; Turn timer 0 on
FLAG:    JNB TF0, FLAG     ; Jump to FLAG if TF0 is not bit (not  equal to 1)
         CLR TR0           ; Turn timer 0 off
```

```
CLR TF0          ; Clear TF0 to zero
RET              ; Return from subroutine
```

4.7 INITIALIZING TIMERS

Now that we've discussed the timer related SFRs we are ready to write code that will initialize the timer and start it running. As you'll recall, we first must decide what mode we want the timer to be in. In this case we want a 16-bit timer that runs continuously; that is to say, it is not dependent on any external pins. We must first initialize the TMOD SFR. Since we are working with timer 0 we will be using the lowest 4 bits of TMOD. The first two bits, GATE0 and C/T0 are both 0 since we want the timer to be independent of the external pins. 16-bit mode is timer mode 1 so we must clear T0M1 and set T0M0. Effectively, the only bit we want to turn on is bit 0 of TMOD. Thus to initialize the timer we execute the instruction:

MOV TMOD, #01h

Timer 0 is now in 16-bit timer mode. However, the timer is not running. To start the timer running, we must set the TR0 bit. We can do that by executing the instruction:

SETB TR0

Upon executing these two instructions timer 0 will immediately begin counting, being incremented once every machine cycle (every 12 crystal pulses).

4.8 READING THE TIMER

There are two common ways of reading the value of a 16-bit timer; which you use depends on your specific application. You may either read the actual value of the timer as a 16-bit number, or you may simply detect when the timer has overflown.

Reading the Value of a Timer

If your timer is in an 8-bit mode; that is, either 8-bit auto reload mode or in split timer mode; then reading the value of the timer is simple. You simply read the 1-byte value of the timer and you're done. However, if you're dealing with a 13-bit or 16-bit timer the job is a little complicated. Consider what would happen if you read the low byte of the timer as 255, then read the high byte of the timer as 15. In this case, what actually happened was that the timer value was 14/255 (high byte 14, low byte 255) but you read 15/255. Why? Because you read the low byte as 255. But when you executed the next instruction a small amount of time passed—but enough for the timer to increment again at which time the value rolled over from 14/255 to 15/0. But in the process you've read the timer as being 15/255. Obviously, there's a problem there. The solution? It's not too tricky, really. You read the high byte of the timer, then read the low byte, then read the high byte again. If the high byte read the second time is not the same as the high byte read the first time, you repeat the cycle. In code, this would appear as:

```
REPEAT:    MOV A, TH0
           MOV R0, TL0
           CJNE A, TH0, REPEAT
```

In this case, we load the accumulator with the high byte of timer 0. We then load R0 with the low byte of timer 0. Finally, we check to see if the high byte we read out of timer 0—which is now stored in the accumulator—is the same as the current timer 0 high byte. If it isn't it means we've just "rolled over" and must reread the timer's value—which we do by going back to REPEAT. When the loop exits, we will have the low byte of the timer in R0 and the high byte in the accumulator. Another much simpler alternative is to simply turn off the timer run bit (i.e., CLR TR0), read the timer value, and then turn on the timer run bit (i.e., SETB TR0). In that case, the timer isn't running so no special tricks are necessary. Of course, this implies that your timer will be stopped for a few machine cycles. Whether or not this is tolerable depends on your specific application.

4.8.1 Detecting Timer Overflow

Often it is necessary to just know that the timer has reset to 0. That is to say, you are not particularly interested in the value of the timer but rather you are interested in knowing when the timer has overflown back to 0. Whenever a timer *overflows* from its highest value back to 0, the microcontroller automatically sets the TFx bit in the TCON register. This is useful since rather than checking the exact value of the timer, you can just check if the TFx bit is set. If TF0 is set it means that timer 0 has overflown; if TF1 is set it means that timer 1 has overflown. We can use this approach to cause the program to execute a fixed delay. As you'll recall, we calculated earlier that it takes the 8051 1/20th of a second to count from 0 to 46,079. However, the TFx flag is set when the timer overflows back to 0. Thus, if we want to use the TFx flag to indicate when 1/20th of a second has passed we must set the timer initially to 65536 less 46079, or 19,457. If we set the timer to 19,457, 1/20th of a second later the timer will overflow. Thus, we come up with the following code to execute a pause of 1/20th of a second:

```
MOV TH0, #76        ; High byte of 19,456 (76 * 256 = 19,456)
MOV TL0, #01        ; Low byte of 19,457 (19,456 + 1 = 19,457)
MOV TMOD, #01       ; Put Timer 0 in 16-bit mode
SETB TR0            ; Make Timer 0 start counting
JNB TF0, $          ; If TF0 is not set, jump back to this same instruction
```

Note: $ is an assembler directive, which is used to move on the same instruction, for example, writing L1: JNB TF0, L1 is equivalent to writing like JNB TF0, $.

In the above code, the first two lines initialize the timer 0 starting value to 19,457. The next two instructions configure timer 0 and turn it on. Finally, the last instruction JNB TF0, $, reads "Jump, if TF0 is not set, back to this same instruction." The "$" operand means, in most assemblers, the address of the current instruction. Thus, as long as the timer has not overflown and the TF0 bit has not been set, the program will keep executing this same instruction. After 1/20th of a second timer 0 will overflow, set the TF0 bit, and program execution will then break out of the loop.

4.8.2 Timing the Length of Events

The 8051 provides another cool toy that can be used to time the length of events. For example, let's say we're trying to save electricity in the office and we're interested in how long a light is turned on each day. When the light is turned on, we want to measure time. When the light is turned off we don't. One option would be to connect the light switch to one of the pins, constantly read the pin, and turn the timer on or off based on the state of that pin. While this would work fine, the 8051 provides us with an easier method of accomplishing this. Looking again at the TMOD SFR, there is a bit called GATE0. So far we've always cleared this bit because we wanted the timer to run regardless of the state of the external pins. However, now it would be nice if an external pin could control whether the timer was running or not. It can. All we need to do is connect the light switch to pin INT0 (P3.2) on the 8051 and set the bit GATE0. When GATE0 is set timer 0 will only run if P3.2 is high. When P3.2 is low (i.e., the light switch is off) the timer will automatically be stopped. Thus, with no control code whatsoever, the external pin P3.2 can control whether or not our timer is running or not.

4.8.3 Using Timers as Event Counters

We've discussed how a timer can be used for the obvious purpose of keeping track of time. However, the 8051 also allows us to use the timers to count events. How can this be useful? Let's say you had a sensor placed across a road that would send a pulse every time a car passed over it. This could be used to determine the volume of traffic on the road. We could attach this sensor to one of the 8051's I/O lines and constantly monitor it, detecting when it pulsed high and then incrementing our counter when it went back to a low state. This is not terribly difficult, but requires some code. Let's say we hooked the sensor to P1.0; the code to count cars passing would look something like this:

```
JNB P1.0,$          ;If a car hasn't raised the signal, keep waiting
JB P1.0,$           ;The line is high which means the car is on the sensor
                    ; right now
INC COUNTER         ;The car has passed completely, so we count it
```

As you can see, it's only three lines of code. But what if you need to be doing other processing at the same time? You can't be stuck in the JNB P1.0,$ loop waiting for a car to pass if you need to be doing other things. Of course, there are ways to get around even this limitation but the code quickly becomes big, complex, and ugly. Luckily, since the 8051 provides us with a way to use the timers to count events we don't have to bother with it. It is actually painfully easy. We only have to configure one additional bit. Let's say we want to use timer 0 to count the number of cars that pass. If you look back to the bit table for the TCON SFR you will see there is a bit called "C/T0"—it's bit 2 (TCON.2). Reviewing the explanation of the bit we see that if the bit is clear then timer 0 will be incremented every machine cycle. This is what we've already used to measure time. However, if we set C/T0 timer 0 will monitor the P3.4 line. Instead of being incremented every machine cycle, timer 0 will count events on the P3.4 line. So in our case we simply connect our sensor to P3.4 and let the 8051 do the work. Then, when we want to know how many cars have passed, we just read the value of timer 0—the value of timer 0 will be the number of cars that have passed. So what exactly is an event? What does timer 0 actually "count"? Speaking at the electrical level, the 8051 counts 1-0 transitions on the P3.4 line. This means that when a car first runs over our sensor, it will raise the input to a high ("1") condition. At that point, the 8051 will not count anything since this is a 0-1 transition. However, when the car has passed the sensor will fall back to a low ("0") state. This is a 1-0 transition and at that instant, the counter will be incremented by 1. It is important to note that the 8051 checks the P3.4 line each instruction cycle (12 clock cycles). This means that if P3.4 is low, goes high, and goes back low in 6 clock cycles it will probably not be detected by the 8051. This also means the 8051 event counter is only capable of counting events that occur at a maximum of 1/24th the rate of the crystal frequency. That is to say, if the crystal frequency is 12.000 MHz it can count a maximum of 500,000 events per second (12.000 MHz * 1/24 = 500,000). If the event being counted occurs more than 500,000 times per second the 8051 will not be able to accurately count.

4.9 STEPS TO PROGRAMMING TIMER IN MODE 1

To generate a time delay or waveforms, using the timer's in mode 1 the following steps are taken:

1. Load TMOD register with the value indicating which timer (timer 0 or timer 1) is to be used and timer mode is selected.
2. Load timer register TLx and THx with the initial count values, according to the time delay value calculated.
3. Start the timer; this may be done either by software control or by hardware control.
4. Keep monitoring the timer flag (TFx) with the instruction "JNB TFx, TARGET" to see and confirm whether the timer flag bit is raised or not. Get out of the loop when timer flag bit TFx becomes high. When TFx is raised indicating the amount of delay is calculated so there is need to stop the timer.
5. Stop the timer. When timer flag bit TFx is raised to high indicating that timer is reached to their maximum limit and overflow back to 0000h and hence to stop the timer.
6. Clear the timer flag bit TFx for the next round, in other words to repeat the process timer flag bit TFx must be cleared. If TFx is not cleared then even if the timer is started for the next round, it is not possible to sense whether the timer has calculated the desired delay or not.
7. Go back to step 2 to load the values of timer register TLx and THx again.

To calculate the exact time delay and square wave frequency generation, we need to know the crystal frequency. Assume that the 8051 microcontroller's crystal frequency is 11.0592 MHz, and the delay calculation in hexadecimal and decimal is as follows:

(*a*) In hexadecimal

$$\text{Time delay} = (\text{FFFFh} - \text{YYXX} + 1) * 1.085\ \mu\text{s}$$

where YYXX are initial count values to be loaded in THx and TLx respectively. Values of YYXX are in hexadecimal format.

(*b*) In decimal

$$\text{Time delay} = (65535 - \text{NNNNN} + 1) * 1.085\ \mu\text{s}$$

or

$$\text{Time delay} = (65536 - \text{NNNNN}) * 1.085\ \mu\text{s}$$

where NNNNN indicates initial count value to be loaded in the timer register TLx and THx respectively.

4.10 STEPS TO PROGRAMMING TIMER IN MODE 2

To generate a time delay or waveform using timer's mode 2 take the following steps:

1. Load TMOD register with the value indicating which timer, timer 0 or timer 1 is to be used and select the timer mode.
2. Load the THx register with initial count value, according to the delay value calculated.
3. Start the timer; this may be done either by using software control or by hardware control. The timer of 8051 microcontroller is supported by both.
4. Keep monitoring the timer flag (TFx) with the instruction "JNB TFx, TARGET" to see and confirm whether timer flag bit is raised or not. Get out of the loop when timer flag bit TFx becomes high.
5. Clear the timer flag bit TFx for the next round, in other words to repeat the process timer flag bit TFx must be cleared. If TFx is not cleared then even if the timer is started for the next round, it is not possible to sense whether the timer has calculated the desired delay or not.
6. Go back to step 4, since mode 2 is auto-reload mode, therefore no need to load the timer register THx again. If we go back to step 3 and start the timer again then there is no issue, it will also produce the same result, but as we are not stopping the timer in step 5, it is advisable to stop the timer first and then go back to step 3.

To calculate exact time delay and square wave frequency generation, we need to know the crystal frequency. Assume that the 8051 microcontroller's crystal frequency is 11.0592 MHz, and the delay calculation in hexadecimal and decimal is as follows:

(*a*) In hexadecimal

$$\text{Time delay} = (\text{FFh} - \text{YY} + 1) * 1.285\ \mu\text{s}$$

where YY are initial count values to be loaded in THx, value YY are in hexadecimal format.

(*b*) In decimal

$$\text{Time delay} = (255 - \text{NNN} + 1) * 1.085\ \mu s$$

or

$$\text{Time delay} = (256 - \text{NNN}) * 1.085\ \mu s$$

where NNN indicates initial count value to be loaded in the timer register THx.

Example 4.1

A 1 kHz square wave signal is to be generated from port pin P1.7. The microcontroller clock frequency is 11.0592 MHz.

(*a*) Determine the required time delay

(*b*) Using timer 0 determine the base number that must go into the TH0 and TL0 registers.

Solution: The required waveform is shown in Figure 4.4.

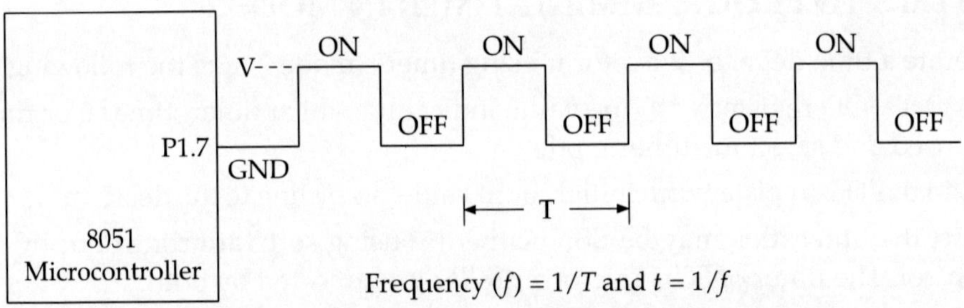

Fig. 4.4 *Square wave signal to be generated at port 1, pin 7*

(*a*) One cycle time T of the required – square wave signal equals 1/frequency

$$T = \frac{1}{f} = \frac{1}{1000} = 1\ \text{ms}$$

For the square wave TON = TOFF

Therefore,

$$\text{Delay timer} = \frac{T}{2} = \frac{1\ \text{ms}}{2} = 0.5\ \text{ms}$$

(*b*) Timer clock = Microclock/12

$$\text{Timer clock} = \frac{11.0592\ \text{MHz}}{12} = 921.6\ \text{kHz}$$

$$\text{Timer cycle time} = \frac{1}{f} = \frac{1}{921.6\ \text{kHz}} = 1.085\ \mu s$$

Method – I

$$\text{Delay count} = \frac{\text{Delay time}}{\text{Timer cycle time}}$$

$$\text{Delay count} = \frac{0.5\text{ ms}}{1.085\ \mu\text{s}}$$

$$= 461 \text{ (Nearest whole mumber)}$$

$$\text{Base number} = 65536 - \text{delay count}$$

$$\text{Base number} = 65536 - 461 = 65075$$

$$65075 \text{ Decimal} = \text{FE33h hexadecimal}$$

Therefore, $\quad \text{TH0} = \text{FEh and TL0} = 33\text{h}$

Method – II

$$\text{Time delay} = (\text{FFFFh} - \text{YYXX} + 1) * 1.085\ \mu\text{s}$$

$$\text{Time delay} = 0.5\text{ ms}$$

$$0.5\text{ ms} = (\text{FFFh} - \text{YYXX} + 1) * 1.085\ \mu\text{s}$$

$$\text{FFFFh} - \text{YYXX} + 1 = \frac{0.5\text{ ms}}{1.085\ \mu\text{s}}$$

$$\text{FFFh} - \text{YYXX} + 1 = 461$$

$$461 \text{ decimal} = \text{01CD hexadecimal}$$

$$\text{YYXX} = \text{FFFFh} + 1 - \text{01CDh}$$

$$\text{YYXX} = \text{FE33h}$$

Therefore, $\quad \text{TH0} = \text{FEh and TL0} = 33\text{ h}$

Example 4.2

Repeat example 3.1 to produce a 2 kHz square wave at port pin P1.7. Assume that the clock frequency remains at the same value.

(*a*) Determine the required time delay.

(*b*) Using timer 0 determine the base number that must go into the TH0 and TL0 registers.

Example 4.3

A 8051 microcontroller having an 11.0592 MHz clock is to be used to generate a 1 kHz square wave signal from pin 7 of port 1. Write a suitable assembly program to achieve this.

Solution:

$$\text{Timer clock} = \frac{11.0592\text{ MHz}}{12} = 921.6\text{ kHz}$$

$$\text{Timer cycle time} = \frac{1}{f} = \frac{1}{921.6\text{ kHz}} = 1.085\ \mu\text{s}$$

Method – I

$$\text{Time count} = \frac{\text{Delay time}}{\text{Timer cycle time}}$$

$$\text{Delay count} = \frac{0.5 \text{ ms}}{1.085\ \mu\text{s}}$$

$$= 461 \text{ (nearest whole number)}$$

$$\text{Base number} = 65536 - \text{delay count}$$

$$\text{Base number} = 65536 - 461 = 65075$$

$$65075 \text{ decimal} = \text{FE33h hexadecimal}$$

Therefore, $$\text{TH0} = \text{FEh and TL0} = 33\text{h}$$

Method – II

$$\text{Time delay} = (\text{FFFFh} - \text{YYXX} +) * 1.085\ \mu\text{s}$$

$$\text{Time delay} = 0.5 \text{ ms}$$

$$0.5 \text{ ms} = (\text{FFFh} - \text{YYXX} + 1) * 1.085\ \mu\text{s}$$

$$\text{FFFFh} - \text{YYXX} + 1 = \frac{0.5 \text{ ms}}{1.085\ \mu\text{s}}$$

$$\text{FFFh} - \text{YYXX} + 1 = 461$$

$$461 \text{ decimal} = \text{01CD hexadecimal}$$

$$\text{YYXX} = \text{FFFFh} + 1 - \text{01CDh}$$

$$\text{YYXX} = \text{FE33h}$$

Therefore, $$\text{TH0} = \text{FEh and TL0} = 33 \text{ h}$$

Assembly Language Program:

```
        ORG 0000H          ; Reset address
        SJMP START         ; Short jump over reserved area
        ORG 0040H          ; Program starts address at 40h
START:  MOV TMOD, #01H     ; Put timer 0 in mode 1
AGAIN:  SETB P1.7          ; Set P1.7 high
        ACALL DELAY        ; Go to 0.5 ms delay
        CLR P1.7           ; Set P1.7 low
        ACALL DELAY        ; Go to 0.5 ms delay
        SJMP AGAIN         ; Repeat
DELAY:  MOV TH0, #0FEH     ; Load higher byte of timer 0
        MOV TL0, #33H      ; Load lower byte of timer 0
        SETB TR0           ; Turn timer 0 on
```

```
FLAG:     JNB TF0, FLAG       ; Repeat until rollover when TF0 = 1
          CLR TR0             ; Timer 0 turn off
          CLR TF0             ; Clear TF0 back to 0
          RET                 ; Return from subroutine
          END                 ; No more assembly language after here
```

Example 4.4

A 8051 microcontroller having an 11.0592 MHz clock is to be used to generate a 4 kHz square wave signal from pin 7 of port 1. Write a suitable assembly program to achieve this. Use timer 1 in auto-reload mode.

Solution: Square wave cycle time $= \frac{1}{4\text{ kHz}} = \frac{1}{4000} = 0.25\ \mu s$

Delay required for a square wave = Half the cycle time (ON Time = OFF Time)

$= 0.125\ \mu s = 125\ \mu s$

In mode 2, count value is calculated as

$$\text{Delay count} = (256 - NNN) * 1.085\ \mu s$$

where NNN is the count value

$$125\ \mu s = (256 - NNN) * 1.085\ \mu s$$

$$256 - NNN = \frac{125\ \mu s}{1.085\ \mu s} = 115$$

$$NNN = 256 - 115 = 141$$

$$141 \text{ decimal} = 5D \text{ hexadecimal}$$

Therefore,

8DH is to go into TH1

8Dh is to go into TL1

TL1 could start with its default value of 00h since the first half cycle would not be seen on the oscilloscope screen. The line MOV TL1, #8Dh can be left out of the program since after the first half cycle TL1 will automatically be reloaded with 8Dh from TH1.

Assembly Language Program:

```
          ORG 0000H           ; Reset address
          SJMP START          ; Short jump over reserved area
          ORG 0040H           ; Program starts address at 40h
START:    MOV TMOD, #20H      ; Put timer 1 in mode 2
          MOV TH1, #8DH       ; Load TH1 byte with count value
AGAIN:    SETB P1.7           ; Set P1.7 high
```

```
        ACALL DELAY     ; Go to 0.5 ms delay
        CLR P1.7        ; Set P1.7 low
        ACALL DELAY     ; Go to 0.5 ms delay
        SJMP AGAIN      ; Repeat
DELAY:  SETB TR1        ; Turn timer 1 on
FLAG:   JNB TF1, FLAG   ; Repeat until rollover when TF1 =1
        CLR TR1         ; Rimer 1 turn off
        CLR TF1         ; Clear TF0 back to 1
        RET             ; Return from subroutine
        END             ; No more assembly language after here
```

Example 4.5

Write an assembly language program to generate pulse width of 5 μs (microsecond) on port pin P0.1 using timer 0. Assume that the crystal frequency of 8051 microcontroller is 11.0592 MHz.

Solution:

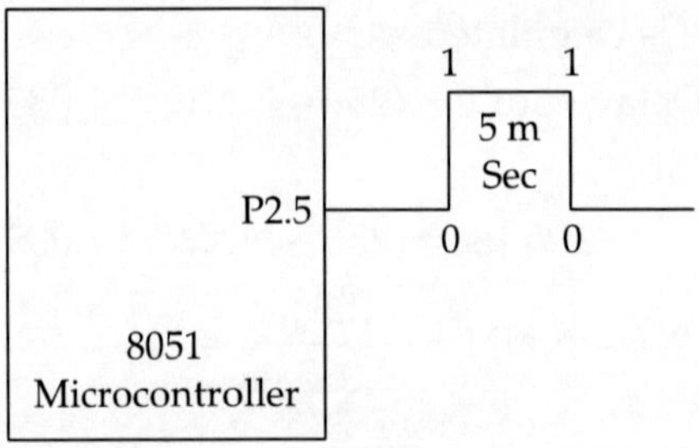

Fig. 4.5

$$\text{Crystal frequency} = 11.0592 \text{ MHz}$$

$$\text{Internal frequency} = \frac{11.0592 \text{ MHz}}{12} = 921.6 \text{ kHz}$$

$$\text{Internal clock time} = \frac{1}{921.6 \text{ kHz}} = 1.085 \text{ μs}$$

$$\text{Time delay} = (\text{FFFh} - \text{YYXX} + 1) * 1.085 \text{ μs}$$

$$\text{Time delay} = 5 \text{ μs (given)}$$

Setting values in TMOD register or configuring TMOD register with initialization of timer 0 in mode 1

TMOD register

D7	D6	D5	D4	D3	D2	D1	D0	
GATE	C/T	M1	M0	GATE	C/T	M1	M0	
0	0	0	0	0	0	0	1	= 01 H

Assembly Language Program:

```
        CLR P2.5            ; Clear P2.5
        MOV TMOD, #01H      ; Set timer 0 in mode 1
        MOV TL0, #00H       ; Load lower byte of timer 0 register
        MOV TH0, #0EEH      ; Load higher byte of timer 0 register
        SETB P0.1           ; Set P0.1 high
        SETB TR0            ; Start timer 0
L1:     JNB TF0, L1         ; Keep monitoring TF0 flag bit until timer 0 rolls
                            ; over from FFFFh to 0000h
        CLR P0.1            ; Clear P0.1
        CLR TR0             ; Stop timer 0
        CLR TF0             ; Clear timer 0 flag
```

Example 4.6

Write an assembly language program to generate a square wave of 2 kHz frequency on the port pin P1.0, use timer 1. Assume that the crystal frequency of 8051 microcontroller is 11.0592 MHz.

Solution:

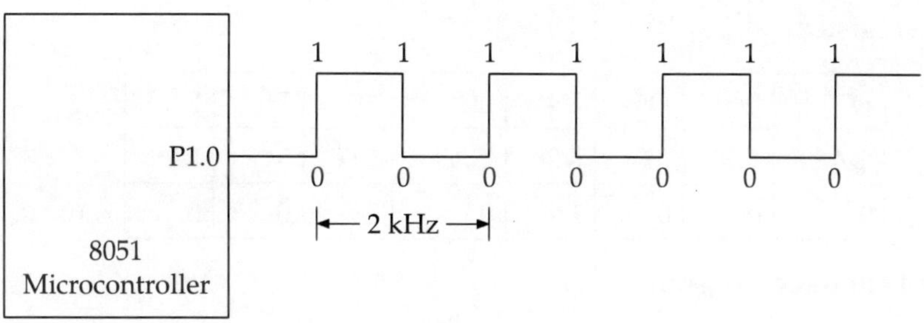

Fig. 4.6

$$\text{Crystal frequency} = 11.0592 \text{ MHz}$$

$$\text{Internal machine cycle frequency} = \frac{11.0592 \text{ MHz}}{12} = 921.6 \text{ kHz}$$

$$\text{Internal clock time} = \frac{1}{921.6 \text{ kHz}} = 1.085\ \mu\text{s}$$

$$\text{Time delay of given square wave, T} = \frac{1}{f} = \frac{1}{2 \text{ kHz}} = 500\ \mu\text{s}$$

$$TON = TOFF = \frac{500\ \mu s}{2} = 250\ \mu s$$

$$\text{Time delay} = (\text{FFFFh} - \text{YYXX} + 1) * 1.085\ \mu s$$

$$\text{Time delay} = 250\ \mu s$$

$$250\ \mu s = (\text{FFFFh} - \text{YYXX} + 1) * 1.085\ \mu s$$

$$\text{FFFFh} - \text{YYXX} + 1 = \frac{250\ \mu s}{1.085\ \mu s}$$

$$\text{FFFFh} - \text{YYXX} + 1 = 230 \text{ decimal}$$

$$230 \text{ Decimal} = \text{E6 hexadecimal}$$

$$\text{YYXX} = \text{FFFFh} + 1 - \text{E6h}$$

$$\text{YYXX} = \text{FE1Ah}$$

Therefore, $\text{TH1} = \text{FEh and TL1} = \text{1Ah}$

or

$$65536 - \text{NNNNN} = 230$$

$$\text{NNNNN} = 65536 - 230$$

$$\text{NNNNN} = 65306$$

$$65306 \text{ Decimal} = \text{FF1A hexadecimal}$$

Setting values in TMOD register or configuring TMOD register with initialization of timer 0 in mode 1

TMOD register

D7	D6	D5	D4	D3	D2	D1	D0	
GATE	**C/T**	**M1**	**M0**	**GATE**	**C/T**	**M1**	**M0**	
0	0	0	1	0	0	0	0	= 10 H

Assembly Language Program:

Method – I

```
          CLR P1.0          ; Clear P1.0
          MOV TMOD, #10H    ; Set timer 1 in mode 1
BACK:     MOV TL1, #1AH     ; Load lower byte of timer 1 register
          MOV TH1, #0FFH    ; Load higher byte of timer 1 register
          SETB P1.0         ; Set P1.0 high
          SETB TR1          ; Start timer 1
L1:       JNB TF1, L1       ; Keep monitoring TF1 flag bit until timer 1 rolls
                            ; over from FFFFh to 0000h
```

```
        CLR P1.0          ; Clear P1.0
        CLR TR1           ; Stop timer 1
        CLR TF1           ; Clear timer 1 flag bit
        MOV TL1, #1AH
        MOV TH1, #0FFH
        SETB TR1
L2:     JNB TF1, L2
        CLR TR1
        CLR TF1
        SJMP BACK
```

Method – II

```
        CLR P1.0
        MOV TMOD, #10H
BACK:   MOV TL1, #1AH
        MOV TH1, #0FFH
        SETB TR1
L1:     JNB TF1, L1
        CLR TR1
        CPL P1.0          ; Complement P1.0 to get alternate high and low
        CLR TF1
        SJMP BACK
```

Example 4.7

Write an assembly language program to generate a square wave of 100 Hz frequency on the port pin P1.1, use timer 1. Assume that the crystal frequency of 8051 microcontroller is 11.0592 MHz.

Solution:

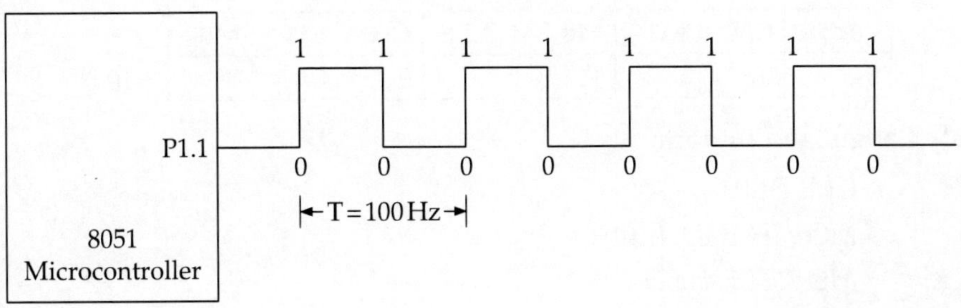

Fig. 4.7

$$\text{Crystal frequency} = 11.0592 \text{ MHz}$$

$$\text{Internal frequency} = \frac{11.0592 \text{ MHz}}{12} = 921.6 \text{ kHz}$$

$$\text{Internal clock time} = \frac{1}{921.6 \text{ kHz}} = 1.085 \text{ ms}$$

$$\text{Time delay of given square wave T} = \frac{1}{f} = \frac{1}{100 \text{ Hz}} = 10 \text{ ms}$$

$$\text{TON} = \text{TOFF} = \frac{10 \text{ ms}}{2} = 5 \text{ ms}$$

$$\text{Time delay} = (\text{FFFFh} - \text{YYXX} + 1) * 1.085 \ \mu\text{s}$$

$$\text{Time delay} = 5 \text{ ms}$$

$$5 \text{ ms} = (\text{FFFFh} - \text{YYXX} + 1) * 1.085 \ \mu\text{s}$$

$$\text{FFFFh} - \text{YYXX} + 1 = \frac{5 \text{ ms}}{1.085 \text{ ms}}$$

$$\text{FFFFh} - \text{YYXX} + 1 = 4608 \text{ decimal}$$

$$4608 \text{ decimal} = 1200 \text{ hexadecimal}$$

$$\text{YYXX} = \text{FFFFh} + 1 - 1200\text{h}$$

$$\text{YYXX} = \text{FE00h}$$

Therefore, $\text{TH1} = \text{FEh and TL1} = 00\text{h}$

or $65536 - \text{NNNNN} = 4608$

$$\text{NNNNN} = 65536 - 4608$$

$$\text{NNNNN} = 60928$$

$$60928 \text{ Decimal} = \text{EE00 hexadecimal}$$

Setting values in TMOD register or configuring TMOD register with initialization of timer 0 in mode 1

TMOD register

D7	D6	D5	D4	D3	D2	D1	D0	
GATE	**C/T**	**M1**	**M0**	**GATE**	**C/T**	**M1**	**M0**	
0	0	0	1	0	0	0	0	= 10 H

Assembly Language Program:

```
        CLR P1.0
        MOV TMOD, #10H
BACK:   MOV TL1, #00H
        MOV TH1, #0EEH
```

```
        SETB TR1
L1:     JNB TF1, L1
        CLR TR1
        CPL P1.1        ; Complement P1.1 to get alternate high and low
        CLR TF1
        SJMP BACK
```

Example 4.8

Write an assembly language program to generate a square wave of 1.8 kHz frequency on the port pin P1.0, use timer 1 in auto-reload mode. Assume that the crystal frequency of 8051 microcontroller is 11.0592 MHz.

Solution:

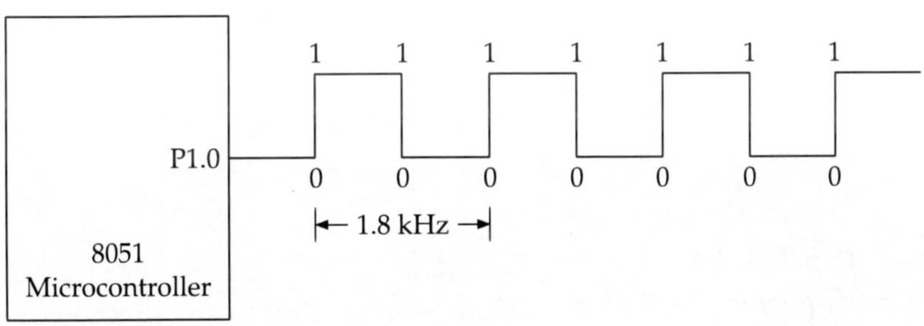

Fig. 4.8

$$\text{Crystal frequency} = 11.0592 \text{ MHz}$$

$$\text{Machine cycle frequency} = \frac{11.0592 \text{ MHz}}{12} = 321.6 \text{ kHz}$$

$$\text{Timer clock time} = \frac{1}{921.6 \text{ kHz}} = 1.085 \text{ μs}$$

$$\text{Time delay of given square wave T} = \frac{1}{f} = \frac{1}{1.8 \text{ kHz}} = 555.52 \text{ μs}$$

$$\text{TON} = \text{TOFF} = \frac{555.52 \text{ μs}}{2} = 277.76 \text{ μs}$$

Since in this problem auto-reload mode is to be used then timer register value is calculated as

$$\text{Time delay} = (\text{FFh} - \text{YY} + 1) * 1.085 \text{ μs}$$

$$\text{Time delay} = 277.76 \text{ μs}$$

$$277.76 \text{ μs} = (\text{FFh} - \text{YY} + 1) * 1.085 \text{ μs}$$

$$FFh - YY + 1 = \frac{277.76}{1.085}\mu s$$

$$FFh - YY + 1 = 256 \text{ decimal}$$

$$256 \text{ decimal} = 0100 \text{ hexadecimal}$$

$$YY = FFh + 1 - 0100\text{ h}$$

$$YY = 00h$$

Setting values in TMOD register or configuring TMOD register with initialization of timer 1 in mode 2

TMOD register

D7	D6	D5	D4	D3	D2	D1	D0	
GATE	**C/T**	**M1**	**M0**	**GATE**	**C/T**	**M1**	**M0**	
0	0	1	0	0	0	0	0	= 20 H

Assembly Language Program:

```
        CLR P1.0
        MOV TMOD, #20H     ; Set timer 1 in auto-reload mode (mode 2)
        MOV TH1, #00H
        SETB TR1
L1:     JNB TF1, L1
        CPL P1.0           ; Complement P1.0 to get alternate high and low
        CLR TF1
        SJMP L1
```

Example 4.9

Assuming that the clock pulses are fed to the pin P3.5, write an assembly language program for counter 1 in auto-reload mode to count the pulses and display the state of timer 1 lower byte register TL1 count on port P2.

Solution:

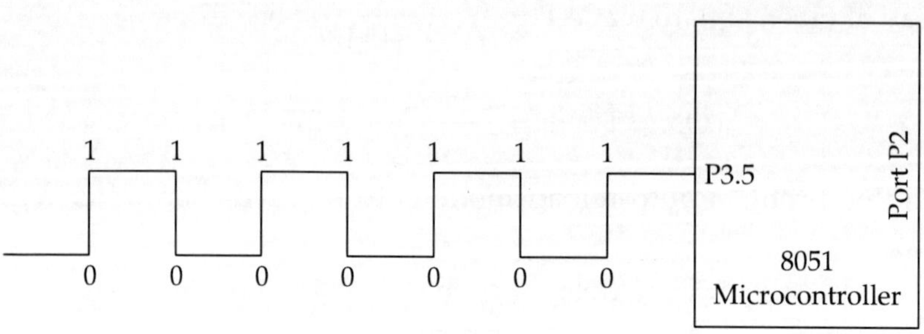

Fig. 4.9

TMOD register

GATE	C/T	M1	M0	GATE	C/T	M1	M0	
0	1	1	0	0	0	0	0	= 60H

Assembly Language Program:

```
        MOV TMOD, #60H    ; Timer 1 in auto-reload mode for counting pulses
        MOV TH1, #00H
        SETB P3.5         ; Make port P3.5 as an input port
BACK:   SETB TR1
L1:     MOV A, TL1
        MOV P2, A
        JNB TF1, L1
        CLR TR1
        CLR TF1
        SJMP BACK
```

Example 4.10

Write an assembly language program to generate a square wave with 33% duty cycle of 2 kHz frequency on the port pin P1.0, use timer 1. Assume that the crystal frequency of 8051 microcontroller is 11.0592 MHz.

Solution:

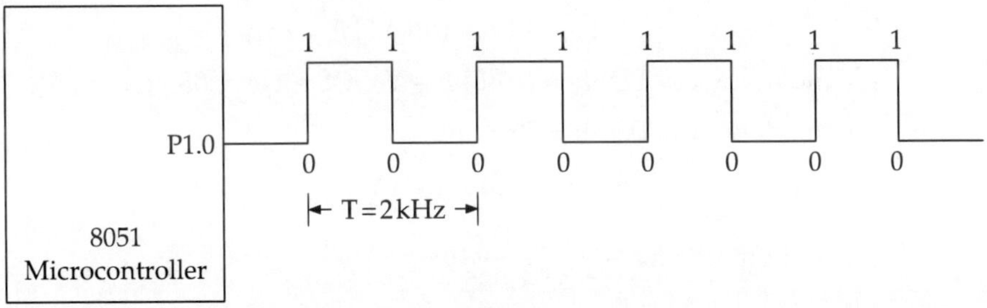

Fig. 4.10

Duty cycle of square wave = 33%, it means that the ON time of the square wave is 33% of T and OFF time is 67% of T.

$$\text{Crystal frequency} = 11.0592 \text{ MHz}$$

$$\text{Internal machine cycle frequency} = 11.0592 \text{ MHz}/12 = 921.6 \text{ kHz}$$

$$\text{Internal clock time} = \frac{1}{921.6 \text{ kHz}} = 1.085 \ \mu\text{s}$$

$$\text{Time delay of given square wave T} = \frac{1}{f} = \frac{1}{2\,\text{kHz}} = 500\ \mu\text{s}$$

$$\text{TON} = 33 * \frac{500}{100}\ \mu\text{s} = 165\ \mu\text{s}$$

$$\text{TOFF} = 67 * \frac{500}{100}\ \mu\text{s} = 335\ \mu\text{s}$$

Or TOFF may also be calculated as

$$\text{T} = \text{TON} + \text{TOFF}$$

$$\text{TOFF} = \text{T} - \text{TON}$$

$$= 50\ \mu\text{s} - 165\ \mu\text{s} = 335\ \mu\text{s}$$

$$\text{Time delay for TON} = (\text{FFFFh} - \text{YYXX} + 1) * 1.085\ \mu\text{s}$$

$$\text{Time delay of TON} = 165\ \mu\text{s}$$

$$165\ \mu\text{s} = (\text{FFFFh} - \text{YYXX} + 1) * 1.085\ \mu\text{s}$$

$$\text{FFFFh} - \text{YYXX} + 1 = \frac{165}{1.085}\mu\text{s}$$

$$\text{FFFFh} - \text{YYXX} + 1 = 152\ \text{decimal}$$

$$152\ \text{Decimal} = 98\ \text{hexadecimal}$$

$$\text{YYXX} = \text{FFFFh} + 1 - 98\text{h}$$

$$\text{YYXX} = \text{FF68h}$$

Therefore, for TON (ON time)

$$\text{TH1} = \text{FFh} \quad \text{and} \quad \text{TL1} = 68\text{h}$$

$$\text{Time delay for TOFF} = (\text{FFFFh} - \text{YYXX} + 1) * 1.085\ \mu\text{s}$$

$$\text{Time delay for TOFF} = 335\ \mu\text{s}$$

$$335\ \mu\text{s} = (\text{FFFFh} - \text{YYXX} + 1) * 1.085\ \mu\text{s}$$

$$\text{FFFFh} - \text{YYXX} + 1 = \frac{335}{1.085}\mu\text{s}$$

$$\text{FFFFh} - \text{YYXX} + 1 = 308\ \text{decimal}$$

$$308\ \text{decimal} = 134\ \text{hexadecimal}$$

$$\text{YYXX} = \text{FFFFh} + 1 - 134\text{h}$$

$$\text{YYXX} = \text{FECCh}$$

Therefore, for TOFF (OFF time)

$$\text{TH1} = \text{FEh} \quad \text{and} \quad \text{TL1} = \text{CCh}$$

Setting values in TMOD register or configuring TMOD register with initialization of timer 0 in mode 1

TMOD register

D7	D6	D5	D4	D3	D2	D1	D0	
GATE	**C/T**	**M1**	**M0**	**GATE**	**C/T**	**M1**	**M0**	
0	0	0	1	0	0	0	0	= 10 H

Assembly Language Program:

```
        CLR P1.0            ; Clear P1.0 (OFF duration)
        MOV TMOD, #10H      ; Set timer 1 in mode 1
BACK:   MOV TL1, #68H       ; Load lower byte of timer 1 register
        MOV TH1, #0FFH      ; Load higher byte of timer 1 register
        SETB P1.0           ; Set P1.0 high (ON duration)
        SETB TR1            ; Start timer 1
L1:     JNB TF1, L1         ; Keep monitoring TF1 flag bit until timer 1 rolls
                              over from FFFFh to 0000h
        CLR P1.0            ; Clear P1.0 (OFF duration)
        CLR TR1             ; Stop timer 1
        CLR TF1             ; Clear timer 1 flag bit
        MOV TL1, #0CCH
        MOV TH1, #0FEH
        SETB TR1
L2:     JNB TF1, L2
        CLR TR1
        CLR TF1
        SJMP BACK
```

Example 4.11

Write an assembly language program to generate a square wave with 66% duty cycle of 4 kHz frequency on the port pin P1.0, use timer 1. Assume that the crystal frequency of 8051 microcontroller is 11.0592 MHz. Generate 200 pulses.

Solution:

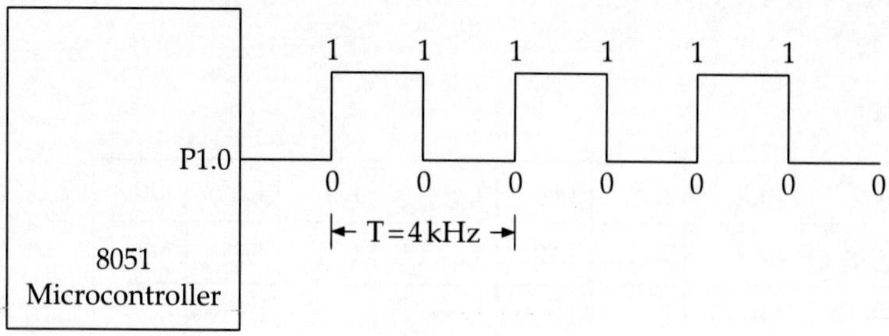

Fig. 4.11

Duty cycle of square wave = 33%, it means that the ON time of the square wave is 66% of T and OFF time is 34% of T.

$$\text{Crystal frequency} = 11.0592 \text{ MHz}$$

$$\text{Internal machine cycle frequency} = 11.0592 \text{ MHz}/12 = 921.6 \text{ kHz}$$

$$\text{Internal clock time} = \frac{1}{921.6 \text{ kHz}} = 1.085 \text{ µs}$$

$$\text{Time delay of given square wave, } T = \frac{1}{f} = \frac{1}{4 \text{ kHz}} = 250 \text{ µs}$$

$$\text{TON} = 66 * \frac{250}{100} \text{ µ sec} = 165 \text{ µs}$$

$$\text{TOFF} = 34 * \frac{250}{100} \text{ µ sec} = 85 \text{ µs}$$

Or TOFF may also be calculated as

$$T = \text{TON} + \text{TOFF}$$

$$\text{TOFF} = T - \text{TON}$$

$$= 250 \text{ µs} - 165 \text{ µs} = 335 \text{ µs}$$

$$\text{Time delay for TON} = (\text{FFFFh} - \text{YYXX} + 1) * 1.085 \text{ µs}$$

$$\text{Time delay of TON} = 165 \text{ µs}$$

$$165 \text{ µs} = (\text{FFFFh} - \text{YYXX} + 1) * 1.085 \text{ µs}$$

$$\text{FFFFh} - \text{YYXX} + 1 = \frac{165 \text{ µs}}{1.085 \text{ µs}}$$

$$\text{FFFFh} - \text{YYXX} + 1 = 152 \text{ decimal}$$

$$152 \text{ decimal} = 98 \text{ hexadecimal}$$

$$\text{YYXX} = \text{FFFFh} + 1 - 98\text{h}$$

$$\text{YYXX} = \text{FF68h}$$

Therefore, for TON (ON time)

$$TH1 = FFh \quad \text{and} \quad TL1 = 68h$$

$$\text{Time delay for TOFF} = (FFFFh - YYXX + 1) * 1.085\ \mu s$$

$$\text{Time delay for TOFF} = 85\ \mu s$$

$$85\ \mu s = (FFFFh - YYXX + 1) * 1.085\ \mu s$$

$$FFFFh - YYXX + 1 = \frac{335\ \mu s}{1.085\ \mu s}$$

$$FFFFh - YYXX + 1 = 78 \text{ decimal}$$

$$78 \text{ decimal} = 4 \text{ hexadecimal}$$

$$YYXX = FFFFh + 1 - 4h$$

$$YYXX = FEB2h$$

Therefore, for TOFF (OFF time)

$$TH1 = FEh \quad \text{and} \quad TL1 = B2h$$

Setting values in TMOD register or configuring TMOD register with initialization of timer 0 in mode 1

TMOD register

D7	D6	D5	D4	D3	D2	D1	D0	
GATE	**C/T**	**M1**	**M0**	**GATE**	**C/T**	**M1**	**M0**	
0	0	0	1	0	0	0	0	= 10 H

Assembly Language Program:

```
         CLR P1.0            ; Clear P1.0 (OFF duration)
         MOV R2, #200        ; Counter for 200 pulses
         MOV TMOD, #10H      ; Set timer 1 in mode 1
REPEAT:  MOV TL1, #68H       ; Load lower byte of timer 1 register
         MOV TH1, #0FFH      ; Load higher byte of timer 1 register
         SETB P1.0           ; Set P1.0 high (ON duration)
         SETB TR1            ; Start timer 1
L1:      JNB TF1, L1         ; Keep monitoring TF1 flag bit until timer 1 rolls
                               over from FFFFh to 0000h
         CLR P1.0            ; Clear P1.0 (OFF Duration)
         CLR TR1             ; Stop timer 1
         CLR TF1             ; Clear timer 1 flag bit
         MOV TL1, #0B2H
```

```
        MOV TH1, #0FFH
        SETB TR1
L2:     JNB TF1, L2
        CLR TR1
        CLR TF1
        DJNZ R2, REPEAT
BACK:   SJMP BACK
```

REVIEW QUESTIONS

1. Write an assembly language program to generate square wave of 2 kHz frequency. Use timer 1 in mode 1. Generate square wave on the port pin P2.5 and generate 200 pulses only. Assume that crystal frequency of 8051 is 11.0592 MHz.
2. Write an assembly language program to generate square wave of 1.5 ms time duration. Use timer 1 in mode 2. Use port pin P2.0 and assume that the crystal frequency of 8051 is 11.0592 MHz.
3. Write an assembly language program to generate square wave of 7.2 Hz frequency on the port pin P1.2 using timer 1 in auto-reload mode. Assume crystal frequency of 8051 is 11.0592 MHz.
4. Write an assembly language program to generate square wave of 100 μs time duration using timer 0 in auto-reload mode with 66% duty cycle. Assume that crystal frequency of 8051 is 11.0592 MHz.
5. Write an assembly language program to generate square wave of 200 μs time duration using timer 1 in auto-reload mode. Assume that crystal frequency of 8051 is 11.0592 MHz.
6. Write an assembly language program to generate 200 square pulses of 2 kHz frequency on the port pin P1.7. Use timer 0 in mode 1, assume crystal frequency as 12 MHz.
7. Write an assembly language program to generate square pulse of 2 kHz frequency on port pin P2.7, use timer 1 in mode 1. The duty cycle of the square wave is 66 %. Assume crystal frequency as 11.0592 MHz.
8. Write an assembly language program to generate square wave of 50 Hz frequency on port pin P2.7, use timer 0 in auto-reload mode. Assume that crystal frequency is 12 MHz.
9. Write an assembly language program to generate 1000 square wave of 100 Hz frequency on the port pin P1.7, use timer 0 in mode 1. Assume that crystal frequency is 12 MHz.
10. Write an assembly language program to generate delay of using timer 1 in mode 1. Assume that crystal frequency is 12 MHz.

5

Chapter

Serial Communication of Microcontroller 8051

One of the 8051's most powerful futures is its integrated UART (universal asynchronous receiver transmitter), also known as a serial port. Integrated serial port provides the flexibility to read and write easily. This chapter deals with the concept of intergrated serial ports and their futures. After this chapter we will be familiarized with the concept of how data will be received and transmitted one bit at a time, how the characters or numbers are displayed or communicated serially via serial communication network.

5.1 UNIVERSAL ASYNCHRONOUS RECEIVER TRANSMITTER (UART)

UARTs are used for serial communication between systems; they can be either half duplex (send or receive but at different intervals) or full duplex (send and receive at the same time). Also known as an RS232 connection, the microcontroller UART can provide the connection with a PC or another microcontroller-based system. Figure 5.1 illustrates possible connection arrangements. In a minimum connection there could be only two transmission lines, transmit (Tx) and receive (Rx) as shown in Figure 5.2. The data is conveyed as a bit stream, either transmit or receive, and the speed is defined by the baud rate, i.e., bits per second.

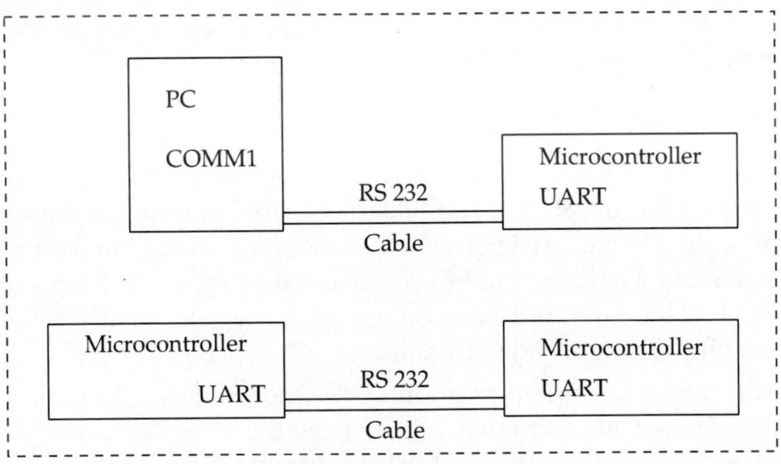

Fig. 5.1 *Use of RS232 interface between PC and microcontroller or between two microcontrollers*

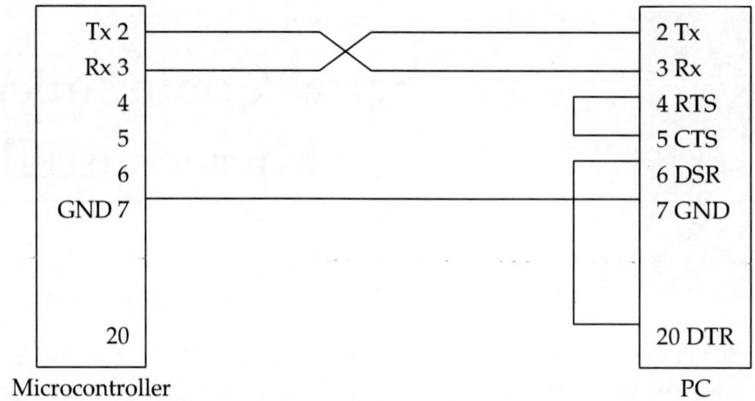

Fig. 5.2 *RS 232 transmit (Tx) and receive (Rx) connections between a PC and microcontroller*

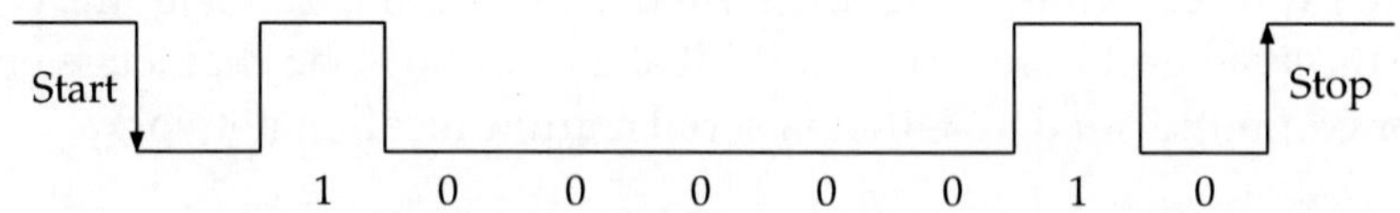

Fig. 5.3 *ASCII bit pattern 01000001 (41h) for character 'A'*

The UART has four modes of operation, 0 to 3. Modes 0 and 2 have fixed baud rates, mode 0 is one-sixth of the oscillator frequency, and mode 2 is 1/16 or 1/32 of the oscillator frequency. For modes 1 and 3 the baud rate can be selected, a typical range is:

75; 150; 300; 600; 1200; 2400; 4800; 9600; 19200; 38400

Modes 0 and 1 are used for connection between two devices. Modes 2 and 3 are used for master slave multiprocessor systems, in principle there could be one master microcontroller and up to 255 slave microcontrollers. In mode 1 ten bits are used to specify an RS232 frame consisting of 1 start bit (logic 0), 8 data bits and 1 stop bit (logic 1). For example, the ASCII bit pattern 0100 0001 (hex 41) represents the character 'A' and is transmitted as shown in Figure 5.3; least significant bit (LSB) first.

5.2 SERIAL COMMUNICATION

One of the 8051's most powerful features is its integrated UART, otherwise known as a serial port. The fact that the 8051 has an integrated serial port means that you may easily read and write values to the serial port. If it were not for the integrated serial port, writing a byte to a serial line would be a rather complicated process requiring turning ON and OFF one of the I/O lines in rapid succession to properly "clock out" each individual bit, including start bits, stop bits, and parity bits.

Actually, if the serial communication port is not integrated within the chip then proper synchronization between the transmitter and the receiver is a complicated process. However, we do not have to do this. Instead, we simply need to configure the serial port's operation mode and baud rate. This configuration can be done by writing very simple code, may require only 5 to 8 lines of code. Once configured, all we have to do is write to an

SFR to write a value to the serial port or read the same SFR to read a value from the serial port. The 8051 will automatically let us know when it has finished sending the character we wrote and will also let us know when it has received a byte so that we can process it. We do not have to worry about transmission at the bit level which saves us quite a bit of coding and processing time.

For example, like in the microprocessor based system where the serial communication facility is implemented using interfacing of additional peripheral devices like UART 8250 or USART 8251 and then these devices need to program properly by writing their command words for proper synchronization between the microprocessor and the peripheral device, this process requires to write sometimes a very long coding simply for synchronization and hence involves a complicated and time consuming process.

5.3 TYPES OF TRANSMISSION

Serial communication is capable of processing or transmitting information by two types: Simplex communication and duplex communication. Further, the duplex communication is also of two types: half duplex communication and full duplex communication.

5.3.1 Simplex Communication

In simplex communication, the serial data is transmitted only in one direction, i.e., from transmitter to receiver.

The simplex is a one-way transmission, data is transferred only in one direction over a single communication channel. For example, microprocessor transmits data to the printer, and to the CRT display unit.

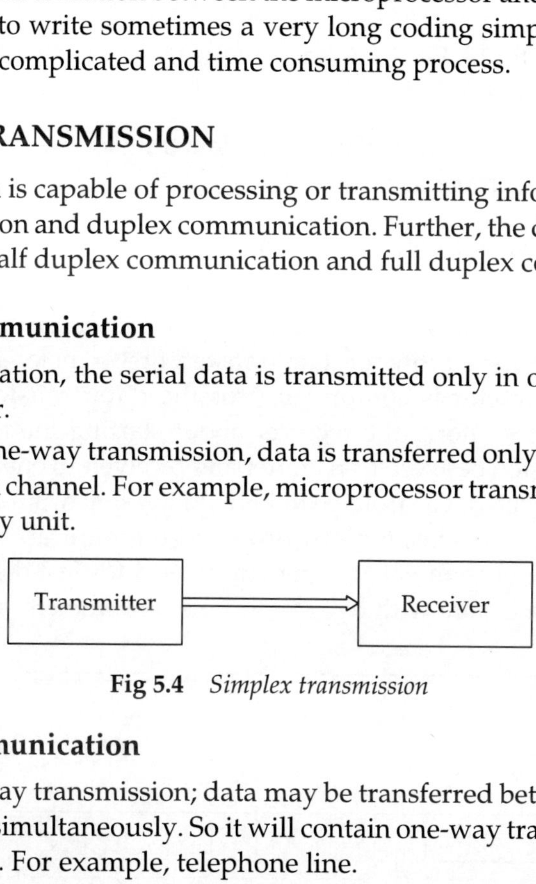

Fig 5.4 *Simplex transmission*

5.3.2 Duplex Communication

The duplex is a two-way transmission; data may be transferred between two transceivers in both the directions simultaneously. So it will contain one-way transmission or two-way transmission at a time. For example, telephone line.

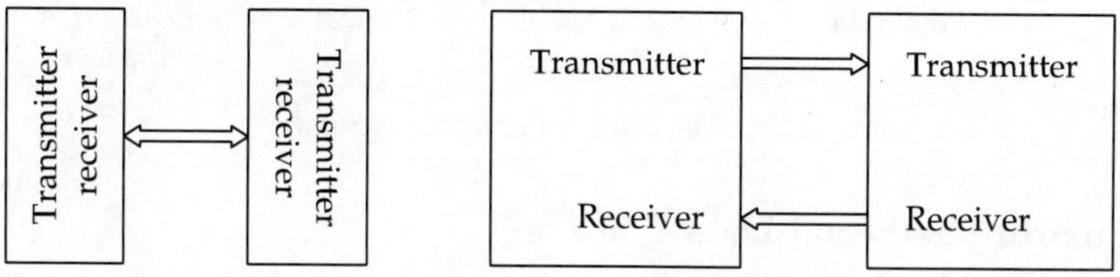

Fig. 5.5 *Duplex transmission*

Half Duplex

The half duplex is a two-way transmission, but in such a way that data may travel in only one direction at a time. For example, walky talky (Wireless System).

Full Duplex

The full duplex is a two-way transmission (communication), but in such a way that the data may travel in both the directions at the same time. For example, telephone line, mobile communication, etc.

5.4 METHODS OF DATA TRANSMISSION / FORMATS OF SERIAL DATA TRANSFER

In modern data communication system for the data transfer, two types of format are used: synchronous and asynchronous.

5.4.1 Synchronous Serial Data Transfer

In this method, block of data bytes are transferred serially at a time. The number of data bytes is not limited. The data bytes are transferred one after the other and so on up to the end. In this method at the starting of data transfer, first transfer synchronous characters one synchronous character is compulsory for the data transfer and 2nd is optional. Synchronous characters inform the receiver about starting the transfer of data, i.e., its provides synchronization between transmitter and receiver. In one synchronous character, 256 receivers are connected, i.e., from 00 to FFh. As the synchronization is done only once and before data transfer begins, the synchronous communication is faster compared to asynchronous communication. So, it is generally used for fast data transfer at high rate. The synchronous data format with two synchronous characters is shown in Figure 5.6.

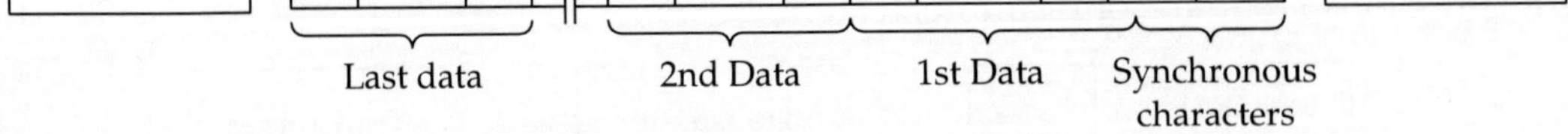

Fig. 5.6 *Synchronous data format*

5.4.2 Asynchronous Serial Data Transfer

In this method, one byte of data is transferred serially at a time. Normally, the output line of transmitter is maintained high and before transferring the data the output line is made low (Logic 0) for one clock pulse to indicate start of data byte, so it is called the start bit. After the data bits are transferred serially D_0 to D_7 (D_7 last), the output is maintained high for specific period to indicate end of data and these bits are called stop bits. These stop bits used are logical high signals so after completion of stop bits, the same logical level is maintained on output line. At the end of every data bytes parity bit will also send, this is optional and only used to maintain specific parity of the data bits sent. In the communication system especially for information theory, parity bit plays an important role in order to maintain the integrity of the information or data communicated between the transmitter and the receiver. In the asynchronous communication, before transmission of any data, the data must be packed between start and stop bits, called frame, and hence asynchronous communication is slower compared to synchronous communication. So it is generally used for slow data transfer at low rate, but having significant advantages over synchronous communication. The asynchronous data format with one start, one stop and one parity bit is shown in Figure 5.7.

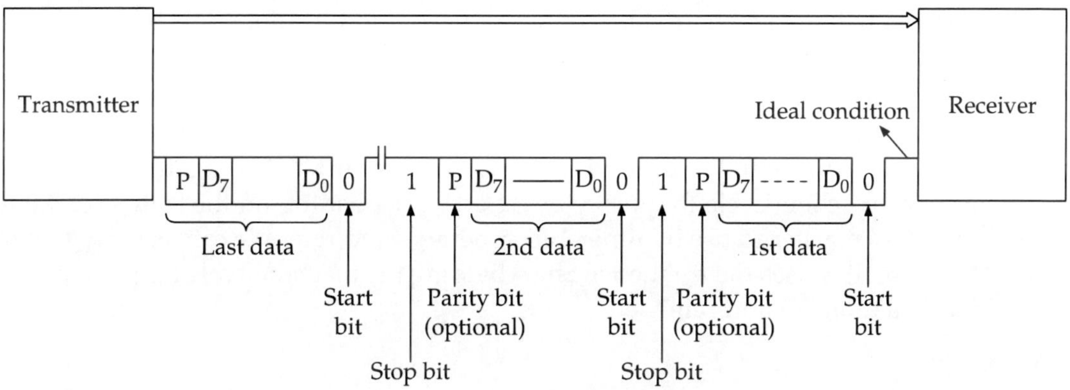

Fig. 5.7 *Asynchronous data format*

5.4.3 Differences between Synchronous and Asynchronous Serial Data Transfer

Synchronous data transfer	Asynchronous data transfer
1. It is used to transfer a group of characters at a time.	It is used to transfer one character at a time.
2. Used for high data transfer rates ≥ 20 K bits/second.	Used for data transfer rates ≤ 20 K bits/second.
3. Synchronous characters are transmitted along with the group of characters.	Synchronous characters are not transmitted along with the characters.
4. No start and stop bits for each character is used.	Start and stop bit for each character is present which forms a frame.
5. One clock is used for both transmitter and receiver.	Two separate clock inputs can be used for transmitter and receiver.
6. Since synchronization is involved, this can be implemented by using hardware only.	No synchronization is required hence hardware and software implementation is possible.

5.5 DATA TRANSFER RATE / BAUD RATE

Data transfer rate is defined as the number of bits transmitted or received in one second. Usually, it is referred as bits per second (BPS) and sometimes it is also called baud rate (BR). Transmission of the serial data requires configuring baud rate; actually at the baud rate it was defined as the number of bits to be transferred in one second. It is not possible to configure data transfer serially without configuring baud rate. It is important to note that it is not necessary that bits per second and baud rate are same. Because the baud rate is used in reference to MODEM (modulator-demodulator), where number of signal changes indicate bits transmission and single signal change may indicate more data bits transmitted, whereas if conductor based transmission is concerned then bits per second and baud rate are same.

Baud rate in 8051 microcontroller is programmable, i.e., different baud rates can be easily achieved without any hardware configuration or modifications; this is simply done by writing a set of instructions.

The 8051 microcontroller divides crystal frequency by 12 to get the machine cycle frequency; in other words we say that internal frequency is $1/12^{th}$ of crystal frequency (external frequency). Serial communication UART circuitry once again divides this frequency, i.e., machine cycle frequency by 32 or 16, depending on the status of SMOD bit in the PCON register to use by timer 1 to generate baud rate. We can also say that on chip UART circuitry uses the frequency after dividing the machine cycle frequency by 32 or 16 for calculation of baud rate.

SMOD = 0

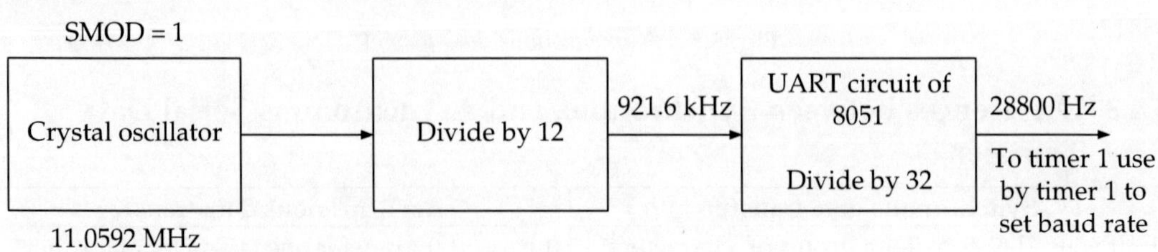

Fig. 5.8 *Setting baud rate for SMOD = 0*

SMOD = 1

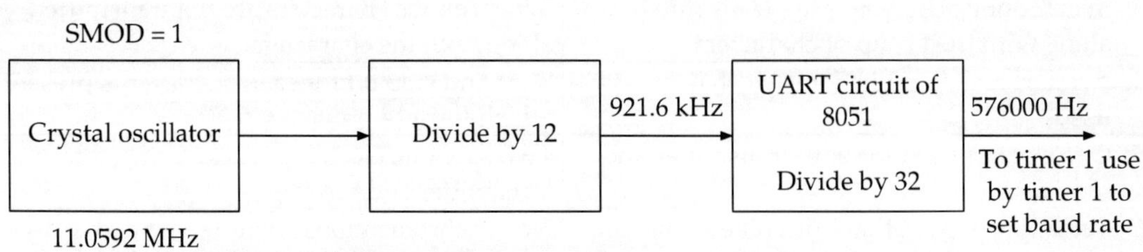

Fig. 5.9 *Setting baud rate for SMOD = 1*

To set baud rate compatible with IBM PC's timer 1 must be operated in auto-reload mode. The reason to use the auto-reload mode is that, it is possible to get maximum overflows (ticks) at very high rate only in auto-reload mode.

The values to be loaded in TH1 register and their corresponding baud rate compatible with IBM PC's are shown in Table 5.1.

Table 5.1 *Timer 1 TH1 register values for various baud rates*

TH1 (Decimal)	TH1 (Hex)	Baud Rate
– 3	FDh	9600
– 6	FAh	4800
– 12	F4h	2400
– 24	E8h	1200
– 48	D0h	600

Programming Tips:

The baud rate indicated in this table is doubled if PCON.7 (SMOD) is set.

Roughly baud rate is calculated as follows:

Baud rate = Timer 1 frequency / TH1

After calculation of TH1 values from the above formula take the inversion of this and convert into hexadecimal equivalent, this is the required value to be loaded in the TH1 register to get the specified baud rate.

Example: Assume that crystal frequency is 11.0592 MHz, then timer 1 frequency becomes

Timer 1 frequency = 11.0592 MHz / 12 * 32

Timer 1 frequency = 28800 Hz

So to set baud rate 9600 then

9600 = 28800 / TH1

Therefore, H1 = 28800 / 9600 = 3

Therefore, TH1 = – 3 = FDh

5.5.1 Calculation of Baud Rate

In modes 1 and 3, the baud rate is determined by how frequently timer 1 overflows. The more frequently timer 1 overflows, the higher the baud rate, i.e., how fast the timer flag bit overflows that decides baud rate. In other words, we can say that baud rate is directly related with timer flag rollover. There are many ways one can cause timer 1 to overflow at a rate that determines a baud rate but the most common method is to put timer 1 in 8-bit auto-reload mode (timer mode 2) and set a reload value (TH1) that causes timer 1 to overflow at a frequency appropriate to generate a baud rate.

To determine the value that must be loaded in TH1 register to generate a given baud rate, we may use the following equation:

$$TH1 = 256 - \frac{2^{SMOD} * \text{Oscillator frequency}}{12 * 32 * \text{Baud rate}}$$

SMOD is a bit 7 of PCON register. PCON register is not a bit addressable register.

$$TH1 = 256 - \frac{2^{SMOD} * \text{Oscillator frequency}}{384 * \text{Baud rate}}$$

If SMOD = 0, then

$$TH1 = 256 - \frac{\text{Oscillator frequency}}{384 * \text{Baud rate}}$$

If SMOD = 1, then

$$TH1 = 256 - \frac{2 * \text{Oscillator frequency}}{384 * \text{Baud rate}}$$

$$TH1 = 256 - \frac{\text{Oscillator frequency}}{192 * \text{Baud rate}}$$

For example, to set the baud rate 9600, for SMOD = 0, the value of TH1 register is calculated as

$$TH1 = 256 - \frac{2^{SMOD} * \text{Oscillator frequency}}{384 * \text{Baud rate}}$$

Assume that oscillator frequency = 11.0592 MHz

$$TH1 = 256 - \frac{2^0 * 11.0592 \text{ MHz}}{384 * 9600}$$

$$TH1 = 256 - 3 = 253$$

$$253 \text{ Decimal} = \text{FD hexadecimal}$$

Therefore, $TH1 = FDh$

5.6 SCON REGISTER (SERIAL CONTROL)/SETTING THE SERIAL PORTS

The first things we must do when using the 8051's integrated serial port is, obviously, configure it. This configuration is done by writing to the SCON register, once the configuration is done then data communication between the transmitter and the receiver is just a fun. This lets us tell the 8051 how many data bits we want, the baud rate we will be using, and how the baud rate will be determined. First, let's present the "Serial Control" (SCON) SFR and define what each bit of the SFR represents:

D7	D6	D5	D4	D3	D2	D1	D0
SM0	SM1	SM2	REN	TB8	RB8	TI	RI

Table 5.2 *Explanation of SCON register*

Bit	Name	Bit address	Function
7	SM0	9Fh	Serial port mode bit 0.
6	SM1	9Eh	Serial port mode bit 1.
5	SM2	9Dh	Multiprocessor communication enable bit.
4	REN	9Ch	Receiver enables. This bit must be set in order to receive characters serially.
3	TB8	9Bh	Transmit bit 8. The 9th bit to transmit in mode 2 and mode 3.
2	RB8	9Ah	Receive bit 8. The 9th bit received in mode 2 and mode 3.
1	TI	99h	Transmit flag. Set when a byte has been transmitted completely.
0	RI	98h	Receive flag. Set when a byte has been received completely.

Additionally, it is necessary to define the function of SM0 and SM1 by an additional table shown in Table 5.3.

Table 5.3 *Explanation of serial mode bits SM0 and SM1*

SM0	SM1	Serial mode	Explanation	Baud rate
0	0	0	8-bit shift register	Oscillator/12
0	1	1	8-bit UART	Set by timer 1
1	0	2	9-bit UART	Oscillator/32
1	1	3	9-bit UART	Set by timer 1

The SCON SFR allows us to configure the serial port. Thus, we'll go through each bit and review its function. The first four bits (bits 4 through 7) are configuration bits.

Bits **SM0** and **SM1** let us set the *serial mode* to a value between 0 and 3, inclusive. The four modes are defined in the chart immediately above. As you can see, selecting the serial mode selects the mode of operation (8-bit/9-bit, UART or shift register) and also determines how the baud rate will be calculated. In modes 0 and 2 the baud rate is fixed based on the oscillator's frequency. In modes 1 and 3 the baud rate is variable based on how often timer 1 overflows.

We'll talk more about the various serial modes in a moment. The next bit, **SM2**, is a flag for "multiprocessor communication". Multiprocessor communication is used to expand the serial communication to the next level like communication between more numbers of devices using multiple microcontrollers. Generally, whenever a byte has been received the 8051 will set the "RI" (receive interrupt) flag. This lets the program know that a byte has been received and that it needs to be processed. However, when **SM2** is set the "RI" flag will only be triggered if the 9th bit received was a "1". That is to say, if SM2 is set and a byte is received whose 9th bit is clear, the RI flag will never be set. This can be useful in certain advanced serial applications. For now it is safe to say that you will almost always want to clear this bit so that the flag is set upon reception of any character. The next bit,

REN, is "receiver enable." This bit is very straightforward: If you want to receive data via the serial port, set this bit. You will almost always want to set this bit. The last four bits (bits 0 through 3) are operational bits. They are used when actually sending and receiving data—they are not used to configure the serial port. The TB8 bit is used in modes 2 and 3. In modes 2 and 3, a total of nine data bits are transmitted. The first 8 data bits are the 8 bits of the main value, and the ninth bit is taken from TB8. If TB8 is set and a value is written to the serial port, the data's bits will be written to the serial line followed by a "set" 9th bit. If TB8 is clear the ninth bit will be "clear." The RB8 also operates in modes 2 and 3 and functions essentially the same way as TB8, but on the reception side. When a byte is received in modes 2 or 3, a total of nine bits are received.

In this case, the first eight bits received are the data of the serial byte received and the value of the ninth bit received will be placed in RB8. TI means "transmit interrupt." When a program writes a value to the serial port, a certain amount of time will pass before the individual bits of the byte are "clocked out" of the serial port. If the program were to write another byte to the serial port before the first byte was completely out, the data being sent would be garbled. Thus, the 8051 lets the program know that it has "clocked out" the last byte by setting the TI bit. When the TI bit is set, the program may assume that the serial port is "free" and ready to send the next byte. Finally, the RI bit means "receive interrupt". It functions similarly to the "TI" bit, but it indicates that a byte has been received. That is to say, whenever the 8051 has received a complete byte it will trigger the RI bit to let the program know that it needs to read the value quickly, before another byte is read.

Bits D7 and D6: SM0 and SM1

Serial Mode Bit 0 and 1

Of the 4 serial modes, only mode 1 was widely used. When mode 1 is chosen then data framing consists of 8-bit data, 1 start bit and 1 stop bit, which makes it compatible with COM (communication) port of IBM compatible PC's. More importantly, serial mode 1 allows the baud rate to be variable and is set by timer 1 of the 8051, so it can be easily programmed to get the different baud rate. In the serial mode 1, for each character a total of 10 bits are transferred where the 1st bit is always the start bit, followed by 8-bit data and finally 1 stop bit.

Mode 0: Shift register

SM0 SM1 = 00

Mode 0: In this mode, the serial port works like a shift register and the data transmission works synchronously with a clock frequency of $f_{osc}/12$. Serial data is received and transmitted through RXD. 8 bits are transmitted/ received at a time. Pin TXD outputs the shift clock pulses of frequency $f_{osc}/12$, which is connected to the external circuitry for synchronization. The shift frequency or baud rate is always 1/12 of the oscillator frequency. The data transmission and reception scheme in mode 0 is shown in Figure 5.10.

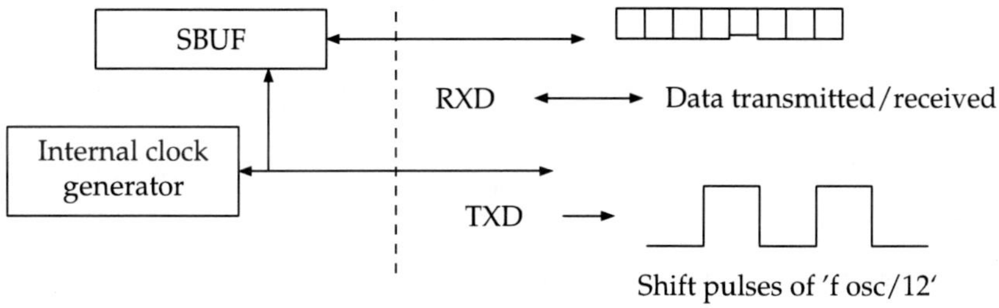

Fig. 5.10 *Data transmission/reception in mode 0*

Mode 1: Standard UART Mode

SM0 SM1 = 01

In mode 1, the serial port functions as an UART mode. 10 bits are transmitted through TXD or received through RXD. The 10 bits consist of one start bit (which is usually '0'), 8 data bits (LSB is sent first/received first), and a stop bit (which is usually '1'). Once received, the stop bit goes into RB8 in the special function register SCON. The baud rate is variable.

Figure 5.11 shows the way the bits are transmitted/ received.

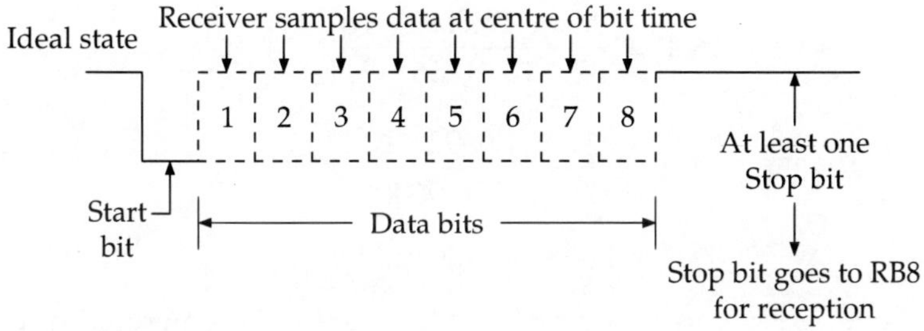

Fig. 5.11 *Data transmission format in UART mode*

Bit time= $1/f_{baud}$

In receiving mode, data bits are shifted into the receiver at the programmed baud rate. The data word (8 bits) will be loaded to SBUF if the following conditions are true.

1. RI must be zero (i.e., the previously received byte has been cleared from SBUF)
2. Mode bit SM2 = 0 or stop bit = 1

   After the data is received and the data byte has been loaded into SBUF, RI becomes one.

Baud rate generation in Mode 1

Timer 1 is used to generate baud rate for mode 1 serial communication by using overflow flag of the timer to determine the baud frequency. Timer 1 is used in timer mode 2 as an auto-reload 8-bit timer. The data rate is generated by timer 1 using the following formula

$$f_{\text{baud}} = \frac{2^{\text{SMOD}}}{32} * \frac{f_{\text{OSC}}}{12 * [256 - \text{TH1}]}$$

where, SMOD is the 7th bit of PCON register, f_{osc} is the crystal oscillator frequency of the microcontroller.

It can be noted that $f_{osc}/(12 \times [256 - (\text{TH1})])$ is the timer overflow frequency in timer mode 2, which is the auto-reload mode.

If timer-1 is not run in mode-2, then the baud rate is

$$f_{\text{baud}} = \frac{2^{\text{SMOD}}}{32}$$

Timer 1 can be run using the internal clock, $f_{osc}/12$ (timer mode) or from any external source via pin T1 (P3.5) (counter mode).

Example: If standard baud rate is desired, then 11.0592 MHz crystal could be selected. To get a standard 9600 baud rate, the setting of TH1 is calculated as follows:

Assuming SMOD to be '0'

$$9600 = \frac{2^0}{32} * \frac{11.0592 * 10^6}{12 * (256 - \text{TH1})}$$

or

$$256 - \text{TH1} = \frac{1}{32} * \frac{11.0592 * 10^6}{12 * 9600} = 3$$

or

$$\text{TH1} = 256 - 3 = 253 = \text{FDh}$$

In mode 1, if SM2 is set to 1, no receive interrupt (RI) is generated unless a valid stop bit is received.

Mode 2: Multiprocessor mode

SM0 SM1 = 10

$$f_{\text{baud}} = \frac{2^{\text{SMOD}}}{64} * f_{\text{OSC}}$$

Mode 3: Multiprocessor mode with variable baud rate

SM0 SM1 = 11

In this mode 11 bits are transmitted through TXD or received through RXD. The various bits are: a start bit (usually '0'), 8 data bits (LSB first), a programmable 9th bit and a stop bit (usually '1').

Mode 3 is same as mode 2, except the fact that the baud rate in mode 3 is variable (i.e., just as in mode 1).

$$f_{baud} = \frac{2^{SMOD}}{32} * \frac{f_{OSC}}{12 * [256 - TH1]}$$

This baud rate holds when timer 1 is programmed in Mode 2.

Operation in Multiprocessor Mode

8051 operates in multiprocessor mode for serial communication mode 2 and mode 3. In multiprocessor mode, a master processor can communicate with more than one slave processor. The connection diagram of processors communicating in multiprocessor mode is given in Figure 5.12

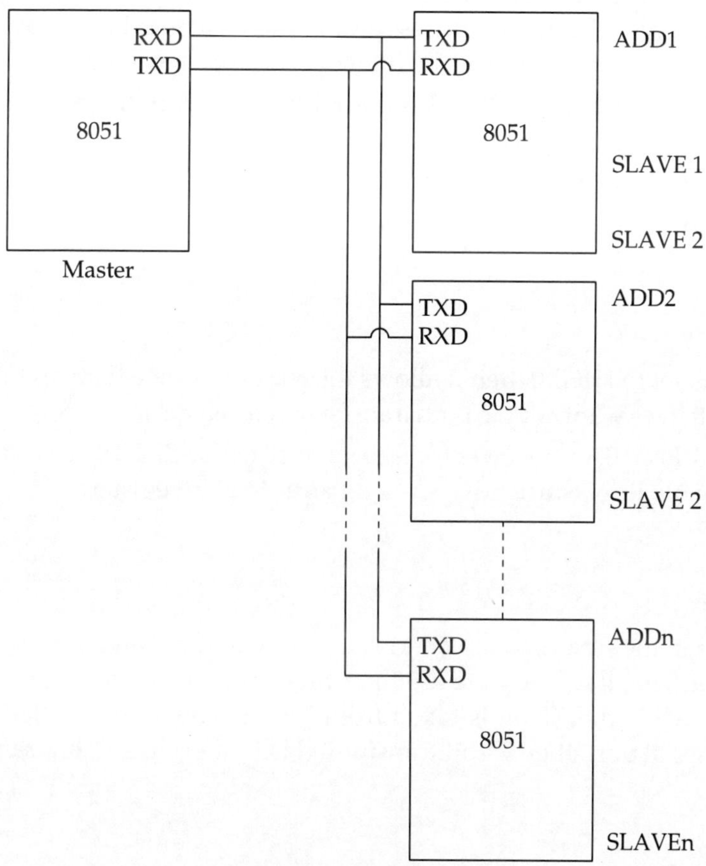

Fig. 5.12 *Microcontroller 8051 in multiprocessor communication*

After the communication between the master and a slave has been established, the data bytes are sent by the master with TB8=0. Hence, other slaves do not respond or get interrupted by this data as their SM2 is pulled high (1).

Bit D5: SM2

Multiprocessor bit

SM2 bit is a flag bit for the "multiprocessor communication". When this bit is set to high then it enables the future of multiprocessing capability of the 8051 microcontroller.

The master communicates with one slave at a time. 11 bits are transmitted by the master, viz, one start bit (usually '0'), 8 data bits (LSB first), TB8 and a stop bit (usually '1'). TB8 is '1' for an address byte and '0' for a data byte.

If the master wants to communicate with a certain slave, it first sends the address of the slave with TB8=1. This address is received by all the slaves. Slaves initially have their SM2 bit set to '1'. All slaves check this address and the slave who is being addressed, responds by clearing its SM2 bit to '0' so that the data bytes can be received.

It should be noted that in modes 2 and 3, receive interrupt flag RI is set if REN=1, RI=0 and the following condition is true.

1. SM2=1 and RB8=1 and a valid stop bit is received, or
2. SM2=0 and a valid stop bit is received.

Bit D4: REN

Receiver enable

When this bit is set to HIGH, then it allows the 8051 to receive data on the RXD pin of the 8051. As a result if we want 8051 to both transfer and receive data, REN must be set to HIGH.

By making REN = 0, the receiver is disabled. Instruction SETB SCON.4 and CLR SCON.4 enable and disable REN, since SCON is a bit addressable register.

Bit D3: TB8

Transmit bit 8

This bit is used in the serial modes 2 and 3. In serial modes 2 and 3, a total of 9 bits of data are transmitted. The first 8 bit data are the 8 bits of the main value and it is stored in the SBUF register, while the 9th bit is taken from TB8. If TB8 is set to HIGH, and a value is written to the serial port, then 9th bit transmitted is 1 otherwise 9th bit transmitted is LOW.

Bit D2: RB8

Receive bit 8

In serial mode 1, this bit gets a copy of the stop bit when an 8-bit data is received. Like TB8, RB8, also used in serial modes 2 and 3 and functions essentially the same way as TB8. When a byte is received in serial modes 2 and 3, a total of 9 bits are received. In this case, the first 8 bits are the main value of the serial byte data and copied to the SBUF register and 9th bit received is copied to RB8.

Bit D1: TI

Transmitter interrupt

This is an extremely important flag bit in the SCON register. When the 8051 microcontroller finishes the transfer of the 8-bit character, it raises the TI flag bit to indicate that serial communication is ready to transfer another byte. The TI flag bit is raised at the beginning of the stop bit.

Bit D0: RI

Receiver interrupt

This is another extremely important flag bit in the SCON register, when the 8051 receives data serially via RXD line, it gets rid of the start and stop bits and places the received byte in the SBUF register. Then it raises the RI flag bit to indicate that a byte has been received completely and should be picked up before it is lost by new incoming byte. RI flag bit is raised halfway through the stop bit.

5.6.1 Importance of TI (Transmitter Interrupt) FLAG Bit

The following steps are observed when the character byte is transferred serially via TXD line:

1. The start bit is transmitted; indicate that data transmitted is going to be stared.
2. The 8-bit main value of the character byte start transmission with LSB (bit D0) first.
3. The stop bit is transmitted. The stop bit is the last bit to be transmitted if the parity bit is not considered. It is during the transmission of stop bit, TI flag bit is raised to one (logic high) by CPU, indicating that transmission is completed. So 8051 is ready to transfer a new character byte.
4. By monitoring TI flag bit we ensure that the previous byte is transmitted completely or not. Suppose TI flag bit doesn't raise and we are trying to send a new character, so there may be chances that some bits of the previous byte are lost and new character gets transmitted. In other words, TI flag bit is raised by CPU in order to ensure proper transmission of data bits.
5. When TI rises, it must be forced to zero by software program in order to transmit next character or byte.

   From the above points it is clear that when 8051 microcontroller finishes transmission, it will raise TI flag bit automatically, gives the information that transmission is completed in all respects.

5.6.2 Importance of RI (Receiver Interrupt) FLAG Bit

In receiving bits via its RXD line, the 8051 microcontroller goes through the following steps:

1. It receives the start bit, indicating that the next bit is the first bit of the character byte it is about to receive.
2. The 8-bit character is received one bit at a time. The LSB, i.e., D0 bit receives first. When the last bit (bit D7) is received, a byte is formed and placed in the SBUF register.
3. The stop bit is received, when receiving the stop bit the 8051 makes RI = 1, indicating that an entire character byte has been received and must be picked up before it gets overwritten by an incoming character.
4. By checking the RI flag bit when it is raised, we know that the character has been received and it is available in the SBUF register. We must copy the SBUF register contents to a safe place in some other register or in memory before it is lost.
5. After the SBUF contents are copied into a safe place, the RI flag bit must be forced to 0 by the software instruction CLR RI in order to allow the next received character byte to be placed in SBUF. Failure to do this causes loss of the received character.

   From the above, we conclude that by checking the RI flag bit, we know whether or not the 8051 has received a character byte. If we fail to copy SBUF into a safe place, we risk the loss of the received byte. More importantly, it must be noted that RI flag bit is raised by the 8051, but it must be cleared by the programmer with an instruction such as CLR RI. It must also be noted that if we copy SBUF into a safe place before the RI flag bit is raised, we risk copying the garbage.

5.7 SETTING THE SERIAL PORT BAUD RATE

Once the serial port mode has been configured, as explained above, the program must configure the serial port's baud rate. This only applies to serial port modes 1 and 3. The baud rate is determined based on the oscillator's frequency when in modes 0 and 2. In mode 0, the baud rate is always the oscillator frequency divided by 12. This means if the crystal is 11.0592 MHz, mode 0 baud rate will always be 921,583 bauds. In mode 2 the baud rate is always the oscillator frequency divided by 64, so a 11.059 MHz crystal speed will yield a baud rate of 172,797. In modes 1 and 3, the baud rate is determined by how frequently timer 1 overflows. The more frequently timer 1 overflows, the higher the baud rate. There are many ways one can cause timer 1 to overflow at a rate that determines a baud rate, but the most common method is to put timer 1 in 8-bit auto-reload mode (timer mode 2) and set a reload value (TH1) that causes timer 1 to overflow at a frequency appropriate to generate a baud rate. To determine the value that must be placed in TH1 to generate a given baud rate, we may use the following equation (assuming PCON.7 is clear). TH1 = 256 - ((Crystal / 384) / Baud). If PCON.7 is set then the baud rate is effectively doubled, thus the equation becomes:

$$\text{TH1} = 256 - ((\text{Crystal} / 192) / \text{Baud})$$

For example, if we have an 11.0592 MHz crystal and we want to configure the serial port to 19,200 bauds, we try plugging it in the first equation:

TH1 = 256 – ((Crystal / 384) / Baud)

TH1 = 256 – ((11059000 / 384) / 19200)

TH1 = 256 – ((28,799) / 19200)

TH1 = 256 – 1.5 = 254.5

As you can see, to obtain 19,200 bauds on a 11.0592 MHz crystal we'd have to set TH1 to 254.5. If we set it to 254 we will have achieved 14,400 bauds and if we set it to 255 we will have achieved 28,800 bauds. Thus, we're stuck... But not quite... to achieve 19,200 baud we simply need to set PCON.7 (SMOD). When we do this, we double the baud rate and utilize the second equation mentioned above. Thus, we have:

TH1 = 256 – ((Crystal / 192) / Baud)

TH1 = 256 – ((11059000 / 192) / 19200)

TH1 = 256 – ((57699) / 19200)

TH1 = 256 – 3 = 253

Here we are able to calculate a nice, even TH1 value. Therefore, to obtain 19,200 baud with an 11.059 MHz crystal we must:

1. Configure serial port mode 1 or 3.
2. Configure timer 1 to timer mode 2 (8-bit auto-reload).
3. Set TH1 to 253 to reflect the correct frequency for 19,200 bauds.
4. Set PCON.7 (SMOD) to double the baud rate.

5.8 WRITING TO THE SERIAL PORT

Once the serial port has been properly configured as explained above, the serial port is ready to be used to send data and receive data. If you thought that configuring the serial port was simple, using the serial port will be a breeze. To write a byte to the serial port, one must simply write the value to the SBUF (99h) SFR. For example, if you wanted to send the letter "A" to the serial port, it could be accomplished as easily as:

MOV SBUF, #'A'

Upon execution of the above instruction, the 8051 will begin transmitting the character via the serial port. Obviously, transmission is not instantaneous—it takes a measureable amount of time to transmit. And since the 8051 does not have a serial output buffer, we need to be sure that a character is completely transmitted before we try to transmit the next character. The 8051 lets us know when it is done transmitting a character by setting the TI bit in SCON. When this bit is set we know that the last character has been transmitted and that we may send the next character, if any. Consider the following code segment:

```
CLR TI              ; Be sure the bit is initially clear
MOV SBUF, #'A'      ; Send the letter 'A' to the serial port
JNB TI, $           ; Pause until the RI bit is set.
```

The above three instructions will successfully transmit a character and wait for the TI bit to be set before continuing. The last instruction says "Jump if the TI bit is not set to $"-- $, in most assemblers, means "the same address of the current instruction". Thus the 8051 will pause on the JNB instruction until the TI bit is set by the 8051 upon successful transmission of the character.

5.9 READING THE SERIAL PORT

Reading data received by the serial port is equally easy. To read a byte from the serial port, one just needs to read the value stored in the **SBUF** (99h) SFR after the 8051 has automatically set the RI flag in SCON. For example, if your program wants to wait for a character to be received and subsequently read it into the accumulator, the following code segment may be used:

```
JNB RI, $           ; Wait for the 8051 to set the RI flag
MOV A, SBUF         ; Read the character from the serial port
```

The first line of the above code segment waits for the 8051 to set the RI flag; again, the 8051 sets the RI flag automatically when it receives a character via the serial port. So as long as the bit is not set, the program repeats the "JNB" instruction continuously. Once the RI bit is set upon character reception, the above condition automatically fails and program flow falls through to the "MOV" instruction which reads the value.

Programming Tips:

For reading and writing to or from serial port, a common register is used called SBUF (serial buffer). The on-chip address of SBUF register is 99h. The SBUF register is not a bit addressable register. Putting a value to SBUF register, then after configuring serial port mode, then SBUF is packed with start and stop bits and then it starts transmission one bit at a time. Similarly, when the bits are received serially then microcontroller gets rid of start and stop bits and stores 8 bits of value in the SBUF register.

5.10 PCON (POWER CONTROL) REGISTER DOUBLING THE BAUD RATE IN THE 8051 MICROCONTROLLER

There are two ways to increase the baud rate of data transfer in the 8051 microcontroller:

1. Use a higher frequency crystal
2. Change a bit in the PCON register, shown below

D7	D6	D5	D4	D3	D2	D1	D0
SMOD	-	-	-	GF1	GF0	PD	IDL

Option 1 is not feasible in many situations, since the system crystal is fixed on most of the systems.

The PCON register is an 8-bit register, of these 8 bits, some are unused and some are used for power control capability of the 8051. The bit that is used for the serial communication is bit D7, the SMOD bit.

When the 8051 is powered up, D7 bit of the PCON register is zero, we can set this bit to HIGH by software and thereby double the baud rate. The following sequence of the instructions must be used to set bit D7 HIGH of the PCON register, since PCON is not a bit addressable register. There may be more methods to set SMOD = 1, without affecting the other bits of PCON register, so one may use any sequence of the program with ensuring that the other bits of the PCON register is unaffected and only SMOD gets high.

```
MOV A, PCON      ; Get original or old value of PCON register in the
                   accumulator
SETB ACC.7       ; Make D7 (SMOD) = 1
MOV PCON, A      ; Now SMOD = 1 without modifying other bits of
                   the PCON register alternatively,
MOV A, PCON      ; Get original or old value of PCON register in the
                   accumulator
OR A, #0FFh      ; Make D7 (SMOD) = 1
MOV PCON, A      ; Now SMOD = 1 without modifying other bits of
                 ; The PCON register
```

Baud rate can be calculated as follows:

$$TH1 = 256 - \frac{2^{SMOD} * \text{Oscillator frequency}}{384 * \text{Baud rate}}$$

Baud rate for SMOD = 0

$$TH1 = 256 - \frac{\text{Oscillator frequency}}{384 * \text{Baud rate}}$$

Baud rate for SMOD = 1,

$$TH1 = 256 - \frac{2 * \text{Oscillator frequency}}{384 * \text{Baud rate}}$$

$$TH1 = 256 - \frac{\text{Oscillator frequency}}{192 * \text{Baud rate}}$$

Baud rate comparisons for SMOD = 0 and SMOD =1 is shown in Table 5.4. It shows that for the same value of TH1, the baud rates are doubled when SMOD = 1.

Table 5.4 *Baud rate comparisons for SMOD = 0 & SMOD = 1*

TH1 (Decimal)	TH1 (Hex)	Baud Rate	
		SMOD = 0	SMOD = 1
– 3	FDh	9600	19200
– 6	FAh	4800	9600
– 12	F4h	2400	4800
– 24	E8h	1200	2400
– 48	D0h	600	1200

Bit D3 & D2: GF1 & GFO

General-purpose flag bit 1 & 0

These bits are used for general purposes; it means that its operation and use is decided by the programmer. When these bits are not used or the role of these bits is not defined for any purpose then these bits are generally set to 0.

Power Saving Modes of 8051 Microcontroller

8051 has two power saving modes. They are

1. Power down mode
2. Idle mode

The two power saving modes are entered by setting two bits IDL and PD in the special function register (PCON) respectively.

The structure of PCON register is as follows:

PCON SFR Address 87H

D7	D6	D5	D4	D3	D2	D1	D0
SMOD	-	-	-	GF1	GF0	PD	IDL

The schematic diagram for 'power down' mode and 'Idle' mode is shown in Figure 5.13.

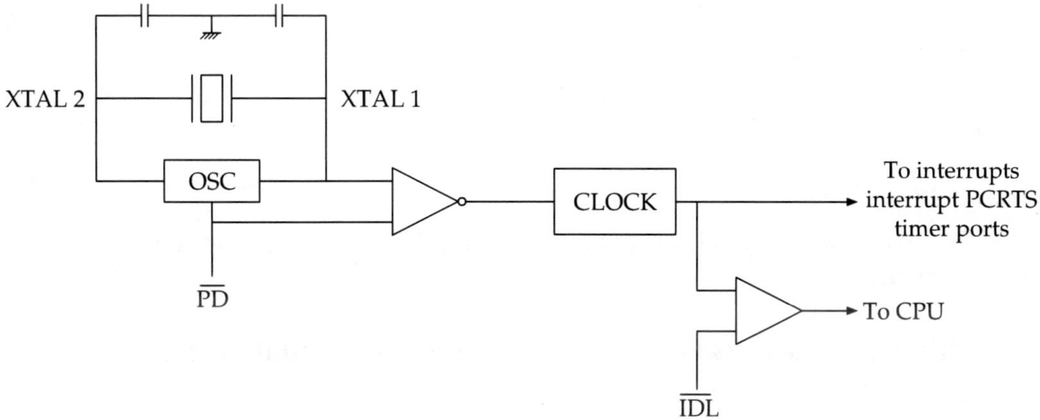

Fig. 5.13 *Schematic diagram for power down and idle mode implementation*

Bit D1: PD

Power down

When this bit is set then power down operation in 80C51BH microcontroller is activated. This facility is available only in CHMOS versions of the microcontrollers.

The power down mode is entered by setting the PD bit to 1. The internal clock to the entire microcontroller is stopped (frozen). However, the program is not dead. The power down mode is exited (PCON.1 is cleared to 0) by hardware reset only. The CPU starts from the next instruction where the power down mode was invoked. Port values are not changed/overwritten in power down mode. V_{cc} can be reduced to as low as 2V in power down mode. However, V_{cc} has to be restored to normal value before power down mode is exited.

Bit D0: IDL

Ideal mode

When this bit is set, then it activates the ideal mode operation in 80C51BH microcontroller. This facility is also available only in CHMOS versions of microcontrollers.

Idle mode is entered by setting IDL bit to 1 (i.e., $\overline{IDL} = 0$). The clock signal is gated off to CPU, but not to the interrupt, timer and serial port functions. The CPU status is preserved entirely. SP, PC, PSW, the accumulator and other registers maintain their data during IDLE mode. The port pins hold their logical states they had at the time IDLE was initiated. ALE and $\overline{PSEN}$ are held at logic high levels.

Ways to exit Idle Mode:

1. Activation of any enabled interrupt will clear PCON.0 bit and hence the idle mode is exited. The program goes to the interrupt service routine (ISR). After RETI is executed at the end of the ISR, the next instruction will start from the one following the instruction that enabled the idle mode.
2. A hardware reset exits the idle mode. The CPU starts from the instruction following the instruction that invoked the 'idle' mode.

Programming Tips:

If 1's (HIGH) are written to PD (power down) and IDL (idle mode) at the same time, PD (power down) takes precedence.

5.11 PROGRAMMING OF TRANSMISSION BYTE SERIALLY

The following steps are taken to transmit a character or byte serially through the TXD line of 8051 microcontroller.

1. The TMOD register must be initialized with 20h indicating timer 1 in auto-reload mode (mode 2), to set the baud rate. The baud rate is configured only using auto-reload mode, because only in auto-reload mode it is possible to get maximum overflow at very high rate.
2. Load TH1 register with a value corresponding to the specified baud rate.
3. Initialize SCON register with value 50h indicating serial mode 1, i.e., 8-bit data, 1 start bit and 1 stop bit. In most of the system use only serial mode 1, because this mode is IBM compatible.
4. Set TR1 = 1 to start the timer 1, to get the baud rate.
5. The character byte to be transmitted serially must be put in the SBUF register.
6. Keep monitoring TI (transmitter interrupt) flag bit using the instruction "CHECK: JNB TI, CHECK", in order to ensure that transmission is completed or not. When TI rises, indicate completion of the transmission of character them go out of the loop.
7. Clear TI flag bit using instruction CLR TI
8. In order to repeat the process, i.e., to transmit another character or byte go to step 5. No need to start the timer again, as once the baud rate is configured; it is available throughout the completion of the program. The baud rate is changed again by loading another value in TH1 register and start the timer or it may lapse by resetting the system.

5.12 PROGRAMMING OF RECEPTION OF BYTE SERIALLY

The following steps help to receive characters or bytes serially through the RXD line of 8051 microcontroller.

1. The TMOD register must be initialized with 20h indicating that timer 1 in auto-reload mode (mode 2), to set the baud rate. The baud rate is configured only using auto-reload mode, because only in auto-reload mode it is possible to get maximum overflow at very high rate.
2. Load TH1 register with a value corresponding to the specified baud rate.
3. Initialize SCON register with value 50h indicating serial mode 1, i.e., 8-bit data, 1 start bit and 1 stop bit. In most of the system use only serial mode 1, because this mode is IBM compatible.
4. Set TR1 = 1 to start the timer 1, to get the baud rate.
5. Keep monitoring RI (receiver interrupt) flag bit using the instruction "CHECK: JNB RI, CHECK", in order to ensure that reception is completed or not. When RI rises, indicate completion of the reception of character then go out of the loop.
6. When RI flag bit rises then there is a byte, i.e., received byte in SBUF register, so it must be picked up before a new incoming character is received. In other words, when RI rises then store the SBUF contents in safe place.
7. Clear RI flag bit using instruction CLR RI.
8. In order to repeat the process, i.e., to receive another character or byte go to step 5. No need to start the timer again, as once the baud rate is configured; it is available throughout the completion of the program. The baud rate is change again by loading another value in TH1 register and start the timer or it may lapse by resetting the system.

5.13 SOLVED EXAMPLES

Example 5.1

8051 microcontroller has an oscillator frequency of 11.0592 MHz. Using timer 1 and configuring the UART in mode 1 write an assembly language program that transmits the ASCII character "N" at a baud rate of 9600.

Solution:

$$\text{TH1} = 256 - \frac{\text{Crystal frequency}}{384 * \text{Baud rate}}$$

By putting baud rate 9600, we get

$$\text{TH1} = \text{FDh}$$

TMOD register

GATE	C/T	M1	M0	GATE	C/T	M1	M0	
0	0	1	0	0	0	0	0	= 20 H

SCON register

SM0	SM1	SM2	REN	TB8	RB8	TI	RI	
0	1	0	1	0	0	0	0	= 50 H

Assembly Language Program:

```
        ORG 0               ; Reset address
        SJMP START          ; Jump over reserved area
        ORG 40H             ; Program start address
START:  MOV SCON, #50H      ; Serial mode 1
        MOV TMOD, #20H      ; Timer 1 mode 2
        MOV TH1, #0FDH      ; Baud rate 9600
        SETB TR1            ; Turn timer 1 on
AGAIN:  MOV SBUF, #'N'      ; ASCII of N into SBUF
HERE:   JNB TI, HERE        ; Stay here till TI is set
        CLR TI              ; Clear TI
        SJMP AGAIN          ; Repeat
        END                 ; End of assembly language
```

Example 5.2

8051 microcontroller has an oscillator frequency of 11.0592 MHz. Using timer 1 and configuring the UART in mode 1, write an assembly language program that transmits the ASCII character "N" at a baud rate of 19200.

Solution:

$$TH1 = 256 - \frac{\text{Crystal frequency}}{384 * \text{Baud rate}}$$

By putting baud rate 19200, we get

$$TH1 = FEh$$

TMOD register

GATE	C/T	M1	M0	GATE	C/T	M1	M0	
0	0	1	0	0	0	0	0	= 20 H

SCON register

SM0	SM1	SM2	REN	TB8	RB8	TI	RI	
0	1	0	1	0	0	0	0	= 50 H

Assembly Language Program:

```
         ORG 0              ; Reset address
         SJMP START         ; Jump over reserved area
         ORG 40H            ; Program start address
START:   MOV SCON, #50H     ; Serial mode 1
         MOV TMOD, #20H     ; Timer 1 mode 2
         MOV TH1, #0FEH     ; Baud rate 19200
         SETB TR1           ; Turn timer 1 on
AGAIN:   MOV SBUF, #'N'     ; ASCII of N into SBUF
HERE:    JNB TI, HERE       ; Stay here till TI set
         CLR TI             ; Clear TI
         SJMP AGAIN         ; Repeat
         END                ; End of assembly language
```

Example 5.3

8051 microcontroller has an oscillator frequency of 11.0592 MHz. Write an assembly language program that receives a character into the serial buffer (SBUF) of the UART and writes the hex value of the character onto port 1. The UART should be configured as mode 1. Use 9600 baud rate.

Solution:

$$\text{TH1} = 256 - \frac{\text{Crystal frequency}}{384 * \text{Baud rate}}$$

By putting baud rate 9600, we get

$$\text{TH1} = \text{FDh}$$

TMOD register

GATE	C/T	M1	M0	GATE	C/T	M1	M0	
0	0	1	0	0	0	0	0	= 20 H

SCON register

SM0	SM1	SM2	REN	TB8	RB8	TI	RI	
0	1	0	1	0	0	0	0	= 50 H

The TI bit in the serial control (SCON) can be left at its default value of zero but the receive bit (REN) must be set.

Thus S0CON is 50H. When the character byte is received RI will set but this must be cleared by the program to enable other receive interrupts to occur. In the simulation, we should expect RB8 to set when the character byte has been received. TB8 and RB8 are mainly used in modes 2 and 3 but in mode 1 RB8 is set by the stop bit, which is the last bit of the 10-bit mode 1 data frame.

Assembly Language Program:

```
        ORG 0               ; Reset address
        SJMP START          ; Jump over reserved area
        ORG 40H             ; Program start address
START:  MOV SCON, #50H      ; Mode 1, REN enabled
        MOV TH1, #0FDH      ; 9600 baud
        MOV TMOD, #20H      ; Timer 1 mode 2
        SETB TR1            ; Turn timer 1 on
STAY:   JNB RI, STAY        ; Stay here, until RI = 1
        CLR RI              ; Clear RI
        MOV A, SBUF         ; Move character from buffer to A
        MOV P1, A           ; Hex value onto port 1
        RETI                ; Return from interrupt
        END                 ; No more assembly language
```

Example 5.4

8051 microcontroller has an oscillator frequency of 11.0592 MHz. Write an assembly language program to transfer the message "BHEL" serially at 4800 baud rate, 8-bit data, 1 start bit and 1 stop bit. Send the message continuously.

Solution:

$$\text{TH1} = 256 - \frac{\text{Crystal frequency}}{384 * \text{Baud rate}}$$

By putting baud rate 4800, we get

$$\text{TH1} = 250 \text{ decimal} = \text{FAh}$$

TMOD register

GATE	C/T	M1	M0	GATE	C/T	M1	M0	
0	0	1	0	0	0	0	0	= 20 H

SCON register

SM0	SM1	SM2	REN	TB8	RB8	TI	RI	
0	1	0	1	0	0	0	0	= 50 H

Assembly Language Program:

Method – I

```
         ORG 0
         SJMP START
         ORG 0040H
START:   MOV TMOD, #20H
         MOV TH1, #-6 / MOV TH1, # 0FAH
         MOV SCON, #50H
         SETB TR1
AGAIN:   MOV A, #'B'
         ACALL SEND
         MOV A, #'H'
         ACALL SEND
         MOV A, #'E'
         ACALL SEND
         MOV A, #'L'
         ACALL SEND
         SJMP AGAIN
SEND:    MOV SBUF, A
HERE:    JNB TI, HERE
         CLR TI
         RET
```

Method – II

```
         ORG 0
         SJMP START
         ORG 0040H
START:   MOV TMOD, #20H
         MOV TH1, #-6 / MOV TH1, # 0FAH
         MOV SCON, #50H
         SETB TR1
LAST:    MOV DPTR, #MSG
AGAIN:   CLR A
         MOVC A, @A+DPTR
         JZ LAST             ; If last character gets out and repeats
                               again
```

```
        ACALL SEND
        INC DPTR
        SJMP AGAIN
SEND:   MOV SBUF, A
HERE:   JNB TI, HERE
        CLR TI
        RET
MSG:    DB 'BHEL', 0
        END
```

Example 5.5

Assume that the 8051 microcontroller's serial port is connected to the COM port of the IBM PC, and on the PC we are using the hyper terminal program to send and receive data serially. Port P1 and P2 of the 8051 are connected to the LED's and switches respectively. Write an assembly language program to perform the following:

1. Send to the PC the message "microcontroller is ready"
2. Receive any data sent by the PC and put it on LED's connected to the port P1.
3. Get data on switches connected to port P2 and send it to the PC serially. The program should perform in part (1) once but part (2) and (3) continuously. Assume the baud rate 4800 with oscillator frequency 11.0592 MHz.

Solution:

$$TH1 = 256 - \frac{\text{Crystal frequency}}{384 * \text{Baud rate}}$$

By putting baud rate 4800, we get

$$TH1 = FAh$$

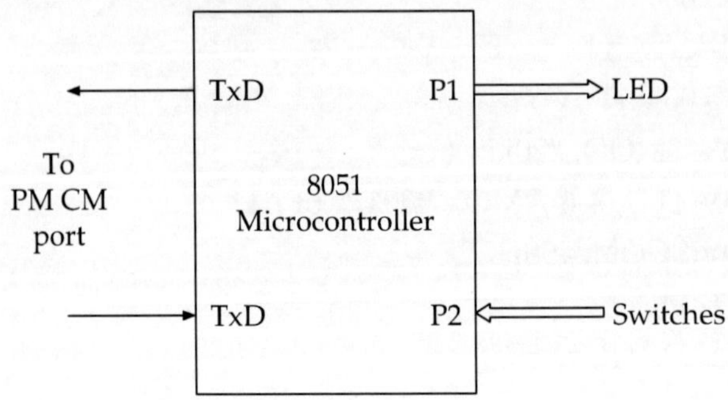

Fig. 5.14 *Example 3.7*

TMOD register

GATE	C/T	M1	M0	GATE	C/T	M1	M0	
0	0	1	0	0	0	0	0	= 20 H

SCON register

SM0	SM1	SM2	REN	TB8	RB8	TI	RI	
0	1	0	1	0	0	0	0	= 50 H

Assembly Language Program:

```
          ORG 0             ; Reset address
          SJMP START        ; Jump over reserved area
          ORG 40H           ; Program start address
START:    MOV P2, #0FFH     ; Port P2 configured as an input port
          MOV SCON, #50H    ; Serial mode 1
          MOV TMOD, #20H    ; Timer 1 mode 2
          MOV TH1, #0FAH    ; Baud rate 19200
          SETB TR1          ; Turn timer 1 on
          MOV DPTR, #MSG    ; Load pointer for message
TR_MSG:   CLR A
          MOVC A, @A+DPTR   ; Get the character
          JZ READ_SW        ; Check for last character, get out if last
          ACALL SEND        ; Otherwise transfer message serially
          INC DPTR          ; Next character
          SJMP TR_MSG       ; Stay in loop until last character is sent
READ_SW:  MOV A, P2         ; Read switches value from port P2
          ACALL SEND        ; Transfer it serially
          ACALL RECEV       ; Get the serial data
          MOV P1, A         ; Display it on LED's
          SJMP READ_SW      ; Repeat the process
SEND:     MOV SBUF, A       ; Load the data into SBUF
HERE:     JNB TI, HERE      ; Stay here till TI set
          CLR TI            ; Clear TI
          RET
RECEV:    JNB RI, RECEV
          MOV A, SBUF
          CLR RI
          RET
```

```
MSG:    DB 'Microcontroller is ready', 0
        END                           ; End of assembly language
```

Example 5.6

Assume a switch is connected to pin P0.3, write an assembly language program to monitor its status and send two messages to the serial port continuously as follows:

SW = 0 send "NO'"

SW = 1 send "YES"

Assume that crystal frequency of the 8051 is 11.0592 MHz, use 19200 baud rate and 8-bit UART. Use SMOD =1

Solution:

$$TH1 = \frac{2^{SMDO} * \text{Crystal frequency}}{384 * \text{Baud rate}}$$

$$TH1 = \frac{2 * \text{Crystal frequency}}{384 * \text{Baud rate}}$$

$$TH1 = 256 - \frac{\text{Crystal frequency}}{384 * \text{Baud rate}}$$

By putting baud rate 19200, we get

$$TH1 = 250 \text{ decimal} = FDh$$

TMOD register

GATE	C/T	M1	M0	GATE	C/T	M1	M0	
0	0	1	0	0	0	0	0	= 20 H

SCON register

SM0	SM1	SM2	REN	TB8	RB8	TI	RI	
0	1	0	1	0	0	0	0	= 50 H

Assembly Language Program:

```
        ORG 0               ; Reset address
        SJMP START          ; Jump over reserved area
        ORG 40H             ; Program start address
START:  MOV A, PCON
        SETB ACC.7
        MOV PCON, A
        MOV SCON, #50H      ; Serial mode 1
```

```
         MOV TMOD, #20H      ; Timer 1 mode 2
         MOV TH1, #0FDH      ; Baud rate 19200
         SETB TR1            ; Turn timer 1 on
         SETB P0.3           ; Make port pin P0.3 as an input port
CHECK:   JB P0.3, NEXT
         MOV DPTR, #MSG1     ; If SW = 0 then send message NO
REP1:    CLR A
         MOVC A, @A+DPTR     ; Read the message
         JZ CHECK            ; Check for last character
         ACALL SEND          ; Send message serially
         INC DPTR            ; Go for next byte
         SJMP REP1
NEXT:    MOV DPTR, #MSG2     ; If SW = 1 then send message YES
REP2:    CLR A
         MOVC A, @A+DPTR     ; Read the message
         JZ CHECK            ; Check for last character
         ACALL SEND          ; Send message serially
         INC DPTR            ; Go for next byte
         SJMP REP2
SEND:    MOV SBUF, A         ; Copy A into SBUF for serial transmission
HERE:    JNB TI, HERE        ; Stay here till TI set
         CLR TI              ; Clear TI
         RET
MSG1:    DB                  'NO', 0
MSG2:    DB                  'YES', 0
         END                 ; End of assembly language
```

Example 5.7

8051 microcontroller has an oscillator frequency of 11.0592 MHz. Using timer 1 and configuring the UART in mode 1, write an assembly language program that repeatedly sends the line of text 'BIT RAIPUR' at a baud rate of 9600.

Solution:

$$TH1 = 256 - \frac{\text{Crystal frequency}}{384 * \text{Baud rate}}$$

By putting baud rate 9600, we get

$$TH1 = FDh$$

TMOD register

GATE	C/T	M1	M0	GATE	C/T	M1	M0	
0	0	1	0	0	0	0	0	= 20 H

SCON register

SM0	SM1	SM2	REN	TB8	RB8	TI	RI	
0	1	0	1	0	0	0	0	= 50 H

Assembly Language Program:

```
         ORG 0                  ; Reset address
         SJMP START             ; Jump over reserved area
         ORG 40H                ; Program start address
START:   MOV SCON, #50H         ; Serial mode 1
         MOV TMOD, #20H         ; Timer 1 mode 2
         MOV TH1, #0FDH         ; Baud rate 9600
         SETB TR1               ; Turn timer 1 on
TEXT:    MOV DPTR, #MSG1        ; Data pointer to message address
NEXTCH:  MOV A, #00h / CLR A    ; Zero the previous character
         MOVC A,@A + DPTR       ; Character into A
         CJNE A, #7EH, TRXCH    ; Checking end of message, _ (ASCII
                                ; Value of _ ; is 7EH)
         MOV A, #0DH            ; Carriage return 0DH
         ACALL SEND             ; Call up send routine
         MOV A, #0AH            ; Line feed ¼0AH
         ACALL SEND             ; Call up send routine
         SJMP TEXT              ; Repeat line of text
TRXCH:   ACALL SEND             ; Send text character
         INC DPTR               ; Increment data pointer
         SJMP NEXTCH            ; Prepare to send next character
SEND:    JNB TI, SEND           ; Check SBUF clear to send
         CLR TI                 ; Clear TI
         MOV SBUF, A            ; Send contents of A
         RET                    ; Return from subroutine
MSG1:    DB    'BIT RAIPUR_'    ; Text message
         END                    ; No more assembly language
```

REVIEW QUESTIONS

1. Explain SCON and PCON registers in detail.
2. Explain baud rate. Also explain how to double the baud rate.
3. Explain multiprocessor communication environment of 8051 microcontroller.
4. Explain serial mode 0 and mode 1 in details.
5. Explain power mode operation of 8051 microcontroller.
6. Explain synchronous and asynchronous serial communication.
7. Explain the differences between synchronous and asynchronous data transfer.
8. Explain asynchronous communication, also explain with the schematic to transfer ASCII code "ABC" serially with one start bit and one stop bit.
9. Explain baud rate calculation in mode 0 and mode1 of serial communication.
10. Explain why only timer 1 in auto-reload mode is used for the calculation of baud rate in serial communication.
11. Write an assembly language program to receive bytes of data serially and put them on port P1, set the baud rate at 4800, use serial mode 1.
12. Write an assembly language program to transfer the message "CSVTU" serially at the baud rate 4800, use SMOD =1. Assume that the crystal of 8051 is 11.0592 MHz.
13. Write an assembly language program to transfer the message "WE ARE READY" serially at the baud rate 19200, use SMOD = 0. Assume that the crystal frequency of 8051 is 12 MHz.
14. Write an assembly language program to transfer the message "NB BAHADURE" serially at 38400 baud rate with 8-bit data, 1 start bit and 1 stop bit.
15. Interface two microcontrollers, one acting as a transmitter and other as a receiver. Write an assembly language program to transmit message "YES" from transmitter microcontroller to the receiver microcontroller serially.

6 Chapter

Interrupts of 8051 Microcontroller

Interrupt is a routine program to break or stop the current execution and branch to another program. Interrupt of microcontroller 8051 consists of two hardware interrupts as well as generated interrupts using timers and serial communication. After going through this chapter reader will get the information about how to configure the interrupts, how the interrupts are recognized and how they can be stopped at the end of the routine.

6.1 8051 MICROCONTROLLER INTERRUPTS

Interrupts are signals that cause the microcontroller to suspend what it is doing and transfer to a special program called an interrupt handler. The interrupt handler is responsible for determining the cause of interrupt, servicing the interrupt and returning the control to the point from where the interrupt was caused.

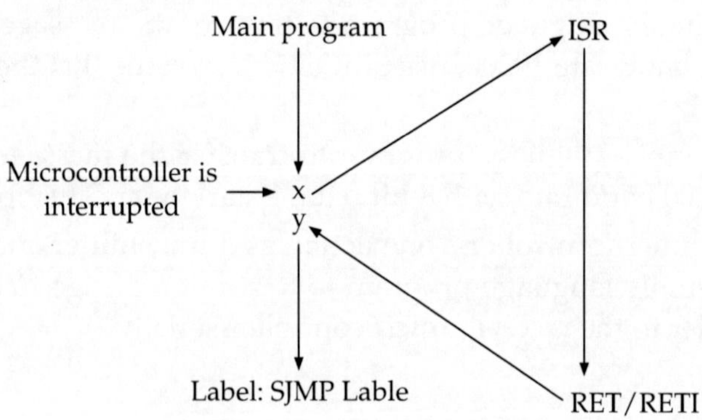

Fig. 6.1 *Interrupt structure*

When a microcontroller executes any program it is called the main program. In between the main program, microcontroller is interrupted then microcontroller will branch from main program to subprogram; this subprogram is called interrupt service routine (ISR). After executing this ISR then RETI/RET instruction comes then microcontroller returns from ISR to the main program from where the microcontroller has left.

As stated earlier in the previous sections, program flow is always sequential, being altered only by those instructions which expressly cause program flow to deviate in some way. However, interrupts give us a mechanism to "put on hold" the normal program flow, execute a subroutine, and then resume normal program flow as if we had never left it. This subroutine, called an interrupt handler, is only executed when a certain event (interrupt) occurs. The event may be one of the timers "overflowing," receiving a character via the serial port, transmitting a character via the serial port, or one of the two "external events." The 8051 may be configured so that when any of these events occur, the main program is temporarily suspended and control passed to a special section of code which presumably would execute some function related to the event that occurred. Once complete, control would be returned to the original program. The main program never even knows it was interrupted.

The ability to interrupt normal program execution when certain events occur makes it much easier and much more efficient to handle certain conditions. If it were not for interrupts we would have to manually check our main program whether the timers had overflown, whether we had received another character via the serial port, or if some external event had occurred. Besides making the main program ugly and hard to read, such a situation would make our program inefficient since we'd be burning precious "instruction cycles" checking for events that usually don't happen.

For example, let's say we have a large 16k program executing many subroutines performing many tasks. Let's also suppose that we want our program to automatically toggle the P3.0 port every time timer 0 overflows. The code to do this isn't too difficult:

```
                JNB TF0, SKIP_TOGGLE
                CPL P3.0
                CLR TF0
SKIP_TOGGLE:    ...
```

Since the TF0 flag is set whenever timer 0 overflows, the above code will toggle P3.0 every time timer 0 overflows. This accomplishes what we want, but is inefficient. The JNB instruction consumes 2 instruction cycles to determine that the flag is not set and jump over the unnecessary code.

Luckily, this isn't necessary. Interrupts let us forget about checking for the condition. The microcontroller itself will check for the condition automatically and when the condition is met will jump to a subroutine (called an interrupt handler), execute the code, then return. In this case, our subroutine would be nothing more than:

```
CPL P3.0
RETI
```

First, you'll notice the CLR TF0 command has disappeared. That's because when the 8051 executes our "timer 0 interrupt routine," it automatically clears the TF0 flag. You'll also notice that instead of a normal RET instruction we have a RETI instruction. The RETI instruction does the same thing as a RET instruction, but tells the 8051 that an interrupt routine has finished. You must always end your interrupt handlers with RETI.

6.1.1 What events can trigger interrupts, and where do they go? Interrupts of 8051 Microcontroller

We can configure the 8051 so that any of the following events will cause an interrupt:

- Timer 0 overflow.
- Timer 1 overflow.
- Reception/transmission of serial character.
- External event 0.
- External event 1.

In other words, we can configure the 8051 so that when timer 0 overflows or when a character is sent/received, the appropriate interrupt handler routines are called. Obviously we need to be able to distinguish between various interrupts and executing different codes depending on what interrupt was triggered. This is accomplished by jumping to a fixed address (vector location) when a given interrupt occurs. Apart from these interrupt sources, many manufactures include RESET as also one of the interrupt sources. Actually, RESET is considered an exception; exceptions are also similar to the interrupts but they are automatically recognized and execute the program associated with the exception when they occur. For example, if we try to divide the number by 0 then, it may be possible that the divide by 0 errors is detected and the microcontroller will switch to another program automatically to display the error message on the display device.

The six interrupts in the 8051 are allocated as follows:

1. **RESET:** When the reset pin is activated, the microcontroller 8051 jumps to the address location 0000h. This is also referred as power on reset.
2. **Timers:** Two interrupts are set aside for the timers: timer 0 and timer 1. Memory location addresses 000Bh and 001Bh in the interrupt vector table belongs to timer 0 and timer 1, respectively.
3. **Hardware external interrupts:** Two interrupts are set aside for hardware external interrupts. Pin number 12 (P3.2) and 13 (P3.3) are for the external hardware interrupts, also referred as EX1 and EX2. Memory location addresses 0003h and 0013h in the interrupt vector table are assigned to INT0 and INT1 external hardware interrupts respectively.
4. **Serial communication:** Serial communication has a single interrupt that belongs to both transmit and receive. The interrupt vector table location 0023h belongs to the serial communication interrupt.

The interrupt sources of 8051 with their routine addresses are shown in Table 6.1.

Table 6.1 *Interrupts with their handler addresses*

Interrupt	Flag	Interrupt handler address / Vector location
External 0	IE0	0003h
Timer 0	TF0	000Bh
External 1	IE1	0013h
Timer 1	TF1	001Bh
Serial	RI/TI	0023h

By consulting Table 6.1 we see that whenever timer 0 overflows (i.e., the TF0 bit is set), the main program will be temporarily suspended and control will jump to 000BH. It is assumed that we have code at address 000BH that handles the situation of timer 0 overflowing.

6.2 INTERRUPTS VERSUS POLLING

A single microcontroller can serve several devices. There are two ways to do that: Interrupts and polling.

Interrupts

In the interrupt method, whenever any device needs its service, the device notifies the microcontroller by sending an interrupt signal; upon receiving an interrupt signal, the microcontroller interrupts whatever it is doing and serves the device. The program associated with the interrupt is called the interrupt service routine (ISR) or interrupt handler.

Polling

In polling, the microcontroller continuously monitors the status of a several devices. When the status condition is met, it performs the service, after that, it moves on to monitor the next device until each one is serviced. Although polling can monitor the status of several devices and serve each of them as certain conditions are met, it is not an efficient use of microcontroller.

The advantage of the interrupts is that the microcontroller can serve many devices, not all at the same time; each device can get the attention of the microcontroller, based on the priority assigned to it. The polling method cannot assign priority since it checks all the devices in a round robin fashion as shown in Figure 6.2. Also in the interrupt method, the microcontroller can ignore or mask or reset a device request for service. This is not possible in polling.

The most important reason that the interrupt method is preferable over the polling is that the polling method wastes the time of the microcontroller in order to search the devices one by one, so in order to avoid tying down the microcontroller, interrupts are used.

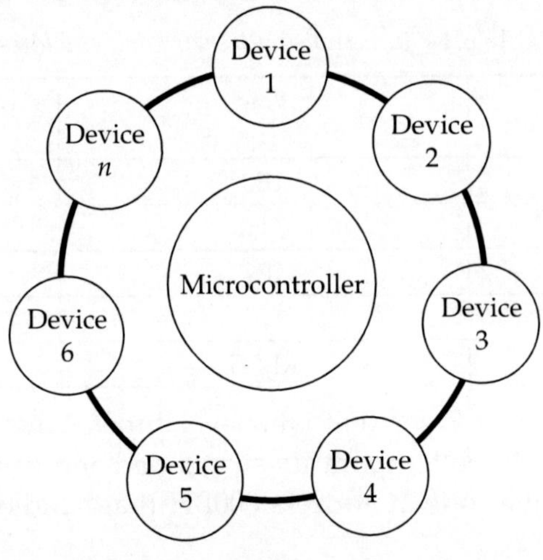

Fig. 6.2 *Polling method to access several devices*

For example, the instruction ***LABEL: JNB TFx, LABEL***, waited until the timer rollover and while we were waiting we could not do anything else, i.e., a waste of this microcontroller's time that could have been used to perform some other useful tasks. In this case, the microcontroller can go about doing other tasks, and when the timer flag bit is raised, the timer will interrupt the microcontroller in whatever it is doing.

6.3 THE INTERRUPT ENABLE (IE) REGISTER / SETTING UP INTERRUPTS

By default at power-up, all interrupts are disabled. This means that even if, for example, the TF0 bit is set, the 8051 will not execute the interrupt. User's program must specifically tell the 8051 that it wishes to enable interrupts and specifically which interrupts it wishes to enable. User's program may enable and disable interrupts by modifying the IE SFR (A8h register):

D7	D6	D5	D4	D3	D2	D1	D0
EA	--	--	ES	ET1	EX1	ET0	EX0

Table 6.2 *Explanation of function of IE register*

Bit	Name	Bit address	Function
7	EA	AFh	Global interrupt enable/disable
6	--	AEh	Undefined (don't care)
5	--	ADh	Undefined (don't care)
4	ES	ACh	Enable serial communication interrupt

3	ET1	ABh	Enable timer 1 interrupt
2	EX1	AAh	Enable external 1 hardware interrupt
1	ET0	A9h	Enable timer 0 interrupt
0	EX0	A8h	Enable external 0 hardware interrupt

As you can see, each of the 8051's interrupts has its own bit in the IE SFR. You enable a given interrupt by setting the corresponding bit. For example, if you wish to enable timer 1 interrupt, you would execute either:

MOV IE, #08h / SETB IE.3

Both of the above instructions set bit 3 of IE, thus enabling timer 1 interrupt. Once timer 1 interrupt is enabled, whenever the TF1 bit is set, the 8051 will automatically put "on hold" the main program and execute the timer 1 interrupt handler at address 001Bh. However, before timer 1 interrupt (or any other interrupt) is truly enabled, you must also set bit 7 of IE. Bit 7, the global interrupt enable/disable, enables or disables all interrupts simultaneously. That is to say, if bit 7 is cleared then no interrupts will occur, even if all the other bits of IE are set. Setting bit 7 will enable all the interrupts that have been selected by setting other bits in IE. This is useful in program execution if you have time-critical code that needs to execute. In this case, you may need the code to execute from start to finish without any interrupt getting in the way. To accomplish this you can simply clear bit 7 of IE (CLR EA) and then set it after your time critical code is done. So, to sum up what has been stated in this section, to enable the timer 1 interrupt the most common approach is to execute the following two instructions:

```
SETB IE.3
SETB EA
```

Thereafter, the timer 1 interrupt handler at 001Bh will automatically be called whenever the TF1 bit is set (upon timer 1 overflow).

8051 provides 5 vectored interrupts. They are:

1. $\overline{\text{NTO}}$
2. TF0
3. $\overline{\text{NT1}}$
4. TF1
5. RI/TI

Out of these, $\overline{\text{NTO}}$ and $\overline{\text{NT1}}$ are external interrupts whereas timer and serial port interrupts are generated internally. The external interrupts could be negative edge triggered or low level triggered. All these interrupts,when activated, set the corresponding interrupt flags. Except for serial interrupt, the interrupt flags are cleared when the processor branches to the interrupt service routine (ISR). The external interrupt flags are cleared on branching to interrupt service routine (ISR), provided the interrupt is negative edge triggered. For low level triggered external interrupt as well as for serial interrupt, the corresponding flags have to be cleared by software by the programmer.

The schematic representations of the interrupt are as follows:

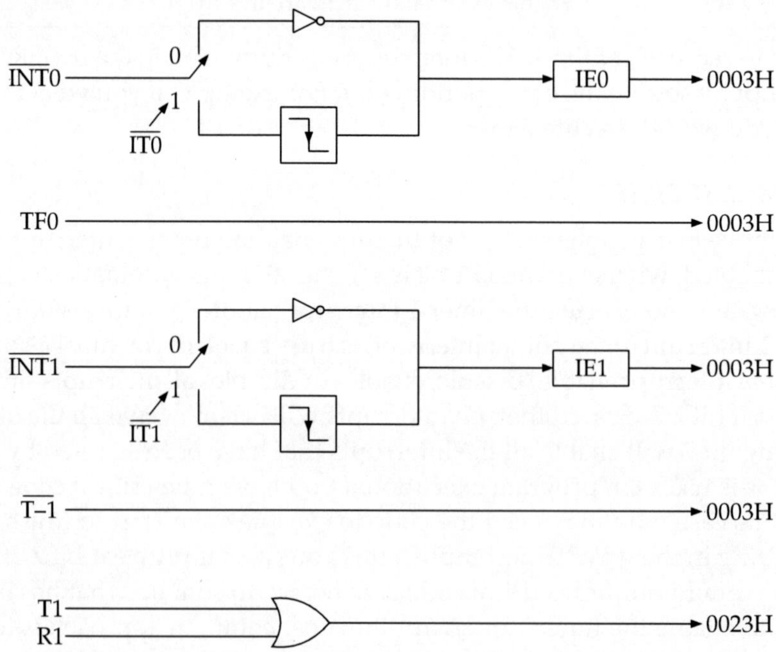

Fig. 6.3 *Interrupt structure of 8051 microcontroller*

6.4 THE INTERRUPT PRIORITY (IP) REGISTER

The 8051 offers two levels of interrupt priority: high and low. By using interrupt priorities, you may assign higher priority to certain interrupt conditions. For example, you may have enabled timer 1 interrupt which is automatically called every time timer 1 overflows. Additionally, you may have enabled the serial interrupt which is called every time a character is received via the serial port. However, you may consider that receiving a character is much more important than the timer interrupt. In this case, if timer 1 interrupt is already executing you may wish that the serial interrupt itself interrupts the timer 1 interrupt. When the serial interrupt is complete, control passes back to timer 1 interrupt and finally back to the main program. You may accomplish this by assigning a high priority to the serial interrupt and a low priority to the timer 1 interrupt. Interrupt priorities are controlled by the IP SFR (B8h). The IP SFR has the following format:

D7	D6	D5	D4	D3	D2	D1	D0
--	--	PT2	PS	PT1	PX1	PT0	PX0

Table 6.3 *Explanation of function of IP register*

Bit	Name	Bit address	Function
7	--	BFh	Undefined (don't care)
6	--	BEh	Undefined (don't care)
5	PT2	BDh	Timer 2 interrupt priority (8052 only)
4	PS	BCh	Serial communication interrupt priority
3	PT1	BBh	External 1 hardware interrupt priority
2	PX1	Bah	Timer 1 interrupt priority
1	PT0	B9h	External 0 hardware interrupt priority
0	PX0	B8h	Timer 0 interrupt priority

When considering interrupt priorities, the following rules apply:

- Nothing can interrupt a high-priority interrupt—not even another high-priority interrupt.
- A high-priority interrupt may interrupt a low-priority interrupt.
- A low-priority interrupt may only occur if no other interrupt is already executing.
- If two interrupts occur at the same time, the interrupt with higher-priority will execute first. If both interrupts are of the same priority, the interrupt which is serviced first by polling sequence will be executed first.

6.4.1 What happens when an interrupt occurs?

When an interrupt is triggered, the following actions are taken automatically by the microcontroller:

- The current program counter is saved on the stack, low-byte first.
- Interrupts of the same and lower priority are blocked.
- In the case of timer and external interrupts, the corresponding interrupt flag is set.
- Program execution transfers to the corresponding interrupt handler vector address.
- The interrupt handler routine executes.

Take special note of the third step: If the interrupt being handled is a timer or external interrupt, the microcontroller automatically clears the interrupt flag before passing control to the interrupt handler routine.

6.4.2 What happens when an interrupt ends?

An interrupt ends when your program executes the RETI instruction. When the RETI instruction is executed, the following actions are taken by the microcontroller:

- Two bytes are popped off the stack into the program counter to restore normal program execution.
- Interrupt status is restored to its pre-interrupt status.

6.5 PROGRAMMING TIMER INTERRUPT ROLLOVER TIMER FLAG AND INTERRUPT

We know that timer flag is raised when the timer rolls over, i.e., overflows and the overflow conditions are checked by monitoring TF bit using the instruction "JNB TFx, Label". In polling method, we have to wait until the timer flag bit TF is raised. The problem with this method is that the microcontroller is tied down while waiting for TF flag bit to be raised and cannot do anything else. Using interrupts this problem is solved and avoids tied down to the microcontroller. If the timer interrupt in IE (interrupt enable) register is enabled, whenever the timer rolls over, timer flag TF bit is raised automatically and the microcontroller is interrupted in whatever it is doing, and jumps to the interrupt vector table to service the interrupt service routine (ISR) of the timer. In this way, the microcontroller can do other things until it is notified that the timer has rolled over. Whenever the timer flag bit is raised and if the corresponding bit of the timer interrupt flag bit is set in the IE register, the microcontroller automatically switches or jumps to the vector location of timers.

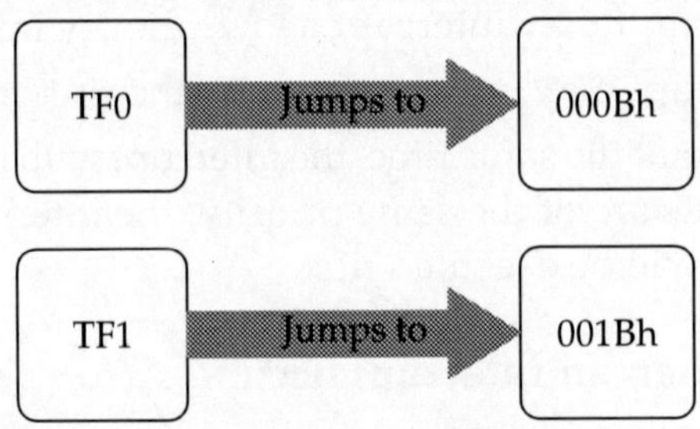

Fig. 6.4 *Timer 0 and Timer 1 interrupt invoked by TF0 and TF1*

6.6 SERIAL INTERRUPTS

The working of serial interrupts is slightly different than the rest of the interrupts. This is due to the fact that there are two interrupt flags: RI and TI and only one interrupt flag bit is associated for both. If either flag is set, a serial interrupt is triggered. As you will recall from the previous section on the serial port, the RI bit is set when a byte is received by the serial port and the TI bit is set when a byte has been sent. This means that when your serial interrupt is executed, it may have been triggered because the RI flag was set or because the TI flag was set or because both flags were set. Thus, your routine must check the status of these flags to determine what action is appropriate. Also, since the 8051 does not automatically clear the RI and TI flags you must clear these bits in your interrupt handler.

```
INT_SERIAL: JNB RI, CHECK_TI     ; If the RI flag is not set, we jump to
                                   check TI
            MOV A, SBUF          ; If we got to this line, it's because the
                                   RI bit was set
```

```
            CLR RI              ; Clear the RI bit after we've processed
                                  it
CHECK_TI:   JNB TI, EXIT_INT    ; If the TI flag is not set, we jump to the
                                  exit point
            CLR TI              ; Clear the TI bit before we send
                                  another character
            MOV SBUF, #'A'      ; Send another character to the serial
                                  port
EXIT_INT:   RETI
```

As you can see, our code checks the status of both the interrupt flags. If both flags were set, both sections of the code will be executed. Also note that each section of the code clears its corresponding interrupt flag. If you forget to clear the interrupt bits, the serial interrupt will be executed over and over until you clear the bit. Thus, it is very important that you always clear the interrupt flags in a serial interrupt.

6.7 PROGRAMMING THE SERIAL COMMUNICATION INTERRUPTS

Fig. 6.5 *Serial interrupts invoked by RI and TI*

We know that TI (transmit interrupt) is raised when the last bit of the framed data, the stop bit, is transferred, indicating that the SBUF register is ready to transfer the next byte. RI (received interrupt) is raised when the entire frame of data, including the stop bit, is received. In other words, when the SBUF register has a byte, RI is raised to indicate that the received byte needs to be picked up before it is lost by new incoming serial data. As far as serial communication is concerned, all the above concepts apply equally when using either polling or an interrupt. The only difference is in how the serial communication needs are served. In the polling method, we wait for the flag (TI or RI) to be raised; while we wait we cannot do anything else. In the interrupt method, we are notified when the 8051 has received a byte or is ready to send the next byte; we can do other things while the serial communication needs are served.

In the 8051 only one interrupt is set aside for serial communication. This interrupt is used to both send and receive data. If the interrupt bit in the IE register is enabled, when RI or TI is raised the 8051 gets interrupted and jumps to memory address location 0023h to execute ISR. In that ISR we must check or examine the RI and TI flags to see which one caused the interrupt and respond accordingly.

6.8 PROGRAMMING EXTERNAL HARDWARE INTERRUPTS

There are only two external hardware interrupts in the 8051 microcontroller: INT0 and INT1. They are located on pin P3.2 and P3.3 of port 3 respectively. The interrupt vector table locations 0003h and 0013h are set aside for INT0 and INT1 respectively. They are enabled and disabled using IE register. There are two types of activation for the external hardware interrupts, they are:

1. level triggered, and
2. edge triggered

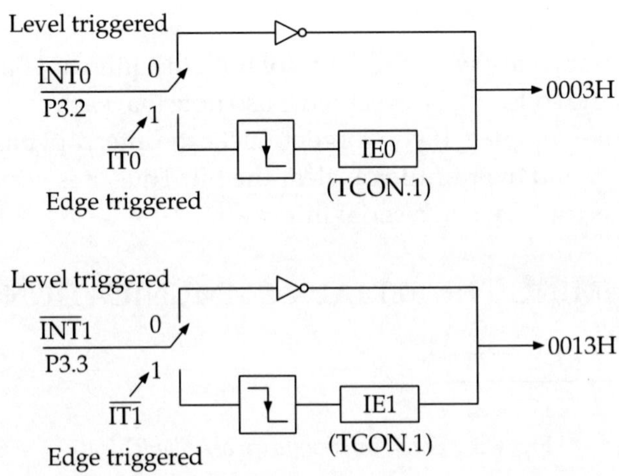

Fig. 6.6 *Level & Edge triggered for INT0 and INT1*

6.8.1 Level Triggered Interrupt

In the level triggered mode, INT0 and INT1 pins are normally high and if a low level signal (as they are low level active pins) is applied to them, it triggers the interrupt. Upon receiving the interrupt signal, microcontroller stops whatever it is doing and jumps to the interrupt vector table to service the interrupt. This is called level triggered or level activated interrupt and is the default mode of activation upon reset of the 8051. The low level signal at the 8051 pin must be removed before the execution of the last instruction of interrupt service routine, i.e., RETI, otherwise, another interrupt will be generated.

6.8.2 Edge Triggered Interrupt

Upon reset the 8051 makes INT0 and INT1 low level triggered interrupts. To make them edge triggered interrupts, we must program the bits of the TCON register. The TCON register consists of IT0 and IT1 flag bits that determine level or edge triggered mode of the hardware interrupts. IT0 and IT1 are bits D0 and D2 of the TCON register respectively. Since TCON is a bit addressable register, these bits are also referred as TCON.0 and TCON.2.

Upon reset TCON.0 and TCON.2 are both 0's i.e. external hardware interrupts of INT0 and INT1 pins are low level triggered. By making TCON.0 and TCON.2 bits high with instruction such as SETB TCON.0 and SETB TCON.2 the external hardware interrupts INT0 and INT1 become edge triggered interrupts.

For example, by applying high to low pulse on INT1 pin, then INT1 hardware interrupt becomes edge triggered hardware interrupts.

The TCON register and their bits related to interrupts are shown below:

TCON (Timer Control) register

D7	D6	D5	D4	D3	D2	D1	D0
TF1	TR1	TF0	TR0	IE1	IT1	IE0	IT0
Timer Bits				Interrupt Bits			

Bits	Name	Function
3	IE1	External hardware interrupts 1 edge flag. Set by CPU when the external interrupt edge (high to low transition) is detected. Cleared by CPU when the interrupt is processed.
2	IT1	Interrupt 1 type control bit. Set/cleared by software to specify falling edge (edge triggered)/level triggered external interrupts.
1	IE0	External hardware interrupts 0 edge flag. Set by CPU when the external interrupt edge (high to low transition) is detected. Cleared by CPU when the interrupt is processed.
0	IT0	Interrupt 0 type control bit. Set/cleared by software to specify falling edge (edge triggered)/level triggered external interrupts. (IT0 = 0, then low level triggered mode is set IT0 = 1, then edge triggered mode is set)

6.9 IMPORTANT INTERRUPT CONSIDERATION: REGISTER PROTECTION

One very important rule applies to all interrupt handlers: Interrupts must leave the processor in the same state as it was in when the interrupt initiated. Remember, the idea behind interrupts is that the main program isn't aware that they are executing in the "background." However, consider the following code:

```
CLR C               ; Clear carry
MOV A, #25h         ; Load the accumulator with 25h
ADDC A, #12h        ; Add 12h, with carry
```

After the above three instructions are executed, the accumulator will contain a value of 37h. But what would happen if right after the MOV instruction an interrupt occurred. During this interrupt, the carry bit was set and the value of the accumulator was changed to 40h. When the interrupt finished and control was passed back to the main program, the ADDC would add 12h to 40h, and add an additional 1h because the carry bit is set. In this case, the accumulator will contain the value 53h at the end of execution.

In this case, the main program has seemingly calculated the wrong answer. How can 25h + 12h yield 53h as a result? It doesn't make sense. A programmer that was unfamiliar with interrupts would be convinced that the microcontroller was damaged in some way, provoking problems with mathematical calculations. What has happened, in reality, is the interrupt did not protect the registers it used.

Restated: An interrupt must leave the processor in the same state as it was in when the interrupt initiated.

What does it mean? It means if your interrupt uses the accumulator, it must ensure that the value of the accumulator is the same at the end of the interrupt as it was at the beginning. This is generally accomplished with a PUSH and POP sequence. For example:

```
PUSH ACC
PUSH PSW
MOV A, #0FFh
ADD A, #02h
POP PSW
POP ACC
```

The *guts* of the interrupt are the MOV instruction and the ADD instruction. However, these two instructions modify the accumulator (the MOV instruction) and also modify the value of the carry bit (the ADD instruction will cause the carry bit to be set). Since an interrupt routine must guarantee that the registers remain unchanged by the routine, the routine pushes the original values onto the stack using the PUSH instruction. It is then free to use the registers it protected to its heart's content. Once the interrupt has finished its task, it pops the original values back into the registers. When the interrupt exits, the main program will never know the difference because the registers are exactly the same as they were before the interrupt executed.

In general, your interrupt routine must protect the following registers:

- PSW
- DPTR (DPH/DPL)
- PSW
- ACC
- B
- Registers R0-R7

Remember that PSW consists of many individual bits that are set by various 8051 instructions. Unless you are absolutely sure of what you are doing and have a complete understanding of what instructions set what bits, it is generally a good idea to *always* protect PSW by pushing and popping it off the stack at the beginning and end of your interrupts. Note also that most assemblers will not allow you to execute the instruction:

PUSH R0

This is due to the fact that depending on which register bank is selected, R0 may refer to either internal ram address 00h, 08h, 10h, or 18h. R0, in and of itself, is not a valid memory address that the PUSH and POP instructions can use. Thus, if you are using any "R" register in your interrupt routine, you will have to push that register's absolute address onto the stack instead of just saying PUSH R0. For example, instead of PUSH R0 you would execute:

PUSH 00h

Of course, this only works if you've selected the default register set. If you are using an alternate register set, you must PUSH the address which corresponds to the register you are using.

6.10 COMMON PROBLEMS WITH INTERRUPTS

Interrupts are a very powerful tools available to the 8051 developer, but when used incorrectly they can be a source of a huge number of debugging hours. Errors in interrupt routines are often very difficult to diagnose and correct. If you are using interrupts and your program is crashing or does not seem to be performing as you would expect, always review the following interrupt-related issues:

- **Register protection**: Make sure you are protecting all your registers, as explained above. If you forget to protect a register that your main program is using, strange results may occur. In our example above, we saw how failure to protect registers caused the main program to apparently calculate that 25h + 12h = 53h. If you witness problems with registers changing values unexpectedly or operations producing "incorrect" values, it is very likely that you've forgotten to protect registers. *It is suggested that always protect the registers used in the main program.*
- **Forgetting to restore protected values**: Another common error is to push registers onto the stack to protect them, and then forget to pop them off the stack before exiting

the interrupt. For example, you may push ACC, B, and PSW onto the stack in order to protect them and subsequently pop only ACC and PSW off the stack before exiting. In this case, since you forgot to restore the value of "B", an extra value remains on the stack. When you execute the RETI instruction, the 8051 will use that value as the return address instead of the correct value. In this case, your program will almost certainly crash.

- **Using RET instead of RETI:** Remember that interrupts are always terminated with the RETI instruction. It is easy to inadvertently use the RET instruction instead. However, the RET instruction will not end your interrupt. Usually, using a RET instead of a RETI will cause the illusion of your main program running normally, but your interrupt will only be executed once. If it appears that your interrupt mysteriously stops executing, verify that you are exiting with RETI.

6.11 SOLVED EXAMPLES

Example 6.1

Show the instruction to

(*a*) Enable serial interrupt, timer 0 interrupt and hardware external interrupt 0.

(*b*) Disable timer 0 interrupt

(*c*) Show how to disable all the interrupts with a single instruction.

Solution: Interrupt enable (IE) register:

EA	--	ET2	ES	ET1	EX1	ET0	EX0	
1	0	0	1	0	0	1	1	= 93H

(*a*) MOV IE, #93H

(*b*) Since IE register is a bit addressable register, we can use the following instructions to access individual bits of register

```
SETB IE.7        ; EA = 1, global enable
SETB IE.4        ; enable serial interrupt
SETB IE.1        ; enable timer 0 interrupt
SETB IE.0        ; enable external hardware interrupt 0
```

(*c*) CLR IE.1 ; disable timer 0 interrupt only

(*d*) CLR IE.7 ; disable all interrupt

Example 6.2

Program the interrupt priority (IP) register to perform the following:

(*a*) Assign the highest priority to timer 0 and show the new priority order.

(*b*) Discuss what happens if external hardware interrupts 0 and timer 0 interrupts are set at the same time.

Solution: Interrupt priority (IP) register:

--	--	PT2	PS	PT1	PX1	PT0	PX0	
0	0	0	0	0	0	1	0	= 02 H

(*a*) MOV IP, #02H

Or

SETB IP.2

Old priority order	New priority order
External hardware interrupt 0 (INT0)	Timer 0 interrupt (TF0)
Timer 0 interrupt (TF0)	External hardware interrupt 0 (INT0)
External hardware interrupt 1 (INT1)	External hardware interrupt 1 (INT1)
Timer 1 interrupt (TF1)	Timer 1 interrupt (TF1)
Serial communication interrupt (RI/TI)	Serial communication interrupt (RI/TI)

(*b*) Timer 0 is assigned the highest priority so the order of the interrupts are

New priority order
Timer 0 interrupt (TF0)
External hardware interrupt 0 (INT0)
External hardware interrupt 1 (INT1)
Timer 1 interrupt (TF1)
Serial communication interrupt (RI/TI)

According to the new priority order of the interrupts if the signal is applied on timer 0 and INT0 at the same time then timer 0 is executed first followed by external hardware interrupt 0 (INT0).

Example 6.3

Assume that after reset, the IP register is programmed with the instruction MOV IP, #0Bh. Discuss the sequence in which the interrupts are serviced.

Solution: Interrupt priority (IP) register:

--	--	PT2	PS	PT1	PX1	PT0	PX0	
0	0	0	0	1	0	1	1	= 0B H

This instruction sets external hardware interrupt 0 (INT0), timer 0 (TF0) and timer 1 (TF1) to the higher priority level compared with rest of the interrupts. So according to program bits, old and new priority level of the interrupts are shown below:

Old priority order	New priority order
External hardware interrupt 0 (INT0)	External hardware interrupt 0 (INT0)
Timer 0 interrupt (TF0)	Timer 0 interrupt (TF0)

External hardware interrupt 1 (INT1)	Timer 1 interrupt (TF1)
Timer 1 interrupt (TF1)	External hardware interrupt 1 (INT1)
Serial communication interrupt (RI/TI)	Serial communication interrupt (RI/TI)

Example 6.4

Write an assembly language program that continuously gets 8-bit data from port P0 and send it to port P1 while simultaneously creating a square wave of 250 μs period on port pin P2.1. Use timer 0 in mode 2 to create the square wave, assume that crystal frequency of 8051 is 11.0592 MHz.

Solution: Calculations of count value in TH0 register

Method - I

$$TON = TOFF = 125\ \mu s$$

$$THO = \frac{125\ \mu s}{1.085\ \mu s} = 115$$

$$-115 \text{ decimal} = 8Dh \text{ hexadecimal (consider only 8-bit value)}$$

Therefore, $TH0 = -115d = 8Dh$

Method – II

$$\text{Time delay} = (FF - YY + 1) * 1.085\ \mu s$$

$$125\ \mu s = (FF - YY + 1) * 1.085\ \mu s$$

$$FF - YY + 1 = 115 \text{ decimal}$$

$$FF - YY + 1 = 73 \text{ hex}$$

$$YY = FF + 1 - 73$$

TMOD Register

GATE	C/T	M1	M0	GATE	C/T	M1	M0	
0	0	0	0	0	0	1	0	= 02H

Interrupt enable (IE) register:

EA	--	ET2	ES	ET1	EX1	ET0	EX0	
1	0	0	0	0	0	1	0	= 82H

Assembly Language Program:

```
ORG 0000H
SJMP MAIN        ; bypass vector location
ORG 0030H
```

```
MAIN:    MOV P0, #0FFH     ; Make port P0 as an input port
         MOV TMOD, #02H    ; Timer 0 in Mode 2
         MOV TH0, #8DH     ; Load TH0 with auto-reload value
         MOV IE, #82H      ; Enable timer 0
         SETB TR0          ; Start timer 0
BACK:    MOV A, P0         ; Get data from port P0 until TF0 raised
         MOV P1, A         ; Send data from P0 to port P1
         SJMP BACK         ; Repeat the process
         SJMP $            ; Stop
         ORG 000BH
         CPL P2.1          ; Interrupt routine of timer 0
         RETI              ; Return from interrupt
```

Example 6.5

Repeat example 5.4 using timer 0 in mode 1.

Solution: Calculations of count value in TH0 – TL0 register

Method – I

$$\text{TON} = \text{TOFF} = 125\ \mu s$$

$$\text{THO} = \frac{125\ \mu s}{1.085\ \mu s} = 115$$

$$-115 \text{ decimal} = \text{FF8Dh hexadecimal (consider 16-bit value)}$$

Method – II

$$\text{Time delay} = (\text{FFFF} - \text{YYXX} + 1) * 1.085\ \mu s$$

$$125\ \mu s = (\text{FFFF} - \text{YYXX} + 1) * 1.085\ \mu s$$

$$\text{FFFF} - \text{YYXX} + 1 = 115 \text{ Dec}$$

$$\text{FFFF} - \text{YYXX} + 1 = 73 \text{ Hex}$$

$$\text{YYXX} = \text{FFFF} + 1 - 73$$

$$\text{YYXX} = \text{FF8D h}$$

Therefore, $\text{TH0} = \text{FFh}$ and $\text{TL0} = \text{8Dh}$

Assembly Language Program:

```
         ORG 0000H
         SJMP MAIN         ; Bypass vector location
         ORG 0030H
```

```
MAIN:   MOV P0, #0FFH      ; Make port P0 as an input port
        MOV TMOD, #01H     ; Timer 0 in mode 1
        MOV TH0, #0FFH     ; Load timer 0 higher byte
        MOV TL0, #8DH      ; Load timer 0 lower byte
        MOV IE, #82H       ; Enable timer 0
        SETB TR0           ; Start timer 0
BACK:   MOV A, P0          ; Get data from port P0 until TF0 raised
        MOV P1, A          ; Send data from P0 to port P1
        SJMP BACK          ; Repeat the process
        SJMP $             ; Stop
        ORG 000BH
        CPL P2.1           ; Interrupt routine of timer 0
        MOV TH0, #0FFH
        MOV TL0, #0A4H
        CLR TR0
        CLR TF0
        RETI               ; Return from interrupt
```

Example 6.6

Write an assembly language program to generate square wave of 50 Hz frequency on port pin P1.2 using interrupt for timer 0. Assume that crystal frequency of 8051 is 11.0592 MHz.

Solution: Calculation of count value

$$T = \frac{1}{f} = \frac{1}{50} = 0.02$$

$$\text{Ton} = \text{Toff} = 0.01 \text{ sec}$$

$$\text{THO} = \frac{125\ \mu s}{1.085\ \mu s} = 9216 > 255$$

Thus, we can use timer 0 in mode 1 (count value exceeds 8-bit number, so mode 2, i.e., auto-reload mode is not possible).

–9216 dec = DC00 hex

TMOD Register

GATE	C/T	M1	M0	GATE	C/T	M1	M0	
0	0	0	0	0	0	0	1	= 01H

Interrupt enable (IE) register:

EA	--	ET2	ES	ET1	EX1	ET0	EX0	
1	0	0	0	0	0	1	0	= 82H

Assembly Language Program:

```
        ORG 0000H
        SJMP MAIN           ; Bypass vector location
        ORG 0030H
MAIN:   MOV TMOD, #01H      ; Timer 0 in mode 1
        MOV TH0, #0DCH      ; Load timer 0 higher byte
        MOV TL0, #00H       ; Load timer 0 lower byte
        MOV IE, #82H        ; Enable timer 0
        SETB TR0            ; Start timer 0
BACK:   SJMP BACK           ; Stop/stays in the loop until timer flag
                            ; Bit is high
        ORG 000BH
        CPL P2.1            ; Interrupt routine of timer 0
        MOV TH0, #0DCH      ; Load timer 0 higher byte
        MOV TL0, #00H       ; Load timer 0 lower byte
        CLR TR0             ; Stop timer 0
        CLR TF0             ; Clear timer 0 flag bit
        RETI                ; Return from interrupt
```

Example 6.7

Write an assembly language program to get data from port P1 and send it to port P2 continuously, while giving a copy to the serial port to be transferred serially. Assume that crystal frequency of 8051 is 11.0592 MHz and baud rate for the serial communication is 9600.

Solution:

TMOD Register (Timer 1 in Mode 2 for baud rate)

GATE	C/T	M1	M0	GATE	C/T	M1	M0	
0	0	1	0	0	0	0	0	= 20H

SCON Register: (for Serial Mode 1)

SM0	SM1	SM2	REN	TB8	RB8	TI	RI	
0	1	0	1	0	0	0	0	= 50H

Interrupt Enable (IE) Register:

EA	--	ET2	ES	ET1	EX1	ET0	EX0	
1	0	0	1	0	0	0	0	= 90H

Assembly Language Program:

```
            ORG 0000H
            SJMP MAIN
            ORG 0030H
MAIN:       MOV P1, #0FFH        ; Make port P1 as an input port
            MOV TMOD, #20H       ; Timer 1 in mode 2 for baud rate
            MOV TH1, #0FDH       ; To set baud rate 9600
            MOV SCON, #50H       ; Serial mode 1
            MOV IE, #90H         ; Enable serial communication interrupt
            SETB TR1             ; Start timer 1 for baud rate
L1:         MOV A, P1            ; Get data from port P1
            MOV SBUF, A          ; Send data serially
            MOV P2, A            ; Send data to the port P2
            SJMP L1              ; Repeat process
            ORG 0023H
            JB TI, TRANS         ; Check for TI flag bit
            CLR RI
            RETI
TRANS:      CLR TI
            RETI
            END
```

Example 6.8

Write an assembly language program to get data from port P1 and send it to port P2 continuously, while sending; receive data from serial port and transfer to port P0. Assume that crystal frequency of 8051 is 11.0592 MHz; baud rate for the serial communication is 9600.

Solution:

TMOD Register: (to set timer 1 in mode 2 for baud rate)

GATE	C/T	M1	M0	GATE	C/T	M1	M0	
0	0	1	0	0	0	0	0	= 20h

SCON Register: (for serial mode 1)

SM0	SM1	SM2	REN	TB8	RB8	TI	RI	
0	1	0	1	0	0	0	0	= 50H

Interrupt Enable (IE) Register:

EA	--	ET2	ES	ET1	EX1	ET0	EX0	
1	0	0	1	0	0	0	0	= 90H

Assembly Language Program:

```
          ORG 0000H
          SJMP MAIN
          ORG 0030H
MAIN:     MOV P1, #0FFH      ; make port P2 as an input port
          MOV TMOD, #20H     ; timer 1 in mode 2 for baud rate
          MOV TH1, #0FDH     ; to set baud rate 9600
          MOV SCON, #50H     ; serial mode 1
          MOV IE, #90H       ; enable serial communication interrupt
          SETB TR1           ; start timer 1 for baud rate
L1:       MOV A, P1          ; get data from port P1
          MOV P2, A          ; send data to the port P2
          SJMP L1            ; repeat the process
          ORG 0023H
          JB RI, RECEV
          CLR TI
          RETI
RECEV:    MOV A, SBUF
          MOV P0, A
          CLR RI
          RETI
          END
```

Example 6.9

Assume that INT0 pin is connected to a switch that is normally high. Whenever the switch goes low, it should turn ON LED. The LED is connected to port pin P1.4 and is normally OFF. When it is turned ON it should stay ON for a fraction of seconds. As long as switch is pressed low, the LED should stay ON.

Solution:

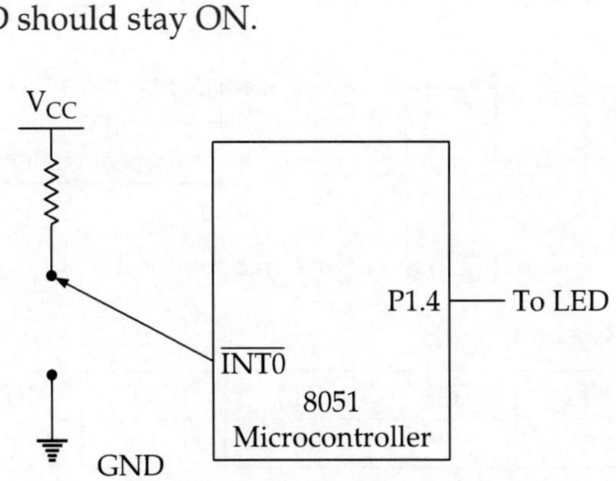

Fig. 6.7 *Figure of example 6.9*

Interrupt Enable (IE) Register:

EA	--	ET2	ES	ET1	EX1	ET0	EX0	
1	0	0	0	0	0	0	1	= 81H

Assembly Language Program:

```
        ORG 0000H
        SJMP MAIN
        ORG 0030H
MAIN:   MOV IE, #81H
HERE:   SJMP HERE
        END
        ORG 0003H
        SETB P1.4
        MOV R2, #0FFH
REP:    DJNZ R2, REP
        CLR P1.4
        RETI
```

Example 6.10

Assume that INT1 pin is connected to a pulse generator; write a program in which the falling edge of the pulse will send a high to port pin P1.4, which is connected to an LED, i.e. LED is turned ON and OFF at the same rate as the pulses are applied to the INT1 pin.

Solution:

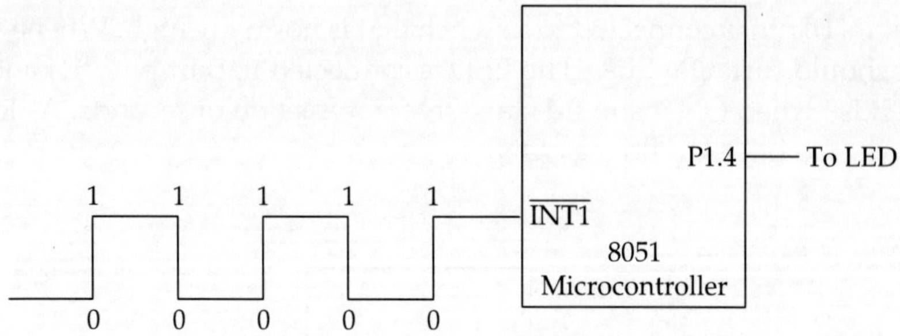

Fig. 6.8 *Figure of example 6.10*

Interrupt Enable (IE) Register:

EA	--	ET2	ES	ET1	EX1	ET0	EX0	
1	0	0	0	0	1	0	0	= 84H

Assembly Language Program:

```
            ORG 0000H
            SJMP MAIN
            ORG 0030H
MAIN:       SETB TCON.2 ; initialize INT1 edge triggered interrupt
            MOV IE, #84H
HERE:       SJMP HERE
            END
```

Interrupt service routine (ISR) for hardware interrupt INT1 to turn ON LED

```
            ORG 0013H
            SETB P1.4
            MOV R2, #0FFH
REP:        DJNZ R2, REP
            CLR P1.4
            RETI
```

Example 6.11

Using interrupt, write an assembly language program to do the following:

(*a*) Receive data serially and send it to port P0.

(*b*) Get data from port P1 and transfer serially and give a copy to port P2.

(*c*) Make timer 0 in auto-reload mode to generate a square wave of 250 μs frequency on the port pin P0.7.

Assume crystal frequency 11.0592 MHz and set baud rate for the serial communication is 4800.

Solution: Calculations of TH1 register for baud rate

$$TH1 = 256 - \frac{\text{Crystal frequency}}{384 * \text{Baud rate}}$$

$$TH1 = 256 - \frac{11.0592 \text{ MHz}}{384 * 4800}$$

$$TH1 = 250 \text{ dec} = \text{FA hex}$$

Calculations of count value in TH0 register for generation of square wave of 250 μs

Method - I

$$TON = TOFF = 125 \ \mu s$$

$$THO = \frac{125\ \mu s}{1.085\ \mu s} = 115$$

–115 decimal = 8Dh hexadecimal (consider only 8-bit value)

Therefore, TH0 = –115d = 8Dh

Method – II

$$\text{Time delay} = (FF - YY + 1) * 1.085\ \mu s$$
$$125\ \mu s = (FF - YY + 1) * 1.085\ \mu s$$
$$FF - YY + 1 = 115 \text{ decimal}$$
$$FF - YY + 1 = 73 \text{ hex}$$
$$YY = FF + 1 - 73$$
$$YY = TH0 = 8Dh$$

TMOD Register:

Timer 1 for generation of baud rate and timer 0 for generation of square wave of 200 μs

GATE	C/T	M1	M0	GATE	C/T	M1	M0	
0	0	1	0	0	0	1	0	= 22H

SCON register:

For transmitting data serially

SM0	SM1	SM2	REN	TB8	RB8	TI	RI	
0	1	0	1	0	0	0	0	= 50H

Interrupt Enable (IE) Register:

For enable serial communication interrupt and timer 0 interrupt

EA	--	ET2	ES	ET1	EX1	ET0	EX0	
1	0	0	1	0	0	1	0	= 92H

Assembly Language Program:

```
        ORG 0000H
        SJMP MAIN          ; Bypass vector location
        ORG 0030H
MAIN:   MOV P1, #0FFH      ; Make port P1 as an input port
        MOV TMOD, #22H     ; Timer 0 in mode 2 for square wave
                             generation and timer 1 in mode 2 for
                             baud rate generation
        MOV TH0, #8DH      ; Load TH0 with auto-reload value
```

```
          MOV TH1, #0FAH     ; Load TH1 value for 4800 baud rate
          MOV SCON, #50H     ; Set serial mode 1
          MOV IE, #92H       ; Enable serial communication and
                               timer 0 interrupts
          SETB TR1           ; Start timer 1 for baud rate
          SETB TR0           ; Start timer 0 for square wave
BACK:     MOV A, P1          ; Get data from port P1 until TF0 raised
          MOV P2, A          ; Send data from P1 to port P2
          MOV SBUF, A        ; Transfers data serially
          SJMP BACK          ; Repeat the process
          SJMP $             ; Stop
                             Interrupt service routine (ISR) for timer 0 to
                             generate square wave
          ORG 000BH
          CPL P0.7           ; Interrupt routine of timer 0
          RETI               ; Return from interrupt
                             Interrupt service routine (ISR) for serial port to
                             check whether data is transmitted or received
                             completely or not
          ORG 0023H
          JB TI, TRANS
          MOV A, SBUF
          MOV P0, A
          CLR RI
          RETI
TRANS:    CLR TI
          RETI
          END
```

Programming Tips: In the above program, if timer 0 is set in Mode 1 for the generation of square wave, then in the main program initialize TH0 = FFh and TL0 = 8Dh, also initialize same values of TH0 and TL0 under interrupt service routine of timer 0.

REVIEW QUESTIONS

1. What are interrupts? Explain with suitable examples.
2. Explain what is interrupt and polling.
3. Explain interrupt enable and interrupt priority registers of 8051 microcontroller.

4. Explain in detail serial communication interrupt of 8051 microcontroller.
5. Explain why only one vector location is reserved for the serial communication interrupt.
6. Draw and explain interrupt structure of 8051 microcontroller in detail.
7. Show the instruction to enable serial communication interrupt and timer 0 interrupt.
8. Assume that, after reset interrupt priority register is programmed with MOV IP, #0Dh. Discuss the sequence in which interrupts are serviced.
9. Show the instruction to assign highest priority to serial communication interrupt and then discuss the order of execution of interrupts.
10. Write an assembly language program to generate square wave of 200 μs duration on port pin P2.1, simultaneously transfer the value of port P0 to port P1, after adding port P0 contents with 03h.
11. Write an interrupt service routine using external hardware interrupt 0 to generate square wave of 50% duty cycle on the port pin P1.7.
12. Write an interrupt service routine using external hardware interrupt 1 to generate square wave of 66% duty cycle on the port pin P1.7.
13. Write an assembly language program to transfer a block of 10 bytes from internal RAM location 30h to internal RAM location 40h using interrupt of timer 0.
14. Design the real time clock showing hour: minutes: seconds using hardware interrupt 1; assume that crystal frequency of 8051 is 12 MHz.
15. Write an interrupt service routine for the multiplication of X * Y and divide X / Y operation. Multiplication program is called under the service routine of hardware external interrupt 0 and division program is called under the service routine of hardware external interrupt 1. Where X and Y are 8-bit unsigned numbers present in the internal RAM location at 30h and 31h.

Chapter 7

Memory Interfacing with Microcontroller 8051

Memory is simply a device that can be used to store the information (data). The semiconductor memories are extensively used because of their small size, low cost, high speed, high reliability and ease of expansion of the memory size. It consists mainly of flip-flops, one flip-flop can hold one bit of data and some additional circuitry such as buffers.

Microcontroller 8051 is equipped with inbuilt or on-chip RAM and ROM of fixed amount in size. For writing small programs and small applications in the microcontroller this fixed amount of memory is sufficient but for writing high end applications especially industrial applications, more amount of memory is required for storing the programs as well as the data. In such situations additional requirement of memory is fulfilled by connecting external memory with the 8051 microcontroller. This technique is actually referred as memory interfacing techniques. Microcontroller 8051 consists 16 address lines, so it is possible to interface maximum of 64 KB of RAM or ROM with 8051 microcontroller.

This chapter illustrates the memory interfacing of different types and sizes with microcontroller 8051.

7.1 SEMICONDUCTOR MEMORY FUNDAMENTALS

In the design of all computers, semiconductor memories are used as primary storage for data and code.

- They are connected directly to the CPU and they are the memory that the CPU asks for information (code or data).
- Among the most widely used are RAM and ROM.
- **Memory Capacity**
    - o The number of bits that a semiconductor memory chip can store is called its chip capacity (bits or bytes).
- **Memory Organization**
    - o Each memory chip contains 2^N locations where N is the number of address pins on the chip.
    - o Each location contains M bits, where M is the number of data pins on the chip.
    - o The entire chip will contain $2^N \times M$ bits.

- o Ex. Memory organization of 4k × 4, 2^{12} = 4096 locations, each location holding 4 bits (so N=12 and M=4)
- Memory speed (access time)

7.2 MEMORY TYPES

The semiconductor memories play a very important role in the microprocessor or microcontroller or in computer based system. Two main categories of semiconductor memories are read only memory (ROM) and random access memory or read/write memory (RAM/RWM). The read only memory as the name suggests, the contents or information is read only and no writing is permitted. The information is entered in read only memory by using a special programmer caller ROM/EPROM burner. The advantage of the read only memory is that the content of the memory is fixed and not alterable by writing programs and also when the power is lost the contents of these memories will be retained as they are.

On the other hand, random access memory (RAM) is generally used to read or write information during program execution. The contents of this memory are not fixed and may be altered through programming; also the contents of this memory are lost when the power is cut off.

The data from the read only memory or from random access memory are accessed randomly by using the corresponding address of the memory location and hence both of these types are called random memories. In other words, any memory location can be accessed directly, by providing the memory address of that memory location.

7.2.1 Read Only Memory Types

1. **ROM (Read Only Memory)**
    - ROM is the type of memory that does not lose its contents when power is turned off. It is also called nonvolatile memory.
2. **PROM (Programmable Memory)**
    - User programmable (one-time programmable) memory.
    - If the information burnt into PROM is wrong, it needs to be discarded since internal fuses are blown permanently.
    - Special equipment needed: ROM burner or ROM programmer.
3. **EPROM (Erasable Programmable ROM) 2000 times**
    - Allows making changes in the contents of PROM after it is burnt.
    - One can program the memory chip and erase it.
    - Erasing its contents can take up to 20 minutes; the entire chip is erased.

- All EPROM chips have a window that is used to shine ultraviolet (UV) radiation to erase its contents.
- Also referred to as UV-EPROM.

4. **EEPROM (Electrically Erasable ROM) 500,000 times**
   - Method of erasure is electrical.
   - Moreover, one can select which byte to be erased.
   - Cost per bit is much higher than for UV-EPROM.
5. **Flash Memory EPROM**
   - First, the process of erasure of the entire content takes less than a second, or one might say in a flash, hence its name: flash memory.
   - When flash memory's contents are erased, the entire device is erased.
   - Even though flash memories are writeable, like EPROMs they find their widest use in microcomputer systems for storage of firmware.
6. **RAM (Random Access Memory) infinite times**
   - RAM is called volatile memory since cutting off the power to the IC will mean loss of data.
   - Also referred to as R/WM (Read And Write Memory)

7.2.2 Read/Write Memory Types (RAM Memory Types)

1. **SRAM (Static RAM)**
   - Storage cells are made of flip-flops and therefore they do not require refreshing to keep their data.
   - A cell handling one bit requires 6 or 4 transistors each, which is too many.
   - SRAMs are widely used for cache memory and battery-backed memory systems.
2. **DRAM (Dynamic RAM)**
   - Uses MOS capacitors to store a bit.
   - Requires constant refreshing due to leakage (every 2 μs – 4 μs).
   - **Advantages:**

     High density (capacity)

     Cheaper cost per bit.
3. **Lower power consumption.**
   - **Disadvantages:**

     While it is being refreshed, data cannot be accessed

     Larger access times

     Too many pins due to large capacity.

Standard EPROM ICs:

Table 7.1 *Standard EPROM ICs*

EPROM	Density (Bits)	Capacity (Bytes)
2716	16 K	2K × 8
2732	32 K	4K × 8
27C64	64 K	8K × 8
27C128	128 K	16K × 8
27C256	256 K	32K × 8
27C512	512 K	64K × 8
27C010	1 M	128K × 8
27C020	2 M	256K × 8
27C040	4 M	512K × 8

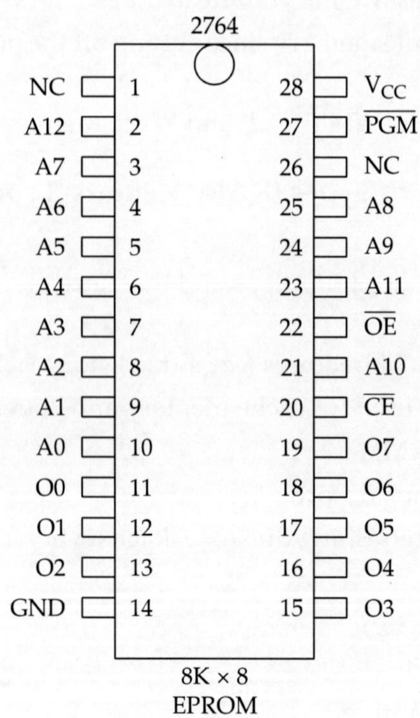

Fig. 7.1 *2764 EPROM IC*

Standard SRAM ICs:

Table 7.2 *Standard SRAM ICs*

EPROM	Density (Bits)	Organization
4361	64 K	64 K × 1
4363	64 K	16 K × 4
4364	64 K	8 K × 8
43254	256 K	64 K × 4
43256 A	256 K	32 K × 8
431000 A	1 M	128 K × 8

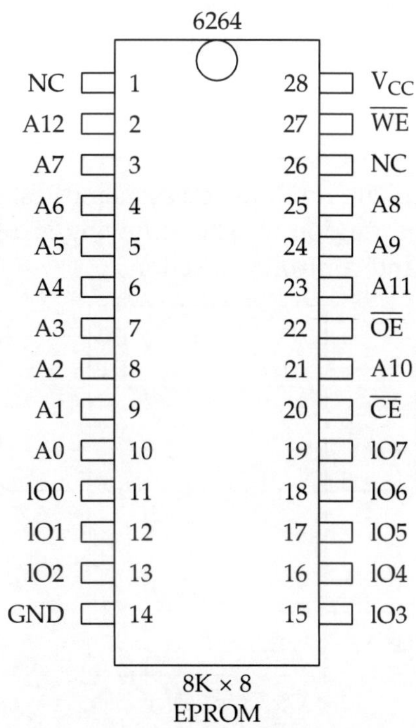

Fig. 7.2 *6264 SRAM IC*

Standard DRAM ICs:

Table 7.3 *Standard DRAM ICs*

DRAM	Density (Bits)	Organization
2164	64 K	64 K × 1
21256	256 K	256 K × 1
21464	256 K	64 K × 4
421000	1 M	1 M × 1
424256	1 M	256 K × 4
44100	4 M	4 M × 1
44400	4 M	1 M × 4
44160	4 M	256 K × 16
416800	16 M	8 M × 2
416400	16 M	4 M × 4
416160	16 M	1 M × 16

In fact all of these devices are random access memories, i.e., each byte or word has an address and can be accessed randomly. The following is a pseudo memory component showing the address, data and control connections.

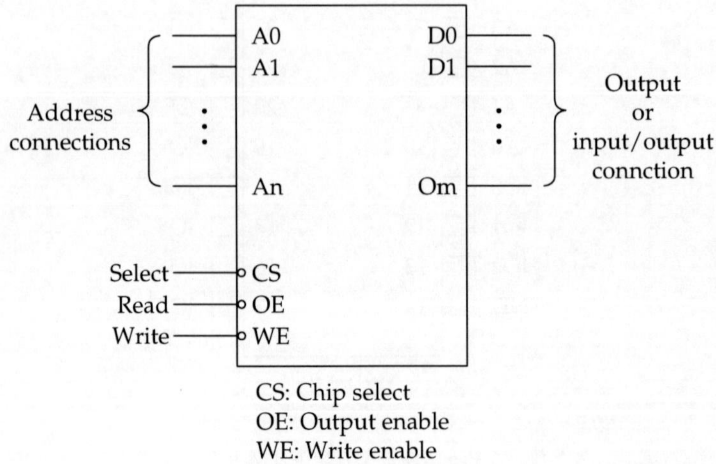

Fig. 7.3 *Memory device*

For ROM and SRAM, the above device has N × M bits. Today the capacity of one chip SRAM memories is less than 1MB, however, there are 128Mbits DRAM memories. In most of the RAM devices, the same pins are used for both read and write data. So at one time, only a read or write operation is possible. Some other devices (e.g. some DRAM devices) have separate data pins for input and output, but still they can do only read or write at one time.

Most of the memory devices produced have an input, sometimes more than one, that selects or enables the memory device. It is most often called a chip select ($\overline{CS}$), chip enable ($\overline{CE}$) or simply select (S) input. The other input(s) are used to retrieve or store data.

An example of a function table:

Table 7.4 *Function of chip select*

CS	OE	WE	D0-Dm
1	x	x	Z (high impedance)
0	x	0	Data in
0	0	1	Data out
0	1	1	Z (high impedance)

7.3 MEMORY STRUCTURE AND ITS REQUIREMENT

A read/write memory consists of an array of registers, in which each register has unique address. The size of the memory is N × M as shown Figure 7.4, where N is the number of registers and M is the word length in number of bits.

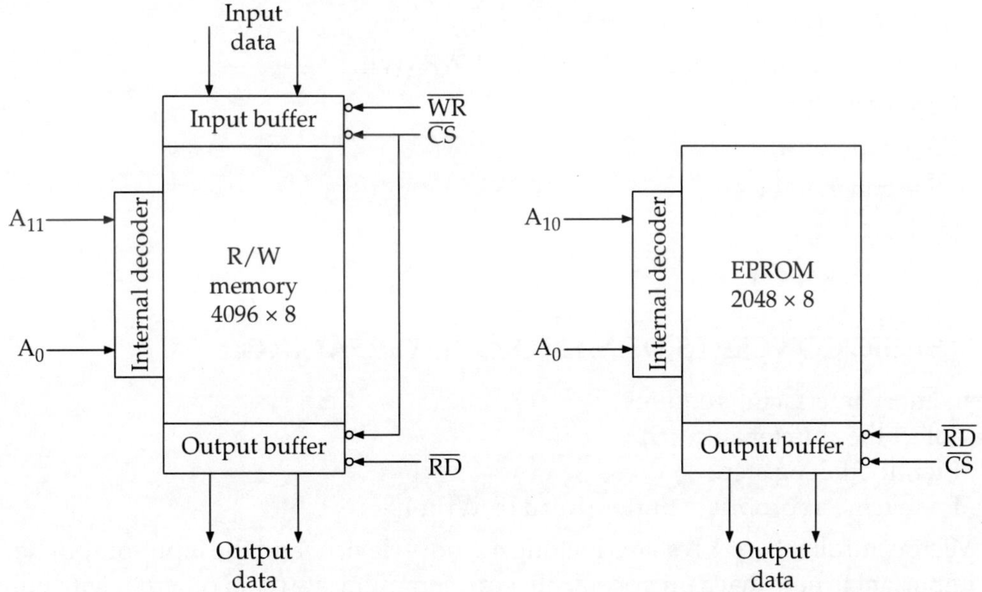

Fig 7.4 *Logic diagram for RAM and EPROM memory organization*

Physical structure of practical memory ICs:

1. **Address Pins:**

No. of Address Pins	No. of Memory Location
8	2^8 = 256 location
9	2^9 = 512 location
10	**2^{10} = 1024 = 1K location**
11	2^{11} = 2048 = 2K
12	2^{12} = 4K
13	2^{13} = 8K
14	2^{14} = 16K
15	2^{15} = 32K
16	**2^{16} = 64K**
17	2^{17} = 128K
18	2^{18} = 256K
19	2^{19} = 512K
20	**2^{20} = 1024K = 1 Mega location**

2. **Data pins:**
   Number of flip-flops in each location is 4/8 then data pins 4/8
3. **Control pins:**
   ROM / EPROM consists of only $\overline{RD}$ ($\overline{OE}$)
   RAM / RWM have control pins $\overline{RD}$ and $\overline{WR}$ ($\overline{WE}$)
4. **Common pins:**
   $\overline{CS}$ (Chip select)
   $\overline{CS}$ is generated using
   (*a*) NAND gate
   (*b*) 3 to 8 decoder

7.4 BASIC CONCEPTS OF MEMORY INTERFACING

The memory interfacing requires:

- Select the chip (enable $\overline{CS}$).
- Identify the register.
- Enable the appropriate buffer (Read or Write)

Microcontroller based systems include memory devices and the input-output devices. It is important to note that a microcontroller can communicate (read or write) with only one device at a time, since the address, data and control lines are common for all the devices.

In order to communicate with memory or IO device, it is necessary to decode the address from the microcontroller.

7.5 MEMORY DECODING

In general, all the memory locations are not implemented, hence all the addresses are not used by the memory devices to select particular memory locations. The remaining lines or unused lines are used to decode to generate chip select signals.

Basically two techniques are used to decode the address:

- Absolute or full decoding
- Linear or partial decoding

7.5.1 Absolute or Full Decoding

In this technique, all the higher address lines are decoded to select the memory chip, and the memory chip is selected only for the specified logic levels on these high order address lines, no other logic levels can select the chip so each location has fixed address. During read, write, output, other locations are not affected. This technique is expensive because it needs more hardware than partial decoding. Figure 7.5 shows the memory interface with absolute decoding

Memory IC	Hex Address	Binary address															
		A15	A14	A13	A12	A11	A10	A9	A8	A7	A6	A5	A4	A3	A2	A1	A0
RAM	0000h	0	0	0	0	0	0	0	0	0	0	0	0	0	0	0	0
1K X 8	03FFh	0	0	0	0	0	0	1	1	1	1	1	1	1	1	1	1
EPROM	0400h	0	0	0	0	0	1	0	0	0	0	0	0	0	0	0	0
1K X 8	07FFh	0	0	0	0	0	1	1	1	1	1	1	1	1	1	1	1
		Enable of Decoder			I/P to the Decoder			To Memory IC									

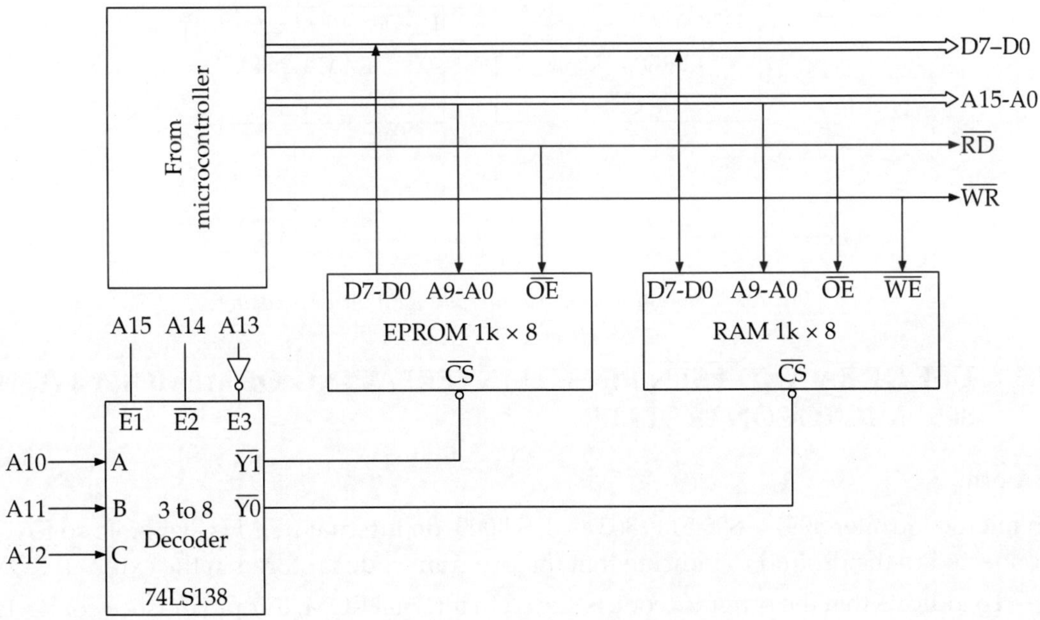

Fig. 7.5 *Absolute decoding technique*

7.5.2 Linear or Partial Decoding

In this technique, all the address lines are not used to generate chip select. This technique is basically used in small systems. In this, individual high order address lines are used to decode the chip select for the memory chips so it reduces the cost of decoding circuit but the drawback is that address of location is not fixed, so each location may have one or more un-decoded lines or multiple addresses (shadow address). The decoding technique using partial or linear decoding is shown in Figure 7.6.

Memory IC	Hex Address	Binary Address A15	A14	A13	A12	A11	A10	A9	A8	A7	A6	A5	A4	A3	A2	A1	A0
EPROM 1K × 8	0000h	0	X	X	X	X	X	0	0	0	0	0	0	0	0	0	0
	03FFh	0	X	X	X	X	X	1	1	1	1	1	1	1	1	1	1
RAM 1K × 8	8000h	1	X	X	X	X	X	0	0	0	0	0	0	0	0	0	0
	83FFh	1	X	X	X	X	X	1	1	1	1	1	1	1	1	1	1
		To $\overline{CS}$	Not Used (Don't Care)					To Memory IC									

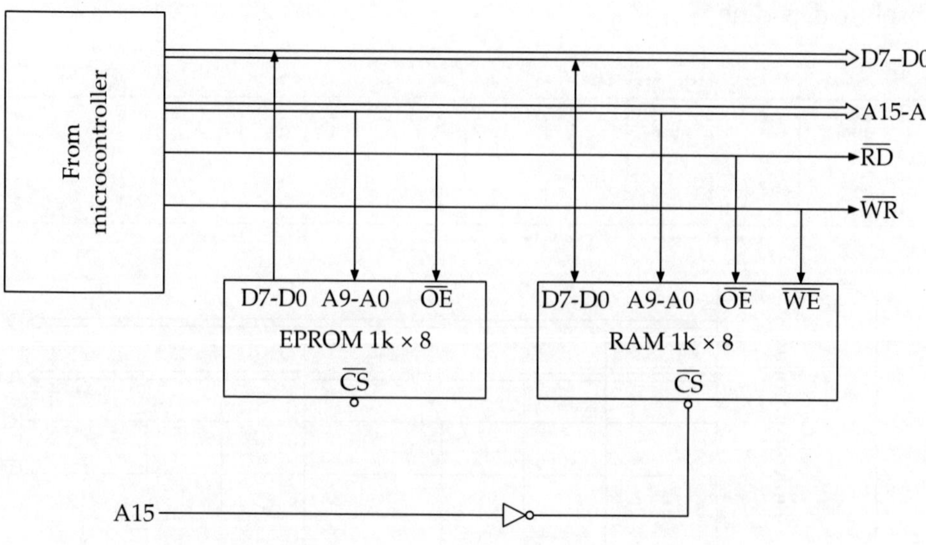

Fig 7.6 *Shows the memory interface with linear decoding*

7.6 USE OF $\overline{EA}$ AND $\overline{PSEN}$ PINS IN INTERFACING OF MEMORY WITH 8051 MICROCONTROLLER

$\overline{EA}$ pin:

In microcontroller 8751 / 89C51 / 8031 / DS 5000, no internal ROM is available so EA# is connected to the ground to indicate that the program code is stored in the external ROM.

To indicate that the program code is stored in internal ROM, this pin must be connected to the V_{CC}.

Microcontroller 8051 consists of on-chip ROM as well as it is supported with external ROM up to 64 KB, in this case

EA# = 0 ; indicates external memory access (# is used as a bar, indicating low level signal)

EA# = 1 ; indicates internal memory access

PSEN# pin:

Another important signal for the 8031/8051 microcontroller is PSEN# (program store enable) signal. PSEN# is an output signal for the 8051/8031 microcontroller and must be connected to the OE# / RD# pin of a ROM containing the program code. In other words, to access external ROM containing program code or opcodes, the 8051/8031 microcontroller uses the PSEN signal. It is important to emphasize the role of EA# and PSEN# when connecting or interfacing the 8051/8031 microcontroller to the external ROM. When EA# pin is connected to ground (GND), the 8051/8031 microcontroller fetches opcode from external ROM by using PSEN# signal.

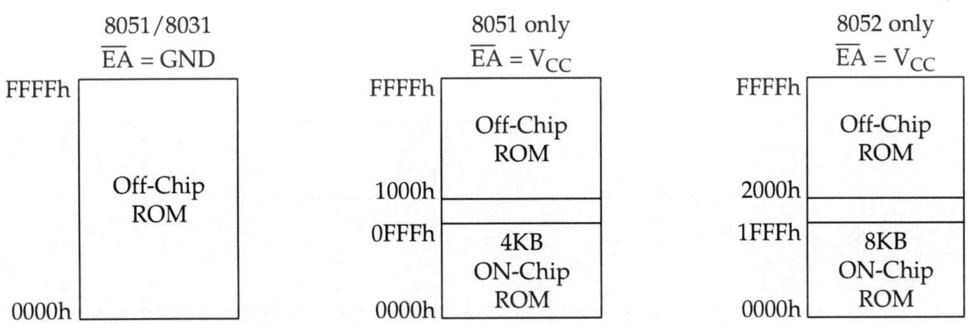

Fig 7.7 *ON and OFF chip program code access*

7.7 SOLVED EXAMPLES

Example 7.1

Show the interfacing of an 8051 microcontroller with 8K bytes of data ROM.

Solution: To access 8K bytes of memory requires 13 address lines, i.e., from A_0 to A_{12}, remaining lines are used for the generation of chip select using either NAND gate or 3 to 8 decoder.

Step – I: Address decoding table or address map

Memory IC	Hexadecimal address	A15	A14	A13	A12	A11	A10	A9	A8	A7	A6	A5	A4	A3	A2	A1	A0
8KB DATA ROM	0000H	0	0	0	0	0	0	0	0	0	0	0	0	0	0	0	0
	1FFFH	0	0	0	1	1	1	1	1	1	1	1	1	1	1	1	1

(Binary address: A15–A13 → To design chip select; A12–A0 → To memory IC)

Step – II: Generation of chip select

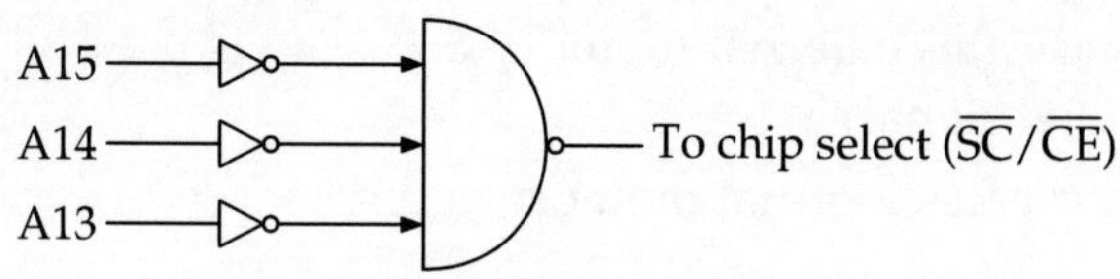

Step – III: Interfacing of 8KB data ROM with 8051 Microcontroller

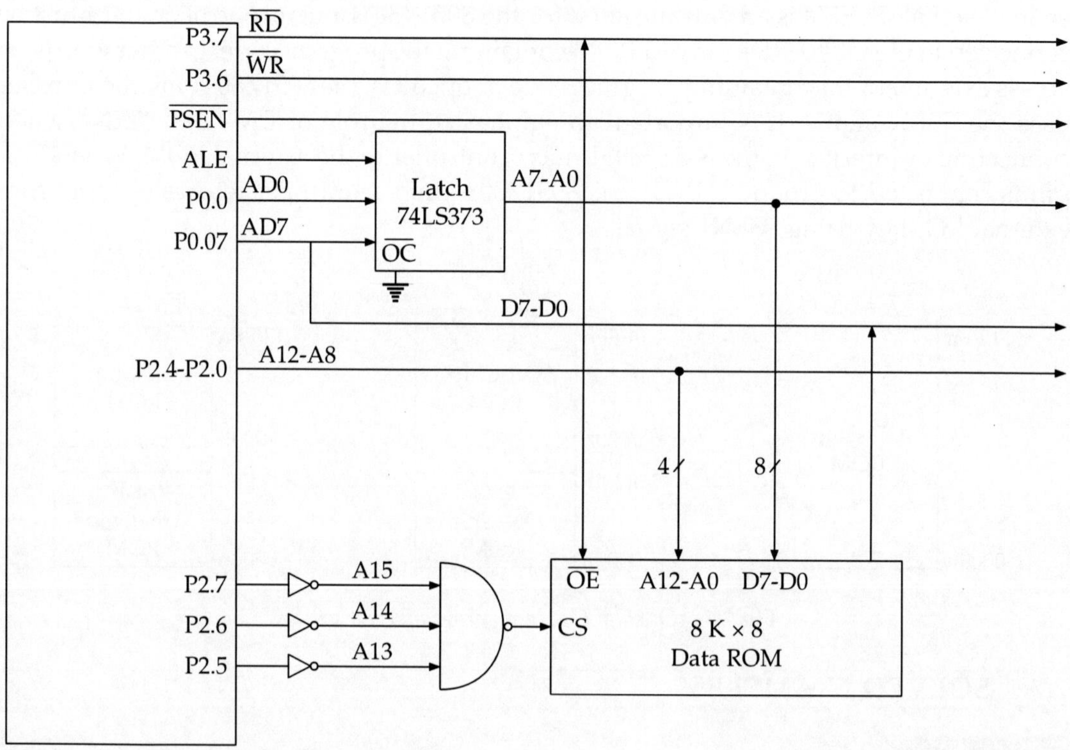

Programming Tips:

When only one memory IC is connected (DATA ROM or DATA RAM) then chip select will be designed either by using NAND gate or 3 to 8 decoder. If more number of memory ICs are used to be interfaced with microcontroller 8051 then only 3 to 8 decoder is used for the implementation of chip select logic.

Example 7.2

Show the interfacing of an 8051 microcontroller with 8K bytes of data ROM and 8K bytes of program ROM.

Solution: For interfacing program ROM (external ROM consisting program opcodes), PSEN# is used to activate both OE# (RD#) and CS#.

For data ROM, use RD# signal to activate OE#, while CS# is activated by simple NAND gate or by 3 to 8 decoder.

For 8K bytes of data ROM, to access 8K bytes of memory requires 13 address lines, i.e., from A_0 to A_{12}.

Step – I: Address decoding table / address map

Memory IC	Hexadecimal address	Binary address															
		A15	A14	A13	A12	A11	A10	A9	A8	A7	A6	A5	A4	A3	A2	A1	A0
8KB DATA	0000H	0	0	0	0	0	0	0	0	0	0	0	0	0	0	0	0
ROM	1FFFH	0	0	0	1	1	1	1	1	1	1	1	1	1	1	1	1

A15–A13: To design chip select; A12–A0: To memory IC

Step – II: Generation of chip select

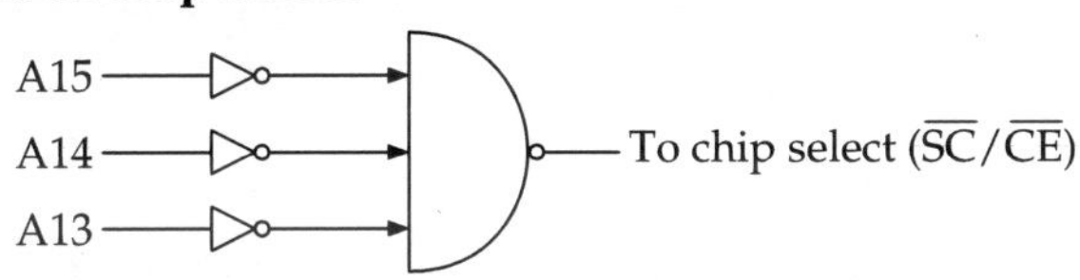

Step – III: Interfacing of 8K bytes of Data ROM and 8K bytes of program ROM with 8051 Microcontroller

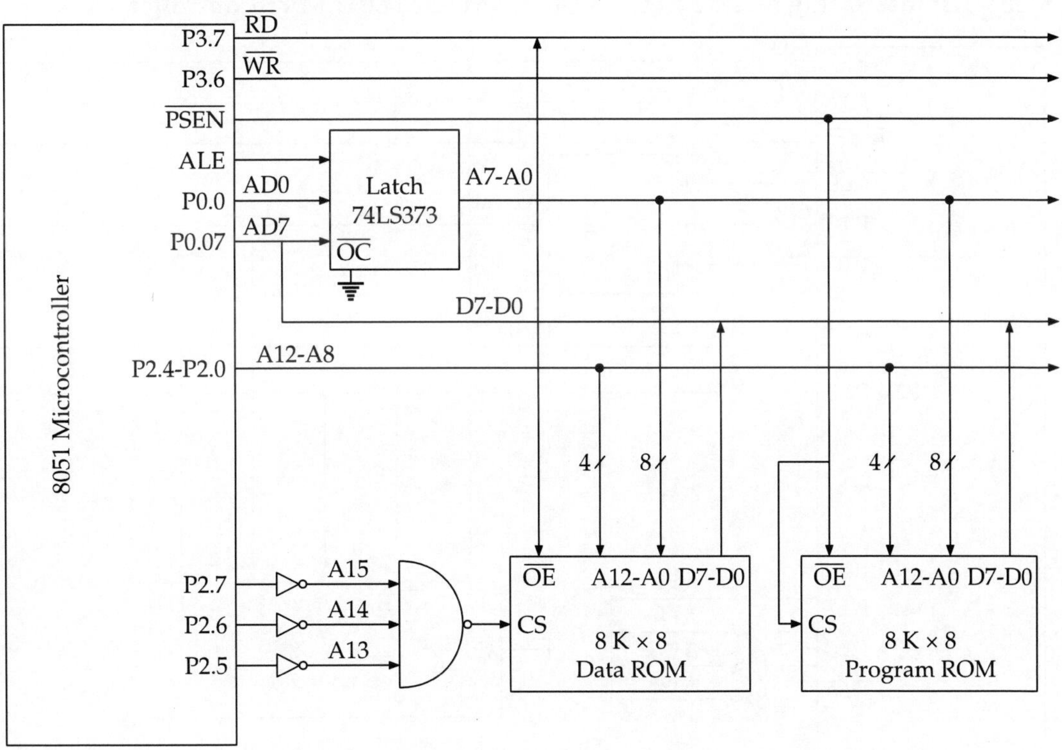

Example 7.3

Show the interfacing of an 8051 microcontroller with 4 K bytes of data RAM with 8051 microcontroller.

Solution: To access 4 K bytes of data RAM requires 12 address lines i.e. from A_0 to A_{11}

Step – I: Address decoding table / address map

Memory IC	Hexadecimal address	A15	A14	A13	A12	A11	A10	A9	A8	A7	A6	A5	A4	A3	A2	A1	A0
4KB DATA RAM	0000H	0	0	0	0	0	0	0	0	0	0	0	0	0	0	0	0
	0FFFH	0	0	0	0	1	1	1	1	1	1	1	1	1	1	1	1

A15–A12: To design chip select; A11–A0: To memory IC

Step – II: Generation of chip select

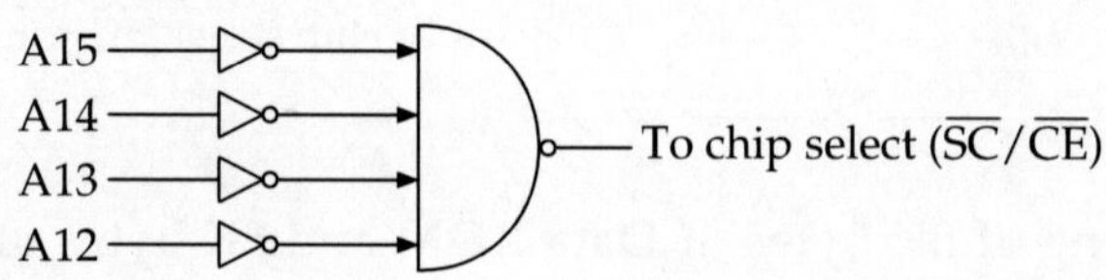

Step – III: Interfacing of 4K bytes of Data RAM with 8051 Microcontroller

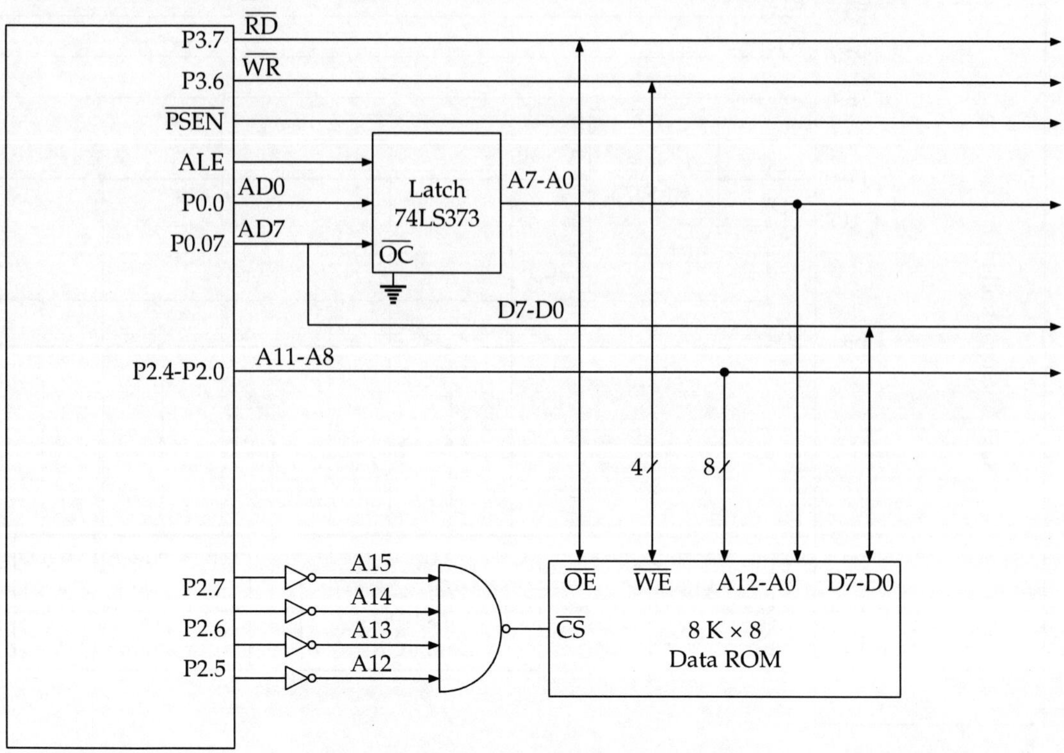

Example 7.4

Show the interfacing of an 8051 microcontroller with 16 K bytes of data RAM with 8051 microcontroller. Assume the starting address of 16 KB data RAM is 8000h.

Solution: To access 16 KB data RAM requires 14 address lines, i.e., from A_0 to A_{13}

Step – I: Address decoding table / Address map

Memory IC	Hexadecimal address	Binary address															
		A15	A14	A13	A12	A11	A10	A9	A8	A7	A6	A5	A4	A3	A2	A1	A0
16KB DATA RAM	8000h	1	0	0	0	0	0	0	0	0	0	0	0	0	0	0	0
	BFFFh	1	0	1	1	1	1	1	1	1	1	1	1	1	1	1	1
		To design chip select		To memory IC													

Step – II: Generation of chip select

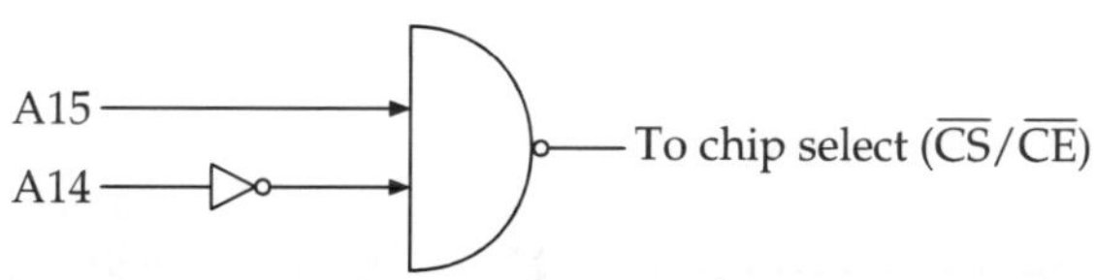

Step – III: Interfacing of 8051 Microcontroller with 16 KB Data RAM

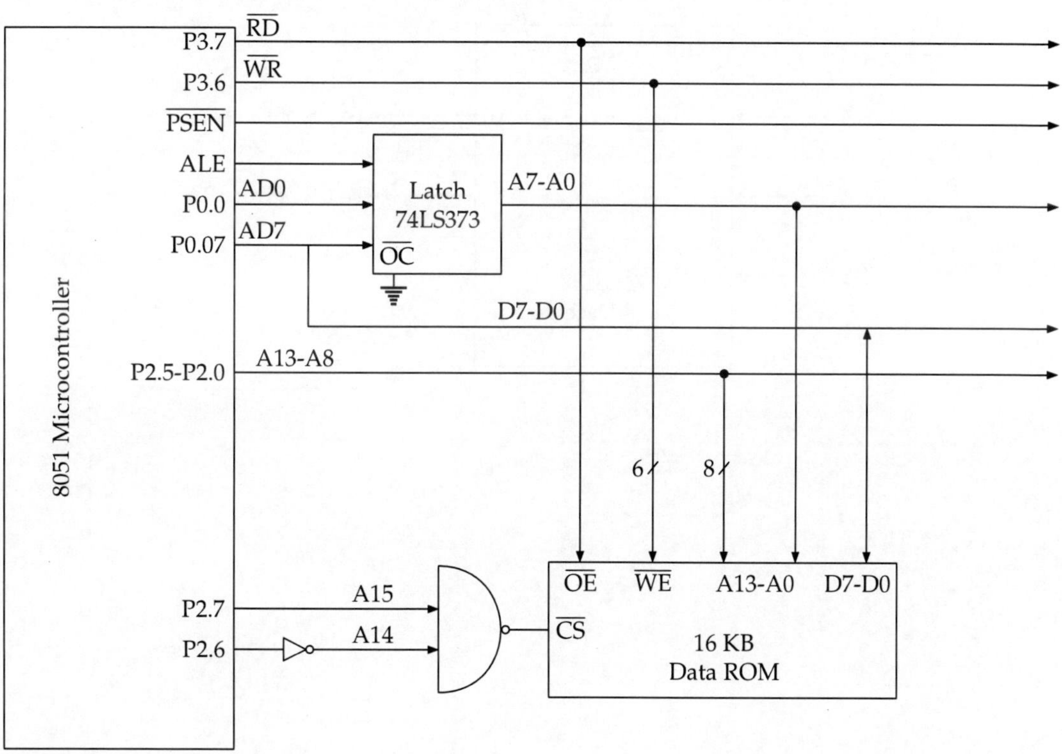

Example 7.5

Show the interfacing of an 8051 microcontroller with 8 KB of program memory, 8K bytes of data ROM and 8K bytes of data RAM. Show the complete address mapping. Use 74LS138 3 to 8 decoder for the generation of chip select.

Solution: To access 8 KB of data ROM and 8 KB of data RAM requires 13 address lines i.e. from $A_0 - A_{12}$

Step – I: Address decoding table / address map

Memory IC	Hexadecimal address	A15	A14	A13	A12	A11	A10	A9	A8	A7	A6	A5	A4	A3	A2	A1	A0
8KB DATA ROM	0000H	0	0	0	0	0	0	0	0	0	0	0	0	0	0	0	0
	1FFFH	0	0	0	1	1	1	1	1	1	1	1	1	1	1	1	1
8KB DATA RAM	2000H	0	0	1	0	0	0	0	0	0	0	0	0	0	0	0	0
	3FFFH	0	0	1	1	1	1	1	1	1	1	1	1	1	1	1	1
		To generate chip select (A15–A13)			To Memory IC (A12–A0)												

(Binary address)

Step – II: Generation of chip select

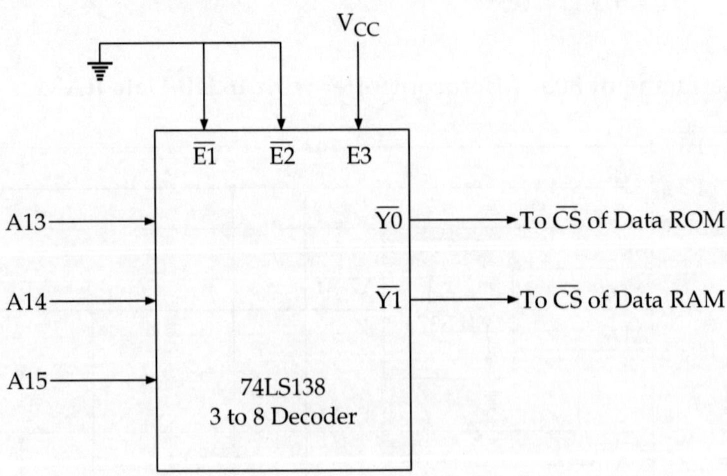

Step – III: Interfacing of 8051 Microcontroller with 8KB Data ROM and 8KB Data RAM and 8KB program ROM

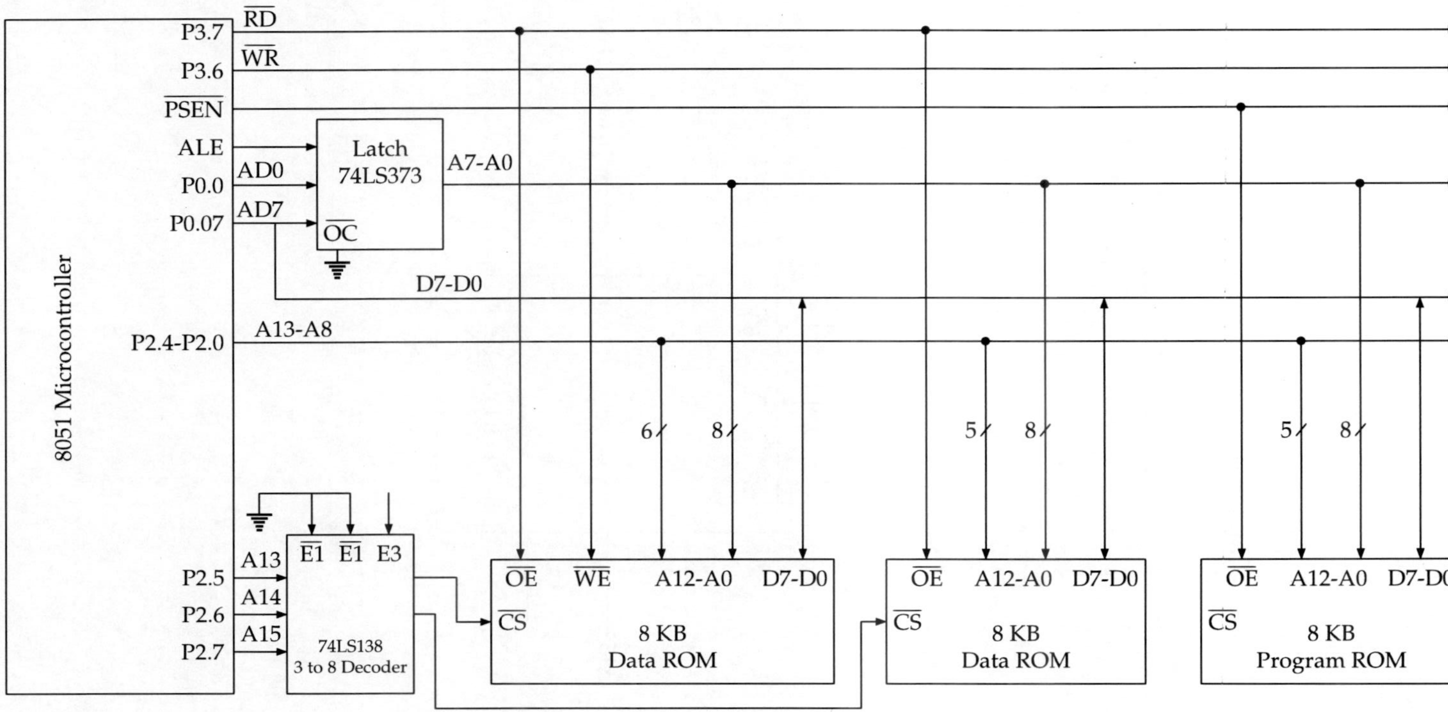

Example 7.6

Show the interfacing of an 8051 microcontroller with 16K bytes of data ROM and 16K bytes of data RAM.

Solution:

Step – I: Address decoding table / address map

Memory IC	Hexadecimal address	A15	A14	A13	A12	A11	A10	A9	A8	A7	A6	A5	A4	A3	A2	A1	A0
16KB DATA ROM	0000H	0	0	0	0	0	0	0	0	0	0	0	0	0	0	0	0
	3FFFH	0	0	1	1	1	1	1	1	1	1	1	1	1	1	1	1
16KB DATA RAM	8000H	1	0	0	0	0	0	0	0	0	0	0	0	0	0	0	0
	BFFFH	1	0	1	1	1	1	1	1	1	1	1	1	1	1	1	1

A15–A14: To generate chip select; A13–A0: To Memory IC

Step – II: Generation of chip select

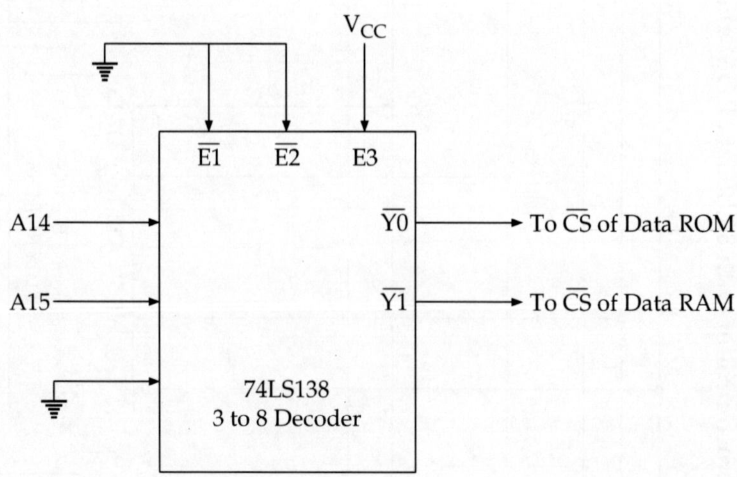

Step – III: Interfacing of 8051 Microcontroller with 16 KB Data RAM, 16 KB Data ROM and 16 KB program ROM

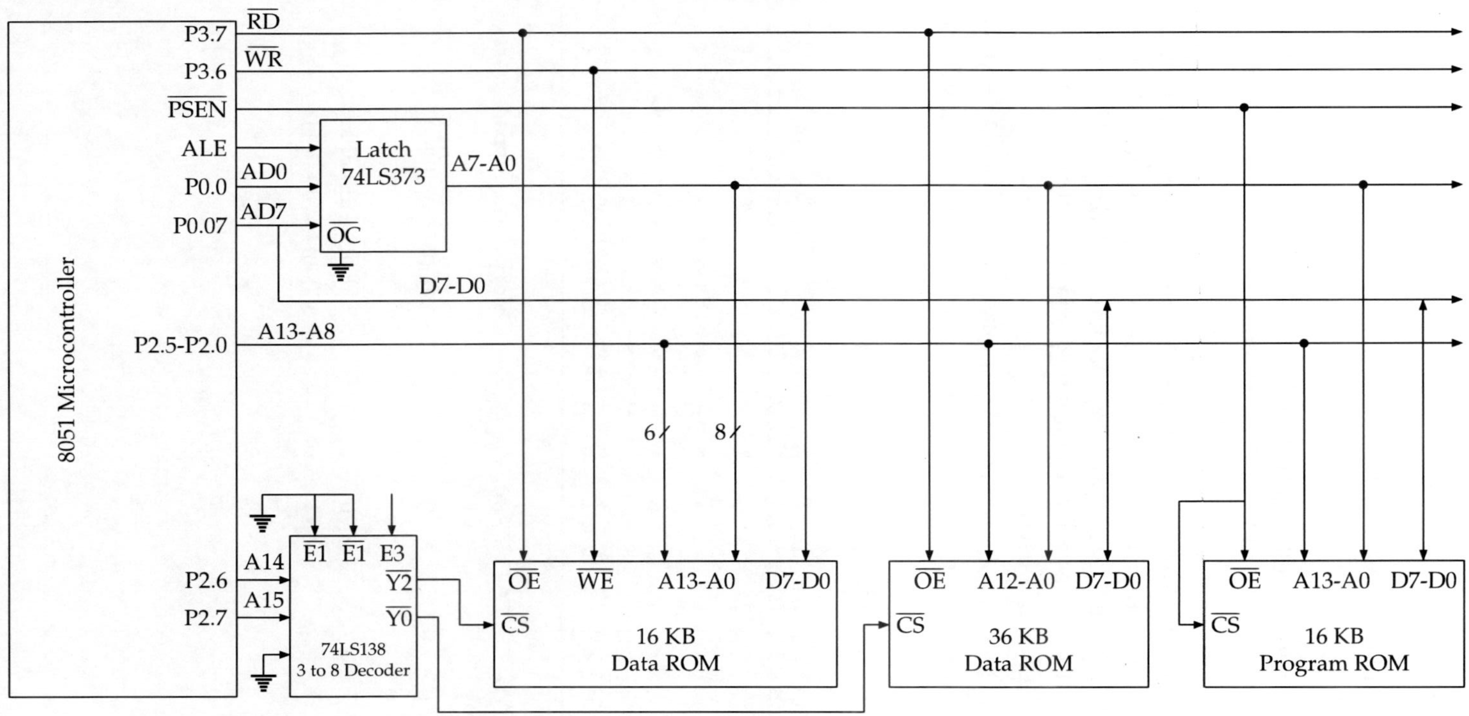

Example 7.7

(*a*) Design the interfacing system for the microcontroller 8051 with 256 KB of SRAM. Assume that available IC is 64 KB of SRAM.

(*b*) Write an assembly language program to read 10 bytes from memory location 20000h and store the contents in the internal memory location from 30h.

Solution:

(*a*) To access 256 KB of memory requires 18 address lines, i.e., from $A_0 - A_{17}$. In the 8051 microcontroller only 16 address lines are available, so in this case 256 KB of memory is generated using the four memory blocks of 64 KB of SRAM. The requirement of A_{16} and A_{17} address lines are provided by port pin P1.0 and P1.1 and for the generation of chip select P1.2 is used.

It may also be possible that instead of using port pin P1.2 for the generation of chip select, the chip select is generated or directly decoded using the port pins P1.0 and P1.1.

Step – I: Address decoding table / address map

Memory IC	Hexadecimal address	Binary address																	
		P1.1 A17	P1.0 A16	P2.7 A15	P2.6 A14	P2.5 A13	P2.4 A12	P2.3 A11	P2.2 A10	P2.1 A9	P2.0 A8	P0.7 A7	P0.6 A6	P0.5 A5	P0.4 A4	P0.3 A3	P0.2 A2	P0.1 A1	P0.0 A0
64KB	00000H	0	0	0	0	0	0	0	0	0	0	0	0	0	0	0	0	0	0
SRAM - I	0FFFFH	0	0	1	1	1	1	1	1	1	1	1	1	1	1	1	1	1	1
64KB	10000H	0	1	0	0	0	0	0	0	0	0	0	0	0	0	0	0	0	0
SRAM - II	1FFFFH	0	1	1	1	1	1	1	1	1	1	1	1	1	1	1	1	1	1
64KB	20000H	1	0	0	0	0	0	0	0	0	0	0	0	0	0	0	0	0	0
SRAM - III	2FFFFH	1	0	1	1	1	1	1	1	1	1	1	1	1	1	1	1	1	1
64KB	30000H	1	1	0	0	0	0	0	0	0	0	0	0	0	0	0	0	0	0
SRAM - IV	3FFFFH	1	1	1	1	1	1	1	1	1	1	1	1	1	1	1	1	1	1

To memory IC

A17	A16	block address space
P1.1	P1.0	
0	0	00000H – 0FFFFH
0	1	10000H – 1FFFFH
1	0	20000H – 2FFFFH
1	1	30000H – 3FFFFH

Step – II: Interfacing of 256KB (4 blocks of 64KB SRAM) memory IC with Microcontroller 8051

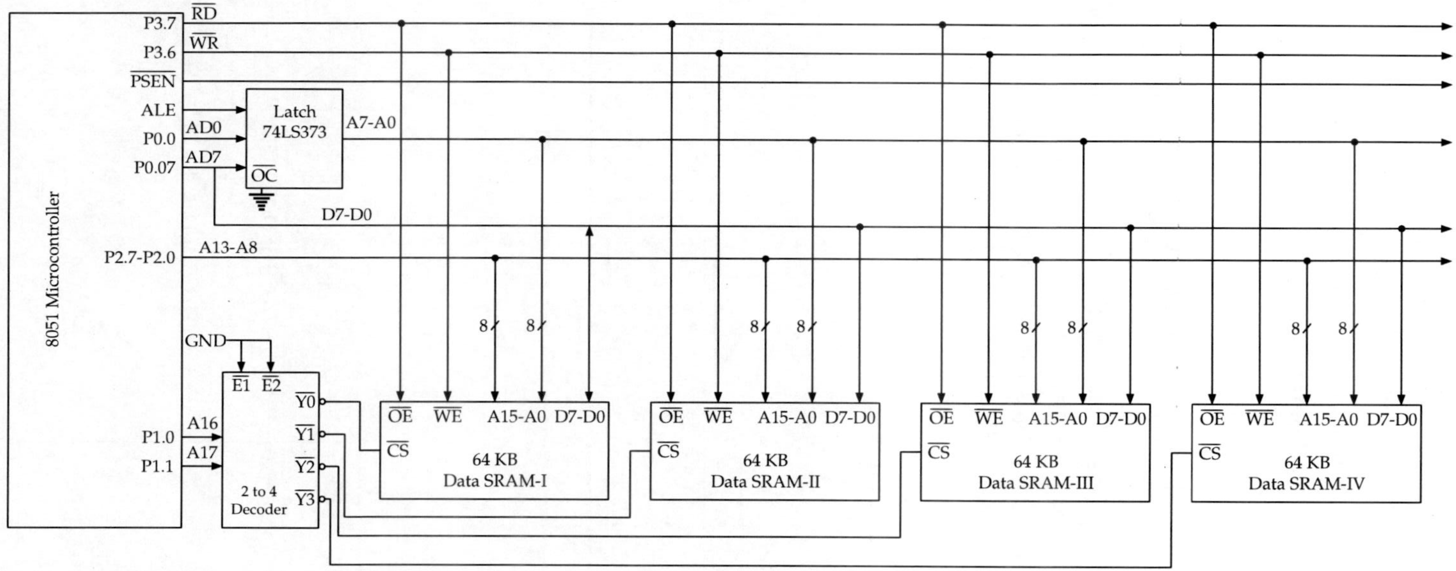

(*b*) Assembly Language Program:

To access memory location 20000h it is necessary to set P1.1 (A17) = 1 and P1.0 (A16) = 0

```
        CLR P1.0            ; A16 = 0
        SETB P1.1           ; A17 = 1
        MOV DPTR, #0000H    ; Starting of 64KB block
        MOV R2, 10          ; Counter for 10 bytes
        MOV R0, #30h        ; Starting internal memory location
L1:     MOVX A, @DPTR       ; Read external memory contents
        MOV @R0, A          ; Store them at internal memory
                              location
```

```
INC R0          ; Next internal memory location
INC DPTR        ; Next external memory location
DJNZ R2, L1     ; Repeat for 10 bytes
L2: SJMP L2     ; End of the program
```

Example 7.8

Show the interfacing of an 8051 microcontroller with 4 KB of data ROM and 8 KB of data RAM.

Solution: To access 4 KB of data ROM requires 12 address lines, i.e., from $A_0 - A_{11}$ and to access 8K bytes of data RAM requires 13 address lines, i.e., from $A_0 - A_{12}$.

In this case address line A_{12} is not used for the 4 KB data ROM but needs to be used for 8 KB data RAM so address line A_{12} is marked as don't care for the 4 KB data ROM. It means that, 4 KB data ROM had a multiple address, it can be accessed from the memory location 0000h to 0FFFh or it can also be accessed from the memory location 1000h to 1FFFh. In other words we say that, 4 KB data ROM having shadow location, this is referred to as partial decoding.

Step – I: Address decoding table

Memory IC	Hexa. address	Binary address A15	A14	A13	A12	A11	A10	A9	A8	A7	A6	A5	A4	A3	A2	A1	A0
4KB DATA ROM	0000H	0	0	0	x	0	0	0	0	0	0	0	0	0	0	0	0
	0FFFH	0	0	0	x	1	1	1	1	1	1	1	1	1	1	1	1
8KB DATA RAM	2000h	0	0	1	0	0	0	0	0	0	0	0	0	0	0	0	0
	3FFFh	0	0	1	1	1	1	1	1	1	1	1	1	1	1	1	1
		to design chip select			To Memory IC												

Step – II: Generation of chip select

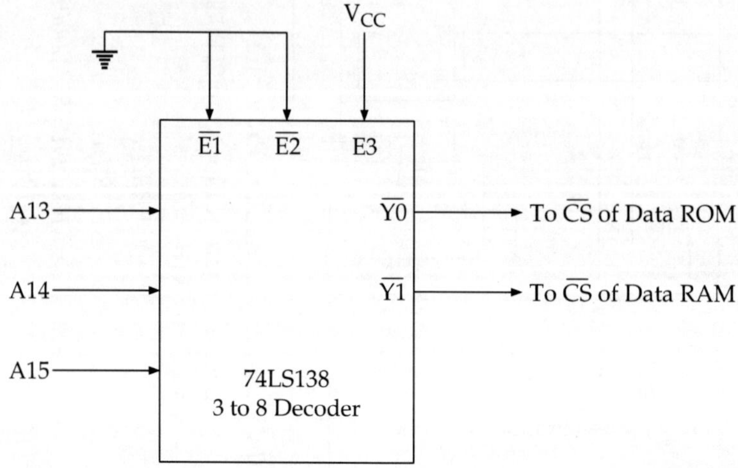

Step – III: interfacing of 4KB Data ROM and 8KB Data RAM with 8051 Microcontroller

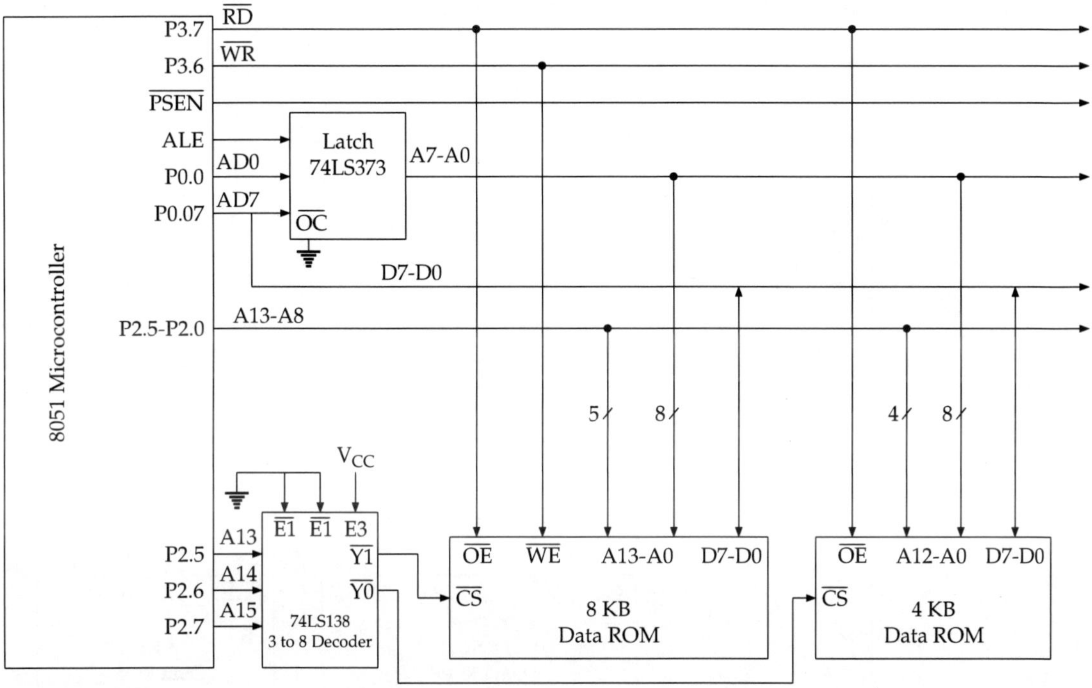

Example 7.9

Show the interfacing of an 8051 microcontroller with 8 KB of data ROM and 8 KB of data RAM using partial decoding techniques.

Solution:

Step – I: Address decoding table

Memory IC	Hexa address	A15	A14	A13	A12	A11	A10	A9	A8	A7	A6	A5	A4	A3	A2	A1	A0
8KB DATA ROM	0000H	0	X	X	0	0	0	0	0	0	0	0	0	0	0	0	0
	1FFFH	0	X	X	1	1	1	1	1	1	1	1	1	1	1	1	1
8KB DATA RAM	8000h	1	X	X	0	0	0	0	0	0	0	0	0	0	0	0	0
	9FFFh	1	X	X	1	1	1	1	1	1	1	1	1	1	1	1	1

(Binary address: A15 – to design chip select; A12–A0 – To Memory IC)

Step – II: Generation of chip select

In this case, step – II is not required. In this problem, address line A_{15} is directly used for the generation of chip select. Address line A_{15} is connected to CS# directly to the 8 KB data ROM and through inverter to the CS# of 8 KB data RAM, so when A15 line is 0 then 8 KB data ROM is selected and when it is 1 then 8 KB data RAM is selected.

In this problem, address lines A_{13} and A_{14} are not used, hence they are marked as don't care, so multiple addresses are possible for accessing the memory location, and hence this decoding technique is called partial decoding technique.

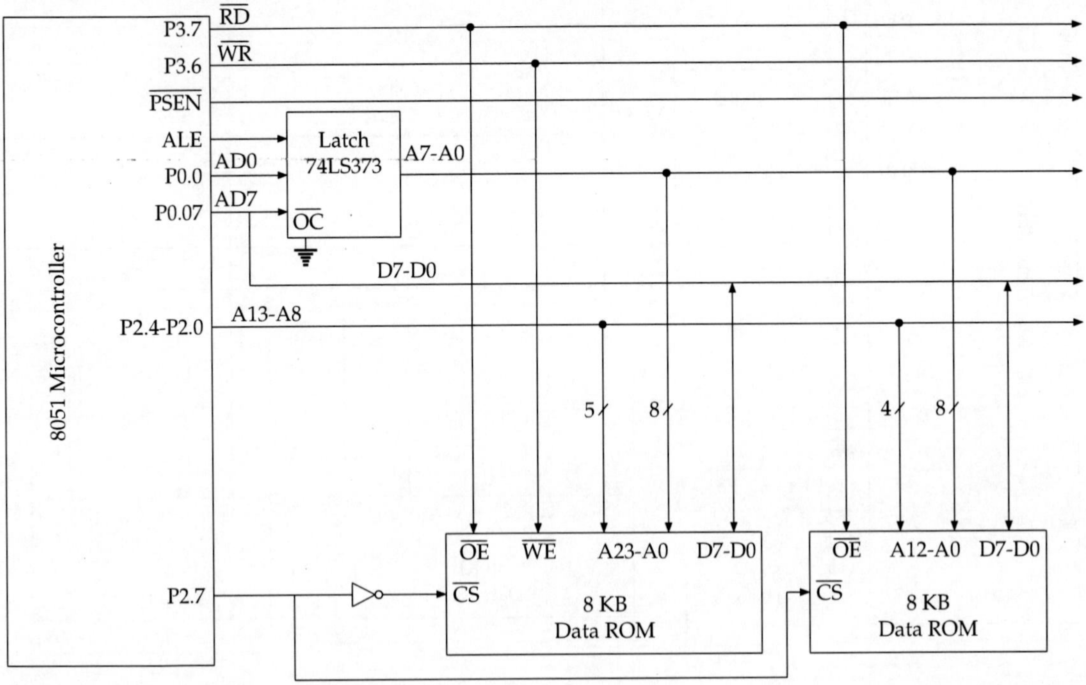

Example 7.10

Show the interfacing of an 8051 microcontroller with 16 KB of data RAM using the available memory IC 6164.

Solution: To access 16 KB data RAM using IC 6164 (8Kx8) requires 4 ICs of 6164.

To access 8KB data RAM requires 13 address lines, i.e., from $A_0 - A_{12}$

Step – I: Address decoding table

Memory IC	Hexa address	A15	A14	A13	A12	A11	A10	A9	A8	A7	A6	A5	A4	A3	A2	A1	A0
		Binary address															
8KB RAM	0000H	0	0	0	0	0	0	0	0	0	0	0	0	0	0	0	0
6164 - I	1FFFH	0	0	0	1	1	1	1	1	1	1	1	1	1	1	1	1
8KB RAM	2000H	0	0	1	0	0	0	0	0	0	0	0	0	0	0	0	0
6164 - II	3FFFH	0	0	1	1	1	1	1	1	1	1	1	1	1	1	1	1
8KB RAM	4000H	0	1	0	0	0	0	0	0	0	0	0	0	0	0	0	0
6164 - III	5FFFH	0	1	0	1	1	1	1	1	1	1	1	1	1	1	1	1
8KB RAM	6000H	0	1	1	0	0	0	0	0	0	0	0	0	0	0	0	0
6164 - IV	7FFFH	0	1	1	1	1	1	1	1	1	1	1	1	1	1	1	1
		To design chip select			To Memory IC												

Step – II: Generation of chip select

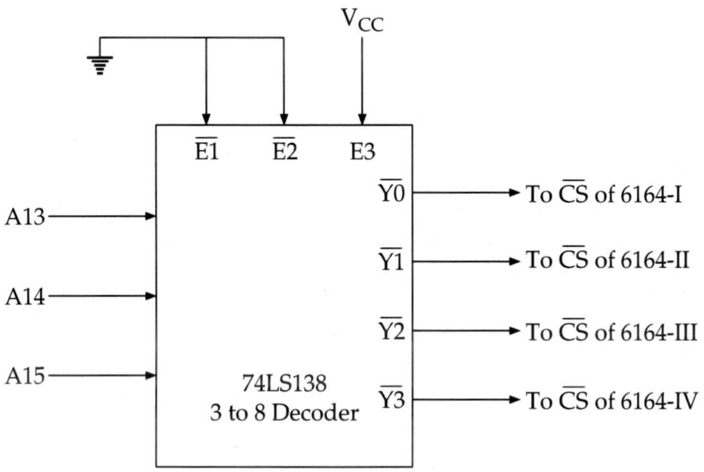

Step – III: Interfacing of 8051 Microcontroller with 4 ICs of 6164

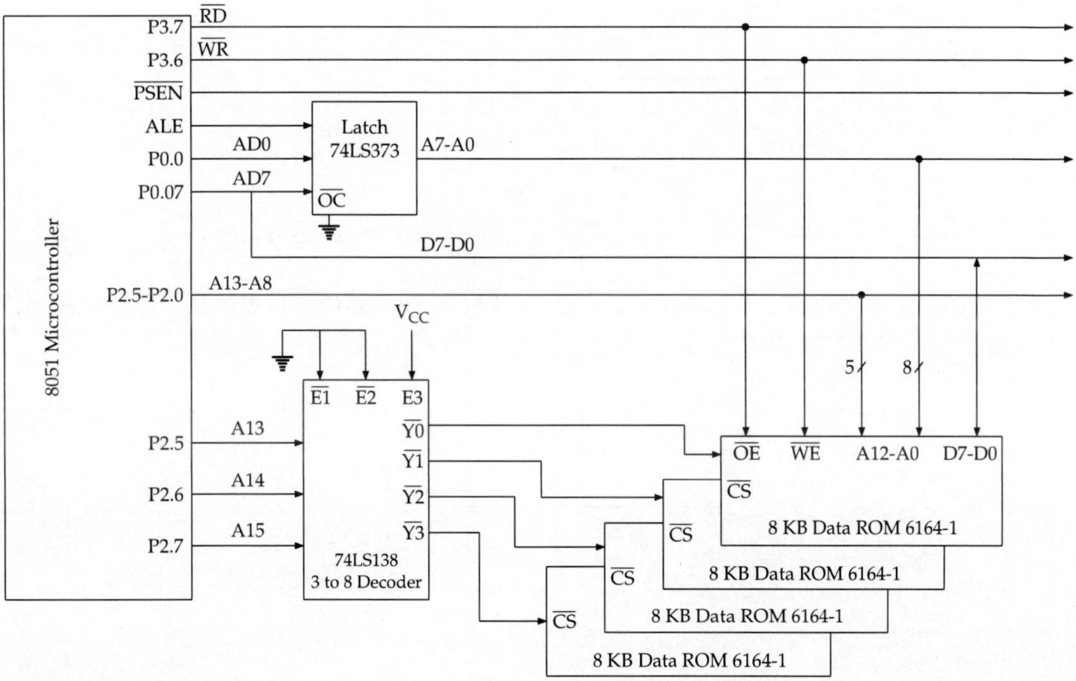

Example 7.11

Show the interfacing of an 8051 microcontroller with 64 KB of data RAM using the available memory IC 6164.

Solution: To interface 64 KB of data RAM using the available IC 6164 requires 8 ICs of 6164 (8 KB).

To access 6164 (8 KB) requires 13 address lines, i.e., from $A_0 - A_{12}$

Step – I: Address decoding table

Memory IC	Hexa address	A15	A14	A13	A12	A11	A10	A9	A8	A7	A6	A5	A4	A3	A2	A1	A0
64KB RAM	0000H	0	0	0	0	0	0	0	0	0	0	0	0	0	0	0	0
6164 - 1	1FFFH	0	0	0	1	0	0	0	0	0	0	0	0	0	0	0	0
64KB RAM	2000H	0	0	1	0	0	0	0	0	0	0	0	0	0	0	0	0
6164 - 2	3FFFH	0	0	1	1	1	1	1	1	1	1	1	1	1	1	1	1
64KB RAM	4000H	0	1	0	0	0	0	0	0	0	0	0	0	0	0	0	0
6164 - 3	5FFFH	0	1	0	1	1	1	1	1	1	1	1	1	1	1	1	1
64KB RAM	6000	0	1	1	0	0	0	0	0	0	0	0	0	0	0	0	0
6164 - 4	7FFFH	0	1	1	1	1	1	1	1	1	1	1	1	1	1	1	1
⋮	⋮		⋮							⋮							
64KB RAM	E000H	1	1	1	0	0	0	0	0	0	0	0	0	0	0	0	0
6164 - 8	FFFFH	1	1	1	1	1	1	1	1	1	1	1	1	1	1	1	1

A15–A13: To generate chip select; A12–A0: To Memory IC

Step – II: Generation of chip select

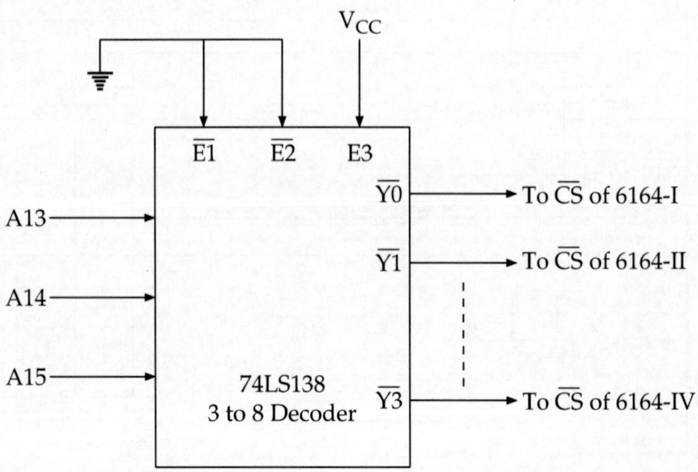

Step – III: Interfacing of 8051 microcontroller with 8 ICs of 6164

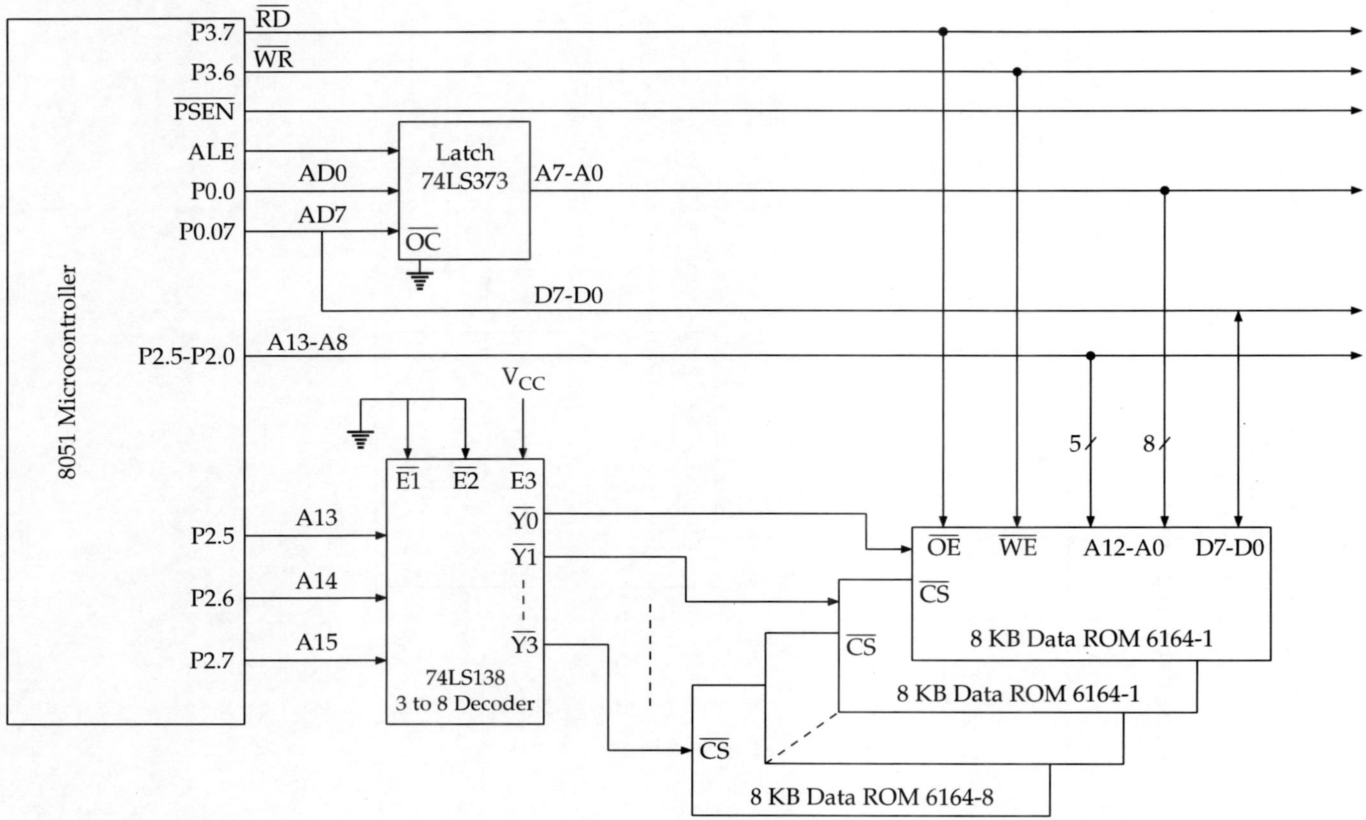

Example 7.12

Show the interfacing of an 8051 microcontroller with 8 KB of data RAM using the available memory IC of 4 KB of data RAM.

Solution: To interface 8 KB of data RAM using the available memory IC of 4 KB of data RAM requires 2 ICs of 4 KB of data RAM.

To access 4KB of data RAM requires 12 address lines $A_0 - A_{11}$.

Step – I: Address decoding table

Memory IC	Hexa. address	A15	A14	A13	A12	A11	A10	A9	A8	A7	A6	A5	A4	A3	A2	A1	A0
4KB DATA RAM - 1	0000H	0	0	0	0	0	0	0	0	0	0	0	0	0	0	0	0
	0FFFH	0	0	0	0	1	1	1	1	1	1	1	1	1	1	1	1
4KB DATA RAM - 2	1000H	0	0	0	1	0	0	0	0	0	0	0	0	0	0	0	0
	1FFFH	0	0	0	1	1	1	1	1	1	1	1	1	1	1	1	1
		to generate chip select				To Memory IC											

Step – II: Generation of chip select

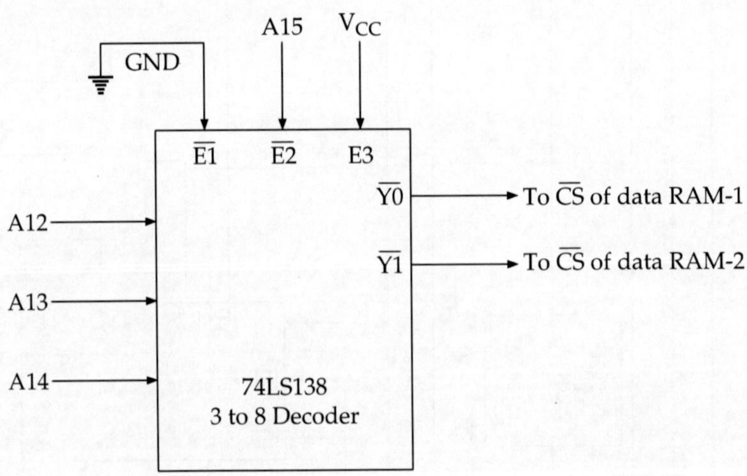

Step – III: Interfacing of 8051 microcontroller with 2 ICs of 4KB Data RAM

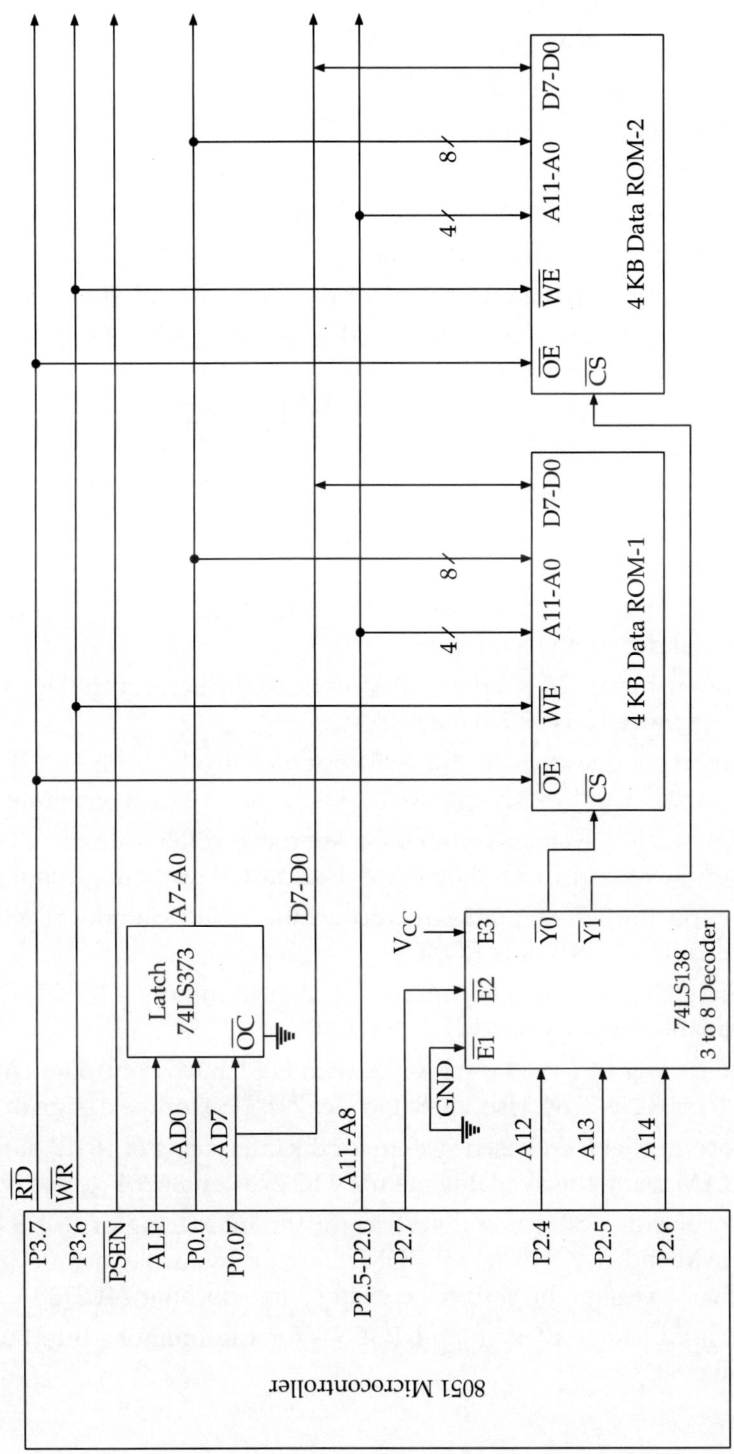

REVIEW QUESTIONS

1. Differentiate between SRAM and DRAM memories.
2. Write short notes on semiconductor memories.
3. What is volatile memory? Explain all types of volatile memory in detail.
4. What is ROM? Explain the families of ROM memories in details.
5. Write short notes on volatile and non-volatile memory.
6. Explain the organization of read/write memory and read only memory.
7. Show the interfacing 2 KB data RAM and 2 KB data ROM with 8051 microcontroller.
8. Show the interfacing of microcontroller based system with 6 KB of data RAM. Assume the available IC is 2 KB of RAM.
9. Show the interfacing of 2 KB of data RAM with 8051 microcontroller using the available memory IC 2114.
10. Show the interfacing of 4 KB data RAM and 8 KB data ROM with 8051 microcontroller.
11. Design the microcontroller 8051 based system with 32 KB data RAM and 32 KB data ROM; use full decoding technique.
12. Design the microcontroller 8051 based system with 32 KB data RAM and 32 KB data ROM; use partial decoding technique.
13. Show the interfacing of 128 KB data RAM with 8051 microcontroller. Assume that the available memory IC is 32 KB data RAM.
14. Design the interfacing system for the 8051 microcontroller with 16 KB data ROM, 16 KB program ROM and 16 KB data RAM. Use NAND gate to generate chip select.
15. Design the interfacing system for the 8051 Microcontroller with 16 KB data ROM, 16 KB program ROM and 16 KB data RAM. Use partial decoding technique.
16. Design the interfacing system for the microcontroller 8051 to interface 6 KB data ROM using the available IC 2 KB data ROM
17. Show the interfacing for the 8051 microcontroller to interface 16 KB of data RAM using the available memory of 4 KB data RAM.
18. Show the interfacing of 64 KB data ROM with 8051 microcontroller. Assume that available memory IC is 2764. Use 3 to 8 decoder 74LS138 for the design of chip select.
19. Design microcontroller 8051 based system for the interfacing of 16 KB data ROM and 16 KB data RAM using the available memory IC 2764 and 6264 respectively.
20. Design microcontroller 8051 based system for the interfacing of 16 KB data ROM, 16 KB data RAM and 16 KB program ROM using the available memory IC 2764 and 6264 respectively. Design the chip select using 3 to 8 decoder 74LS138.
21. Explain the significance of EA# and PSEN# for the memory interfacing in the microcontroller 8051.

22. Find the address range of the memory design in the diagram shown below:

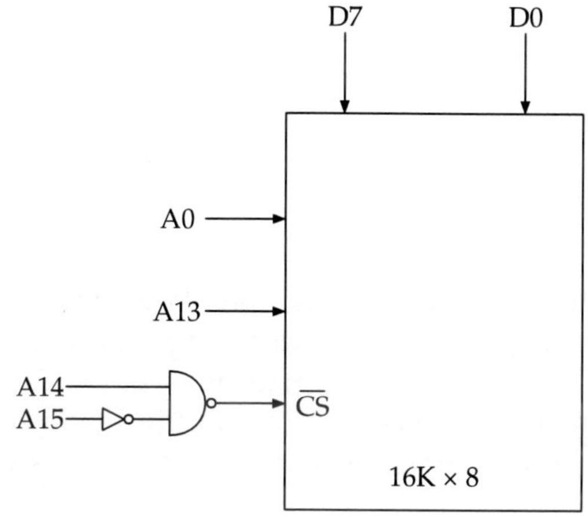

23. Find the address range for Y0, Y2, Y3 and Y6 of the 74LS138 decoder shown below:

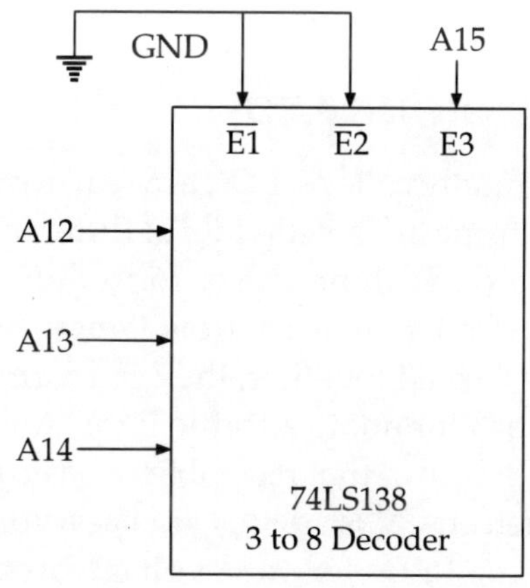

24. Find the address range for Y0, Y2, and Y3 of the 74LS138 decoder shown below:

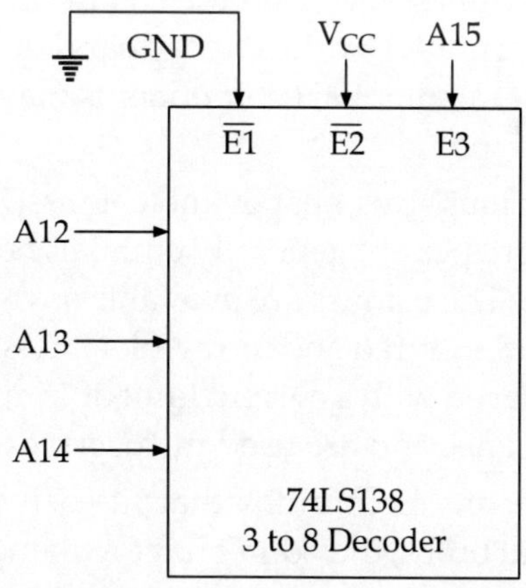

8 Chapter

Applications of 8051 Microcontroller

In this chapter, readers will get information about interfacing of the devices with microcontroller 8051 and program them according to the requirement. In the beginning, we understand how to interface and program small LEDs, 7-segment LEDs, interfacing and programming of some real-time devices, and in the later section, interfacing and programming of some very useful devices like ADCs, DACs, stepper motors, DC motors, keyboards and LCDs.

The advantage of microcontrollers over microprocessors is, microcontrollers are capable to interface and program small devices easily without any external hardware devices.

8.1 LIGHT EMITTING DIODE (LED)

Light emitting diodes, commonly called LEDs, are real unsung heroes from the last many decades in the world of electronics. They do lots of different jobs and are found in almost all kinds of devices. Among other things, they form numbers on digital clocks, display devices, transmit information from remote controls, light up watches and tell you when your appliances are turned on. Collected together, they can form images on a jumbo television screen referred as LED TV or illuminate a traffic light. A light-emitting diode (LED) is a two-lead semiconductor light source that resembles a basic PN-junction diode, except that an LED also emits light. When an LEDs anode lead has a voltage that is more positive than its cathode lead by at least the LEDs forward voltage drop, current flows. Electrons are able to recombine with holes within the device, releasing energy in the form of photons. This effect is called electroluminescence, and the colour of the light (corresponding to the energy of the photon) is determined by the energy band of the semiconductor. An LED is often small in area, and integrated optical components may be used to shape its radiation pattern.

LEDs are available in various sizes, shapes and colours. Standard sizes for the LEDs are 3, 5 and 8 mm in diameter. Shapes are generally circular; however, rectangular shapes for the LED's are also available. The colours are available as red, green, blue, yellow, orange, and white. Bicolour, tricolour, infrared and ultraviolet LEDs are also available. As a normal practice, they are to be interfaced with a current limit of 15mA. The infrared and ultraviolet LEDs have a special significance and are used in special devices.

Basically, LEDs are just tiny light bulbs that fit easily into an electrical circuit. But unlike ordinary incandescent bulbs, they don't have a filament that will burn out, and they don't get especially hot. They are illuminated solely by the movement of electrons in a semiconductor material, and they last just as long as a standard transistor. The lifespan of an LED surpasses the short life of an incandescent bulb by thousands of hours. Tiny LEDs are already replacing the tubes that light up LCD HDTVs to make dramatically thinner televisions. The LED based TV are superior not only because of the clarity they offer but their power consumption is very less and the life span is also great.

Like other ordinary diodes, LEDs also have polarities: anode and cathode. Figure 8.1 shows the symbol of LED. For easy identification of the polarities of the diodes, consider that the lead for anode is provided longer than that of the cathode; also if it is viewed from the top, cathode side is seen as a flat surface and rounded for anode. LEDs are simplest type of indicators and may be used to display two states ON and OFF.

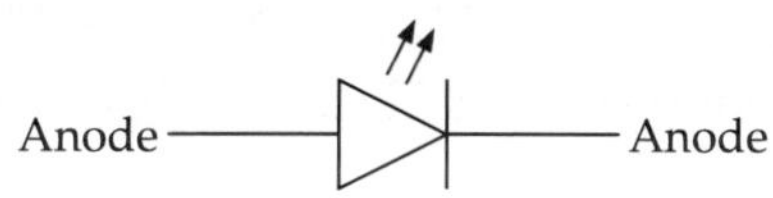

Fig. 8.1 *LEDs symbol*

An LED consists of a chip of semiconducting material doped with impurities to create a P-N junction. As in other diodes, current flows easily from the P-side, or anode to the N-side, or cathode, but not in the reverse direction. Charge-carriers-electrons and holes—flow into the junction from electrodes with different voltages. When an electron meets a hole, it falls into a lower energy level and releases energy in the form of a photon.

Most materials used for LED production have very high refractive indices. This means that much light will be reflected back into the material at the material/air surface interface.

To interface LEDs with microprocessor or microcontroller based system, we must consider the following:

(*a*) LEDs glow only when they are forward biased.

(*b*) Their resistance is almost zero for all practical purpose.

It means that they must be interfaced with proper polarity and with a proper value of resistance in series. This resistance is a current limiting resistor, and its value is to be calculated as per the supplied voltage applied to the LED and current limit of the LED.

Example 8.1

Blinking Light Emitting Diode (LED)

Show the interfacing of LED with port pin P2.1 of Microcontroller 8051 and write an assembly language program to blink LED continuously. The operating frequency of 8051 microcontroller is 11.0592 MHz.

Solution: Hardware arrangement:

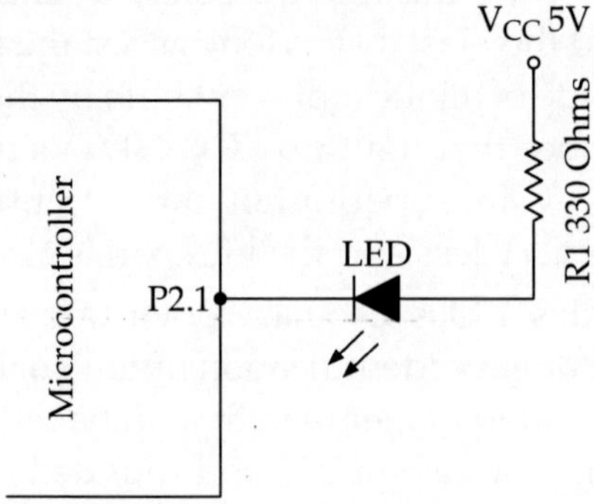

Fig. 8.2 *Interfacing of LED with 8051 microcontroller*

Figure 8.2 shows how to interface the LED to microcontroller. As you can see the anode is connected through a resistor to Vcc and the Cathode is connected to the microcontroller pin. So when the port pin is **HIGH** the LED is **OFF** and when the port pin is **LOW** the LED is turned **ON**.

Flashing LED ALGORITHM

1. Start.
2. Turn ON LED.
3. Turn OFF LED.
4. GO TO 2.

We now want to flash an LED. It works by turning ON a LED & then turning it OFF & then looping back to START. However, the operating speed of microcontroller is very high so the flashing frequency will also be very fast to be detected by human eye, and hence the effect of ON and OFF the LED is not clearly observable by the human eye. By introducing some delay between ON and OFF the LED, it is possible to make the LED ON and stays ON for some fraction of second and then OFF for again some fraction of second, so the effect of ON and OFF may be clearly observed by the programmer.

Modified Flashing LED ALGORITHM with delay

1. Start.
2. Turn ON LED.
3. Wait for some time (delay).
4. Turn OFF LED.
5. Wait for some time (delay).
6. Go to 2.

You can see in the modified algorithm that after turning ON the LED the controller waits for the delay period & then turns OFF the LED & again waits for the delay period & then goes back to the start. This delay allows the user to observe the effect of ON and OFF the LED.

Assembly Language Program:

Method – I

```
          ORG 0000h
LOOP:     CLR P2.1
          ACALL DELAY
          SETB P2.1
          ACALL DELAY
          JMP LOOP
DELAY:    MOV R3, #0FFH
REP:      MOV R2, #0FFH
HERE:     DJNZ R2, HERE
          DJNZ R3, REP
          RET
```

In the above program LED is connected to P2.1. The above program can also be written as follows:

Method – II

```
          ORG 0000h
LOOP:     CPL P2.1
          CALL DELAY
          JMP LOOP
DELAY:    MOV R3, #0FFH
REP:      MOV R2, #0FFH
HERE:     DJNZ R2, HERE
          DJNZ R3, REP
          RET
```

The only drawback of the second program (Method – II) is that the LEDs ON time will be equal to LEDs OFF time. Whereas in the first program if different delay routines are called, the LEDs ON time can be different than that of LEDs OFF time.

Simulation using Protius and IDE 8051 simulation software:

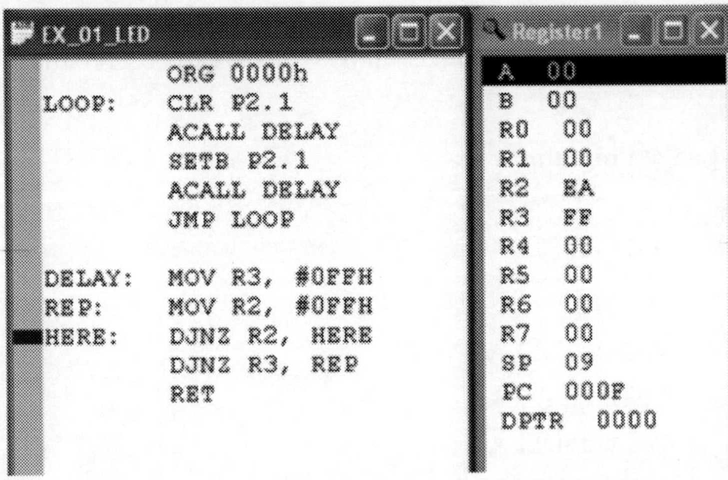

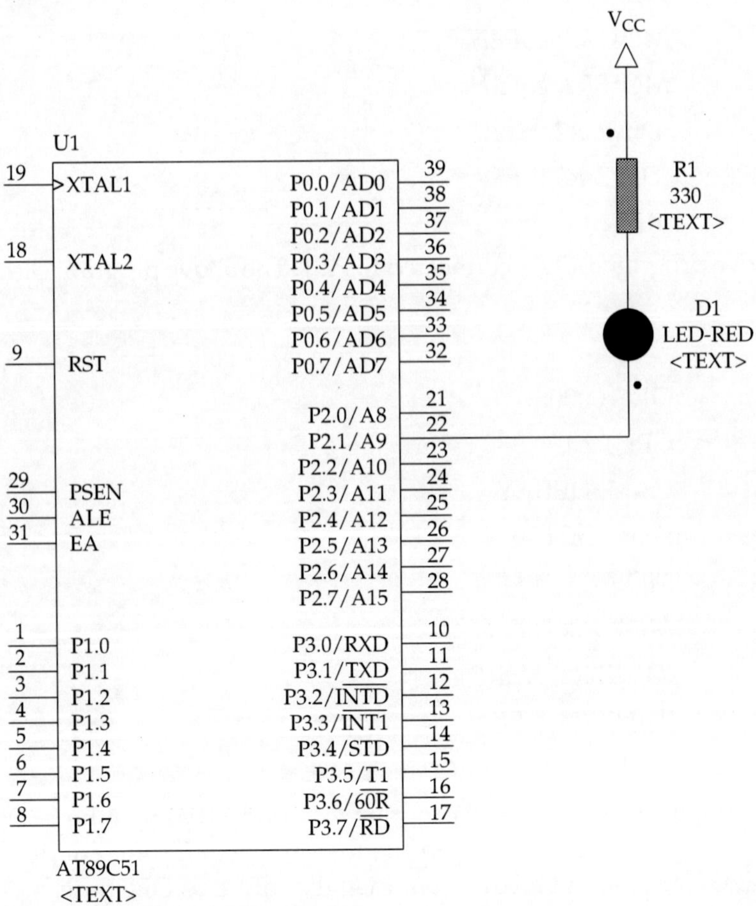

Example 8.2

Interfacing of Multiple LEDs in CA configuration

Show the interfacing of 8-LEDs using common anode arrangement with port 2 pins of 8051 microcontroller. Write an assembly language program to blink the alternate LEDs continuously. The operating frequency of 8051 microcontroller is 11.0592 MHz.

Solution:

Hardware arrangement:

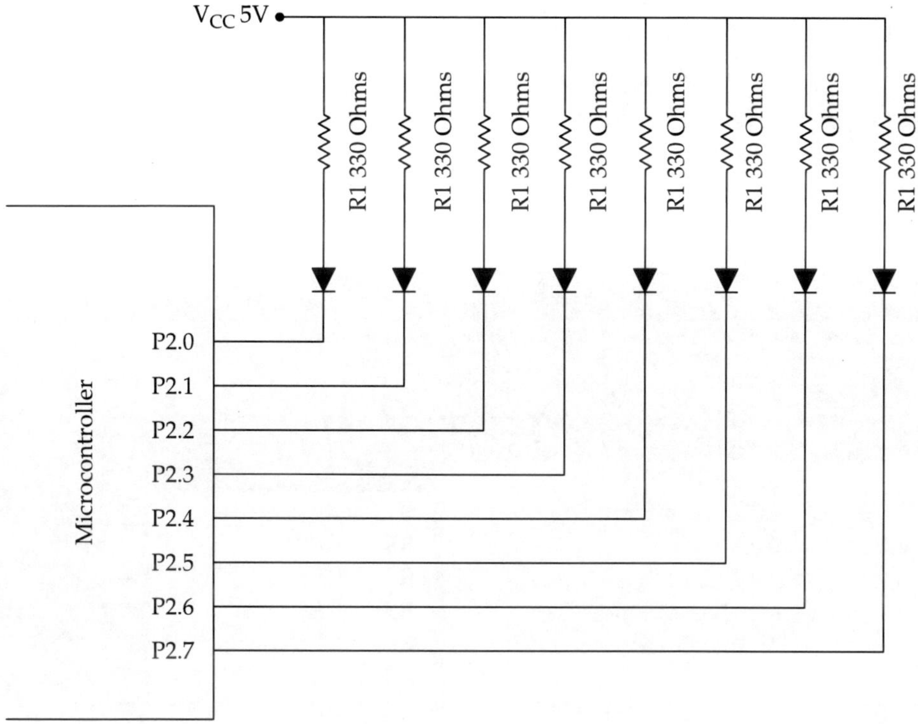

Fig. 8.3 *Interfacing of 8 LEDs with microcontroller 8051*

Figure 8.3 shows how to interface 8 LEDs to microcontroller. As you can see the anode is connected through a resistor to Vcc and the cathode is connected to the microcontroller pin. So when the corresponding port Pin is **HIGH** the LED is **OFF** and when the port pin is **LOW** the LED is turned **ON**. So when alternate HIGH – LOW is transferred on the port pins, then alternate LEDs are showing as OFF – ON.

Assembly Language Program:

```
            ORG 0000h
            MOV A, #55H         ; transfer alternate LOW – HIGH to the
                                  accumulator
LOOP:       CPL A               ; complement accumulator
            MOV P2, A           ; transfers accumulator to port P2
            ACALL DELAY         ; get suitable delay between ON - OFF
            JMP LOOP            ; repeat the process
DELAY:      MOV R3, #0FFH
REP:        MOV R2, #0FFH
HERE:       DJNZ R2, HERE
            DJNZ R3, REP
            RET
```

Simulation using Protius and IDE 8051 simulation software:

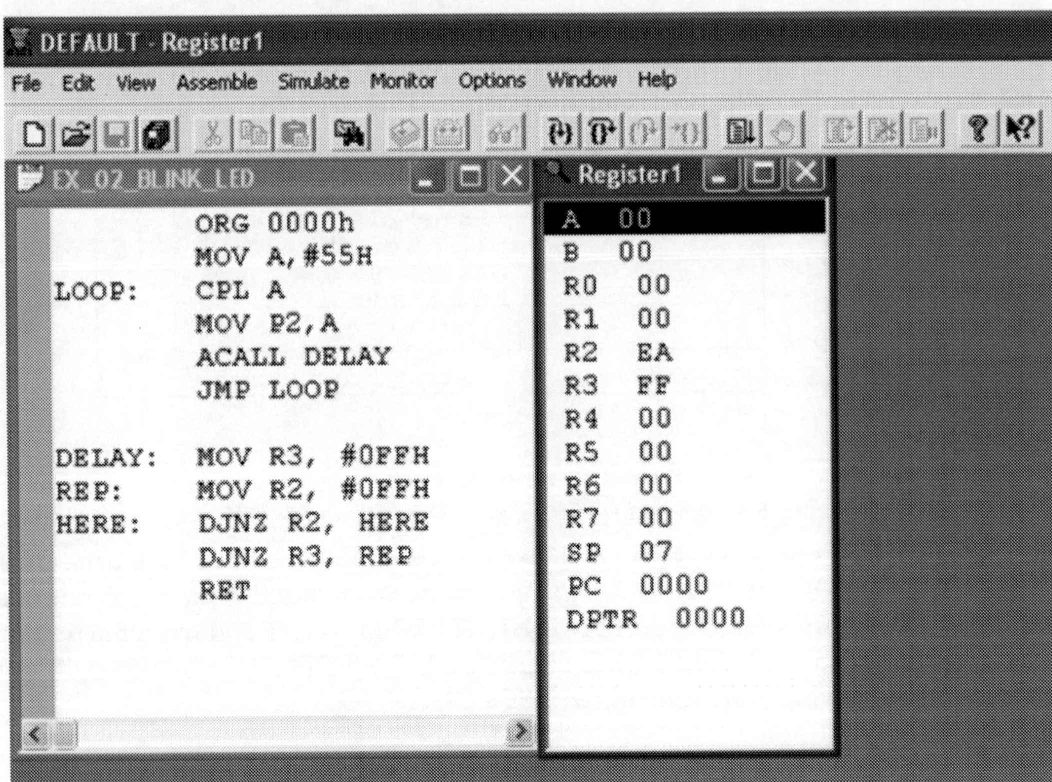

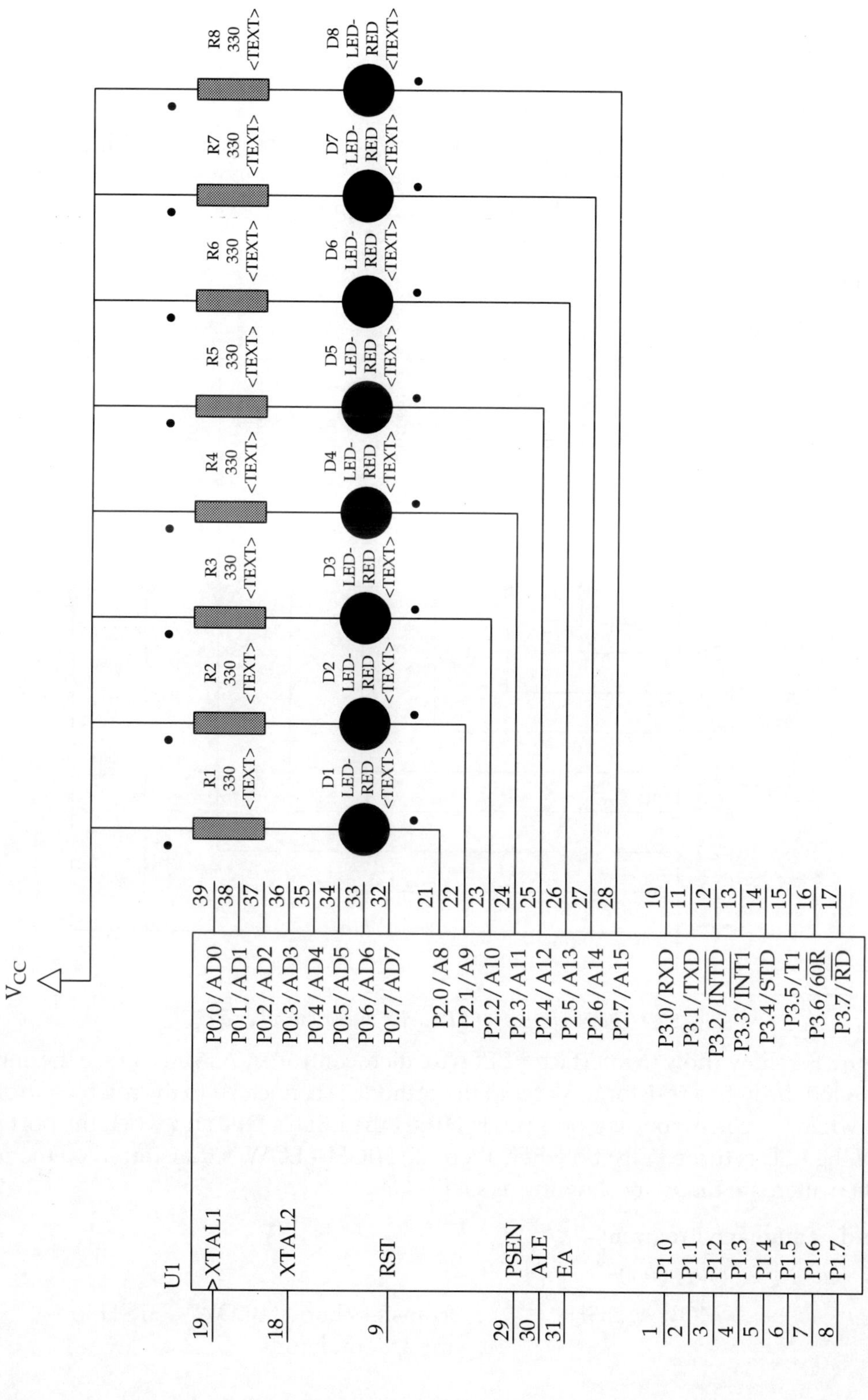
VCC
R1 330 <TEXT>
R2 330 <TEXT>
R3 330 <TEXT>
R4 330 <TEXT>
R5 330 <TEXT>
R6 330 <TEXT>
R7 330 <TEXT>
R8 330 <TEXT>
D1 LED-RED <TEXT>
D2 LED-RED <TEXT>
D3 LED-RED <TEXT>
D4 LED-RED <TEXT>
D5 LED-RED <TEXT>
D6 LED-RED <TEXT>
D7 LED-RED <TEXT>
D8 LED-RED <TEXT>
U1
19 XTAL1
18 XTAL2
9 RST
29 PSEN
30 ALE
31 EA
1 P1.0
2 P1.1
3 P1.2
4 P1.3
5 P1.4
6 P1.5
7 P1.6
8 P1.7
39 P0.0/AD0
38 P0.1/AD1
37 P0.2/AD2
36 P0.3/AD3
35 P0.4/AD4
34 P0.5/AD5
33 P0.6/AD6
32 P0.7/AD7
21 P2.0/A8
22 P2.1/A9
23 P2.2/A10
24 P2.3/A11
25 P2.4/A12
26 P2.5/A13
27 P2.6/A14
28 P2.7/A15
10 P3.0/RXD
11 P3.1/TXD
12 P3.2/INTD
13 P3.3/INT1
14 P3.4/STD
15 P3.5/T1
16 P3.6/60R
17 P3.7/RD
AT89C51 <TEXT>

Example 8.3

Interfacing of Multiple LEDs in CA configuration

Show the interfacing of 8 LEDs using common anode arrangement with port 2 pins of 8051 microcontroller. Write an assembly language program to generate rotating blink on LEDs continuously. Assume that operating frequency of 8051 microcontroller is 11.0592 MHz.

Solution:

Hardware arrangement:

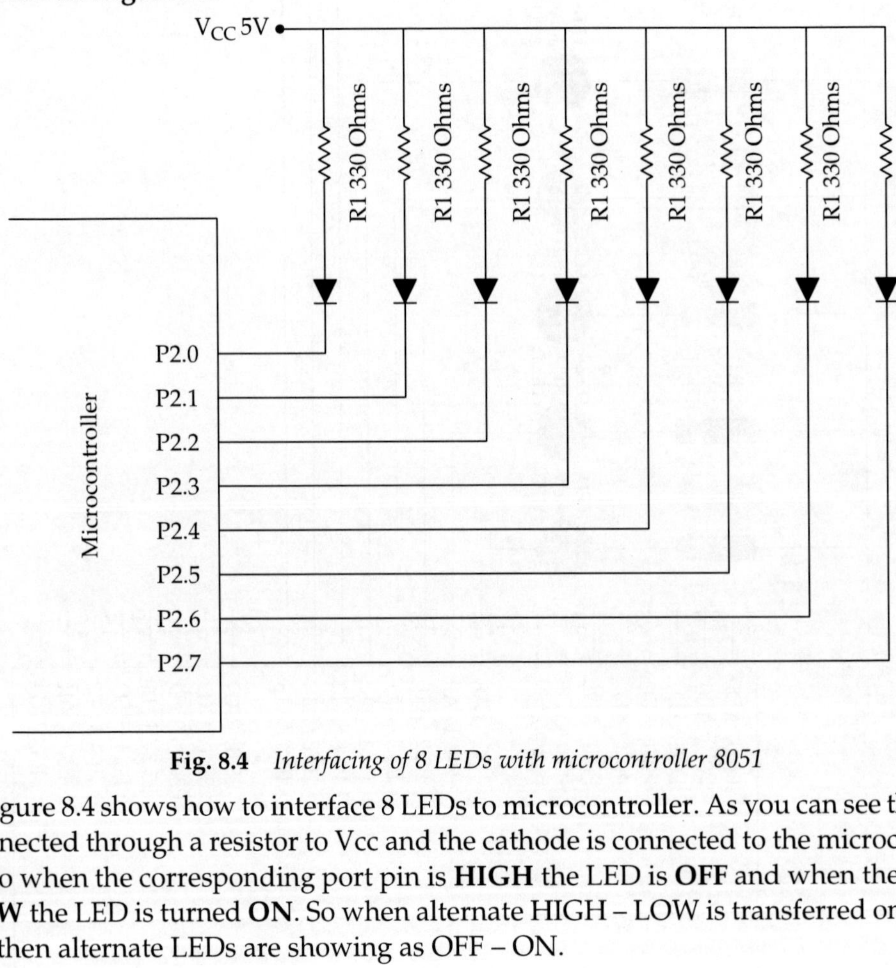

Fig. 8.4 *Interfacing of 8 LEDs with microcontroller 8051*

Figure 8.4 shows how to interface 8 LEDs to microcontroller. As you can see the anode is connected through a resistor to Vcc and the cathode is connected to the microcontroller pin. So when the corresponding port pin is **HIGH** the LED is **OFF** and when the port pin is **LOW** the LED is turned **ON**. So when alternate HIGH – LOW is transferred on the port pins, then alternate LEDs are showing as OFF – ON.

Assembly Language Program:

```
ORG 0000h
MOV A, #55H        ; Transfer alternate LOW – HIGH to
                   the accumulator
```

```
LOOP:     RL A             ; Rotate accumulator left
          MOV P2, A        ; Transfers the accumulator to port P2
          ACALL DELAY      ; Get suitable delay between ON - OFF
          JMP LOOP         ; Repeat the process
DELAY:    MOV R3, #0FFH
REP:      MOV R2, #0FFH
HERE:     DJNZ R2, HERE
          DJNZ R3, REP
          RET
```

Example 8.4

ON and OFF Light Emitting Diode (LED) Using Keys

Show the interfacing of LED with port pin P2.0 of microcontroller 8051 and write an assembly language program to ON the LED when the key K1 is pressed and OFF the LED when key K2 is pressed. Assume that operating frequency of 8051 microcontroller is 11.0592 MHz.

Solution:

Hardware arrangement:

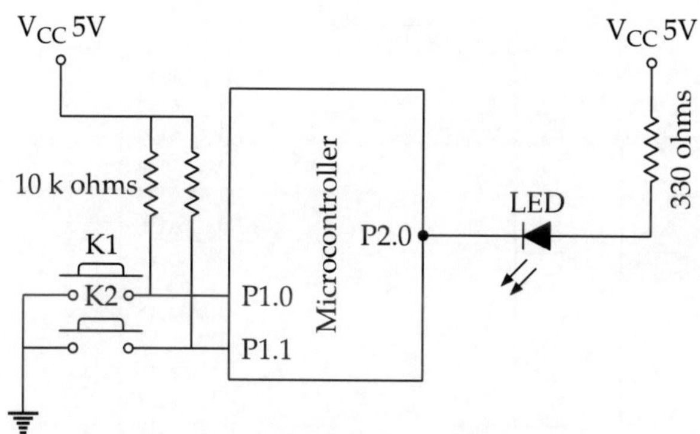

Fig. 8.5 *Interfacing of LED with two keys for ON and OFF*

Assembly Language Program:

```
          ORG 0000h
          SETB P1.0        ; Set port pin P1.0 as input (key K1)
          SETB P1.1        ; Set port pin P1.1 as input (key K2)
LOOP:     MOV A, P1        ; Read key values
          RRC A            ; Rotate towards right to check key K1
```

```
            JC KEY_K2         ; If carry key K1 is not pressed go for key
                                K2
            CLR P2.0          ; If no carry key K1 is pressed ON the
                                LED
            ACALL DELAY
            SJMP LOOP
KEY_K2:     RRC A             ; Rotate towards right to check key K2
            JC LOOP           ; If carry key read the keys again and repeat
            SETB P2.0         ; If no carry key K2 is pressed OFF the
                                LED
            ACALL DELAY
            SJMP LOOP
DELAY:      MOV R3, #0FFH
REP:        MOV R2, #0FFH
HERE:       DJNZ R2, HERE
            DJNZ R3, REP
            RET
```

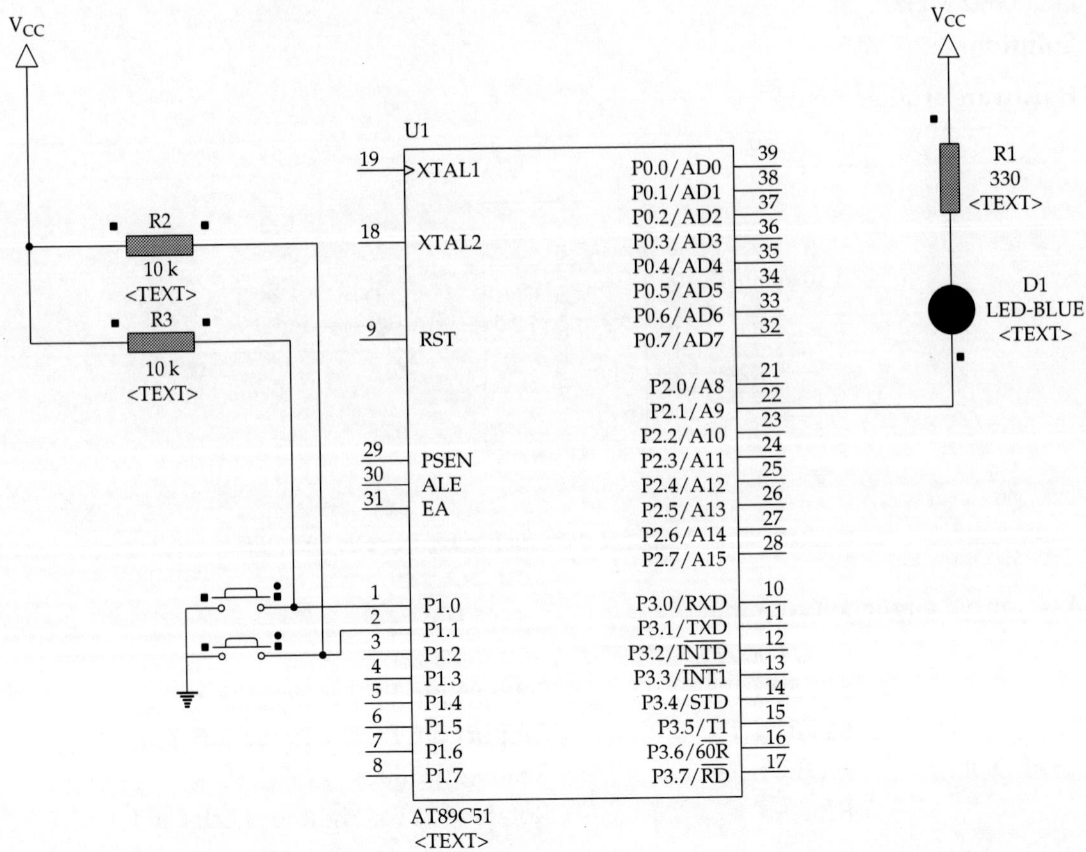

8.2 SEVEN-SEGMENT LEDs

Basics of seven-segment LEDs:

The LED finds its place in many applications in these modern electronic fields. One of them is the seven-segment display. Seven-segment displays contain the arrangement of the LEDs in "eight" (8) passion, and a dot (.) with a common electrode, lead (anode or cathode). The purpose of arranging it in that passion is that we can make any number out of that by switching ON and OFF the particular LED's. Figure 8.6 shows the block diagram of the seven-segment LED arrangement.

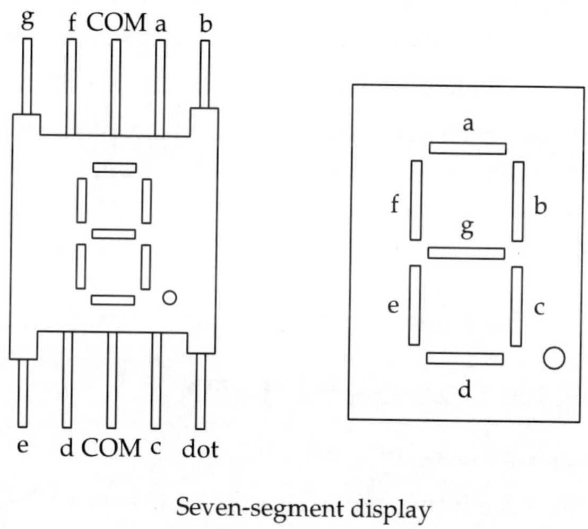

Fig. 8.6 *Pin configuration of Seven-Segment Display*

Seven-segment displays are basically 7 LEDs. There are two types of 7 segment displays: common cathode and common anode.

- Common cathode – where all the segments share the same cathode.
- Common anode – where all the segments share the same anode.

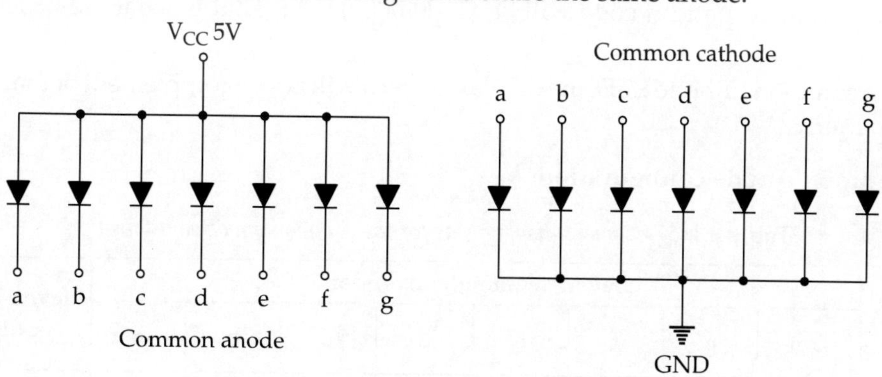

Fig. 8.7 *Configuration of common anode and common cathode of 7-segment LED*

In common anode in order to turn ON a segment, the corresponding pin must be set to 0 and to turn it OFF it is set to 1. Similarly in common cathode in order to turn ON a segment, the corresponding pin must be set to 1 and to turn it OFF it is set to 0. As shown in Figure 8.7 in common anode type, all the anodes of LEDs are connected common with the +5V power supply and hence the name common anode 7-segment LED. Similarly, for common cathode configuration, it shows that all the cathodes of LEDs are connected common to the ground, and hence the name common cathode 7-segment LED.

Two methods of interfacing LED with the Microcontroller Intel 8051/AT89C51.

1. Using lookup table. This uses 7 output pins of microcontroller.
2. Using 7447 decoder. This method uses 4 output pins of microcontroller.

The difference between the two main methods is simple and clear. In both the cases, microcontroller communicates with external world through its ports. But, in the 1st case, we connect all the 8 pins of the port directly to the LED and control the voltage through the ports manually to display the desired number. But, in the second case, we send the BCD of the number that we wanted to display to a middleware IC 7447, the BCD to LED code converter, which by itself gives out the correspondent 7-segment codes to the LED.

8.2.1 Look-up Table For Seven-Segment LED

Calculation of lookup table is as follows:

The lookup table contains the input pattern for the LED legs, to display the corresponding digits. The table shows the seven segment requirement pattern to display the hex number, with the seven segment conversion.

For example, let us consider the display of the number 0, where we need to switch ON all the LEDs which are there at the boundary, i.e., for a COMMON ANODE (CA) LED, we should supply 5 volts to these LEDs. The 6 LEDs ('a' through 'f') should get binary 1, the dot and the (middle) hyphen segment should get 0 volts or the binary zero. Effectively the seven-segment pattern code will be (1100 0000) C0h. That is what we OUT through the port pins.

For a common cathode LED, the display pattern will be the complement of that common anode pattern.

For Common Anode configuration:

Table 8.1 *7-segment display code for the common anode configuration*

Hex No.	Seven-segment conversion								Seven-segment equivalent
	Dot	g	f	e	d	c	b	a	
0	1	1	0	0	0	0	0	0	C0h
1	1	1	1	1	1	0	0	1	F9h

2	1	0	1	0	0	1	0	0	A4h
3	1	0	1	1	0	0	0	0	B0h
4	1	0	0	1	1	0	0	1	99h
5	1	0	0	1	0	0	1	0	92h
6	1	0	0	0	0	0	1	0	82h
7	1	1	1	1	1	0	0	0	F8h
8	1	0	0	0	0	0	0	0	80h
9	1	0	0	1	1	0	0	0	98h

For Common Cathode configuration:

Table 8.2 *7-segment display code for the common cathode configuration*

Hex No.	Seven segment conversion								Seven-segment equivalent
	Dot	g	f	e	d	c	b	a	
0	0	0	1	1	1	1	1	1	3Fh
1	0	0	0	0	0	1	1	0	06h
2	0	1	0	1	1	0	1	1	5Bh
3	0	1	0	0	1	1	1	1	4Fh
4	0	1	1	0	0	1	1	0	66h
5	0	1	1	0	1	1	0	1	6Dh
6	0	1	1	1	1	1	0	1	7Dh
7	0	0	0	0	0	1	1	1	07h
8	0	1	1	1	1	1	1	1	7Fh
9	0	1	1	0	0	1	1	1	67h

8.2.2 Using 7447 Decoder

The IC7447 is a BCD to 7-segment pattern converter. This setup is the advanced form of the previous setup where we entered the patterns manually to display the desired character. Here in this case, the IC7447 takes the binary coded decimal (BCD) as the input and outputs the relevant 7-segment code. We connect first four pins of the microcontroller port 2 to the 7447 and the output 8 pins of 7447 to the 8 legs of the LED as shown in the figure. The circuit diagrams are shown below, the first figure is interfacing the CA LED whereas the second is of CC LED. The number required to display is sent as the lower nibble of the port 2 of the microcontroller. The 7447 converts the four input bits (BCD) to their corresponding 7-segment codes. The outputs of the 7447 are connected to the 7-segment display.

Example 8.5

Show the interfacing of seven segment display device of common anode type with 8051 microcontroller and write an assembly language program to display digit 3 continuously. Assume that operating frequency of 8051 microcontroller is 11.0592 MHz.

Solution:

Hardware arrangement:

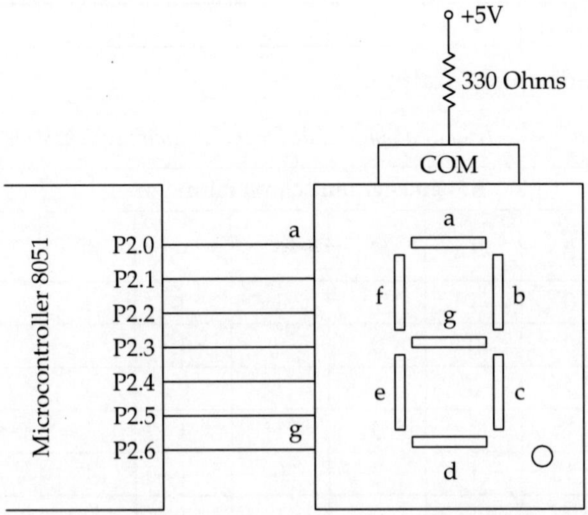

Fig. 8.8 *Interfacing of common anode type 7-segment LED with microcontroller 8051*

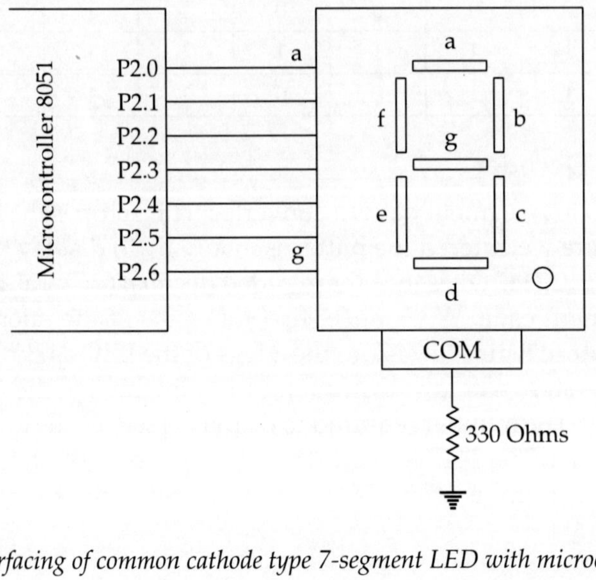

Fig. 8.9 *Interfacing of common cathode type 7-segment LED with microcontroller 8051*

Figure 8.8 and 8.9 shows how to interface 7-segment LED display to a microcontroller. Now according to the lookup table containing the seven-segment pattern to display the corresponding hex digits, e.g., consider we have to display '1'. From the above figure we come to know that turning ON segment B & C will show '1' on the 7-segment display so P2.1 & P2.2 should be LOGIC 0 whereas rest of the pins should be LOGIC 1. Table 8.1 shows the lookup table for COMMON ANODE display.

Assembly Language Program (using common anode configuration):

```
            ORG 0000H
            MOV P2, #0FFH       ; OFF all the LED's
LOOP:       MOV P2, #0B0H       ; Display code for 3
            ACALL DELAY
            SJMP LOOP
DELAY:      MOV R3, #0FFH
REP:        MOV R2, #0FFH
HERE:       DJNZ R2, HERE
            DJNZ R3, REP
            RET
```

Simulation using Protius and IDE 8051 simulation software:

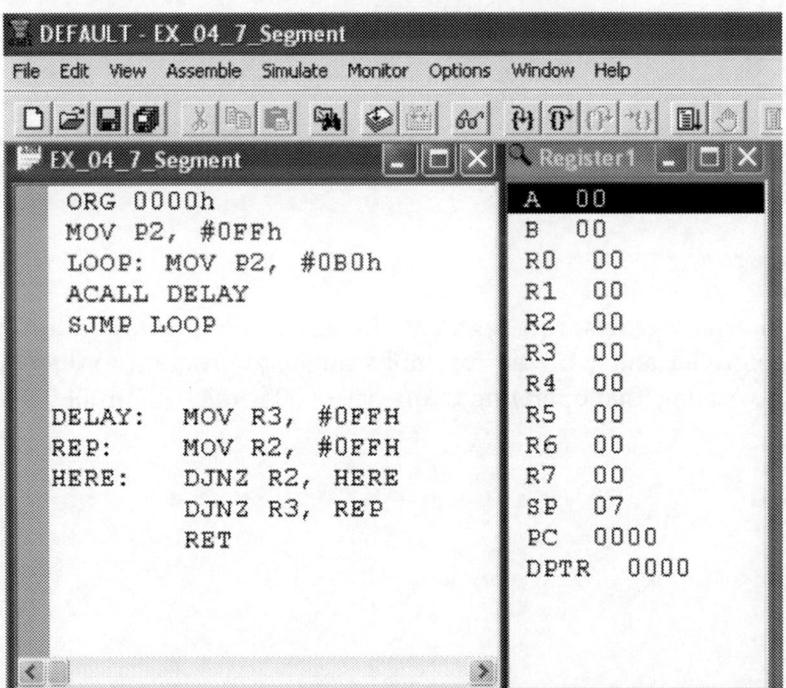

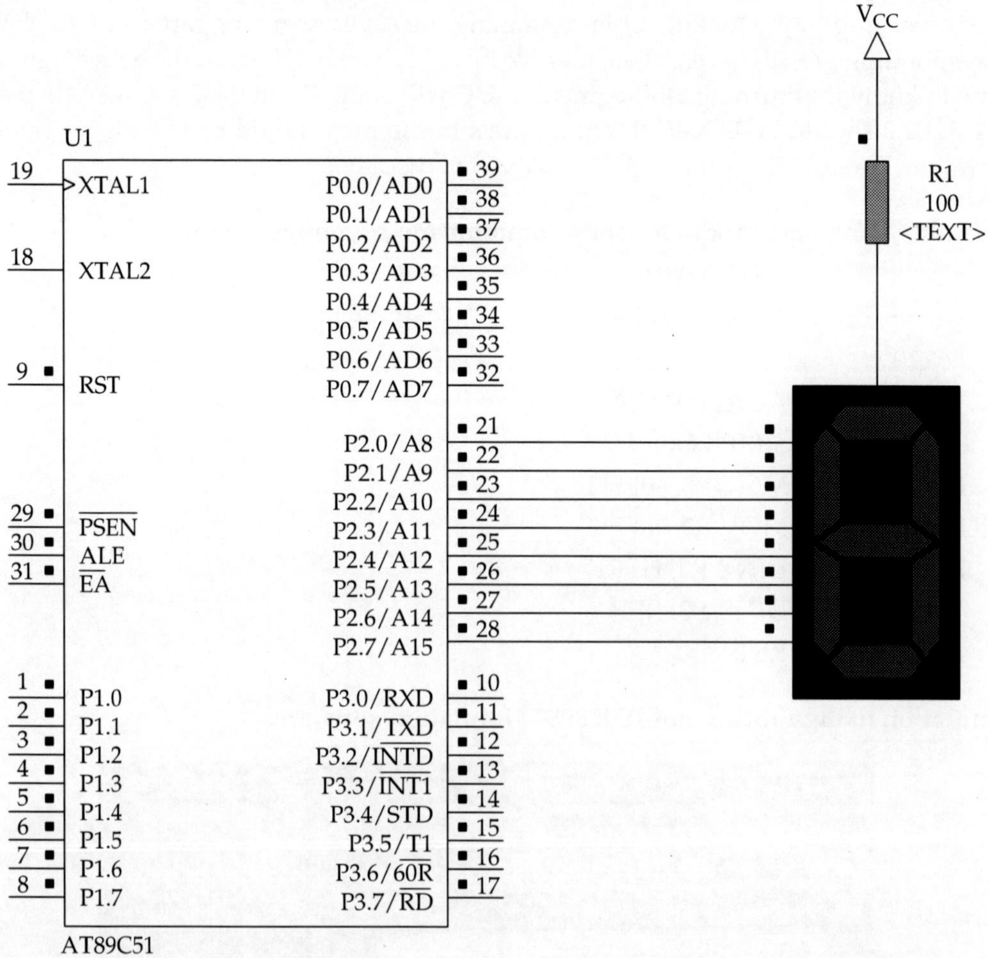

Example 8.6

Show the interfacing of seven-segment display device common anode type with 8051 microcontroller and write an assembly language program to display digit 0 to 9 continuously. Assume that operating frequency of 8051 microcontroller is 11.0592 MHz.

Solution: Hardware arrangement:

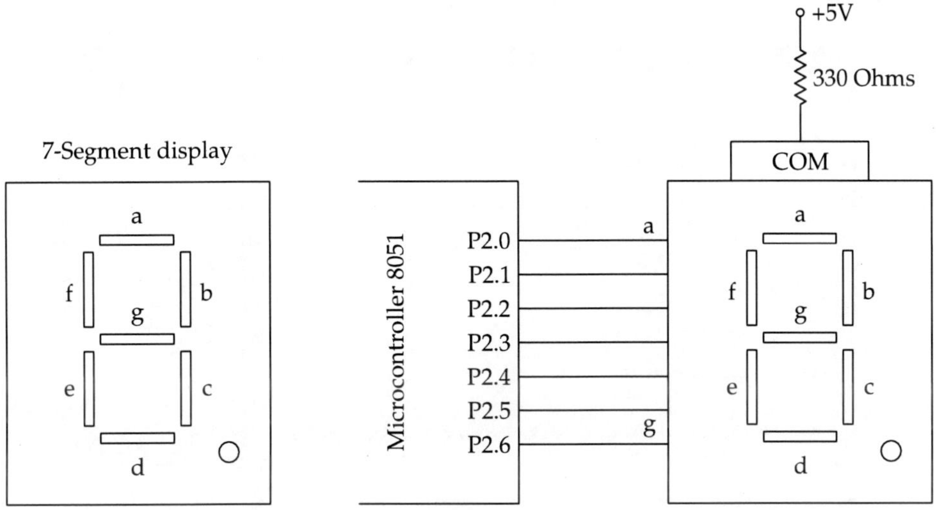

Fig. 8.10 *Interfacing of 7-segment LED (common anode) with 8051 microcontroller*

Assembly Language Program:

```
        ORG 0000h
REPEAT: MOV R2, #10
        MOV DPTR, #MSG
LOOP:   CLR A
        MOVC A, @A+DPTR
        MOV P2,A
        ACALL DELAY
        INC DPTR
        DJNZ R2, LOOP
        LJMP REPEAT
DELAY:  MOV R3, #0FFH
REP:    MOV R4, #0FFH
        MOV R5, #0FFH
HERE:   DJNZ R5, HERE
HERE1:  DJNZ R4, HERE1
        DJNZ R3, REP
        RET
MSG:    DB 0C0H,0F9H,0A4H,0B0H,99H,92H,82H,0F8H,80H,98H
        END
```

Simulation using Protius and IDE 8051 simulation software:

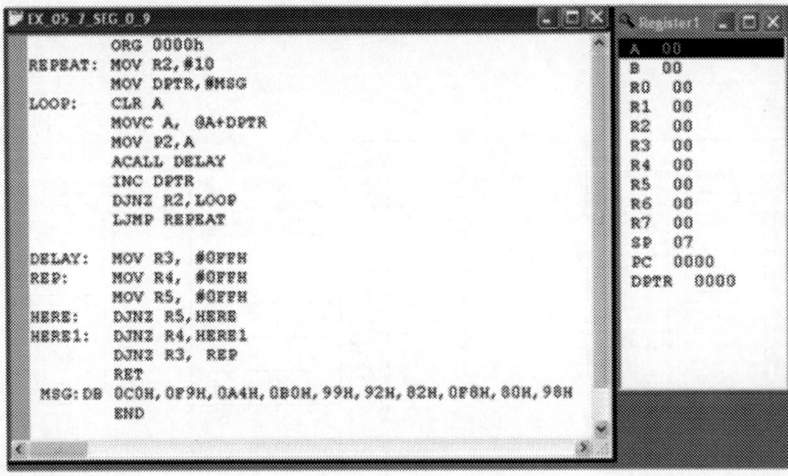

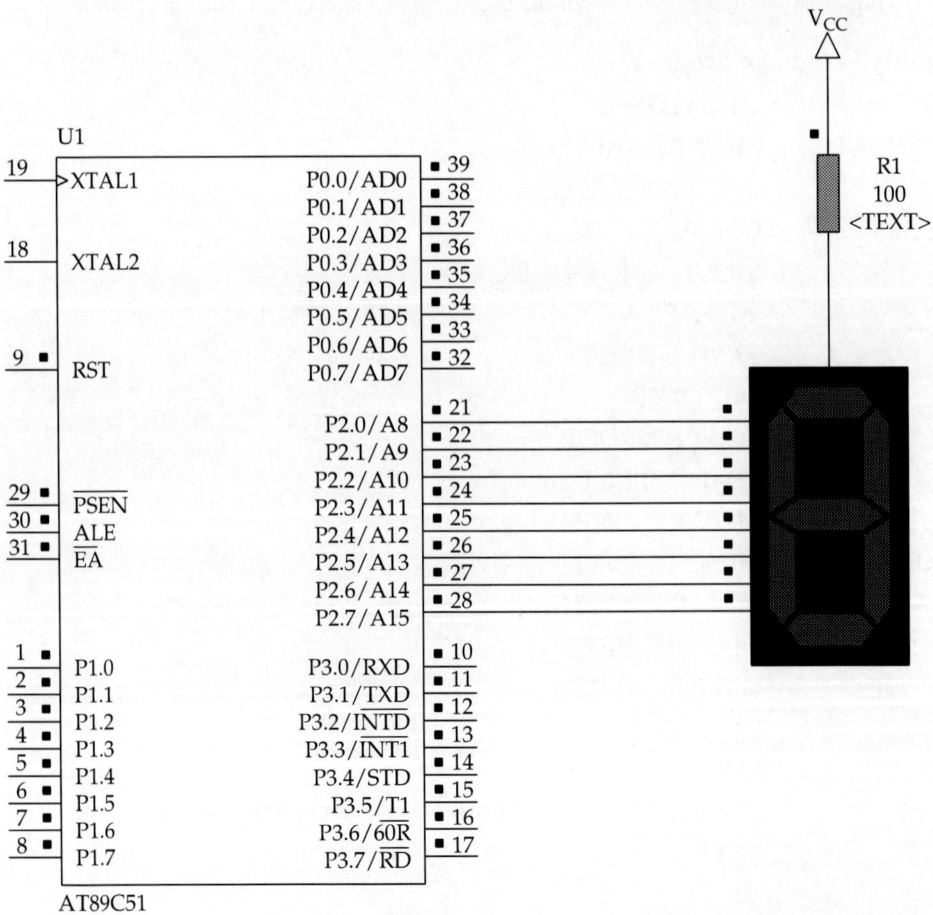

Example 8.7

Show the interfacing of seven-segment display device of common anode type using 7447 decoder with 8051 microcontroller and write an assembly language program to display digits 0 to 9 continuously. Assume that operating frequency of 8051 microcontroller is 11.0592 MHz.

Solution:

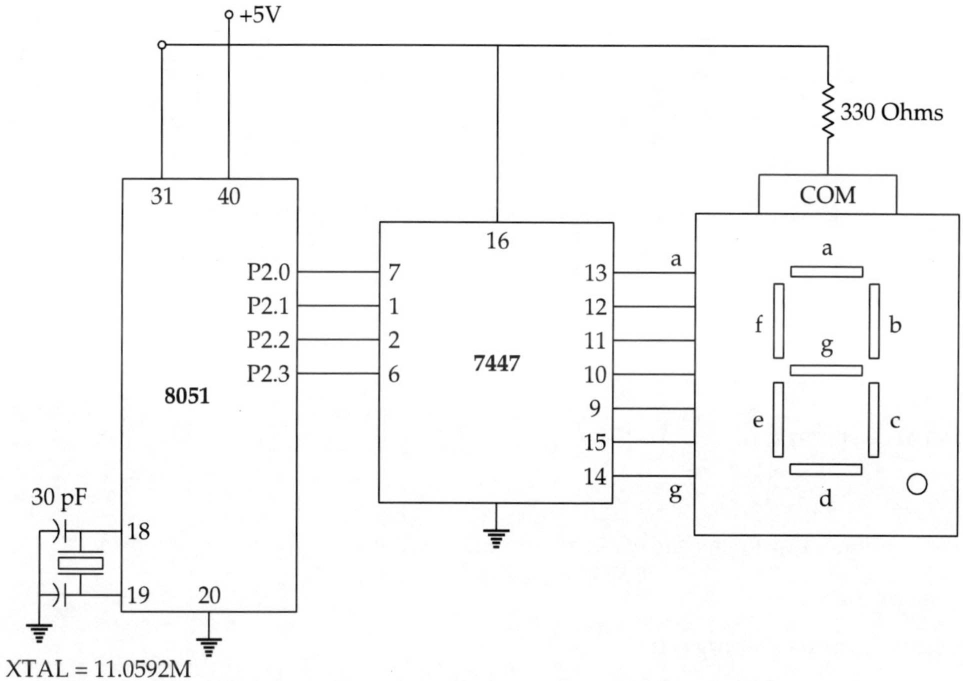

Fig. 8.11 *Interfacing of 7-segment LED in common anode configuration with 8051 microcontroller using IC 7447*

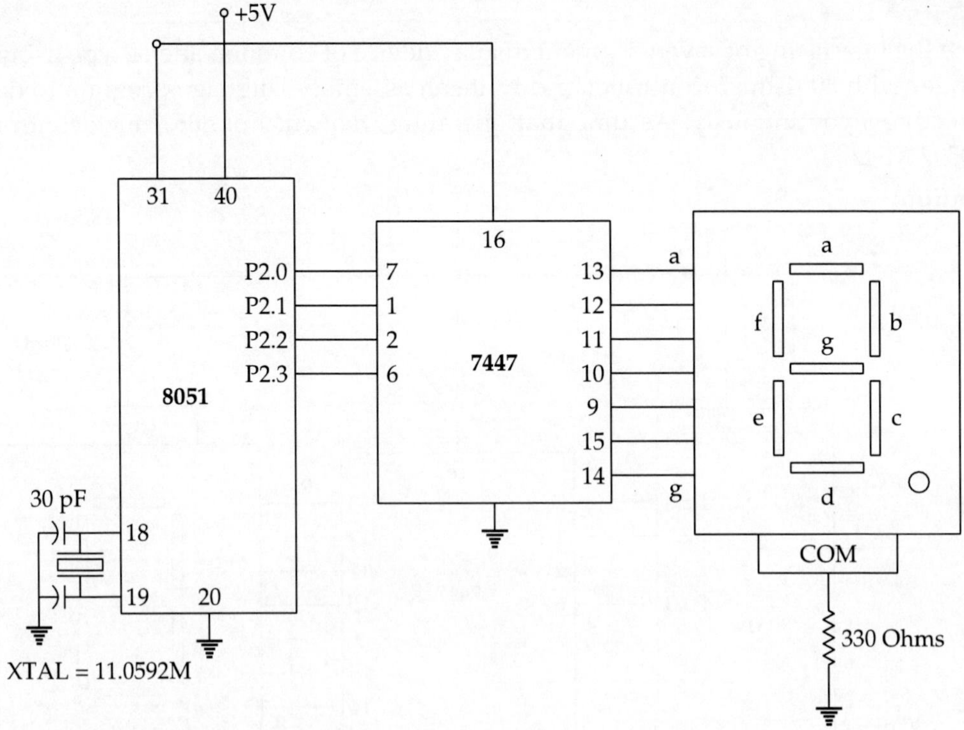

Fig. 8.12 *Interfacing of 7-segment LED in common cathode configuration with 8051 microcontroller using IC 7447*

Assembly Language Program:

```
            ORG 0000h
AGAIN:      MOV A, #00H          ; Start from zero
UP:         MOV P2, A            ; Move to Port 2
            ACALL DELAY
            INC A
            CJNE A, #0AH, UP     ; Send only from 0 to 9
            SJMP AGAIN
DELAY:      MOV R3, #0FFH
REP:        MOV R4, #0FFH
            MOV R5, #0FFH
HERE:       DJNZ R5, HERE
HERE1:      DJNZ R4, HERE1
            DJNZ R3, REP
            RET
```

Simulation using Protius and IDE 8051 simulation software:

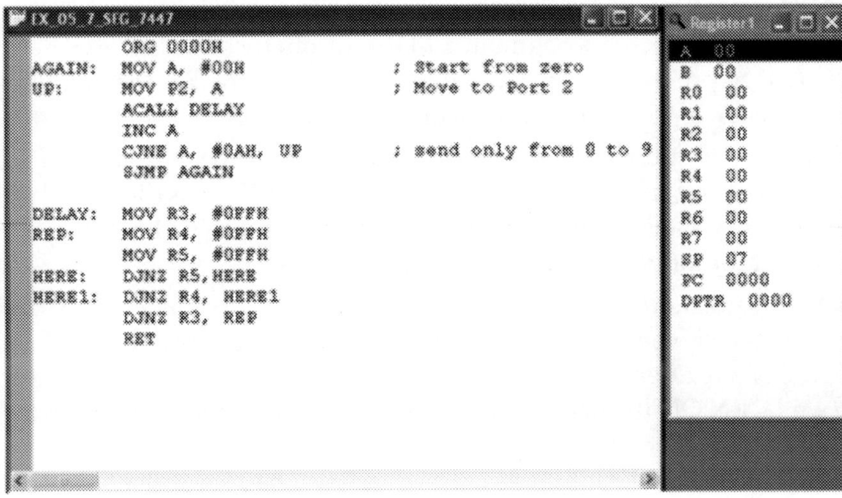

```
         ORG 0000H
AGAIN:   MOV A, #00H            ; Start from zero
UP:      MOV P2, A              ; Move to Port 2
         ACALL DELAY
         INC A
         CJNE A, #0AH, UP       ; send only from 0 to 9
         SJMP AGAIN

DELAY:   MOV R3, #0FFH
REP:     MOV R4, #0FFH
         MOV R5, #0FFH
HERE:    DJNZ R5,HERE
HERE1:   DJNZ R4, HERE1
         DJNZ R3, REP
         RET
```

```
A    00
B    00
R0   00
R1   00
R2   00
R3   00
R4   00
R5   00
R6   00
R7   00
SP   07
PC   0000
DPTR 0000
```

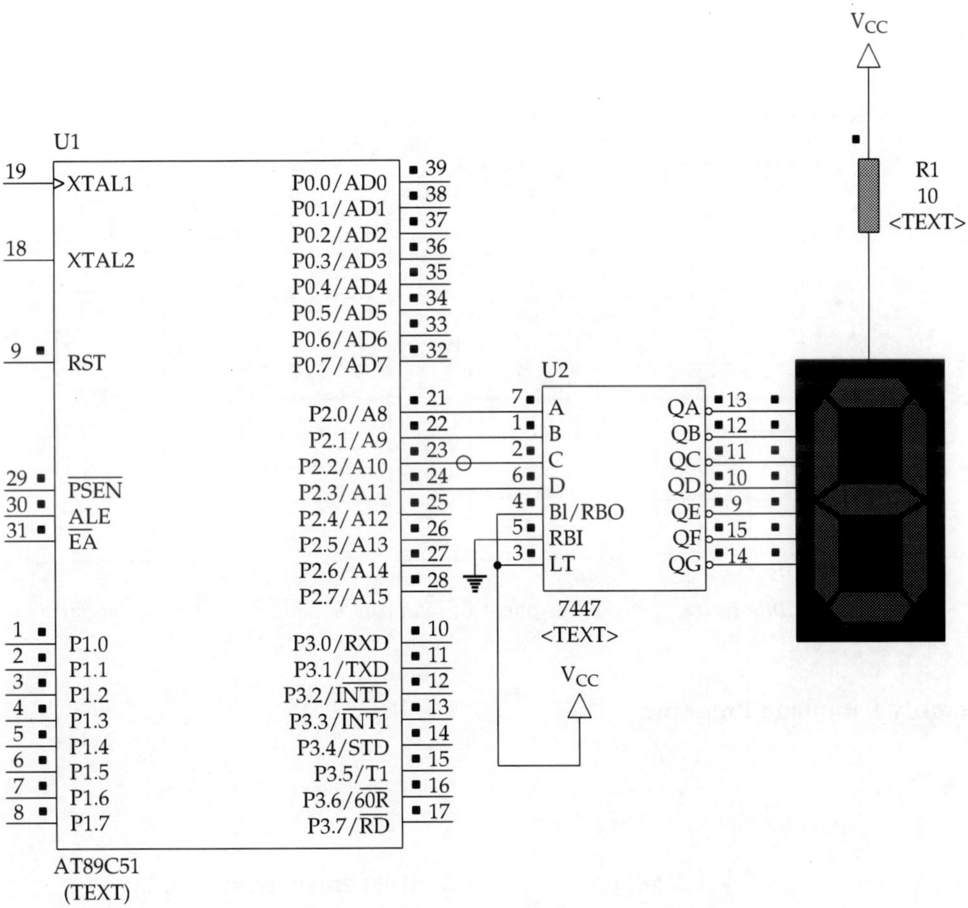

Example 8.8

Show the interfacing of two 7-segment LED with microcontroller 8051, and write an assembly language program to display digit 31 continuously. For the delay, assume that operating frequency of 8051 microcontroller is 11.0592 MHz.

Solution: We can now interface a single 7 segments to the microcontroller but for interfacing multiple 7 segments we use scanning principle where one 7-seg is displayed after another but this process is very fast hence the flickering cannot be seen by human eye. Figure 8.12 shows the circuit for interfacing two 7-segment displays.

When interfacing more than one 7-segment displays the segments (a – g) of all displays are connected together whereas their ANODE (cathode in case of common cathode displays) are switched ON one after another. Consider we have to display '31' on the above 7-seg display so we TURN ON the first transistor by setting its corresponding pin to 1 and then give the 7-seg equivalent code for '3' which is B0h. Then we TURN OFF the first transistor and TURN ON the second & output its corresponding 7-seg equivalent code of '1' i.e. F9h. Then we again go back to display '3'. This is a never ending loop.

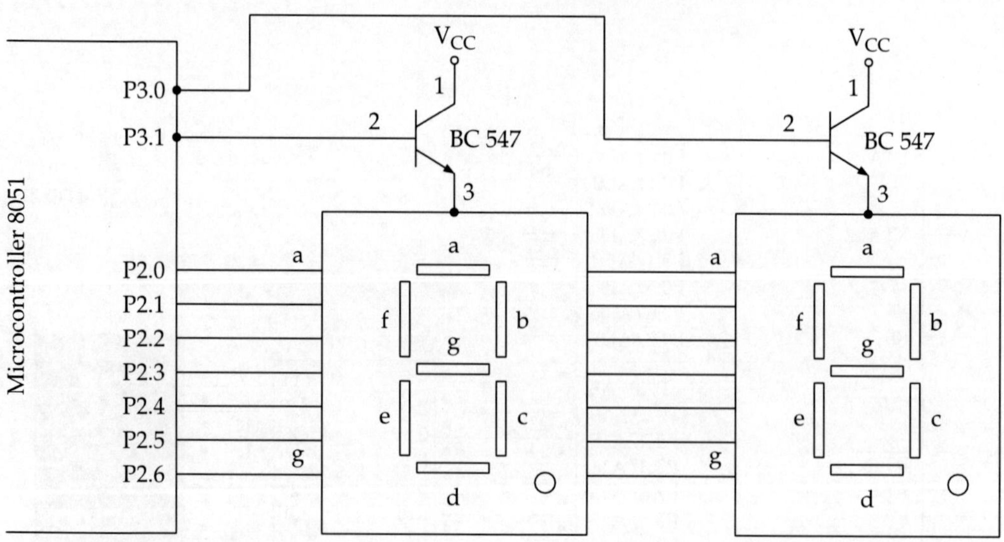

Fig. 8.13 *Interfacing of two 7 – segment LED in common anode configuration with 8051 microcontroller*

Assembly Language Program:

```
ORG 0000h
CLR P3.0
CLR P3.1
MOV P2, #0FFh        ; Off all seven-segment display
```

```
Loop:   CLR P3.0
        SETB P3.1
        MOV P2, #B0h
        CALL delay
        CLR P3.1
        SETB P3.0
        MOV P2, #F9h
        CALL delay
        AJMP loop
```

8.3 STEPPER MOTOR

This section of stepper motors will explain everything that you need to know about stepper motors, that is, how the stepper motors are interfaced with the microcontroller; why the drivers are needed for the interfacing; types of drivers used for the interfacing; how the full step and half step rotation is performed on the stepper motors; what are the types of stepper motors, how we can buy the stepper motors from the market, etc. Stepper motors can be used in various areas of your microcontroller projects such as making robots, robotic arm, automatic door lock system and automation related areas, etc. This section will explain the construction of stepper motors (universal, unipolar and bipolar stepper motors), basic principle, different controlling types (half step and full step), interfacing Techniques (using drivers L293D or ULN2003) and programming your microcontroller in assembly language programming to control stepper motor.

A stepper motor is widely used device that translates electrical pulses into mechanical movement. In applications such as disk drives, dot matrix printers, and robotics, the stepper motor is used for position control. The sequence of the applied pulses is directly related to the direction of motor shafts rotation. The speed of the motor shafts rotation is directly related to the frequency of the input pulses and the length of rotation of input pulses applied. The advantages and disadvantages of stepper motor are given below.

Advantages

1. The rotation angle of the motor is proportional to the input pulse.
2. The motor has full torque at stand still (if the windings are energized).
3. Precise positioning and repeatability of movement.
4. Have an accuracy of 3–5% of a step and this error is non-cumulative from one step to the next.
5. Excellent response to starting, stopping, reversing.

6. Very reliable since there are no contact brushes in the motor. Therefore, the life of the motor is simply dependant on the life of the bearing.
7. The motor's response to digital input pulses provides open-loop control, making the motor simpler and less costly to control.
8. It is possible to achieve very low speed synchronous rotation with a load that is directly coupled to the shaft.
9. A wide range of rotational speed is proportional to the frequency of the input pulses.

(*b*) Disadvantages

1. Resonance can occur if not properly controlled.
2. Not easy to operate at extremely high speeds.

The stepper motors commonly have a permanent magnet called rotor (also called shaft) surrounded by a stator. There are also stepper motors called variable reluctance stepper motors that do not have permanent magnet rotor.

The most common stepper motors have four stator windings that are paired with a center tapped common as shown in Figure 8.13.

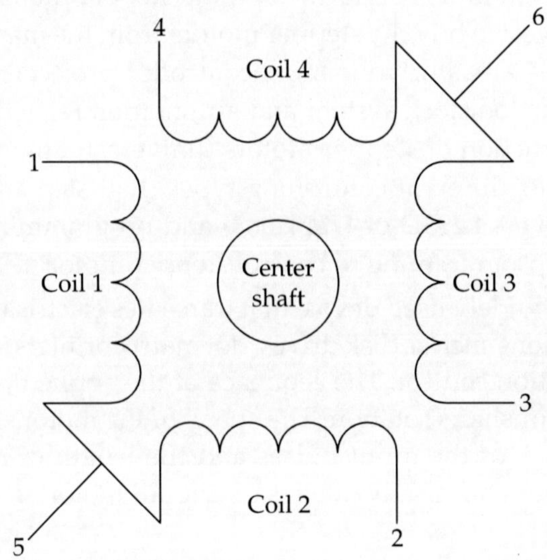

Fig. 8.14 *Stepper motor with four windings*

This type of stepper motor is commonly referred to as a 4-phase or unipolar stepper motor. The center tap allows a change of current direction in each of two coils when a winding is grounded, thereby resulting in a polarity change of the stator.

The stepper motor discussed here has a total of 6 leads, 4 leads representing the four stator windings and 2 commons for the center tapped leads. As the sequence of power is applied to each stator winding, the rotor will rotate. There are several widely used sequences where each has a different degree of precision. Table 8.3 shows a 2-phase, 4-step stepping sequences for the movement of the stepper motor in clockwise and counterclockwise directions.

Table 8.3 *Stepping sequence for stepper motor*

	Step	Winding A	Winding B	Winding C	Winding D	
Clockwise ↓	1	1	0	0	1	Counter clockwise ↑
	2	1	1	0	0	
	3	0	1	1	0	
	4	0	0	1	1	

It is important to note that although we can start with any of the sequence in Table 8.3, once we start we must continue in the proper order. For example, if we start with step-3 we must continue in the sequence of steps 4 – 1 – 2 – 3 – 4 – 1……etc.

8.3.1 Stepper Motor Types

There are three basic stepper motor types:

1. Active rotor: Permanent magnet (PM)
2. Reactive rotor: Variable reluctance (VR)
3. Combination of VR and PM: Hybrid (HB)

These are brushless electrical machines which rotate in fixed angular increments when connected to a sequentially switched DC current. When alternating current is used, the rotation is essentially continuous.

Permanent Magnet (PM)

Often referred as a "tin can" or "canstack" motor the permanent magnet step motor is a low-cost and low-resolution type motor with typical step angles of 7.50 to 150. PM motors as the name implies have permanent magnets added to the motor structure. The rotor no longer has teeth as with the VR motor. Instead the rotor is magnetized with alternating north and south poles situated in a straight line parallel to the rotor shaft. These magnetized motor poles provide increased magnetic flux intensity and because of this the PM motor exhibits improved torque characteristics when compared with the VR type.

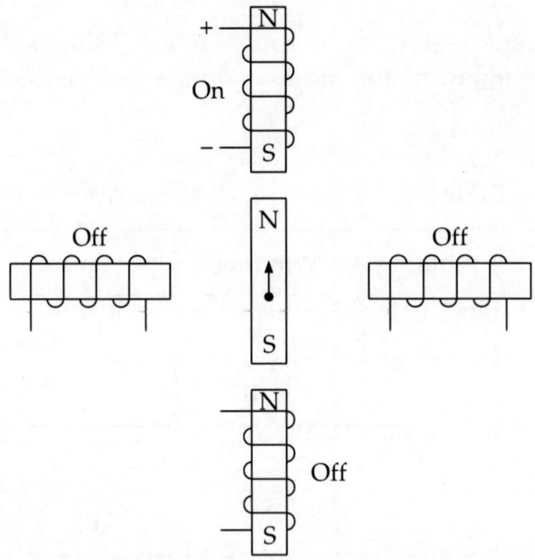

Fig. 8.15 *Permanent magnet type stepper motor*

Variable reluctance (VR)

This type of stepper motor has been around for a long time. It is probably the easiest to understand from a structural point of view. Figure 8.16 shows a cross section of a typical VR stepper motor. This type of motor consists of a soft iron multi-toothed rotor and wound stator. When the stator windings are energized with DC current, the poles become magnetized. Rotation occurs when the rotor teeth are attracted to energized stator poles.

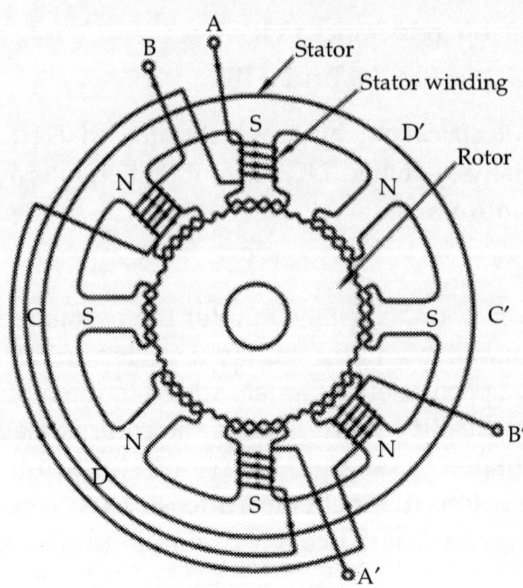

Fig. 8.16 *Cross section view of 1.8° VR motor*

Hybrid (HB)

The hybrid stepper motor is more expensive than the PM stepper motor but provides better performance with respect to step resolution, torque and speed. Typical step angles for the HB stepper motor range from 3.60 to 0.90. The hybrid stepper motor combines the best features of both the PM and VR type stepper motor. The rotor is multi-toothed like the VR motor and contains an axially magnetized concentric magnet around its shaft. The teeth on the rotor provide an even magnetic flux to preferred locations in the air gap. This further increases the detent, holding and dynamic torque characteristics of the motor when compared with both the VR and PM types.

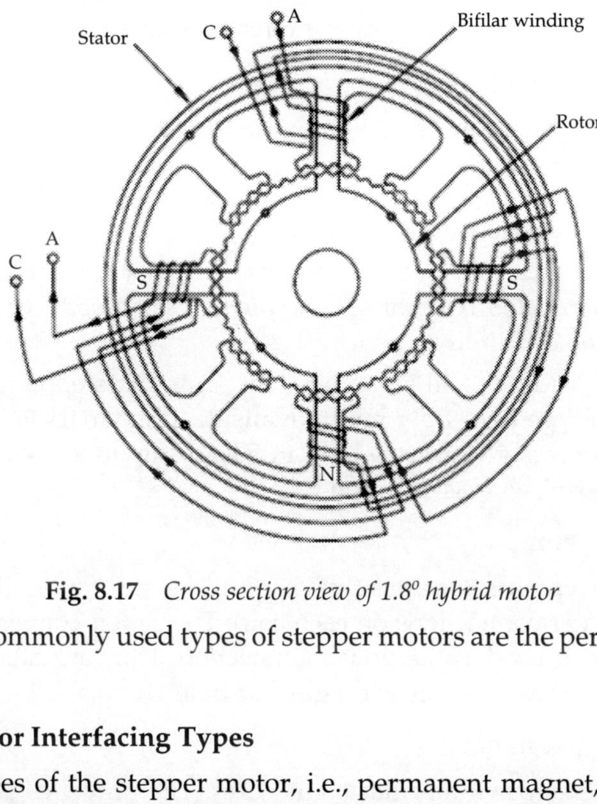

Fig. 8.17 *Cross section view of 1.8° hybrid motor*

The two most commonly used types of stepper motors are the permanent magnet and the hybrid types.

8.3.2 Stepper Motor Interfacing Types

Apart from the types of the stepper motor, i.e., permanent magnet, variable reluctance and hybrid types, the stepper motor had three different types of interfacing also. They are: universal, unipolar and bipolar. The type of the interfacing whether universal, unipolar or bipolar can be identified very easily by the number of connections to the motor. A universal stepper motor has eight, while the unipolar has six and bipolar has four winding connections. The universal stepper motor can be configured for all the three motors, while the unipolar can be either unipolar or bipolar, obviously the bipolar cannot be configured for universal or unipolar mode. Figure 8.18 shows the basic internal connections of all three types of configuration.

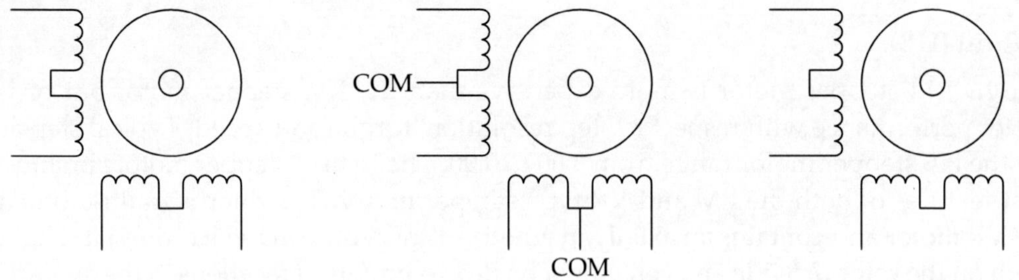

Fig. 8.18 *Types of stepper motor interfacing*

1. **Bipolar stepper motor**

   The bipolar stepper motor usually has four wires coming out of it. Unlike unipolar steppers, bipolar steppers have no common center connection. They have two independent sets of coils instead. You can distinguish them from unipolar steppers by measuring the resistance between the wires. You should find two pairs of wires with equal resistance. If you've got the leads of your meter connected to two wires that are not connected (i.e. not attached to the same coil), you should see infinite resistance (or no continuity).

   As already said, we will talk mostly on "unipolar stepper motors" which is the most common type of stepper motor available in the market. A simple example of 6-lead step motor is given below and in 5-lead step motor wire 5 and 6 are joined together to make 1 wire as common.

2. **Unipolar stepper motor**

   The unipolar stepper motor has five or six wires and four coils (actually two coils divided by center connections on each coil). The center connections of the coils are tied together and used as the power connection. They are called unipolar steppers because power always comes in on this one pole.

3. **Universal stepper motor**

   The universal stepper motor has eight wires. The universal stepper motor can be used as a bipolar or unipolar by using four or six wires out of eight wires. The name universal because, this stepper motor is used to configure as a bipolar stepper motor as well as in unipolar stepper motor.

8.3.3 Connecting Unipolar Stepper Motor

There are actually many ways you can interface a stepper motor to your controller, out of them the most used interfaces are:

1. Interface using L293D-H-bridge motor driver
2. Interface using ULN2003/2004-darlington arrays

We will discuss both connection techniques one by one. The above-mentioned methods need 4 controller pins for interface.

Connecting unipolar stepper using L293D

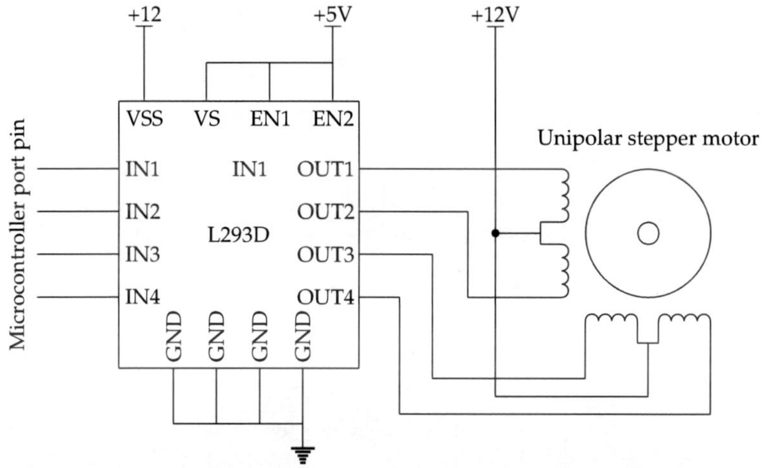

Fig. 8.19 *Interfacing of unipolar Stepper motor using L293D*

As you see in the circuit above the four pins "Controller pin" 1, 2, 3 and 4 will control the motion and direction of the stepper motor according to the step sequence programmed in the controller.

Connecting unipolar stepper using ULN 2003 driver IC:

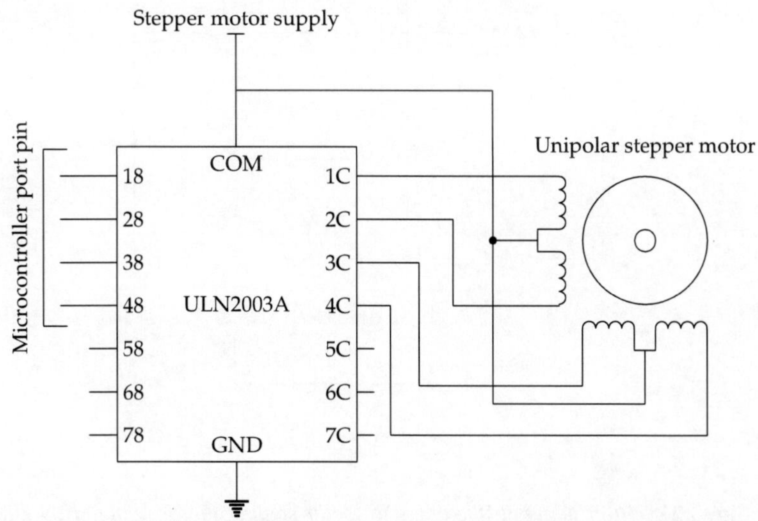

Fig. 8.20 *Interfacing of unipolar stepper motor using ULN 2003*

As already discussed in case of L293D, here in this circuit too the four pins "Controller pin" 1, 2, 3 and 4 will control the motion and direction of the stepper motor according to the step sequence sent by the microcontroller.

2-wire connection for Unipolar Stepper Motor

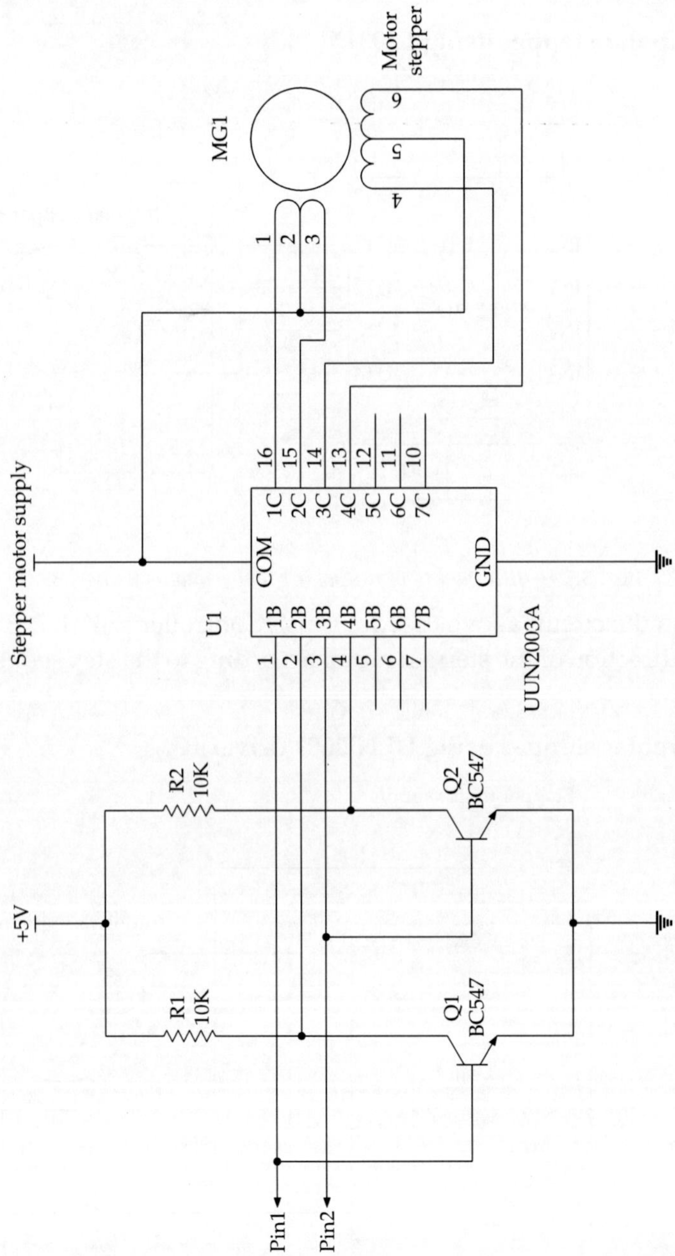

Fig. 8.21 *Interfacing of two wires unipolar stepper motor using ULN 2003*

We have seen the generally used 4-wire connection method for interfacing unipolar stepper motor, but we can simplify the design to make controller use less pins with the help of 2-wire connection method. The circuit for 2-wire connection is shown in Figure 8.21.

8.3.4 Connecting Bipolar Stepper Motor

We have studied that Bi-polar stepper motors have two different coils. The step sequence for bipolar stepper motor is same as that of unipolar stepper motors. The driving circuit for this requires an H-bridge as it allows the polarity of the power applied to be controlled independently. This can be done as shown in Figure 8.22.

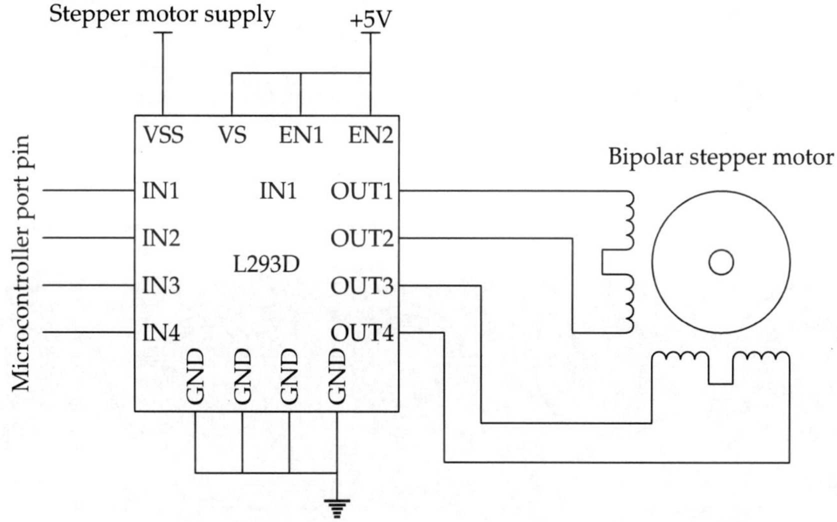

Fig. 8.22 *Interfacing of bipolar stepper motor using L293D*

Now we have seen the methods for connecting stepper motors with microcontroller. So keeping these circuits in mind, we will now look at the programming of microcontroller to control stepper motors.

8.3.5 Step Sequence

Stepper motors can be driven in two different patterns or sequences namely,

1. Full step sequence
2. Half step sequence

We will go through these sequences one by one.

1. Full Step Sequence

In the full step sequence, two coils are energized at the same time and the motor shaft rotates. The order in which coils has to be energized is given in Table 8.4

Table 8.4 *Full mode sequence*

	Full Mode Sequence					
	Step	**A**	**B**	**C**	**D**	
Clockwise ↓	0	1 (ON)	1 (ON)	0 (OFF)	0 (OFF)	Counter Clockwise ↑
	1	0 (OFF)	1 (ON)	1 (ON)	0 (OFF)	
	2	0 (OFF)	0 (OFF)	1 (ON)	1 (ON)	
	3	1 (ON)	0 (OFF)	0 (OFF)	1 (ON)	

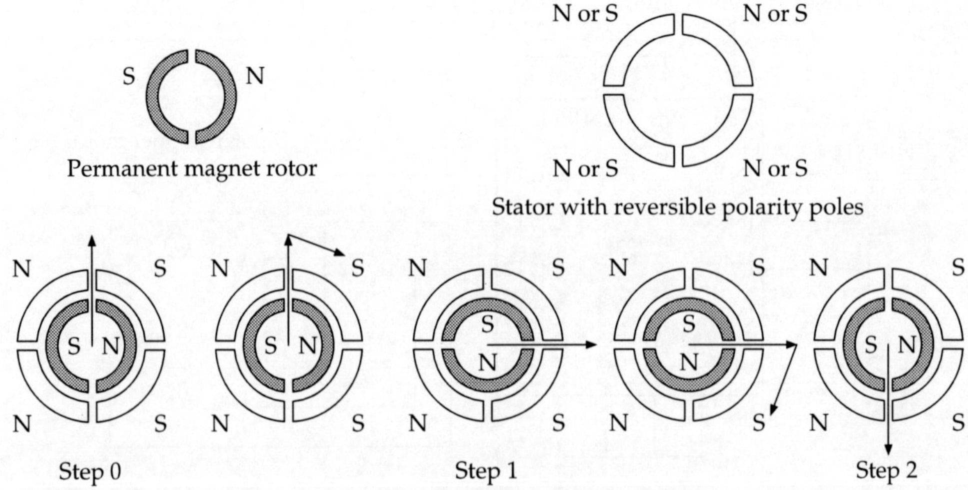

Fig. 8.23 *Stepper motor step sequences (Full)*

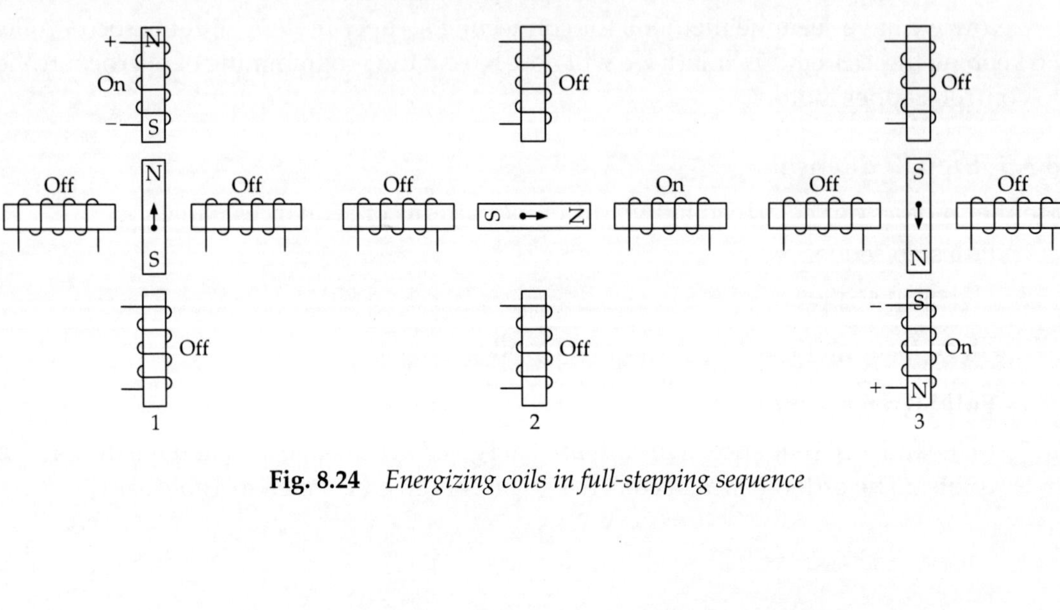

Fig. 8.24 *Energizing coils in full-stepping sequence*

2. Half Step Sequence

In half mode step sequence, motor step angle reduces to half the angle in full mode. So the angular resolution is also increased, i.e., it becomes double the angular resolution in full mode. Also in half mode sequence, the number of steps gets doubled as that of full mode. Half mode is usually preferred over full mode. Table 8.5 shows the pattern of energizing the coils.

Table 8.5 *Half mode sequence*

	Half Mode Sequence					
	Step	**A**	**B**	**C**	**D**	
	0	1	1	0	0	
	1	0	1	0	0	
	2	0	1	1	0	
Clockwise ↓	3	0	0	1	0	**Counter Clockwise** ↑
	4	0	0	1	1	
	5	0	0	0	1	
	6	1	0	0	1	
	7	1	0	0	0	

In the previous illustration, i.e., in full step sequence, only one of the electromagnets was on at a time. We can get finer resolution by sometimes turning on two electromagnets at a time. This technique is called half stepping. The illustration of half-stepping sequence is shown in Figure 8.25.

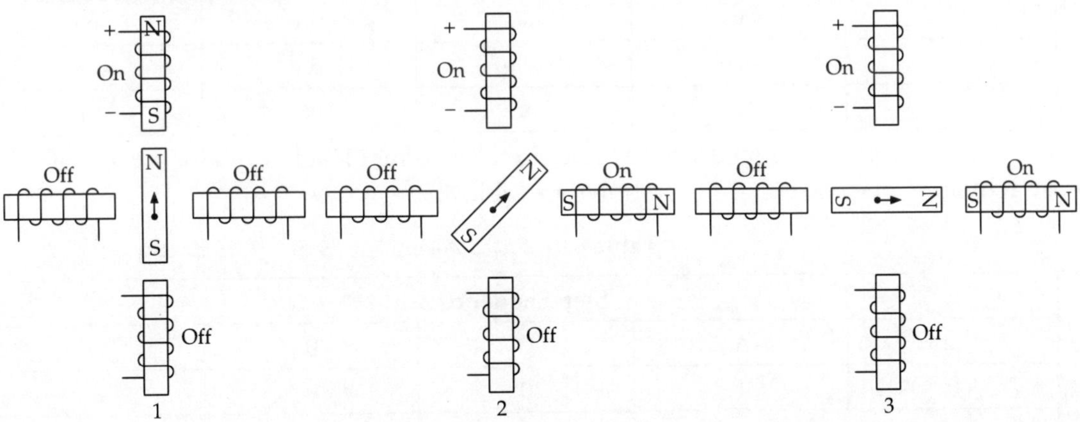

Fig. 8.25 *Energizing coils in half-stepping sequence*

Step Sequence for 2-wire control of unipolar stepper motor

As seen in the above explanation, in every step of the sequence, two wires are always set to opposite polarities. Because of this, it's possible to control steppers with only two wires instead of four, with a slightly more complex circuit. The stepping sequence is the same as it is for the two coils A and B, and the opposite polarity value is given to C and D. The sequence is given in Table 8.6

Table 8.6 *2-wire mode sequence*

2-wire Mode Sequence		
Step	**A**	**B**
0	0	1
1	1	1
2	1	0
3	0	0

Step Sequence for Bipolar stepper motor

Bipolar motor has simpler construction. It has two windings with no center taps and a permanent magnet at the center just like unipolar stepper motors. Being simpler in construction, the stepping sequence is a little complex, as the power for both the coils has to be controlled in such a way that the polarity of the poles gets reversed. This polarity sequence is shown in Table 8.7.

Table 8.7 *Polarity sequence*

Polarity Sequence				
Step	**A**	**C**	**B**	**D**
0	+ve	−ve	−ve	−ve
1	−ve	−ve	+ve	−ve
2	−ve	+ve	−ve	−ve
3	−ve	−ve	−ve	+ve

The above polarity sequence can be interpreted in terms of logic levels for microcontroller by activating one coil at a time as shown in Table 8.8.

Table 8.8 *Step sequence*

Step Sequence				
Step	**A**	**C**	**B**	**D**
0	1	0	0	0
1	0	0	1	0
2	0	1	0	0
3	0	0	0	1

We have now learnt most of the necessary things regarding a stepper motor.

8.3.6 Step Angle

The movement of motor depends on internal construction of motor, in particular the number of teeth on the stator and the rotor. The step angle is the minimum degree of rotation associated with a step angle. Various motors have different step angles. Step angle of the stepper motor is defined as the angle traversed by the motor in one step. To calculate the step angle, simply divide 360 by number of steps a motor takes to complete one revolution. As we have seen that in half mode, the number of steps taken by the motor to complete one revolution gets doubled, so step angle reduces to half.

As in above examples, stepper motor rotating in full mode takes 4 steps to complete a revolution. So step angle can be calculated as

Step angle $Ø = 360°/4 = 90°$ and in case of half mode step angle gets half so 45°.

So this way we can calculate step angle for any stepper motor. Usually, step angle is given in the specification sheet of the stepper motor you are using. Knowing stepper motor's step angle helps you calibrate the rotation of motor also to help you move the motor to correct angular position.

Table 8.9 shows some step angles for various motors. The term steps per revolution is the total number of steps needed to rotate one complete rotation or 360° .

Steps per revolution = 360°/step angle in degrees

Table 8.9 *Stepper motor step angles*

Step angles	Steps per revolution
0.72	500
1.8	200
2.0	180
2.5	144
5.0	72
7.5	48
15	24

Example 8.9

Stepper Motor Interface

Show the interfacing of stepper motor with 8051 microcontroller and write an assembly language program to rotate stepper motor clockwise continuously. Assume the oscillator frequency of 8051 is 11.0592 MHz.

Solution:

Hardware Arrangement:

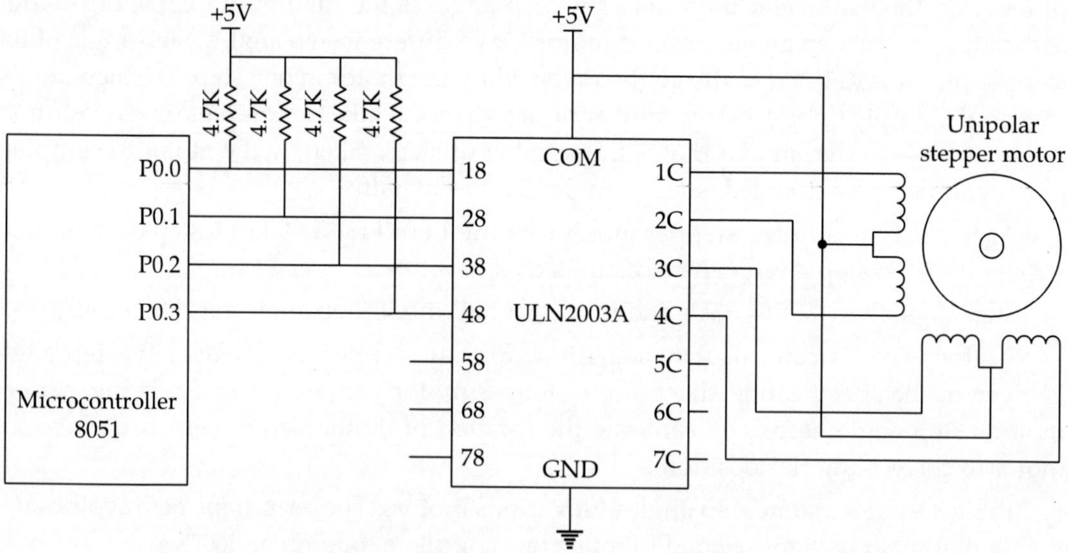

Fig. 8.26 *Interfacing of unipolar stepper motor with 8051 microcontroller using ULN 2003A*

Assembly Language Program:

In this problem, it is assumed that the stepper motor is connected at port 0.0 to port 0.3. Adjusting the delay will increase or decrease the speed of the motor. Here just for demonstration I have taken some delay, you can change it as you want.

Method – I

```
         MOV A, #66h        ; Load step sequence
BACK:    MOV P0, A          ; Issue sequence to motor
         RR A               ; Rotate right (clockwise pattern)
         ACALL DELAY        ; Wait for some time
         SJMP BACK          ; Keep going
DELAY:   MOV R7,#04
WAIT2:   MOV R6,#0FFH
WAIT1:   MOV R5,#0FFH
WAIT:    DJNZ R5,WAIT
```

```
        DJNZ R6,WAIT1
        DJNZ R7,WAIT2
        RET
        END
```

Method – II

```
        ORG 0000H
        STEPPER EQU P0
        SJMP MAIN
        ORG 0030H           ; bypass interrupt vector location
MAIN:   MOV STEPPER, #0CH
        ACALL DELAY
        MOV STEPPER, #06H
        ACALL DELAY
        MOV STEPPER, #03H
        ACALL DELAY
        MOV STEPPER, #09H
        ACALL DELAY
        SJMP MAIN

DELAY:  MOV R7,#04
WAIT2:  MOV R6,#0FFH
WAIT1:  MOV R5,#0FFH
WAIT:   DJNZ R5,WAIT
        DJNZ R6,WAIT1
        DJNZ R7,WAIT2
        RET
        END
```

Simulation using Protius software:

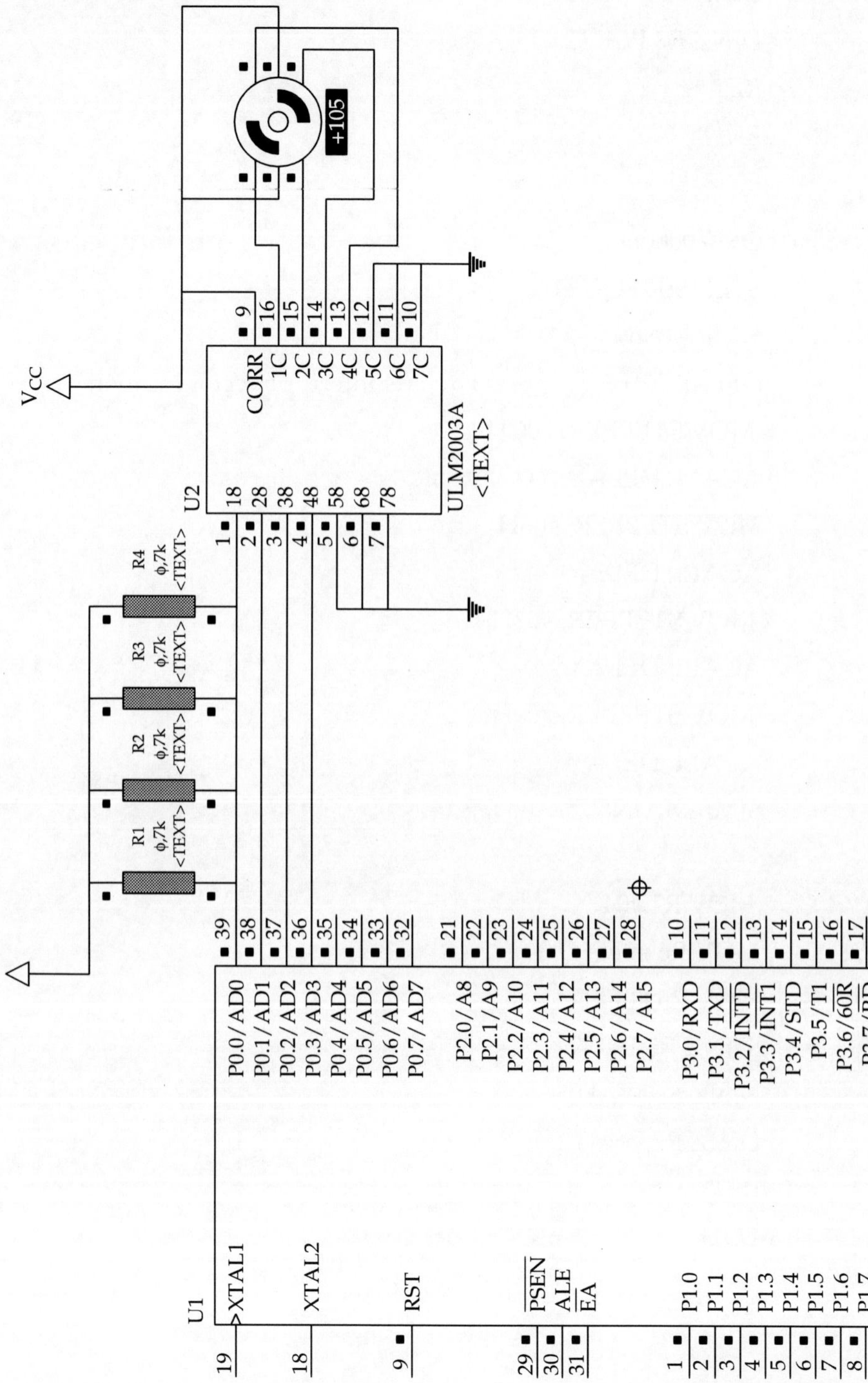

Example 8.10

Stepper Motor Interface

Show the interfacing of stepper motor with 8051 microcontroller and write an assembly language program to rotate stepper motor clockwise continuously in half step sequence. Assume that the oscillator frequency of 8051 is 11.0592 MHz.

Solution:

Hardware arrangement:

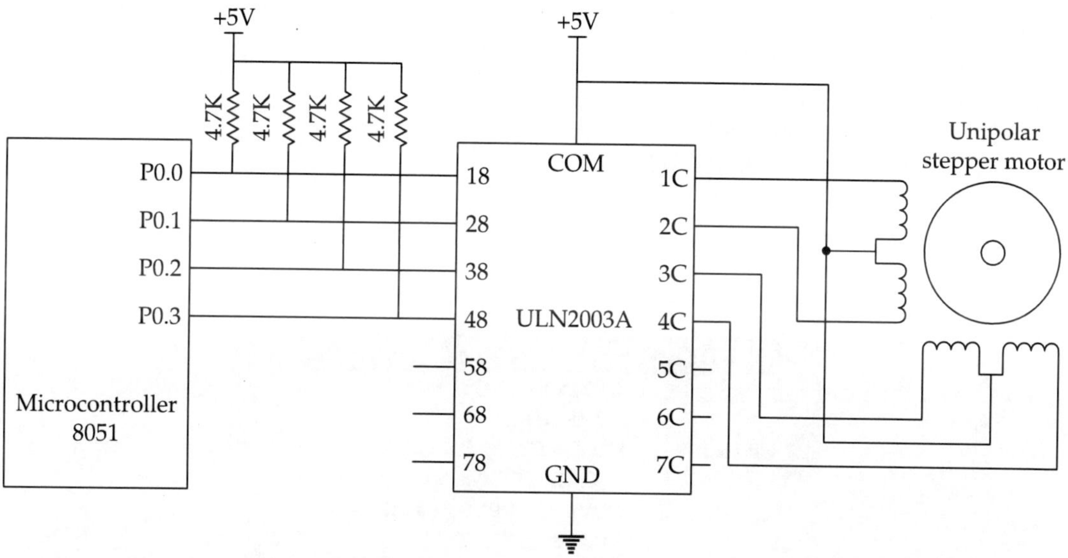

Fig. 8.27 *Interfacing of unipolar stepper motor with 8051 microcontroller using ULN 2003A*

Assembly Language Program:

```
        ORG 0000H
        STEPPER EQU P0
        SJMP MAIN
        ORG 0030H               ; Bypass interrupt vector location
MAIN:   MOV STEPPER, #08H
        ACALL DELAY
        MOV STEPPER, #0CH
        ACALL DELAY
        MOV STEPPER, #04H
        ACALL DELAY
        MOV STEPPER, #06H
        ACALL DELAY
        MOV STEPPER, #02H
        ACALL DELAY
```

```
            MOV STEPPER, #03H
            ACALL DELAY
            MOV STEPPER, #01H
            ACALL DELAY
            MOV STEPPER, #09H
            ACALL DELAY
            SJMP MAIN
DELAY:      MOV R7,#04
WAIT2:      MOV R6,#0FFH
WAIT1:      MOV R5,#0FFH
WAIT:       DJNZ R5,WAIT
            DJNZ R6,WAIT1
            DJNZ R7,WAIT2
            RET
            END
```

Simulation using Protius and IDE 8051 simulation software:

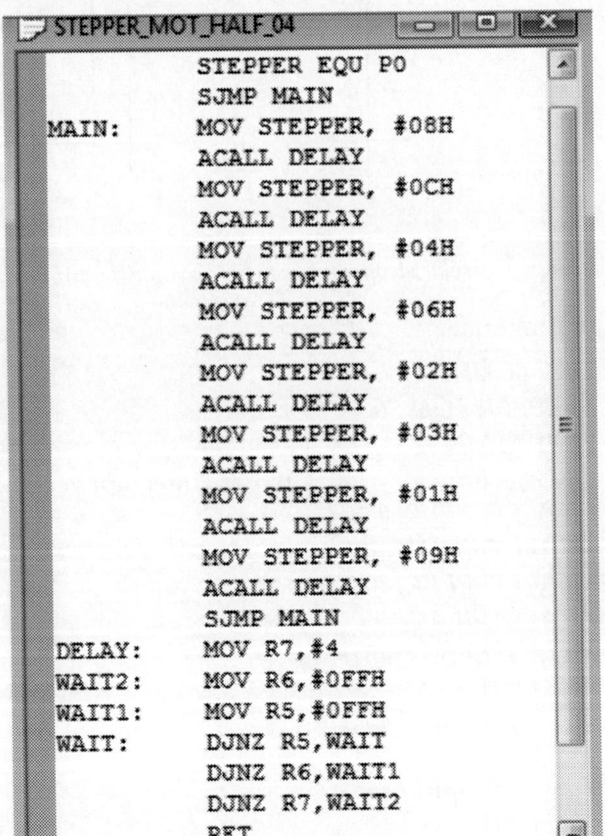

```
STEPPER_MOT_HALF_04
            STEPPER EQU P0
            SJMP MAIN
MAIN:       MOV STEPPER, #08H
            ACALL DELAY
            MOV STEPPER, #0CH
            ACALL DELAY
            MOV STEPPER, #04H
            ACALL DELAY
            MOV STEPPER, #06H
            ACALL DELAY
            MOV STEPPER, #02H
            ACALL DELAY
            MOV STEPPER, #03H
            ACALL DELAY
            MOV STEPPER, #01H
            ACALL DELAY
            MOV STEPPER, #09H
            ACALL DELAY
            SJMP MAIN
DELAY:      MOV R7,#4
WAIT2:      MOV R6,#0FFH
WAIT1:      MOV R5,#0FFH
WAIT:       DJNZ R5,WAIT
            DJNZ R6,WAIT1
            DJNZ R7,WAIT2
            RET
```

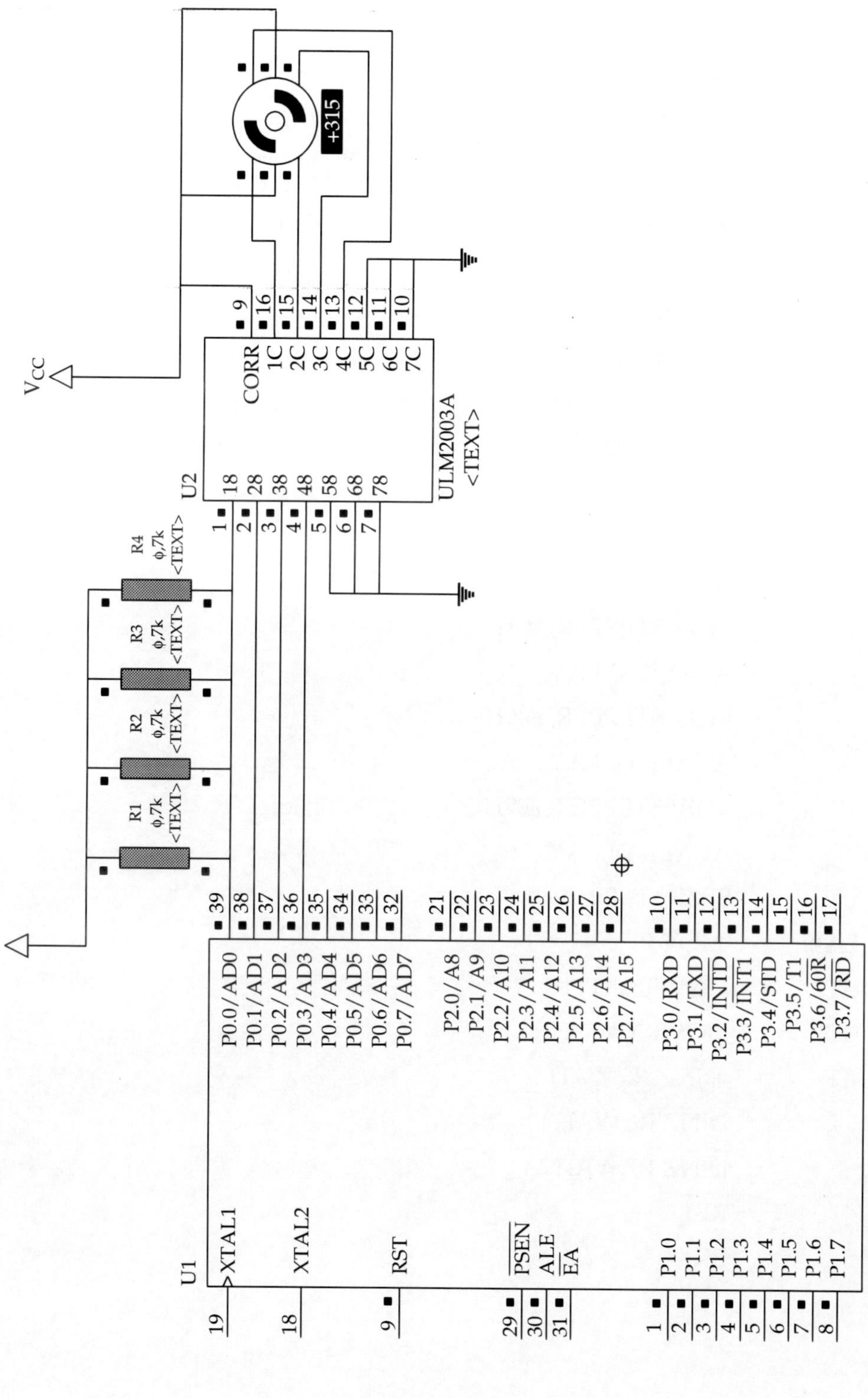
U1
AT89C51
(TEXT)
U2
ULM2003A
<TEXT>
VCC
R1
R2
R3
R4
+315

Example 8.11

Stepper Motor Interface for 2-wire unipolar

Show the interfacing for 2-wire connection of unipolar stepper motor with 8051 microcontroller and write an assembly language program to rotate stepper motor clockwise continuously in half step sequence. Assume that oscillator frequency of 8051 is 11.0592 MHz.

Solution:

Assembly Language Program:

```
            ORG 0000H
            STEPPER EQU P0
            SJMP MAIN
            ORG 0030H              ; bypass interrupt vector location
MAIN:       MOV STEPPER, #03H
            ACALL DELAY
            MOV STEPPER, #01H
            ACALL DELAY
            MOV STEPPER, #00H
            ACALL DELAY
            MOV STEPPER, #02H
            ACALL DELAY
            SJMP MAIN
DELAY:      MOV R7,#04
WAIT2:      MOV R6,#0FFH
WAIT1:      MOV R5,#0FFH
WAIT:       DJNZ R5,WAIT
            DJNZ R6,WAIT1
            DJNZ R7,WAIT2
            RET
            END
```

Simulation using Protius and IDE 8051 simulation software:

STEPPER_MOT_2_WIRE

```
          ORG 0000H
          STEPPER EQU P0
          SJMP MAIN
          ORG 0030H
MAIN:     MOV STEPPER, #03H
          ACALL DELAY
          MOV STEPPER, #01H
          ACALL DELAY
          MOV STEPPER, #00H
          ACALL DELAY
          MOV STEPPER, #02H
          ACALL DELAY
          SJMP MAIN

DELAY:    MOV R7,#4
WAIT2:    MOV R6,#0FFH
WAIT1:    MOV R5,#0FFH
WAIT:     DJNZ R5,WAIT
          DJNZ R6,WAIT1
          DJNZ R7,WAIT2
          RET
```

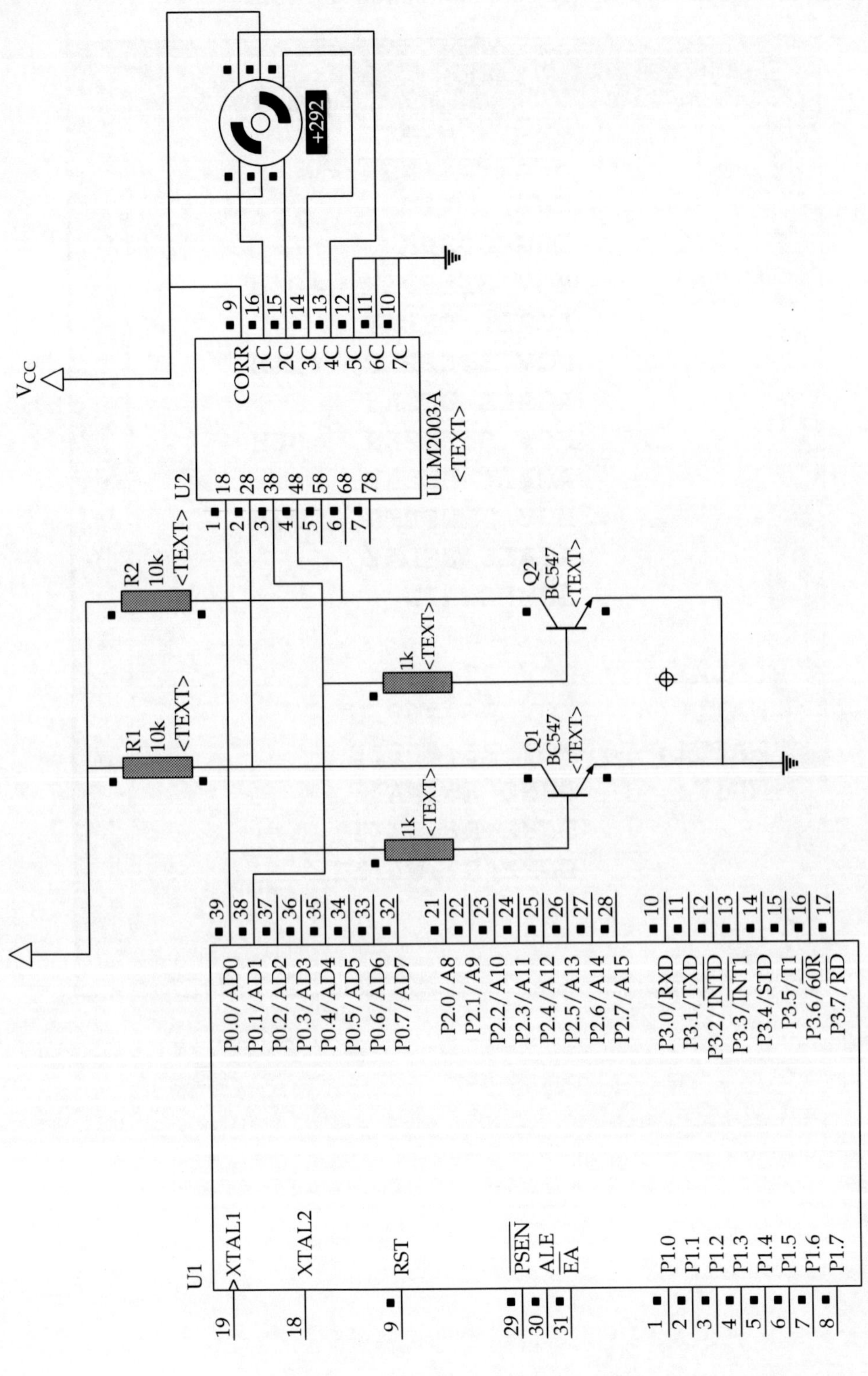
U1
AT89C51
(TEXT)
U2
ULM2003A
R1
10k
R2
10k
Q1
BC547
Q2
BC547
1k
V_{CC}
+292

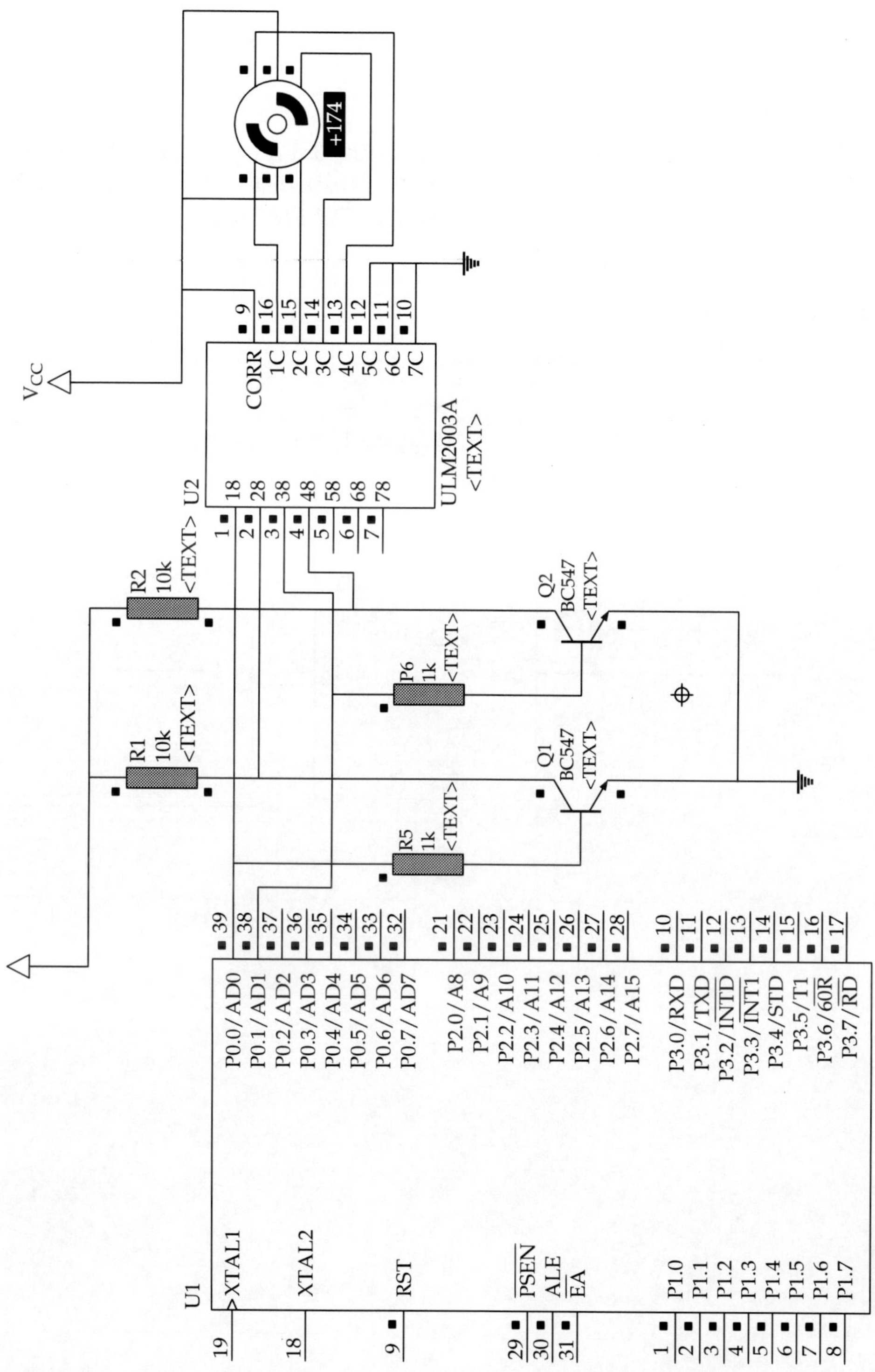
VCC
U1
AT89C51
(TEXT)
U2
ULM2003A
<TEXT>
R1
10k
<TEXT>
R2
10k
<TEXT>
P6
1k
<TEXT>
R5
1k
<TEXT>
Q1
BC547
<TEXT>
Q2
BC547
<TEXT>
+174

Example 8.12

Bipolar Stepper Motor Interface

Show the interfacing for bipolar stepper motor with 8051 microcontroller and write an assembly language program to rotate stepper motor clockwise continuously in full step sequence. Assume that oscillator frequency of 8051 is 11.0592 MHz.

Solution:

Hardware arrangement:

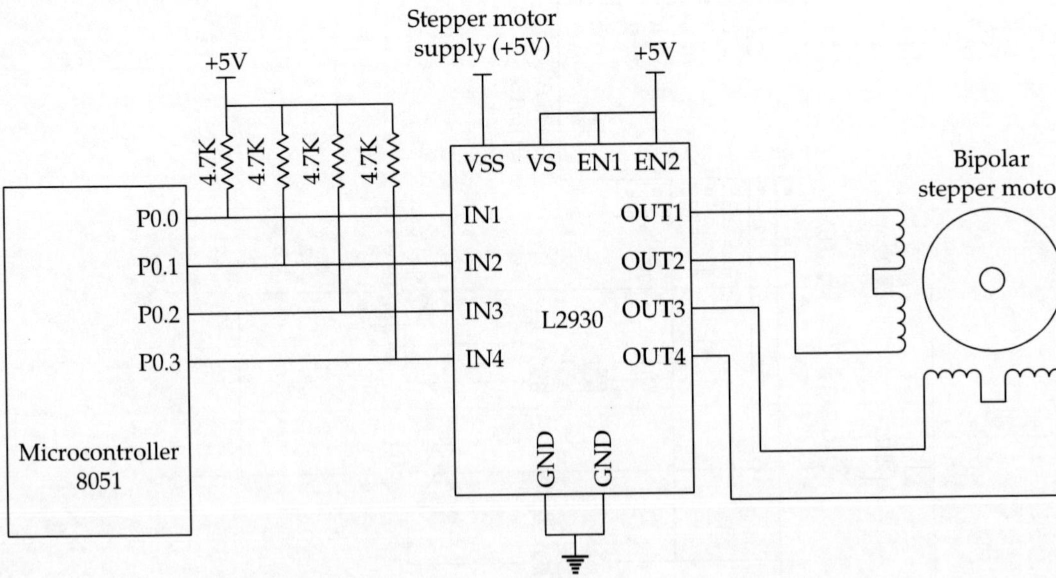

Fig. 8.28 *Interfacing of bipolar stepper motor using L293D*

Assembly Language Program:

```
        ORG 0000H
        STEPPER EQU P0
        SJMP MAIN
        ORG 0030H
MAIN:   MOV STEPPER, #08H
        ACALL DELAY
        MOV STEPPER, #02H
        ACALL DELAY
        MOV STEPPER, #04H
        ACALL DELAY
        MOV STEPPER, #01H
        ACALL DELAY
        SJMP MAIN
DELAY:  MOV R7,#04
WAIT2:  MOV R6,#0FFH
WAIT1:  MOV R5,#0FFH
WAIT:   DJNZ R5,WAIT
        DJNZ R6,WAIT1
        DJNZ R7,WAIT2
        RET
        END
```

Simulation result using Protius simulation software:

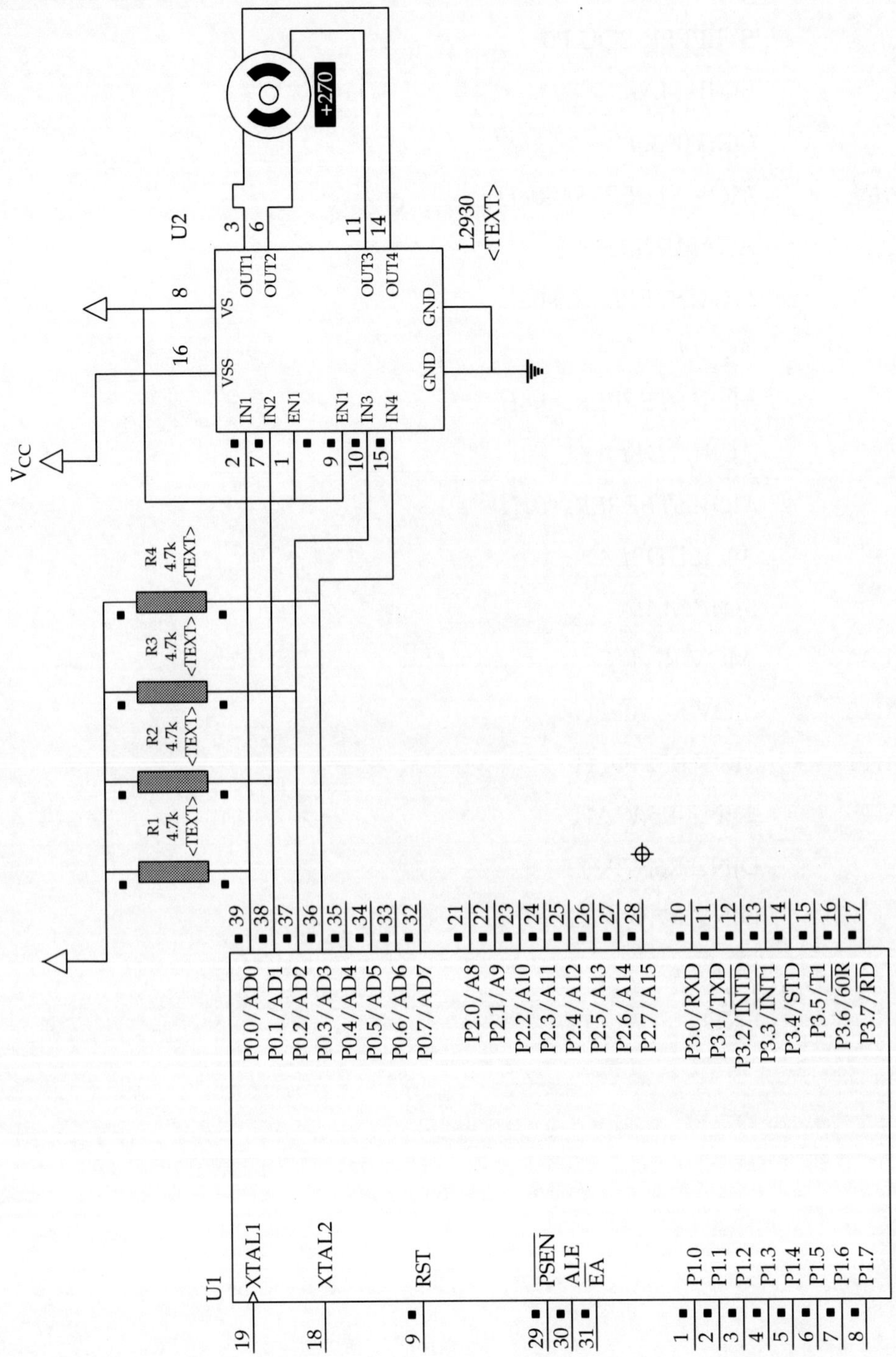

Example 8.13

A switch is connected to port pin P2.0. Write a program to monitor the status of switch and perform the following:

(*a*) If SW = 0 then stepper motor moves clockwise

(*b*) If SW = 1 then stepper motor moves anticlockwise.

Solution:

Hardware arrangement:

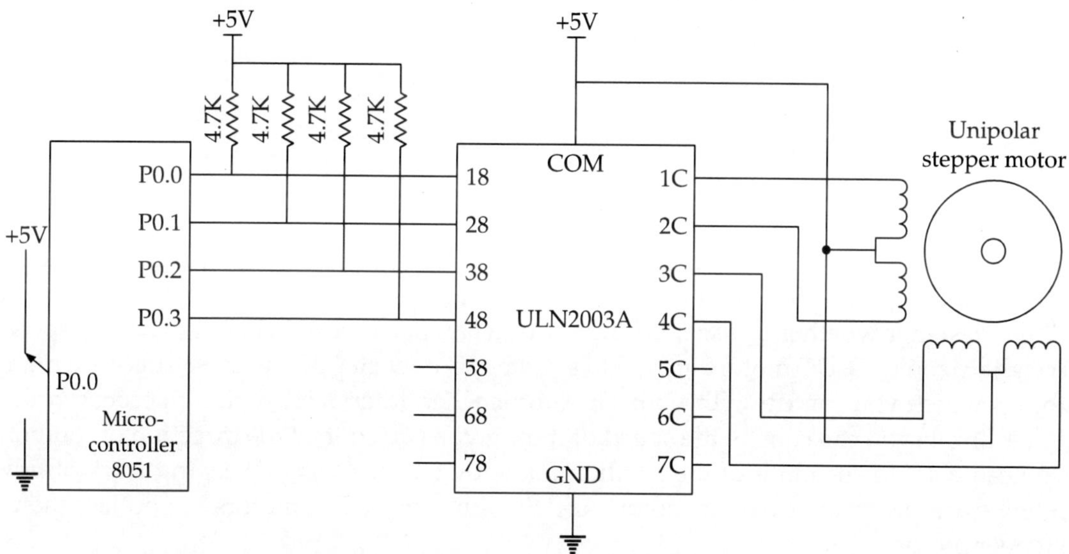

Fig. 8.29 *Interfacing of unipolar stepper motor with 8051 microcontroller with switch*

Assembly Language Program:

```
        SETB P2.0        ; Initialize port pin P2.0 as an input port
        MOV A, #66h
        MOV P0, A
TURN:   JB P2.0, CCW
        RR A
        MOV P0, A
        ACALL DELAY
        SJMP TURN
CCW:    RL A
        MOV P0, A
```

```
        ACALL DELAY
        SJMP TURN
DELAY:  MOV R7, #04
WAIT2:  MOV R6,#0FFH
WAIT1:  MOV R5,#0FFH
WAIT:   DJNZ R5,WAIT
        DJNZ R6,WAIT1
        DJNZ R7,WAIT2
        RET
        END
```

8.4 DC MOTORS

Whenever a robotics hobbyist talks about making a robot, the first thing that comes to his mind is making the robot move on the ground. And there are always two options in front of the designer whether to use a DC motor or a stepper motor. When it comes to speed, weight, size, cost... DC motors are always preferred over stepper motors. There are many things which you can do with your DC motor when interfaced with a microcontroller. For example you can control the speed of motor, you can control the direction of rotation, you can also do encoding of the rotation made by DC motor, i.e., keeping track of how many turns are made by your motors, etc. So you can see DC motors are no less than a stepper motor.

Here we will learn how to interface a DC motor with a microcontroller. Usually, H-bridge is preferred way of interfacing a DC motor. These days many IC manufacturers have H-bridge motor drivers available in the market, like L293D is the most used H-Bridge driver IC. H-bridge can also be made with the help of transistors and MOSFETs, etc. Rather than being cheap, they only increase the size of the design board, which is sometimes not required so using a small 16-pin IC is preferred for this purpose.

Working Theory of H-Bridge

The name "H-bridge" is derived from the actual shape of the switching circuit which controls the motion of the motor. It is also known as "full bridge". Basically, there are four switching elements in the H-bridge as shown in Figure 8.30.

As you can see in the figure, there are four switching elements named as "high side left", "high side right", "low side right", "low side left". When these switches are turned on in pairs, motor changes its direction accordingly. Like, if we switch on high side left and low side right then motor rotates, in forward direction, as current flows from power supply through the motor coil goes to ground via switch low side right. This is shown in Figure 8.31.

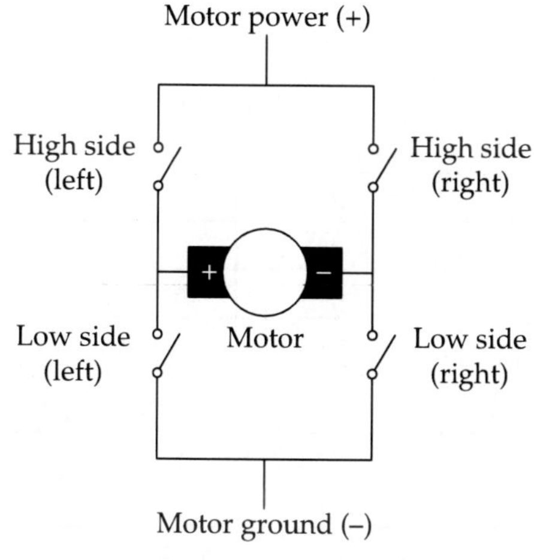

Fig. 8.30 *H-bridge*

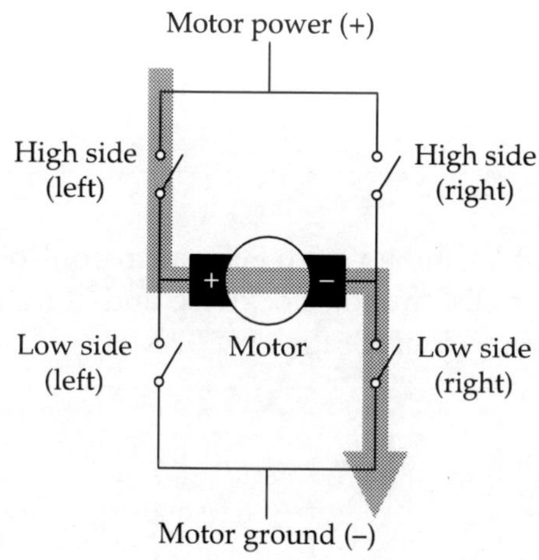

Fig. 8.31 *H-bridge flow of current when high side left and low side right are switched ON*

Similarly, when you switch on low side left and high side right, the current flows in opposite direction and the motor rotates in backward direction. This is the basic working of H-bridge. We can also make a small truth table according to the switching of H-bridge explained above. The sequence of operation of DC motor is shown in Table 8.10 and the truth table for the same is shown in Table 8.11.

As already said, H-bridge can be made with the help of transistors as well as MOSFETs; the only thing is the power handling capacity of the circuit. If motors are needed to run with high current then lot of dissipation is there. So head sinks are needed to cool the circuit.

Table 8.10 *Sequence of operation of DC motor*

Truth Table				
High Left	**High Right**	**Low Left**	**Low Right**	**Description**
On	Off	Off	On	Motor runs clockwise
Off	On	On	Off	Motor runs anticlockwise
On	On	Off	Off	Motor stops or decelerates
Off	Off	On	On	Motor stops or decelerates

Table 8.11 *Truth table of DC motor operation*

Truth Table		
A	**B**	**Operation**
0	0	Motor stops
0	1	Motor runs in one direction (clockwise)
1	0	Motor runs in opp. direction (anticlockwise)
1	1	Motor stops

Example 8.14

DC Motor Interfacing

Show the interfacing of DC motor with 8051 microcontroller and write an assembly language program to rotate DC motor clockwise and counterclockwise continuously. Assume that oscillator frequency of 8051 is 11.0592 MHz.

Solution:

Hardware arrangement:

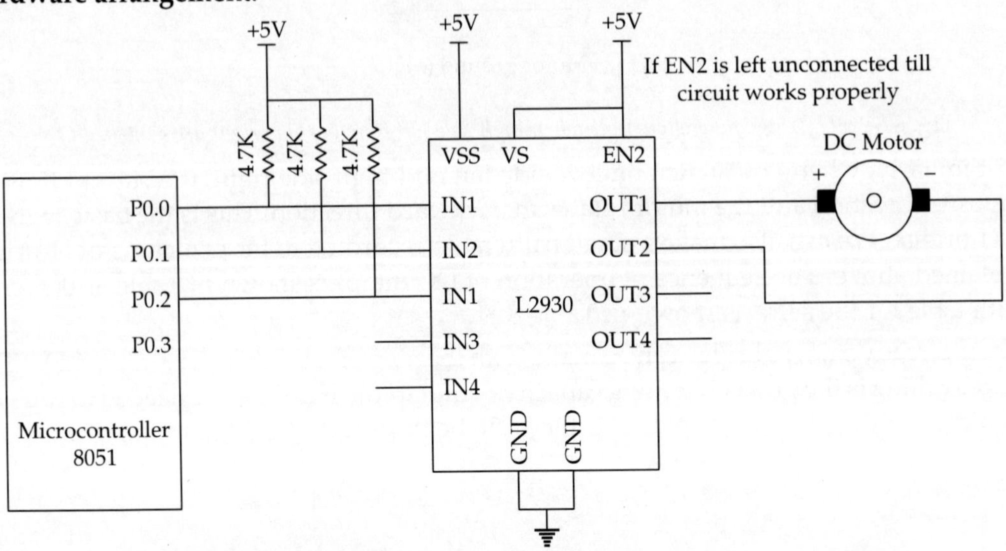

Fig. 8.32 *Interfacing of DC motor with 8051 microcontroller using L293D*

Assembly Language Program:

```
L293D_A    EQU P0.0          ; L293D A - positive of motor
L293D_B    EQU P0.1          ; L293D B - negative of motor
L293D_E    EQU P0.2          ; L293D E - enable pin of IC
           ORG 0000H
           SJMP MAIN
MAIN:      ACALL ROT_F       ; Rotate motor forward
           ACALL DELAY       ; Let the motor rotate
           ACALL BREAK       ; Stop the motor
           ACALL DELAY       ; Wait for some time
           ACALL ROT_B       ; Rotate motor backward
           ACALL DELAY       ; Let the motor rotate
           ACALL BREAK       ; Stop the motor
           ACALL DELAY       ; Wait for some time
           SJMP MAIN         ; Do this in loop
ROT_F:     SETB L293D_A      ; Make positive of motor 1
           CLR L293D_B       ; Make negative of motor 0
           SETB L293D_E      ; Enable to run the motor
           RET               ; Return from routine
ROT_B:     CLR L293D_A       ; Make positive of motor 0
           SETB L293D_B      ; Make negative of motor 1
           SETB L293D_E      ; Enable to run the motor
           RET               ; Return from routine
BREAK:     CLR L293D_A       ; Make positive of motor 0
           CLR L293D_B       ; Make negative of motor 0
           CLR L293D_E       ; Disable the o/p
           RET               ; Return from routine
DELAY:     MOV R7,#20H
BACK:      MOV R6,#0FFH
BACK1:     MOV R5,#0FFH
HERE:      DJNZ R5, HERE
           DJNZ R6, BACK1
           DJNZ R7, BACK
           RET
```

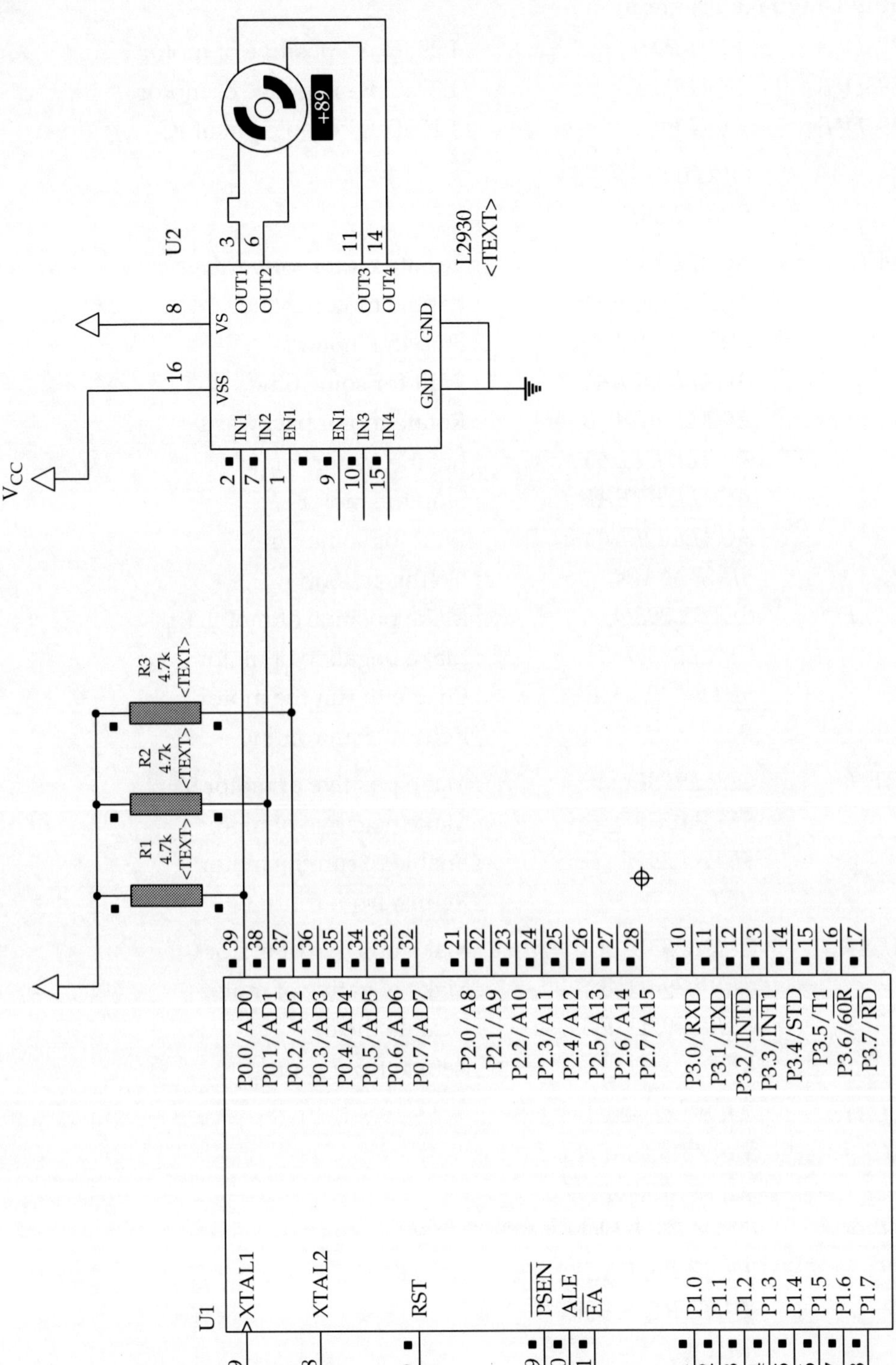
V_CC
U2
L2930
<TEXT>
VSS
VS
OUT1
OUT2
OUT3
OUT4
GND
GND
IN1
IN2
EN1
EN1
IN3
IN4
16
8
3
6
11
14
2
7
1
9
10
15
+89
R1
4.7k
<TEXT>
R2
4.7k
<TEXT>
R3
4.7k
<TEXT>
U1
XTAL1
XTAL2
RST
PSEN
ALE
EA
P1.0
P1.1
P1.2
P1.3
P1.4
P1.5
P1.6
P1.7
AT89C51
(TEXT)
P0.0/AD0
P0.1/AD1
P0.2/AD2
P0.3/AD3
P0.4/AD4
P0.5/AD5
P0.6/AD6
P0.7/AD7
P2.0/A8
P2.1/A9
P2.2/A10
P2.3/A11
P2.4/A12
P2.5/A13
P2.6/A14
P2.7/A15
P3.0/RXD
P3.1/TXD
P3.2/INTD
P3.3/INT1
P3.4/STD
P3.5/T1
P3.6/60R
P3.7/RD

Example 8.15

DC Motor Interfacing

Show the interfacing of DC motor with 8051 microcontroller and write an assembly language program to rotate DC motor clockwise fast and slow at regular intervals using the control of PWM. Assume that oscillator frequency of 8051 is 11.0592 MHz.

Solution: Hardware arrangement:

Same As Previous Problem

Assume the following PWM waveform, for the fast operation (rotation) of DC motor ON time is higher than OFF time and for slower operation of DC motor ON time is lower than OFF time. In ON time DC motor rotates and during OFF time DC motor stops.

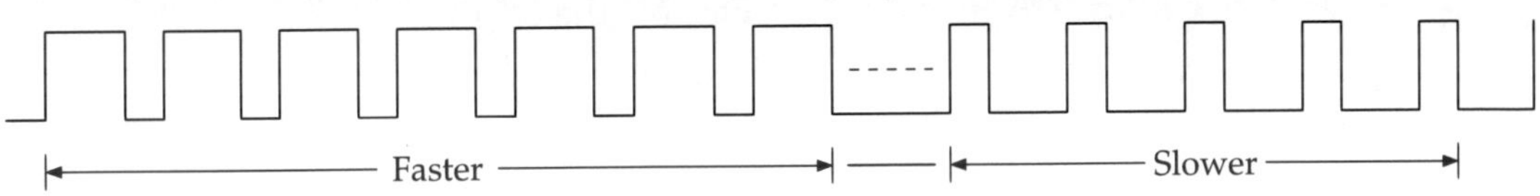

Fig. 8.33 *PWM waveform*

Assembly Language Program:

```
L293D_A EQU P0.0                ; L293D A - positive of motor
L293D_B EQU P0.1                ; L293D B - negative of motor
L293D_E EQU P0.2                ; L293D E - enable pin of ic
        ORG 0000H
        SJMP MAIN
MAIN: MOV R2, #0FFH             ; Count for clockwise rotation (fast)
REP:    ACALL ROT_F             ; Rotate motor forward
        ACALL DELAY             ; Let the motor rotate
        ACALL DELAY
        ACALL DELAY
        ACALL BREAK             ; Stop the motor
        ACALL DELAY
        DJNZ R2, REP
        MOV R2, #0FFH           ; Count for clockwise rotation (slow)
REP1:   ACALL ROT_F
        ACALL DELAY
        ACALL BREAK
```

```
            ACALL DELAY          ; Wait for some time
            ACALL DELAY
            ACALL DELAY
            DJNZ R2, REP1
            SJMP MAIN            ; Do this in loop
ROT_F:      SETB L293D_A         ; Make positive of motor 1
            CLR L293D_B          ; Make negative of motor 0
            SETB L293D_E         ; Enable to run the motor
            RET                  ; Return from routine
BREAK:      CLR L293D_A          ; Make positive of motor 0
            CLR L293D_B          ; Make negative of motor 0
            CLR L293D_E          ; Disable the o/p
            RET                  ; Return from routine
DELAY:      MOV R7,#20H
BACK:       MOV R6,#0FFH
BACK1:      MOV R5,#0FFH
HERE:       DJNZ R5, HERE
            DJNZ R6, BACK1
            DJNZ R7, BACK
            RET
```

8.5 SERVOMOTOR

A **servomotor** is a rotary actuator that allows for precise control of angular position, velocity and acceleration. It consists of a suitable motor coupled to a sensor for position feedback. It also requires a relatively sophisticated controller, like drivers are needed in the stepper motors and the DC motors, often a dedicated module designed specifically for use with servomotors.

Servomotors are not a different class of motors, on the basis of fundamental operating principle, but use servomechanism to achieve closed loop control with a generic open loop motor to control the motor rotation. In other words, a servomotor is just a regular motor with a sensor installed, typically to measure angular position during operation.

A servomotor is an electromechanical device in which the electrical input determines the position of motor armature. It is actually an assembly of four things: a normal DC motor, a gear reduction unit, a position-sensing device and a control circuit. Servomotors are used extensively in robotics industry and radio-controlled cars. The implemented types in modern servosystems are AC servomotors based on induction motor designs, DC servomotors based on DC motor designs, DC brushless motors and AC brushless servomotors based on synchronous motor designs. These motors are working in a closed loop control systems where the programmed positions of motion and velocity feedback controllers are required.

The motor is paired with some type of encoder to provide position and speed feedback. In the simplest case, only the position is measured. The measured position of the output is compared to the command position, the external input to the controller. If the output position differs from that required, an error signal is generated which then causes the motor to rotate in either direction, as needed to bring the output shaft to the appropriate position. As the positions approach, the error signal reduces to zero and the motor stops. Figure 8.35 shows the inside of the servo motor.

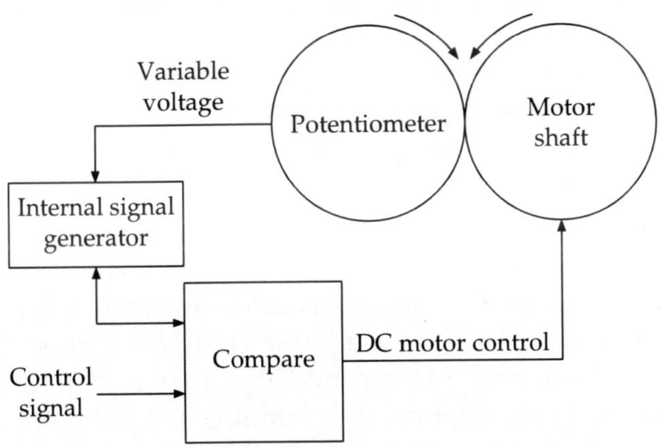

Fig. 8.34 *Inside of the servomotor*

More sophisticated servomotors measure both the position and also the speed of the output shaft. They may also control the speed of their motor, rather than always running at full speed. Both of these enhancements, usually in combination with a PID control algorithm, allow the servomotor to be brought to its commanded position more quickly and more precisely, with less overshooting.

By comparing servomotors and stepper motors, servomotors are generally used as a high performance alternative to the stepper motor. Stepper motors have some inherent ability to control position, as they have built-in output steps. This often allows them to be used as an open-loop position control, without any feedback encoder, as their drive signal specifies the number of steps of movement to rotate. This lack of feedback though limits their performance, as the stepper motor can only drive a load that is well within its capacity, otherwise missed steps under load may lead to positioning errors. The encoder and controller of a servomotor are an additional cost, but they optimize the performance of the overall system (for all of speed, power and accuracy) relative to the capacity of the basic motor. With larger systems, where a powerful motor represents an increasing proportion of the system cost, servomotors have the advantage.

Pulse-width modulation (PWM), or pulse-duration modulation (PDM), is a commonly used technique for controlling power to inertial electrical devices, made practical by modern electronic power switches.

The main advantage of PWM is that power loss in the switching devices is very low. When a switch is off there is practically no current, and when it is on, there is almost no voltage drop across the switch. Power loss, being the product of voltage and current, is thus in both cases close to zero. PWM also works well with digital controls, which, because of their on/off nature, can easily set the needed duty cycle.

PWM has also been used in certain communication systems where its duty cycle has been used to convey information over a communication channel.

Servomechanisms are used in modern automatic machine tools, satellite-tracking antennas, remote control airplanes, automatic navigation systems on boats and planes, and antiaircraft-gun control systems, fly-by-wire systems in aircraft which use servos to actuate the aircraft's control surfaces and radio-controlled models which use RC servos for the same purpose. Many autofocus cameras also use servomechanism to accurately move the lens, and thus adjust the focus. A modern hard disk drive has a magnetic servo system with sub-micrometer positioning accuracy.

How do you communicate the angle at which the servo should turn? The control wire is used to communicate the angle. The angle is determined by the duration of a pulse that is applied to the control wire. This is called pulse width modulation. The servo expects to see a pulse every 20 milliseconds (.02 second/50Hz frequency). The length of the pulse will determine how far the motor turns. A 1.5 millisecond pulse, for example, will make the motor turn to the 90 degree position (often called the neutral position). If the pulse is shorter than 1.5 ms, then the motor will turn the shaft to closer to 0 degree. If the pulse is longer than 1.5 ms, the shaft turns closer to 180 degrees. The standard time versus angle for the PWM based servomotor is shown in Figure 8.36.

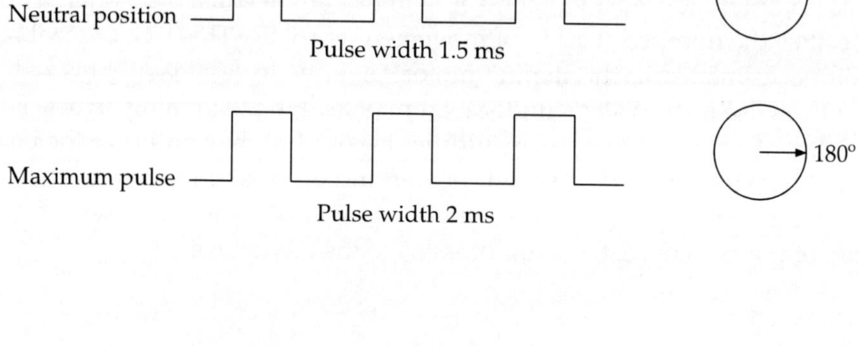

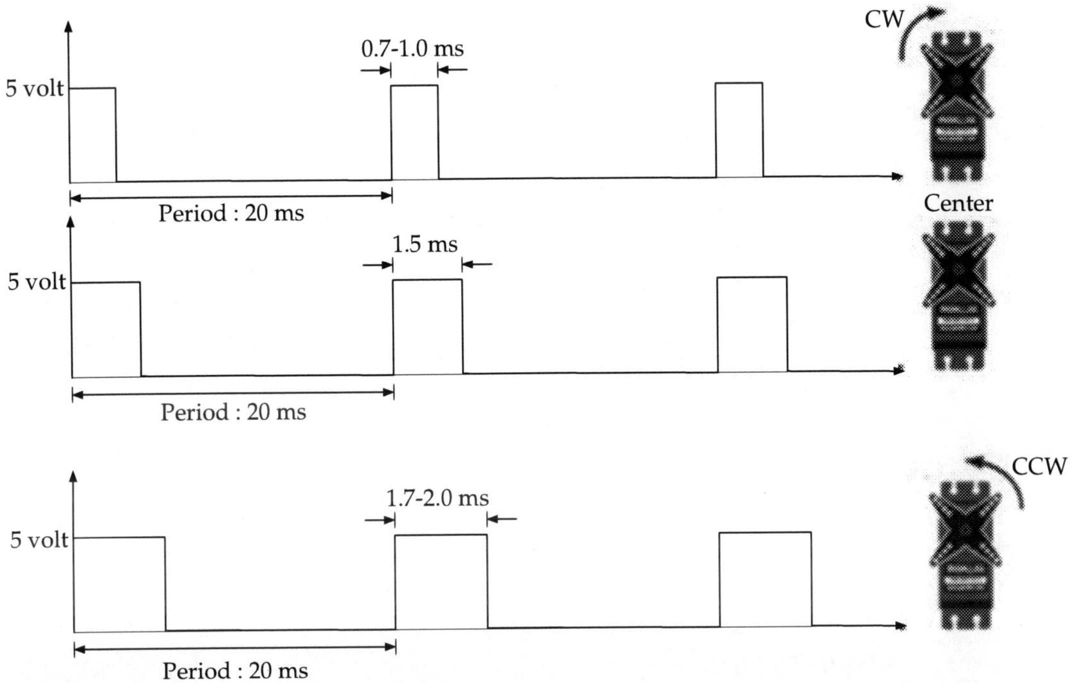

Fig. 8.35 *Controlling of servomotor with PWM*

Example 8.16

Servomotor Interfacing

Show the interfacing of a servomotor with 8051 microcontroller and write an assembly language program to rotate the servomotor clockwise continuously. Assume that oscillator frequency of 8051 is 11.0592 MHz.

Solution:

Hardware arrangement:

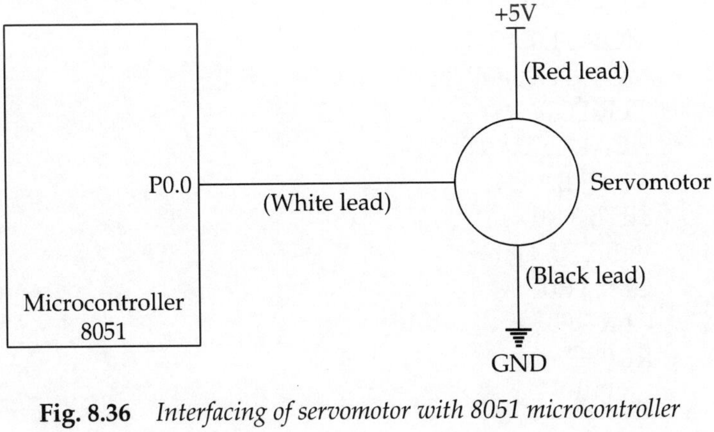

Fig. 8.36 *Interfacing of servomotor with 8051 microcontroller*

Assembly Language Program:

```
        ORG 0000H
        SETB P1.0
        SETB P1.1
BACK:   JNB P1.0, ONE
        JNB P1.1, TWO
        MOV TMOD, #01 ; Center
        MOV TL0, #99H
        MOV TH0, #0FAH
        SETB P0.0
        ACALL START
        MOV TL0,#29H
        MOV TH0,#0F9H
        CLR P0.0
        ACALL START
        SJMP BACK
ONE:    MOV TMOD,#01       ; Left
        MOV TL0,#99H
        MOV TH0,#0F4H
        SETB P0.0
        ACALL START
        MOV TL0,#29H
        MOV TH0,#0F9H
        CLR P0.0
        ACALL START
        SJMP BACK
TWO:    MOV TMOD, #01      ; Right
        MOV TL0, #0B0H
        MOV TH0, #0FCH
        SETB P0.0
        ACALL START
        MOV TL0, #29H
        MOV TH0, #0F9H
        CLR P0.0
        ACALL START
        SJMP BACK
START:  SETB TR0
GO:     JNB TF0, GO
        CLR TR0
        CLR TF0
        RET
        END
```

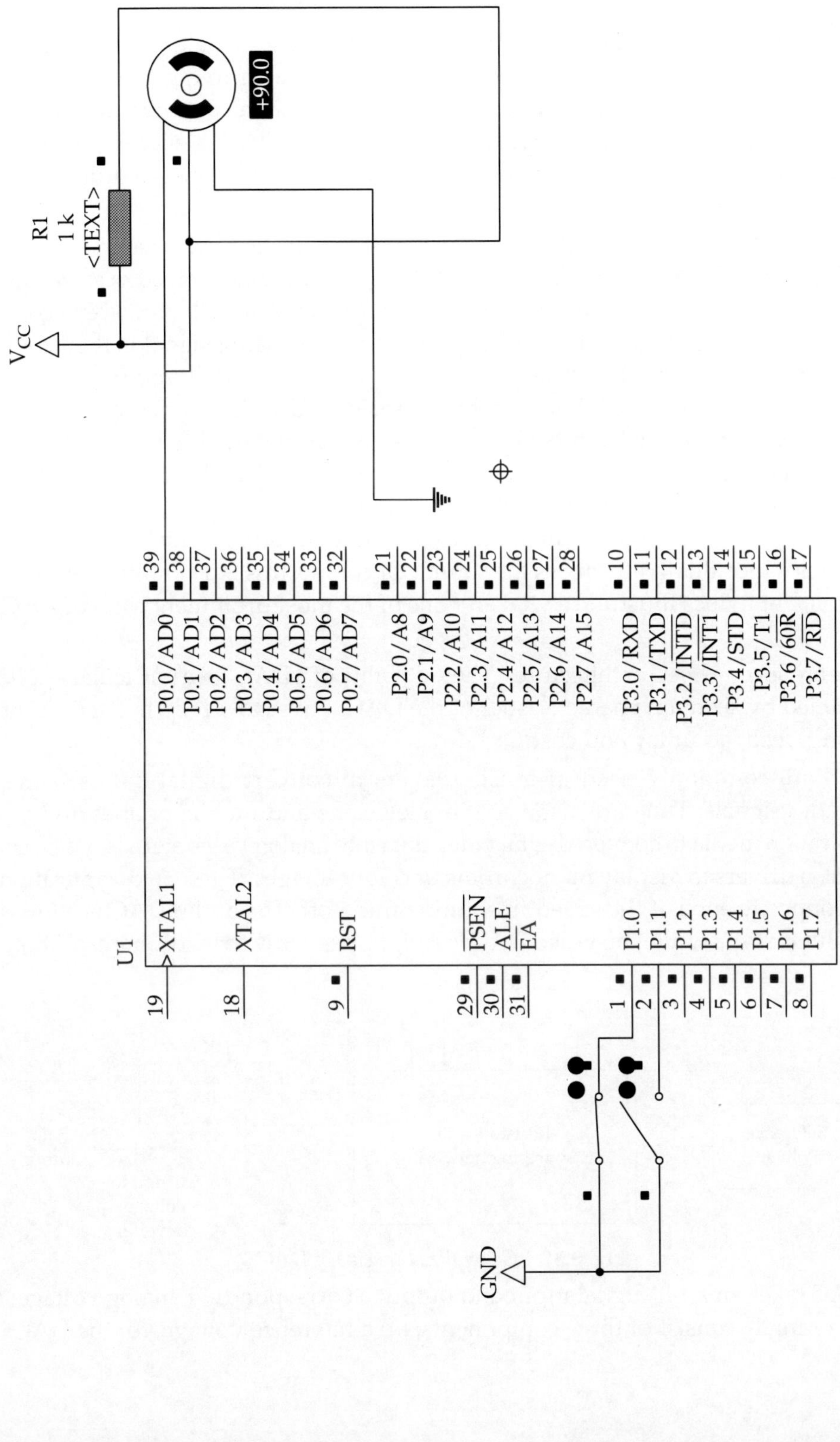
R1
1 k
<TEXT>
V_{CC}
+90.0
GND
U1
XTAL1
XTAL2
RST
PSEN
ALE
EA
P1.0
P1.1
P1.2
P1.3
P1.4
P1.5
P1.6
P1.7
AT89C51
(TEXT)
P0.0/AD0
P0.1/AD1
P0.2/AD2
P0.3/AD3
P0.4/AD4
P0.5/AD5
P0.6/AD6
P0.7/AD7
P2.0/A8
P2.1/A9
P2.2/A10
P2.3/A11
P2.4/A12
P2.5/A13
P2.6/A14
P2.7/A15
P3.0/RXD
P3.1/TXD
P3.2/INTD
P3.3/INT1
P3.4/STD
P3.5/T1
P3.6/60R
P3.7/RD

8.6 DIGITAL-TO-ANALOG CONVERTER (DAC)

A **digital-to-analog converter** is a function that converts digital data (usually binary) into an analog signal (current, voltage, or electric charge). An analog-to-digital converter (ADC) performs the reverse function. Unlike analog signals, digital data can be transmitted, manipulated, and stored without any degradation. But a DAC is needed to convert the digital signal to analog to drive an earphone or loudspeaker amplifier in order to produce sound (analog air pressure waves).

DACs and their inverse, ADCs, are part of an enabling electronics technology that has contributed significantly to the digital revolution. To illustrate, consider a typical long-distance telephone call or mobile communication system. The caller's voice is converted into an analog electrical signal by a microphone and the analog signal is then converted to a digital stream by an ADC. The digital stream is then divided into packets where it will be mixed with other digital data, not necessarily audio. The digital packets are then sent to the destination, but each packet may take a completely different route and may not even arrive at the destination in the correct time order. The digital voice data is then extracted from the packets and assembled into a digital data stream. A DAC converts it into an analog electrical signal which drives an audio amplifier which, in turn, drives a loudspeaker which finally produces the sound. Of course, this is a simplified and stylized description, but it does illustrate a vital and one of the most prominent roles of ADCs and DACs.

There are several DAC architectures. The suitability of a DAC for a particular application is determined by six main characteristics of the DAC: physical size, power consumption, resolution, speed, accuracy and cost.

DACs are commonly used in music players to convert digital data streams into analog audio signals. They are also used in televisions and mobile phones and in other communication media to convert digital video data into analog video signals which connect to the screen drivers to display monochrome or colour images. These two applications use DACs at opposite ends of the speed/resolution trade-off. The audio DAC is a low speed high resolution type while the video DAC is a high speed low to medium resolution type.

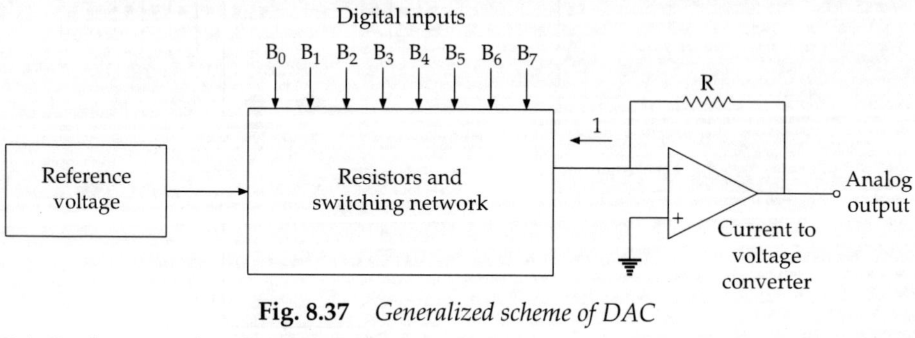

Fig. 8.37 *Generalized scheme of DAC*

A DAC takes an n-bit digital input and output a corresponding analog voltage. DAC systems normally consist of three components: (a) a reference voltage, (b) the DAC itself,

(c) an op amp for output buffering and for the conversion of current output to voltage output, the generalized scheme is shown in Figure 8.38 and the simplified scheme is shown in Figure 8.38.

As per the Nyquist-Shannon sampling theorem, a DAC can reconstruct the original signal from the sampled data provided that its bandwidth meets certain requirements. Digital sampling introduces quantization error that manifests as low-level noise added to the reconstructed signal.

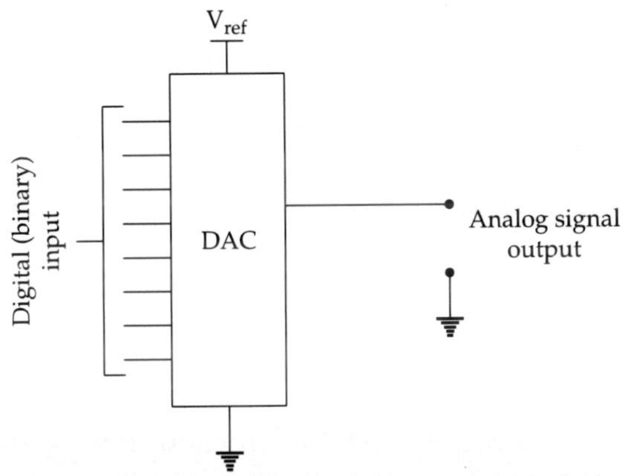

Fig. 8.38 *Simplified scheme of DAC*

Example 8.17

Show the interfacing of DAC0808 with 8051 microcontroller. Write an assembly language program to generate a sawtooth wave form (stair step ramp) on port P1.

Solution:

Hardware arrangement:

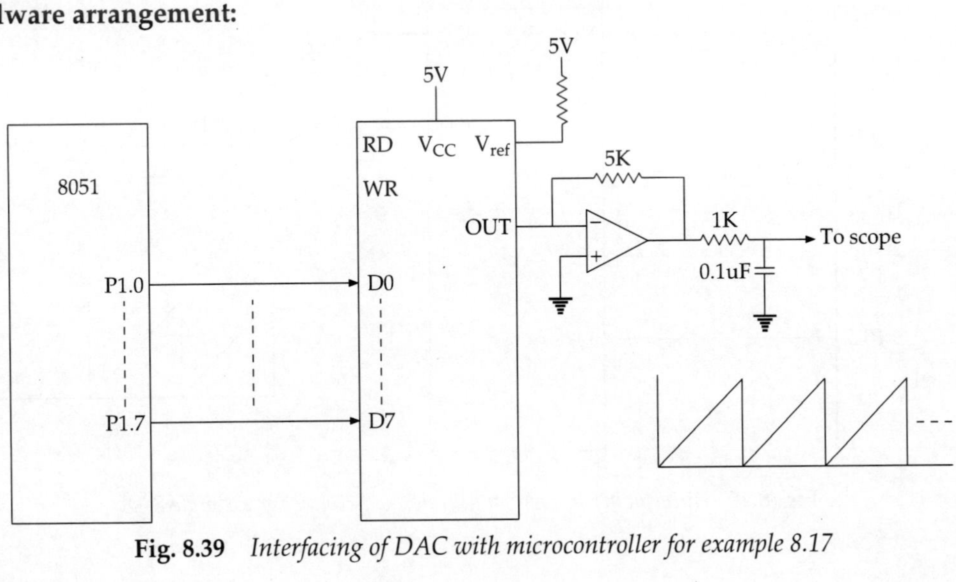

Fig. 8.39 *Interfacing of DAC with microcontroller for example 8.17*

Assembly Language Program:

```
        CLR A / MOV A, #00h
AGAIN:  MOV P1, A          ; Send data to DAC
        INC A              ; Count from 00 to FF
        ACALL DELAY        : Let DAC recover
        SJMP AGAIN
DELAY:  MOV R7,#20H
BACK:   MOV R6,#0FFH
BACK1:  MOV R5,#0FFH
HERE:   DJNZ R5, HERE
        DJNZ R6, BACK1
        DJNZ R7, BACK
        RET
```

Example 8.18

Show the interfacing of DAC0808 with 8051 microcontroller. Write an assembly language program to generate a reverse sawtooth wave form (stair step ramp) on port P1.

Solution:

Hardware arrangement:

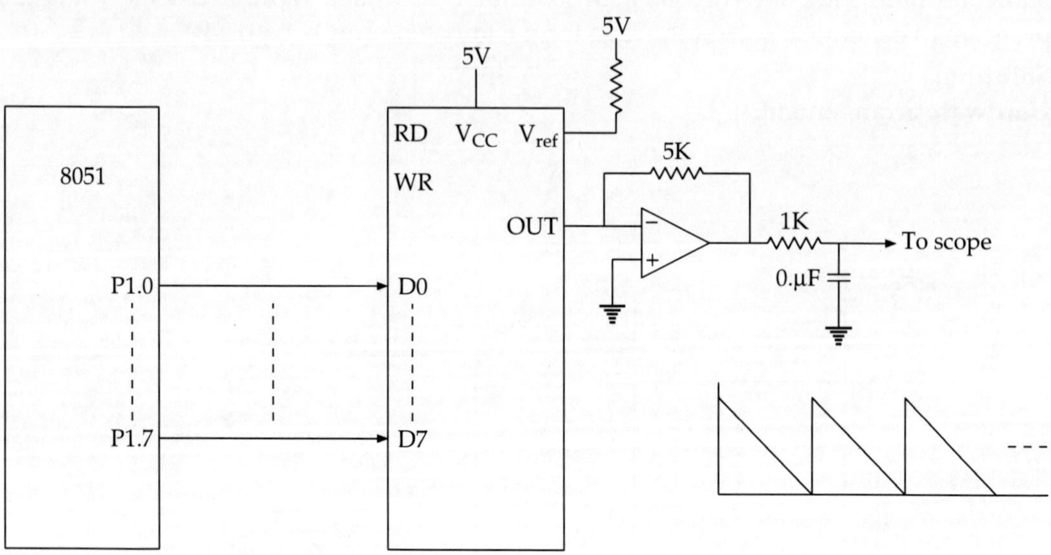

Fig. 8.40 *Interfacing of DAC with microcontroller for example 8.18*

Assembly Language Program:

```
         MOV A, #0FFh
AGAIN:   MOV P1, A          ; Send data to DAC
         DEC A              ; Count from FF to 00
         ACALL DELAY        : Let DAC recover
         SJMP AGAIN
DELAY:   MOV R7,#20H
BACK:    MOV R6,#0FFH
BACK1:   MOV R5,#0FFH
HERE:    DJNZ R5, HERE
         DJNZ R6, BACK1
         DJNZ R7, BACK
         RET
```

Example 8.19

Show the interfacing of DAC0808 with 8051 microcontroller and write an assembly language program to generate triangular waveform at the DAC output using port P1 of the microcontroller. Generate 150 cycles, use P1.

Solution:

Hardware arrangement:

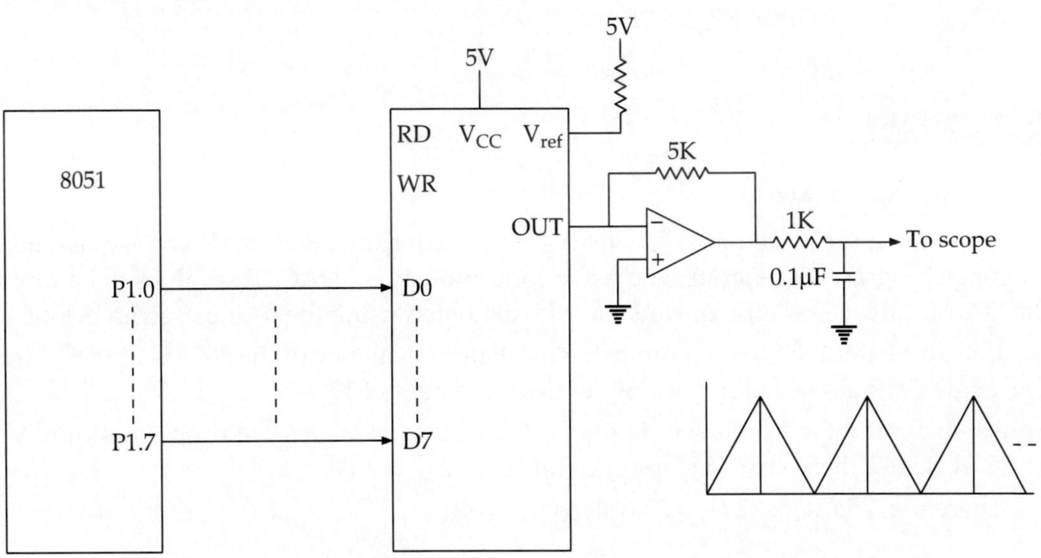

Fig. 8.41 *Interfacing of DAC with microcontroller for example 8.19*

Assembly Language Program:

```
        MOV R2, #150
AGAIN:  MOV R3, #0FEH
        MOV R4, #0FEh
        CLR A
L1:     MOV P1, A
        INC A
        ACALL DELAY
        DJNZ R3, L1
        MOV A, #0FFh
L2:     MOV P1, A
        DEC A
        ACALL DELAY
        DJNZ R4, L2
        DJNZ R2, AGAIN
        END
DELAY:  MOV R7,#20H
BACK:   MOV R6,#0FFH
BACK1:  MOV R5,#0FFH
HERE:   DJNZ R5, HERE
        DJNZ R6, BACK1
        DJNZ R7, BACK
        RET
```

Example 8.20

Generating Sine Wave

Show the interfacing of DAC 0808 with microcontroller 8051 and write an assembly language program to generate sine wave continuously on DAC. The values to be sent to the DAC for the generation of sine wave is calculated using the formula given below.

Use port P1 of the 8051. Assume that oscillator frequency of the 8051 is 11.0592 MHz. The angle θ should be from 0° to 360° in the increment of 15°.

Solution: To find the values sent to the DAC for various angles, we simply multiply V_{out} by 25.6 because there are 256 steps and full scale V_{out} is 10V.

Therefore, 256 steps/10V = 25.6 steps per volt.

Table 8.12 *Values sent to DAC for sine wave generation for example 8.20*

Angle (decimal)	sin θ	V_{OUT} = 5V + (5V*sin θ)	Values sent to DAC = V_{OUT} * 25.6
0	0	5	128
15	0.259	6.295	161
30	0.5	7.5	192
45	0.707	8.535	218
60	0.866	9.33	238
75	0.966	9.83	251
90	1	10	256
105	0.966	9.83	251
120	0.866	9.33	238
135	0.707	8.535	218
150	0.5	7.5	192
165	0.259	6.295	161
180	0	5	128
195	–0.259	3.705	94
210	– 0.5	2.5	64
225	–0.707	1.465	37
240	–0.866	0.67	17
255	–0.966	0.17	4
270	–1	0	0
285	–0.966	0.17	4
300	–0.866	0.67	17
315	–0.707	1.465	37
330	–0.5	2.5	64
345	–0.259	3.705	94
360	0	5	128

Assembly Language Program:

```
REP:     MOV DPTR, #SIN_TABLE
         MOV R2, #25          ; for total 25 values from 0 to
                                360
AGAIN:   CLR A
         MOVC A, @A+DPTR
         MOV P1, A
```

```
            ACALL DELAY
            INC DPTR
            DJNZ R2, AGAIN
            SJMP REP
DELAY:      MOV R7, #20H
BACK:       MOV R6, #0FFH
BACK1:      MOV R5, #0FFH
HERE:       DJNZ R5, HERE
            DJNZ R6, BACK1
            DJNZ R7, BACK
            RET
SIN_TABLE:  DB 128, 161, 192, 218, 238, 251, 256 ....... 128
            END
```

8.7 ANALOG-TO-DIGITAL CONVERETERS (ADCs)

Analog-to-digital converter is a device that converts a continuous physical quantity (usually voltage) to a digital number that represents the quantity's amplitude.

The conversion involves quantization of the input, so it necessarily introduces a small amount of error. Instead of doing a single conversion, an ADC often performs the conversions periodically. The result is a sequence of digital values that have converted a continuous-time and continuous-amplitude analog signal to a discrete-time and discrete-amplitude digital signal.

An ADC is defined by its bandwidth (the range of frequencies it can measure) and its signal-to-noise ratio (how accurately it can measure a signal relative to the noise it introduces). The actual bandwidth of an ADC is characterized primarily by its sampling rate and to a lesser extent by how it handles errors such as aliasing. The dynamic range of an ADC is influenced by many factors, including the resolution (the number of output levels it can quantize a signal to), linearity and accuracy (how well the quantization levels match the true analog signal) and jitter (small timing errors that introduce additional noise).

ADCs are chosen to match the bandwidth and required signal-to-noise ratio of the signal to be quantized. If an ADC operates at a sampling rate greater than twice the bandwidth of the signal, then perfect reconstruction is possible given an ideal ADC and neglecting quantization error. The presence of quantization error limits the dynamic range of even an ideal ADC, however, if the dynamic range of the ADC exceeds that of the input signal, its effects may be neglected resulting in an essentially perfect digital representation of the input signal.

An ADC may also provide an isolated measurement such as an electronic device that converts an input analog voltage or current to a digital number proportional to the magnitude of the voltage or current. However, some non-electronic or only partially electronic devices, such as rotary encoders, can also be considered ADCs. The digital, output may use different coding schemes. Typically, the digital output will be a two's complement binary number that is proportional to the input, but there are other possibilities. An encoder, for example, might output a gray code. The inverse operation is performed by a digital-to-analog converter (DAC).

Table 8.13 shows the various analog-to-digital converters that are most commonly used with the microcontroller 8051 and its family.

Table 8.13 *Analog-to-digital converter ICs*

Name of the IC	Description
ADC 0800	8-bit ADC
ADC 0801	8-bit ADC, 100 μs, 0.25 LSB
ADC 0802	8-bit ADC, 100 μs, 0.5 LSB
ADC 0804	8-bit ADC, 100 μs, 1.0 LSB
ADC 0808	8-bit 8 channel ADC, 100 μs
ADC 0809	8-bit 8 channel ADC (equivalent to ADC 0808)
AD 571	10-bit, ADC, with reference and clock signals
MAX 1204	5V, 8-channel, serial, 10-bit ADC with 3V digital interface
MAX 1202	5V, 8-channel, serial, 12-bit ADC with 3V digital interface
MAX 195	16-bit, self calibrating, 10 μs sampling ADC

8.7.1 Analog-to-digital converter IC ADC 0801/0802/0803/0804/0805

The ADC0801, ADC0802, ADC0803, ADC0804 and ADC0805 are CMOS 8-bit successive approximation A/D converters that use a differential potentiometric ladder — similar to the 256R products. These converters are designed to allow operation with the NSC800 and INS8080A derivative control bus with TRI-STATE output latches directly driving the data bus. These A/Ds appear like memory locations or I/O ports to the microprocessor or microcontroller and no interfacing logic is needed. Differential analog voltage inputs allow increasing the common-mode rejection and offsetting the analog zero input voltage value. In addition, the voltage reference input can be adjusted to allow encoding any smaller analog voltage span to the full 8 bits of resolution.

The features of the ADC 0801/0802/0803/0804/0805 are as follows:

(*a*) Compatible with 8080 μP derivatives, no interfacing logic needed, access time 135 ns
(*b*) Easy interface to all microprocessors, or operates "stand alone"
(*c*) Differential analog voltage inputs
(*d*) Logic inputs and outputs meet both MOS and TTL voltage level specifications
(*e*) Works with 2.5V (LM336) voltage reference
(*f*) On-chip clock generator
(*g*) 0V to 5V analog input voltage range with single 5V supply
(*h*) No zero adjust required
(*i*) 0.3" standard width 20-pin DIP package
(*j*) 20-pin molded chip carrier or small outline package
(*k*) Operates ratiometrically or with 5 VDC, 2.5 VDC, or analog span adjusted voltage reference

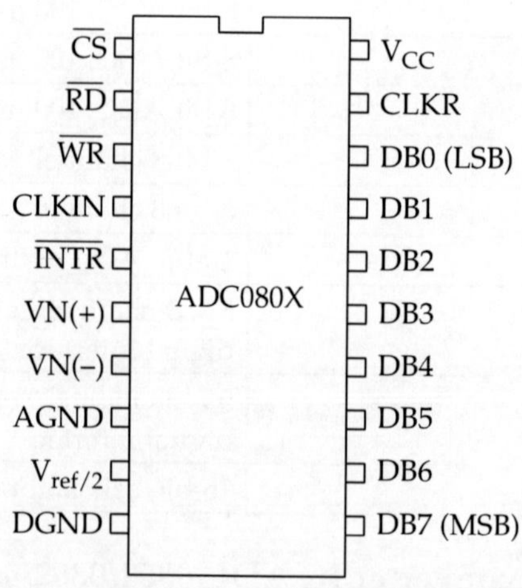

Fig. 8.42 *Pin layout of ADC080X*

8.7.2 Analog-to-digital Converter IC ADC 0808/0809

The ADC0808, ADC0809 data acquisition component is a monolithic CMOS device with an 8-bit analog-to-digital converter, 8-channel multiplexer and microprocessor and microcontroller compatible control logic. The 8-bit A/D converter uses **successive approximation** as the conversion technique. The converter features a high impedance chopper stabilized comparator, a 256R voltage divider with analog switch tree and a successive approximation register. The 8-channel multiplexer can directly access any of 8-single ended analog signals. The device eliminates the need for external zero and full-scale adjustments. Easy interfacing to microprocessors and microcontroller is provided by the latched and decoded multiplexer address inputs and latched TTL TRI-STATE outputs. The design of the ADC0808, ADC0809 has been optimized by incorporating the most desirable aspects of several A/D conversion techniques. The ADC0808, ADC0809 offers high speed, high accuracy, minimal temperature dependence, excellent long-term accuracy and repeatability, and consumes minimal power. These features make this device ideally suited to applications from process and machine control to consumer and automotive applications.

The features of the ADC 0808/0809 are as follows:

(*a*) Easy interface to all microprocessors
(*b*) Operates ratiometrically or with 5 VDC or analog span adjusted voltage reference
(*c*) No zero or full-scale adjust required
(*d*) 8-channel multiplexer with address logic
(*e*) 0V to 5V input range with single 5V power supply
(*f*) Outputs meet TTL voltage level specifications
(*g*) Standard hermetic or molded 28-pin DIP package
(*h*) 28-pin molded chip carrier package

Example 8.21

Show the interfacing of ADC 0804 with microcontroller 8051 and write an assembly language program to read the value and store in internal memory location 30h.

Solution: The pin configuration of ADC 0804 is shown in Figure 8.47 and typical application schematic with microcontroller is shown in Figure 8.48.

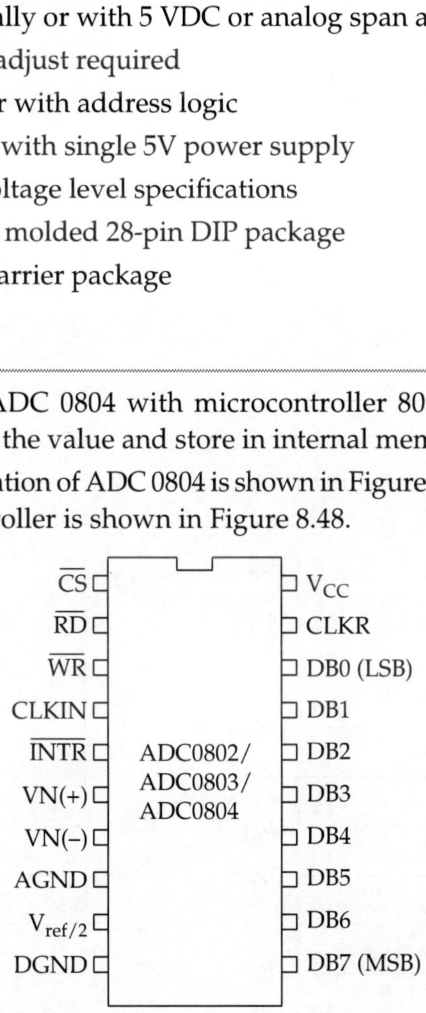

Fig. 8.43 *Pin configuration of ADC 0804*

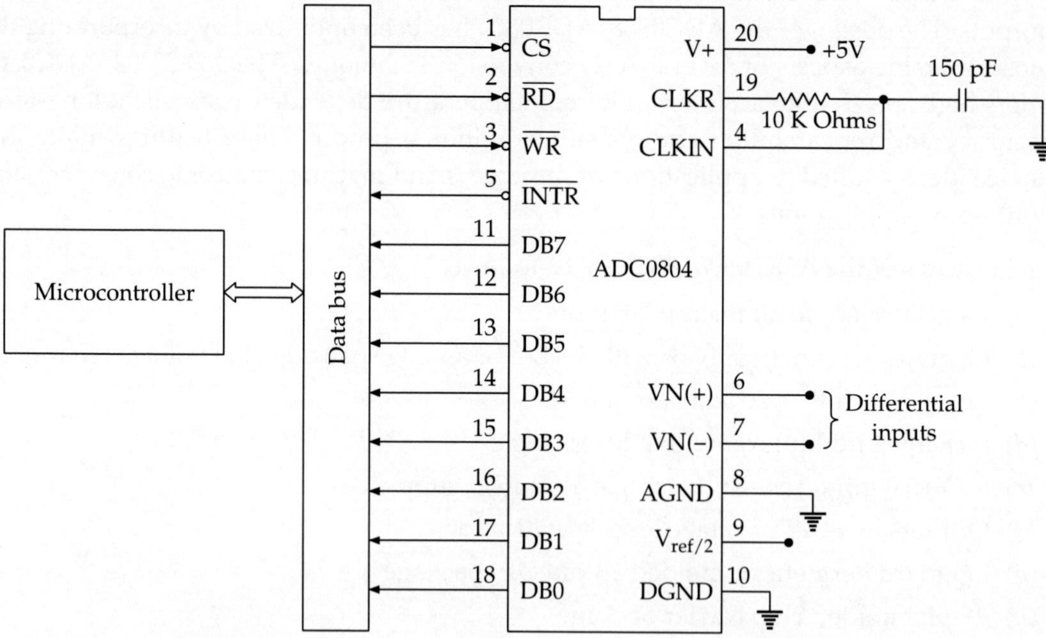

Fig. 8.44 *Schematic interfacing of ADC 0804 with microcontroller*

Hardware arrangement:

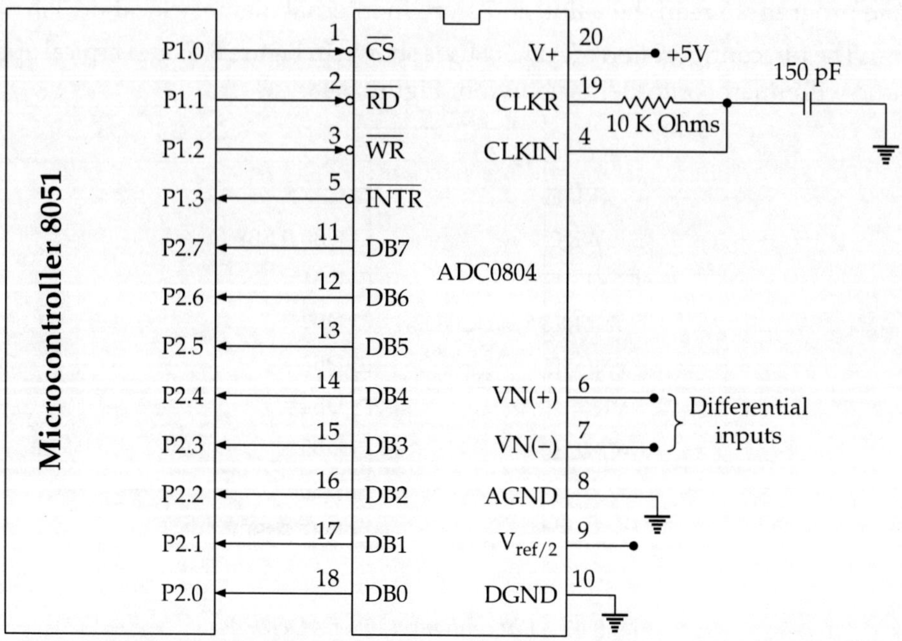

Fig. 8.45 *Interfacing of ADC 0804 with microcontroller for example 8.21*

The following points are observed to start the conversion:

1. Make chip select (CS) signal low to select the ADC 0804
2. Make write (WR) signal low
3. Make chip select high
4. Wait for INTR signal to go low; when it goes low, it indicates conversion ends.

Once the conversion in ADC is done, the data is available in the output latch of the ADC. Data of new conversion is only available for reading after ADC 0804 made INTR signal low or say when the conversion is completed. Below are the steps observed to read the output from the ADC 0804.

1. Make chip select (CS) pin low.
2. Make read (RD) signal low.
3. Read the data from port where ADC is connected.
4. Make read (RD) signal high.
5. Make chip select (CS) high.

Assembly Language Program:

```
RD          EQU              P1.0     ; Read signal P1.0
WR          EQU              P1.1     ; Write signal P1.1
CS          EQU              P1.2     ; Chip select signal P1.2
INTR        EQU              P1.3     ; INTR signal P1.3
ADC_PORT    EQU              P2 ; ADC data port pin P2
            ORG 0000h
START:      ACALL CONV       ; Start ADC Conversion
            ACALL READ       ; Read converted value
            MOV P3, 30h      ; Move the value to the port P3
            SJMP START       ; Repeat the process
CONV:       CLR CS           ; Make CS low
            CLR WR           ; Make WR low
            ACALL DELAY      ; Give small delay
            SETB WR          ; Make WR high
            SETB CS          ; Make CS high
WAIT:       JB INTR, WAIT    ; Wait for INTR signal
            RET              ; Conversion completed
READ:       CLR CS           ; Make CS low
```

```
CLR RD              ; make RD low
MOV A, ADC_PORT     ; Read the converted value
MOV 30h, A          ; Store the converted value in internal
                    ; RAM 30h
SETB RD             ; make RD high
SETB CS             ; Make CS high
RET                 ; Reading completed
DELAY:              MOV R7, #50H
BACK:               DJNZ R7, BACK
                    RET
```

8.8 LIQUID CRYSTAL DISPLAY (LCD)

The most commonly used character based LCDs are based on Hitachi's HD44780 controller or other which are compatible with HD44580. In this application, we will discuss character based LCDs, and their interfacing with microcontroller 8051.

8.8.1 Introduction

The most commonly used LCDs found in the market today are 1-line, 2-line or 4-line LCDs which have only 1 controller and support at most 80 characters, whereas LCDs supporting more than 80 characters make use of 2 HD44780 controllers.

Most LCDs with 1 controller have 14 pins and LCDs with 2 controller has 16 pins (two pins are extra in both for back-light LED connections). Pin description is shown in Table 8.14.

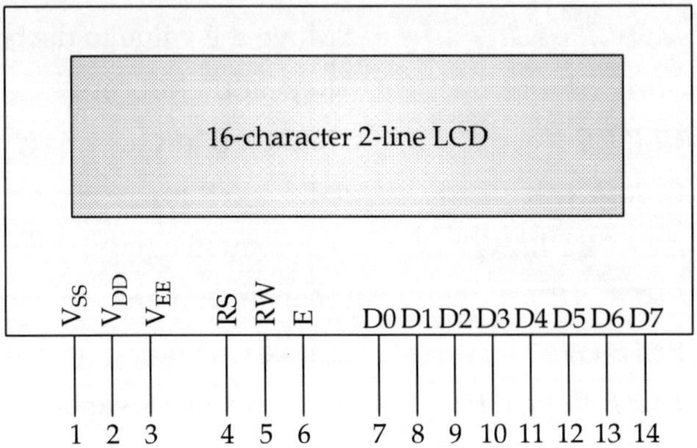

Fig. 8.46 *Character LCD type HD44780 pin diagram*

Table 8.14 *Pin description of character LCD type HD44780*

Pin No.	Name	Description
1	VSS	Power supply (GND)
2	VCC	Power supply (+5V)
3	VEE	Contrast adjust
4	RS	Register select 0 = To select command register 1 = To select data register
5	R/W	Read/Write 0 = Write to the LCD 1 = Read from the LCD
6	EN	Enable signal
7	D0	Data bus line D0 (LSB)
8	D1	Data bus line D1
9	D2	Data bus line D2
10	D3	Data bus line D3
11	D4	Data bus line D4
12	D5	Data bus line D5
13	D6	Data bus line D6
14	D7	Data bus line D7 (MSB)

Pin description of LCD

The pin description of the LCD shown in Figure 8.47 is as follows:

Power supply and contrast adjust signals:

Pin No. 1,2 and 3

Vcc, Vss and VEE

Vcc and Vss signals are used to provide power supply requirement for the LCD. Vcc is connected to the +5V supply and Vss is connected to the ground, whereas VEE is used for controlling LCD contrast.

Pin No. 4

RS, Register Select

There are two very important registers inside the LCD, they are command register and data register. The RS pin is used for their selection as follows:

If RS = 0, then the instruction command register is selected, allowing the user to send command such as clear display, return home, etc., to the LCD.

If RS = 1, then the data register is selected, allowing the user to send data to be displayed on the LCD.

Pin No. 5

R/W, Read / write

This input allows the user to read information from the LCD or to write information to the LCD.

If R/W = 0, then writing operation to the LCD is initiated

If R/W = 1, then reading from the LCD operation is initiated

Pin No. 6

E, Enable

This signal is used by the LCD to latch the information presented to its data bus lines.

When data is applied to the data bus lines, a high-to-low pulse must be applied to this pin in order for the LCD to latch in the data presented at the data bus lines. The minimum pulse duration is 450 ns for latching the information on the data bus lines.

Pin No. 7 to 14

D0 to D7, data bus lines

The 8-bit data pins, D0 to D7, are used to send information to the LCD or read the contents of the LCD's internal registers.

8.8.2 LCD Commands

Only the instruction register (IR) and the data register (DR) of the LCD can be controlled by the MCU. Before starting the internal operation of the LCD, control information is temporarily stored into these registers to allow interfacing with various MCUs, which operate at different speeds, or various peripheral control devices. The internal operation of the LCD is determined by signals sent from the MCU. These signals, which include register selection signal (RS), read/write signal (R/W), and the data bus (DB0 to DB7), make up the LCD instructions. There are four categories of instructions that:

- Designate LCD functions, such as display format, data length, etc.
- Set internal RAM addresses
- Perform data transfer with internal RAM
- Perform miscellaneous functions

The various command codes that can be sent to the LCD to clear the display or force the cursor to the home position or blink the cursor is shown in Table 8.15.

Table 8.15 *LCD command codes*

Command code (Hex)	Command to the LCD instruction register
01	Clear display screen
02	Return home
04	Decrement cursor (shift cursor to left)
06	Increment cursor (shift cursor to right)
05	Shift display right
07	Shift display left
08	Display off, cursor off
0A	Display off, cursor on
0C	Display on, cursor off
0E	Display on, cursor on (cursor blinking)
0F	Display on, cursor on (cursor blinking)
10	Shift cursor position to left
14	Shift cursor position to right
18	Shift the entire display to the left
1C	Shift the entire display to the right
80	Force cursor to beginning on the first line
C0	Force cursor to beginning on the second line
38	Initialization of the LCD / 2 lines and 5 x 7 matrix

8.8.3 Busy Flag of LCD

As discussed in the previous section, there must be some delay for LCD to successfully process the command or data. So this delay can be made either with a delay loop of specified time more than that of LCD process time or we can read the busy flag, which is recommended. The reason to use busy flag is that delay produced is almost for the exact amount of time for which LCD needs to process the time. So is best suited for every application.

Steps to read busy flag

When we send the command, the BF or D7th bit of the LCD becomes 1 and as soon as the command is processed the BF = 0. Following are the steps to be kept in mind while reading the busy flag.

- Select command register
- Select read operation
- Send enable signal
- Read the flag

So following the above steps we can write the code in assembly as below:

```
LCD_BUSY:   SETB LCD_D7          ; Make D7th bit of LCD data
                                   port as input
            SETB LCD_EN          ; Make port pin as output
            CLR LCD_RS           ; Select command register
            SETB LCD_RW          ; read the value
CHECK:      CLR LCD_EN
            SETB LCD_EN          ; Enable High → Low
            JB LCD_D7, CHECK     ; Read busy flag again and
                                   again till it becomes 0
            RET                  ; Return from busy routine
```

The above routine will provide the necessary delay for the instructions to complete. If you don't want to read the busy flag you can simply use a delay routine to provide the specific amount of delay. A simple delay routine for the LCD is given below:

```
LCD_BUSY:   MOV R7, #50H
BACK:       MOV R6, #FFH
L1:         DJNZ R6, L1
            DJNZ R7, BACK
            RET                  ; Return from busy routine
```

Now we are ready with the initialization routine and the busy routine for LCD. In the next section, we will see how to send data and command to the LCD.

Example 8.22

Show the interfacing of 16 × 2 LCD with microcontroller 8051 and write an assembly language program to display text message "NB BAHADURE".

Solution:

Hardware arrangement:

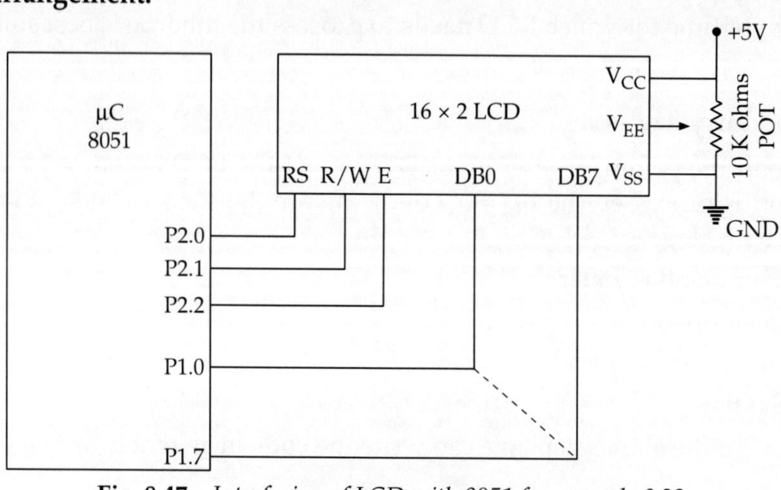

Fig. 8.47 *Interfacing of LCD with 8051 for example 8.22*

Port P1 of the microcontroller is connected to the data bus of LCD, whereas port P2.0, P2.1 and P2.2 are connected to the RS, R/W and E signals of the LCD respectively.

When RS is 0 then command code of the LCD is initialized and when RS =1 then ASCII code of the text is transferred on the LCD. The information transferred on the LCD either the command code or text messages it should be latched on the LCD, this latching operation is initiated when HIGH-to-LOW pulse is transferred on E signal of the LCD.

Assembly Language Program:

Method – I

```
; P2.0            Is connected to the RS pin of LCD
; P2.1            Is connected to the R/W pin of LCD
; P2.2            Is connected to the E pin of LCD
ORG 0000H
MOV A, #38H       ; Initialize 16 × 2 LCD
ACALL COMMAND     ; Send command
ACALL DELAY       ; Delay for some time
MOV A, #0EH       ; Display on cursor on
ACALL COMMAND
ACALL DELAY
MOV A, #01H       ; Clear LCD
ACALL COMMAND
ACALL DELAY
MOV A, #06        ; Shift cursor right
ACALL COMMAND
ACALL DELAY
MOV A, #84h       ; Cursor at line 1, position 4
ACALL COMMAND
ACALL DELAY
MOV A, #'N'       ; Display letter N
ACALL DATA
ACALL DELAY
MOV A, #'B'       ; Display letter B
ACALL DATA
```

```
        ACALL DELAY
        MOV A, #' '        ; Display space
        ACALL DATA
        ACALL DELAY
        MOV A, #'B'        ; Display letter B
        ACALL DATA
        ACALL DELAY
        MOV A, #'A'        ; Display letter A
        ACALL DATA
        ACALL DELAY
        MOV A, #'H'        ; Display letter H
        ACALL DATA
        ACALL DELAY
        MOV A, #'A'        ; Display letter A
        ACALL DATA
        ACALL DELAY
        MOV A, #'D'        ; Display letter D
        ACALL DATA
        ACALL DELAY
        MOV A, #'U'        ; Display letter U
        ACALL DATA
        ACALL DELAY
        MOV A, #'R'        ; Display letter R
        ACALL DATA
        ACALL DELAY
        MOV A, #'E'        ; Display letter E
        ACALL DATA
        ACALL DELAY
AGAIN:  SJMP AGAIN
        COMMAND:           MOV P1, A
        CLR P2.0
        CLR P2.1
```

```
            SETB P2.2
            ACALL DELAY
            CLR P2.2
            RET
DATA:       MOV P1,A
            SETB P2.0
            CLR P2.1
            SETB P2.2
            ACALL DELAY
            CLR P2.2
            RET
DELAY:      MOV R3, #75H
H2:         MOV R4,#0FFH
H1:         DJNZ R4,H1
            DJNZ R3,H2
            RET
```

Method – II

This method uses look-up table for the commands as well as data, so program length is reduced, resulting in saving the memory bytes.

```
            ; P2.0              Is connected to the RS pin of LCD
            ; P2.1              Is connected to the R/W pin of LCD
            ; P2.2              Is connected to the E pin of LCD
            ORG 0000H
            MOV DPTR, #COM_TAB
BACK:       CLR A
            MOVC A, @A+DPTR
            JZ DATA_SEND
            ACALL COMMAND
            ACALL DELAY
            INC DPTR
            SJMP BACK
```

```
DATA_SEND:MOV DPTR, #DATA_TAB
BACK1:    CLR A
          MOVC A, @A+DPTR
          JZ STOP
          ACALL DATA
          ACALL DELAY
          INC DPTR
          SJMP BACK1
STOP:     SJMP STOP
COMMAND:  MOV P1, A
          CLR P2.0
          CLR P2.1
          SETB P2.2
          ACALL DELAY
          CLR P2.2
          RET
DATA:     MOV P1,A
          SETB P2.0
          CLR P2.1
          SETB P2.2
          ACALL DELAY
          CLR P2.2
          RET
DELAY:    MOV R3, #75H
H2:       MOV R4,#0FFH
H1:       DJNZ R4,H1
          DJNZ R3,H2
          RET
COM_TAB:  DB 38H, 01H, 06H, 0EH, 84H, 00H
DATA_TAB: DB 'NB BAHADURE', 00H
          END
```

Simulation using Protius and IDE 8051 simulation software:

```
          ORG 0000H
          MOV DPTR, #COM_TAB
BACK:     CLR A
          MOVC A, @A+DPTR
          JZ DATA_SEND
          ACALL COMMAND
          ACALL DELAY
          INC DPTR
          SJMP BACK
DATA_SEND:  MOV DPTR, #DATA_TAB
BACK1:    CLR A
          MOVC A, @A+DPTR
          JZ STOP
          ACALL DATA
          ACALL DELAY
          INC DPTR
          SJMP BACK1
STOP:     SJMP STOP

COMMAND:MOV P1, A
          CLR P2.0
          CLR P2.1
          SETB P2.2
          ACALL DELAY
          CLR P2.2
          RET
```

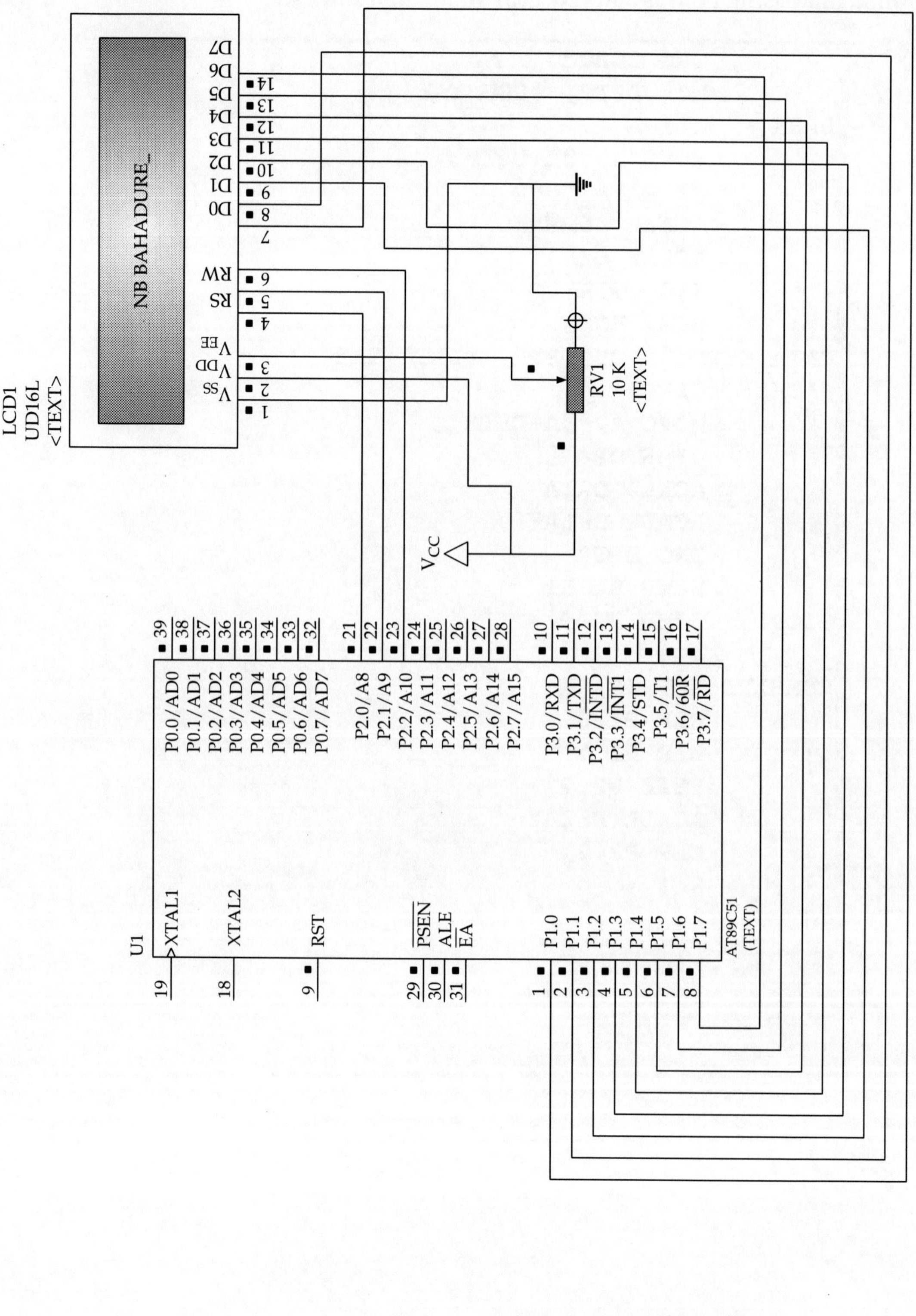
LCD1
UD16L
<TEXT>
NB BAHADURE_
VSS
VDD
VEE
RS
RW
D0
D1
D2
D3
D4
D5
D6
D7
RV1
10 K
<TEXT>
VCC
U1
XTAL1
XTAL2
RST
PSEN
ALE
EA
P0.0/AD0
P0.1/AD1
P0.2/AD2
P0.3/AD3
P0.4/AD4
P0.5/AD5
P0.6/AD6
P0.7/AD7
P2.0/A8
P2.1/A9
P2.2/A10
P2.3/A11
P2.4/A12
P2.5/A13
P2.6/A14
P2.7/A15
P3.0/RXD
P3.1/TXD
P3.2/INTD
P3.3/INT1
P3.4/STD
P3.5/T1
P3.6/60R
P3.7/RD
P1.0
P1.1
P1.2
P1.3
P1.4
P1.5
P1.6
P1.7
AT89C51
(TEXT)

8.9 KEYBOARD / KEYPAD INTERFACING

Keyboards are organized in a matrix of rows and columns. The CPU accesses both rows and columns through ports therefore with two 8-bit ports an 8 × 8 matrix of keys can be connected to the microcontroller. When a key is pressed, a row and a column make a contact; otherwise there is no connection between than.

Scanning and identification of the key:

Figure 8.49 shows a 4 × 4 matrix keyboard connected to two ports. The rows are connected to the output port and the columns are connected to the input port. If no key has been pressed, reading the input port will yield 1's for all columns since they are all connected to the high (Vcc). If all the rows are grounded and a key is pressed, one of the columns will have 0 since the key pressed provides the path to ground. The function of the microcontroller is to scan the keyboard continuously to detect and identify the key pressed.

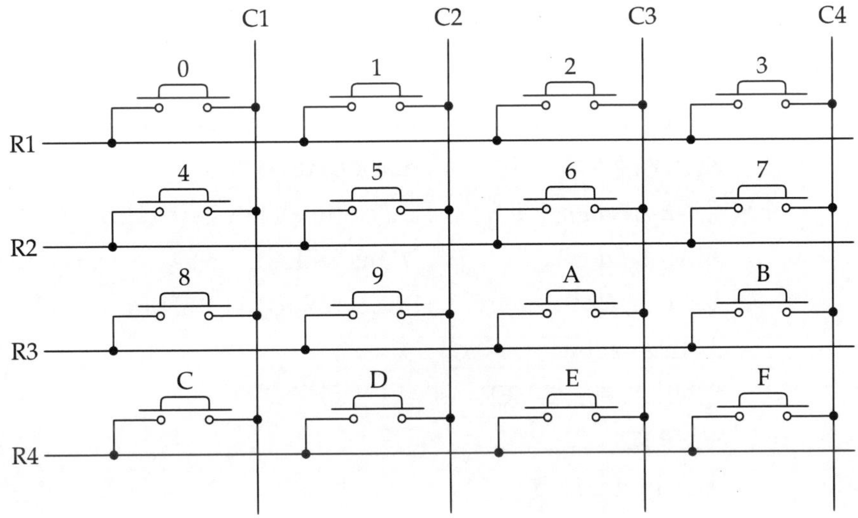

Fig. 8.48 *4 × 4 matrix keyboard*

Grounding rows and reading the column:

To detect a pressed key, the microcontroller grounds all rows by providing 0 to the output latch then it reads the column. If the data read from the column is D3 – D0 = 1111, no key has been pressed and process continues until a key pressed is detected. However, if one of the column bits has a zero, this means that a key has occurred, e.g., if D3 – D0 = 1101, this means that a key in D1 column has been pressed. After a key press is detected, the microcontroller will go through the process of identifying the key. Starting with the top row, the microcontroller grounds it by providing a low to row D0 only; then it reads the column. If the data read is all 1's, no key in that row is activated and the process is moved to the next row. It grounds the next row, reads the column and checks for any zero. This process continues until the row is identified. After identification if the row in which the key has been pressed, the next task is to find out which column the pressed key belongs to. This should be easy since microcontroller knows at any time which row and column are being occurred.

Example 8.23

Interface 4 × 4 matrix keyboards with microcontroller 8051. Write an assembly language program to read the keypad and send the ASCII code of pressed key to port P0.

P1.0 – P1.3 connected to rows (output)

P2.0 – P2.3 connected to the column (input)

Solution:

Assembly Language Program:

```
            MOV P2, #0FFh         ; P2 input
K1:         MOV P1, #00h          ; Grounds all rows at once
            MOV A, P2             ; Reads column ensure all keys open
            ANL A, #0Fh           ; mask unused bits
            CJNE A, #0Fh, K1      ; check till all keys are released
            ACALL DELAY           ; 20 ms DELAY
K2:         MOV A, P2             ; SEE ANY KEY PRESSED
            ANL A, #0F
            CJNE A, #0F, OVER
            SJMP K2
OVER:       ACALL DELAY
            MOV A, P2
            ANL A, #0F
            CJNE A, #0F, OVER1    ; Key pressed find row
            SJMP K2               ; if none keep polling
OVER1:      MOV P1, #11111110B    ; GROUND ROW 0
            MOV A, P2
            ANL A, #0F
            CJNE A, #0F, ROW_0
            MOV P1, #11111101B    ; GROUND ROW 1
            MOV A, P2
            ANL A, #0F
```

```
            CJNE A, #0F, ROW_1
            MOV P1, #11111011B   ; GROUND ROW 2
            MOV A, P2
            ANL A, #0F
            CJNE A, #0F, ROW_2
            MOV P1, #11110111B   ; GROUND ROW 3
            MOV A, P2
            ANL A, #0F
            CJNE A, #0F, ROW_3
            LJMP K2              ; IF NONE FALSE INPUT, REPEAT
ROW_0:      MOV DPTR, #KCODE0
            SJMP FIND
ROW_1:      MOV DPTR, #KCODE1
            SJMP FIND
ROW_2:      MOV DPTR, #KCODE2
            SJMP FIND
ROW_3:      MOV DPTR, #KCODE3
            SJMP FIND
FIND:       RRC A                ; See if any CY bit is low
            JNC ASC              ; If zero get ASCII code
            INC DPTR
            SJMP FIND
; Suppose A= 11111110 it means column 0 is detected so send ASCII
ASC:        CLR A
            MOVC A, @A+DPTR
            MOV P0, A
            LJMP K1
; ASCII Look up table for each row
KCODE0: DB '0', '1', '2', '3'     ; row 0
KCODE1: DB '4', '5', '6', '7'     ; row 1
KCODE2: DB '8', '9', 'A', 'B'     ; row 2
KCODE3: DB 'C', 'D', 'E', 'F'     ; row 3
```

Example 8.24

Show the interfacing of 16 push button keys arranged in 4 × 4 matrixes and common cathode type seven-segment LED display with 8051 microcontroller and write an assembly language program to display the pressed key on the seven-segment LED. Assume that the oscillator frequency of microcontroller is 11.0592MHz.

Solution: Hex key pad is essentially a collection of 16 keys arranged in the form of a 4 × 4 matrix. Hex key pad usually has keys representing numeric's 0 to 9 and characters A to F. The simplified diagram of a typical hex key pad is shown in the figure below.

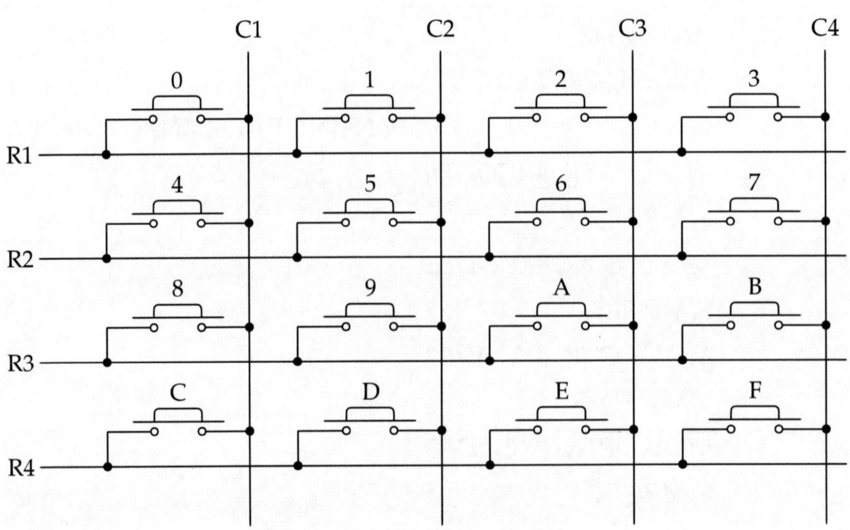

Fig. 8.49 *4 × 4 matrix keyboard*

The hex keypad has 8 communication lines namely R1, R2, R3, R4, C1, C2, C3 and C4. R1 to R4 represents the four rows and C1 to C4 represents the four columns. When a particular key is pressed, the corresponding row and column to which the terminals of the key are connected gets shorted. For example, if key 1 is pressed row R1 and column C1 gets shorted and so on. The program identifies which key is pressed by a method known as column scanning. In this method, a particular row is kept low (other rows are kept high) and the columns are checked for low. If a particular column is found low then it means that the key connected between that column and the corresponding row (the row that is kept low) is being pressed. For example, if row R1 is initially kept low and column C1 is found low during scanning, it means key 1 is pressed.

Hardware arrangement:

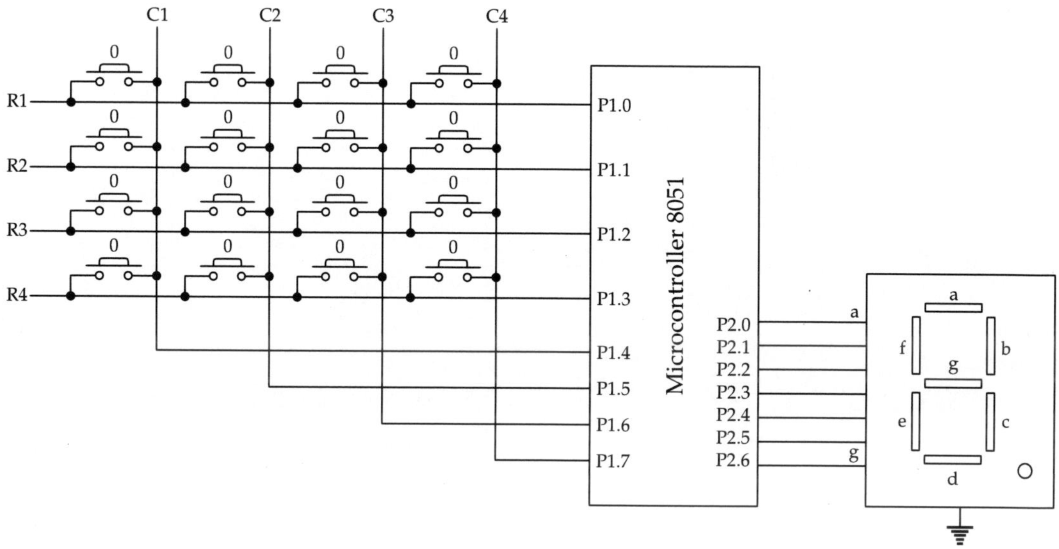

Fig. 8.50 *Interfacing of hex keypad and 7-segment LED with 8051 microcontroller*

Programming Tips: If it is implemented practically using the breadboard or on PCB then depending on the voltage source current limiting resistances of 330 ohms may be connected between each port pin (port P2 in the above circuit) and each seven segment LED terminal.

Assembly Language Program:

```
        ORG 0000H
        MOV DPTR,#LOOK_TAB  ; Moves starting address of look-up
                            table to DPTR
        MOV A,#11111111B    ; Loads A with all 1's
        MOV P2,#00000000B   ; Initializes P0 as output port
BACK:   MOV P1,#11111111B   ; Loads P1 with all 1's
        CLR P1.0            ; Makes row 1 low (ground row 1)
        JB P1.4,NEXT1       ; Checks whether column 1 is low and
                            jumps to NEXT1 if not low
        MOV A,#00           ; Loads A with 00 if column is low (that
                            means key 1 is pressed)
        ACALL DISPLAY       ; Calls DISPLAY subroutine
```

```
NEXT1:    JB P1.5,NEXT2       ; Checks whether column 2 is low and so
                              on.
          MOV A,#01
          ACALL DISPLAY
NEXT2:    JB P1.6,NEXT3
          MOV A,#02
          ACALL DISPLAY
NEXT3:    JB P1.7,NEXT4
          MOV A,#03
          ACALL DISPLAY
NEXT4:    SETB P1.0           ; Make row 1 high
          CLR P1.1            ; Make row 2 low (ground row 2)
          JB P1.4,NEXT5
          MOV A,#04
          ACALL DISPLAY
NEXT5:    JB P1.5,NEXT6
          MOV A,#05
          ACALL DISPLAY
NEXT6:    JB P1.6,NEXT7
          MOV A,#06
          ACALL DISPLAY
NEXT7:    JB P1.7,NEXT8
          MOV A,#07
          ACALL DISPLAY
NEXT8:    SETB P1.1           ; Make row 2 high
          CLR P1.2            ; Make row 3 low (ground row 3)
          JB P1.4,NEXT9
          MOV A,#08
          ACALL DISPLAY
NEXT9:    JB P1.5,NEXT10
          MOV A,#09
          ACALL DISPLAY
```

```
NEXT10:    JB P1.6,NEXT11
           MOV A,#10
           ACALL DISPLAY
NEXT11:    JB P1.7,NEXT12
           MOV A,#11
           ACALL DISPLAY
NEXT12:    SETB P1.2              ; Make row 3 high
           CLR P1.3               ; Make row 4 low (ground row 4)
           JB P1.4,NEXT13
           MOV A,#12
           ACALL DISPLAY
NEXT13:    JB P1.5,NEXT14
           MOV A,#13
           ACALL DISPLAY
NEXT14:    JB P1.6,NEXT15
           MOV A,#14
           ACALL DISPLAY
NEXT15:    JB P1.7,BACK
           MOV A,#15
           ACALL DISPLAY
           LJMP BACK
DISPLAY:   MOVC A,@A+DPTR         ; Gets digit drive pattern for the current
                                  ; Key from look up table
           MOV P2,A               ; Puts corresponding digit drive pattern
                                    into P0
           RET
LOOK_TAB:  DB 3FH, 06H, 5Bh, 4Fh
           DB 66h, 6Dh, 7Dh, 07h
           DB 7Fh, 67h, 77h, 7Ch
           DB 39h, 5Eh, 79h, 71h
           END
```

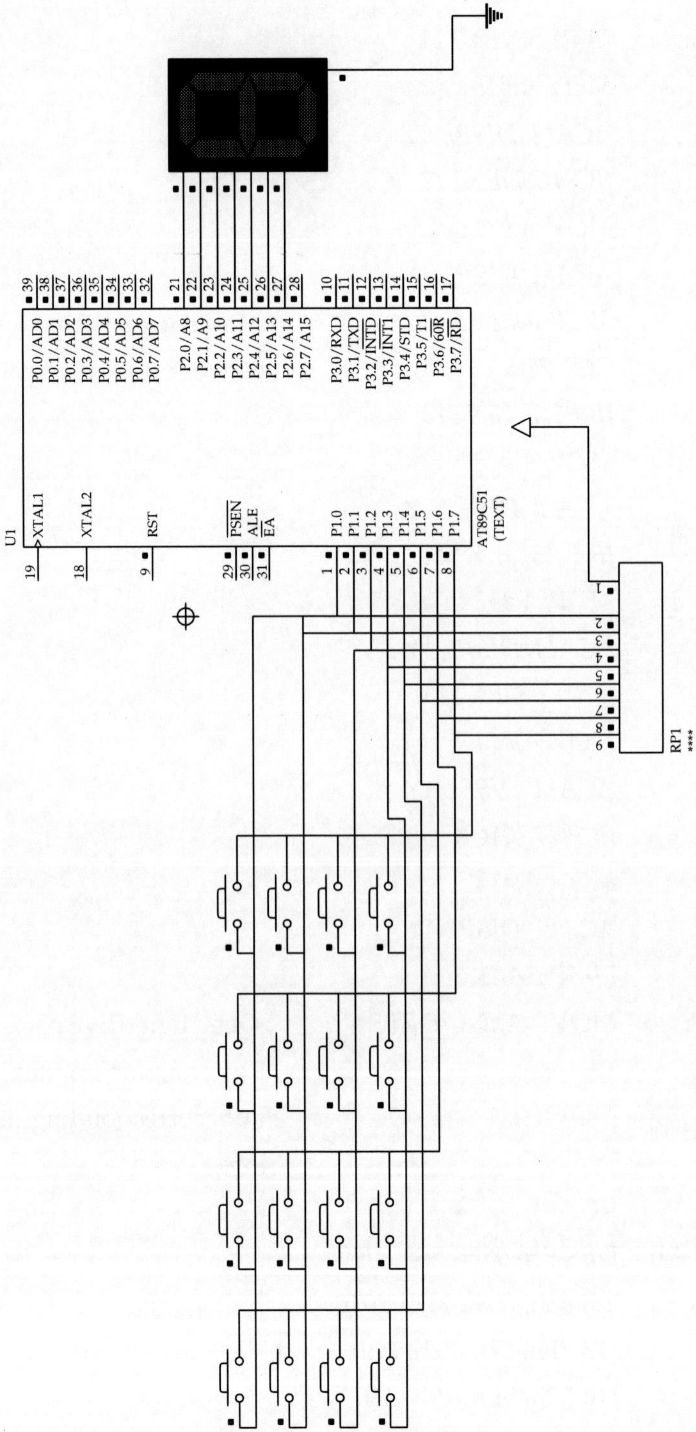
U1
XTAL1
XTAL2
RST
PSEN
ALE
EA
P1.0
P1.1
P1.2
P1.3
P1.4
P1.5
P1.6
P1.7
AT89C51
(TEXT)
P0.0/AD0
P0.1/AD1
P0.2/AD2
P0.3/AD3
P0.4/AD4
P0.5/AD5
P0.6/AD6
P0.7/AD7
P2.0/A8
P2.1/A9
P2.2/A10
P2.3/A11
P2.4/A12
P2.5/A13
P2.6/A14
P2.7/A15
P3.0/RXD
P3.1/TXD
P3.2/INTD
P3.3/INT1
P3.4/STD
P3.5/T1
P3.6/60R
P3.7/RD
RP1
****

Approach of the program:

First, the program initializes port 2 as an output port by writing all 0's to it and port 1 as an input port by writing all 1's to it. Then the program makes row 1 low by clearing P1.0 and scans the columns one by one for low using JB instruction. If column C1 is found low, which means 1 is pressed and accumulator is loaded by zero and DISPLAY subroutine is called. The display subroutine adds the content in A with the starting address of look up table stored in DPTR and loads A with the data to which the resultant address points (using instruction MOVC A,@A+DPTR). The present data in A will be the digit drive pattern for the current key press and this pattern is put to port 2 for display. This way the program scans for each key one by one and puts it on the display if it is found to be pressed.

Example 8.25

Show the interfacing of 16 push button keys arranged in 4 × 4 matrixes with 8051 microcontroller and write an assembly language program to display the pressed key on the 16 × 2 LCD. Assume that oscillator frequency of microcontroller is 11.0592 MHz.

Solution:

Hardware arrangement: (*see* Figure 8.51)

Assembly Language Program:

```
        ORG 0000H
        MOV P1,#0FFH        ; Make port P1 as an input port
        LJMP MAIN
B0:     MOV A,#'0'          ; Key 1 is pressed display 0
        LCALL WRITE_CHR
        LCALL DELAY
        LJMP KEYPAD0
B1:     MOV A,#'1'          ; Key 2 is pressed display 1
        LCALL WRITE_CHR
        LCALL DELAY
        LJMP KEYPAD1
B2:     MOV A,#'2'          ; Key 3 is pressed display 2
        LCALL WRITE_CHR
        LCALL DELAY
        LJMP KEYPAD2
B3:     MOV A,#'3'
        LCALL WRITE_CHR
        LCALL DELAY
        LJMP KEYPAD3
```

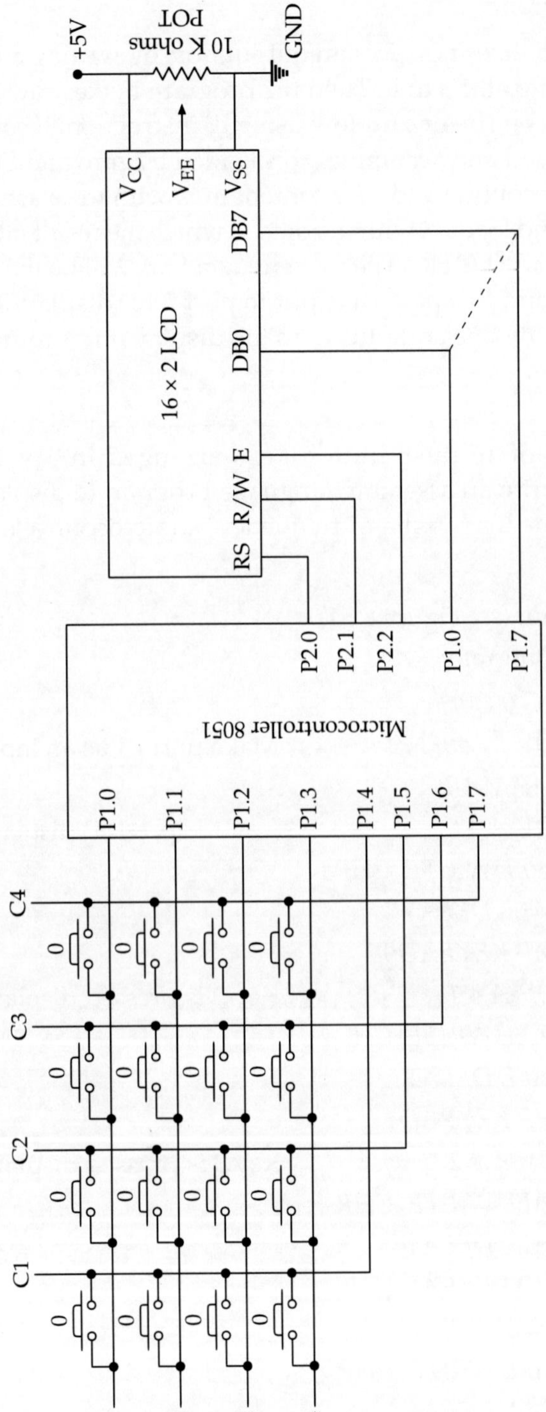

Fig. 8.51 *Interfacing of Hex keypad and LCD with 8051 microcontroller*

```
B4:     MOV A,#'4'
        LCALL WRITE_CHR
        LCALL DELAY
        LJMP KEYPAD4
B5:     MOV A,#'5'
        LCALL WRITE_CHR
        LCALL DELAY
        LJMP KEYPAD5
B6:     MOV A,#'6'
        LCALL WRITE_CHR
        LCALL DELAY
        LJMP KEYPAD6
B7:     MOV A,#'7'
        LCALL WRITE_CHR
        LCALL DELAY
        LJMP KEYPAD7
B8:     MOV A,#'8'
        LCALL WRITE_CHR
        LCALL DELAY
        LJMP KEYPAD8
B9:     MOV A,#'9'
        LCALL WRITE_CHR
        LCALL DELAY
        LJMP KEYPAD9
B10:    MOV A,#'A'
        LCALL WRITE_CHR
        LCALL DELAY
        LJMP KEYPAD10
B11:    MOV A,#'B'
        LCALL WRITE_CHR
        LCALL DELAY
        LJMP KEYPAD11
B12:    MOV A,#'C'
        LCALL WRITE_CHR
        LCALL DELAY
        LJMP KEYPAD12
B13:    MOV A,#'D'
        LCALL WRITE_CHR
        LCALL DELAY
        LJMP KEYPAD13
```

```
B14:        MOV A,#'E'
            LCALL WRITE_CHR
            LCALL DELAY
            LJMP KEYPAD14
B15:        MOV A,#'F'
            LCALL WRITE_CHR
            LCALL DELAY
            LJMP KEYPAD15
B1_0:       LJMP B0
B1_1:       LJMP B1
B2_1:       LJMP B2
B3_1:       LJMP B3
B4_1:       LJMP B4
B5_1:       LJMP B5
B6_1:       LJMP B6
B7_1:       LJMP B7
B8_1:       LJMP B8
B9_1:       LJMP B9
B10_1:      LJMP B10
B11_1:      LJMP B11
B12_1:      LJMP B12
B13_1:      LJMP B13
B14_1:      LJMP B14
B15_1:      LJMP B15
MAIN:       LCALL INI_LCD
            LCALL WRITE_PROG
            LCALL INI_LCD1
KEYPAD:     CLR P1.0            ; Ground row 1
            JNB P1.4,B1_0       ; If key is pressed in column 1 go to
                                  display 0
KEYPAD0:    JNB P1.5,B1_1       ; If key is pressed in column 2 go to
                                  display 1
KEYPAD1:    JNB P1.6,B2_1       ; If key is pressed in column 3 go to
                                  display 2
KEYPAD2:    JNB P1.7,B3_1       ; If key is pressed in column 4 go to
                                  display 3
KEYPAD3:    SETB P1.0           ; Ungrounded row 1
            CLR P1.1            ; Ground row 2
            JNB P1.4,B4_1       ; If key is pressed in column 1 go to
                                  display 4
```

```
KEYPAD4:   JNB P1.5,B5_1      ; If key is pressed in column 1 go to
                                display 5
KEYPAD5:   JNB P1.6,B6_1      ; If key is pressed in column 1 go to
                                display 6
KEYPAD6:   JNB P1.7,B7_1      ; If key is pressed in column 1 go to
                                display 7
KEYPAD7:   SETB P1.1          ; Ungrounded row 2
           CLR P1.2           ; Grounded row 3
           JNB P1.4,B8_1      ; If key is pressed in column 1 go to
                                display 8
KEYPAD8:   JNB P1.5,B9_1      ; If key is pressed in column 1 go to
                                display 9
KEYPAD9:   JNB P1.6,B10_1     ; If key is pressed in column 1 go to
                                display A
KEYPAD10:  JNB P1.7,B11_1     ; If key is pressed in column 1 go to
                                display B
KEYPAD11:  SETB P1.2          ; Ungrounded row 3
           CLR P1.3           ; Grounded row 4
           JNB P1.4,B12_1     ; If key is pressed in column 1 go to
                                display C
KEYPAD12:  JNB P1.5,B13_1     ; If key is pressed in column 1 go to
                                display D
KEYPAD13:  JNB P1.6,B14_1     ; If key is pressed in column 1 go to
                                display E
KEYPAD14:  JNB P1.7,B15_1     ; If key is pressed in column 1 go to
                                display F
KEYPAD15:  SETB P1.3          ; Ungrounded row 4
           LJMP KEYPAD        ; Repeat the process
INI_LCD:   MOV A,#38H
           LCALL WRITE_CMD
           MOV A,#0CH
           LCALL WRITE_CMD
           MOV A,#06H
           LCALL WRITE_CMD
           MOV A,#80H
           LCALL WRITE_CMD
           RET
```

```
INI_LCD1:   MOV A,#0C0H
            LCALL WRITE_CMD
            RET
DELAY:      MOV TMOD,#01H        ; Delay of approx. 1 second
            MOV R1,#20
REP:        MOV TH0,#-49990
            MOV TL0,#-49990
            SETB TR0
            JNB TF0,$
            CLR TF0
            CLR TR0
            DJNZ R1,REP
            RET
WRITE_CMD:  CLR P3.7
            CLR P3.5
            MOV P2,A
            SETB P3.6
            CLR P3.6
            LCALL DELAY
            RET
WRITE_CHR:  SETB P3.7
            CLR P3.5
            MOV P2,A
            SETB P3.6
            CLR P3.6
            LCALL DELAY
            RET
WRITE_PROG: MOV DPTR,#TABLE
            MOV R2,#14
L1:         CLR A
            MOVC A, @A+DPTR
            INC DPTR
            LCALL WRITE_CHR
            DJNZ R2, L1
            RET
TABLE:      DB 'KEYPAD PROGRAM'
LAST:       SJMP LAST
            END
```

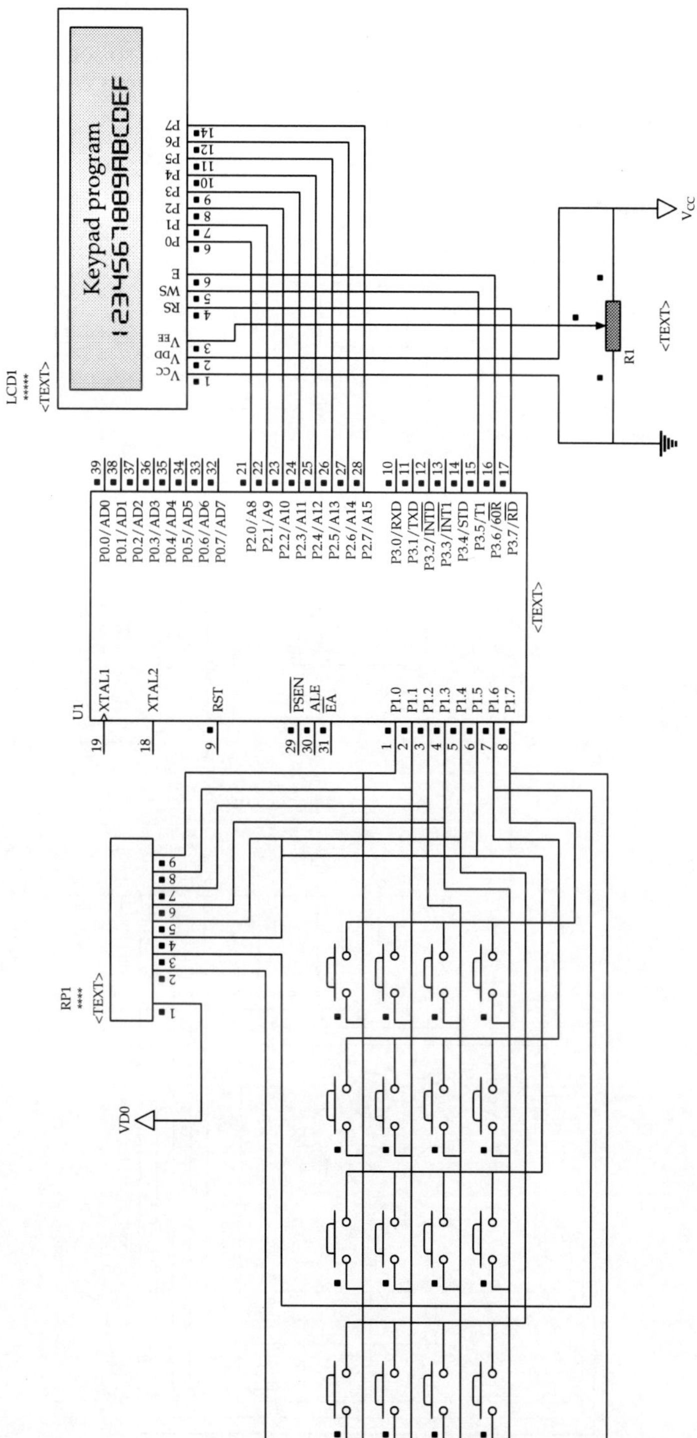

LCD1
Keypad program
123456789ABCDEF
U1
R1
RP1
VDD
VCC

Example 8.26

Show the interfacing of telephone keypad with 8051 microcontroller and write an assembly language program to display the pressed key on the 16 × 2 LCD. Assume that oscillator frequency of microcontroller is 11.0592 MHz.

Solution:

Hardware arrangement:

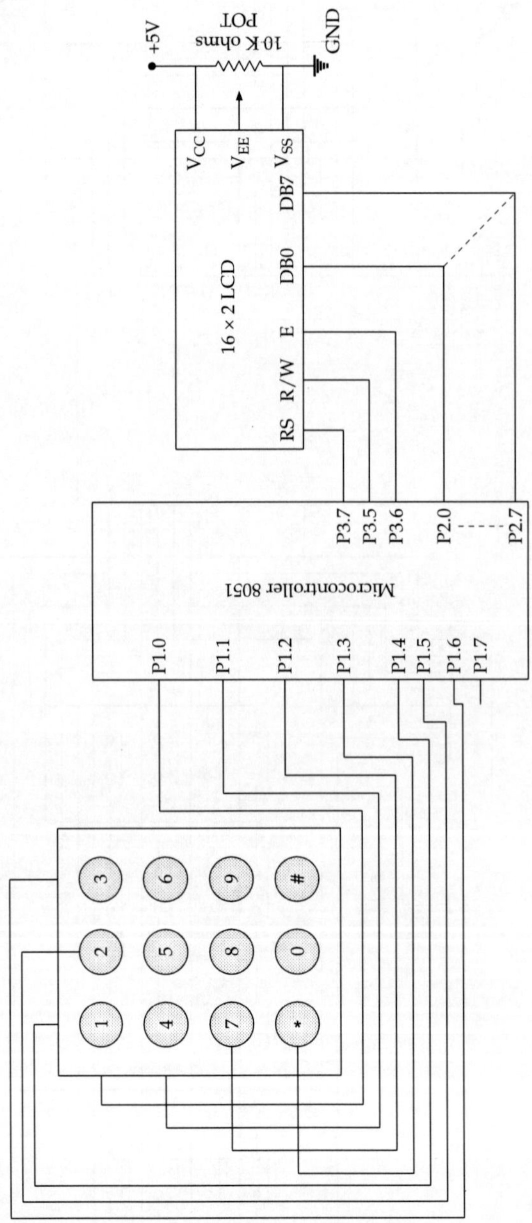

Fig. 8.52 *Interfacing of keypad and LCD with microcontroller 8051*

Assembly Language Program:

```
        ORG 0000H
        MOV P1,#0FFH
        LJMP MAIN
B1:     MOV A,#'1'
        LCALL WRITE_CHR
        LCALL DELAY
        LJMP KEYPAD1
B2:     MOV A,#'2'
        LCALL WRITE_CHR
        LCALL DELAY
        LJMP KEYPAD2
B3:     MOV A,#'3'
        LCALL WRITE_CHR
        LCALL DELAY
        LJMP KEYPAD3
B4:     MOV A,#'4'
        LCALL WRITE_CHR
        LCALL DELAY
        LJMP KEYPAD4
B5:     MOV A,#'5'
        LCALL WRITE_CHR
        LCALL DELAY
        LJMP KEYPAD5
B6:     MOV A,#'6'
        LCALL WRITE_CHR
        LCALL DELAY
        LJMP KEYPAD6
B7:     MOV A,#'7'
        LCALL WRITE_CHR
        LCALL DELAY
        LJMP KEYPAD7
```

```
B8:         MOV A,#'8'
            LCALL WRITE_CHR
            LCALL DELAY
            LJMP KEYPAD8
B9:         MOV A,#'9'
            LCALL WRITE_CHR
            LCALL DELAY
            LJMP KEYPAD9
B10:        MOV A,#'*'
            LCALL WRITE_CHR
            LCALL DELAY
            LJMP KEYPAD10
B11:        MOV A,#'0'
            LCALL WRITE_CHR
            LCALL DELAY
            LJMP KEYPAD11
B12:        MOV A,#'#'
            LCALL WRITE_CHR
            LCALL DELAY
            LJMP KEYPAD12
B1_1:       LJMP B1
B2_1:       LJMP B2
B3_1:       LJMP B3
B4_1:       LJMP B4
B5_1:       LJMP B5
B6_1:       LJMP B6
MAIN:       LCALL INI_LCD
            LCALL WRITE_PROG
            LCALL INI_LCD1
KEYPAD:     CLR P1.0
            JNB P1.4,B1_1
KEYPAD1:    JNB P1.5,B2_1
```

```
KEYPAD2:    JNB P1.6,B3_1
KEYPAD3:    SETB P1.0
            CLR P1.1
            JNB P1.4,B4_1
KEYPAD4:    JNB P1.5,B5_1
KEYPAD5:    JNB P1.6,B6_1
KEYPAD6:    SETB P1.1
            CLR P1.2
            JNB P1.4,B7
KEYPAD7:    JNB P1.5,B8
KEYPAD8:    JNB P1.6,B9
KEYPAD9:    SETB P1.2
            CLR P1.3
            JNB P1.4,B10
KEYPAD10:   JNB P1.5,B11
KEYPAD11:   JNB P1.6,B12
KEYPAD12: SETB P1.3
            LJMP KEYPAD
INI_LCD:    MOV A,#38H
            LCALL WRITE_CMD
            MOV A,#0CH
            LCALL WRITE_CMD
            MOV A,#06H
            LCALL WRITE_CMD
            MOV A,#80H
            LCALL WRITE_CMD
            RET
INI_LCD1:   MOV A,#0C0H
            LCALL WRITE_CMD
            RET
DELAY:      MOV TMOD,#01H
            MOV R1,#20
```

```
REP:        MOV TH0,#-49990
            MOV TL0,#-49990
            SETB TR0
            JNB TF0,$
            CLR TF0
            CLR TR0
            DJNZ R1,REP
            RET
WRITE_CMD:  CLR P3.7
            CLR P3.5
            MOV P2,A
            SETB P3.6
            CLR P3.6
            LCALL DELAY
            RET
WRITE_CHR:  SETB P3.7
            CLR P3.5
            MOV P2,A
            SETB P3.6
            CLR P3.6
            LCALL DELAY
            RET
WRITE_PROG: MOV DPTR,#TABLE
L1:         CLR A
            MOVC A, @A+DPTR
            JZ L2
            INC DPTR
            LCALL WRITE_CHR
            SJMP L1
L2:         RET
TABLE:      DB 'KEY_PROG BY BAHADURE','0'
LAST:       SJMP LAST
            END
```

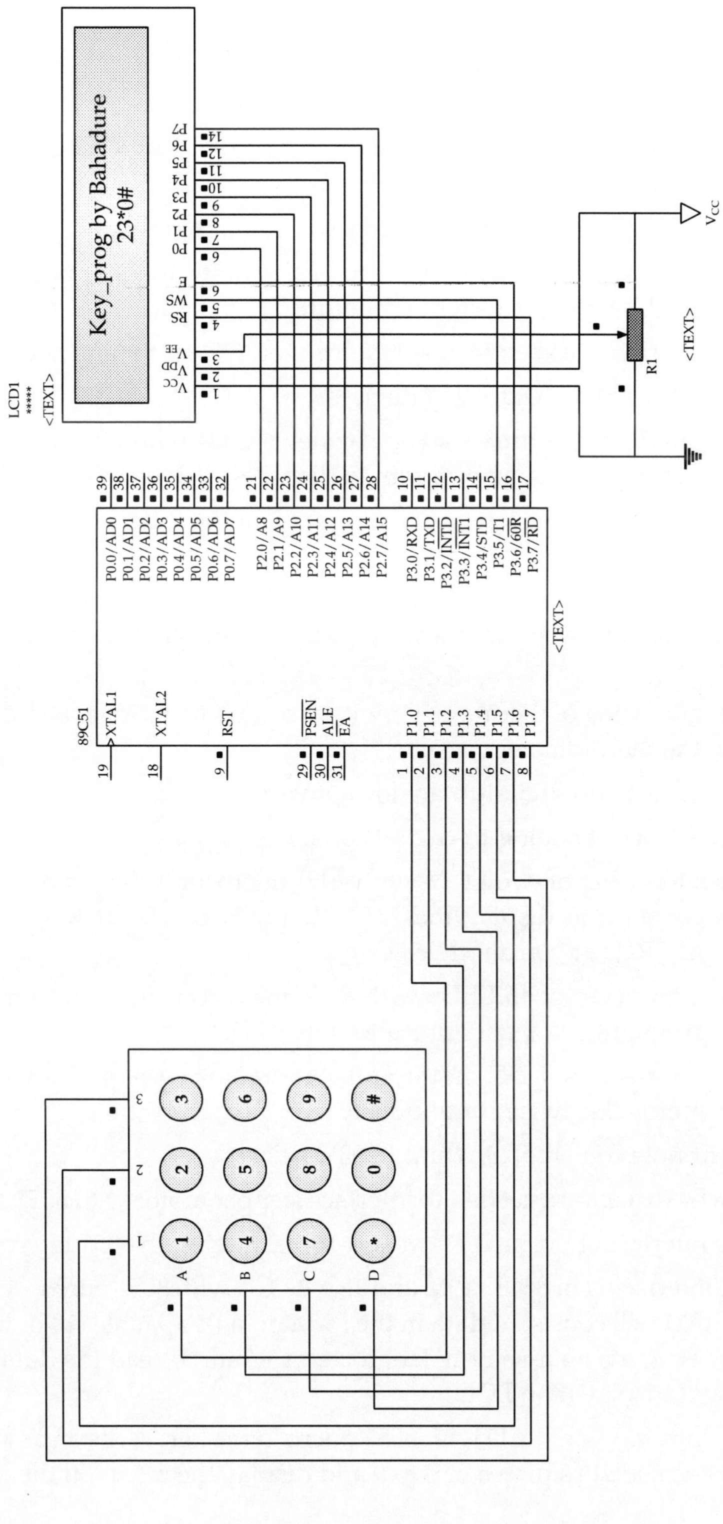
LCD1
Key_prog by Bahadure
23*0#
89C51
R1
V_CC

REVIEW QUESTIONS

1. Write short notes of LEDs and seven-segment LEDs.
2. Explain the common anode and common cathode seven-segment LEDs with data bit pattern for numbers 0 to 9.
3. What is stepper motor? Explain the types of stepper motor.
4. Show the interfacing of stepper motor and rotate the stepper motor clockwise continuously. Use universal stepper motor.
5. Explain the differences between stepper motor, DC motor and servomotor.
6. Write short notes of LCD display devices.
7. Show the interfacing of three seven-segment LEDs with 8051 microcontroller and write an assembly language program to display '1 2 3' on the seven-segment LEDs.
8. Write a short note on the keyboard; also explain the different types of keys used in the interfacing.
9. Show the interfacing of one seven-segment LED and one 16 x 2 LCD with 8051 microcontroller and write an assembly language program to display single number 3 on both.
10. Show the interfacing of thermometer with 8051 microcontroller and display the value read from the thermometer on LCD.
11. Write a short note on digital-to-analog converter.
12. Explain the types of analog-to-digital converter.
13. Show the interfacing of two LCDs with 8051 microcontroller, and write an assembly language program to display message "NILESH BAHADURE" on one LCD and message "BIT Raipur" on another LCD.
14. Show the interfacing of 16 LEDs with 8051 microcontroller and write an assembly language program to ON and OFF alternate LED.
15. Show the interfacing of DC motor and servomotor with 8051 microcontroller and rotate the motor clockwise continuously.
16. Write short notes on ULN2003 and L293D drivers.
17. Explain why drivers are needed to interface stepper motor with 8051 microcontroller.
18. Show the interfacing on 8 KB RAM and 16 × 2 LCD with 8051 microcontroller.
19. Show the interfacing of 8 KB ROM and 16 × 2 LCD with 8051 microcontroller. Assume that the ROM will consist of data in the form of look-up table (10 bytes in equivalent ASCII code). Write an assembly language program to read the data from the ROM location and display it on LCD.
20. Show the interfacing of LCD with 8051 microcontroller. Write an assembly language program to calculate squares of 0 to 9 and display the result on the LCD.

21. Show the interfacing of LED with 8051 microcontroller and write an assembly language program to perform addition of 10 bytes present in the memory location from 2000h, blink the LED if the carry is generated.

22. Show the interfacing of two switches and one LCD with 8051 microcontroller and write an assembly language program as if the switch one is pressed, microcontroller performs addition operation and when switch 2 is pressed, microcontroller performs subtraction operation, display the result on the LCD.

Chapter 9

Embedded Systems

In this chapter, readers will get information about of the embedded systems. Embedded systems are common platforms found in most of the modern electronic devices, i.e., from simple mobile devices to highly advanced aircraft.

9.1 INTRODUCTION

Embedded systems are the platform for the application of microcontrollers. Embedded systems are omnipresent. These are seen in your home, college, office, shopping mall and so on and so forth. Even when you are on the move – either on the most basic two wheeler or on an advanced aircraft – you are still amidst embedded systems. They are a unique combination of computer hardware, software and perhaps additional mechanical or other parts, designed to perform a specific function within a given time frame.

The embedded software is required for all real-time applications and developed using a real-time operating system (RTOS), as it helps to schedule and execute tasks based on the priority in a predictable manner. To cite just a few examples, embedded software allow your washing machine to choose speed according to the type of cloth, confer thinking power to the microwave ovens and propel rocket launchers into space.

Industry analysts estimate that over 54 percent of future software development will be in the embedded technology space. According to the IDC report, the international market as a whole expects product development worth $75 billion.

Embedded system is teamwork of three different sets of people: hardware engineering, software engineering and application domain experts.

As a hardware engineer, you will be responsible for hardware platform and module prototyping, debugging and testing. Also, you have to ensure compliance with standards and product specifications and initiate design changes as necessary.

If you are comfortable in the software domain, you can start as an embedded system programmer. This role will engage you in activities like analysis and optimization of embedded software for the targeted RTOS. Your tasks may include developing installable and built-in device drivers, kernel modifications and embedded applications.

Want to explore the application possibilities of the domain? Give a thought to beginning as a wireless protocol engineer. You will be assigned a task for the development, integration and testing of various protocols within an embedded firmware stack used in mobile devices.

If a particular application is developed and is embedded in the microcontroller IC so that the IC is dedicated to that application only, for example, for the railway system we developed the system to display the message "WELCOME TO THE INDIAN RAILWAY". To design this we create dedicated hardware, assembly or C language program and may be application software using real-time operating system to run this application within the schedule-time. Then this type of applications are called embedded based applications.

Another example includes that in the security system used in a car: air bag, here this air bag must open within the time when its service is required otherwise it is useless. So this type of system, where timing is the ultimate important factor, is always designed on the platform of embedded system on the platform of RTOS.

Who's Hiring?

Almost all industry segments are increasing the use of embedded systems for increased sophistication in the areas of networking, automation and control. As a result, software, telecommunication and electronics companies are heavily diversifying into it. As the application increases, the list of stakeholders can move only one way. The maximum demand is for professionals with 2-5 years of experience. Entry level hiring is around 20 per cent of the total talent acquisition.

The major recruiters, however, are big players like Texas Instruments, STMicroelectronics, Cadence Design, Honeywell, Delphi, TCS, Wipro, HCL, Infosys, Hawaii, Tata Elxsi, Mistral, LG, Portal player, CMC, GDA technologies, future software, ATS, and Serveen.

Entry level recruitment depends on some predefined skill sets. On the software side, the basic knowledge of "C" and Assembly language programming is a must. On the other side, sound knowledge of basic electronics, especially digital and analog design techniques, microprocessors and microcontrollers (e.g., RISC processor or 8051), and system knowledge related to communication protocol and real-time operating system can be helpful in getting a strong foothold in the field.

If we compare the basic skill sets required by any industry segment in India with the knowledge level of a fresher, a significant drawback of our educational system is revealed. The same holds true for the embedded sector too, which suffers from the lack of quality manpower. In fact, people with knowledge of embedded system design are rare despite the fact that this is one of the fastest growing sectors of electronics and designing concepts is a significant challenge.

Some believe that a computer teamed with the right software can design it all – cars, an air plane, even a spacecraft. However, when designing and developing embedded system, hardware and software tools always walk in step, helping each other attain perfection. While software helps design the hardware in fast, easy, graphical ways, hardware helps to test the software. Both go hand in hand through all the stages of designing and developing an embedded system, right from hardware platform creation and verification (simulation) to software platform creation, software application creation and software verification (debugging).

Hardware as well as software tools play a critical role in choosing, putting together and testing the various components of an embedded system, viz., microprocessors, microcontrollers or digital signal processors (DSPs), application specific integrated circuits (ASICs), field programmable gate arrays (FPGAs), analog-to-digital and digital-to-analog converters, communication paths, and memory devices and so on.

The key hardware tools used in the development, testing and debugging of embedded systems include development and evaluation boards, FPGAs, emulators, debuggers, logic analyzers and oscilloscopes. These not only help improve the quality and performance of the targeted embedded system but in many cases also reduce the time-to-market.

As the role of embedded systems in our lives is increasing – with these little, invisible computers powering devices ranging from microwave ovens to modems – the tools used to make these are also constantly evolving with the goal of enabling the developers to create more potent embedded systems within a shorter span of time. The hardware tools used for embedded development also fall in this dynamic category.

Emulators

Emulators are basically pieces of hardware that imitate the behaviour of other pieces of hardware, usually a system in the design stage. Developers can use an emulator in place of a processor or chip, and run the system with live data to see how it works. They can watch the application run step-by-step and easily identify errors. Emulators are important tools in functional verification and debugging of embedded systems where simulation can be slow and costly.

One of the most popular types of hardware emulators used in embedded design and development is the in-circuit emulator, which is used to test embedded software. Basically, it allows the developers to test the embedded software on the target hardware in order to ensure that there is no faulty code.

Earlier, emulators were independent electronic devices that could be plugged into a board in place of the microprocessor or other chips, to emulate their behaviour for testing purposes. However, the current generation of emulators allows the users to test their code on the actual target system itself using special joint test action group (JTAG) based interfaces.

An FPGA is essentially a chip that can be configured for various purposes by the user. As a result, it can be used to prototype system-on-a-chip, ASIC and embedded system designs. These prototypes, in turn, are used for functional verification of the hardware design, as well as the software and firmware being developed for it. In many cases, FPGA based functional verification is considered to be more accurate than software based simulation.

Earlier, it was believed that FPGA based prototyping and verification are cumbersome and time consuming for large systems. However, the situation has changed now. New age FPGAs are used extensively to test and prototype large modules of designs. These can even handle multi-million gate designs. A single high performance ASIC prototyping systems (HAPS) board from Synopsis, for example, can hold up to 8 million ASIC gates. More recent solutions from Synopsis can actually break a large full chip design into partitions, load these partitions into a multi-FPGA board and prototype a full system-on-a-chip in one go. This was not possible until a few years ago when engineers had to manually partition and prototype the design bottom up.

Apart from identifying faults and testing or debugging software developed for embedded systems, FPGA based prototypes have also been found to be useful in kindling customer interest by showing them a close-to-real demonstration of the system features.

Xilinx and Altera are the market leaders in the FPGA space, while companies such as Synopsis and Mentor Graphics provide software based tools for the design and implementation of FPGAs.

9.2 EMBEDDED SYSTEM

There are many definitions, but as a general concept, embedded systems are information processing systems that are embedded into a larger product. Embedded system is a special-purpose computer system that is used for a particular task. The special computer system is usually less powerful than general-purpose systems, although some exceptions do exist where embedded systems are very powerful and complicated. Usually, a low power consumption CPU with a limited amount of memory is used in embedded systems. Many embedded systems use very small operating systems; most of these provide limited operating system capabilities.

Embedded systems are generally not directly visible to the user. They are available almost in all kinds of information processing automated systems. Examples of embedded systems include information processing systems in telecommunication system, in transportation system, in fabrication equipment, in biomedical instrumentation, aircraft system, and in consumer electronics.

9.2.1 Embedded Systems and Power Consumption

Power consumption is also an important issue in embedded systems since most of the devices based on the embedded system are portable devices. This is also related to the moving parts issue in a sense. Components that consume more power need cooling, so some type of fan and heat sink must be attached to the system. For this reason, people always prefer CPUs and other components that use less power and generate less heat. It is also beneficial for systems that run on batteries to use less power. High power consuming components drain batteries quickly.

9.2.2 Embedded Operating Systems

Since embedded systems are usually not general-purpose systems, they need not be as sophisticated as commercial general-purpose systems. These operating systems are smaller in size, prompted in service and quickly bootable. Most of these operating systems also offer real-time capabilities, which are missing in general-purpose systems (microprocessor based systems). Although there are hundreds of operating systems available for the embedded systems, some of the most commonly used ones are:

- VxWorks
- pSOS
- Embedded Linux
- QNX
- Windows CE
- RTOS (Real time operating system)

When you use Linux as an embedded operating system, you can take out many parts of the kernel that are required in the general-purpose Linux systems. Which parts should be excluded from the kernel depends on the requirements of a particular system. For example, you can take out the entire networking code if the embedded system is not going to be networked. Similarly, you can take out most of the secondary storage related drivers and file systems support. Kernel parts that are the most likely for removal from the embedded Linux are:

- Disk drivers
- CD-ROM drivers
- Even if an embedded system is intended to be networked, you may not need all of the routing codes and drivers for all types of network adapters.
- Sound and multimedia related drivers and components. However you may need these components if the embedded system is used as a computer game.
- Most of the file system support.
- Any other thing that is not required. You can go through the kernel configuration process to determine what is not necessary.

The trimmed version of the Linux kernel will be much smaller compared to the kernels you find in most Linux distributions. Many modern operating systems are also used in embedded systems these days. Embedded Linux products are becoming more and more popular. Since embedded systems are used for special purpose tasks, there is usually no need to add new applications to the system once it is built. Therefore, programs and operating systems are usually loaded only once into read-only memory. The read-only memory is available in many forms and includes some type of ROM, flash chips and flash cards. In case an upgrade is required, a new ROM chip replaces the old one. If you look at your surroundings, you will find tiny computers everywhere. They are present even in places where you rarely notice them. Some of the most common embedded systems used in everyday life are:

- Refrigerators
- Washing machines
- Microwave ovens
- TV, VCR, DVD players
- Remote controllers
- Cameras
- Two-wheelers (ignition systems)
- Cars (antilock brakes, engine control, sun roof, climate control, automatic transmission and so on)
- Electronic toys
- Personal digital assistants
- Printers

Complicated and more sophisticated embedded systems include routers and switches where high performance CPUs are used.

9.2.3 Embedded Systems and Moving Parts

Most of the embedded systems are supposed to be used for a long period of time and without any maintenance. This is the very nature of embedded systems applications, to manufacture a system and then leave it to run independently for its intended life. For this reason, most embedded systems usually don't have moving components because any moving or mechanical part experiences wear and tear and needs replacement periodically. Typical examples of moving components include fans, disk drives and so on. Fans are used for cooling in power supplies and CPU heat sinks in desktop computers. Disk drives (both floppy and hard disk) are used as a storage medium for operating systems, applications and data. In embedded systems, ROM is used for storage of operating system and application data to eliminate the need for disk drives. These days, flash cards are also common and you can update a system just by replacing the flash memory card. In embedded systems you have to use special types of power supplies that don't need any fan for cooling. Many different types of power supplies are available in the market that don't heat up when used continuously for long periods of time.

9.2.4 Applications of Embedded Systems

Embedded systems are found in a variety of common electronic devices, such as:

(*a*) **Consumer electronics:** Cell phones, pagers, digital cameras, camcorders, videocassette recorders, portable video games, calculators, electronic toys, digital pen, and personal digital assistants.

(*b*) **Home appliances :** Microwave ovens, answering machines, thermostat, home security, washing machines, and lighting systems.

(*c*) **Office automation:** Fax machines, copiers, printers, and scanners.

(*d*) **Business equipment:** Cash registers, curbside check-in, alarm systems, card readers, product scanners, and automated teller machines (ATMs).

(*e*) **Medical system:** There is a huge potential for improving the medical service by taking advantage of information processing taking place within medical equipment. The latest EEG and ECG systems are already equipped with embedded systems.

(*f*) **Military applications:** Information processing has been used in military equipment for many years. The RADAR and SONAR systems already consist of embedded systems.

(*g*) **Smart buildings:** Information processing system can be used to increase the comfort level in buildings, can reduce the energy consumption within buildings, and can improve safety and security. For example, energy can be saved on cooling, heating and lighting rooms which are empty. Available rooms can be displayed at appropriate places, simplifying ad-hoc meetings and cleaning.

(*h*) **Robotics:** Mechanical aspects are very important for robots. Most of the characteristics used for the other systems are also equally applicable for the robotics control.

(*i*) **Trains:** Safety features contribute significantly to the total value of trains, and dependability is extremely important.

(*j*) **Aircraft electronics:** A significant amount of the total value of airplanes is due to the information processing equipment, including flight control systems, anti-collision systems, pilot information systems and others.

(*k*) **Automobiles:** Transmission control, cruise control, fuel injection, anti-lock brakes, car parking system, speed controller, and active suspension.

Modern cars can be sold only if they contain a significant amount of electronics. These include air bag control systems, engine control systems, anti-braking systems (ABS), air conditioning control, GPS systems, safety features and many more.

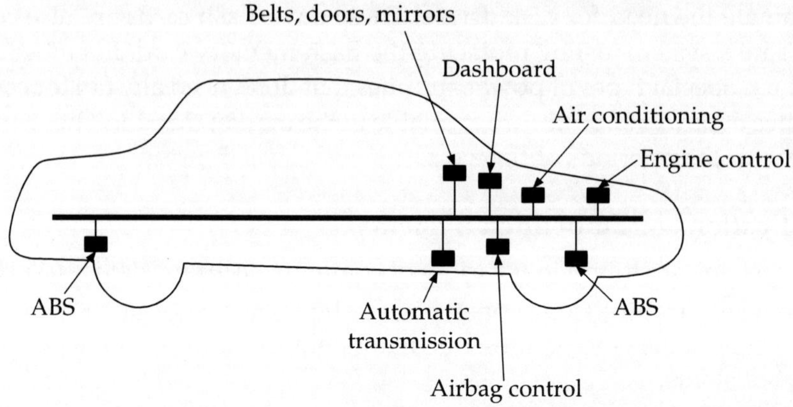

Fig. 9.1 *Use of embedded systems in automobile*

One might popularly say that nearly any device that runs on electricity either already has, or will soon have, a computing system embedded within it.

9.3 CHARACTERISTICS OF EMBEDDED SYSTEMS

The common characteristics of the embedded systems are as follows:

(*a*) Embedded systems have to be dependable

Most of the embedded systems are used for the safety devices and therefore have to be dependable. If it is not used for the safety devices, then also dependability must be ensured for the other systems like two-wheelers, cars, aircraft, rockets, trains, electronic toys, etc.

Dependability encompasses the following aspects of a system:

1. Reliability: Reliability in a sense that a system must not fail.
2. Maintainability: The need to ensure that a failing system can be repaired within a certain time frame. It must be repairable.
3. Availability: The system must be available. To achieve high order of availability, the system must attain the high order of reliability as well as maintainability.
4. Safety: This is to ensure that the system is safe and does not cause any harm, especially a failing system.
5. Security: The system must be secure. Most of the embedded system applications are found in the communication, so it is the need to ensure that confidential data will remain confidential and that authenticate communication must be guaranteed.

(*b*) Embedded systems have to be efficient

The efficiency of the embedded systems are evaluated on the following basis:

1. Energy: Most of the embedded systems are portable and obtain their energy or power from the batteries. To ensure the usability of the embedded system based devices for the longer run, the embedded system must be energy efficient.
2. Code size: Embedded systems are programmable devices, so the program written for the embedded system has to be stored with the system (on chip). Generally, in the embedded system there are no hard discs available on which the code of the program may be stored. Also to maintain the portability of the system, the code size should be as small as possible for the intended application.
3. Runtime efficiency: The hardware requirement for implementing the required functionality should be minimum as possible. The time constraint is met when least amount of hardware resources are used for the system. In order to reduce the power consumption, clock frequencies and supply voltages should be as small as possible. Above all to improve the working efficiency of the embedded systems, only the necessary hardware components should be present.

4. Weight: Again to meet the portability of the system, the system must be light in weight.
5. Cost: Because of the high volume productivity, the embedded system must be cost effective, so it is within the reach of the maximum. Especially in the market of consumer electronics, where competition is high and technology is affected very quickly, proper use of hardware resources and the software development budget are required.

(c) Embedded systems must have single functionality

Embedded systems are generally designed in such a way that it will meet the requirement of specific tasks. For example, a pager is always a pager. In contrast to the general-purpose processor system, where it is possible to run more than one application at a time like word processors, video games, spreadsheets, etc., in the embedded system, only one specific application is run, so it is designed in this constraint only.

For example, in a car system for monitoring the parking system one dedicated embedded system is available and for controlling the airbag another embedded system is available. In short, for each dedicated application there's an embedded system available.

(d) Embedded systems have a dedicated user interface

(e) Embedded systems must satisfy real-time constraints

(f) Embedded systems are hybrid systems

Hybrid system ensures that the system has both analog and digital parts. Analog parts are used for accepting continuous signal, especially from the outside world, but for the processing on the microcontroller based systems, where the values are accepted only in digital form, the embedded system must have hybrid systems.

(g) Embedded systems are reactive systems

Reactive system is a system where the input is accepted continually and some computations are performed on that, so the output forms a new state.

It is not necessary that all the embedded systems will have all the above characteristics, but if the systems meet most of the characteristics listed above they are called embedded systems.

9.4. ARCHITECTURE OF REAL-TIME EMBEDDED SYSTEMS

General-purpose computers or microprocessor systems have a generic architecture, but it cannot be defined for real-time embedded systems. There are as many architectures as the number of manufactures. As embedded systems are specially designed systems to perform particular tasks or operations, generalizing them would severely dilute the purpose of an embedded system.

However, for the sake of understanding, we can discuss some common form of systems at the block diagram level.

Some of these parts used in the real-time embedded systems may be programmable and therefore must have some place to keep these programs. In real-time embedded systems, the on-chip or on-board non-volatile memory does keep these programs. These programs are a part of the real-time operating system (RTOS) and continually run as long as the device is receiving power. A part of the RTOS also executes itself in the stand-by mode while taking a very little power from the battery. This is also called the sleep mode of the system.

An example of proposed architecture of an embedded system is shown in Figure 9.2.

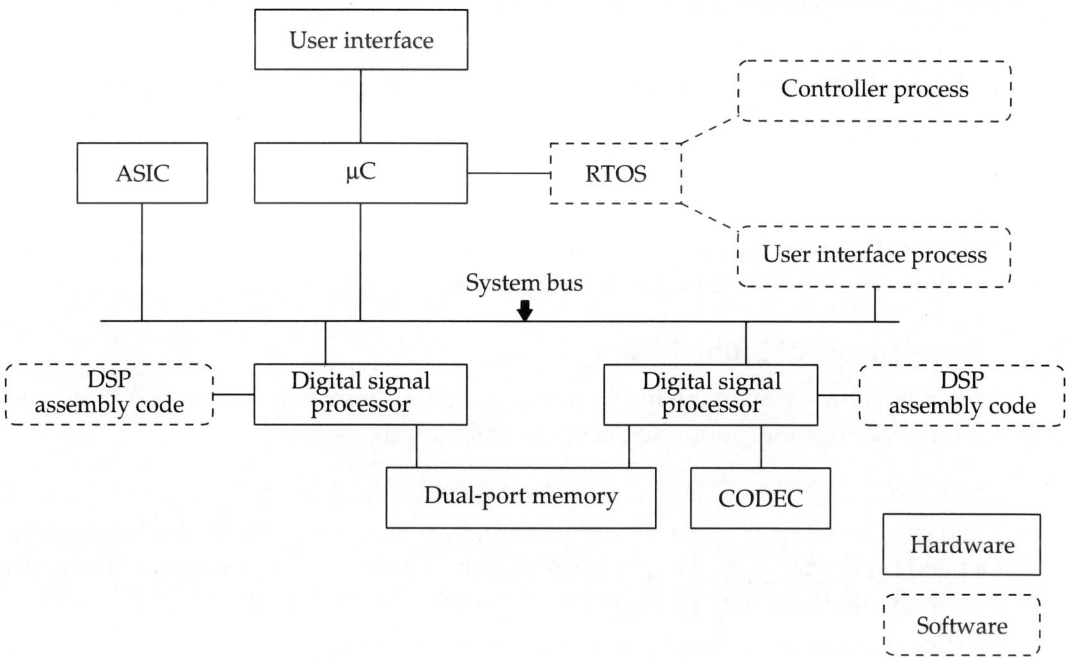

Fig. 9.2 *Architecture of an embedded system*

The explanation of various parts is as follows:

User Interface: For interacting with users, may consist of keyboard, touch pad etc.

ASIC: (Application specific integrated circuit) for specific functions like motor control, data modulation, etc.

Microcontroller (μC): a family of microprocessors

Real Time Operating System (RTOS): Contains all the software for the system control and user interface.

Controller Process: Overall control algorithm for the external process. It also provides timing and control for the various units inside the embedded system.

Digital Signal Processor (DSP): A typical family of microprocessors, used for advanced computing calculations, to process signals, etc. The DSP is a heavy duty processor capable of doing real-time signal processing and control. Both the DSPs along with their operating systems and codes are independent of each other. They share the same memory without interfering with each other. The RTOS controls the timing requirement of all the devices.

DSP Assembly Code: Code for DSP stored in program memory

Dual Ported Memory: Data Memory accessible by two processors at the same time

CODEC: Compressor/decompressor of the data.

User Interface Process: Part of the RTOS that runs the software for user interface activities

Controller Process: The part of the RTOS that runs the software for timing and control amongst the various units of the embedded system

9.4.1 Components of Embedded Systems

From the architecture of an embedded system, we have developed some sense about what type of components are used in the design of embedded systems.

(*a*) Processor

The central processing unit is the most important part of the embedded system. Depending on the type of application or task to perform the processors are broadly classified into three major categories:

1. Microprocessors/general-purpose microprocessors
2. Microcontroller
3. Digital signal processors

For more specific applications, customized processors can also be designed. Unless the demand is high, the design and manufacturing cost of such processors will be high. For example, ARM processors, where the architectural design is not fixed and it offers different varieties based on the need of the applications. Therefore, in most of the applications the design is carried out using already available processors in the market.

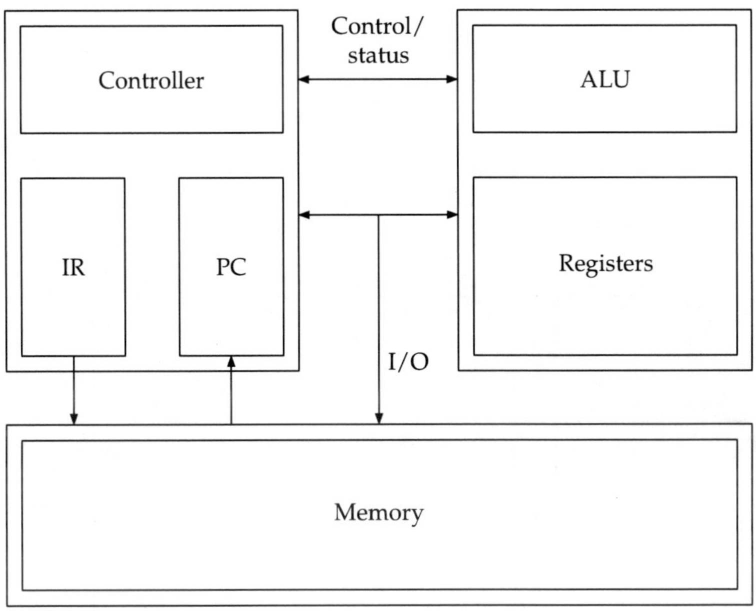

Fig. 9.3 *Block diagram of a general-purpose microprocessor*

General-purpose microprocessor block diagram is shown in Figure 9.3. Microprocessor is generally used to provide low cost solution but the size of the embedded systems is increased in this case, because number of other required devices are not available inside the microprocessor like serial port, interrupt, etc. So these devices need to be connected externally, hence the size of the embedded system is large.

The architecture of typical microcontroller, for example, microcontroller 8051 is shown in Figure 9.4. Microcontroller based embedded system is small in size in contrast to microprocessor based, because the requirement of additional devices is nullified in the microcontroller.

The digital signal processors have been designed based on the modified harvard architecture to handle real-time signals and these processors are faster compared to microprocessors and microcontrollers. The features of these processors are suitable for implementing signal processing algorithms. One of the common operations required in such applications is array multiplication. For example, convolution and correlation require array multiplication. This is accomplished by multiplication followed by accumulation and addition. This is generally carried out by multiplier and accumulator (MAC) units. In digital signal processors, generally all the instructions are executed in single cycle.

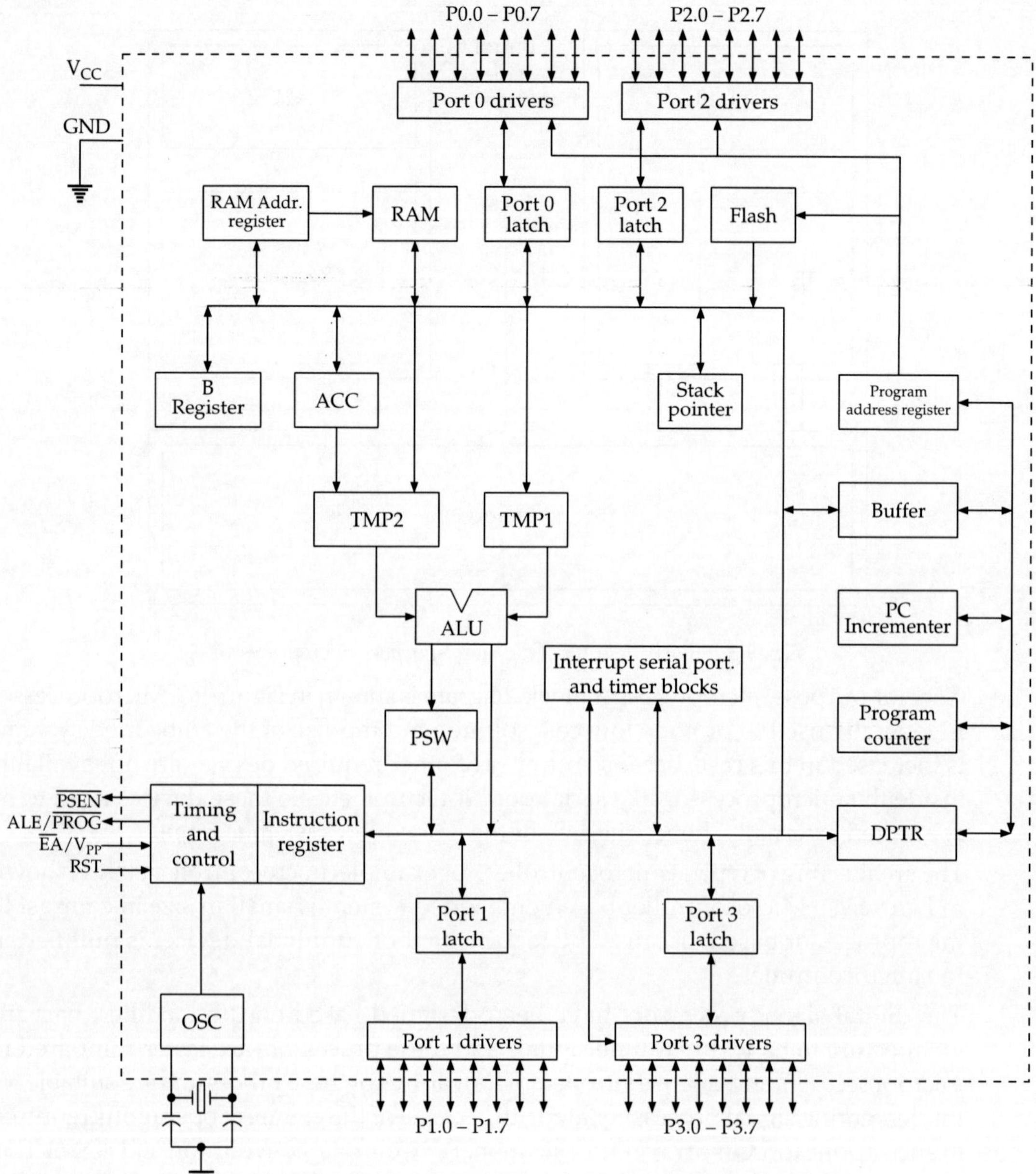

Fig. 9.4 *The architecture of a 8051 microcontroller*

(b) Memory

Compactness, *speed* and *low power* consumption are the characteristics required for the memory to be used in an embedded system. Therefore, very low power semiconductor memories are used in almost all such devices. For housing the operating system, Read Only Memory (ROM) is used. For example, you may like to change the ring tone of your mobile device and keep it for some time. You may like to change the screen colour, etc. In these cases, the memory should be capable of retaining the information even after the power is removed. In other words, the memory should be non-volatile and should be easily programmable too. It is achieved by using Flash memories. Flash memories are also faster in operation compared to other similar kind of memories, so it is highly suited to the embedded systems.

(c) Input/output devices and interfaces

Input/output interfaces are necessary to make the embedded systems interact with the external world. They could be visual display units such as TFT screens in a mobile phone, touch pad keyboard, antenna, microphones, speakers, etc. These embedded systems should also have open interfaces to other devices such as desktop computers, local area network (LAN) and other embedded systems. For example, you may like to download your address book into your personal digital assistant (PDA). Or you may like to download some mp3 songs from your favourite internet site into your mp3 player. These input/output devices along with standard software protocols in the RTOS provide the necessary interface to these standards.

(d) Software as operating system

The embedded system is just the physical body as long as it is not programmed. Whenever you switch on your mobile device you might have marked some activities on the screen. Whenever you move from one city to the other, you might have noticed the changes on your screen. Or sometimes you might have noticed the no-signal sign on the mobile device. These activities are taken care of by the real-time operating system sitting in the non-volatile memory of the embedded systems.

Apart from these, the embedded systems may have various other components and application specific integrated circuits (ASIC) for specialized functions such as motor control, modulation, demodulation, CODEC.

(e) Application software

Application software is designed to perform the specific task or operation on an embedded system. Application software may also be referred to as users program.

9.4.2 Characteristics of Real-Time Embedded Systems

Real-time systems cover such an enormous range of applications and products that a generalization of the characteristics into a set that is applicable to each and every system is difficult. Different categories of real-time systems may exhibit the characteristics that we identify to different extent or may not even exhibit some of the characteristics at all.

1. **Time constraints:** Every real-time task is associated with some time constraints. One form of time constraints that is very common is deadlines associated with tasks. A task deadline specifies the time before which the task must complete and produce the results. Other types of timing constraints are delay and duration. It is the responsibility of the real-time operating system (RTOS) to ensure that all tasks meet their respective time constraints.
2. **New Correctness Criterion:** The notion of correctness in real-time systems is different from that used in the context of traditional systems. In real-time systems, correctness implies not only logical correctness of the results, but the time at which the results are produced is important. If the result is produced after the deadline, then this result is considered the incorrect or inappropriate result.
3. **Safety Criticality:** For traditional non-real-time systems, safety and reliability are independent issues. However, in many real-time systems these two issues are intricately bound together making them *safety-critical*. Note that a safe system is one that does not cause any damage even when it fails. A reliable system on the other hand, is one that can operate for long duration without exhibiting any failures.
4. **Concurrency:** A real-time system usually needs to respond to several independent events within very short and strict time bounds. For instance, if the motor car meets accident then safety air bag must open within a time frame, otherwise it is useless. These systems can be considered to be non-deterministic, since the behaviour of the system depends on the exact timing of its inputs.
5. **Task Criticality:** Task criticality is a measure of the cost of failure of a task. Task criticality is determined by examining how critical are the results produced by the task to the proper functioning of the system. A real-time system may have tasks of very different criticalities. It is therefore natural to expect that the criticalities of the different tasks must be taken into consideration while designing for fault-tolerance. The higher the criticality of a task, the more reliable it should be made. Further, in the event of a failure of a highly critical task, immediate failure detection and recovery are important. However, it should be realized that task priority is a different concept and task criticality does not solely determine the task priority or the order in which various tasks are to be executed.
6. **Reactive:** Real-time systems are often *reactive*. A reactive system is one in which an on-going interaction between the computer and the environment is maintained. Ordinary systems compute functions on the input data to generate the output data.
7. **Stability:** Under overload conditions, real-time systems need to continue to meet the deadlines of the most critical tasks, though the deadlines of non-critical tasks may

not be met. This is in contrast to the requirement of *fairness* for traditional systems even under overload conditions.

8. **Exception Handling:** Many real-time systems work round-the-clock and often operate without human operators. For example, consider a small automated chemical plant that is set up to work non-stop. When there are no human operators, taking corrective actions on a failure becomes difficult. Even if no corrective actions can be immediately taken, it is desirable that a failure does not result in catastrophic situations. A failure should be detected and the system should continue to operate in a degraded mode rather than shutting off abruptly.

9.5 EMBEDDED OPERATING SYSTEMS

Except for very simple systems, scheduling, task switching, and I/O requires the support of an operating system suited for embedded applications. The following are the essential features of real-time and embedded operating systems:

(*a*) Flexible: Due to the large variety of the operating systems, there is also a large variety of requirements for the functionality of embedded operating systems. Hence, we need operating systems which can be flexible towards the application at hand. Configurability is therefore one of the main characteristics of embedded operating systems.

(*b*) Compact in size: There are effectively no devices like hard disk, a keyboard, a screen or a mouse or any other similar peripheral device that needs to be supported by all the versions of the operating system, except may be the system timer.

(*c*) Protection mechanisms are not always necessary, since embedded systems are typically designed for a single or dedicated purpose and untested programs are hardly ever loaded, it means that the program loaded in the embedded systems is already tested, it can be assumed to be reliable.

(*d*) Interrupts can be employed by any process. In desktop computing applications, interrupts would be a serious source of unreliability to allow any process to use interrupts directly, whereas in embedded systems it is possible to let interrupts directly start and stop tasks.

(*e*) Many embedded systems are real-time systems and hence the operating system used in the embedded systems must be a real-time operating system.

9.6 REAL-TIME OPERATING SYSTEM (RTOS)

Real-time operating system is an operating system that supports the construction of real-time systems. A RTOS is an operating system intended to serve real-time application process data as it comes in, typically without any buffering delays. Processing time requirements are measured in tenths of a second or even shorter.

The following are the key characteristics of the real-time operating systems:

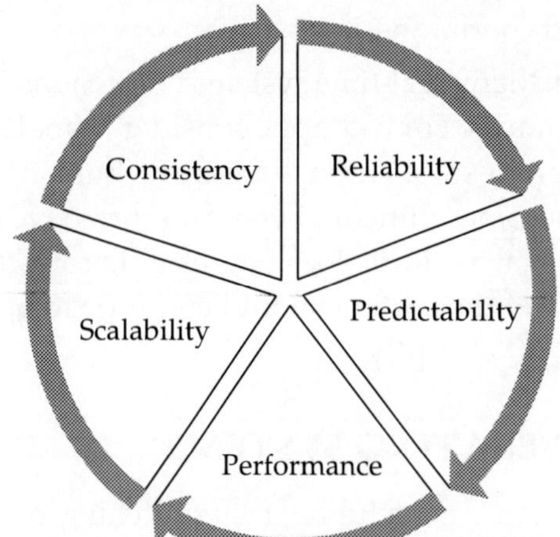

1. The timing behaviour of the operating system (OS) must be predictable. For each service of the OS, an upper bound on the execution time must be guaranteed.
2. The OS must manage the timing and scheduling of tasks. Scheduling can be defined as mapping from the set of tasks to intervals of execution time, including the mapping to start times as a special case. Also, the OS possibly has to be aware of task deadlines so that the OS can apply appropriate techniques.
3. The OS must be fast. In addition to being predictable, the OS must be capable of supporting applications with deadlines that are fractions of a second.

   Each RTOS includes a so-called real-time OS kernel. This kernel manages the resources which are found in every system, including the processor, the memory and the system timer. In fact, kernel will take care of all kinds of resources of the operating system. Protection mechanisms need not be present.

   There are two types of RTOSs: General-purpose OS and real-time kernel.

   (*a*) General-purpose OS type RTOSs: For these operating systems, some drivers, such as disk, network drivers, or audio drivers are implicitly assumed to be present, and they are embedded into the kernel. The application software and middleware are implemented on top of the application programming interface, which is standard for all applications.

   (*b*) Real-time kernel type RTOSs: Since there is hardly any standard device in embedded systems, device drivers are not deeply embedded into the kernel, but are implemented on the top of the kernel. Only the necessary drivers are included. Application software and middleware may be implemented on top of the appropriate drivers, not on top of a standard API of the OS.

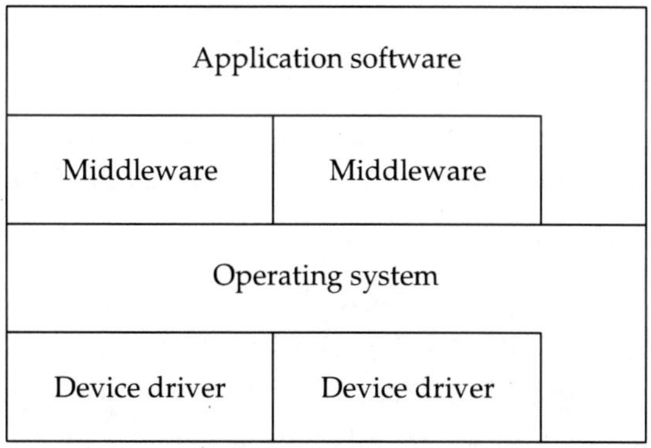

Fig. 9.5 *General-purpose operating system*

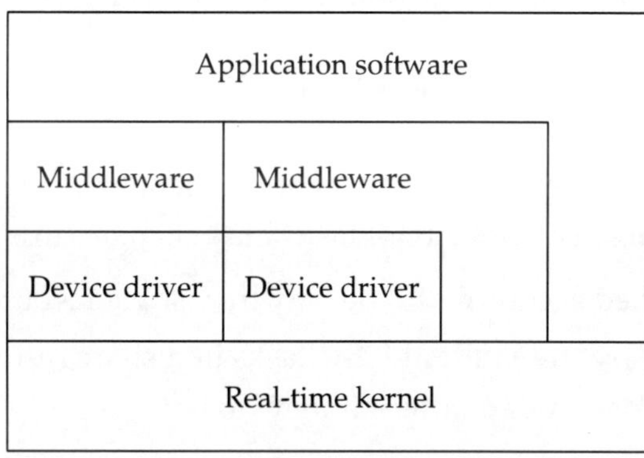

Fig. 9.6 *Real-time operating system*

9.7 TYPES OF REAL-TIME TASKS

"A real-time system is a system whose correctness includes its response time as well as its functional correctness". The system doesn't just run software to plough through the process; it needs to do it in a timely manner. So, real-time systems care about timeliness, but just how timely isn't clear until you add a modifier to "real-time". In other words, real-time is very much a gray region. "Response to what?" To "events", which are either external event, communicated to the system via its input/output interfaces, or to timer-triggered events within the system itself?

Classification of real-time systems

1. The type of system can be either a uniprocessor, multiprocessor or distributed system.
2. Moving to an multiprocessor or distributed real-time systems adds lots of difficulty by requiring real-time communication and synchronization mechanisms between the processors.

There are two different execution models:

1. In a **preemptive** model of execution, a task may be interrupted (preempted) during its execution and another task run in its place.
2. In a **non-preemptive** model of execution after a task that starts executing, no other task may execute until this task concludes or yields the CPU.

The task model for a real-time system is of four types:

1. **Purely periodic**: Every task is cyclic and executes periodically, even IO is polled. The nature of tasks doesn't vary much between cycles.
2. **Aperiodic/asynchronous**: Most tasks in the system aren't periodic.
3. A **static** system is a system where all tasks are known at design time including their release times.
4. A **dynamic** system is a system that can dynamically create and destroy tasks at runtime.

In many real-time systems, the tasks have different priorities. There are two possible models:

1. **Static-priorities:** The priorities of tasks don't change during execution.
2. **Dynamic-priorities:** The priority of tasks may change during execution.

Sets of tasks are classified according to the way they share resources:

1. We will call a task set **dependent** if the tasks in it share resources.
2. We will call a task set **independent** if they do not.

The real-time task specifies the following:

- The challenges
- Can we use existing systems?
- Proposed solutions
- Case studies of real-time operating systems.

A task is the basic unit of execution. Each task has three important properties:

(*a*) **Release time:** The point in time from which the task can be executed.

(*b*) **Deadline:** The point in time by which the task must complete.

(*c*) **Execution time:** The time the task takes to execute.

For example, brake controller

Classification of task timing constraints

There are two types of timing constraints : hard and soft.

- If missing a deadline is a fatal error or has disastrous consequences, it is considered **hard** (missing an occasional deadline is not acceptable).
- Otherwise (if some percentage of missed deadlines is acceptable) it is **soft**.

Classic examples:

Braking systems controller → hard
Multimedia streaming → soft

9.7.1 Hard Real-Time Systems

A hard real-time task is one that is constrained to produce its results within certain predefined time bounds. The system is considered to have failed whenever any of its hard real-time task does not produce its required results before the specified time bound.

An example of a system having hard real-time tasks is a braking system controller, anti-missile system, robot, etc. An example is the robot system which cyclically carries out a number of activities including communication with the host system, logging all completed activities, sensing the environment to detect any obstacles present, tracking the objects of interest, path planning, effecting next move, etc. Now consider that the robot suddenly encounters an obstacle. The robot must detect it and as soon as possible try to escape colliding with it. If it fails to respond to it quickly then it would collide with the obstacle and the robot would be considered to have failed. Therefore, detecting obstacles and reacting to it are hard real-time tasks.

Applications having hard real-time tasks are typically safety-critical. This means that any failure of a real-time task, including its failure to meet the associated deadlines, would result in severe consequences. This makes hard real-time tasks extremely critical. Criticality of a task can range from extremely critical to not so critical. Task criticality therefore is a different dimension from hard or soft characterization of a task. Criticality of a task is a measure of the cost of a failure; the higher the cost of failure, the more critical is the task.

For hard real-time tasks in practical systems, the time bounds usually range from several microseconds to a few milliseconds. It may be noted that a hard real-time task does not need to be completed within the shortest time possible, but it is merely required that the task must complete within the specified time bound. In other words, there is no reward in completing a hard real-time task much ahead of its deadline.

The hard real-time systems have the following parameters:

1. Contains tasks with hard timing constraints.
2. Requires formal verification/guarantees to always meeting its hard deadlines (except for fatal errors).

   Examples: Air traffic control, vehicle subsystems control, medical systems.
3. The scheduler then does one of the two operations :
   (a) The first would be the allowance of the process, guaranteeing that the process will be completed on time.
   (b) The second would be the rejection of the process. This is known as resource reservation. This type of guarantee requires that the scheduler knows the exact length of time each type of process of the operating system requires to perform.

3. Therefore, each operation must be guaranteed to take an optimal amount of time.
4. Hard real-time systems are composed of special purpose software running on hardware dedicated to their critical process.

9.7.2 Soft Real-Time Systems

Soft real-time tasks also have time bounds associated with them. However, unlike hard real-time tasks, the timing constraints on soft real-time tasks are not expressed as absolute values. Instead, the constraints are expressed either in terms of the average response times required.

An example of a soft real-time task is web browsing, and multimedia streaming. Normally, after a URL (Uniform Resource Locater) is clicked, the corresponding web page is fetched and displayed within a couple of seconds on the average. However, when it takes several minutes to display a requested page, we still do not consider the system to have failed, but merely express that the performance of the system has degraded or the speed of the internet system is slow.

The soft real-time systems consist of the following parameters:

1. Contain only tasks with no hard timing constraints.
2. Also known as "best effort" systems.
3. Most modern operating systems can serve as the base for soft real-time systems.
   Examples: Multimedia transmission and reception, networking, telecom (cellular) networks, web sites and services, computer games.
4. Less restrictive time by nature.
5. Requires that critical processes receive priority over less critical ones.
6. Implementation of soft real-time applications requires careful design of the scheduler and other related aspects of the operating system.
7. System interactions must be known as well as the system must have priority scheduling, giving the highest priority to real-time processes. In addition, the dispatch latency must be small.

9.8 FEATURES OF REAL-TIME OPERATING SYSTEMS

In the embedded systems, real-time operation is the responsibility of real-time operating system. These features also help us to make clear differences between a traditional operating system and a real-time operating system. In the following, we identify some important features required of a real-time operating system, and especially those that are normally absent in traditional operating systems.

1. Scheduling
2. Resource allocation
3. Interrupt handling
4. Other issues like kernel size

9.8.1 Scheduling in Real-Time Operating Systems

More information about the tasks is known along with some more additional information as given below:

- Number of tasks
- Resource requirements
- Release time
- Execution time
- Deadlines

Being a more deterministic system, better scheduling algorithms can be devised. Some of the scheduling algorithms are given below:

(*a*) Clock driven scheduling
(*b*) Weighted round robin scheduling
(*c*) Priority scheduling

(*a*) Clock Driven Scheduling

Clock and timer services with adequate resolution are one of the most important issues in real-time programming. Hard real-time application development often requires support of timer services with resolution of the order of a few microseconds. And even finer resolution may be required in case of certain special applications. Clocks and timers are a vital part of every real-time operating system. On the other hand, traditional operating systems often do not provide time services with sufficiently high resolution. For example in the motor car system, if an accident happens then air bag must be open or start in limited or desired time frame, otherwise it may cause a big problem.

The salient parameters required to accomplish clock driven scheduling must satisfy the following points:

(*i*) All parameters about jobs (release time/execution time/deadline) should be known in advance.
(*ii*) Schedule can be computed offline or at some regular time instances.
(*iii*) Minimal runtime overhead.
(*iv*) Not suitable for many applications.

(*b*) Weighted Round Robin Scheduling

A real-time operating system must support static priority levels. When once the programmer assigns a priority value to a task, the operating system does not change it by itself. Static priority levels are also called *real-time priority levels*. Such priority levels that are changed by the operating system dynamically are obviously not static priorities.

(*i*) Jobs scheduled in first in first out (FIFO) manner.

(*ii*) Time quantum given to jobs is proportional to its weight.

Example: Use high speed switching network

(*iii*) Not suitable for precedence constrained jobs.

Example: Job A can run only after Job B. No point in giving time quantum to Job B before Job A.

(*c*) Priority Scheduling

In real-time embedded system, it is needed to ensure that the processor never leaves any job assigned to the embedded systems. So to ensure proper execution of all the tasks assigned to the embedded system, the best possible way is to allocate priority scheduling to each and every task in the systems.

The constraints of the priority scheduling are given below:

(*i*) Processor should never be left idle when there are ready tasks.

(*ii*) Processor should be allocated to processes according to priorities.

(*iii*) Priorities should be static at design time and dynamic at runtime

9.8.2 Resource Allocation

If real-time tasks are allowed to share critical resources among themselves using the traditional resource sharing techniques, then the response times of tasks can become unbounded leading to deadline misses. This is one important reason as to why every commercial real-time operating system should at the minimum provide the basic priority inheritance mechanism. Support of priority ceiling protocol is also desirable, if large and moderate sized applications are to be supported. The resource allocation can also be accomplished using clock driven scheduling, weighted round robin scheduling and priority scheduling mechanisms. Some resources are non-preemptible, for example, semaphores.

9.8.3 Interrupt Handling

An interrupt is a signal from a device attached to a computer or from a program within a computer that causes the main program that is operating system to stop and figure out what to do next.

Interrupts cause the processor to suspend the operations whatever it is doing; instead execute the code that will respond to the event whatever caused the interrupt.

Interrupt latency is defined as the time delay between the occurrence of an interrupt and the running of the corresponding ISR (interrupt service routine). In real-time operating systems, the upper bound on interrupt latency must be bounded and is expected to be less than a few microseconds. The way low interrupt latency is achieved, is by performing bulk of the activities of ISR in a deferred procedure call (DPC). A DPC is essentially a task that performs most of the ISR activity. A DPC is executed later at a certain priority value.

Further, supports for nested interrupts are usually desired. That is, a real-time operating system should not only be preemptive while executing kernel routines, but should be preemptive during interrupt servicing as well. This is especially important for hard real-time applications with sub-microsecond timing requirements.

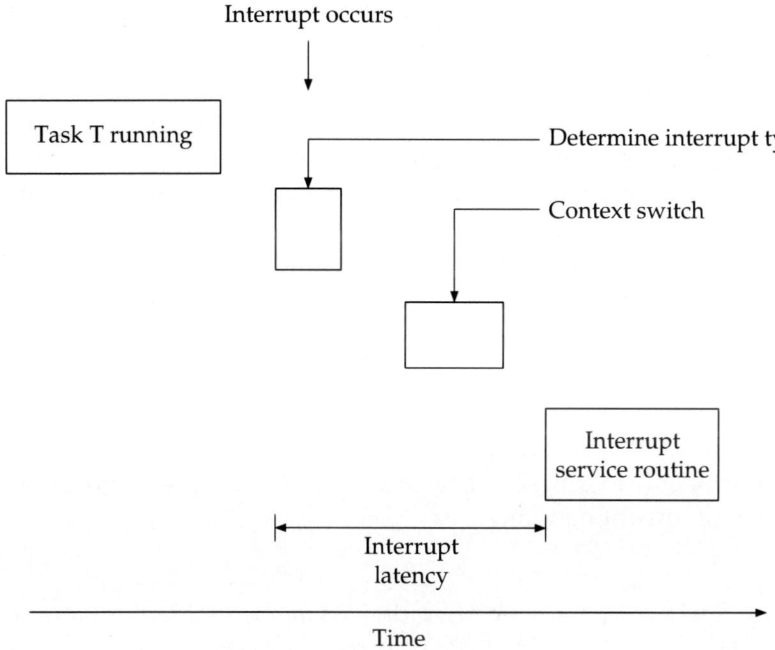

Fig. 9.7 *Interrupt latency in task execution*

9.8.4 Other issues like Kernel Size

Most real-time systems do not provide the features found in a standard desktop system. Reasons include: Real-time systems are typically single-purpose. Real-time systems often do not require interfacing with a user. Features found in a desktop PC require more substantial hardware than what is typically available in a real-time system. For timely perfect real-time embedded systems, the following parameters should be met:

1. Interrupt latency should be very small
   - Kernel has to respond to real-time events
   - Interrupts should be disabled for minimum possible time
2. For embedded applications kernel size should be small, so it should fit in the ROM.
3. Sophisticated features can be removed, so
   - No virtual memory
   - No protection

9.9 OTHER REAL-TIME OPERATING SYSTEMS

1. PSOS

PSOS is a popular real-time operating system that is being primarily used in embedded applications. It is available from Wind River Systems, a big player in the real-time operating system arena. It is a host-target type of real-time operating system. PSOS is being used in several commercial embedded products. An example application of PSOS is in the base stations of the cellular systems.

PSOS consists of 32 priority levels. In the minimal configuration, the foot print of the operating system is only 12 Kbytes. For sharing critical resources among real-time tasks, it supports priority inheritance and priority ceiling protocols. It supports segmented memory management. It allocates tasks to memory regions. A memory region is a physically contiguous block of memory. A memory region is created by the operating system in response to a call from an application.

In most modern operating systems, the control jumps to the kernel when an interrupt occurs. PSOS takes a different approach. The device drivers are outside the kernel and can be loaded and removed at the runtime. When an interrupt occurs, the processor jumps directly to the interrupt service routine pointed to by the vector table. The intention is not only to gain speed, but also to give the application developer complete control over interrupt handling.

2. VRTX

VRTX is a POSIX-RT compliant operating system from Mentor Graphics. VRTX has been certified by the US FAA (Federal Aviation Agency) for use in mission and life critical applications such as avionics. VRTX has two multitasking kernels: VRTXsa and VRTXmc. VRTXsa is used for large and medium applications. It supports virtual memory. It has a POSIX-compliant library and supports priority inheritance. Its system calls are deterministic and fully preemptible. VRTXmc is optimized for power consumption and ROM and RAM sizes. It has therefore a very small foot print. The kernel typically requires only 4 to 8 Kbytes of ROM and 1 Kbyte of RAM. It does not support virtual memory. This version is targeted for cell phones and other small hand-held devices

3. VxWorks

VxWorks is a product from Wind River Systems, who also developed PSOS operating system. It is host-target system. The host can be either a Windows or a UNIX machine. It supports most POSIX-RT functionalities. VxWorks comes with an integrated development environment (IDE) called Tornado. In addition to the standard support for program development tools such as editor, cross-compiler, cross-debugger, etc. Tornado contains VxSim and WindView. VxSim simulates a VxWorks target for use as a prototyping and testing environment. WindView provides debugging tools for the simulator environment. VxMP is the multiprocessor version of VxWorks.

VxWorks was deployed in the Mars Pathfinder which was sent to Mars in 1997. Pathfinder landed in Mars, responded to ground commands, and started to send science and engineering data. However, there was a hitch: it repeatedly reset itself. Remotely using trace generation, logging, and debugging tools of VxWorks, it was found that the cause was unbounded priority inversion. The unbounded priority inversion caused real-time tasks to miss their deadlines, and as a result, the exception handler reset the system each time. Although VxWorks supports priority inheritance, using the remote debugging tool, it was found to have been disabled in the configuration file. The problem was fixed by enabling it.

4. QNX

QNX is a product from QNX Software System Ltd. QNX Neutrino offers POSIX-compliant APIs and is implemented using microkernel architecture.

Because of the fine grained scalability of the microkernel architecture, it can be configured to a very small size – a critical advantage in high volume devices, where even a 1% reduction in memory costs can return millions of dollars in profit.

5. μC/OS – II

μC/OS-II is a free RTOS, easily available on the internet. It is written in ANSI C and contains small portion of assembly code and hence also named as μC (microcontroller) operating system. The assembly language portion has been kept to a minimum to make it easy to port it to different processors. To date, μC/OS-II has been ported to over 100 different processor architectures ranging from 8-bit to 64-bit microprocessors, microcontrollers, and DSPs. Some important features of μC/OS-II are highlighted in the following:

- μC/OS-II was designed so that the programmer can use just a few of the offered services or select the entire range of services. This allows the programmer to minimize the amount of memory needed by μC/OS-II on a per-product basis.
- μC/OS-II has a fully preemptive kernel. This means that μC/OS-II always ensures that the highest priority task that is ready would be taken up for execution.
- μC/OS-II allows up to 64 tasks to be created. Each task operates at a unique priority level. There are 64 priority levels. This means that round-robin scheduling is not supported. The priority levels are used as the PID (process identifier) for the tasks.
- μC/OS-II uses a partitioned memory management. Each memory partition consists of several fixed sized blocks. A task obtains memory blocks from the memory partition and the task must create a memory partition before it can be used. Allocation and de-allocation of fixed-sized memory blocks is done in constant time and is deterministic. A task can create and use multiple memory partitions, so that it can use memory blocks of different sizes.

- μC/OS-II has been certified by Federal Aviation Administration (FAA) for use in commercial aircraft by meeting the demanding requirements of its standard for software used in avionics. To meet the requirements of this standard, it was demonstrated through documentation and testing that it is robust and safe.

6. RT Linux

Linux is by and large a free operating system. It is robust, feature rich, and efficient. Several real-time implementations of Linux (RT-Linux) are available. It is a self-host operating system and its structure is shown in Figure 9.8.

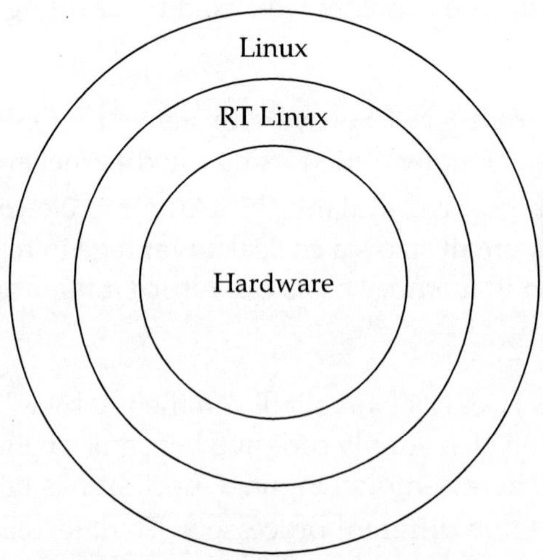

Fig. 9.8 *Structure of RT Linux*

RT Linux runs along with a Linux system. The real-time kernel sits between the hardware and the Linux system. The RT kernel intercepts all interrupts generated by the hardware. If an interrupt is to cause a real-time task to run, the real-time kernel preempts Linux, if Linux is running at that time, and lets the real-time task run. Thus, in effect Linux runs as a task of RT Linux. The real-time applications are written as loadable kernel modules. In essence, real-time applications run in the kernel space. In the approach taken by RT Linux, there are effectively two independent kernels: real-time kernel and Linux kernel. Therefore, this approach is also known as the *dual kernel approach* as the real-time kernel is implemented outside the Linux kernel. Any task that requires deterministic scheduling is run as a real-time task. These tasks preempt Linux whenever they need to execute and yield the CPU to Linux only when no real-time task is ready to run.

7. Lynx

Lynx is a self host system. The currently available version of Lynx (Lynx 3.0) is a microkernel-based real-time operating system, though the earlier versions were based on monolithic design. Lynx is fully compatible with Linux. With Lynx's binary compatibility, a Linux program's binary image can be run directly on Lynx. On the other hand, for other Linux compatible operating systems such as QNX, Linux applications need to be recompiled in order to run on them. The Lynx microkernel is 28 KB in size and provides the essential services in scheduling, interrupt dispatch, and synchronization. The other services are provided as kernel plug-ins (KPIs). By adding KPIs to the microkernel, the system can be configured to support I/O, file systems, sockets, and so on. With full configuration, it can function as a multipurpose Unix machine on which both hard and soft real-time tasks can run. Unlike many embedded real-time operating systems, Lynx supports memory protection.

8. **Windows CE**

Windows CE is a stripped down version of Windows, and has a minimum footprint of 400 KB only. It provides 256 priority levels. To optimize performance, all threads are run in the kernel mode. The timer accuracy is 1 msec for sleep and wait related APIs. The different functionalities of the kernel are broken down into small non-preemptive sections. So, during system call preemption is turned off for only short periods of time. Also, interrupt servicing is preemptable. That is, it supports nested interrupts. It uses memory management unit (MMU) for virtual memory management.

Windows CE uses a priority inheritance scheme to avoid priority inversion problem present in Windows NT. Normally, the kernel thread handling the page fault (i.e., DPC) runs at priority level higher than NORMAL. When a thread with priority level NORMAL suffers a page fault, the priority of the corresponding kernel thread handling this page fault is raised to the priority of the thread causing the page fault. This ensures that a thread is not blocked by another lower priority thread even when it suffers a page fault.

REVIEW QUESTIONS

1. Define embedded systems.
2. Explain the features of the embedded systems.
3. Explain the applications of the embedded systems.
4. Write short notes on the operating system used for the embedded system.
5. Explain the differences between the operating systems used in the large computers (desktop system) and those used in the embedded system.
6. Explain why there is a need of real-time operating systems for the development of the embedded systems.
7. Write short notes on interrupt handling and resource allocation.
8. Explain in detail the scheduling in real-time operating systems.
9. Write short notes on hard real-time system and soft real-time system.

10. Explain the types of real-time tasks.
11. Explain the characteristics of real-time system.
12. Explain the characteristics of embedded systems.
13. Draw and explain the architecture of the embedded systems. Explain the roles of digital signal processor in the embedded systems.
14. Write a short note on emulator.
15. Explain what kind of microcontroller or processor is best suited for the development of the embedded systems and why.
16. Write short notes on the power consumption of the embedded system devices.
17. Most of the handheld devices like mobile, tabs, etc., use embedded system inside. Explain why.
18. Write short notes on the RISC architecture used for the development of the embedded systems.
19. Draw and explain structure of RT Linux.
20. Explain the significance of real-time system in the embedded systems.
21. Explain how combination of functions reduces memory requirements in the embedded systems.

PIC, 8096 and ARM Microcontrollers

10 Chapter

In this chapter readers will get insights of other important 8-bit as well as 16- and 32-bit microcontrollers used for the small as well as in the large systems. This chapter also provides valuable information about ARM microcontroller, which is now a part of almost all modern electronic handheld devices.

SECTION – A

PERIPHERAL INTERFACE CONTROLLER (PIC) MICROCONTROLLER

10.1 BASIC DIFFERENCES BETWEEN 8051 MICROCONTROLLER AND PIC MICROCONTROLLER

(*a*) 8051 MICROCONTROLLER

1. The **Intel 8051** is based on Harvard architecture.
2. Instruction Set : CISC
3. Large number of instructions (usually 250)
4. Adopted by number of semiconductor manufacturing companies like ATMEL, NxP (Philips), Silicon, Cirrus Logic, etc.
5. Instruction cycles varies from 1 to 3 cycles.

(*b*) PIC Microcontroller

1. The PIC is also based on Harvard architecture.
2. Instruction set : RISC
3. Small number of instructions (less than 35 for basic PIC microcontroller)
4. Manufactured by one and only Microchip Technology Incorporation.
5. All single-cycle instructions except for program branches, which are two-cycles.
6. In-Circuit Debug feature via two pins.

10.2 CLASSIFICATION AND SELECTION GUIDE OF PIC MICROCONTROLLER

PIC microcontroller can be classified on the basis of memory type (Flash, OTP, ROM, ROMless etc.), I/O pin count (4-18 pins, 20-28 pins, 32-44 pins, 45 and above pins), Memory size (0.5-1K, 2-4K, 8-16K, 24-32K, 48-64K, 96-128K, etc), special features (CAN, USB, LCD, motor control, radio frequency).

The selection of PIC microcontroller can be made on the following basis:

(*a*) **By product family**

PIC 10 MCU
PIC 12 MCU
PIC 16 MCU
PIC 18 MCU

(*b*) **By application features**

CAN
Ethernet
LCD
Extreme low power
USB
Motor control

(*c*) **By memory size**

< 1 KB
2 KB – 4 KB
8 KB – 16 KB
24 KB – 32 KB
48 KB – 64 KB
96 KB – 128 KB

(*d*) **By port pin count**

6, 8 and 14 pins
18, 20 pins
28, 40, 44 pins
64, 80, 100 pins

(*e*) **By memory types**

Flash
OTP
ROM
ROMless

10.3 INTRODUCTION TO PIC MICROCONTROLLER

PIC stands for Peripheral Interface Controller given by Microchip Technology to identify its single-chip microcontrollers. These devices have been very successful in 8-bit microcontrollers. The main reason is that Microchip Technology has continuously upgraded the device architecture and added needed peripherals to the microcontroller to suit customers' requirements.

The architecture of various PIC microcontrollers can be divided as follows:

1. **Low end PIC Microcontrollers**

   Microchip PIC microcontrollers are available in various types. When PIC microcontroller MCU was first available from General Instruments in early 1980s, the microcontroller consisted of a simple processor executing 12-bit wide instructions with basic I/O functions. These devices are known as low-end architectures. They have limited program memory and are meant for applications requiring simple interface functions, small program and data memories. Some of the low-end device numbers are

   12C5XX

   16C5X

   16C505

2. **Mid end PIC Microcontrollers**

   Mid range PIC architectures are built by upgrading low-end architectures with more number of peripherals, more number of registers and more data/program memory. Some of the mid-range devices are:

   16C6X

   16C7X

   16F87X

   Program memory type is indicated by an alphabet.

   C = EPROM

   F = Flash

   RC = Mask ROM

10.4 ADVANTAGES OF THE PIC MICROCONTROLLER

The PIC microcontrollers are gaining high popularity because of the following advantages over other microcontrollers.

1. The PIC microcontroller is based on harvard architecture.
2. It is based on RISC instruction set.
3. Speed: 1 instruction cycle = 4 clock cycles.

4. Instruction set simplicity: The instruction set consists of just 35 instructions (as opposed to 111 instructions for 8051).
5. Power-on-reset and brown-out reset. Brown-out-reset means when the power supply goes below a specified voltage (say 4V), it causes PIC to reset; hence malfunction is avoided.
6. A watch dog timer (user programmable) resets the processor if the software/program ever malfunctions and deviates from its normal operation.
7. PIC microcontroller has four optional clock sources.
   - Low power crystal
   - Mid range crystal
   - High range crystal
   - RC oscillator (low cost).
8. Programmable timers and on-chip ADC.
9. Up to 12 independent interrupt sources.
10. Powerful output pin control (25 mA (max.) current sourcing capability per pin).
11. EPROM/OTP/ROM/Flash memory option.
12. I/O port expansion capability.
13. Free assembler and simulator support from the manufacturers of the PIC chip, i.e., from Microchip.

10.5 CPU ARCHITECTURE AND MEMORY ORGANIZATION OF PIC MICROCONTROLLER

The CPU uses Harvard architecture with separate program and variable (data) memory interface. This facilitates instruction fetch and the operation on data/accessing of variables simultaneously.

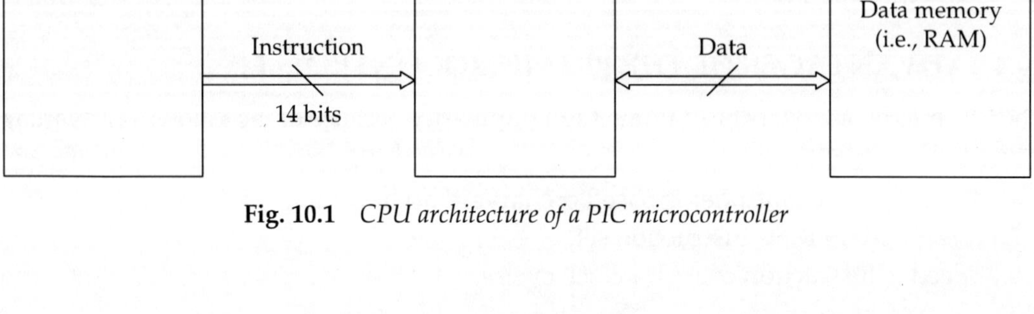

Fig. 10.1 *CPU architecture of a PIC microcontroller*

10.5.1 Memory organization of PIC Microcontroller

(a) Program Memory:

PIC microcontroller has 13 bits of program memory address. Hence, it can address up to 8k of program memory. The program counter is 13-bit. PIC 16C6X or 16C7X program memory is 2k or 4k. While addressing 2k of program memory, only 11-bits are required. Hence, two most significant bits of the program counter are ignored. Similarly, while addressing 4k of memory, 12 bits are required. Hence, the MSB of the program counter is ignored. Figure 10.2 shows the program memory map of PIC16C74A PIC microcontroller.

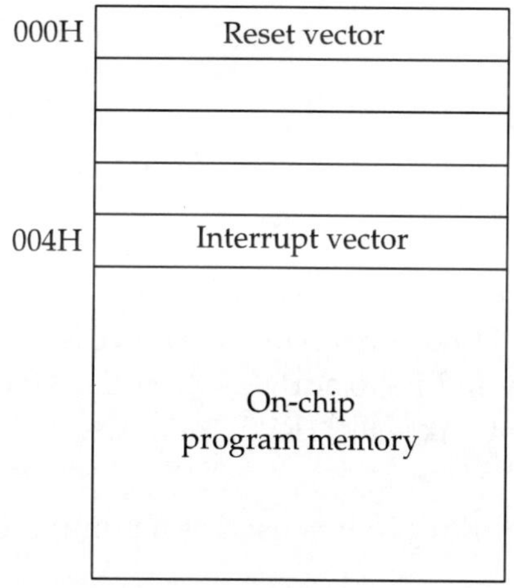

Fig. 10.2 *Program memory map*

(b) Data Memory (Register File):

On reset, the program counter is cleared and the program starts at 00H. Here a 'goto' instruction is required that takes the processor to the mainline program.

When a peripheral interrupt, that is enabled, is received, the processor goes to 004H. A suitable branching to the interrupt service routine (ISR) is written at 004H.

Data memory is also known as register file. Register File consists of two components:

1. General purpose register file (same as RAM).
2. Special purpose register file (similar to SFR in 8051).

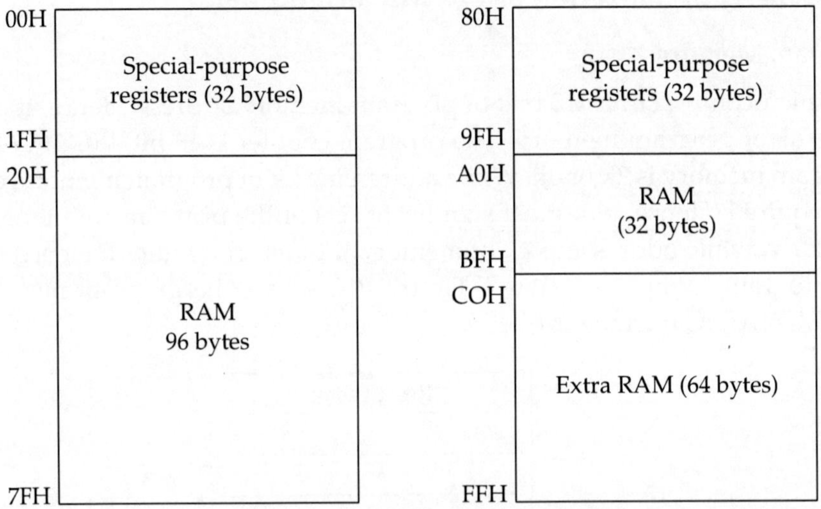

Fig. 10.3 *Data memory map*

The special purpose register file consists of input/output ports and control registers. Addressing from 00H to FFH requires 8 bits of address. However, the instructions that use direct addressing modes in PIC to address these register files use 7 bits of instruction only. Therefore, the register bank select (RP0) bit in the STATUS register is used to select one of the register banks.

In indirect addressing, FSR register is used as a pointer to anywhere from 00H to FFH in the data memory.

10.5.2 Specification of PIC Microcontrollers

Some 12-bit PIC Microcontrollers

Microcontroller	Program memory	Data memory	Max. speed (MHz)	I/O ports	A to D Converter
12C508	512 X 12	25	4	6	--
16C54	384 X 12	25	20	12	--
16C57	2048 X 12	72	20	20	--
16C505	1024 X 12	41	4	12	--
16C58A	2048 X 12	73	20	12	--

Some 16-bit PIC Microcontrollers

Microcontroller	Program memory	Data memory	Max. speed (MHz)	I/O ports	A to D Converter
16C554	512 X 14	80	20	13	--

16C64	2014 X 14	128	20	33	--
16C74A	4096 X 14	192	20*	33	8 bits 8 channels
16F84	1024 X 14	36	10	13	--
16F627	1024 X 14	224	20`	16	--
16F628	2048 X 14	224	20	16	--
16F676	1024 X 14	64	20	12	8
16F73	4096 X 14	192	20	22	5
16F876	8192 X 14	368	20	22	5
16F877	8192 X 14	368	20	33	8 (10 bits 8 channels)

PIC Microcontroller clock

Most of the PIC microcontrollers can operate up to 20 MHz. One instruction cycle (machine cycle) consists of four clock cycles.

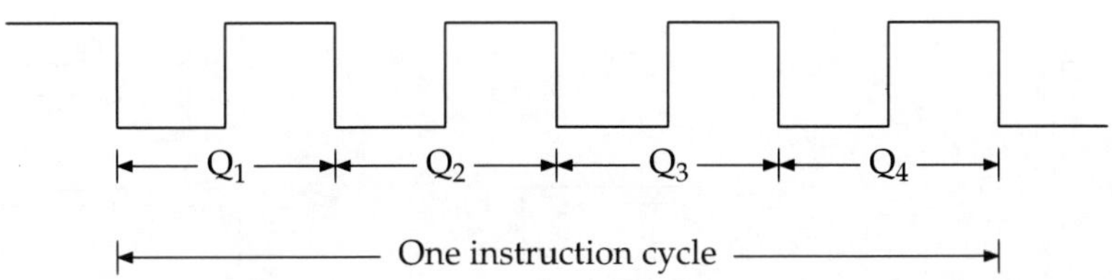

Fig. 10.4 *Relation between instruction cycles and clock cycles for PIC microcontroller*

Instructions that do not require modification of program counter content get executed in one instruction cycle.

Although the architectures of various mid range 8-bit PIC microcontroller are not the same, the variation is mostly interns of addition of memory and peripherals. We will discuss here the architecture of a standard mid-range PIC microcontroller, 16C74A. Unless mentioned otherwise, the information given here is for a PIC 16C74A Microcontroller Chip.

10.6 ARCHITECTURE OF 16C74A PIC MICROCONTROLLER

The basic architecture of PIC16C74A is shown in Figure 10.5. The architecture consists of program memory, file registers and RAM, ALU and CPU registers. It should be noted that the program counter is 13-bit and the program memory is organized as 14-bit word. Hence, the program memory capacity is 8k × 14 bit. Each instruction of PIC16C74A is 14-bit long. The description of registers of PIC16C74A, program counter – stack and register file is explained as follows:

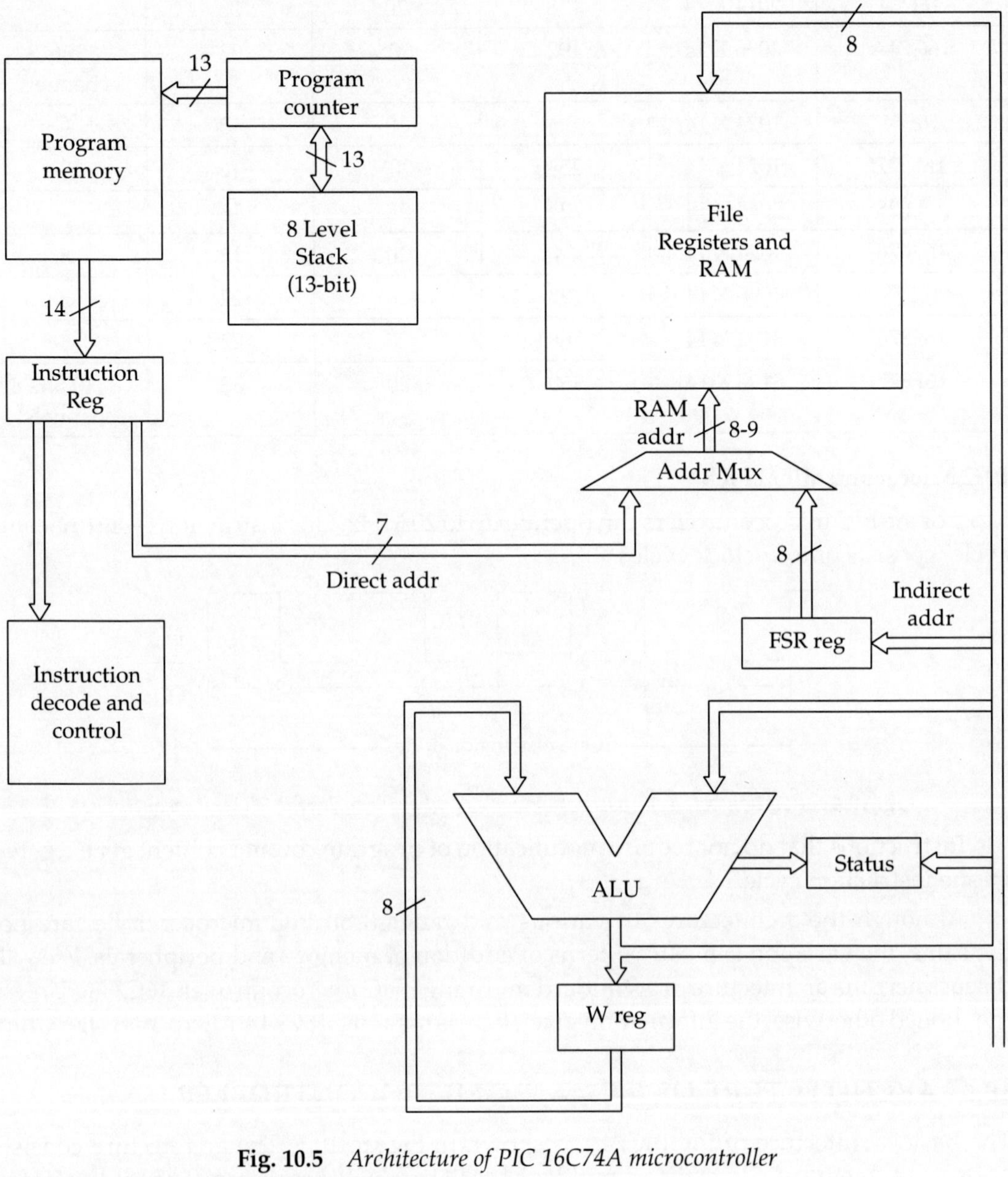

Fig. 10.5 *Architecture of PIC 16C74A microcontroller*

10.6.1 Registers of PIC16C74A

CPU registers (registers commonly used by the CPU). The various CPU registers are discussed here.

(*a*) Working Register:

W, the working register (also called accumulator register), is used by many instructions as the source of an operand. This is similar to accumulator in 8051. It may also serve as the destination for the result of the instruction execution. It is an 8 - bit register.

7   0

W, Working register

(*b*) Status Register:

The STATUS register is an 8-bit register that stores the status of the processor. This also stores carry, zero and digit carry bits. STATUS - address 03H, 83H

7	6	5	4	2	2	1	0
0	0	RP0	NOT_TD	NOT_PD	Z	DC	C

C = Carry bit

DC = Digit carry (same as auxiliary carry)

Z = Zero bit

NOT_TO and NOT_PD - Used in conjunction with PIC's sleep mode

RP0- Register bank select bit used in conjunction with direct addressing mode.

(*c*) FSR Register:

(File Selection Register, address = 04H, 84H)

FSR is an 8-bit register used as data memory address pointer. This is used in indirect addressing mode.

(*d*) INDF Register:

(INDirect through FSR, address = 00H, 80H)

INDF is not a physical register. Accessing INDF access is the location pointed to by FSR in indirect addressing mode.

(*e*) PCL Register:

(Program Counter Low Byte, address = 02H, 82H)

PCL is actually the lower 8 bits of the 13-bit program counter. This is both a readable and writable register.

(*f*) PCLATH Register:

(Program Counter Latch, address = 0AH, 8AH)

PCLATH is an 8-bit register which can be used to decide the upper 5 bits of the program counter. PCLATH is not the upper 5 bits of the program counter. PCLATH can be read from or written to without affecting the program counter. The upper 3 bits of PCLATH remain zero and they serve no purpose. When PCL is written to, the lower 5 bits of PCLATH are automatically loaded to the upper 5 bits of the program counter, as shown in Figure 10.6.

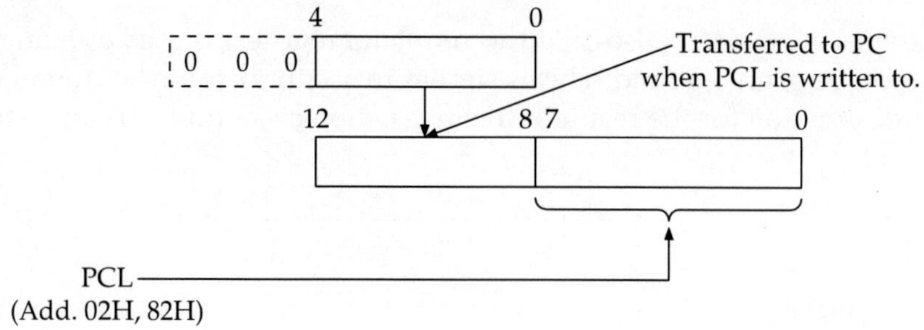

Fig. 10.6 *Schematic presentation of PCL being loaded from PCLATH*

10.6.2 Program Counter Stack

An independent 8-level stack is used for the program counter. As the program counter is 13 bit, the stack is organized as 8x13 bit registers. When an interrupt occurs, the program counter is pushed onto the stack. When the interrupt is being serviced, other interrupts remain disabled. Hence, other 7 registers of the stack can be used for subroutine calls within an interrupt service routine or within the mainline program.

10.6.3 Register File Map

It can be noted that some of the special purpose registers are available both in Bank-0 and Bank-1. These registers have the same value in both banks. Changing the register content in one bank automatically changes its content in the other bank. The complete map for the register file is shown in Figure 10.7.

10.7 PORT STRUCTURE AND PIN CONFIGURATION OF PIC16C74A

As mentioned earlier, there is a large variety of PIC microcontrollers. However, the midrange architectures are widely used. Our discussion will be mainly confined to PIC16C74A whose architecture has most of the required features of a mid-range PIC microcontroller. Study of any other mid-range PIC microcontroller will not cause much variation from the basic architecture of PIC16C74A.

PIC16C74A has 5 I/O ports. Each port is a bidirectional I/O port. In addition, they have the following alternate functions:

Port	Alternative uses of I/O pins	No. of I/O pins
Port A	A/D converter inputs	6
Port B	External interrupt inputs	8
Port C	Serial port, Timer I/O	8
Port D	Parallel slave port	8
Port E	A/D converter inputs	3
	Total I/O pins	33
	Total pins	40

Address	Bank-0	Bank-1	Address
00	INDF	INDF	80
01	TMRO	OPTION	81
02	PCL	PCL	82
03	STATUS	STATUS	83
04	FSR	FSR	84
05	PORT 1	TRISA	85
06	PORT B	TRISB	86
07	PORT C	TRISC	87
08	PORT D	TRISD	88
09	PORT E	TRISE	89
0A	PCLATH	PCLATH	8A
0B	INTCON	INTCON	8B
0C	PIR1	PIE1	8C
0D	PIR2	PIE2	8D
0E	TMR1L	PCON	8E
0F	TMR1H		8F
10	T1CON		90
11	TRM2		91
12	T2CON	PR2	92
13	SSPBUF	SSPADD	93
14	SSPCON	SSPSTAT	94
15	CCPR1L		95
16	CCPR1H		96
17	CCP1CON		97
18	RCSTA	TXSTA	98
19	TXREC	SPBRC	99
1A	RCREG		9A
1B	CCPR2L		9B
1C	CCPR2H		9C
1D	CCP2CON		9D
1E	ADRES		9E
1F	ADCONO	ADCON1	9F
20 – 7F	General purpose RAM	General purpose RAM	A0 – BFH

Fig. 10.7 *Register file map*

In addition to I/O pins, there is a master clear pin (MCLR) which is equivalent to reset in 8051. However, unlike 8051, MCLR should be pulled low to reset the microcontroller. Since PIC16C74A has inherent power-on reset, no special connection is required with MCLR pin to reset the microcontroller on power-on. There are two V_{DD} pins and two V_{SS} pins. There are two pins (OSC1 and OSC2) for connecting the crystal oscillator/ RC oscillator. Hence, the total number of pins with a 16C74A is 33+7=40. This IC is commonly available in a dual-in-pin (DIP) package.

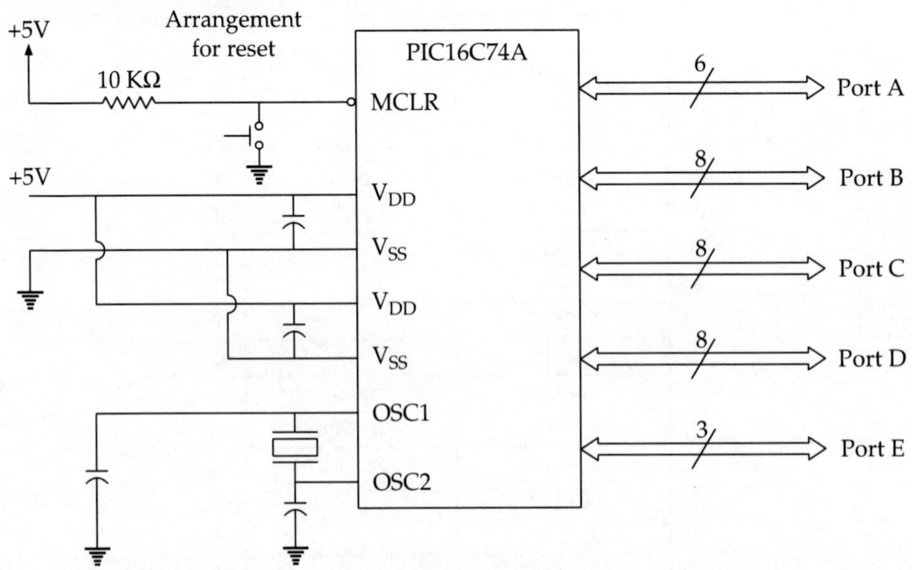

Fig. 10.8 *Pin configuration of PIC16C74A microcontroller*

SECTION – B

8096 16-BIT MICROCONTROLLER

10.8 BASICS OF 8096 16-BIT MICROCONTROLLER

INTEL 8096, a second generation processor belongs to MCS 96 family. This is a high performance 16-bit microcontroller with register to register architecture. This is designed to handle high speed calculations and fast input/output operations which are preferred in high speed modern control applications. The 8096 with 16-bit CPU horse power, high speed math processing and high speed I/O is ideal for complex motor control and axis control systems. Hence, it is used in 3-phase large horse power AC motors and robotics. The 10-bit ADC option makes it most suitable option for data acquisition systems and closed loop analog controllers.

The 8096 is a 16-bit microcontroller designed for high speed control functions. The microcontroller 8096 is fabricated using Intel's HMOS – III process. The 8096 microcontroller supports bit, byte, and word operations. With a 12 MHz input frequency, the 8096 microcontroller can perform a 16-bit addition in 1 μs and a 16 x 16 bit multiplication and 32/16 bit division operation in 6.25 μs. Instruction execution times average is 1 to 2 μs in typical applications.

Four high speed trigger inputs are provided to record the times at which external events occur. Six high speed pulse generator outputs are provided to trigger external events at present times. The high speed output unit can simultaneously perform software timer functions. 8096 microcontrollers support up to four 16-bit software timers that can be in operation at once. The on-chip ADC includes a sample and hold, and converts up to 8 multiplexed analog input channels to 10-bit digital values. With a 12 MHz crystal frequency, each conversion takes 22 μs time.

8096 can be configured in two modes: (i) Single chip mode, and (ii) expanded mode. In the single chip mode, the internal ROM or EPROM is accessed by making the pin EA (active low) HIGH. For ROM-less chip to access the external memory, the pin EA is made low. In the expanded mode both internal and external (OFF CHIP) memory can be accessed using the multiplexed bus architecture. It has nearly 230 bytes of on-chip RAM and one 10-bit A/D converter with sample hold circuit. There are five on chip I/O ports each of 8-bit width. The 8096 microcontroller has a complete set of 16-bit arithmetic instructions including multiply and divide operations. It has pulse width modulation output with dedicated baud rate generator. It has one on-chip full duplex serial port. There are 20 interrupt sources and 8 interrupt vectors on 8096. It has two 16-bit timers, timer 1 and timer 2 and one 16-bit watchdog timer. This 8096 is available as 48 pin DIP (dual in-line package) and 68 pin PLCC and also 68 pin leadless chip carrier IC. It is also available as a 68 pin PGA (pin grid array) package.

Typical applications using the 8096 microcontroller include closed loop control, mid-range digital signal processing, modems, motor control systems, printers, engine control systems, photocopiers, anti-lock brakes, air-conditioners control system, disk drives, and medical instrumentation systems, etc.

10.8.1 Features of 8096 Microcontroller

1. 232 8-bit register file
2. Register to register architecture
3. 10-bit analog-to-digital converter (ADC) with sample and hold.
4. Five 8-bit input/output ports
5. 20 interrupt sources
6. Full duplex serial port
7. Dedicated baud rate generator

8. Pulse width modulated (PWM) output
9. 16 × 16 bit multiplication and 32/16 bit division operation perform in 6.25 μs.
10. 16-bit watchdog timer
11. Four 16-bit software timers
12. Two 16-bit timers/counters
13. ROM/EPROM lock
14. Run-time programmable EPROM
15. High speed input/output subsystem
16. Extended temperature available
17. Extended burn-in facility available

10.9 ARCHITECTURE OR BLOCK DIAGRAM OF 8096 MICROCONTROLLER

The 8096 CPU has a 16-bit ALU which operates on a 256-byte register file instead of an accumulator. Any of the locations in the register file can be used for sources or destinations for most of the instructions. Hence, this is called a ***register to register architecture***. Many of the instructions can also use bytes or words from anywhere in the 64 KB address space as operands. Locations 00h through 17H are the I/O control registers or special function registers (SFRs). Locations 18H and 19H contain the stack pointer which serves as general-purpose RAM when not performing the stack operations. The remaining bytes of the register file serves as general purpose RAM, accessible as bytes, words or double-words. Calculations performed by the CPU take place in the RALU. This RALU contains a 17-bit ALU, the program status word (PSW), the program counter (PC), a loop counter and three temporary registers. The RALU operates directly on the register files, thus eliminating the problem with the accumulator and providing direct control of I/O operations through the SFRs.

The functions of the different parts of the architecture are explained below.

10.9.1 PSW Register

PSW register is a 16-bit register, whose lower byte can act as INTERRUPT MASK register. Z is the zero flag, N is the negative flag, V is the overflow flag, VT is the overflow trap flag, C is the carry flag, I is the global interrupt enable flag and ST is the sticky bit, it is set during a right shift.

D 15	D 14	D 13	D 12	D 11	D 10	D 09	D 08	D 07	D 06	D 05	D 04	D 03	D 02	D 01	D 00
Z	N	V	VT	C	–	I	ST	INTERRUPT MASK							

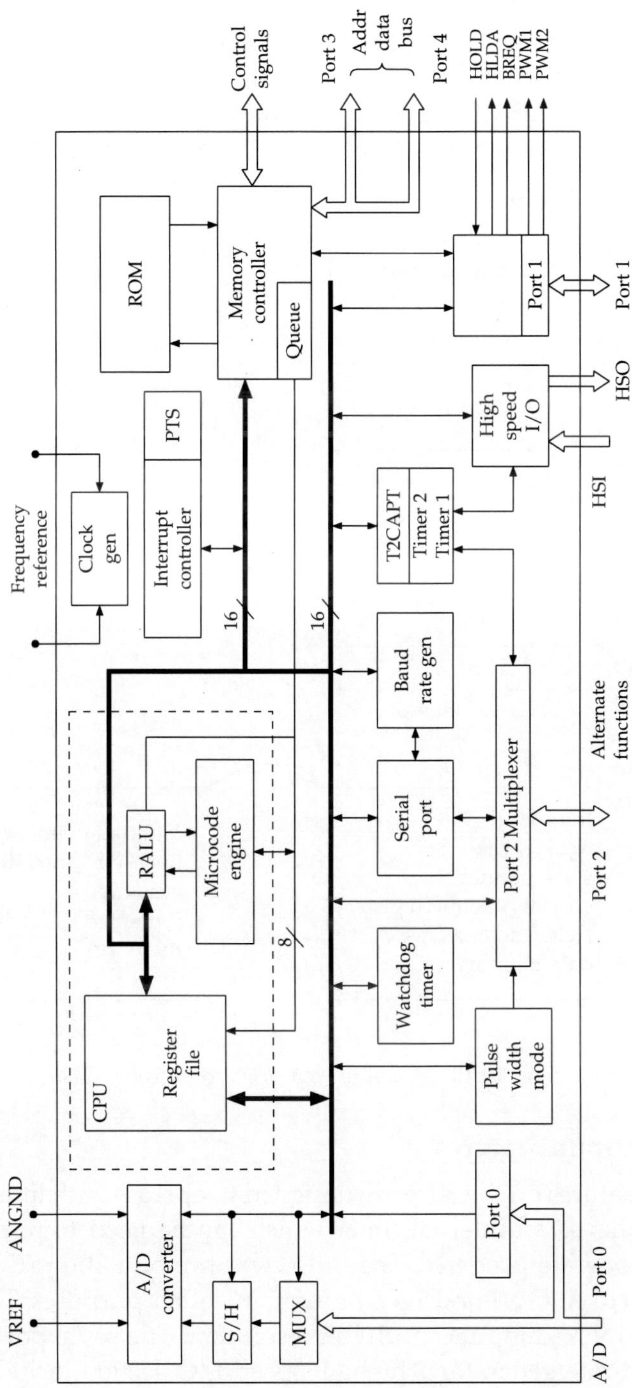

Fig. 10.9 *Block diagram of 8096 microcontroller*

10.9.2 Memory Mapping

The addressable memory space on the 8096 is 64 KB, most of which are available to the user for program or data memory storage. The locations which have special purpose are 0000H through 00FFH and 1FFFH through 2010H. All other locations can be used either for program or for data storage or for memory mapped peripherals.

The 64 KB memory space mapping is shown in the diagram.

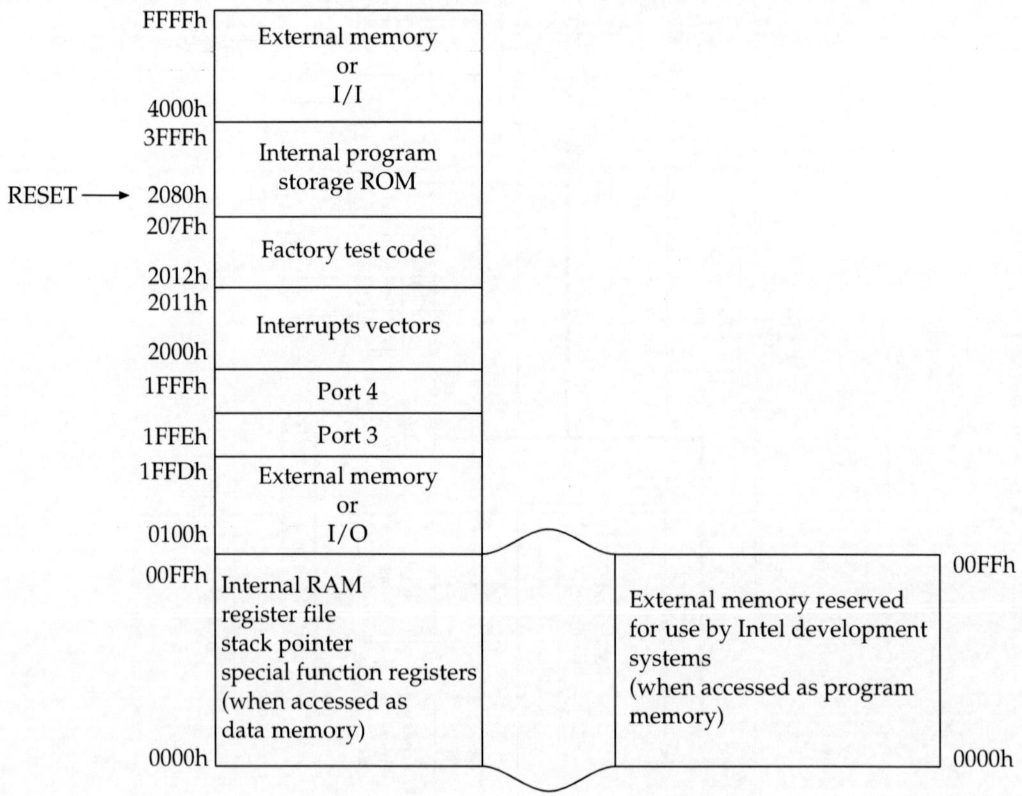

Fig. 10.10 *Memory map of microcontroller 8096*

10.9.3 Input/output Features:

Most of the I/O features on the 8096 are designed to operate with little CPU intervention. The watchdog timer is an internal timer which can be used to reset the system if the software fails to operate properly. The pulse width modulation (PWM) output can be used as a rough D to A, a motor driver, or for many other purposes. The A to D converter (ADC) has 8 multiplexed inputs and 10-bit resolution. The serial port has several modes and its own baud rate generator. The high speed I/O section includes a 16-bit timer, a 16-bit counter, a 4-input programmable edge detector, 4 software timers, and a 6-output programmable event generator.

10.9.4 PWM Output

Analog outputs are just as important as analog inputs when connecting to a peripheral device. Digital to analog converters are difficult to make on a microcontroller because of all the digital noise and the necessity of providing an on-chip, relatively high current, rail to rail driver. They also take up a fair amount of silicon area which can be better used for other features. The A to D converter does use a D to A, but the currents involved are very small. The PWM signal is a variable duty cycle, fixed frequency waveform that can be integrated to provide an approximation to an analog output. The frequency is fixed at a period of 64 microseconds for a 12 MHz clock speed. Controlling the PWM simply requires writing the desired duty cycle value (an 8-bit value) to the PWM register.

10.9.5 General purpose I/O ports

There are five 8-bit I/O ports namely port 0, port 1, port 2, port 3 and port 4.

(*a*) PORT 0

Port 0 is only an input port which is also used as the analog input port for the on-chip ADC. So, if the analog input features are not used, Port 0 can be used as input port only. Port 0 is shown in Figure 10.11

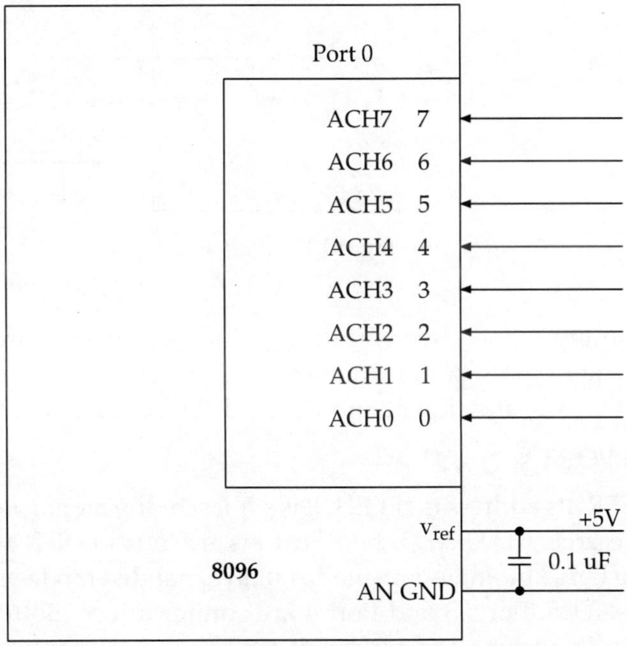

Fig. 10.11 *Port 0*

(*b*) PORT 1

This is a quasi-bidirectional port which can be used either as input or as the out port. It is mapped at the memory address 0FH. If any of the port 1 pins is to be used as input port, the corresponding pin must be made high by writing the data 1. The weak internal pull-up is designed to be overridden by the external device which drives the line. When the output drive capability is sufficient to drive a 74LSxx Input, a CMOS device driven by port1 will require a pull-up resistor of around 10K to +5V in order to bring the output well above the normal CMOS threshold voltage of 2.5 volts. Port 1 inputs can be driven by either CMOS or TTL devices with no extra parts. Port 1 is shown in Figure 10.12.

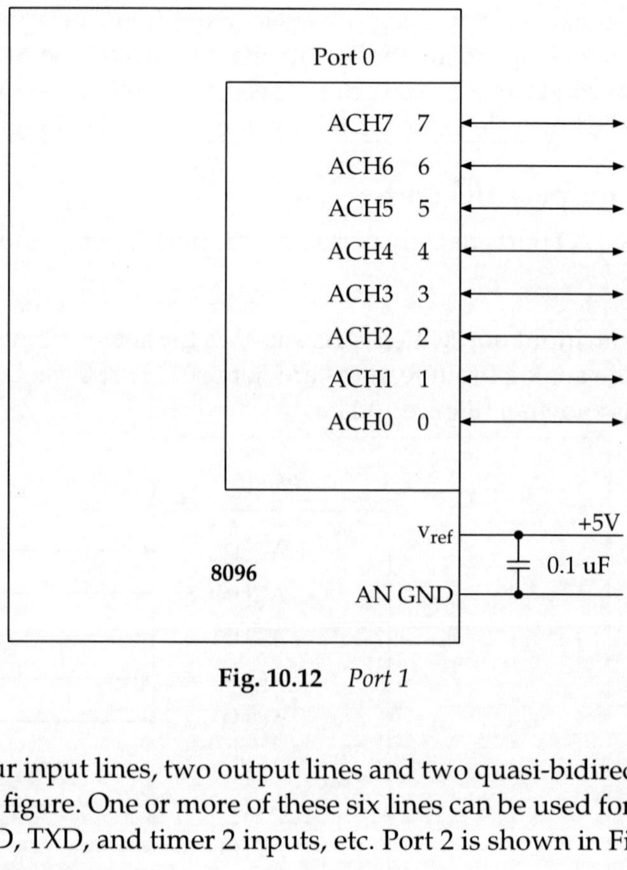

Fig. 10.12 *Port 1*

(*c*) PORT 2

Port 2 has four input lines, two output lines and two quasi-bidirectional I/O lines as shown in the figure. One or more of these six lines can be used for alternate purpose of PWM, RXD, TXD, and timer 2 inputs, etc. Port 2 is shown in Figure 10.13.

(*d*) PORT 3 and PORT 4

It is outside SFR. Its address is 1FFEH. It is a quasi bidirectional port. During extended operation, it carries AD00-AD07 bits. Port 4 is also outside SFR; its address is 1FFFH. It is similar to Port 3 in single chip mode but in expanded mode of operation, it carries signals AD8-AD15. Ports 3 and Port 4 are similar in use. Both of them have open drain outputs. By writing 1 to any line it can be used as an input and other lines can serve as output lines. Each output line requires a pull-up resistor of about 15 K. In the expansion mode, the bus lines can gain the ability to drive both high and low, forming the expansion bus without the need of pull-up resistors. Ports 3 and 4 are shown in Figure 10.14.

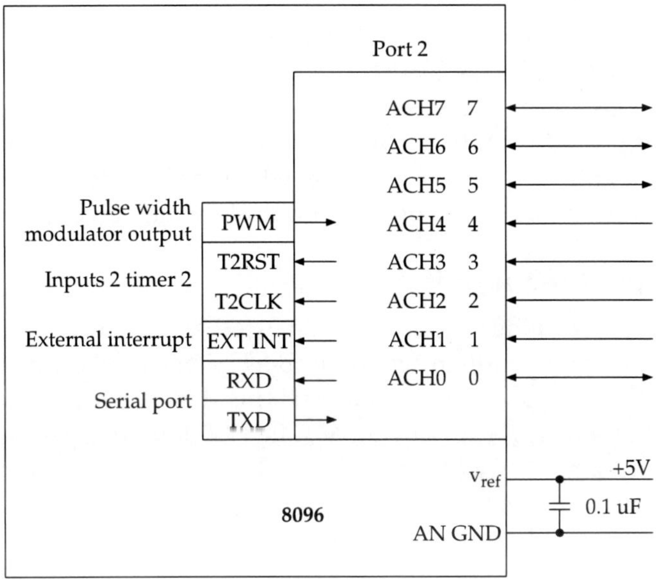

Fig. 10.13 *Port 2*

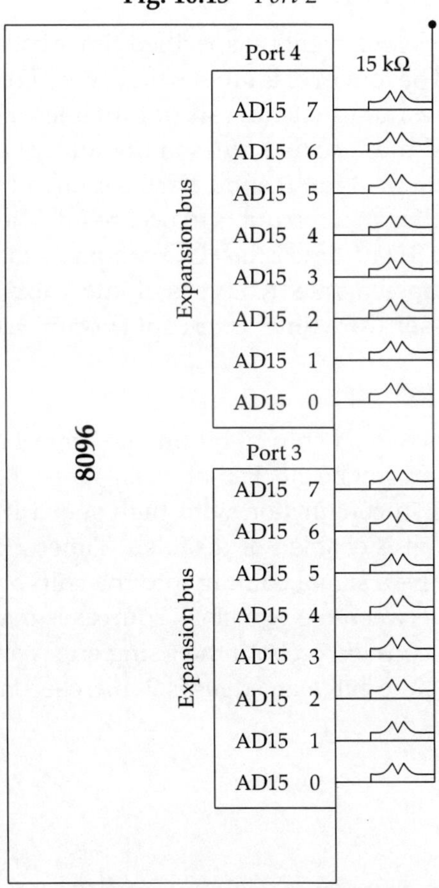

Fig. 10.14 *Port 3 and Port 4*

10.9.6 Reset and Self Protection Options

When the reset pin is driven to low the 8096, regardless of what it is doing will reset and start executing from the address 2080H. The reset pin is a bidirectional line with a strong internal pull-up. This line may also be driven by internal watchdog timer. The INTEL 8096 microcontroller is provided with on-chip self protection circuitry, to protect the chip from large currents. The chip is automatically reset when the V_{DD} deviates from the prescribed levels. Diode circuits are provided on the chip itself, which gives self protection.

10.9.7 High Speed Input Unit (HSI)

This unit is used to record the time at which an external event occurs with respect to timer 1. It can monitor four independently configurable HSI lines and capture the value of Timer 1 when an event takes place. There are 4 lines (HIS.0-HIS.3) available in this unit. The HIS unit can store up to 8 entries (Timer 1 values). The HSI unit can generate an interrupt when loading an entry into the HIS holding register or loading the sixth entry into the FIFO. The four types of events that can trigger captures include "rising edges only", "falling edges only", "rising or falling edges" or "every eighth rising edge".

10.9.8 High Speed Output Unit (HSO)

This HSO unit is used to trigger events at specified times based on timer 1 or timer 2 with minimal CPU overhead. The four events are starting an A/D conversion, resetting of timer 2 setting four software flags, and switching six output lines (HSO.0-HSO.5). The HSO unit stores pending events and the specified times in the content addressable memory (CAM) file. This CAM file stores up to 8 commands. Each command specifies the action time, the nature of the action, whether an interrupt is to occur and whether timer 1 or timer 2 is the reference timer. For every 8 state times, the HSO compares the CAM locations for the time matches. The HSO unit can generate two types of interrupts (i) HSO execution interrupt (Vector = 2006H) and (ii) software timer interrupt (Vector =200BH).

10.9.9 Timers

INTEL 8096 controller has two on-chip 16-bit timers Timer 1 and Timer 2. Timer 1 can act as a 16-bit counter also and can be clocked at every 2 ms., i.e., for every 8 internal clock cycles. This timer is used in conjunction with high speed I/O system. When timer 1 is overflown, the interrupt bit is enabled or disabled. Timer 2 can be used as a 16-bit even counter which is clocked by a signal coming into the chip on either of the two pins Port 2.3 or HIS.1. The choice between the two clock sources is made by setting or clearing bit 7 of the IOC.0. Timer 2 is counted on both the rising edges and falling edges of the input signal and the minimum time between edges is 2 microseconds.

10.9.10 Watchdog Timer

An on-chip 16-bit watchdog timer is available in 8096 which helps to recover the controller from the software upsets. This 16-bit WDT is a counter which is incremented every state time. This counter is cleared by program after periodic interval and not allowed to overflow. However, if the program does not progress properly by any reason such as electrostatic discharge (ESD) or due to any hardware related problems, the overflow occurs. And the hardware reset is initiated to restart the microcontroller. This process avoids the system from having a malfunction for longer than 16 ms under 12 MHz frequency operation. When the WDT overflows, it pulls down the RESET pin for at least two state times resetting 8096 and also any other devices connected to the RESET line.

10.9.11 Serial Port

The 8096 has an on-chip full duplex serial port to reduce the total number of chips required in the system. The serial port is similar to that on the 8051 controller. It has one synchronous and three asynchronous modes. In the asynchronous modes, baud rates of up to 187.5 K baud can be used, while in the synchronous mode rates up to 1.5 M baud are available. The chip has a baud rate generator which is independent of Timer 1 and Timer 2, so using the serial port does not take away any of the HSI, HSO or timer flexibility or functionality. Control of the serial port is provided through the SPCON/SPSTAT (Serial Port Control /Serial Port Status) register. The serial port is configured in four modes. The four modes of the serial port are referred to as modes 0, 1, 2 and 3. Mode 0 is the synchronous mode, and is commonly used to interface to shift registers for I/O expansion. In this mode, the port outputs a pulse train on the TXD pin and either transmits or receives data on the RXD pin. Mode 1 is the standard asynchronous mode, 8 bits plus a stop and start bit are sent or received. Modes 2 and 3 handle 9 bits plus a stop and start bit. The difference between the two is that, in Mode 2 the serial port interrupt will not be activated unless the ninth data bit is a one; in Mode 3 the interrupt is activated whenever a byte is received. These two modes are commonly used for inter-processor communication. Baud rates for all of the modes are controlled through the baud rate register. This is a byte wide register which is loaded sequentially with two bytes, and internally stores the value as a word. The least significant byte is loaded to the register followed by the most significant. The most significant bit of the baud value determines the clock source for the baud rate generator. If the bit is a one, the XTAL1 pin is used as the source, if it is a zero, the T2 CLK pin is used.

10.9.12 Interrupt Structure

There are 20 different interrupt sources that can be used on the 8096. The 20 sources vector through 8 locations or interrupt vectors. All these interrupts are enabled or disabled using the 9th bit of PSW register. If this bit is set to 1 all the interrupts are enabled and disabled when reset to zero. Control of the interrupts is handled through the interrupt pending register (INT_PENDING), the interrupt mask register (INT_MASK), and the I bit in the PSW (PSW.9). The content of the interrupt mask register determines whether the pending interrupt is serviced or not. If it is to be serviced, the CPU pushes the contents of the program counter on to the stack and reloads it with the vector corresponding to the desired interrupt. When the hardware detects one of the 8 interrupts, it sets the corresponding bit in the interrupt pending register. This register can be read or modified as a byte register. Individual interrupts can be enabled or disabled by setting or clearing the bits in the Interrupt mask register. The INT_MASK register can be accessed as the lower bits of the PSW register.

10.9.13 A to D Converter

Analog inputs are frequently required in every microcontroller application. The 8096 controller has one 10-bit on-chip A/D converter that can use any one of eight input channels. The conversions are done using the successive approximation method, and require 168 state times (42 microseconds with a 12 MHz clock.). The main components of the A/D converter are 8-analog input channels, one 8 to 1 multiplexer, a sample and hold capacitor and resistor ladder. The A/D converter performs a conversion in 88 time states. Upon the completion of each conversion, the converter can generate a conversion complete interrupt.

10.10 WINDOW SELECTION REGISTER (WSR)

WSR is an SFR which is used to select horizontal windows and vertical windows.

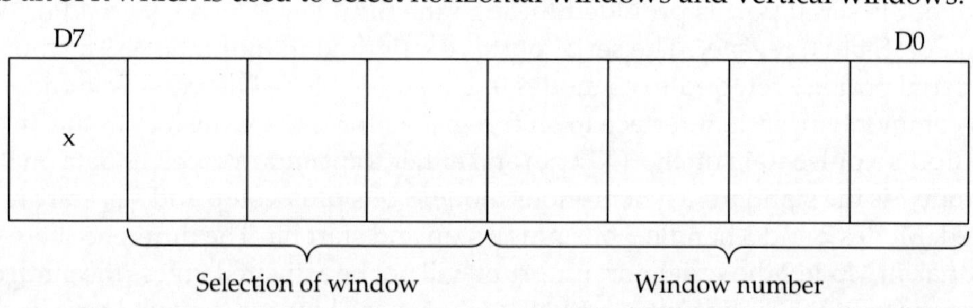

Fig. 10.15 *Window selection register*

The 4 LSBs of WSR are used to select the horizontal window. The four-bit value will indicate window number. 3 bits of WSR, i.e., BIT 4, 5 and 6 will decide the vertical window configuration. If BIT 4 is set it will select 4 vertical windows, if BIT 5 is set it will select 8 vertical windows and if BIT 6 is set it will select 16 vertical windows.

Once a vertical window is selected, the memory location within a window will be selected by the instruction directly. Figure 10.16 shows the breakup of internal RAM as windows.

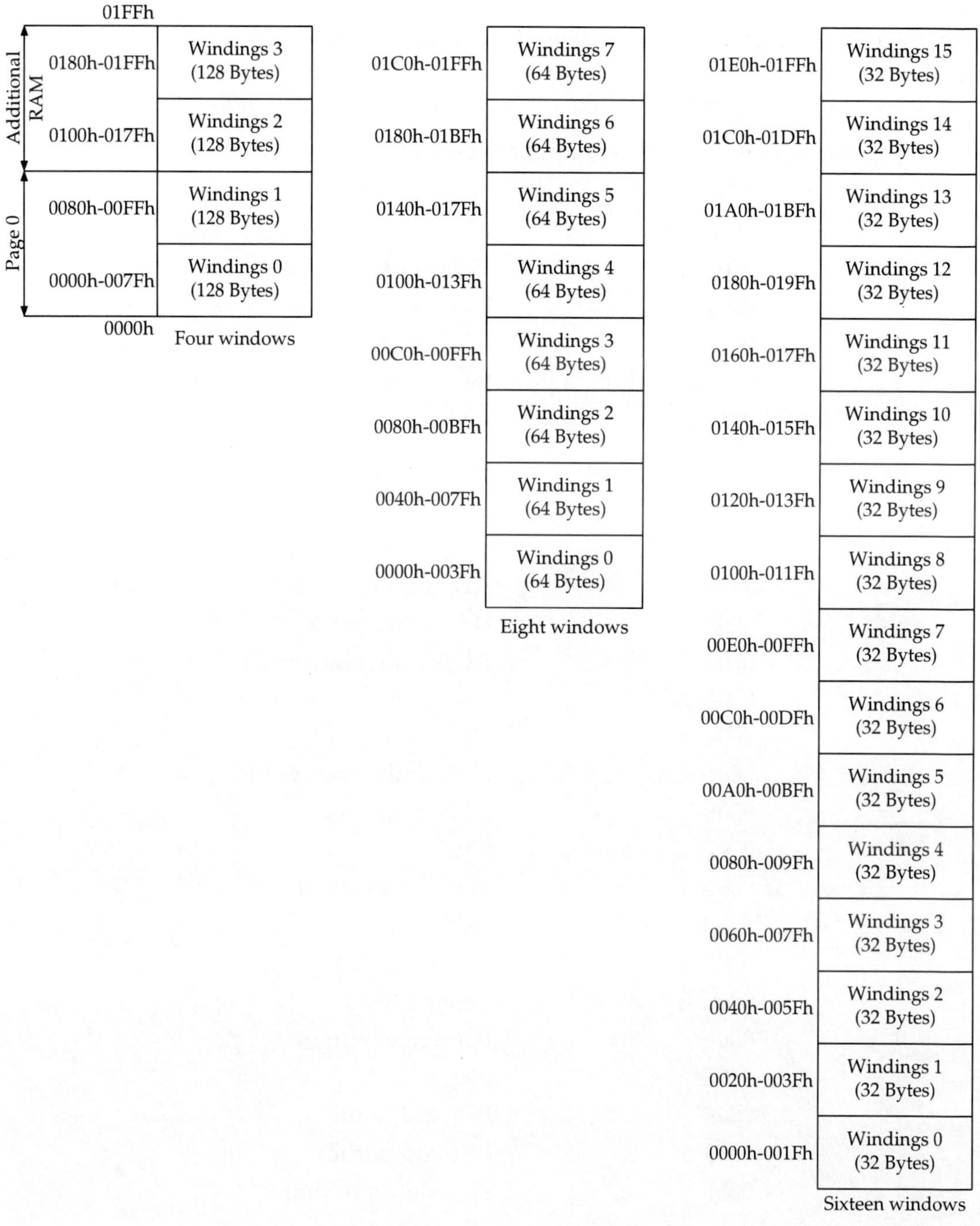

Fig. 10.16 *Break-up of internal RAM*

10.11 SFRs OF MICROCONTROLLER 8096

Special function registers are available from address 0000H-0017H, i.e., 24 addresses. Out of these 24 addresses, 9 addresses are common to horizontal window 0 (read), horizontal window 0 (write), horizontal window 1 (read) and horizontal window 15.

1. Registers common to all windows are given below.

ADD (Hex)	SFR
0000-0001H	R0
0008H	INT_mask
0009H	INT_pending
0012H	INT_pend1
0013H	INT_mask1
0014H	WSR

2. Registers of horizontal window 0 (read) and window 0 (write)

ADD (Hex)	SFR
0002	AD_command (write)
0002	AD_Result_L0 (read)
0003	HSI_mode (write)
0003	AD_result_high (read)
0004-0005	HS0_Time (write)
0004-0005	HS0_Time (read)
0006	HS0_Command (write)
0006	HS0_status (read)
0007	SBUF
000A	Watchdog (write)
000B	I0C2 (write)
000A-000B	Timer 1 (read)
000C-000D	Timer 2 (read)
000E	Baud rate (write)
000E	P0 (read)
000F	P1 (read/write)
0010	P2 (read/write)
0011	SP_Control (write)
0011	SP_stat (read)
0015	10C0 (write)
0015	10S0 (read)

0016	10C1 (write)
0016	10S1 (read)
0017	10S2 (read/write)

3. Horizontal window 1 (read/write)

ADD (Hex)	SFR
0003	AD_Time
0004	PTSSEL_L0
0005	PTSSEL_H0
0006	PTSSRV_L0
0007	PTSSRV_H0
000C	10C3
0016	PWM2 control
0017	PWM1 control

Unused addresses are reserved for future expansion. Common addresses are same as earlier.

4. SFRs at horizontal window 15

ADD (Hex)	SFR
000C	T2 Capture_L0
0016	T2 Capture_L0

10.12 ADDRESSING MODES

To perform any operation in the microcontroller we have to send corresponding instruction to the microcontroller, along with the instruction, we have to provide three pieces information to the microcontroller.

(*a*) Operation to be performed or opcode

(*b*) Address of source of data

(*c*) Address of destination of result

The method by which the address of source of data and address of destination of result is given along with the instruction are called addressing modes.

The 8096 instruction set supports six addressing modes. They are:

1. Immediate addressing mode
2. Register direct addressing mode
3. Indirect addressing mode
4. Indirect with auto increment mode

5. Short indexed mode
6. Long indexed mode.

These addressing modes increase the flexibility and overall execution speed of 8096 controller. Each instruction uses at least one of the addressing modes. The register, direct and immediate addressing modes execute faster than the other addressing modes. Both of the indirect addressing modes use the value in a word register as the address of the operand. The indirect auto increment mode, increments a word address by one after a byte operation and two after a word operation.

The explanation of each addressing modes is given below with examples.

(*a*) Immediate addressing mode

The data required for the execution of the instruction is given directly along with the instruction. Such instructions are called immediate addressing mode.

For example:

1. LD AX, #55AH
2. ADD AX,NUM, #1234H

(*b*) Register direct addressing mode

The data required for the execution of the instruction is present in the register and the name of the register is given along with the instruction. Such instructions are called register addressing mode or register direct addressing mode.

For example:

1. SUB BX, AX, CX
2. INC CX
3. LD AX, CX
4. ST AX, CX
5. CLR BX
6. ADDB DL, BL, BL
7. POP AX

(*c*) Indirect addressing mode

The data required for the execution of the instruction is present in the memory location and the address of this memory location is present in the register, and the name of this register is given along with the instruction. Such instructions are called indirect addressing mode.

For example:

1. LD AX, [SI]
2. ST [SI], CX

(*d*) **Indirect with auto increment mode**

The data required for the execution of the instruction is present in the memory location and the address of this memory location is present in the register, and the name of this register is given along with the instruction with auto-increment of the memory location using + sign along with the instruction. Such instructions are called indirect addressing with auto increment mode.

For example:

1. INC [CX]+
2. ADD AX, CX, [BX]+

(*e*) **Short indexed mode**

The data required for the execution of the instruction is present in the memory location and the address of this memory location is calculated by adding the number present in the register with short displacement or offset given along with the instruction. Such instructions are referred as short indexed mode addressing.

For example:

1. LD AX, DISP [SI]

(*f*) **Long indexed mode**

The data required for the execution of the instruction is present in the memory location and the address of this memory location is calculated by adding the number present in the register with long displacement or offset given along with the instruction. Such instructions are referred as long indexed mode addressing.

For example:

1. LD AX, [SI+offset]

10.13 INSTRUCTIONS OF 8096

Maximum number of operands given along with the instruction can be up to 3. The memory locations are treated as registers and each memory location used should be given a name. These names are to be used in the instructions. The instructions of 8096 microcontroller are classified as:

(*a*) **Data Transfer Instructions**

1. LD/LDB
2. LDBSE
3. LDBZE
4. ST/STB
5. PUSH
6. PUSHF

7. POP
8. POPF

(b) Arithmetic Instructions

1. ADD/ADDB
2. ADDC/ADDCB
3. SUB/SUBB
4. SUBC/SUBCB
5. CMP/CMPB
6. MUL/MULU
7. MULB/MULUB
8. DIV/DIVB
9. DIVU/DIVUB

(c) DATA Manipulation Instructions

1. CLR
2. EXT
3. NOT

(d) BIT Manipulation Instructions

1. SETC
2. CLRC
3. EIDI

(e) Logical Instructions

1. AND/ANDB
2. OR/ORB
3. XOR/XORB

(f) Program Control Flow Instructions

1. CALL
2. SCALL
3. LCALL
4. JMP
5. SJMP
6. LJMP
7. RET
8. BR

(g) Conditional Jump Based on Flags

1. JC
2. JNC
3. JE
4. JNE
5. JGE
6. JLT
7. JGT
8. JLE
9. JH
10. JNH
11. JV
12. JNV
13. JVT
14. JNVT
15. JST
16. JNST
17. JBS
18. JBC

(h) Miscellaneous Instructions

1. RST
2. TRAP

SECTION – C

ARM MICROCONTROLLER

10.14 INTRODUCTION

ARM: Advanced RISC Machine

Initially developed at acorn computers limited of cambridge, england, between 1983 and 1985 hence named as acorn RISC machine from acorn computers limited. In 1990, ARM Ltd. was established and ARM was renamed as advanced RISC machines.

ARM is a 16/32-bit RISC processor architecture currently being developed by the ARM corporation. ARM is the industry's leading provider of 16/32 bit embedded RISC microprocessor solutions.

The business model behind ARM is based on licensing the ARM architecture to companies who want to manufacture ARM-based CPU's or system-on-a-chip products. ARM does not make integrated circuits (ICs), they only provide prototype architecture to the manufacturers. The best part is that ARM processors can be used in any domain. The ARM provides two main types of licenses:

1. Implementation license
2. Architecture license.

1. The implementation license provides complete information required to design and manufacture integrated circuits containing an ARM processor core. ARM provides hard and soft views to licensees. Soft views include gate level netlists (RTL source code), meaning synthesizable from licensees using a suitable gate library. Hard cores are the final layout given to the customer.

   A hard core is optimised for a specific manufacturing process, while a soft core can be used in any process but is less optimised.
2. The architecture license enables the licensee to develop their own processors compliant with the ARM ISA (ISA: instruction set architecture).

ARM provides comprehensive support required in developing a complete system. It also develops technologies to assist with the designing of the ARM architecture.

ARM processors possess a unique combination of features that makes ARM the most popular embedded architecture and processing system today. There are numerous reasons why ARM architecture is so popular and gaining high demand since its inception. First, ARM cores are very simple compared to most other general-purpose processors, which means that they can be manufactured using a comparatively small number of transistors, and other devices leaving plenty of space on the chip for application-specific macrocells. A typical ARM chip can contain several peripheral controllers, a digital signal processor, and some amount of on-chip memory, along with an ARM core.

Second, both ARM ISA and pipeline design are aimed at minimising energy consumption — a critical requirement in mobile embedded systems and small handheld devices. Third, the ARM architecture is highly modular: the only mandatory component of an ARM processor is the integer pipeline; all other components, including caches, MMU, floating point and other co-processors are optional, which gives a lot of flexibility in building application-specific ARM-based processors.

Finally, while being small, and low-power, ARM processors provide high performance for embedded applications. For example, the ARM Cortex processor running at 400MHz provides performance comparable to Pentium 2 at 300MHz, while using thirty times less energy.

10.15 FEATURES OF ARM

1. ARM is a RISC processor.
2. It consists of large register file R0 to R15.
3. Architectural simplicity which allows very small implementations, which results in very low power consumption.

4. It supports load and store register, so data processing is only in register contents.
5. Uniform and fixed length instructions.
6. 32-bit processor.
7. It supports 32-bit ARM instruction sets and 16-bit THUMB instruction sets.
8. Latest ARM provides mixture of 32-bit and 16-bit instruction sets.
9. Good speed and power consumption ratio
10. High code density
11. Mostly single cycle execution
12. Speed 1 MHz to 1.25 GHz
13. ARM supports JAVA Jazelle direct byte code execution
14. DSP enhanced instructions
15. Conditional execution of all instructions
16. 32-bit barrel shifter
17. In-built circuit for debugging
18. ARM has a good support for RTOS like Linux, QNX, VxWorks, and FreeRTOS etc.

10.16 HISTORY, VERSIONS AND DEVELOPMENT OF THE ARM PROCESSOR

The history of ARM started in 1983, when a company named acorn computers was looking for a 16-bit microprocessor for their next desktop machine. They quickly discovered that existing commercial microprocessors did not satisfy their requirements. These processors were slower than standard memory parts available at the time. The processor available had complex instruction sets that included instructions taking hundreds of cycles to execute, leading to high interrupt latencies and slower execution speed. The reason for these deficiencies was that early integrated microprocessors were modelled after processors of minicomputers, which consisted of many chips, were driven by microcode and had very complex instruction sets, also for the processing they required large number of supporting devices to be interfaced with the processor.

Therefore, acorn engineers considered designing their own microprocessor to suite their requirement. However, resources required for such a project were well beyond what the company could afford. In early 1980s, microprocessor architectures became so complex that it took many years even for large companies with significant expertise in the area to design a new processor.

Fortunately, just two years before this, the berkeley RISC 1 project had shown that it had actually been possible to build a very simple processor with performance comparable to the most advanced CISC processors of the time. Acorn decided to pick up the berkeley approach, and two years later, in 1985, they released their first 26-bit acorn RISC machine (ARM) processor, which also became the first commercial RISC processor in the world. It used less than 25,000 transistors — a very small number even for 1985, but still performed comparably to or even outperformed the Intel 80286 processor that came out at about the same time. This first ARM architecture is now referred to as the ARM version 1 architecture. It was followed by the second processor in 1987. The main new feature introduced in ARM version 2 was coprocessor support. Later, the ARM v2 architecture was implemented with on-chip cache in the ARM3 processor.

In 1990, Apple made a strategic decision to use an ARM processor in their newton PDA. A joint venture was co-founded by acorn and apple to design a new processor. The company was called ARM, standing for Advanced RISC Machines. The third version of the ARM architecture was developed by this company and featured 32-bit addressing, MMU support and 64-bit multiply-accumulate instructions. It was implemented in ARM6 and ARM7 cores. The release of these processors and the apple newton PDA in 1992 marked ARM's move to the embedded market. Later versions will now be equipped with many important features like introduction of digital signal processor, trust zone virtualization technology, java byte code extension to the ARM instruction set, improved thumb instruction set, improved floating point supports, etc. Table 10.1 summarizes the history of the ARM architecture.

10.16.1 ARM Versions

(*a*) Version 1

1. The first ARM processor developed at Acron Computers Limited between 1983-1985
2. Supported only 26-bit addressing and had no multiply or coprocessor support

(*b*) Version 2

It was still a 26-bit address machine but included the 32-bit result multiply instruction and coprocessor support

(*c*) Version 2a

1. The ARM3 chip was the first ARM with an on-chip cache
2. Added the atomic load and store (SWP) instruction
3. Introduced the use of coprocessor 15 as the system control coprocessor to manage the cache

(*d*) Version 3

1. The first ARM processor designed by ARM limited following their establishment as a separate company in 1990 was the ARM6
2. It had 32-bit addressing capability
3. Separate CPSR and SPSRs

4. Added the undefined and abort modes to allow coprocessor emulation and virtual memory support in supervisor mode
5. Backward compatible with version 2a

(*e*) Version 3G

Version 3 without backwards compatibility to version 2a

(*f*) Version 3M

Introduced the signed and unsigned multiply and multiply-accumulate instructions that generate the full 64-bit result

(*g*) Version 4

1. Add the signed and unsigned half-word and signed byte load and store instructions
2. Reserve some of the SWI space for architecturally defined operations
3. Introduced the system mode
4. A system mode that uses the user registers

(*h*) Version 4T

Introduced the 16-bit thumb compressed form of the instruction set

(*i*) Version 5T

A superset of architecture version 4T, adding the BLX, CLZ and BPK instructions

(*j*) Version 5TE

Add the signal processing instruction set extensions and java byte code engine (Jazelle)

(*k*) Version 6

1. Supports for SIMD by adding media instructions
2. Improved thumb 2 provides the use of mixture of 16- and 32-bit instructions.
3. Enhanced support for trust zone virtualization technology, this make this core ideal for audio-video applications

(*l*) Version 6M

1. Targeted for low cost high performance device
2. Used in cortex M0 and cortex M2 processors

(*m*) Version 7

1. All cortex processors except M series have ARM 7 core.
2. NEON technology support
3. Optimized thumb 2 core
4. Enhanced floating operation for 3D graphics

Table 10.1 *Summary of ARM processor*

Core	Architecture	Performance functionality
ARM 1	V1	--
ARM 2	V2	--
ARM 2aS, ARM 3	V2a	--
ARM 6, ARM 600, ARM 610	V3	Basic ARM processors
ARM 7, ARM 700, ARM 710	V3	Basic ARM processors
ARM 7TDMI, ARM 710T, ARM 720T, ARM 740T, SecureCore - 100	V4T	Classic ARM processors
StrongARM, ARM 8, ARM 810	V4	Classic ARM processors
ARM 9TDMI, ARM 920T, ARM 940T	V4T	Classic ARM processors
ARM 9E-S	V5TE	Classic ARM processors
ARM 10TDMI, ARM 1020E, Xscale	V5TE	Classic ARM processors
ARM 11MP, ARM 176JZ, ARM 1136J, ARM1156T2	V6	Classic ARM processors
Cortex – A5, Cortex – A8, Cortex – A9,	V7	Application cortex processors
Cortex – R4	V7	Embedded cortex processors
Cortex – M0, Cortex – M1	V6M	Embedded cortex processors
Cortex – M3, SecureCore - 300	V7M	Embedded cortex processors

10.16.2 ARM Nomenclature

The nomenclature used for the identification of the type of ARM processor core is as follows:

ARMxyzTDMIEJFS

1. x – Series

   y – MMU

   It indicates integrated memory management unit.

3. z – Cache

   It indicates on-chip cache memory supports for the ARM core.

4. T – Thumb

   It indicates thumb instruction i.e. 16 – bit instructions available in the ARM core.

5. D – Debugger (JTAG Debug)

   Inbuilt debugging facility used for debugging the software code.

6. M – Multiplier

   ARM processor has inbuilt hardware multiplier unit for doing multiplication operation.
7. I – Embedded ICE macrocel

   This is the hardware circuit which is used to generate trace information and this feature is used in advance debugging and very useful in bug fixing.
8. E – Enhanced instructions

   Enhanced instruction sets for digital signal processor.
9. J – Java acceleration by jazelle

   Hardware circuit which is used to run JAVA byte code.
10. F – Vector floating point

    This is the hardware implementation of floating operations.
11. S – Synthesizable version

    It means ARM architecture can be modified, because it will come in terms of soft processor core, so manufacturers of the IC can decide addition or deletion of some parts as per requirement.

    Example:

    1. ARM7TDMI

       This is ARM 7 family core processor, which has the following parameters:

       T = Thumb instruction set, D = Debugging unit, M = Memory management unit, I = Trace circuit is inside the core (embedded trace macrocel)

       ARM7TDMI is the basic core and all cores after this have TDMI by default.
    2. ARM946E – S

       ARM9xx core, it consists of enhanced instruction set for digital signal processor and is also synthesizable so manufacturers can modify the core according to the requirements.

10.17 DATA SIZES AND INSTRUCTION SETS

The standard ARM word is 32 bits long. ARM is a 32-bit load/store RISC architecture. The word may be divided into four 8-bit bytes. ARM allows addresses to be 32 bits long; an address refers to a byte, not a word, word 0 is at location 0 and word 1 is at location 4. PC is incremented by 4 in sequential access.

ARM supports both little-endian and big endian. ARM can be configured at power-up to address the bytes in a word in either *little-endian* or *big-endian* mode.

(*a*) Big-endian

The most significant byte of a word is stored at the lowest numbered byte.

(*b*) Little-endian

The least significant byte of a word is stored at the lowest numbered byte.

When used in relation to ARM cores

(*a*) Half word means 16 bits (two bytes)

(*b*) Word means 32 bits (four bytes)

(*c*) Double word means 64 bits (eight bytes)

ARM cores implement two basic instruction sets

(*a*) ARM instruction set – instructions are all 32 bits long

(*b*) Thumb instruction set – instructions are a mix of 16 and 32 bits

Thumb-2 technology added many extra 32- and 16-bit instructions to the original 16-bit Thumb instruction set

Depending on the core, may also implement other instruction sets as given

(*a*) VFP instruction set – 32 bit (vector) floating point instructions

(*b*) NEON instruction set – 32 bit SIMD instructions

(*c*) Jazelle-DBX provides acceleration for Java VMs (with additional software support)

(*d*) Jazelle-RCT provides support for interpreted languages

10.18 ARM OPERATING MODES

ARM has seven basic operating modes; they are User, FIQ, IRQ, Supervisor, Abort, Undef, and System.

- Each mode has access to its own stack space and a different subset of registers
- Some operations can only be carried out in a privileged mode

1. **User mode:** User mode is the main execution mode. By running application software in user mode, the operating system can achieve protection and isolation. User mode is the only unprivileged mode whereas all other modes are privileged modes and therefore only used to run system software.
2. **FIQ Mode:** Fast interrupt request processing mode is entered whenever the processor receives an interrupt signal from the designated fast interrupt source.
3. **IRQ:** Normal interrupt request processing mode is entered whenever the processor receives an interrupt signal from any other interrupt source.
4. **Supervisor mode:** Software interrupt mode is entered when the processor encounters a software interrupt instruction. Software interrupts are a standard way to invoke operating system services on ARM, so supervisor mode is a protected mode for the operating system.
5. **Undef mode:** Undefined instruction mode is entered when the processor attempts to execute an instruction that is supported neither by the main integer core nor by one of the coprocessors. This mode can be used to implement coprocessor emulation.

6. **System Mode:** System mode is used for running privileged operating system tasks. System mode is the privileged user mode for the operating system.
7. **Abort Mode:** Abort mode is entered in response when a data or instruction pre-fetch is aborted.

The operating mode of the ARM core changes can be controlled by software and can also be caused by external interrupts or exception processing.

Most of the application programs execute in user mode; user mode is the only unprivileged mode. Privileged modes (that is, all modes other than User mode) are entered to service interrupts or exceptions, or to access protected resources.

The summary of the ARM operating modes is given in Table 10.2.

Table 10.2 *Summary of ARM operating modes*

Mode	Description	Operation
User	A normal program execution state, under this mode most of the applications/OS task runs.	Unprivileged Mode
FIQ	Entered when a high priority interrupt is raised.	Privileged Mode
IRQ	Entered when a normal priority interrupt is raised.	Privileged Mode
Supervisor	Entered on reset and when a Supervisor call instruction (SVC) is executed.	Privileged Mode
Abort	Entered in response when a data or instruction pre-fetch is aborted.	Privileged Mode
Undef	This mode can be used to implement coprocessor emulation.	Privileged Mode
System	System mode is the privileged user mode for the operating system	Privileged Mode

10.19 BLOCK DIAGRAM OF ARM CORE

The prototype core diagram of ARM architecture is shown in Figure 10.17. This diagram is only prototype and the devices available in the core are adjustable according to the requirement of the manufacturer.

The main blocks of the ARM core are as follows:

1. Register bank
2. Arithmetic logic unit (ALU)
3. Instruction decoder and control unit
4. Instruction pipeline
5. Address register and address incrementer
6. Barrel shifter
7. 32 x 8 Multiplier

8. Scan control
9. Write data register

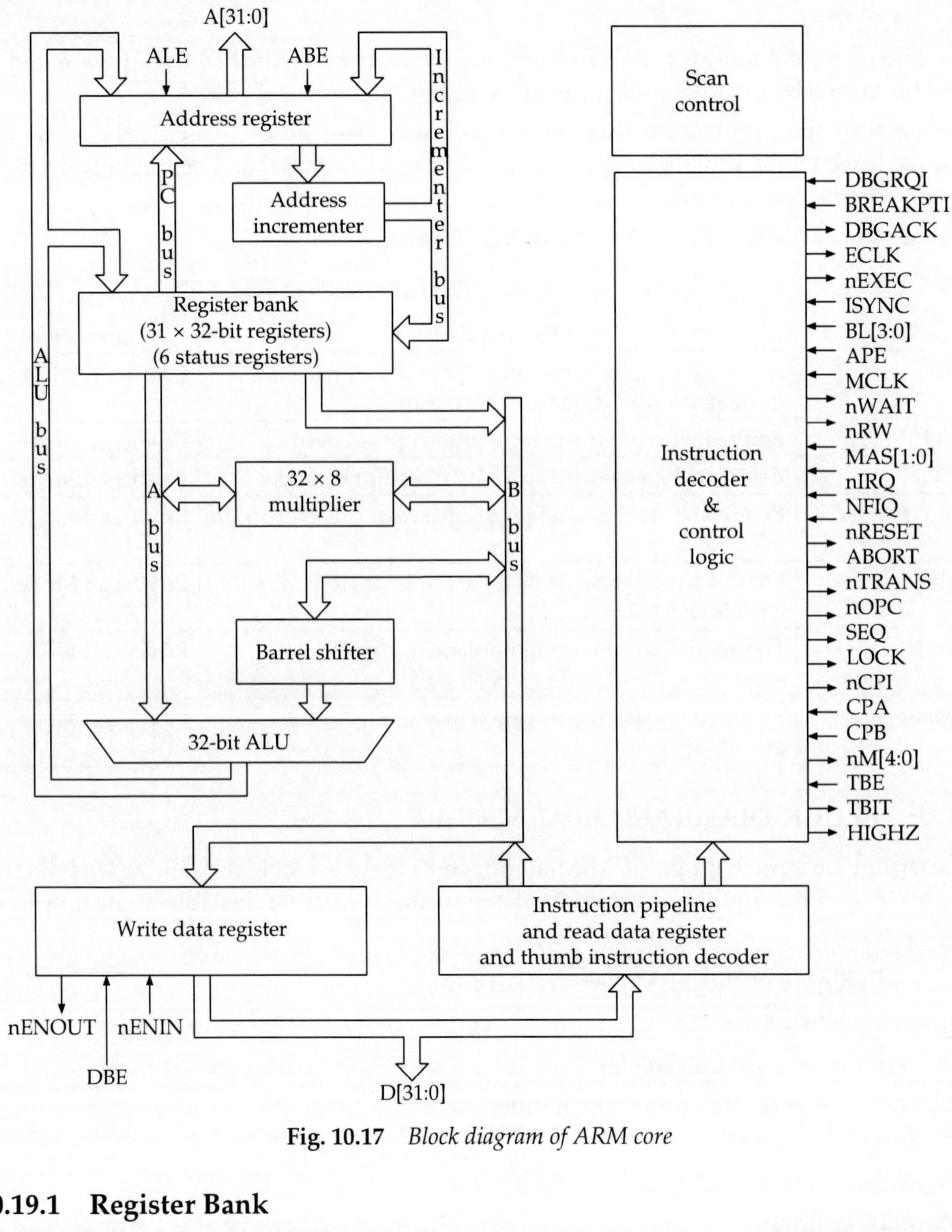

Fig. 10.17 *Block diagram of ARM core*

10.19.1 Register Bank

The ARM has a total of 37 registers. The ARM architecture provides 31 general-purpose 32-bit registers and 6 program status registers (one current program status register [CPSR] and five saved program status registers [SPSR]). Not all of these registers are always available, it depends on the operating modes of the ARM processor. The ARM architecture provides 16 general-purpose registers in the user mode. The complete register structure of the ARM architecture is shown in Figure 10.18. Register 15 is the program counter, but can be manipulated as a general-purpose register. The general-purpose register number 14 is used as a link register by the branch-and-link instruction. Register 13 is typically used as stack pointer, although this is not mandated by the architecture.

ARM state general purpose and program counter registers

System and user mode	FIQ mode	Supervisor mode	Abort mode	IRQ mode	Undefined mode
R0	R0	R0	R0	R0	R0
R1	R1	R1	R1	R1	R1
R2	R2	R2	R2	R2	R2
R3	R3	R3	R3	R3	R3
R4	R4	R4	R4	R4	R4
R5	R5	R5	R5	R5	R5
R6	R6	R6	R6	R6	R6
R7	R7	R7	R7	R7	R7
R8	R8_fiq	R8	R8	R8	R8
R9	R9_fiq	R9	R9	R9	R9
R10	R10_fiq	R10	R10	R10	R10
R11	R11_fiq	R11	R11	R11	R11
R12	R12_fiq	R12	R12	R12	R12
R13	R13_fiq	R13_svc	R13_abt	R13_irq	R13_und
R14	R14_fiq	R14_svc	R14_abt	R14_irq	R14_und
R15(PC)	R15(PC)	R15(PC)	R15(PC)	R15(PC)	R15(PC)

ARM state program status registers

CPSR	CPSR	CPSR	CPSR	CPSR	CPSR
	SPSR_fiq	SPSR_fiq	SPSR_fiq	SPSR_fiq	SPSR_fiq

(shaded) = Banked registers

Fig. 10.18 *ARM registers*

The current program status register (CPSR) contains four 1-bit condition flags (negative, carry, zero and overflow) and four fields reflecting the execution state of the processor. The 'T' field is used to switch between ARM and Thumb instruction sets. The 'I' and 'F' flags enable normal and fast interrupts respectively. Finally, the 'mode' field selects one of seven execution modes, as discussed in Section 10.22.

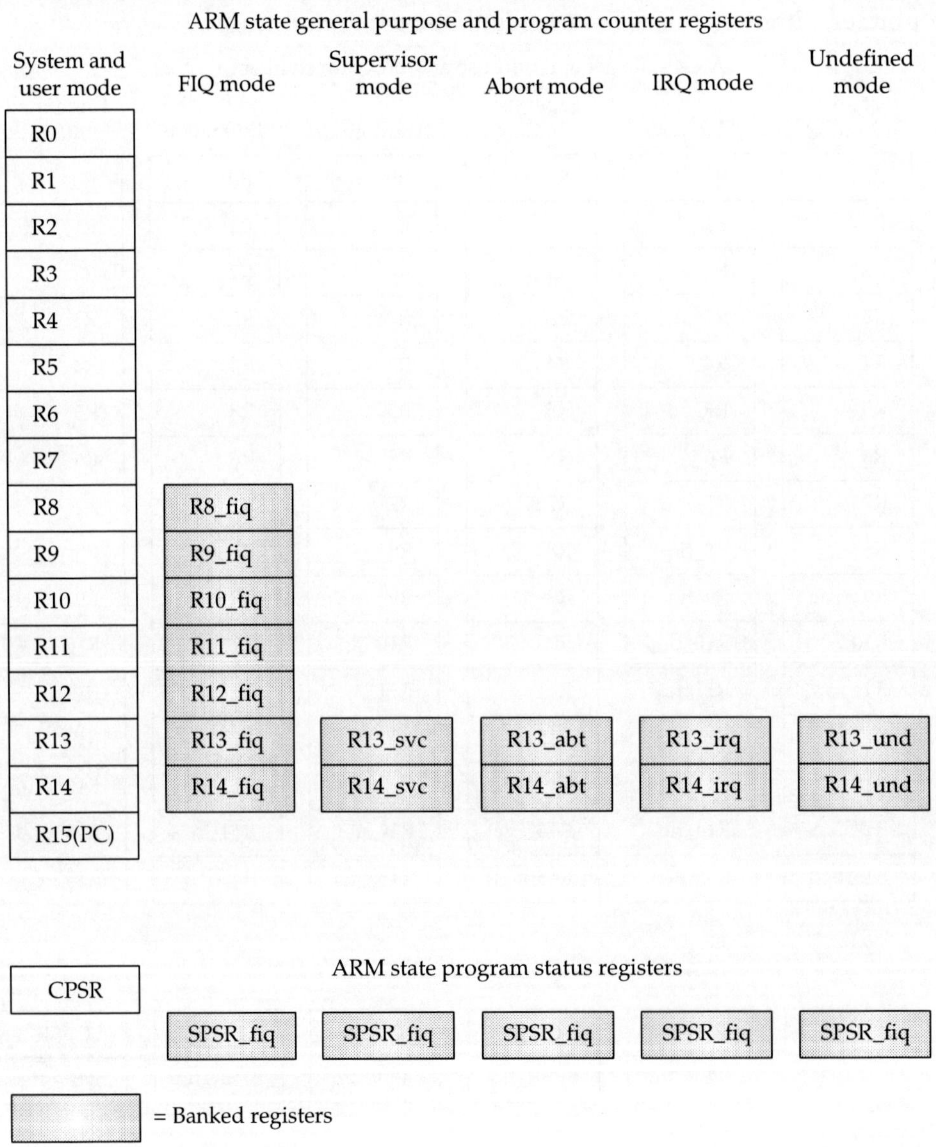

Fig. 10.19 *Effective ARM registers in different operating modes*

In addition to user-visible registers, ARM architecture provides several registers available in privileged modes only (shaded registers in Figure 10.18). Effective portion of the registers area available in different operating modes of ARM architecture is shown in Figure 10.19. SPSR registers are used to store a copy of the value of the CPSR register before an exception was raised. Those privileged modes that are activated in response to exceptions have their own R13 and R14 registers, which allows avoiding saving the corresponding user registers on every exception.

In order to further reduce the amount of state that has to be saved during handling of fast interrupts, ARM provides 5 additional registers available in the fast interrupt processing mode only.

By convention r13 = SP (stack pointer) and r14=LR (link register). In User mode: r14 is used as a link register to store the return address when a call to a subroutine is made. In the exception handling modes, r14 holds the return address for the exception.

By convention r15 = PC (program counter) which is incremented by one word for each instruction. To return from a subroutine MOV PC, LR instruction is used.

10.19.2 Program Status Register/Flag Register

In the ARM core 6 registers of 32-bit long are available as current program status register and saved program status register, generally referred as program status register. One CPSR and five SPSRs are available in the ARM core and both of these follow the same format. CPSR copies the arithmetic logic unit's status flag and the current processor operating mode and the status of interrupt disable flag.

The SPSR is used to store the CPSR when an exception is taken. One SPSR is accessible in each of the exception-handling mode, so five SPSRs are available for each exception handler. *User mode* and *System mode* do not have an SPSR because they are not exception handling modes.

The program status register for the ARM core ARM7TDMI is shown in Figure 10.20.

N	Z	C	V		Undefined	I	F	F	MODE
31	30	29	28	27 24	23 8	7	6	5	4 0

Fig. 10.20 *Current program status register*

In the ARM7TDMI only four bits from bits 31-24 are used as a flag and they are bits 31, 30, 29 and 28.

Bit 31 = N, defines the negative status flag; this flag bit is set when the result is negative in two's complement form.

Bit 30 = Z, defines zero status flag; whenever the result of any arithmetic or logical operation is zero, then this flag indicates logic one, informing the status of the result.

Bit 29 = C, defines carry flag; whenever the result of any arithmetic operation exceeds than its limits of the number used in the register then this flag is raised, indicating carry generation in the result.

Bit 28 = V, defines overflow status flag; whenever the result of any arithmetic or logical operation for the signed numbers exceeds than their limit value then this flag is automatically raised, indicating the result produced is incorrect.

Bits 23-8 are undefined in the ARM7TDMI core.

Bits 7-0 are called controlling flags as they are used to control certain operations in the ARM processor.

Bit 7 = I, if set to logic 1, then disable interrupt (IRQ disable)

Bit 6 = F, if set to logic 1, then disable fast interrupt (FIQ disable)

Bit 5 = T, indicating processor is in ARM state or Thumb state. If T = 0, then processor is in ARM state and when T =1, then processor is in Thumb state.

Bits 4 – 0 = MODE, these five bits filed indicate processor operating mode, i.e., whether processor is operating in user mode, system mode, IRQ mode, FIQ mode, supervisor mode, abort mode or in undefined mode.

Table 10.3 lists the various processor operating modes and the associated 5-bit binary patterns.

Table 10.3 *Processor operating modes*

Mode	Abbreviation	Operation	Mode (bits 4 – 0)
User	usr	Unprivileged	10000
System	sys	Privileged	11111
Interrupt request (IRQ)	irq	Privileged	10010
Fast Interrupt request (FIQ)	fiq	Privileged	10001
Supervisor	svc	Privileged	10011
Abort	abt	Privileged	10111
Undefined	und	Privileged	11011

To access the program status register i.e. CPSR and SPSR one must use MSR and MRS instructions. When MRS instruction is executed then it moves the program status register (PSR) to the general-purpose register, and when the MSR instruction is executed then it moves register contents to the program status register (PSR).

10.19.3 Arithmetic Logic Unit

This unit is used to perform all the arithmetic and logical operations supported by ARM core architecture.

10.19.4 Instruction Decoder and Control Unit

This unit is used to read the opcode of the instruction from the memory and provides the required signal for execution of the same instruction. Most of the instructions of ARM core are executed in single cycle.

10.19.5 Instruction Pipeline

One of the important properties of the ARM architecture is its simplicity. Without register windows, register renaming, out-of-order execution, multiple instruction issue, speculation, and other complex optimisations found in modern microprocessors and microcontrollers, even the latest ARM pipelines remain quite simple.

A pipeline is the mechanism used to execute instructions. Using pipeline when the current instruction is being executed, the next instruction is in the pipeline for the fetch operation, so in terms of this multiple operations are performed simultaneously, and hence the result in speed up of operation.

The different stages of the pipeline supported by ARM core is as follows:

(*a*) 3-stage pipeline
(*b*) 5-stage pipeline
(*c*) 6-stage pipeline
(*d*) 8-stage pipeline

The ARM7TDMI uses 3-stage pipeline in order to increase the speed of the flow of instructions to the processor. The more number of stages in the pipeline will reflect more execution speed of the instructions. Pipeline structure allows several operations to be performed simultaneously, rather than serially. The program counter register points to the instruction being fetched, not executed. Figure 10.21 shows the simplified way of 3-stage pipeline execution.

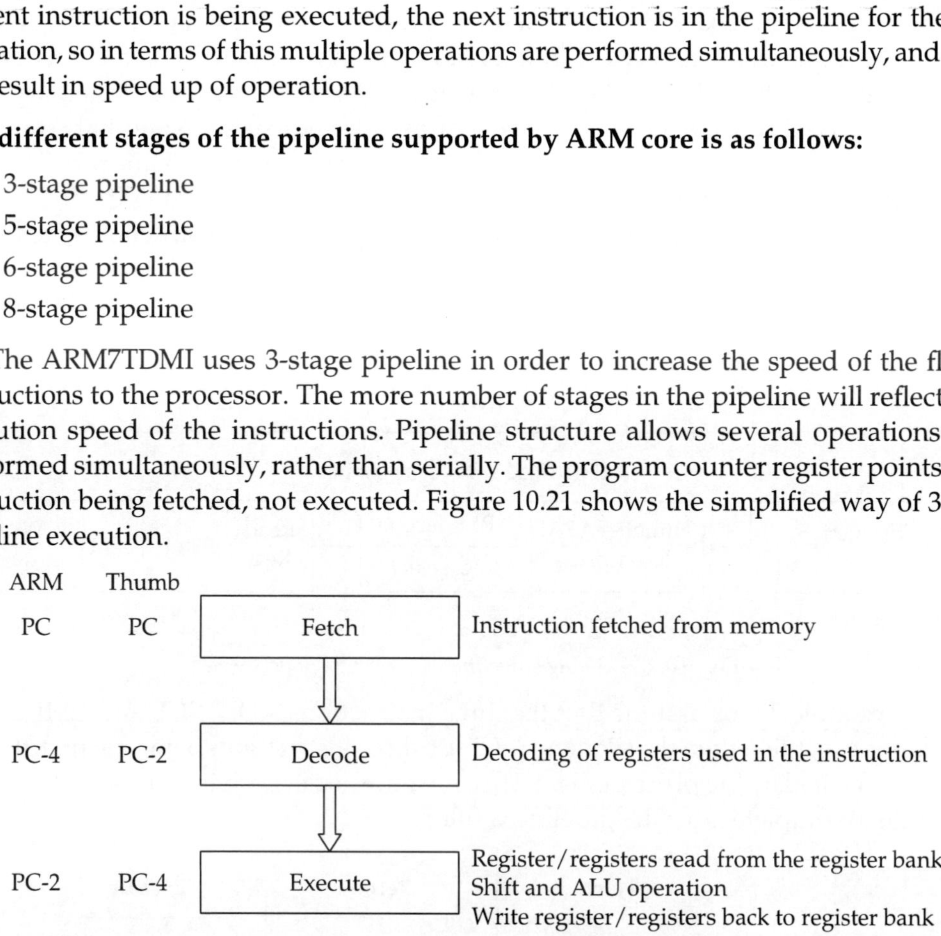

Fig. 10.21 *Example of 3-stage pipeline*

(*a*) 3-stage Pipeline

Figure 10.23 shows the original 3-stage ARM pipeline that remained essentially unchanged from the first ARM processor to the ARM7TDMI core. It is a classical fetch-decode-execute pipeline, which, in the absence of pipeline hazards and memory accesses, completes one instruction per cycle. The first pipeline stage, i.e., fetch stage reads an instruction from memory and increments the value of the instruction address register, which stores the value of the next instruction to be fetched. This value is also stored in the PC register. The next stage, i.e., decode stage decodes the instruction and prepares control signals required to execute it. The third stage, i.e., execute stage does all the actual work, it reads operands from the register file, performs ALU operations, reads or writes memory, if necessary, and finally writes back modified register values. In case the instruction being executed is a data processing instruction, the result generated by the ALU is written directly to the register file and the execution stage completes in one cycle. If it is a load or store instruction, the memory address computed by the ALU is placed on the address bus and the actual memory access is performed during the second cycle of the execute stage. Note that since PC is accessible as a general-purpose register, there are several places in the pipeline where the next instruction address can be issued. Under normal conditions, it is incremented on every cycle during the fetch stage. If a data processing instruction specifies R15 as its destination operand, then the result of the ALU operation is used as the next instruction address. Finally, a load to R15 has a similar effect. Figure 10.22 shows the 3-stage pipeline structure.

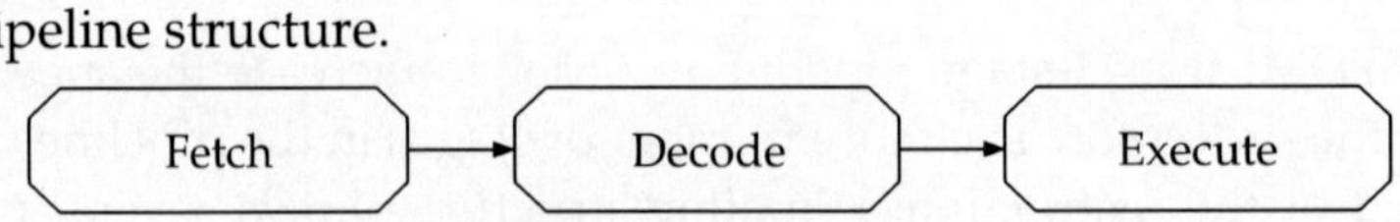

Fig. 10.22 *3-stage pipeline*

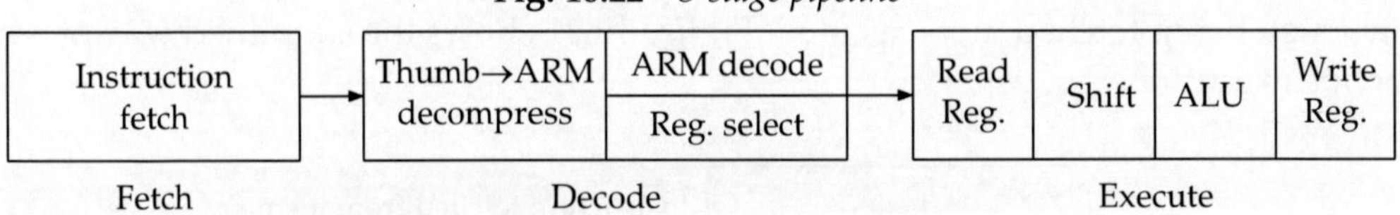

Fig. 10.23 *3-stage pipeline with operation performed*

For example, let us assume that the three instructions ADD, SUB and CMP is given for the execution then the sequence of these three instructions being fetched, decoded and executed by the processor is shown in Figure 10.24. Each instruction takes a single cycle to complete after the pipeline is filled.

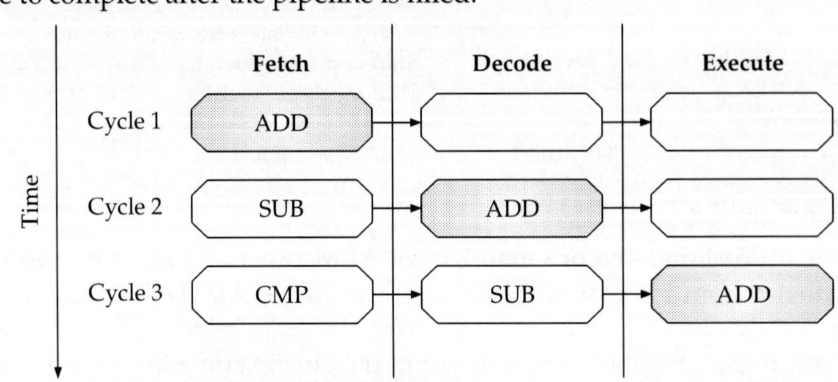

Fig. 10.24 *3-stage pipeline execution steps*

The three instructions are placed into the pipeline sequentially. In the first cycle, the core fetches the first instruction, i.e., ADD instruction from memory. In the second cycle the core fetches the second instruction, i.e., SUB instruction from memory, at the same time it will also decode the first instruction ADD. In the third cycle the first instruction, i.e., ADD instruction is executed, the second instruction SUB is decoded and the third instruction CMP is fetched from the memory. This pipeline procedure is actually referred as filling the pipeline. The pipeline allows the ARM core to execute an instruction every cycle.

(*b*) 5-stage Pipeline

The pipeline structure shown in Figure 10.25 assumes availability of only one memory port, which means that every data transfer instruction causes a pipeline stall, because the next instruction cannot be fetched while memory is being read or written. One way around this problem, which was used in ARM9TDMI and later microarchitectures, is to use separate instruction and data caches. This allows to modify the pipeline to avoid stalls on data transfer instructions. The different operations performed in the 5-stage pipeline is outlined in Figure 10.26.

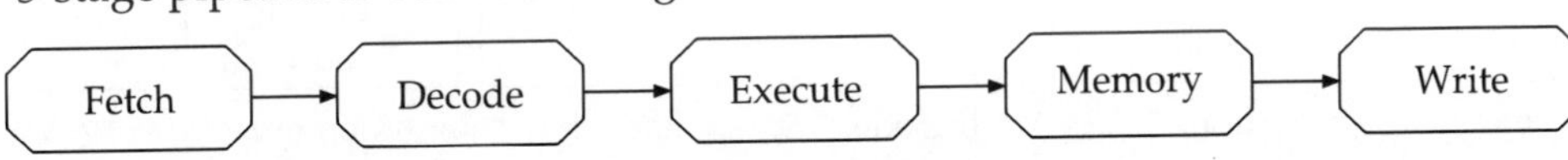

Fig. 10.25 *5-stage pipeline*

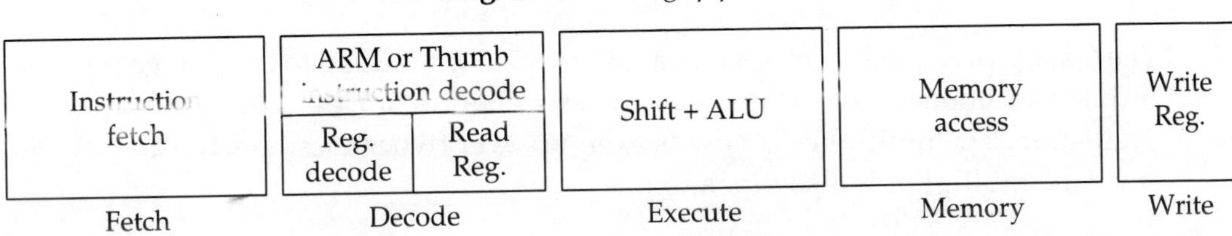

Fig. 10.26 *5-stage pipeline with operation performed*

By comparing 3-stage pipeline and 5-stage pipeline, it was clear that a lot of work was carried out in the execute stage of the 3-stage pipeline. Now in 5-stage pipeline, the execute stage has been split into three more stages to allow greater throughput. This then means the CPI is about 1.5 compared against 1.9 for 3-stage pipeline, and the operating frequency is approximately double for 5-stage pipeline over 3-stage pipeline on the same fabrication process. Therefore, at least double the processing power is available.

(*c*) 6-stage Pipeline

The ARM10 core made further improvements to the pipeline structure which uses 6-stage pipeline as shown in Figure 10.27. Designers realised that the pipeline performance was mainly constrained by memory bandwidth. Therefore, they made both instruction and data buses 64-bit wide. In the fetch stage, this allowed to fetch two instructions on each cycle, which enabled the introduction of a static branch prediction unit. The branch predictor tries to look ahead of the instruction stream and predicts backward pointing branches as taken and forward pointing branches as not taken. This trivial algorithm eliminates the penalty of branches for loops that execute many times. In the execution section of the pipeline, 64-bit data bus allowed to improve the performance of multiple-register transfer instructions by transferring two registers at a time. The different operations performed in different stages of the pipeline is shown in Figure 10.28.

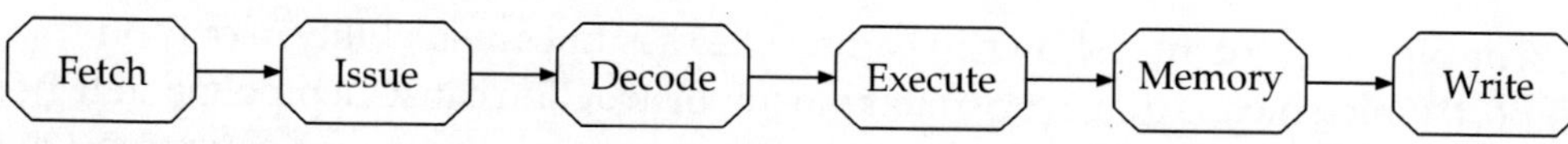

Fig. 10.27 *6-stage pipeline*

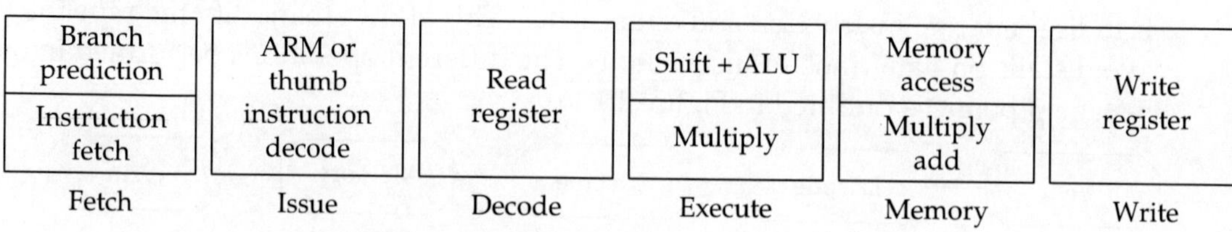

Fig. 10.28 *6-stage pipeline with operation performed*

The 6-stage pipeline illustrates that another stage was added to the 5-stage pipeline to provide additional time to handle coprocessor instruction decode and handle branch prediction. The multiplier is now broken up over two stages, execute and memory, since the multiplier is also pipelined.

(*d*) 8 – stage Pipeline

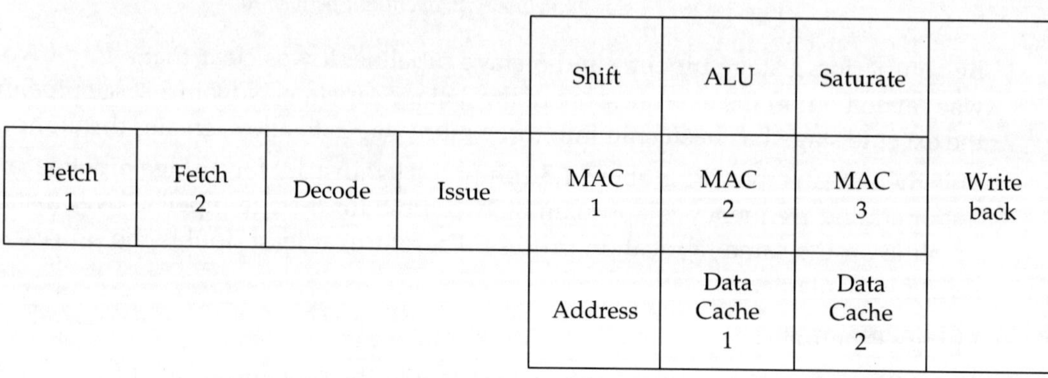

Fig. 10.29 *8-stage pipeline with operation performed*

The ARM11 core which uses 8-stage pipeline structure introduced two main changes to the pipeline architecture as shown in Figure 10.29. First, the shift operation was separated into a separate pipeline stage. Second, both instruction and data cache accesses are now distributed across 2 pipeline stages. Note that the instruction execution section of the resulting 8-stage pipeline is split into three separate pipelines that can operate concurrently in some situations and can commit instructions out-of-order. Still, fetch and decode stages are executed in order.

In the 8-stage pipeline, processor is a single issue processor, meaning that only one instruction per cycle can be issued from the issue stage to one of the 3 backend pipeline stages.

While the instructions are issued in order, they may complete out of order. This will depend on the availability of data, length of execution and memory access times.

10.19.6 Barrel Shifter

Barrel shifter is actually used for multiple rotation and shifting operation in single machine cycle. When the operand 2 is given along with the instruction, then Barrel shifter for operand 2 allows full 32-bit shift within instruction cycle. For example, assume that the contents of the register is rotating towards left by one position then it requires one machine cycle, then for rotation by two times requires more time or more machine cycles compared to one time rotation, but with the help of inbuilt barrel shifter, the rotation by multiple bits requires the same time as the time required by one time rotation.

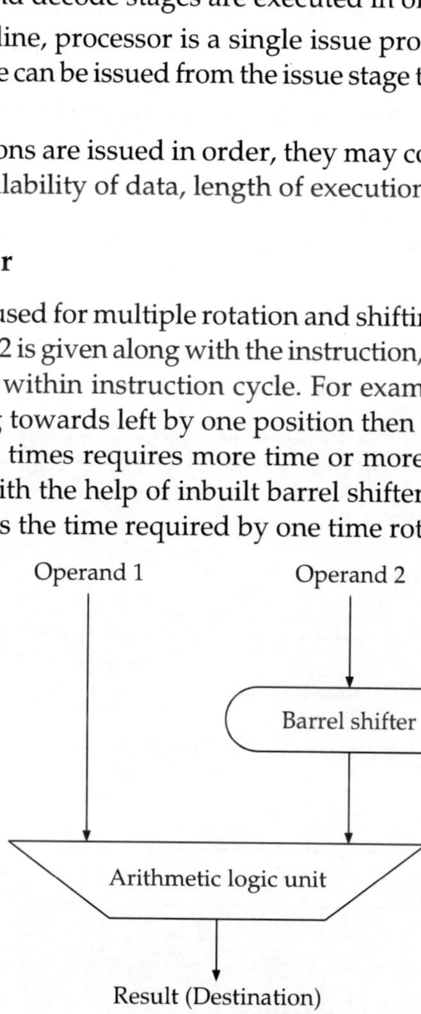

Fig. 10.30 *Barrel shifter and ALU*

To illustrate the role of barrel shifter, we can take the example as shown in Figure 10.30 and add a shift operation to the move instruction. Suppose the operand 2 is the value of register r0 which transferred to the register using the instruction ***LDR r0, [r1, r2, LSL #2]***. In this instruction the value of register r2 is multiplied by 4, this is because of the rotation of the bits of r2 register towards left by two positions using the instruction *LSL*, then the contents of the memory location whose address is calculated as r1+r2*4 is copied to the register r0. Even the instruction consists of the rotation operation cascaded in the data transfer operation, this instruction completes in a single cycle only because of the barrel shifter, remember r0 is the operand 2, which are given to the ALU through barrel shifter. The operand 1 is given to the ALU without any preprocessing of registers and then finally some arithmetic or logical operations are performed on the operand 1 and operand 2.

The barrel shifter can be used in the following instructions of the ARM core, it is specially used when multiple bits are shifted or rotated. Preprocessing or shift occurs within the cycle time of the instruction. This is particularly useful for loading constants into register and achieving fast multiplication or division by a power of 2.

Logical Shift Left (LSL)

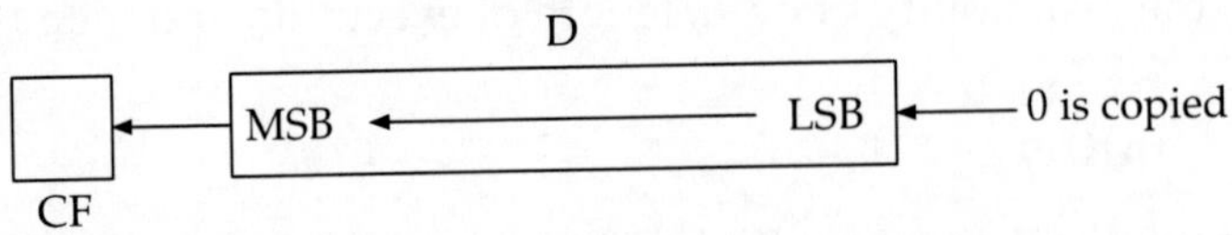

Multiplication by a power of 2

Logical Shift Right (LSR)

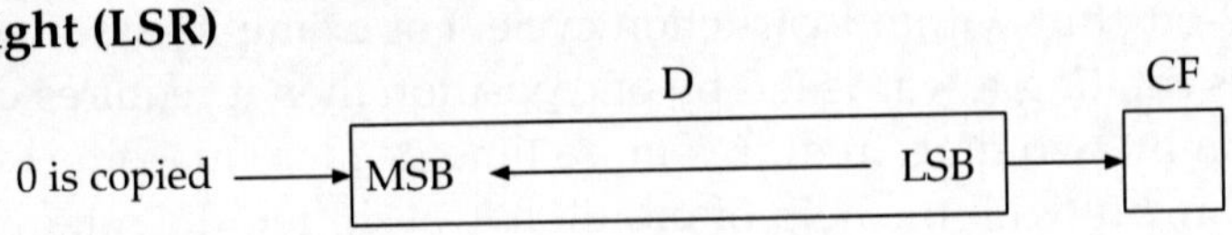

Division by a power of 2

Arithmetic Shift Right (ASR)

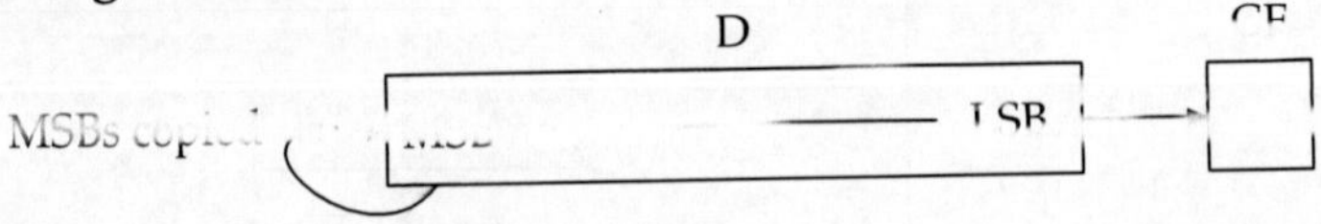

Division by a power of 2, preserving the sign bit

Rotate Right (ROR)

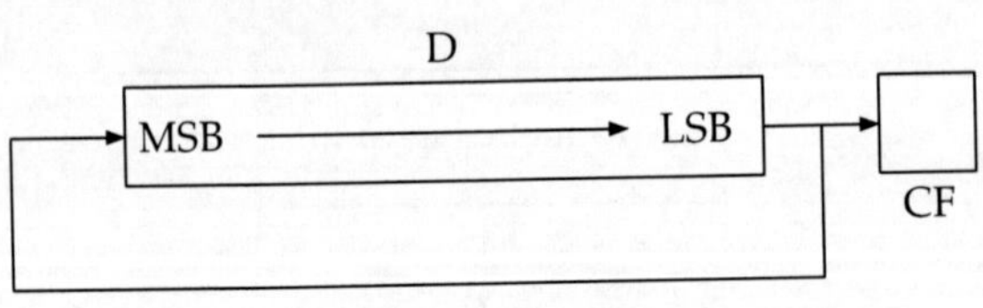

Bit rotate with wrap around from MSB to LSB

Rotate Right Extended (RRX)

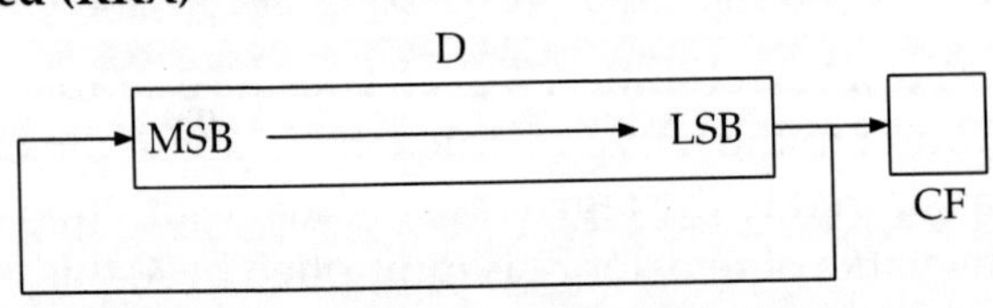

Single bit rotate with wrap around from CF to MSB and then from MSB to LSB.

For example:

1. ADD r2, r3, r4, LSL #2 ; r2 = r3 + (r4 * 4)

   The above instruction rotates the contents of register r4 towards left by two times and hence it was multiplied by $2^2 = 4$, this type of operation is done in no time just because of the barrel shifter.

2. LSL r2, #3 ; r2 = r2 * 8

   The above instruction rotates/shifts the contents of register r2 towards left by three times and hence it was multiplied by $2^3 = 8$.

3. LDR r0, [r1, r2, LSL #2] ; load r0 with data at address ; {r1 + r2 * 4}

   The above instruction rotates/shifts the contents of register r2 towards left by two positions and hence it was multiplied by $2^2 = 4$.

10.20 COPROCESSORS

Coprocessors are actually additional processors, which are normally used to perform additional mathematical or other operations, specially floating point operations. Coprocessors can be attached to the ARM processors. A coprocessor extends the processing features of an existing core by extending the additional instruction set or by configuration registers. It is possible to use more than one coprocessor with the ARM core via the coprocessor interface.

To access coprocessor, one must use the dedicated instructions available in the ARM core. The new instructions of the coprocessors are processed in the decode stage of the ARM pipeline. If the decode stage detects the coprocessor instruction in the decode stage, then it offers it to the relevant coprocessor. But suppose that the coprocessor is not present and the instruction of the coprocessor is given for the execution, then the ARM takes an undefined instruction exception, which allows the user to emulate the behaviour of the coprocessor in software.

10.21 ARM EXCEPTION

An exception occurs when the normal flow of program execution is interrupted, like some out of the range operation is performed, or the system reset, or some non-available instruction is given for execution. When an exception or interrupt sequence occurs, the processor sets the program counter register to a specific memory address to service the program associated with the exception or interrupt. The address where the exception program is present is within a special address range called the vector table, and the vector address is the default address where the exception program is available. Whenever the exception or interrupt occurs the processor state just prior to handling the exception must be preserved, the program flow will automatically back from where the exception or interrupt occurs.

The memory map address 0x00000000 to 0x0000001C is reserved for the vector table, a set of 32 bit word. On some of the processors the vector table can be optionally located at the higher address in memory; in the higher address of the memory the reserved area is in the range of 0xFFFF0000 to 0xFFFF001C.

To process the exception, the ARM core uses the banked core registers to save the current state, the old program counter value (return address) are copied to the r14 (LR) register and the CPSR contents are copied to the SPSR registers.

The diffrent ARM exception types are listed below:

1. **Reset**

   Reset vector is the starting location from where the first instruction is executed by the processor upon system reset. When the signal is applied on the reset pin then the program counter register is set to the reset vector location and hence the execution begins from this location.

2. **Undefined instruction**

   Undefined instruction exception is generated when trying to decode an instruction that is supported neither by the main integer core nor by one of the coprocessors.

3. **Software interrupt**

   Software interrupt exception is raised by a special instruction, i.e., SWI instruction. The SWI instruction is executed to request an operating system service.

4. **Prefetch abort**

   Prefetch abort exception is raised by memory management hardware when memory access permissions are violated during instruction fetch. The actual abort exception occurs in the decode stage.

5. **Data abort**

   Data abort exception is similar to the Prefetch abort exception, but is raised by memory management hardware when a load or store instruction violates memory access permissions. So whenever an attempt is made to access data memory without the correct access permissions, the data abort exception is generated.

6. **Interrupt request (IRQ)**

   Interrupt request exception is raised whenever the processor receives an interrupt signal from any non-fast interrupt source. Interrupt request vector is used by the external hardware circuits to interrupt the normal execution flow of the processor. It can only be raised if IRQ's are not masked in the CPSR.

7. **Fast interrupt request (FIQ)**

   Fast interrupt exception is raised whenever the processor receives an interrupt signal from the designated fast interrupt source.

Table 10.4 shows the exception types along with their vector location.

Table 10.4 *Vector table*

Exception	Abbreviation	Address	High address	Priority
Reset	RESET	0X00000000	0xFFFF0000	1 (Highest)
Undefined instruction	UNDEF	0X00000004	0xFFFF0004	6
Software interrupt	SWI	0X00000008	0xFFFF0008	6 (Lowest)
Prefetch abort	PABT	0X0000000C	0xFFFF000C	5
Data abort	DABT	0X00000010	0xFFFF0010	2
Reserved	--	0X00000014	0xFFFF0014	--
Interrupt request	IRQ	0X00000018	0xFFFF0018	4
Fast interrupt request	FIQ	0X0000001C	0xFFFF001C	3

Except for the reset exception, all other exceptions are handled in a similar way: The processor switches to the corresponding execution mode and saves the address of the instruction following the exception entry instruction in R14 of the new mode, saves the old value (return address) of CPSR to SPSR of the new mode, disables IRQ (in case of a fast interrupt, FIQ is also disabled), and starts execution from the relevant exception vector.

10.22 ASSEMBLER AND ASSEMBLER DIRECTIVES

Assembler is a software program used to convert assembly language program to the equivalent machine language program (binary language program). Here, the source code is converted into object code with the help of the assembler. We used to call assembler as Asmblr.

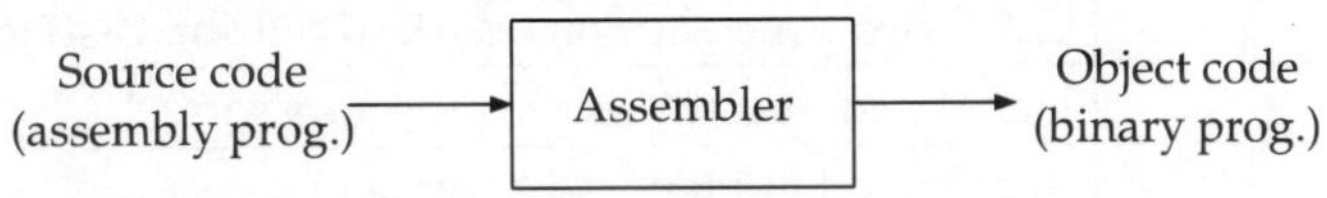

Assembler Directive

It is the instruction used to inform the assembler about conversion of assembly program into machine language program. It is also called pseudo or dummy instruction because these instructions are not executed by ARM processor (microcontroller). Rather, they are executed by the assembler. In other words, we say that assembler directives are used for assembling and they are not consuming any memory location because they are not executed by microcontroller.

Different types of assembler directives supported by ARM processor are shown in Table 10.5.

Table 10.5 *Summary of assembler directives of ARM core*

Assembler directives	Meaning
$	Includes a file. Assembler control
/*comment*/	C-style comment delimiter. Assembler control
//	C++ style comment delimiter. Assembler control
#define	Assigns a value to a label. C-style preprocessor
ARM	Interprets subsequent instructions as 32-bit (ARM) instructions.
THUMB	Interprets subsequent instructions as Thumb execution-mode instructions.
CODE16	Interprets subsequent instructions as 16-bit (Thumb) instructions.
CODE32	Interprets subsequent instructions as 32-bit (ARM) instructions.
DATA	Defines an area of data within a code section. Mode control
ELSE	Assembles instructions if a condition is false. Conditional assembly
ELSEIF	Specifies a new condition in an IF…ENDIF block. Conditional assembly
END	Terminates the assembly of the last module in a file. Module control
ENDIF	Ends an IF block. Conditional assembly
#else	Assembles instructions if a condition is false. C-style preprocessor
#endif	Ends a #if, #ifdef, or #ifndef block. C-style preprocessor
#if	Assembles instructions if a condition is true. C-style pre-processor
#ifdef	Assembles instructions if a symbol is defined. C-style pre-processor
#ifndef	Assembles instructions if a symbol is undefined. C-style preprocessor
#include	Includes a file. C-style preprocessor
#message	Generates a message on standard output. C-style preprocessor
#undef	Undefined a label. C-style preprocessor
#pragma	Recognized but ignored. C-style preprocessor
EQU	Assigns a permanent value local to a module.
EXTERN	Imports an external symbol.
EXTRN	Imports an external symbol.
IF	Assembles instructions if a condition is true.
IMPORT	Imports an external symbol.
INCLUDE	Includes a file.
LIBRARY	Begins a module; an alias for PROGRAM and NAME.
NAME	Begins a program module. Module control
PUBLIC	Exports symbols to other modules. Symbol control

RADIX	Sets the default base. Assembler control
SECTION	Begins a section. Section control
SECTION_TYPE	Sets ELF type and flags for a section. Section control
VAR	Assigns a temporary value. Value assignment

10.23 ARM INSTRUCTION SETS

The ARM core provides two types of instruction sets:

1. ARM instruction sets, which are standard 32-bit instruction set.
2. Thumb instruction sets, which are in 16-bit compressed form, have code density better than most of the complex instruction set computing (CISC) processors.

ARM core supports a number of instructions which are categorized according to the operation performed by the instruction.

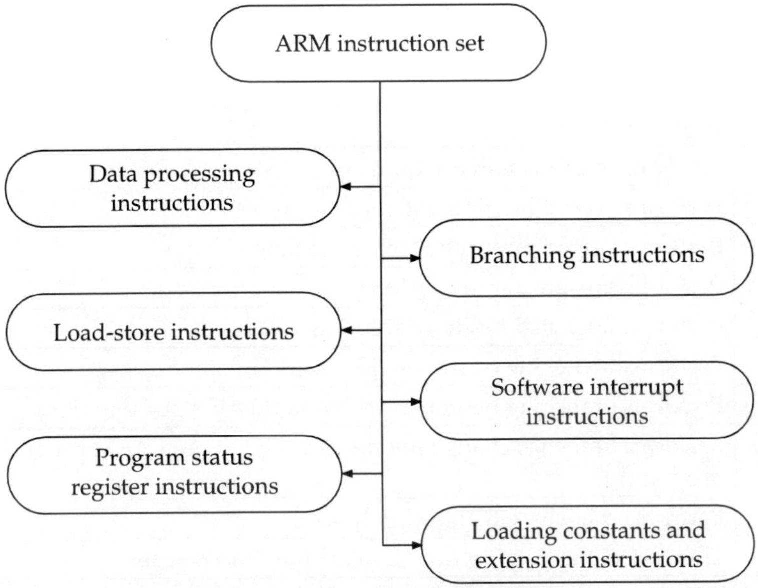

Table 10.6 gives the summary of the ARM instruction sets.

Table 10.6 *Summary of ARM instruction set*

Mnemonics	Description
MOV	Move a 32-bit value into a register
MVN	Move the logical NOT of 32-bit value into a register.
ADD	Add two 32-bit values
ADC	Add two 32-bit values with carry from the previous operation if any.
SUB	Subtract two 32-bit values

SBC	Subtract two 32-bit values with carry (borrow)
RSB	Reverse subtract two 32-bit values
RSC	Reverse subtract two 32-bit values with carry (borrow)
AND	Performs bitwise ANDing of two 32-bit values
ORR	Performs bitwise ORing of two 32-bit values
EOR	Performs exclusive ORing of two 32-bit values
BIC	Performs logical bit clear (AND NOT)
LSL	Performs logical shifting towards left
LSR	Performs logical shifting towards right
ASR	Performs arithmetic right shift
ROR	Performs rotate towards right
RRX	Performs rotate right extended (through carry)
CMP	Performs comparison on two 32-bit values
CMN	Performs comparison of negative values
TST	Test bits of a 32-bit values
TEQ	Test for equality of two 32-bit values
MUL	Performs multiply
MLA	Performs multiply and accumulate
SMULL	Performs signed multiply long
SMLAL	Performs singed multiply accumulate long
UMULL	Performs unsigned multiply long
UMLAL	Performs unsigned multiply accumulate long
B	Performs branching operation (like JUMP operation)
BL	Performs branching operation with link (like CALL operation)
BX	Performs branch exchange (used to switch between ARM program and Thumb program)
BLX	Performs branch exchange with link
LDR	This instruction loads word (32-bit value) into register
STR	This instruction saves word from a register (register to memory)
LDRB	This instruction loads byte into a register
STRB	This instruction saves byte from a register (register to memory)
LDRH	This instruction loads halfword (16-bit value) into a register
STRH	This instruction saves halfword (16-bit value) from a register (register to memory)
LDRSB	This instruction loads signed byte into a register
LDRSH	This instruction loads signed halfword (16-bit value) into a register
LDM	This instruction loads multiple registers (from memory to registers)

STM	This instruction saves multiple registers into memory (from register to memory)
SWP	This instruction swaps a word (32-bit number) between memory and a register
SWPB	This instruction swaps a byte between memory and a register
SWI	This instruction allows execution of software interrupt exception
MRS	This instruction copies program status register (CPSR or SPSR) to a general purpose register
MSR	This instruction copies a general purpose register to a program status register (CPSR or SPSR)
CDP	Coprocessor data processing. This instruction performs an operation in a coprocessor
MRC	This instruction moves from coprocessor to ARM7TDMI-S register (L=1)
MCR	This instruction moves from ARM7TDMI-S register to coprocessor (L=0)
LDC	This instruction loads from memory to coprocessor
STC	This instruction stores from coprocessor to memory
LDR	This instruction loads constant value to the register
ADR	This instruction loads address into the register
QSUB	Signed saturated 32-bit subtract
QDSUB	Singed saturated double 32-bit (64- bit) subtract
SMLAxy	Signed multiply and accumulate 32-bit
SMLALxy	Signed multiply and accumulate 64-bit
SMLAWy	Signed multiply and accumulate 32-bit
SMULxy	Signed multiply
SMULWy	Signed multiply

10.24 DATA PROCESSING INSTRUCTIONS

Data processing instructions manipulate data within registers. They are move instructions, arithmetic instructions, logical instructions, comparison instructions, multiply instructions. Most of the data processing instructions can process one of their operand given along with the instruction using barrel shifter for faster operation. These instructions generally operate only on the registers and they do not affect memory.

The status flags, i.e., negative (N), carry (C), zero (Z) and overflow (V) flags are affected depending on the operation performed by instructions. If the suffix 'S' is used on the data processing instructions, then it updates the status flags in the CPSR register.

Syntax:

<Instruction> {<cond>} {S} Rd, Rn, Operand2

{cond} two-character condition mnemonic for conditional execution. See Table 10.8 for summary on conditional execution

Rd, Rn are expressions evaluating valid register numbers

10.24.1 Data Movement Instruction

Data movement instruction does not specify Rn in the above syntax. It copies the operand 2 into the destination register Rd, where the operand 2 is a register or immediate value given along with the instruction.

MOV	Move a 32-bit value into a register
MVN	Move the logical NOT of 32-bit value into a register

Examples:

1. MOV r7, r3 ; r3 is moved to r7
2. MVN r6, #0 ; two's complement of 0 is moved to r6
3. MOVS r7, r6 ; set the flags
4. MOV r0, r1, LSL #4 ; r0 = r1 * 16
5. MOV r0, r1, LSL r2 ; r0 =

10.24.2 Arithmetic Instructions

The arithmetic instruction performs addition and subtraction of 32-bit signed and unsigned numbers with and without updation of flags.

Syntax:

<Instruction> {<cond>} {S} Rd, Rn, Operand2

ADD	Add two 32-bit values
ADC	Add two 32-bit values with carry from the previous operation if any
SUB	Subtract two 32-bit values
SBC	Subtract two 32-bit values with carry (borrow)
RSB	Reverse subtract two 32-bit values
RSC	Reverse subtract two 32-bit values with carry (borrow)

Examples:

```
ADD r0, r1, r2          ; r0 = r1+r2 and don't update flags
ADDS r0, r1, r2         ; r0 = r1+r2 and update flags
ADD r5, #0x24           ; r5 = r5+ 36
ADD r0, r1, r2, LSL #2  ; r0 = r1 + (r2 * 4)
SUB r0, r1, r2          ; r0 = r1 – r2
RSB r0, r1, r2          ; r0 = r2 – r1
RSB r1, r2, #3          ; r1 = 0x3 – r2
SUBS r2, r2, #2         ; r2= r2 – 1 and update flags
```

10.24.3 Logical Instructions

Logical instructions perform bitwise logical operations on the two source operands.

Syntax:

<Instruction> {<cond>} {S} Rd, Rn, Operand2

AND	Performs bitwise ANDing of two 32-bit values
ORR	Performs bitwise ORing of two 32-bit values
EOR	Performs exclusive ORing of two 32-bit values
BIC	Performs logical bit clear (AND NOT)

Examples:

1. AND r0, r1, r2 ; r0:= r1 AND r2
2. ORR r0, r1, r2 ; r0 = r1 OR r2
3. BIC r0, r1, r2 ; r0 = r1 AND NOT (r2)
   Assume r1= 0b1111 and r2 = 0b1100 (b stands for binary number)
   Then r0 = 0b1111 AND NOT (0b1100) = 0b0011

10.24.4 Barrel Shifter

Barrel shifter is used for multiple shifting within a single machine cycle.

LSL	Performs logical shifting towards left
LSR	Performs logical shifting towards right
ASR	Performs arithmetic right shift
ROR	Performs rotate towards right
RRX	Performs rotate right extended (through carry)

See the section 10.23.6 for more information on the above instruction.

Examples:

1. MOVS r1, r2, LSL #1 ; r1 = r2 * 2
   Assume r1 = 0x00000002 and r2 = 0x00000004 (x stands for hexadecimal number)
   r1 = r2 * 2 = 0x00000008
2. ADD r0, r1, r2, LSL # 2 ; r0 = r1 + r2 * 4
   Assume r1 = 0x00000002 and r2 = 0x00000004 (x stands for hexadecimal number)
   R0 =r1+ r2 * 4 =0x00000002 + 0x00000008 = 0x0000000A

10.24.5 Comparison Instructions

The comparison instructions are used to compare the two 16-bit values. They update the CPSR flag bits according to the result, but do not affect the other registers, i.e., the registers used in the comparison instructions including the destination register is unaffected. In comparison instruction addition of suffix 'S' for updating the status flags are not required, it will be affected automatically depending on the relation of the two numbers used for the comparison.

Syntax:

<Instruction> {<cond>} {S} Rn, Operand2

CMP	Performs comparison on two 32-bit values
CMN	Performs comparison of negative values
TST	Tests bits of 32-bit values
TEQ	Tests for equality of two 32-bit values

Examples:

1. CMP r0, r7 ; r0 – r7 = flags affected
   Assume r0 = 0x0000FFFF and r7 = 0x0000FFFF then the status flags NZCV = 0100 (numbers equal, so zero flag = 1)
2. CMP r4, #0 ; r4 – 0 = flags affected
   Assume r4 = 0x0000000F then NZCV = 0000 (r4 is greater than the immediate value so no borrow is required for the subtraction).

10.24.6 Multiply Instructions

The multiply instructions are multiply the two operand of 32-bit value given along with the instruction and depending upon the instruction, accumulate the results in another register. The long multiplies adjust the result onto a pair of registers representing a 64-bit value.

Syntax:

<Instruction> {<cond>} {S} Rd, Rm, Rn, Operand2

<Instruction> {<cond>} {S} Rd, Rm, Rn

<Instruction> {<cond>} {S} Rd_Lo, Rd_Hi, Rm, Rn

MUL	Performs multiply
MLA	Performs multiply and accumulate
SMULL	Performs signed multiply long
SMLAL	Performs singed multiply accumulate long
UMULL	Performs unsigned multiply long
UMLAL	Performs unsigned multiply accumulate long

Examples:

1. MUL r0, r1, r2 ; r0 = r1*r2

   Assume r1 = 0x00000002 and r2 = 0x00000004 then

   r0 = r1 * r2 = 0x00000008
2. MUL r4, r3, r2 ; r4 = r3*r2
3. MLA r0, r1, r2, r3 ; r0 = (r1 * r2) + r3

   Assume r1 = 0x00000002, r2 = 0x00000004 and r3 = 0x00000002 then

   r0 = r1 * r2 + r3= 0x0000000A
4. UMULL r0, r1, r2, r3 ; [r1, r0] = r2 * r3

   This instruction produces long result, i.e., when the two 32-bit numbers are multiplied the result will be represented in the 64-bit form and the higher order of result (bits 63 – 32) is copied to the higher byte of the register and lower order of result (bits 31 – 0) is copied to lower byte of the resister.

   Assume r2 = 0x00000004 and r3 = 0x00000002 then

   [r1, r0] = r2 * r3= 0x00000008

   So, r1 = 0x00000000 and r0 = 0x00000008

10.25 BRANCHING INSTRUCTIONS

Branching instructions allow to change the program flow or are used to call a subprogram. Branching instructions are allowed to change the value of the program counter, i.e., a new address is copied to the program counter and hence the execution begins from the new location whose address is present in the program counter register.

Syntax:

<Instruction> {<cond>} label

<Instruction> {<cond>} label | Rm

<Instruction> {<cond>} Rm

{cond} two-character condition mnemonic for conditional execution. See Table 10.8 for summary on conditional execution

Rm is expressions evaluating valid register number

B	Performs branching operation (like JUMP operation)
BL	Performs branching operation with link (like CALL operation), return address saved in link register (LR)
BX	Performs branch exchange (used to switch between ARM program and Thumb program)
BLX	Performs branch exchange with link, return address saved in link register (LR)

Examples:

```
1. B loop
2. BL loop1
3. BL next_program
   ...
   ...
   next_program
   MOV pc, lr (or MOV r15, r14)      ; return to the main program
4. B goto
   MOV r0, #5
   ADD r0, r1, #4
goto
   SUB r9, r8,#2
   MOV r10, r9
5. goto2
      MOV r0, #5
      ADD r0, r1, #4
   SUB r9, r8,#2
   B goto2
6. BL subprogram
   MOV r0, #5
   ADD r0, r1, #4
   SUB r9, r8,#2
   MOV r10, r9
   .
   .
   .
   subprogram
      LDR r0, [r2]
      STR r0, [r3]
      .
      .
      .
   MOV pc, lr       ; return to the main program by moving pc = lr. This is the last
                      instruction written under the subprogram.
```

10.26 LOAD AND STORE INSTRUCTIONS

The ARM is classified as a load and store architecture and this functionality is fundamental to its operation. The instruction set allows data processing to occur only on the core's registers, so there must be a way to get data in and out of the core from RAM or ROM. There are four load-store instructions that enable you to accomplish that very task.

Load-store instructions are used to perform data transfer between memory and processor registers. There are three types of load-store instructions: single register transfer, multiple register transfer and swap.

10.26.1 Single Register Transfer

Syntax:

<Instruction> {<cond>} {B} Rd, index addressing

<Instruction> {<cond>} {B} Rd, mode and index addressing

{cond} two-character condition mnemonic for conditional execution. See Table 10.8 for summary on conditional execution

Rd is expressions evaluating valid register number

LDR	This instruction loads word (32-bit value) into register
STR	This instruction saves word from a register (register to memory)
LDRB	This instruction loads byte into a register
STRB	This instruction saves byte from a register (register to memory)
LDRH	This instruction loads halfword (16-bit value) into a register
STRH	This instruction saves halfword (16-bit value) from a register (register to memory)
LDRSB	This instruction loads signed byte into a register
LDRSH	This instruction loads signed halfword (16-bit value) into a register

Examples:

1. LDR r0, [r2] ; [r2, #0] → r0

   This instruction loads register r0 with the contents of the memory location pointed by the register r1. #0 indicates the offset from the register r2 is zero.
2. STR r0, [r1] ; r0 → [r1, #0]

   This instruction saves or stores the contents of register r0 to the memory location pointed by r1.
3. LDR r0, [r1]

   Assume that

   r0 = 0x00000000

r1 = 0x00001000
mem32[0x00001000] = 0x000000FF
mem32[0x00001002] = 0x000000AA
Then after execution of LDR r0, [r1] gives
r0 = 0x000000FF and r1 = 0x00001000

4. LDR r0, [r1, #2]
   Assume that
   r0 = 0x00000000
   r1 = 0x00001000
   mem32[0x00001000] = 0x000000FF
   mem32[0x00001002] = 0x000000AA
   Then after execution of LDR r0, [r1, #2] gives
   r0 = 0x000000AA and r1 = 0x00001000
5. LDR r0, [r1, #2]!
   Assume that
   r0 = 0x00000000
   r1 = 0x00001000
   mem32[0x00001000] = 0x000000FF
   mem32[0x00001002] = 0x000000AA
   Then after execution of LDR r0, [r1, #2]! gives
   r0 = 0x000000AA and r1 = 0x00001002
6. LDR r0, [r1], #2
   Assume that
   r0 = 0x00000000
   r1 = 0x00001000
   mem32[0x00001000] = 0x000000FF
   mem32[0x00001002] = 0x000000AA
   Then after execution of LDR r0, [r1], #2 gives
   r0 = 0x000000FF and r1 = 0x00001002
7. STR r1, [r2,r4]! ; Store r1 at r2+r4 (both of which are registers) and
   ; write back address to r2.
   STR r1,[r2],r4 ; Store r1 at r2 and write back r2+r4 to r2.

10.26.2 Multiple Register Transfer

Load-store multiple instructions can allow to transfer multiple registers between memory and the processor registers in a single instruction. These instructions allow up to 16 registers to be transferred. Multiple register transfer instructions are more efficient than a single register transfer for moving blocks of data between registers and the memory. Though this does not happen in a single instruction cycle (it takes about 2 cycles per register), it is invoked with a single instruction so there is potentially a great saving in code space. Note that the multiple load/stores must take place on 32-bit words only.

Syntax:

<Instruction> {<cond>} <addressing mode of load store> Rn {!}, <registers> {^}

<Instruction> {<cond>} <FD | ED | FA | EA | IA | IB | DA | DB>> Rn {!}, <registers> {^}

{cond} two-character condition mnemonic for conditional execution. See Table 10.8 for summary on conditional execution

Rn is an expression evaluating a valid register number

<register> is a list of registers and register ranges enclosed in {}

{!} if present requests write back (w=1), otherwise w=0

{^} if present set S bit to load the CPSR along with the program counter (PC), or force transfer of user back when in privileged mode

LDM	This instruction loads multiple registers (from memory to registers)
STM	This instruction saves multiple registers into memory (from register to memory)

Transferring multiple words implies a pointer must be moving through memory adding or subtracting offsets from a base register and indeed this is the case. So to the root of the instruction, a two-letter prefix has to be added. The prefix tells the base register to increment after each load (IA), increment before each load (IB), decrement after each load (DA), or decrement before each load (DB). Load/store multiples are also conditionally executed like all the other ARM instructions with one of the post-fixed condition codes that you have already seen.

Example:

1. LDMIA r0!, {r1 – r3}

   Assume that

   mem32[0x00001000] = 0x00000001

   mem32[0x00001004] = 0x00000002

   mem32[0x00001008] = 0x00000003

   r0 = 0x00001000

   r1 = r2 = r3 = 0x00000000

   So after execution of the instruction, the contents of the registers is as follows:

   r0 = 0x0000100C

   r1 = 0x00000001, r2 = 0x00000002 and r3 = 0x00000003

10.26.3 SWAP Instructions

The SWAP instructions are used to swap the contents of the register with the contents of the memory. Swap instructions behaviour is dependent on whether the memory region is cachable or noncachable. Swap instructions to cachable regions of memory are useful for implementing semaphores or other synchronization primitives in multithreaded uniprocessor software systems. Swap instructions to noncachable memory regions are useful for synchronization between two bus masters in a multi-master bus system. This can be two processors, or one processor and a DMA controller.

Syntax:

<Instruction> {<cond>} Rd. Rm, [Rn]

{cond}	two-character condition mnemonic for conditional execution. See Table 10.8 for summary on conditional execution
Rd, Rm, Rn	are expressions evaluating valid register numbers

SWP	This instruction swaps a word (32-bit number) between memory and a register
SWPB	This instruction swaps a byte between memory and a register

Examples:

1. SWP r0, r1, [r2]

   This instruction loads register r0 with the word (32-bit number) addressed by register r2, and stores r1 at r2.
2. SWPB r2, r3, [r4]

   This instruction loads register r2 with the byte addressed by register r4, and stores the byte present in the register r3 at r4
3. SWPEQ r0, r0, [r1]

   This instruction conditionally swaps the contents of the word addressed by r1 with r0.
4. SWP r0, r1, [r2]

   Assume that

   mem32[0x00001000] = 0x00000008

   r0 = 0x00000000, r1 = 0x00000002 and r2 = 0x00001000

   Then after execution of the instruction the contents of the registers is as follows:

   mem32[0x00001000] = 0x00000002

   r0 = 0x00000008, r1 = 0x00000002 and r2 = 0x00001000

10.27 SOFTWARE INTERRUPT INSTRUCTION

Software interrupt instruction causes a software interrupt exception, which provides a mechanism for applications to call operating system routines (programs).

Syntax:

<Instruction> {<cond>} SWI_Number

{cond} two-character condition mnemonic for conditional execution. See Table 10.8 for summary on conditional execution

When the processor executes SWI instruction, then program counter (PC) register is automatically loaded with the vector address of software interrupt 0x00000008 in the vector table. This instruction also forces the processor mode to SVC, which allows an operating system routine to be called in a privileged mode.

SWI	This instruction allows execution of software interrupt exception

10.28 PROGRAM STATUS REGISTER INSTRUCTIONS

The ARM instruction set provides two instructions MRS and MSR to directly control a program status register (PSR). The MRS and MSR instructions are formed from a subset of the data processing operations and are implemented using the TEQ, TST, CMN and CMP instructions without the S flag set.

These instructions allow access to the CPSR and SPSR registers. The MRS instruction allows the contents of the CPSR or SPSR register to be moved to a general register. The MSR instruction allows the contents of a general register to be moved to the CPSR or SPSR register.

N	Z	C	V		Undefined	I	F	T	MODE
31	30	29	28	27 24	23 8	7	6	5	4 0

Fig. 10.31 *Program status register*

Syntax:

MRS {cond} Rd,<psr>

MSR {cond} <psr>,Rm

MSR {cond} <psrf>,Rm

MSR {cond} <psrf>,<#expression>

{cond} two-character condition mnemonic for conditional execution. See Table 10.7 for summary on conditional execution

Rd, Rm are expressions evaluating valid register numbers

<psr> is CPSR or SPSR

<psrf> is CPSR_flg or SPSR_flg

<#expression> where this is used, the assembler will attempt to generate a shifted immediate 8-bit field to match the expression. If this is impossible, it will give an error.

MRS	This instruction copies program status register (CPSR or SPSR) to a general purpose register
MSR	This instruction copies a general purpose register to a program status register (CPSR or SPSR)

Examples:

1. MRS r0, cpsr ; cpsr [bits 31 – 0] $\rightarrow$ r0 [bits 31 – 0]
2. MSR cpsr_flg, r0 ; r0 [bits 31 - 28] $\rightarrow$ cpsr [bits 31 – 28]
3. MSR cpsr_flg,#0xA0000000 ; 0xA (binary bits 1010) $\rightarrow$ cpsr [bits 31 - 28]
4. This instruction clears flags Z and V and sets N and C.
5. MSR cpsr_flg,#0x50000000 ; 0x5 (binary bits 0101) $\rightarrow$ cpsr [bits 31 – 28]
6. This instruction sets flags Z and V and clear N and C.
7. MRS r1, cpsr ; cpsr [bits 31 – 0] $\rightarrow$ r1 [bits 31 – 0]
8. MRS r1, spsr ; spsr [bits 31 – 0] $\rightarrow$ r1 [bits 31 – 0]

10.29 COPROCESSOR INSTRUCTIONS

Coprocessor instructions are used to extend the instruction set by allowing use of additional instructions supported by coprocessor with the ARM instruction set. This class of instructions is used to tell a coprocessor to perform some internal operation. No result is communicated back to ARM7TDMI-S, and it will not wait for the operation to complete. The coprocessor could contain a queue of such instructions awaiting execution, and their execution can overlap other activities, allowing the coprocessor and ARM7TDMI-S to perform independent tasks in parallel.

Syntax:

Instruction (CDP) {cond} p#,<expression1>,cd,cn,cm{,<expression2>}

Instruction (LDC or STC) {cond}{L} p#,cd,<Address>

Instruction (MRC or MCR){cond} p#,<expression1>,Rd,cn,cm{,<expression2>}

{cond} two-character condition mnemonic for conditional execution. See Table 10.8 for summary on conditional execution

p# the unique number of the required coprocessor

<expression1> evaluated to a constant and placed in the CP Opc field

cd, cn and cm evaluate to the valid coprocessor register numbers CRd, CRn and CRm respectively

<expression2> where present is evaluated to a constant and placed in the CP field

{L} when present perform long transfer (N=1), otherwise perform short transfer (N=0)

Rd is an expression evaluating a valid ARM7TDMI-S register number

CDP	Coprocessor data processing. This instruction performs an operation in a coprocessor
MRC	This instruction moves from coprocessor to ARM7TDMI-S register (L=1)
MCR	This instruction moves from ARM7TDMI-S register to coprocessor (L=0)
LDC	This instruction loads from memory to coprocessor
STC	This instruction stores from coprocessor to memory

Examples:

1. CDP p1,10,c1,c2,c3 ; Request coprocessor 1 to do operation 10 on CR2 and CR3, and put the result in CR1.
1. CDPEQ p2,5,c1,c2,c3,2 ; If Z flag is set request coprocessor 2 to do operation 5 (type 2) on CR2 and CR3, and put the result in CR1.
1. LDC p1,c2,table ; Load c2 of coprocessor 1 from address table, using a PC relative address.
1. STCEQLp2,c3,[r5,#24]! ; Conditionally store c3 of coprocessor 2 into an address 24 bytes up from r5, write this address back to r5, and use long transfer option.

10.30 LOADING CONSTANTS INSTRUCTION

Actually, there is no instruction available in ARM core to perform data transfer of constant value or immediate number to the register, but to aid programming there are two pseudo (dummy) instructions to move a 32-bit value into a register.

Syntax:

LDR Rd, = constant

ADR Rd, label

ADR instruction is used to write a relative address into a register, which will be encoding using a PC relative expression.

LDR	This instruction loads constant value to the register
ADR	This instruction loads address into the register

Examples:

1. LDR r1, =0xAA

   This instruction moves constant AAh to the register r1, alternative to this instruction is

   MOV r1, #0xAA
2. LDR r2, =0xF000FF00

10.31 ARM EXTENSION INSTRUCTIONS

The ARMv5E extensions provide many new instructions, the summary of the extension instruction is given in Table 10.7. One of the most important additions in the extension instruction is the signed multiply accumulate instructions that operate on 16-bit data. These instructions are executed in single cycle on many of the ARMv5E core.

Table 10.7 *Instructions introduced in ARMv5E extensions*

Instruction	Description	Syntax
CLZ	Count leading zero	CLZ {cond} Rd, Rm
QADD	Signed saturated 32-bit add	QADD {cond} Rd, Rm, Rn
QDADD	Singed saturated double 32-bit (64-bit) add	QDADD {cond} Rd, Rm
QSUB	Signed saturated 32-bit subtract	QSUB {cond} Rd, Rm
QDSUB	Singed saturated double 32-bit (64-bit) subtract	QDSUB {cond} Rd, Rm
SMLAxy	Signed multiply and accumulate 32-bit	SMLAxy {cond} Rd, Rm, Rs, Rn
SMLALxy	Signed multiply and accumulate 64-bit	SMLALxy {cond} Rd_Lo, Rd_Hi, Rm, Rs
SMLAWy	Signed multiply and accumulate 32-bit	SMLAWy {cond} Rd, Rm, Rs, Rn
SMULxy	Signed multiply	SMULxy {cond} Rd, Rm, Rs
SMULWy	Signed multiply	SMULWy {cond} Rd, Rm, Rs

Example:

1. CLZ r0, r1

   This instruction counts the number of zeros present in register r1. It will count upto the first 1 detected in the register r1 and counting is done from MSB to LSB.

   Suppose r1 = 0b00000000000000010000000000000000

   So after execution of the instruction CLZ r0, r1, it gives

   r0 = 15

10.32 CONDITIONAL EXECUTION INSTRUCTIONS

In ARM architecture, all instructions are conditionally executed according to the state of the CPSR condition codes and the instruction's condition field. This field (bits 31-28 of CPSR) determines the circumstances under which an instruction is to be executed. If the state of the C, N, Z and V flags fulfils the conditions encoded by the field, the instruction is executed, otherwise it is ignored. By using conditional execution instructions, you can increase performance and code density. There are sixteen possible conditions, the condition field is a two-letter mnemonic appended to the instruction mnemonic of the ARM instructions.

For example, a branch (B in assembly language) becomes BEQ for "Branch if Equal", which means the branch will only be taken if the Z flag is set.

In practice, fifteen different conditions may be used; these are listed in Table 10.8. The sixteenth (1111) is reserved, and must not be used.

In the absence of a suffix, the default mnemonic is AL, or always execute. This means the instruction will always be executed regardless of the CPSR condition codes.

Table 10.8 *Summary of condition code*

Mnemonics	Condition	Meaning	Code
EQ	Zero (Z) flag set	Equal	0000
NE	Zero (Z) flag clear	Not equal	0001
CS	Carry (C) flag set	Unsigned higher or same	0010
CC	Carry (C) flag clear	Unsigned lower	0011
MI	Negative (N) set	Negative	0100
PL	Negative (N) clear	Positive or zero	0101
VS	Overflow (V) set	Overflow	0110
VC	Overflow (V) clear	No overflow	0111
HI	Carry (C) flag set and zero (Z) flag clear	Unsigned higher	1000
LS	Carry (C) flag clear and zero (Z) flag set	Unsigned lower or same	1001
GE	Negative (N) equals overflow (V)	Greater or equal	1010
LT	Negative (N) not equals overflow (V)	Less than	1011
GT	Zero (Z) clear AND (Negative (N) equals overflow (V))	Greater than	1100
LE	Zero (Z) set OR (Negative (N) not equals overflow (V))	Less than or equal	1101
AL	--	Always	1110

Examples:

1. ADDEQ r0, r1, r2 ; r0 = r1+r2 if zero flag is set
2. MOVEQ r1, #0 ; r1=0x0 if zero flag is set
3. BEQ goto ; Jump to the branch address goto if zero flag is ; set
4. BNE goto ; Jump to the branch address goto if zero flag is ; clear

REVIEW QUESTIONS

1. Explain the difference between 8051 microcontroller and PIC microcontroller.
2. Explain the features of PIC microcontroller
3. Draw and explain the architecture of PIC microcontroller.
4. Explain program and data memory of PIC microcontroller.
5. Explain the classification of program memory of PIC microcontroller.
6. Explain the register bank of PIC microcontroller.
7. Explain the advantages of PIC microcontroller.
8. Explain the selection criterion of PIC microcontroller.
9. Draw and explain the architecture of 8096 microcontroller.
10. Explain the features of 8096 microcontroller.
11. What is watchdog timer. Explain in detail.
12. Draw and explain PSW of 8096 microcontroller.
13. Explain memory mapping of 8096 microcontroller.
14. Explain the significance of general-purpose input output ports of 8096 microcontroller.
15. Write short notes on high speed input and high speed output of 8096 microcontroller.
16. Draw and explain window selection register in detail.
17. Write a short note on special function registers of 8096 microcontroller.
18. What is an addressing mode? Explain the addressing modes of 8096 microcontroller.
19. Explain the register bank of the ARM core.
20. Assume the following data:

Address	Contents
0x20	0x05
0x21	0xF0
0x22	0x03
0x23	0xFA

    If r0 has the value 0x20, what is the content of r5 after executing the following instruction?

    LDRB r5, [r0], #2
21. Write short notes on load-store instructions of ARM core, also explain what is the significance of "!" in the load-store instructions.
22. Draw and explain program status register of the ARM core.
23. Write a short note on the coprocessor used with the ARM core.
24. What is an instruction pipeline. Explain how the speed of operation is increased using instruction pipeline.

25. Draw and explain the internal working of the 5-stage instruction pipeline.
26. What is exception? Explain ARM core exceptions in detail.
27. Outline the difference between the interrupt and the exception.
28. What is barrel shifter? Explain with suitable example.
29. Explain the operating modes of the ARM core.
30. Draw and explain the block diagram of the ARM core.
31. Write short notes on implementation and architectural licensees.
32. Explain the features of the ARM core.
33. Write a short note on the history and versions of the ARM core.
34. Write a short note on the data sizes of the ARM core.
35. Explain with suitable example, why the ARM processor is so popular in the embedded systems.
36. Explain why ARM processor is best suited for embedded applications.
37. Explain what are the characteristics required for development of the embedded application and what are the characteristics suited in the ARM core for the same.
38. Explain all the conditional prefix instructions.
39. Explain the following instructions of the ARM core.
    (*a*) ADD r0, #0x25
    (*b*) MOV r0, r1, LSL #4
    (*c*) SWI
    (*d*) LDR r0, [r1+4]
    (*e*) MSR cpsr_flg ,#0x50000000
    (*f*) MRS r2, cpsr
    (*g*) BL
    (*h*) BNE label
    (*i*) MOVEQ r1, #0
    (*j*) CLZ r0, r1
    (*k*) RSB r0, r1, r2
    (*l*) LDC p1,c2,table

Chapter 11

Applications of Single Board Computer

In today's hectic world of competition, the indispensable tool which can't be avoided is a computer. As we all know with the advent of technology in the silicon area and the decreasing cost of integrated technology, people are moving into the usage of processors which exceptionally give performance of real time control, monitoring and protection in a cost effective manner thus proving their worthiness in the area of real-time operating system (RTOS).

11.1 INTRODUCTION

The RTOS may be new to you but it is a multitasking operating system intended for multiple real-time applications. It is mainly used in embedded applications. It facilitates the creation of a real-time system. It is a tool for the software developer which provides a layer abstraction between the hardware and the software. RTOS will continue the evolution of microcontroller design. In simple terms, if the OS responds to the amount of work quickly in a given period of time, then the OS can be valued to a real-time operating system. A system is said to be real-time if it is required to complete its work and deliver its services on time. The most practical example that we can see in today's world is the flight control system and why not because all the tasks in that system must execute on time so as to avoid cases of flight crashing while landing, takeoff as life is the most valuable of all. It is a hard real-time system meaning failure to meet the deadlines is fatal. So, it is of prime importance to execute the work on time. In a hard real-time system, validation is done by probably correct procedures or extensive simulation that the system always meets the timing constraints. Unlike hard real-time system, in the soft real-time system the late completion of jobs is undesirable but not fatal. RTOS is totally a different subject of which its concepts are not covered in this book but this chapter will give you a list of porting the RTOS into the processor and the usage in the practical world. The processor choice varies from application to application. Since we are now aware about processors and controllers, it is the correct time to introduce the term single board computer (SBC). SBC is a combination of single circuit board with microprocessor, memory IO devices which not only reduce the number of circuit boards thus cost but also eliminate bus driver circuits, connectors, etc., depending on several parameters each discussed below. SBCs are most commonly used for process control in the rack mount format or for control and interfacing on embedding with the other devices. SBCs can be connected with the analog input/analog output cards, digital input/digital output cards (relay cards) through a FRC cable along with the power supply thus making it functional for control, monitoring and protection applications mounted on the rack. As we all know, motherboards are made by millions of vendors for the consumers and can be easily replaced if damaged at low cost whereas the SBC requires an expertise market thus increase in the cost. So, proper care and precautions must be taken while handling the SBCs. As compared to the multiboard computers, SBCs are lighter in weight and power efficient because of high levels of integration and reduced component and connector counts. SBCs can be booted either from the hard disk or through the compact flash disk having space little more than the space of the OS installed. The user can even synchronize the bus voltages and currents to a standard time reference of one pulse-per-second (1 PPS), conforming to IEEE C 37.118 and IEC61850-90-5 standards. These measurements taken at high speed of 10/25/50 observations per second are time-stamped according to a common time reference provided by global positioning system (GPS). There are various parameters that determine the choice of the SBC as discussed in the sections ahead.

11.2 CLOCK FREQUENCY

Clock frequency is the rated speed of the internal logic elements and is the frequency at which chips like CPU or a core of multi-core processor is running which is generally measured in hertz. It can be in kilo (K), mega (M) or in giga (G) Hz. The speed of the very first microprocessor is 375 kHz. Practically, if the processor's clock is faster, the speed of the calculations and the speed of the processing of functions will be faster. You might be thinking sometimes how important is the frequency of core in any processor. The speed of the CPU generally depends on two things. First is the instruction set which will tell you in a single cycle how much amount of the work can be done. Last but not the least will be the clock frequency, i.e., how many cycles are performed per second. The product of the two will give you the perceivable speed. So, if the user scales down a 4GHz processor into 1GHz, the perceivable speed will get reduced by a quarter. If you put your mind more into it you would want to know why such high frequency cores are needed and why not go for low frequency cores as they use too much power. Well if we copy a file from one HDD say Transcend into the Seagate HDD, the hard drives would not be able to keep up with the processor anyways as the user will gain least benefit from the fast processor. The above statement depicts that if the CPU frequency is faster, then the actual benefits can be met for intensive applications like games, file compressors, etc. which will perform better as compared to the slower frequency CPU. That's why modern laptops scale the CPU frequency. The CPU runs at a slow frequency when the fast CPU is not needed thus consuming less power. The CPU runs at the highest possible frequency, when the fast CPU is needed thus making the computer more responsive. So, many applications require a higher clock speed. So, we can come to the conclusion that the clock speed and the frequency are inversely related because the CPU is waiting for the data so as to take/give data. Hence, if the CPU is working and waiting for the bus, then it is consuming more power and gaining little benefit. Let's solve some simple problems in order to brush up the mind.

Example 11.1

If the user is having the clock multiplier of 2 with a 1 GHz bus, then what is the clock frequency?

Solution: Clock frequency = Clock multiplier * Processor's bus frequency.

Given: Clock multiplier = 2 and the processor's bus frequency is 1GHz. So, the clock frequency is the product of the 2.e. 2*1 = 2 GHz. Thus, the clock frequency with 1 GHz bus having a multiplier of 2 will have the clock frequency of 2 GHz.

Example 11.2

If the user is having the clock multiplier of 3 with a 2 GHz CPU, then what is the processor's bus frequency?

Solution: As explained earlier the formula for calculating the clock bus frequency, the processor's bus frequency will be 2/3 =.666 GHz or 666 MHz.

Note: The quoted CPU speed is the speed at which the *internal* clock runs. This is faster than the external clock (the FSB (front side bus) clock) which determines how fast the resources can be accessed by the CPU on the motherboard. The CPU can process faster if it is a fast internal clock, but a slower FSB means it takes a long time to access memory.

11.3 CACHE

One of the most significant events in the history of computing is the caches and the caching. The caches are used from the ultra-low power chips to the highest Intel end core. Earlier the CPUs were not fast due to the slow pace of the main memory. In order to increase the memory access time, a new type of fast memory was needed. So the concept of cache came which stores the copies of the data from the main memory locations. The cache's dependency varies from CPU to CPU. L1 and L2 cache levels are organized as data cache levels. Have you ever come across a notion that if any read/write operation is done by the processor first from the cache (just to check whether the data is present in the cache) instead of the main memory, will the faster operation of read/write happen. Well the answer is yes. There are three independent caches:

1. Instruction cache: It's a catalyst for executing fetching instruction.
2. Data cache: It's a catalyst for speeding up fetching of the data.
3. Translation look aside buffer: It's a catalyst for both data and executable instructions for speeding up virtual to physical address translation.

During the operation of read/write at a particular location in the main memory, the cache memory is first checked for the contents of the memory location in the cache lines hoping for the presence of the address. The CPU if immediately reads or writes the data in the cache lines then the processor makes the **Cache hit.** There are chances that the processor may not find the memory location looking in the cache, hence the processor makes a **Cache miss**. This is where L2 cache comes into play. The L2 cache is on the same processor chip and uses the same die as the CPU, however, it is still not a part of the core of the CPU. Access proportion of hitting the cache hit is the hit rate. A computer can have different levels of cache memory. The level number will indicate the distance from the CPU. The cache closest to CPU is faster and stores less number of data than the other levels of cache. There are different levels of cache:

(*a*) L1 cache is the fastest cache and it usually comes within the processor chip. The L1 cache typically ranges in size from 8KB to 64KB and uses the high-speed SRAM (static RAM) instead of the slower and cheaper DRAM (dynamic RAM) used for main memory. It is the first level of memory the processor speeds then comes the other levels.

(*b*) L2 caches come between L1 and RAM (processor-L1-L2-RAM) and are bigger than the primary cache (typically 64 KB to 4 MB).

(*c*) L3 cache is not found nowadays as its function has been replaced by L2 cache. L3 caches are found on the motherboard rather than the processor. It is kept between RAM and L2 cache.

So if your system has L1, L2 and L3 cache, data fetching will be

$$L1 \longrightarrow L2 \longrightarrow L3 \longrightarrow RAM$$

i.e., if the data is not there in L1 it will check L2 then L3 then RAM.

Note: In the case of multicore processors, each core shares a common L3 cache but has its own dedicated L1 and L2 cache.

11.4 SYSTEM CHIPSET

A chipset is a combination of electronic components in an integrated circuit which is designed to perform one or more related functions. It plays a vital role in determining the performance of the system. The communication between the processor and the external devices is controlled by the chipset. We can view the chipset in the heart of the CPU, i.e., the motherboard. As already stated it is designed to perform various functions like CPU, functions of a computer, modem functions, etc.

Newer chipset is being replaced with two or more physical chips. The basic function of a chipset is the communication between the components of a modern PC. The chipset makes various interfaces for having connections with the additional components like USB, IDE, PCI-E, etc. Due to the difficulty of integration of all the components into a single chip, it is made of two units which allow different combinations of Northbridge and Southbridge chips. The high speed channels such as communication with the RAM and cache is controlled by Northbridge part whereas the Southbridge part manages relatively slower devices such as USB, etc. The only company that manufactures its own chipsets is the Intel. You can view the list of Intel chipsets in the above link http://en.wikipedia.org/wiki/List_of_Intel_chipsets.

11.5 BIOS

Bios stands for basic input/output system. On all the motherboards and SBCs, there is a chip containing instructions and setup of how the system will operate and should boot. On each computer, all the devices like ports, CPU, hard drives functions together. There is a code in the BIOS controlling the keyboard through PS2 or USB, ports for serial or ethernet, drives whether want to run the operating system through flash or hard disk and number of miscellaneous functions. Bios will come up with every desktop or laptop devices. On booting up of the SBC, the Bios will first determine whether all the attachments are perfectly connected or not and are in operation, then it will load the operating system into the RAM from the hard disk or through the flash. The sizes of the drives are independent and may vary depending on their storage. The Bios will always be found in the ROM chip ensuring its availability all the time irrespective of the disk failures. Bios are copied from ROM to RAM each time on booting of computer since RAM is faster than ROM. This phenomenon is known as **shadowing**. The Bios supports services such as RTC (real-time clock) date and time. Bios sets up interrupt vectors in the main memory for accessing the routines and then loading the operating system thus passing control to it. PCs having flash Bios means Bios has been recorded on a flash memory chip which can be updated whenever required. The microprocessor gets the instructions from the Bios on startup of the computer since it cannot get from the operating system since it is located onto the hard disk and the processor cannot get it without some instructions which tell it how. The programs and the operating system don't have to know the hardware addresses about the input-output devices attached to the PC. Bios loads basic computer hardware and also includes a power on self-test (POST) verifying the computer to meet with the requirements so as to boot up properly. We know that if there is any malfunctioning with the computer it is recommended to run the POST test first so that depending upon the combination of beeps it can be easily known what is malfunctioning within the computer. Functions of BIOS are as follows:

(*a*) For custom settings the CMOS setup is checked. The user can configure hardware settings from the configuration program which includes settings such as date, time, passwords, etc.

(*b*) Will initiate the bootstrap sequence and locate the operating system and will pass control to it.

(*c*) Bios drivers will give the computer basic operational control over computer's hardware.

(*d*) It initializes the registers and manages the power.

(*e*) POST is done for testing the hardware to make sure no error exists before loading the operating system.

(*f*) Loading of the interrupt handlers.

To access the BIOS hold down the key as soon as the computer begins to start up. For example, for Hercules II SBC board press F2. The above function key may vary from SBC to SBC. The beep code of Dell machines is as follows (Table 11.1).

Table 11.1 *Dell BIOS beep codes*

Beep Codes	Possible Causes	Corrective Action
1 – 2	No video card detected	Reseat the video card
1 – 2 – 2 – 3	BIOS ROM checksum error	---------
1 – 3 – 1 – 1	DRAM refresh error	Reseat the memory modules
1 – 3 – 1 – 3	8742 keyboard controller error	Reseat the keyboard connector
1 – 3 – 3 – 1	Memory defective or not present	Reseat the memory modules
1 – 3 – 4 – 1	RAM failure on line xxx	Reseat the memory modules
1 – 3 – 4 – 3	RAM failure on data bits xxx of low byte on memory bus	Reseat the memory modules
1 – 4 – 1 – 1	RAM failure on data bits xxx of high byte on memory bus	Reseat the memory modules

Note: *This is only for reference. For other laptops you can Google on internet.*

11.6 SYSTEM MEMORY

We should know what memory is, what are the different types of memories, what are the differences between the memories and what is their hierarchy. There are internal storage areas in the computer. Memories are the actual chip which holds the data. It is the holding place for the data and the instructions where the processor can reach quickly. Memory plays a vital role in the different aspects of the computer system:

(*a*) **Reliability and stability:** System performance can be poor due to bad memory. So, it is important to have the memory of high quality so as to exhibit fewer problems. But make sure the memory should not be even high-quality if using the wrong one.

(*b*) **Software Support:** More memory is required for the newer programs than the older ones as it will give access to programs which we cannot use with the lesser amount.

(*c*) **Upgradability:** There are some universal memories which can be migrated to future systems or can be used continuously if the motherboard is upgraded.

(*d*) **Performance:** Insufficient memory can cause the processor to work less than half of its full capacity which is far below its performance potential. Thus, the type and the amount of memory is an important contributing factor to the overall performance.

Every computer has main memory or RAM (random access memory) which is a certain amount of physical memory. The RAM can be thought of as an array of boxes each storing one byte of information. A computer that has 1GB of memory can hold up to 1GB of information. In RAM, the information is accessed randomly unlike CDs or hard drives where the information is accessed sequentially, thus making data access faster by the computer. Whenever a program such as any internet browser is open, it is first placed into the RAM from the hard drive thus allowing the programs to communicate with the processor at higher speeds.

Deciding factors for the RAM

- The computer user.
- Proper planning of the programs to run on the computer.
- The number of programs opened at a given time.
- System requirements of the programs running on the computer.
- The memory requirements of the programs used most.

General Recommendations of the RAM

- Minimum: 1GB
- Recommended: 4-6GB
- Excellent: 8GB or more.

There are two kinds of memory: primary memory and secondary memory. Primary memory is the main memory accessible by the CPU. RAM is the best example. Secondary memory is located outside the computers. Examples include magnetic disk, floppy disk. Primary memory is more expensive, thus less in size as compared to the secondary memory. Unlike ROM, RAM is volatile that is when the power is cutoff, all the data contained in the memory is lost. So, the memory which loses the data when the power supply is lost is called volatile memory. We can both write data into RAM and read data from the RAM.

Difference between RAM and ROM

Table 11.2 *Differences between RAM and ROM*

Features	RAM	ROM
Full Form	Random access memory	Read Only Memory
Work	Information present in the RAM can be both read and write operated.	Information present in the ROM can only be read operated.
Definition	Data storage form that can be accessed from any memory location at any time allowing quick accessibility.	Data storage form once program cannot be altered.

Volatility	Contents are lost once the device is off.	Contents are retained even if the device is off.
Storage	Used to store temporary information.	Used to store permanent operation.
Speed	Speed is faster as it can assist the processor to boost up the speed.	Speed is slower since it cannot assist the processor to boost up the speed.
Storage of information	For storage of information, electricity is needed.	Electricity is not needed to preserve the information.
Size of chip	Larger than ROM.	Smaller than RAM.
Cost	High	Low
Types	SRAM (Static RAM) and DRAM (Dynamic RAM).	PROM (programmable read only memory), EPROM (erasable PROM), EEPROM (electronically erasable PROM).

Registers: The registers are needed to transfer the processed speed at high speed since the CPU processes data and instructions at high speed. Here the data or the information is stored temporarily and is passed on as directed by the control unit.

The memory hierarchy of different contents is dependent upon the speed and the size. It is shown in Table 11.3.

Table 11.3 *Memory hierarchy*

Content	Size	Speed
Registers	<1KB	5ns
Cache Memory	<4MB	10ns
Main Memory	<4GB	100ns
Disk Storage	>1GB	5ms

11.7 POWER MANAGEMENT

The power management is a critical factor because it keeps on monitoring the system temperature and provides alerts if exceeded unexpectedly. Advanced power management (APM) facility provides for power management. APM is an application program interface allowing the power management to be included into the BIOS by the computer and the BIOS manufacturers. APM was developed by Microsoft and Intel. The power usage of a system based on its activity is controlled by APM. APM will reduce the power to unused system resources as the system activity decreases until the system is brought into a suspended state. It was difficult to manage the thermal properties and the power usage by the operating system. BIOS managed the hardware and the user was having limited configurability into the power system management settings. The APM Bios is specific to the hardware platform and is supplied by the vendor. We all know that if the desktop or laptop is idle and not performing any application, the system is in low power sleep state instead of shutting down so that the system can be resumed faster on pressing any button or hovering of the mouse or any physical contact with the CPU/keyboard thus ensuring that the current work session is over. The power management event includes standby, suspend requests, resume requests, OEM events, low battery notifications. These events can be sent to the operating system from the APM Bios. There are two different power states: system power states and the device power states. The system power states include Full On, APM Enabled, APM Standby, APM Suspend, and off. All these system states have differences in the latency, i.e., the system to come up into the working state. The small explanations of the system power states are given below:

1. **Full On:** Whenever the power management is not done by the system, the default mode is the Full On Mode.
2. **APM Enabled:** The work is not being done by the system thus un-powering the unused devices.
3. **APM Standby:** After a **short** period of inactivity, the system enters into the APM standby state. The recovery from the APM Standby state to the APM Enabled state is instantaneous.
4. **APM Suspend:** After a **long** period of inactivity, the system enters into the APM Suspend state. The recovery from the APM Suspend state to the APM Enabled state takes a long time.
5. **Off:** Off state is the state when the computer is turned off.

Device Power States

Either by the system Bios or by the operating system, the devices may be power managed. The operating system takes the power management of the device by telling the Bios to stop power managing a particular device using the disabled device power management function. With the help of APM Bios or APM Driver using APM function calls devices are controlled in the background.

APM Device Mode	Features
Device on	All device features are available as the device is fully powered and able to perform work.
Device power managed	Some features work at low power level while the device is working. Here the power is maintained and the device is having variable number of device power sub-modes. Power and operational parameters are retained.
Device low power	Here the device is not working and the devices may have variable number of power sub-modes. Power and operational parameters are retained.
Device off	Here the device is not working. Power is off and operational parameters are not retained.

The successor to APM is the advanced configuration and power interface is the standard written by the vendors namely Intel, Microsoft and Toshiba and later joined by HP and Phoenix to provide the interface for the hardware resources and power management. It provides more control and flexibility to the operating system.

11.8 SOLID STATE DRIVE (SSD)

Solid state drive is a data storage device that stores the data permanently using integrated circuit assemblies as memory. SSD can be thought of an oversized version of the humble USB memory stick. The information is stored in microchips. A typical SSD is called NAND based flash gate memory. The above memory is non-volatile as even if the power is lost the information is retained. There is no mechanical arm to read and write data in SSD as it relies on controller to perform operations related to read and write. The above controller will decide the speed of the SSD. The speed of the drive depends on the following factors:

- How to store the data
- How to retrieve the data
- How to cache the data
- How to clean up the data.

The form factor of the SSD comes in the standard sizes of 1.8″, 2.5″ or 3.5″ which can be fitted into the connectors like SATA, m-SATA. The bottom view figure of a SSD is shown in Figure 11.1.

Fig. 11.1 *SSD bottom view*

In SSD, the power drawn is less averaging 2-3 watts resulting in more than 30 minutes battery boost. It is more expensive than the hard disk drive (HDD). Its capacity is not larger than 1TB. Its booting time is less and has no moving parts. When it is running, it produces no vibration. Since it is drawing less power, heat produced is less. Here, the file opening speed is faster and is safe from any effects of magnetism. The first SSD was developed in 1970s in the IBM in supercomputers. But due to the greater cost per GB of storage information, these drives will not be able to completely replace the hard drives. Thus, SSD is the choice by the user under two suitable conditions:

- If there is a requirement for faster operation,
- Based on its usage, if we can accept that it has limited storage and can work around that.

Unlike SDD, there is HDD. The HDD was first introduced by IBM in 1956. It will store data on a rotating platter using the concept of magnetism. The faster the platter spins, the faster the HDD can perform. Here, the data can be stored largely and at cheap rate. HDD's generally use IDE or SATA interface and the power supply to charge it. HDD is available in the size of 2.5″ Form factor or 3.5″ form factor for desktop applications. The bottom view of the HDD is shown in Figure 11.2.

Fig. 11.2 *HDD (Bottom side)*

Here are some aspects about HDD:

- The power drawn averaging 6-7 watts is more and hence uses more battery.
- HDD is cheaper and costs around $.06 per gigabyte.
- Its capacity is more than the SDD and is exceeded to more than 4TB.
- The boot up time is more and averages 30-40 sec.
- The spinning can be felt during boot up and is clearly audible to human ear.
- Plattering spinning can be resulted in the vibration.
- A measurable amount of heat is produced during spinning.
- The file opening speed is slower.
- The data can be erased though magnets, thus magnetism affects the data storage.

HDD is the best choice of the user under the following requirements:

- If the storage capacity is large in TBs then the best choice is HDD only.
- If the speed is not in the table of how fast the program is responding.
- If the money spent appears economical.

Some Important Terms

- **Form Factor:** The physical arrangement of the computer hardware object is called form factor. This term is commonly heard in describing the size of the SBC or the computer case etc.
- **Sata:** It stands for serial ATA and is an IDE standard which connects devices like hard drives, optical drives to the SBCs. It refers to the types of connections and the cables that follow these standards. The main connection types for SSD and HDD drives is Sata.

Fig. 11.3 *SATA Cable*

- **m-Sata:** m-Sata is the Mini Sata which a low profile interface connector enables more effective Sata integration with small Form Factor. m-Sata enables compact integration in a wide variety of applications for the hard drives and the solid state drives which have data transfer rate of 1.5 Gb/s and 3Gb/s for high performance. The SATA-IO released the m-Sata specification in the year 2009. The available manufacturers include Intel, Toshiba, Samsung, HP etc. m-Sata Flash disk has features like:
  - ✓ It is a compatible interface to Sata.
  - ✓ Comes as MLC's and SLC's. SLC or single-layered cells are used when the cell endurance is higher, speed required is faster and the power consumption is low. Ex: FDMM-16G-XT, etc. Whereas MLC is the multi-layered cell used when the cost is less and the storage density is more, e.g., FDMS-16G-XT, etc.

✓ It can burst read/write up to 300MBps.
✓ It is SMART device, i.e., Self Monitored Analysis and Reporting Technology.
✓ It has low power consumption.
✓ It can operate between -40 to 85 degree Celsius.
✓ It is RoHS compliant, i.e., Restriction of Hazardous Substance.

11.9 WATCHDOG TIMER

A watchdog timer is a simple operation hardware timing device triggering the system reset after elapsing a designated amount of time. A watchdog timer can be built into the processor itself or can be a standalone hardware component. It is not possible to cope up with all the software and hardware anomalies by the developer. So, the developer can mitigate the risks by employing watchdog timer. For the embedded systems to be robust, it should be intelligent enough to cope up with both software and hardware anomalies. In most of the scenarios, embedded systems are not accessible to operator and operate in total isolation. Whenever the software hangs manually, resetting the software is not possible always. In such cases, this can result in damaged hardware and incur significant cost impact. If the humans are unable to handle the faults in a timely manner, then the CPU cannot depend on humans to reboot. It must be self-reliable and intelligent enough to do the reset operation.

To avoid reset, the application must "kick the dog", i.e., the application must periodically reset the watchdog timer before the interval elapses. There are two types of watchdog timers:

11.9.1 External Watchdog Timer

These timers are ICs which will physically assert the reset pin of the processor. An output pin must be asserted by the processor in some fashion to reset the timing mechanism of the watchdog. This is the most appropriate watchdog because of the complete dependence by the processor. Some external watchdog timers come up with the "Windowed" state. It will minimize the likelihood of errant software resetting the watchdog.

11.9.2 Internal Watchdog Timer

Many microcontrollers and processors have built-in watchdog circuitry available to the programmer. A memory-mapped counter will trigger a non-mask able interrupt or will reset when the counter reaches a pre-defined value. Instead of issuing a reset via an I/O assertion, an internal counter of reset of an initial value is done. Watchdog configuration is controlled user software. Watchdog may even be used as a general-purpose timer in some cases. But there are some problems in the internal watchdog timers as compared to the external watchdog timers as they are not "safe". The problems with the internal timers are as follows:

- Watchdog issuing a non-maskable interrupt instead of a reset may not properly re-initialize the system.
- Watchdog control registers may be inadvertently overwritten by runaway code thus disabling the watchdog altogether.
- There are no outside peripherals as reset is limited to the processor itself.
- Most built-in watchdogs have extra safety-steps designed to prohibit errant code from interfering with the operation of the watchdog timer in order to circumvent the issues.

Thus the watchdog timers can be of great deal of reliability to embedded systems if used properly. The overall design must focus on resetting the watchdog timer. The operation integrity of the system must be verified and used as a criterion for resetting the watchdog timer. Validation of the software must be done in the time expected. Enough debugging information must be added to help the debug situation whenever the experience handled by the software is through hardware malfunction or software fault.

11.10 EXPANSION INTERFACE

Before explaining something about the expansion interface, it is important to know a few terms.

1. **Bus:** A bus is a collection of wires transmitted from one computer part to another. In computer terms, it is the internal bus connecting the computer components to the CPU and the main memory. There are two types of buses: Data bus which transfers the data and the address bus which will transfer the information of the data location. How much data can be transmitted at a time is the size of the bus or actually the width. The 16-bit data can be transmitted through 16-bit bus, the 32-bit data can be transmitted through 32-bit bus and so on. The applications can run faster if the data transferred is fast. The old ISA bus is replaced by the PCI bus on personal computers.
2. **ISA bus:** The term ISA stands for industry standard architecture. It is a type of computer bus allowing the expansion cards to be connected to the PCs motherboard. Suppose we want to add a video card, an extra serial port or the network card, the ISA slot can be used. For IBM's first generation PC, the Intel's 8088 microprocessor was supported by the ISA bus. With the CPU clock, the synchronicity was included by the ISA bus. Using multiple expansion cards on a memory channel, the ISA bus provides direct memory access allowing separate interrupt request transactions for each card. ISA remains the most common expansion bus for most of the 1980s and 1990s. The ISA bus also supports IBM AT with Intel 80286 processor and extended ISA. The ISA bus was replaced by the PCI and the AGP slots by the end of the twentieth century. Nowadays, a computer only supports PCI and AGP expansion cards.
3. **Expansion bus:** An expansion bus is a collection of wires allowing computer expansion by inserting PCBs. As we have discussed, an expansion bus called the ISA bus has been utilized by the PCs.

4. **Peripheral Component Interconnect Bus:** A PCI Bus is a bus used to attach the hardware to the computer. It was introduced by Intel in 1992. PCI follows the PnP specification. The PnP is the Plug and Play where without the physical interference of the user configures the hardware device with jumpers; the computer detects and configures a piece of hardware automatically. Peripheral devices have their own memory. The general standard for a PC expansion bus is the PCI local bus. It has replaced the ISA bus and the Video Electronics Standard Association bus. The Peripheral Component Interconnect Special Interest Group standardizes the PCI specifications. The operating system searches for all the PCI buses during the system start up so as to attain the information about the resources which are needed by each device. A system resource assigned by the operating system includes interrupt requests, memory and allotting input/output space. The main features of the PCI bus are:
   - ✓ Support ISA/EISA/MCA (micro-channel architecture) bus.
   - ✓ Can configure through software and registers.
   - ✓ Specification is processor independent.
   - ✓ Support of 5V and 3.3V power supplies.
   - ✓ Support of Multi-processor systems.
   - ✓ The 32-bit standard bus width has a transfer rate of 133 Mbps.
5. **PCI-Express:** It is a superset of PCI and ISA bus. It was previously known as the "3GIO". The base transfer rate is 2.5 Gbps and performs serial data transfers. Here, the system bus throughput is high and I/O pin count is less.

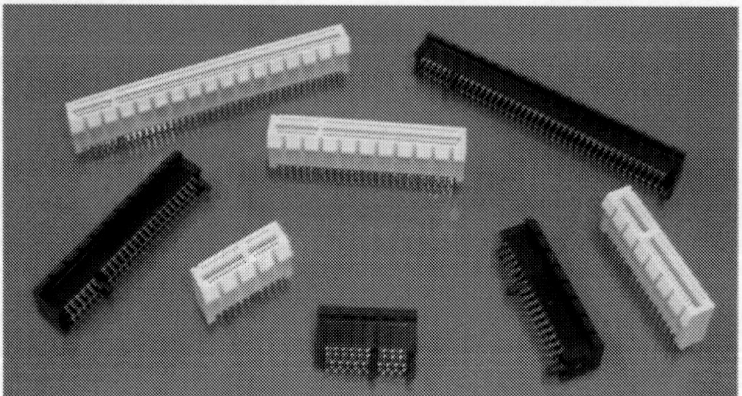

Fig. 11.4 *PCI-express connectors*

Here the data transferred in packets is bidirectional such that the data can flow to and fro from the device simultaneously. For 8-bit parallel streams the parallelized serial data transfers are referred to as "PCI-Express x8". For 16-bit parallel streams the parallelized serial data transfers are referred to as "PCI-Express x16". The PCI-e Product is available in 4 sizes:

- 1x 36 lines
- 4x 64 lines
- 8x 98 lines
- 16x 164 lines

11.10.1 Comparison Between PCI and PCI-e

1. PCI is slower than the PCI-e if speed is the parameter.
2. The interface in PCI is parallel whereas PCI-e has serial interface.
3. Each PCI-e lane is capable of delivering up to 1Gbps.
4. The number of slots in PCI are standardized whereas in PCI-e it varies depending on the number of lanes.
5. Most manufacturers still use PCI rather than PCI-e due to the simplicity and for most of the devices speed is not the issue.

The expansion interface supports PC-104, PC-104 plus, PCI slot x1, mini-PCI-e, etc.

11.10.2 PC-104

The desktop ISA and PCI bus repackaged and hardened for industrial use is the PC-104. Defined by its stacking bus and compact footprint structure, PC-104 is an embedded computer standard which is a ruggedized and modular version of the PC. The sizes of PC-104 modules are small, 3.6 × 3.8 inches. PC-104 modules mate together through stackable PCI, ISA and PCI-e connectors. Due to the following mainstream PC bus development, the PC-104 leverages the hardware and software markets. It minimizes the development costs. Due to the universally completely defined pin outs, interchangeability and interoperability is enabled. (Interchangeability comes into picture whenever two or more items are similar in function and physical characteristics considering equivalence in durability and in performance. Each can be replaced with the other without causing the need for adjustment to fulfill the same requirement. On the other hand, interoperability is the ability of the computer system to run application program from different vendors and interact with the other systems across LAN or WAN independent of the operating system and the physical architecture. It conforms to open standards such as the internet feasibility through hardware and software components.) Users can choose a wide array of PC/104 modules to tailor a system fitting into their project requirements. A standard PC/104 architecture will include the ISA bus connector at the bottom edge of the board. The total number of pins in the connector is 104, thus comprising its name as PC/104. PC 104 P1 bus has 64 pins just like the PC-XT and is combined with 40 pins on P2 with full AT compatibility. Thus, the sum of 64 and 40 is 104 pins. One of the important features of PC/104 is the reliability of pin and socket connector. Each connector has been designed in such a way that more than 1 module can be stacked. PC 104 boards can be stacked one on top of another to form a system with larger I/O requirements. A cross platform technology of how an 8-bit PC/104 module, a 16-bit PC/104 module, and two 32-bit PC/104-*Plus* expansion modules can plug together in the same stack is shown in Figure 11.5. How a standard PC-104 module looks is shown in Figure 11.6

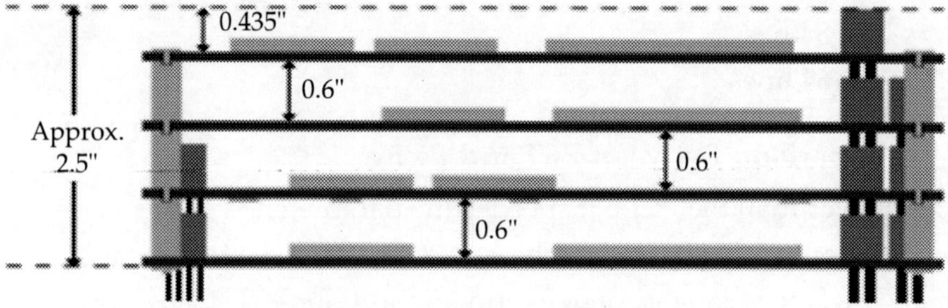

Fig. 11.5 *Module PC – 104 Stack*

Fig. 11.6 *PC-104*

The advantages of PC/104 modules are:

1. Stackable.
2. Compact.
3. Rugged.
4. Inter-operable.
5. Inter-changeable.
6. Based on PC technology.

11.10.3 PC/104 Plus

PCI/104 plus is the PCI implementation of PC/104 for I/O functions which requires higher data transfer speeds. It is the 120-pin PCI connected located at the top edge of the board for higher bandwidth applications and for high speed calculations. It supports a 32-bit multiplexed address/data path operating at PCI local bus speeds of up to 33 MHz yielding a peak bandwidth of 132 MB/sec. These boards are used for high function data rates available on the PCI bus. PC/104 plus boards retain the PC/104 connector so that another board can be inserted on top of it so as to maintain compatibility with PC/104. How a standard PC-104 plus connector and PCI-104 connector looks is shown in Figures 11.7 and 11.8 respectively.

Fig. 11.7 *PC – 104 Plus*

Fig. 11.8 *PCI 104 connector stacked on top*

Thus, the above connectors are needed based on their requirements and applications wherever applied on different projects. Also see how ISA, PCI and PCI-E connectors look like in Figure 11.9.

Fig. 11.9 *Left: ISA, Middle: PCI and Right: PCI-Express*

Note: PCI-104 connector will retain the PCI connector only thus omitting the ISA connector at the bottom edge. This connector is compatible with a PC/104-Plus or PCI-104 CPU. This connector is not compatible with PC/104 CPU board.

11.11 BATTERY

Every SBC or a PC motherboard has a battery. Battery serves some important purposes of feeding the Real Time Clock of the computer (showing data and time) and feeding the configuration memory. Battery is often removable and a new battery can be inserted but make sure to turn off the computer and use anti-static gloves so as to take care of static electricity. The BIOS is powered by an on-board battery for several reasons.

1. It's not a hard disk drive or normal memory, so if the power cuts off, the BIOS memory will be "flashed" and all the variable data would be lost.
2. It provides a stable and reliable power source, which helps "buffer" the BIOS from any damage due to a power surge.

11.12 INPUT–OUTPUT AND INTERFACES

In a SBC, there are I/O devices which are connected to it. These devices have their own purpose. The different I/O devices connected to the SBC are as follows:

1. **Compact Flash –** Booting of the OS through flash disk.
2. **Sata Connector –** Booting of the OS through hard disk.
3. **IDE Connector –** Booting of the OS through hard disk.
4. **Serial Ports –** For the serial communication with the display panel. Can be either RS-232 or Rs-485 communication.
5. **Ethernet Ports –** For the TCP/IP communication with the PC panel.
6. **VGA Port –** For viewing the BIOS and doing the coding in the OS.
7. **Power Cable –** For giving the supply to the SBC.

8. **USB Port** – For taking the data and also for booting depending on the SBC to SBC. Also can be used for keyboard and mouse inputs.
9. **PS/2** – Used for keyboard and mouse inputs.
10. **Utility Board Pin** – Having the utility for both power reset and the ON/OFF switch.
11. **Analog Connector** – For the analog input and analog output channels. Analog input channels are those which take the inputs from the field. The inputs may be in the form of voltages, currents, active power, reactive power, total power, a standard input of 4-20 mA. Some analog parameters to be recorded must be in the form of 4-20 mA. For example, if the operator wants to measure the SF6 gas pressure filled inside the circuit breaker, he/she has to connect a 4-20 mA sensor mounted into the breaker which must be hardware in loop with the monitoring system relay, thus output in the form of 4-20 mA can be seen in the front panel on the monitoring system relay or to the control room HMI. The other analog parameters which are recorded in the same form are hydraulic pressure, dew point measurement, ambient temperature, local control cabinet temperature, etc. We know that the microprocessor systems work in the digital domain. So, the analog parameters like voltage and current must be digitized before feeding into the relay. The analog input modules are sampled periodically at prefixed sampling intervals which is interrupt controlled. The rms computation will be done from the instantaneous values of the analog inputs after resolving them into sine and cosine phasors; the values will be scaled to a limit and the above values will be stored in the memory. In order to avoid errors in the final computed value, precise transformation of the field inputs needs to be done. From the composite signal, the fundamental component is extracted after subjecting the inputs to the filtering stage due to the non-ideality of filtering action as some amount of harmonics may be present. The filtering stage consists of buffer stage so as to minimize the loading effect, the differential amplifier stage so as to raise the voltage level (amplitude correction), the all-pass filter stage so as to adjust the phase shift. We know that the all-pass filter has the gain of 1 and the phase shift can be varied from 0 to 180 degree. Last stage is the chebyshev 2nd order low pass filter with the sallen and key circuit design with proper choice of resistors and capacitors thus extracting the fundamental component of 50 Hz. The last stage can also give us n*50 harmonics with the proper choice of resistors and capacitors. Analog output is the DAC part. The digital value will be computed and will give us in the form of analog voltage values.
12. **Digital I/O:** There can be *n* number of digital input or digital output channels. There can be 24V or 220V DC inputs which are optically isolated with the help of an opto-isolator and are brought down to the microprocessor reading level of +5V. But the most important point to remember is that "there must be ohmic isolation between the 24V/220V DC in the field and +5V at the microprocessor input". This can be taken while taking the consideration of ground. For digital outputs, commands like trip/close operation is performed. You will get clearer about these inputs and outputs scenario after viewing a short example in the latter half of this chapter. Figure 11.10 shows the interfaces and the cables whereas Figure 11.11 displays the utility board.

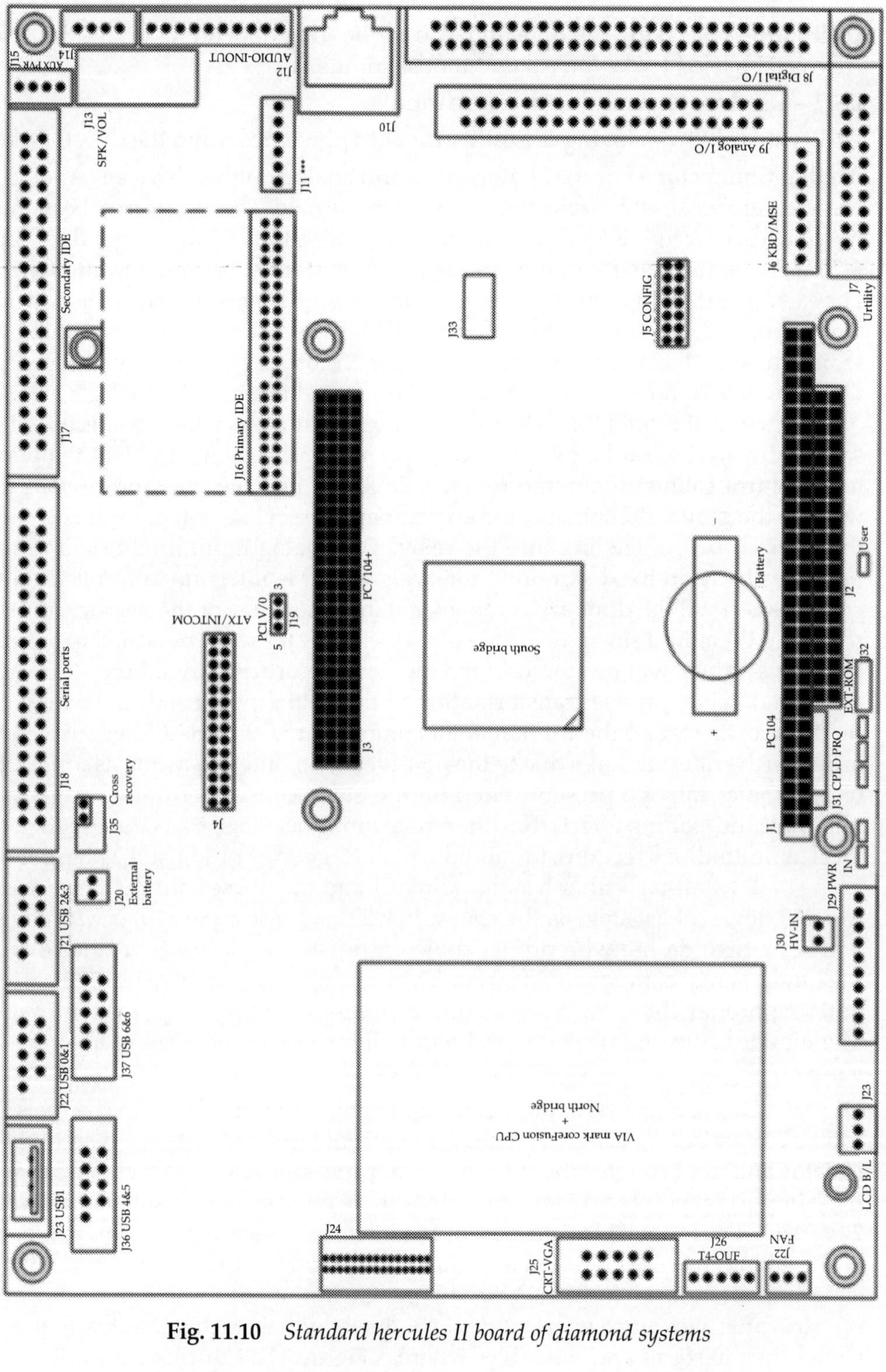

Fig. 11.10 *Standard hercules II board of diamond systems*

Table 11.4 *Pin description of Hercules II Board*

Connector	Description	Manufacturer Part No.
J1	PC/104, ISA bus A,B	EPT 962-60323-12
J2	PC/104, ISA bus C,D	EPT 962-60203-12
J3	PC/104-Plus PCI bus connector	-
J6	PS/2 Mouse and keyboard	Digikey 640456-8
J7	Standard button/LED utility connector	Phyco 2120-20S
J8	Data acquisition digital I/O, 50-pin	Phyco 2120-50S
J9	Data acquisition analog I/O, 40-pin	Phyco 2120-40S
J10	RJ-45 Ethernet	Capsco Sales, Inc. GD-PNS-88
J11	Ethernet header	Digikey 640456-6
J12	Audio I/O	Digikey A1925
J13	Speaker	Standard 2x5, 0.1″ Box header
J14	CD Input	Molex 70543-0003
J15	External auxiliary power (output)	Digikey 640456-4
J16	Primary IDE	All American Semiconductor 2115-2X22GDP/PPTB
J17	Secondary IDE	Phyco 2120-40S (with pin 20 removed)
J18	RS232/RS485/RS422 serial I/O	Phyco 2120-40S
J20	External battery	Digikey A1921
J21	USB 2/3 (USB 1.1)	Standard 2x5, 0.1″ header (with pin 1 removed)
J22	USB 0/1 (USB 1.1)	Standard 2x5, 0.1″ header (with pin 1 removed)
J23	USB1 (USB 1.1)	Capsco Sales, Inc. KUSB-AS-1-N-WHT
J24	LCD panel (LVDS interface)	JST Part No.: BM30B-SRDS-G-TF
J25	VGA	Standard 2x5, 0.1″ Box header
J26	Video/TV out	Digikey 640456-5
J27	CPU fan	Heilind Electronics 89400-0320
J28	LCD backlight	Digikey A19470
J29	Low-voltage power input	Digikey A1925
J30	High-voltage power input (optional)	Digikey A1921
J34	Compact Flash slot (on bottom of board)	-
J36	Non-functional	Standard 2x5, 0.1″ header (with pin 1 removed)

From Figure 11.10, there are various I/O devices, interfaces and connectors as listed in Table 11.4

Fig. 11.11 *Diamond systems utility board having both reset and power switch buttons*

11.13 DISPLAY

Whatever the calculations of the analog and digital parameters for input and output channels that have been done at the application side (SBC side) in any operating system, we will need to get it displayed to the user so that it can communicate with the application side where the interaction between the board and the humans can occur. The best example in today's environment is the mobile where the user gives the command and the application side responds irrespective of the OS. Say, if want to shut down the mobile; the shutdown application pops out thus the user can talk to the mobile. Now, a short example will be demonstrated with Hercules II SBC board of the Diamond Systems in QNX RTOS along with the external Adon Module Connectors of diamond systems mated into the SBC board communicating to the humans RS-232, RS-485 communication media and TCP/IP protocol of analog I/O and digital I/O channels. The Hercules II board contains a 50-pin digital connector consisting of 5 ports and 40-pin analog I/O connector. For more information you can refer to the Hercules II EBX manual from the website http://www.diamondsystems.com/

11.13.1 RS-485 Communication on Hercules Board

The above program will give you a feel of RS-485 communication media using Modbus protocol in QNX Real Time Operating System. Modbus protocol has a standard data format for requesting the data from the slave. The above protocol can be implemented in both Master and Slave Mode. If you are more interested to learn the Modbus protocol, you can learn the Modbus protocol from the Internet. It is an open source protocol and is not proprietary. If the port is in the Master Mode, the Master will query the slave in the above packet format:

1. Slave ID (1 byte)
2. Function Code (1 byte)
3. Starting Address (2 bytes)
4. Count Value (2 bytes) - This indicates the end address. This will be more clear after observing the above example.
5. Last will be the CRC of the first 6 bytes (2 bytes)

11.13.2 Source Code for RS 485 Communication Media using Hercules Board with Modbus Protocol

```
// Header Files required
#include <stdlib.h>
#include <stdio.h>
#include <string.h>
#include <sched.h>
#include <unistd.h>
#include <pthread.h>
#include <err.h>
#include <sys/netmgr.h>
#include <sys/neutrino.h>
#include <math.h>
#include <termios.h>                /* POSIX terminal control definitions */
#include <sys/types.h>
#include <sys/stat.h>
#include <fcntl.h>
#include <sys/mman.h>
#include <time.h>
#include <sys/io.h>
#include <signal.h>
#include <hw/inout.h>
#include "dscud.h"
/* For "devctl()" Header File Required for RTS/CTS handle in QNX Libraries*/
#include <devctl.h>
#include <sys/dcmd_chr.h>
```

```
// Com1 as Master
int sp,i,j,t=0,lm = 0;
unsigned char chr;
unsigned short crc,poly,crc2,crc3,tmp,respcrc,respcrc1,resptmp,resptmp1;
unsigned char respframe[17],respframe1[17];
unsigned char sendframe[17];
unsigned char frame[8];
int chk=0;
// computing the crcdata
int crcdata;
int tempdata;
// getting the current crc data
int current1;
int bytecount = 0; int frrecv = 0;
int local = 1;
// Function declaration for Serial Port Communication
int initport(int fd1);
unsigned char tx_header[6];
int data = 0, fd, error_flides;
union
{
char ss[1000],ss2[1000];
float fls[500],fls2[1000];
float V1data[1000],V1data_inc[1000],V2data[1000],V2data_inc[1000];
} u2;
int initport(int fd)
/* Function to initialise the serial port*/
{
struct termios options;
/* Terminal control structure */
tcgetattr(fd, &options);
// get the parameters associated with the terminal
cfsetispeed(&options, B19200);    // Input baud rate
cfsetospeed(&options, B19200);    // Output baud rate
options.c_cflag |= (CLOCAL | CREAD);
// Ignoring the Modbus Status Lines and Enabling the Receiver
options.c_cflag &= ~PARENB;    // Disabling the Parity Bot
options.c_cflag &= ~CSTOPB;    // No Stop Bits
```

```
options.c_cflag &= ~CSIZE;
// Disabling the number of data bits per character
options.c_cflag |= CS8;                  // 8 data bits
tcsetattr(fd, TCSANOW, &options);
//sets the current terminal control settings for the opened device indicated by fd
return 1;
tcflush(fd,TCIFLUSH);
/* Throw away all input data */
}
static void *Thread1( void *arg )
{
// packet being sent
                                          while(1)
{
rs_flush(sp);
/* Let's turn ON RTS now. */
data = _CTL_RTS_CHG | _CTL_RTS;
if (error = devctl (sp, DCMD_CHR_SERCTL, &data,size of (data), NULL))
// The devctl() function
// sends the device-specific command dcmd to the process managing the device
{
fprintf(stderr, "Error setting RTS: %s\n", strerror ( error ));
exit(EXIT_FAILURE);
}
/* RTS should now be ON. */
check_RTS(sp);                   // Function Call
// sending the data to Com3
// packet ready for Query
frame[0]=0x01;                   // slave address
frame[1]=0x03;                   // function code
frame[2]=0x00;                   // starting address HB
frame[3]=0x00;                   // starting address LB
frame[4]=0x00;                   // No. of bytes HB
frame[5]=0x06;                   // No. of bytes LB
crc=0xffff;                      // Initially make ccrc to be oxffff
poly=0xa001;                     // The above polynomial is 0xa001 for CRC
calculation
for (j=0;j<6;j++)                // Loop for 6 since calculating for first 6 bytes
```

```
{
tmp=0x0000;
crc= frame[j] ^ crc;                    // XORing
for (i=0;i<8;i++)
{
tmp=crc & 0x0001;
crc=crc>>1;                             // Bit-Wise Right shifting by 1
if (tmp == 0x0001)
{
crc = crc ^ poly;
}
}
}                                       // end of for
/*************Reversing the last 2 bytes after CRC calculation for sending the query
to the Slave*************/
frame[6]=crc;
printf("frame6 crc is %02x\n",frame[6]);
crc2=crc>>8;
frame[7]=crc2;
printf("frame7 crc2 is %02x\n",frame[7]);
for(i=0;i<8;i++)
{
rs_putch(sp,frame[i]);                  // sending the Modbus Data Packet Serially to the
Slave
}
printf("\n\nMaster Query Sent by sp: ");
for(i=0;i<8;i++)
{
printf(" %x ",frame[i]);
}
rs_flush(sp);
/* Now let's turn RTS OFF. */
data = _CTL_RTS_CHG | 0;
if (error = devctl (sp, DCMD_CHR_SERCTL, &data, size of (data), NULL))
{
fprintf(stderr, "Error setting RTS: %s\n",strerror ( error ));
exit(EXIT_FAILURE);                     // exiting from the program if failed
}
```

```
usleep (1000*1000);              // Sleep for 1 sec
/* RTS should now be OFF. */
check_RTS(sp); // Function Call
printf("\n Master Query Received by sp: ");
if(tcischars(sp) != 0)
//number of characters waiting to be read, or -1 if an error occurred.
{
while(t<17)
// Indicating 16 bytes have been received from the slave
{
respframe1[t]=rs_getch(sp);       // getting the bytes one by one
t +=1; // Incrementing by one for getting the next byte
}
}
for (t=0;t<17;t++) /// Printing the Bytes received
{
printf("\n respframe1[%i] %x ",t,respframe1[t]);
}
/*****************************Checking of CRC for the Bytes Received****************/
respcrc1 = 0xFFFF;
for (j=0;j<17;j++)                // Since 17 bytes have been received from the slave
{
resptmp1=0x0000;
respcrc1= respframe1[j] ^ respcrc1;
for (i=0;i<8;i++) // since each byte is of 8-bit
{
resptmp1=respcrc1 & 0x0001;
respcrc1=respcrc1>>1;
if (resptmp1 == 0x0001)
{
respcrc1 = respcrc1 ^ poly;
}
}
}                                 // end of for
printf("\n respcrc1 = %x ",respcrc1);
// Checking if the data bytes received serially from the Slave are correct or not. If
found correct will be 0 at the end
if( respcrc1==0x0000 )
```

```
{ //// Taking the data bytes in a union so as to convert into Floating values
for(i=0;i<3;i++)
{
u2.ss2[4*i+3] = respframe1[3 + 4*i];
u2.ss2[4*i+2] = respframe1[4 + 4*i];
u2.ss2[4*i+1] = respframe1[5 + 4*i];
u2.ss2[4*i+0] = respframe1[6 + 4*i];
}
// Converting the data bytes received into Float
for(i=0;i<3;i++)
{
u2.V2data_inc[i] = u2.fls2[i];
}
// Printing the Floating Point Vales
for(i=0;i<3;i++)
{
u2.V2data[i] = u2.V2data_inc[i];
}
printf("\n V2data[0] = %f ", u2.V2data[0]);
printf("\n V2data[1] = %f", u2.V2data[1]);
printf("\n V2data[2] = %f", u2.V2data[2]);
printf("\n received correctly\n");
}
Else                              // If wrong data received from the slave.
{
printf ("\nCRC CHECKSUM FAIL\n");
}
rs_flush(sp);                     // Flushing the data from the port
}                                 // end of while1
return arg;
}                                 // end of thread
int main(int argc, char *argv[])
{
void *Thread_rc;                  // void pointer for return value of Thread
// Parameters for Thread of 1 sec
unsigned long period[1] = {1000};
unsigned long base_period = 1000000;
sp = open("/dev/ser4", O_RDWR ); //Open the serial port 4
```

```
if (sp == -1) // If unable to open the serial port
{
perror("open_port: Unable to open /dev/ttyS4 - ");
return 1;
}
else
{
fcntl(sp, F_SETFL, 0);                    // Provide control over an open file
printf("\n Port 4 opened successfully");
}
// F_SETFL - Set the file status flags
initport(sp);                             // Initialise the port
rs_flush(sp);
printf(" Main code started \n");
/************Thread Creation of 1 sec**************/
period[0] *= base_period;
pthread_create( NULL, NULL, Thread1, &period[0] );
while (1) {};
// Joining the thread
pthread_join((pthread_t)Thread1,&Thread_rc);
// Closing the serial Port
rs_close(sp);
return (int)Thread_rc;
}
int check_RTS(int fd)
{
// int data = 0, error;
/* Let's see if RTS is set, tell devctl() we're requesting line status information and
devctl() then assigns data the line status information for us. */
if (error = devctl (fd, DCMD_CHR_LINESTATUS, &data,sizeof(data), NULL))
{
fprintf(stderr, "\n Error setting RTS: %s", strerror ( error ));
exit(EXIT_FAILURE);
}
if (data & _LINESTATUS_SER_RTS)
{
printf("\n RTS is SET!");
}
```

```
else
{
printf("\n RTS is NOT set ");
}
return(1);
}
```

Explanation of check_RTS function

The two main areas of interest are the setting of data and the devctl() call. The data variable is used for both sending and receiving data.

Data is assigned a value that is sent to the device via devctl() when setting RTS.

If data is equivalent to:	RTS TURNS:
_CTL_RTS_CHG | _CTL_RTS	ON
_CTL_RTS_CHG	OFF

We call devctl() with dcmd set to the DCMD_CHR_LINESTATUS macro and data containing any value (zero is clean) so as to check if RTS is set or not. The devctl() function returns with data containing the Line Status value. This is used to determine what lines are set on that device. In our example, we check against _LINESTATUS_SER_RTS.

Explanation of the Source Code

1. First define the header files for the functions to be used. Some of the functions mentioned here are not actually required but it will at least make you aware of the libraries.
2. There is one special header file which has been highlighted in black letters, namely, "dscud.h" which is the header file for initializing the board which will be used when we will use the analog and digital I/O ports of the board. Another is the source file "rs_lib.c" which is the file for the communication with the external devices. The above file will be used for Receiving(rs_getch), Sending(rs_putch), Flushing(rs_flush), Closing(rs_close) etc. The actual code inside the rs_lib.c file is shown in section 11.13.2.

11.13.3 Source Code for Sending Serially Data to the External Device

```
// Function for sending serially data to the external device
void rs_putch(const int fd, const int character)
{
char b[1];
b[0]=(char)character;
write(fd,b,1);
```

```
}
// Function for receiving serially data from the external device
char rs_getch(const int fd)
{
char b[1];
if( read(fd,b,1)!=0)
return b[0];
}
// Flushing the data from the serial port device
void rs_flush(const int fd)
{
tcflush(fd,TCIOFLUSH);
}
// Closing of the Serial Port
void rs_close(const int fd)
{
struct termios prevt;
tcsetattr(fd,TCSANOW,&prevt);
close(fd);
}
```

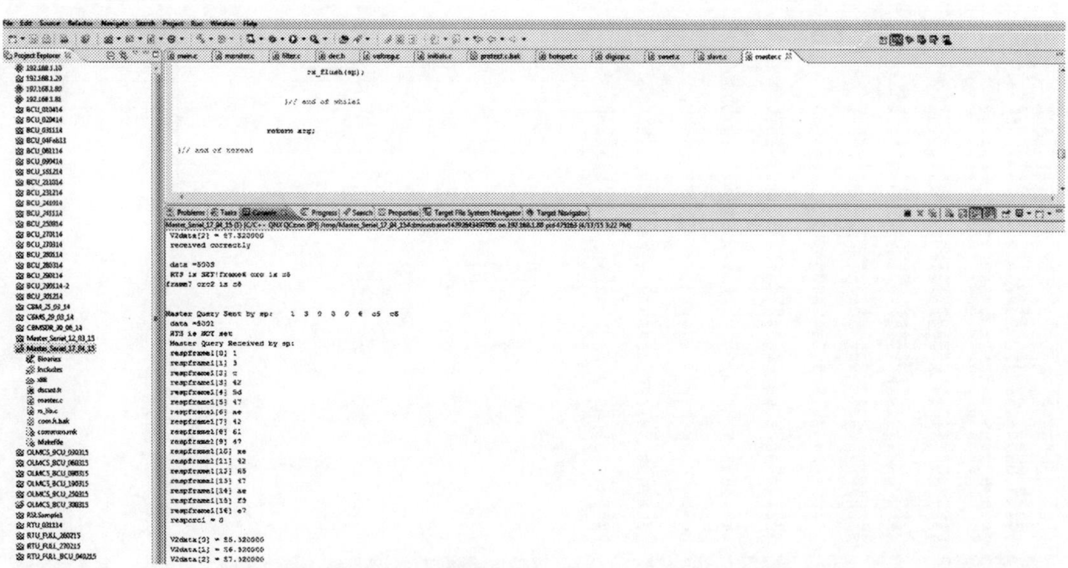

Fig. 11.12 *Console Window of QNX using Momentics IDE*

1. The sequence to transmit data on a shared RS-485 cable is as follows:
   (*a*) To enable transmit Set RTS high.
   (*b*) Send the data which is echoed to the receive buffer of the same port.
   (*c*) To disable transmit set RTS low.
2. The serial port COM4 is used for RS-485 communication.
3. The modbus query sent to the slave is 1 3 0 0 0 6 C5 C8.
   (*a*) Slave ID is 1.
   (*b*) Function code is 3.
   (*c*) Starting address is 0 × 00000. That's why 0 0.
   (*d*) Count is 0 × 0006. That's why 0 6.
   (*e*) The CRC computed from the first 6 bytes is C8C5. Reversing the bytes for sending it to the slave. The slave having slave ID 1 will see that from the starting address 0 × 0000 with the function code 3, the total number of floating values requested will be 3 since the count is 6 and each floating value occupies 4 bytes, i.e., at count of 2. The master can also query for the integer values but here we are **assuming** that the master queries for floating bytes from the slave. The console window as seen from the QNX OS is shown in Figure 11.12.
4. Here we are assuming the slave from the open source testing tool MOD_RSSim which can be easily downloaded from the Internet as it is an open source. We are feeding the data as 55.32 , 56.32 and 57.32 in the simulator and can see that in Figure 11.13.

   From Figure 11.12, we can see that the data bytes received as a response from the slave by the master is as follows:

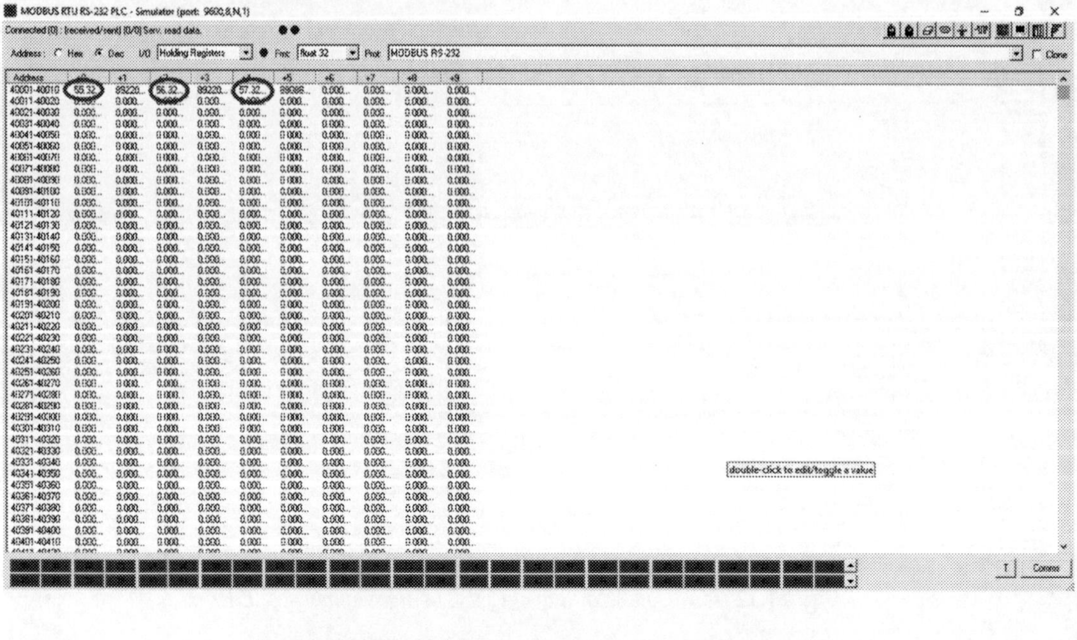

Fig. 11.13 *Mod_RS Sim Simulator where values are fed*

1 3 c 42 5d 47 ae 42 61 47 ae 42 65 47 ae f3 e7. The explanation of the above data bytes is explained below:

(*a*) Slaved ID is 1. It is of 1 byte.

(*b*) Function code is 3. It is of 1 byte.

(*c*) The total number of data bytes (not including the CRC bytes, the last 2 bytes) are 12. Thus, its hexadecimal representation is C. Since the count value was 6, so the total number of actual data bytes sent as response by the slave is 12 (2*6).

(*d*) 42 5d 47 AE is the hexadecimal representation of the floating value 55.32. So it is of 4 bytes.

(*e*) 42 61 47 AE is the hexadecimal representation of the floating value 56.32. So it is of 4 bytes.

(*f*) 42 65 47 AE is the hexadecimal representation of the floating value 57.32. So it is of 4 bytes.

(*g*) F3 E7 is the CRC data bytes of the first 15 data bytes. It is of 2 bytes.

5. You can see from Figure 11.12 that the data bytes received is displayed.

   The communication between the Master and the Slave is depicted in Figure 11.14.

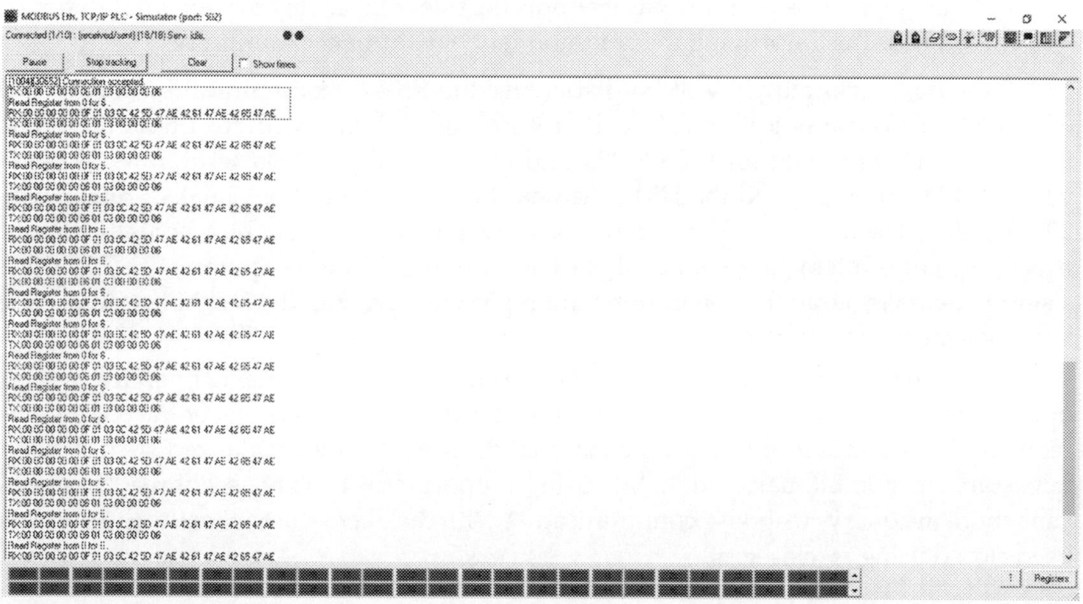

Fig. 11.14 *Communication Log File between the Master and Slave*

6. Last, the data bytes received have been converted into the floating values with the help of union. The data bytes which were sent from the Slave has been received correctly by the Master and is displayed as

   V2data[0] = 55.32

   V2data[1] = 56.32

   V2data[2] = 57.32

   Thus, we can see that the SBC Board can communicate with the external devices through RS-485 communication media. The RS-485 communication media is preferred for multi-point application whereas RS-232 communication media is used for point to point application. The other benefits of RS-485 is as follows:

   1. The mode of operation is differential.
   2. Up to 256 transceivers are allowed to be connected to the bus.
   3. Maximum cable length is up to 4000 ft.
   4. The maximum data transmission speed is 10Mbps.
   5. The level is compatible with the TTL thus facilitating the connection with the TTL circuit.
   6. It has good noise immunity as it adopts the role of balancing driver and differential receiver thus improving the common mode disturbance immunity.

In the next example we will be displaying the RS-232 communication media and ethernet based communication (TCP/IP protocol) of the board with the human machine interface form. The code for the RS-232 and the TCP/IP has been written in QNX OS using POSIX threading and the HMI form has been written in the visual studio C# IDE 2010 version. The main aim of the above example is to give a feel of how actually the end user/operator is interacting with the board according to its own requirement. This is just a simple demo program but on understanding, it will give you the feel of how real-time scenarios are being handled.

Inside the main function, first the serial port has been opened using open function as open("/dev/ser4", O_RDWR) and the port has been initialized. Whatever the parameters required for communication with the external devices has been initialized such as baud rate, start bit, stop bit, parity, data bits using initport (sp). Then a 1 sec thread has been run where the query/response communication with the Slave starts. Finally, the port has been closed using rs_close(sp).

11.13.4 RS-232 and Ethernet Based Communication with the Hercules Board

The above program demonstrates the RS-232 communication media and the TCP/IP communication protocol. The source code at the application side (SBC Board side) for the above has been written in POSIX C in QNX RTOS and the HMI has been designed in visual studio 2010 version IDE. The above HMI will give you a feel of how the communication with the SBC board is taking place and the behaviour of the board w.r.t. their analog input and analog output channels. Kindly observe the above demo carefully. The source code at the application side consists of the following source and header files:

1. **Structs.h:** For initializing the variables used.
2. **Dec.h:** The above header file will store the function declarations used in the main program.
3. **Filtering.c:** For the sampling and the rms computation.
4. **Dscud.h:** The header file for using the SBC. This header file comes with the library.
5. **Initials.c:** For initializing the board, for assigning the analog channels and freeing up the board at the end.
6. **Rs_lib.c/RS_Lib.h:** Files used for communication as explained in the previous RS-485 demo. The header file will include the function declarations required for the communication over the serial port.
7. **Main.c:** Main file for communication with the HMI.

11.13.5 Source code for Structs.h Header File

```
// Header Libraries
#include"dscud.h"
#include<math.h>
#include<stdio.h>
#include<stdlib.h>
#include<pthread.h>
#include<unistd.h>
#include<sys/types.h>
#include<sys/stat.h>
#include <fcntl.h>
#include <time.h>
// Variables initialization
float fvrs1,fvrc1,firs1,firc1;
float Vpr1,Ipr1,Vr1,Ir1;            //These values are for 1st Feeder
float Vrs1,vsm_old[40],Irs1,ism_old[40];
float fvys1,fvyc1,fiys1,fiyc1,Vpy1,Ipy1,Vy1,Iy1;
float Vys1,vsm_old2[40],Iys1,ism_old2[40];
float fvbs1,fvbc1,fibs1,fibc1,Vpb1,Ipb1,Vb1,Ib1;
float Vbs1,vsm_old3[40],Ibs1,ism_old3[40];
float fins1,finc1,Ipn1,In1;
float Ins1,ism_old4[40];
float Vbbs1,fvbbs1,fvbbc1,Vpbb1,Vbb1,vsm_old4[40];
```

```
float fvrs2,fvrc2,firs2,firc2;
float Vpr2,Ipr2,Vr2,Ir2;               //These values are for 2nd Feeder
float Vrs2,Irs2,vsm_old1[40],ism_old1[40];
float fvys2,fvyc2,fiys2,fiyc2,Vpy2,Ipy2,Vy2,Iy2;
float Vys2,Iys2,vsm_old21[40],ism_old21[40];
float fvbs2,fvbc2,fibs2,fibc2,Vpb2,Ipb2,Vb2,Ib2;
float Vbs2,Ibs2,vsm_old31[40],ism_old31[40];
float fins2,finc2,Ipn2,In2;
float Ins2,ism_old41[40];
float Vbbs2,fvbbs2,fvbbc2,Vpbb2,Vbb2,vsm_old41[40];
float fvrs3,fvrc3,firs3,firc3,fvys3,fvyc3,fiys3,fiyc3,fvbs3,fvbc3,fibs3,fibc3,fins3,finc3,f
vbbs3,fvbbc3;
float Vpr3,Ipr3,Vr3,Ir3;               //These values are for 3rd Feeder
float Vrs3,vsm_old20[40],Irs3,ism_old20[40];
float Vpy3,Ipy3,Vy3,Iy3;
float Vys3,vsm_old22[40],Iys3,ism_old22[40];
float Vpb3,Ipb3,Vb3,Ib3;
float Vbs3,vsm_old32[40],Ibs3,ism_old32[40];
float Ipn3,In3;
float Ins3,ism_old42[40];
float Vbbs3,Vpbb3,Vbb3,vsm_old42[40];
float fvrs4,fvrc4,firs4,firc4,fvys4,fvyc4,fiys4,fiyc4,fvbs4,fvbc4,fibs4,fibc4,fins4,finc4,f
vbbs4,fvbbc4;
float Vpr4,Ipr4,Vr4,Ir4;               //These values are for 4th Feeder
float Vrs4,Irs4,vsm_old13[40],ism_old13[40];
float Vpy4,Ipy4,Vy4,Iy4;
float Vys4,Iys4,vsm_old23[40],ism_old23[40];
float Vpb4,Ipb4,Vb4,Ib4;
float Vbs4,Ibs4,vsm_old33[40],ism_old33[40];
float Ipn4,In4;
float Ins4,ism_old43[40];
float Vbbs4,Vpbb4,Vbb4,vsm_old43[40];
```

11.13.6 Source Code for Dec.h Header File

```
// Header Files
#include"dscud.h"
#include<math.h>
```

```
#include<stdio.h>
#include<stdlib.h>
#include<pthread.h>
#include<unistd.h>
#include<sys/types.h>
#include<sys/stat.h>
#include <fcntl.h>
#include <time.h>
void sampling(void);
// Sampling is the method where a complete wave is sampled
void measurand(void);          // for rms computation
void initials(void);           // For initializing the board
void adcinit(void);            // For initializing the analog input channels
void dacinit(void);            // for initializing the analog output channels
void finally(void);            // free the Board
void serial_timer(char ch2);   // running the serial communication
void metering();
// Calling the metering function for mapping the rms values computation to the variables. The above variables mapped will be used for communication.
```

11.13.7 Source Code for Filtering.c Header File

```
// Header Files
#include"dscud.h"
#include"dec.h"
#include"structs.h"
#include<math.h>
#include<stdio.h>
#include<stdlib.h>
#include<pthread.h>
#include<unistd.h>
#include<sys/types.h>
#include<sys/stat.h>
#include <fcntl.h>
#include <time.h>
// extern is used as the structure variable has been declared in initials.c
extern DSCADSCAN adscan;
```

```
extern DSCB dscb;
int i=0;
extern float sin_value[40],cos_value[40];
// Array Initialization for Sampling at 500micro sec
DSCSAMPLE smpl[32];
/// DFT Computation function
void sampling()
{
float diffsmpl;
// For Data Acqusition of 32 variables collecting 40 samples since sampling at 500micro sec. (20ms/500 micro sec = 40  // as one sinusoidal wave is of 50 Hz(20 msec) )
dscADScan(dscb,&adscan,smpl);
Vrs1=smpl[0]*2.0/32768; // R_Phase Voltage of first feeder
Irs1=smpl[1]*20.0/32768; // R_Phase Current of first feeder
Vys1=smpl[2]*2.0/32768; // Y_Phase Voltage of first feeder
Iys1=smpl[3]*20.0/32768; // Y_Phase Current of first feeder
Vbs1=smpl[4]*2.0/32768; // B_Phase Voltage of first feeder
Ibs1=smpl[5]*20.0/32768; // B_Phase Current of first feeder
Vbbs1=smpl[6]*2.0/32768; // Neutral Voltage of first feeder
Ins1=smpl[7]*20.0/32768; // Neutral Current of first feeder
// Similarily R,Y, B and Neutral Phase Voltages and Currents of second feeder
Vrs2=smpl[8]*2.0/32768; //samples of 2nd feeder
Irs2=smpl[9]*20.0/32768; //samples of 3rd feeder
Vys2=smpl[10]*2.0/32768;
Iys2=smpl[11]*20.0/32768;
Vbs2=smpl[12]*2.0/32768;
Ibs2=smpl[13]*20.0/32768;
Vbbs2=smpl[14]*2.0/32678;
Ins2=smpl[15]*20.0/32768;
// Similarily R,Y, B and Neutral Phase Voltages and Currents of third feeder
Vrs3=smpl[16]*2.0/32768;//samples of 3rd feeder
Irs3=smpl[17]*20.0/32768;
Vys3=smpl[18]*2.0/32768;
Iys3=smpl[19]*20.0/32768;
Vbs3=smpl[20]*2.0/32768;
```

```
Ibs3=smpl[21]*20.0/32768;
Vbbs3=smpl[22]*2.0/32768;
Ins3=smpl[23]*20.0/32768;
// Similarily R,Y, B and Neutral Phase Voltages and Currents of fourth feeder
Vrs4=smpl[24]*2.0/32768; //samples of 4th feeder
Irs4=smpl[25]*20.0/32768;
Vys4=smpl[26]*2.0/32768;
Iys4=smpl[27]*20.0/32768;
Vbs4=smpl[28]*2.0/32768;
Ibs4=smpl[29]*20.0/32768;
Vbbs4=smpl[30]*2.0/32678;
Ins4=smpl[31]*20.0/32768;
// Sample Collections for the difference between the 2 samples for 4 feeders of 3 phases of each analog parameters and
// Converting each Analog sampled value into sine and cosine
diffsmpl=Vrs1-vsm_old[i]; // Computation for 1st feeder
fvrs1=fvrs1+(diffsmpl)*sin_value[i];
fvrc1=fvrc1+(diffsmpl)*cos_value[i];
diffsmpl=Irs1-ism_old[i];
firs1=firs1+(diffsmpl)*sin_value[i];
firc1=firc1+(diffsmpl)*cos_value[i];
diffsmpl=Vys1-vsm_old2[i];
fvys1=fvys1+(diffsmpl)*sin_value[i];
fvyc1=fvyc1+(diffsmpl)*cos_value[i];
diffsmpl=Iys1-ism_old2[i];
fiys1=fiys1+(diffsmpl)*sin_value[i];
fiyc1=fiyc1+(diffsmpl)*cos_value[i];
diffsmpl=Vbs1-vsm_old3[i];
fvbs1=fvbs1+(diffsmpl)*sin_value[i];
fvbc1=fvbc1+(diffsmpl)*cos_value[i];
diffsmpl=Ibs1-ism_old3[i];
fibs1=fibs1+(diffsmpl)*sin_value[i];
fibc1=fibc1+(diffsmpl)*cos_value[i];
diffsmpl=Ins1-ism_old4[i];
fins1=fins1+(diffsmpl)*sin_value[i];
```

```
finc1=finc1+(diffsmpl)*cos_value[i];
diffsmpl=Vbbs1-vsm_old4[i];
fvbbs1=fvbbs1+(diffsmpl)*sin_value[i];
fvbbc1=fvbbc1+(diffsmpl)*cos_value[i];
diffsmpl=Vrs2-vsm_old1[i]; //Computation for 2nd feeder
fvrs2=fvrs2+(diffsmpl)*sin_value[i];
fvrc2=fvrc2+(diffsmpl)*cos_value[i];
diffsmpl=Irs2-ism_old1[i];
firs2=firs2+(diffsmpl)*sin_value[i];
firc2=firc2+(diffsmpl)*cos_value[i];
diffsmpl=Vys2-vsm_old21[i];
fvys2=fvys2+(diffsmpl)*sin_value[i];
fvyc2=fvyc2+(diffsmpl)*cos_value[i];
diffsmpl=Iys2-ism_old21[i];
fiys2=fiys2+(diffsmpl)*sin_value[i];
fiyc2=fiyc2+(diffsmpl)*cos_value[i];
diffsmpl=Vbs2-vsm_old31[i];
fvbs2=fvbs2+(diffsmpl)*sin_value[i];
fvbc2=fvbc2+(diffsmpl)*cos_value[i];
diffsmpl=Ibs2-ism_old31[i];
fibs2=fibs2+(diffsmpl)*sin_value[i];
fibc2=fibc2+(diffsmpl)*cos_value[i];
diffsmpl=Ins2-ism_old41[i];
fins2=fins2+(diffsmpl)*sin_value[i];
finc2=finc2+(diffsmpl)*cos_value[i];
diffsmpl=Vbbs2-vsm_old41[i];
fvbbs2=fvbbs2+(diffsmpl)*sin_value[i];
fvbbc2=fvbbc2+(diffsmpl)*cos_value[i];
diffsmpl=Vrs3-vsm_old20[i]; // Computation for 3rd feeder
fvrs3=fvrs3+(diffsmpl)*sin_value[i];
fvrc3=fvrc3+(diffsmpl)*cos_value[i];
diffsmpl=Irs3-ism_old20[i];
firs3=firs3+(diffsmpl)*sin_value[i];
firc3=firc3+(diffsmpl)*cos_value[i];
diffsmpl=Vys3-vsm_old22[i];
```

```
fvys3=fvys3+(diffsmpl)*sin_value[i];
fvyc3=fvyc3+(diffsmpl)*cos_value[i];
diffsmpl=Iys3-ism_old22[i];
fiys3=fiys3+(diffsmpl)*sin_value[i];
fiyc3=fiyc3+(diffsmpl)*cos_value[i];
diffsmpl=Vbs3-vsm_old32[i];
fvbs3=fvbs3+(diffsmpl)*sin_value[i];
fvbc3=fvbc3+(diffsmpl)*cos_value[i];
diffsmpl=Ibs3-ism_old32[i];
fibs3=fibs3+(diffsmpl)*sin_value[i];
fibc3=fibc3+(diffsmpl)*cos_value[i];
diffsmpl=Ins3-ism_old42[i];
fins3=fins3+(diffsmpl)*sin_value[i];
finc3=finc3+(diffsmpl)*cos_value[i];
diffsmpl=Vbbs3-vsm_old42[i];
fvbbs3=fvbbs3+(diffsmpl)*sin_value[i];
fvbbc3=fvbbc3+(diffsmpl)*cos_value[i];
diffsmpl=Vrs4-vsm_old13[i]; //Computation for 4th feeder
fvrs4=fvrs4+(diffsmpl)*sin_value[i];
fvrc4=fvrc4+(diffsmpl)*cos_value[i];
diffsmpl=Irs4-ism_old13[i];
firs4=firs4+(diffsmpl)*sin_value[i];
firc4=firc4+(diffsmpl)*cos_value[i];
diffsmpl=Vys4-vsm_old23[i];
fvys4=fvys4+(diffsmpl)*sin_value[i];
fvyc4=fvyc4+(diffsmpl)*cos_value[i];
diffsmpl=Iys4-ism_old23[i];
fiys4=fiys4+(diffsmpl)*sin_value[i];
fiyc4=fiyc4+(diffsmpl)*cos_value[i];
diffsmpl=Vbs4-vsm_old33[i];
fvbs4=fvbs4+(diffsmpl)*sin_value[i];
fvbc4=fvbc4+(diffsmpl)*cos_value[i];
diffsmpl=Ibs4-ism_old33[i];
fibs4=fibs4+(diffsmpl)*sin_value[i];
fibc4=fibc4+(diffsmpl)*cos_value[i];
```

```
diffsmpl=Ins4-ism_old43[i];
fins4=fins4+(diffsmpl)*sin_value[i];
finc4=finc4+(diffsmpl)*cos_value[i];
diffsmpl=Vbbs4-vsm_old43[i];
fvbbs4=fvbbs4+(diffsmpl)*sin_value[i];
fvbbc4=fvbbc4+(diffsmpl)*cos_value[i];
// Assigning the new computed values with the old ones of each phase for each feeder
vsm_old[i]=Vrs1; //Computation for 1st feeder
ism_old[i]=Irs1;
vsm_old2[i]=Vys1;
ism_old2[i]=Iys1;
vsm_old3[i]=Vbs1;
ism_old3[i]=Ibs1;
ism_old4[i]=Ins1;
vsm_old4[i]=Vbbs1;
vsm_old1[i]=Vrs2; //Computation for 2nd feeder
ism_old1[i]=Irs2;
vsm_old21[i]=Vys2;
ism_old21[i]=Iys2;
vsm_old31[i]=Vbs2;
ism_old31[i]=Ibs2;
vsm_old41[i]=Vbbs2;
ism_old41[i]=Ins2;
vsm_old20[i]=Vrs3;//Computation for 3rd feeder
ism_old20[i]=Irs3;
vsm_old22[i]=Vys3;
ism_old22[i]=Iys3;
vsm_old32[i]=Vbs3;
ism_old32[i]=Ibs3;
ism_old42[i]=Ins3;
vsm_old42[i]=Vbbs3;
vsm_old13[i]=Vrs4;//Computation for 4th feeder
ism_old13[i]=Irs4;
vsm_old23[i]=Vys4;
ism_old23[i]=Iys4;
```

```
vsm_old33[i]=Vbs4;
ism_old33[i]=Ibs4;
vsm_old43[i]=Vbbs4;
ism_old43[i]=Ins4;
i=(i+1)%40; // Doing the same for 40 samples
}
// For rms computation of each phase of 32 Analog parameters
void measurand()
{
Vpr1 = ((fvrs1*fvrs1)+(fvrc1*fvrc1));
Ipr1 = ((firs1*firs1)+(firc1*firc1));
Vpy1 = ((fvys1*fvys1)+(fvyc1*fvyc1));
Ipy1 = ((fiys1*fiys1)+(fiyc1*fiyc1));
Vpb1 = ((fvbs1*fvbs1)+(fvbc1*fvbc1));
Ipb1 = ((fibs1*fibs1)+(fibc1*fibc1));
Ipn1 = ((fins1*fins1)+(finc1*finc1));
Vpbb1= ((fvbbs1*fvbbs1)+(fvbbc1*fvbbc1));
Vpr2 = ((fvrs2*fvrs2)+(fvrc2*fvrc2));
Ipr2 = ((firs2*firs2)+(firc2*firc2));
Vpy2 = ((fvys2*fvys2)+(fvyc2*fvyc2));
Ipy2 = ((fiys2*fiys2)+(fiyc2*fiyc2));
Vpb2 = ((fvbs2*fvbs2)+(fvbc2*fvbc2));
Ipb2 = ((fibs2*fibs2)+(fibc2*fibc2));
Ipn2 = ((fins2*fins2)+(finc2*finc2));
Vpbb2= ((fvbbs2*fvbbs2)+(fvbbc2*fvbbc2));
Vpr3 = ((fvrs3*fvrs3)+(fvrc3*fvrc3));
Ipr3 = ((firs3*firs3)+(firc3*firc3));
Vpy3 = ((fvys3*fvys3)+(fvyc3*fvyc3));
Ipy3 = ((fiys3*fiys3)+(fiyc3*fiyc3));
Vpb3 = ((fvbs3*fvbs3)+(fvbc3*fvbc3));
Ipb3 = ((fibs3*fibs3)+(fibc3*fibc3));
Ipn3 = ((fins3*fins3)+(finc3*finc3));
Vpbb3= ((fvbbs3*fvbbs3)+(fvbbc3*fvbbc3));
Vpr4 = ((fvrs4*fvrs4)+(fvrc4*fvrc4));
```

```
Ipr4 = ((firs4*firs4)+(firc4*firc4));
Vpy4 = ((fvys4*fvys4)+(fvyc4*fvyc4));
Ipy4 = ((fiys4*fiys4)+(fiyc4*fiyc4));
Vpb4 = ((fvbs4*fvbs4)+(fvbc4*fvbc4));
Ipb4 = ((fibs4*fibs4)+(fibc4*fibc4));
Ipn4 = ((fins4*fins4)+(finc4*finc4));
Vpbb4= ((fvbbs4*fvbbs4)+(fvbbc4*fvbbc4));
Vr1 += sqrt(Vpr1);
Ir1 += sqrt(Ipr1);
Vy1 += sqrt(Vpy1);
Iy1 += sqrt(Ipy1);
Vb1 += sqrt(Vpb1);
Ib1 += sqrt(Ipb1);
In1 += sqrt(Ipn1);
Vbb1 += sqrt(Vpbb1);
Vr2 += sqrt(Vpr2);
Ir2 += sqrt(Ipr2);
Vy2 += sqrt(Vpy2);
Iy2 += sqrt(Ipy2);
Vb2 += sqrt(Vpb2);
Ib2 += sqrt(Ipb2);
In2 += sqrt(Ipn2);
Vbb2 += sqrt(Vpbb2);
Vr3 += sqrt(Vpr3);
Ir3 += sqrt(Ipr3);
Vy3 += sqrt(Vpy3);
Iy3 += sqrt(Ipy3);
Vb3 += sqrt(Vpb3);
Ib3 += sqrt(Ipb3);
In3 += sqrt(Ipn3);
Vbb3 += sqrt(Vpbb3);
Vr4 += sqrt(Vpr4);
Ir4 += sqrt(Ipr4);
Vy4 += sqrt(Vpy4);
```

```
Iy4 += sqrt(Ipy4);
Vb4 += sqrt(Vpb4);
Ib4 += sqrt(Ipb4);
In4 += sqrt(Ipn4);
Vbb4 += sqrt(Vpbb4);
//
printf("\n rms computation done");
}
```

11.13.8 Source Code for Dscud.h Header File

// Dscud.h is a huge file but the functions and structures used in the above program is as shown below. This will give you the feel // of structures used, how the functions are used and parameters assigned.

```
typedef struct
{
// These members are filled automatically by dscInitBoard //
// THE USER DOES NOT HAVE TO FILL THESE IN MANUALLY //
BYTE boardtype;                /* Contains the boardtype constant */
DSCB boardnum;                 /* The handle to the board */
// The user must fill in everything below that pertains to board before the call to
dscInitBoard //
WORD base_address;             /* Base address of the board */
// Use int_level for boards with only one interrupt //
BYTE int_level;                /* Interrupt level of the board */
} DSCCB;
typedef struct
{
BYTE current_channel;
/* INPUT: Which channel to read AD samples from */
BYTE gain;
/* INPUT: 0 = gain of 1, 1 = gain of 2, 2 = gain of 4, 3 = gain of 8 */
BYTE range;                    /* INPUT: 0 = 5V range, 1 = 10V range */
BYTE polarity;                 /* INPUT: 0 = bipolar, 1 = unipolar */
BYTE load_cal;                 /* INPUT: 0 = do not load calibration values, 1
                                  = load */
BYTE scan_interval;
```

```
/* For DMM32, Hercules, DMM48 only. Use constants SCAN_INTERVAL_X */
BYTE addiff;
/* For Hercules and Elektra only, 0 = SINGLE_ENDED or 1 = DIFFERENTIAL */
} DSCADSETTINGS;
typedef struct
{
BYTE low_channel;            /* INPUT: Low channel of the scan */
BYTE high_channel;           /* INPUT: High channel of the scan */
DSCSAMPLE *sample_values;
/* I/O: The user must point this to memory that has been allocated and ready to
accept (high - low + 1) DSCADSAMPLE values into it */
BYTE gain; /* Obsolete */
} DSCADSCAN;
typedef struct
{
BYTE polarity;               /* INPUT: 0 = BIPOLAR, 1 = UNIPOLAR */
BYTE load_cal;
/* INPUT: 0 = FALSE, 1= TRUE. Loads calibration from eeprom */
FLOAT range;                 /* INPUT: Absolute value of maximum voltage
*/
} DSCDASETTINGS;
/*//////////////////
// A/D Functions //
///////////////// */
BYTE DSCUDAPICALL dscADSetSettings(DSCB board, DSCADSETTINGS* settings);
BYTE DSCUDAPICALL dscADScan(DSCB board, DSCADSCAN* dscadscan,
DSCSAMPLE* sample_values);
/*//////////////////
// D/A Functions //
///////////////// */
BYTE DSCUDAPICALL dscDASetSettings( DSCB board, DSCDASETTINGS *
dasettings);
BYTE DSCUDAPICALL dscDAConvert(DSCB board, BYTE channel, DSCDACODE
output_code);
/*//////////////////
// DIO Functions //
///////////////// */
```

```
BYTE DSCUDAPICALL dscDIOSetConfig(DSCB board, BYTE* config_bytes);
BYTE DSCUDAPICALL dscDIOOutputByte(DSCB board, BYTE port, BYTE digital_
value);
```

11.13.9 Source Code for Initials.c Header File

```
// Header Files Declaration
#include"dec.h"
#include"dscud.h"
#include <string.h>
#include <stdlib.h>
#include <stdio.h>
DSCADSCAN adscan;                    // Structure used for assigning the channels of
the board
DSCB dscb;                           // Handler to the board
extern float sin_value[40],cos_value[40];
// Array Initialization for Sampling at 500micro sec so that at one cycle 40 samples
can be collected
// Function used for Initializing the board
void initials()
{
int j;
DSCCB dsccb;                         // Structure for handling the board
BYTE confbyte;                       // /* Contains the confbyte constant */
dscInit(DSC_VERSION);                // Initializing function
memset(&dsccb,0,sizeof(DSCCB));      // Set memory to the value
dsccb.base_address = 0x240;          // Base address of the board
dsccb.int_level = 5;                 // Interrupt level of the board
dscInitBoard DSC_HERCEBX, & dsccb, &dscb);
confbyte= 0x1f;
// For making 5 ports. i.e. Port A, B, C, D and E as the DI Ports as there are 5 ports
in the SBC Board
dscDIOSetConfig (dscb,&confbyte); // For Digital Input /Output Functions
// Initializing for 40 channels scanning
for (j=0;j<40;j++)
{
sin_value[j] = 0.05*sin (j*0.157142858);
```

```
cos_value[j] = 0.05*cos (j*0.157142858);
}
// Each Port is of 8 bits
 dscDIOOutputByte(dscb,0,0); // Initially Making Port 0 as 00
 dscDIOOutputByte(dscb,1,0); // Initially Making Port 1 as 00
 dscDIOOutputByte(dscb,2,0); // Initially Making Port 2 as 00
 dscDIOOutputByte(dscb,3,0); // Initially Making Port 3 as 00
 dscDIOOutputByte(dscb,4,0); // Initially Making Port 4 as 00
}
// For ADScanning channel parameters
void adcinit()
{
DSCADSETTINGS adsettings;
// Structure for ADScanning parameters like Gain, Range, Loading of Calibration
values
memset(&adsettings, 0, sizeof(DSCADSETTINGS));      // Set memory to the value
adsettings.range = RANGE_10;        // /* INPUT: 0 = 5V range, 1 = 10V range */
adsettings.polarity = BIPOLAR;      // /* INPUT: 0 = bipolar, 1 = unipolar */
adsettings.gain = GAIN_1;
// /* INPUT: 0 = gain of 1, 1 = gain of 2, 2 = gain of 4, 3 = gain of 8 */
adsettings.load_cal = TRUE;
// /* INPUT: 0 = do not load calibration values, 1 = load */
adsettings.scan_interval = SCAN_INTERVAL_9;
// 4 /* For DMM32, Hercules, DMM48 only. Use constants SCAN_INTERVAL_X.
It is for scan interval in microseconds constant */
dscADSetSettings(dscb, &adsettings);    // For A/D Function
memset(&adscan, 0, sizeof(DSCADSCAN));      // Set memory to a given value
adscan.low_channel = 0;
// /* INPUT: Low channel of the scan */
// For making 32 channels as the Analog Input from
adscan.high_channel = 31;
// /* INPUT: High channel of the scan */
// total 40 Analog Channels
}
// For DA Scan Channel parameters
```

```
void dacinit()
{
DSCDASETTINGS dasettings;
// Structure for D/A parameters like Polarity, Load Calibration etc.
memset(&dasettings, 0, sizeof(DSCDASETTINGS));        // Set memory to the value
dasettings.polarity = UNIPOLAR;
/* INPUT: 0 = BIPOLAR, 1 = UNIPOLAR */
dasettings.load_cal = TRUE;
/* INPUT: 0 = FALSE, 1= TRUE. Loads calibration from eeprom */
dscDASetSettings(dscb, &dasettings);      // For D/A function
// Forcefully writing the output_code to 2048. The DAC is of 12-bit. Maximum upto
//4096 which corresponds to +5Vand if the output code is 0, it is 0V. Assigning the
same to all the channels i.e. from 0 to 3. In total, there are 4 Analog Output channels.
dscDAConvert ( dscb, 0, 2048 );
dscDAConvert ( dscb, 1, 2048 );
dscDAConvert ( dscb, 2, 2048 );
dscDAConvert ( dscb, 3, 2048 );
}
// for freeing of the board
void finally()
{
dscFreeBoard(dscb);
dscFree();
}
```

11.13.10 Source Code for Main.C Header File

```
// Header files required for the main program
#include "rs_lib.h"
#include "dscud.h"
#include "dec.h"
#include "structs.h"
#include <errno.h>
#include <fcntl.h>
#include <stdio.h>
#include <stdlib.h>
#include <string.h>
```

```
#include <unistd.h>
#include <sys/iofunc.h>
#include <sys/dispatch.h>
#include <sys/neutrino.h>
#include <sys/resmgr.h>
#include <sys/syspage.h>
#include <sched.h>
#include <pthread.h>
#include <err.h>
#include <sys/netmgr.h>
#include <math.h>
#include <termios.h>                 /* POSIX terminal control definitions */
#include <sys/types.h>
#include <sys/stat.h>
#include <sys/mman.h>
#include <time.h>
#include <sys/io.h>
#include <signal.h>                  // For timer create
#include <pthread.h>
#include <inttypes.h>
#include <hw/inout.h>
#include <stdint.h>
#include<sys/socket.h>
#include<arpa/inet.h>
#include<netinet/in.h>               // Header files required for the socket connection
#define PORT 1234                    // Macro Port No.
#define MAXPENDING 1                 // For listening of one slave
#define BUFLEN 500                   // For allocation of memory for 500 bytes
int z,inp1,inp2,inp3,inp4,inp5,inp6,inp7,inp8;     // For Digital output
int a;
int sp,,flag=0,laststgp;
char str[1000],ch,ch1='1';
float sin_value[40],cos_value[40];
#define MY_PULSE_CODE_PULSE_CODE_MINAVAIL
// Macro for receiving the Pulse bit
```

```
pthread_mutex_t mutex = PTHREAD_MUTEX_INITIALIZER;
// Macro for Mutex
////// TCP/IP Communication variables
// sockaddr_in is the structure variable for the a structure to contain an internet address
struct sockaddr_in echoServAddr;
struct sockaddr_in echoClntAddr;
unsigned int clntlen;
int clntSock;
int servSock;
unsigned char buffer[BUFLEN];
char c[1]; // handshaking variable
typedef struct
//Structure for metering values so as to communicate with the HMI
{
float Vr1,Ir1,Vy1,Iy1,Vb1,Ib1,Vbb1,In1,Vr2,Ir2,Vy2,Iy2,Vb2,Ib2,Vbb2,In2;
  float Vr3,Ir3,Vy3,Iy3,Vb3,Ib3,Vbb3,In3,Vr4,Ir4,Vy4,Iy4,Vb4,Ib4,Vbb4,In4;
} rx_meter;
rx_meter rx_meter1;
unsigned char *addr;
typedef union {
  struct _pulse pulse;
  /* your other message structures would go here too */
} my_message_t;
unsigned long ii,j,k,l,m,n,o,p,q,r,s,t; // Digital Input Variables initialization
int initport(int fd)
/* Function to initialize the serial port*/
  {
struct termios options;
/* Terminal control structure */
  tcgetattr(fd, &options);
  cfsetispeed(&options, B19200); // Input baud rate
  cfsetospeed(&options, B19200); // Output baud rate
  options.c_cflag |= (CLOCAL | CREAD);
  options.c_cflag &= ~PARENB;
```

```
  options.c_cflag &= ~CSTOPB;
  options.c_cflag &= ~CSIZE;
  options.c_cflag |= CS8;
  tcsetattr(fd, TCSANOW, &options);
  return 1;
  tcflush(fd,TCIFLUSH);
  }
static void *Thread1( void *arg ) {      //Every 500usec for sampling etc
   struct itimerspec itime;
   struct sigevent event;
   unsigned long *period = arg;
   my_message_t msg;
   timer_t timer_id;
   int chid, rcvid;
   chid = ChannelCreate( 0 );
   event.sigev_notify = SIGEV_PULSE;
// Connect Attach will establish a connection between a process and a channel
event.sigev_coid = ConnectAttach( ND_LOCAL_NODE, 0, chid, _NTO_SIDE_
CHANNEL, 0 );
   event.sigev_priority = 26;
   event.sigev_code = MY_PULSE_CODE;
// CLOCK_REALTIME is the machine's best-guess as to the current wall-clock,
time-of-day time
   timer_create( CLOCK_REALTIME, &event, &timer_id );
   itime.it_value.tv_sec = 0;
   itime.it_value.tv_nsec = *period;
   itime.it_interval = itime.it_value;
   timer_settime( timer_id, 0, &itime, NULL );
   while(1)
   {
// MsgReceive will Wait for a message or pulse on a channel
   if ( 0 == ( rcvid = MsgReceive( chid, &msg, sizeof msg, NULL ) ) )
   {
   if(( msg.pulse.code )==MY_PULSE_CODE)
// will check if the pulse has been received or not
   {
```

```
// Locking the mutex a piece of code then no other thread can execute the same
region //until it is unlocked by the thread who locked it.
   pthread_mutex_lock (&mutex);
   // Sampling function is called
   sampling();
   // Unlocking the mutex
   pthread_mutex_unlock (&mutex);
   }
   }
   }
   return arg;
}
static void *Thread2( void *arg ) {          //      1msec thread for Measurand
   struct itimerspec itime;
   struct sigevent event;
   unsigned long *period = arg;
   my_message_t msg;
   timer_t timer_id;
   int chid, rcvid;
   chid = ChannelCreate( 0 );
   event.sigev_notify = SIGEV_PULSE;
event.sigev_coid = ConnectAttach( ND_LOCAL_NODE, 0, chid, _NTO_SIDE_
CHANNEL, 0 );
   event.sigev_priority = 26;
   event.sigev_code = MY_PULSE_CODE;
   timer_create( CLOCK_REALTIME, &event, &timer_id );
   itime.it_value.tv_sec = 0;
   itime.it_value.tv_nsec = *period;
   itime.it_interval = itime.it_value;
   timer_settime( timer_id, 0, &itime, NULL );
   while(1)
   {
   // Same a waiting for the pulse unlike previous thread
   if ( 0 == ( rcvid = MsgReceive( chid, &msg, sizeof msg, NULL ) ) )
```

```
    {
    if(( msg.pulse.code )==MY_PULSE_CODE)
    {
    // Call of the Measurand function for rms computation
    measurand();
    }
    }
    }
    return arg;
}
static void *Thread3( void *arg ) {          //     500msec thread for Metering,
Communication etc
    struct itimerspec itime;
    struct sigevent event;
    unsigned long *period = arg;
    my_message_t msg;
    timer_t timer_id;
    int chid, rcvid;
    chid = ChannelCreate( 0 );
    event.sigev_notify = SIGEV_PULSE;
event.sigev_coid = ConnectAttach( ND_LOCAL_NODE, 0, chid, _NTO_SIDE_
CHANNEL, 0 );
    event.sigev_priority = 26;
    event.sigev_code = MY_PULSE_CODE;
    timer_create( CLOCK_REALTIME, &event, &timer_id );
    itime.it_value.tv_sec = 0;
    itime.it_value.tv_nsec = *period;
    itime.it_interval = itime.it_value;
    timer_settime( timer_id, 0, &itime, NULL );
    while(1)
    {
    // Waiting for the pulse
    if ( 0 == ( rcvid = MsgReceive( chid, &msg, sizeof msg, NULL ) ) )
    {
    if(( msg.pulse.code )==MY_PULSE_CODE)
    {
```

```
// metering function is called for storing the rms computed values in the structure variables
    metering();
    // Making the 32 Analog values to be 0 for taking the samples of the next cycle
    Vr1=0;Ir1=0;Vy1=0;Iy1=0;Vb1=0;Ib1=0;Vbb1=0;In1=0;
    Vr2=0;Ir2=0;Vy2=0;Iy2=0;Vb2=0;Ib2=0;Vbb2=0;In2=0;
    Vr3=0;Ir3=0;Vy3=0;Iy3=0;Vb3=0;Ib3=0;Vbb3=0;In3=0;
    Vr4=0;Ir4=0;Vy4=0;Iy4=0;Vb4=0;Ib4=0;Vbb4=0;In4=0;
    }
    }
    }
    return arg;
}
void *ThreadMain( void *threadArgs ) {// 1 sec thread for TCP/IP comunication
char ch1;
int msglen=0;
    ii = in8(0x330);
// Reading of 1-8 DI channels of 96 channel Adon Module JP1 from the address 0x330
    j = in8(0x331);
// Reading of 9-16 DI channels of 96 channel Adon Module JP1 from the address 0x331
    k = in8(0x332);
// Reading of 17-24 DI channels of 96 channel Adon Module JP1 from the address 0x332
    l = in8(0x334);
// Reading of 25-32 DI channels of 96 channel Adon Module JP1 from the address 0x334
    m = in8(0x335);
// Reading of 33-40 DI channels of 96 channel Adon Module JP1 from the address 0x335
    n = in8(0x336);
// Reading of 41-48 DI channels of 96 channel Adon Module JP1 from the address 0x336
    o = in8(0x338);
// Reading of 1-8 DI channels of 96 channel Adon Module JP2 from the address 0x338
    p = in8(0x339);
```

```
// Reading of 9-16 DI channels of 96 channel Adon Module JP2 from the address 0x339
   q = in8(0x33a);
// Reading of 17-24 DI channels of 96 channel Adon Module JP2 from the address 0x33a
   r = in8(0x33c);
// Reading of 25-32DI channels of 96 channel Adon Module JP2 from the address 0x33c
   s = in8(0x33d);
// Reading of 33-40 DI channels of 96 channel Adon Module JP2 from the address 0x33d
   t = in8(0x33e);
// Reading of 41-48 DI channels of 96 channel Adon Module JP2 from the address 0x33e
// Running at every 1 sec
   while(1)
   {
           usleep(1000*1000);
   printf("\n Serial Communication Thread");
   // Waiting to receive the data on the serial port sp. If received will communicate
if(tcischars(sp) != 0)
//number of characters waiting to be read, or -1 if an error occurred.
   {
ch1 = rs_getch(sp);printf("\n1 %c",ch1);
   // Calling the serial_timer function
   serial_timer(ch1);}
   else
   {
rs_flush(sp);}
   printf("\n TCP Thread");
memset(buffer,'\0',500);
   // Reading from the HMI in the socket clntSock
if((msglen = read(clntSock,buffer,BUFLEN)) <0)
   {
printf("Failed to receive\n"); }
   else
   {
```

```
// Reading the first string stored in the buffer
ch1 = buffer[0];
// Printing the first string stored in the buffer
printf("\n buffer = %c \n",buffer[0]);
// Checking if the character is M or not. If Yes get inside the loop else check for the
if statement
if(ch1 == 'M')
{
memset(str,'\0',500);
// fills the first 500 bytes of the memory area
// pointed to by str with the constant '\0'
// sprint will write the formatted data into string str
sprintf(str,"m%02x%02x%02x%02x%02x%02x%02x%02x%02x%02x%02x%02x%04d
%04d%04d%04d%04d%04d%04d%04d%04d%04d%04d%04d%04d%04d%04d%04d
%04d%04d%04d%04d%04d%04d%04d%04d%04d%04d%04d%04d%04d%04d%04d
%04d",ii,j,k,l,m,n,o,p,q,r,s,t,(int)(rx_meter1.Vr1*100.0),(int)(rx_meter1.Ir1*100.0),(int)
(rx_meter1.Vy1*100.0),(int)(rx_meter1.Iy1*100.0),(int)(rx_meter1.Vb1*100.0),(int)
(rx_meter1.Ib1*100.0),(int)(rx_meter1.Vbb1*100.0),(int)(rx_meter1.In1*100.0),(int)
(rx_meter1.Vr2*100.0),(int)(rx_meter1.Ir2*100.0),(int)(rx_meter1.Vy2*100.0),(int)
(rx_meter1.Iy2*100.0),(int)(rx_meter1.Vb2*100.0),(int)(rx_meter1.Ib2*100.0),(int)
(rx_meter1.Vbb2*100.0),(int)(rx_meter1.In2*100.0),(int)(rx_meter1.Vr3*100.0),(int)
(rx_meter1.Ir3*100.0),(int)(rx_meter1.Vy3*100.0),(int)(rx_meter1.Iy3*100.0),(int)
(rx_meter1.Vb3*100.0),(int)(rx_meter1.Ib3*100.0),(int)(rx_meter1.Vbb3*100.0),(int)
(rx_meter1.In3*100.0),(int)(rx_meter1.Vr4*100.0),(int)(rx_meter1.Ir4*100.0),(int)
(rx_meter1.Vy4*100.0),(int)(rx_meter1.Iy4*100.0),(int)(rx_meter1.Vb4*100.0),(int)
(rx_meter1.Ib4*100.0),(int)(rx_meter1.Vbb4*100.0),(int)(rx_meter1.In4*100.0));
if((msglen=write(clntSock,str,strlen(str))) < 0)
{
  printf("Failed to send data\n");
  continue;
}
}
// If the character received is 'p' from the HMI
else if(ch1 == 'p')// 40 O/P Board
{
c[0] = 'P';
// Sending the character 'P' to the HMI as a means of handshaking. If failed to send,
then will generate the error
```

```
if(write(clntSock,c,1)<0)
  {
  perror("Error Writing to socket of settings page");
  continue;
  }
// Reading the ASCII value from the HMI and stores in the buffer
if(read(clntSock,buffer,80)<0)
  {
  perror("Read Error");
  continue;
  }
// Converting the ASCII value into decimal form
  inp1 = 10.0*(int)(buffer[0]-'0') + 1.0*(int)(buffer[1]-'0');
  printf("\n inp1 = %d",inp1);
// Checking if the value sent from the HMI is 01 or not
  if(inp1 == 01)
  {
// Send the byte value 0x00 to 8 Digital Output channels starting from the address
0x250
  out8(0x240+0x10,0x00);
  }
// Checking if the value sent from the HMI is 02 or not
  else if(inp1 == 02)
  {
// Send the byte value 0xaa to 8 Digital Output channels starting from the address
0x250
  out8(0x240+0x10,0xaa);
  }
// Checking if the value sent from the HMI is 03 or not
  else if(inp1 == 03)
  {
// Send the byte value 0x55 to 8 Digital Output channels starting from the address
0x250
  out8(0x240+0x10,0x55);
  }
// Checking if the value sent from the HMI is 04 or not
```

```
else if(inp1 == 04)
{
// Send the byte value 0xff to 8 Digital Output channels starting from the address
0x250
out8(0x240+0x10,0xff);
}
}// end of P
// If the character received is 'o' from the HMI
else if(ch1 == 'o')// 40 O/P Board
{
c[0] = 'O';
// Sending the character 'O' to the HMI as a means of handshaking. If failed to send,
then will generate the error
if(write(clntSock,c,1)<0)
{
perror("Error Writing to socket of settings page");
continue;
}
// Reading the ASCII value from the HMI and stores in the buffer
if(read(clntSock,buffer,80)<0)
{
perror("Read Error");
continue;
}
// Converting the ASCII value into the decimal form
inp2 = 10.0*(int)(buffer[0]-'0') + 1.0*(int)(buffer[1]-'0');
// printing the decimal value
printf("\n inp2 = %d",inp2);
// Checking if the value received from the HMI is 01
if(inp2 == 01)
{
// Send the byte value 0x00 to 8 Digital Output channels starting from the address
0x251
out8(0x240+0x11,0x00);
}
// Checking if the value received from the HMI is 02
```

```
else if(inp2 == 02)
{
// Send the byte value 0xaa to 8 Digital Output channels starting from the address
0x251
out8(0x240+0x11,0xaa);
}
// Checking if the value received from the HMI is 03
else if(inp2 == 03)
{
// Send the byte value 0x55 to 8 Digital Output channels starting from the address
0x251
out8(0x240+0x11,0x55);
}
// Checking if the value received from the HMI is 04
else if(inp2 == 04)
{
// Send the byte value 0xff to 8 Digital Output channels starting from the address
0x251
out8(0x240+0x11,0xff);
}
}// end of O
// Check if the character received is 'r'
else if(ch1 == 'r')// 40 O/P Board
{
c[0] = 'R';
// Sending 'R' to the HMI as a means of handshaking. Will generate error if failed
to send
if(write(clntSock,c,1)<0)
{
perror("Error Writing to socket of settings page");
continue;
}
// Reading the ASCII values from the HMI and stores into the buffer
if(read(clntSock,buffer,80)<0)
{
perror("Read Error");
```

```
  continue;
  }
// Converting the ASCII value into the decimal
  inp3 = 10.0*(int)(buffer[0]-'0') + 1.0*(int)(buffer[1]-'0');
// Printing the decimal value
  printf("\n inp3 = %d",inp3);
// Checking if the value received from the HMI is 01
  if(inp3 == 01)
  {
// Send the byte value 0x00 to 8 Digital Output channels starting from the address
0x252
  out8(0x240+0x12,0x00);
  }
// Checking if the value received from the HMI is 02
  else if(inp3 == 02)
  {
// Send the byte value 0xaa to 8 Digital Output channels starting from the address
0x252
  out8(0x240+0x12,0xaa);
  }
// Checking if the value received from the HMI is 03
  else if(inp3 == 03)
  {
// Send the byte value 0x55 to 8 Digital Output channels starting from the address
0x252
  out8(0x240+0x12,0x55);
  }
// Checking if the value received from the HMI is 04
  else if(inp3 == 04)
  {
// Send the byte value 0xff to 8 Digital Output channels starting from the address
0x252
  out8(0x240+0x12,0xff);
  }
  }// end of R
// Receiving character 'q' from the HMI
```

```
else if(ch1 == 'q')// 40 O/P Board
{
c[0] = 'Q';
/// Sending 'Q' to the HMI on receiving q. Will Generate error if failed to send.
if(write(clntSock,c,1)<0)
{
perror("Error Writing to socket of settings page");
continue;
}
// Will read the ASCII values and store it in the buffer
if(read(clntSock,buffer,80)<0)
{
perror("Read Error");
continue;
}
// Converting the ASCII value into the decimal
inp4 = 10.0*(int)(buffer[0]-'0') + 1.0*(int)(buffer[1]-'0');
// Printing the decimal value
printf("\n inp4 = %d",inp4);
// Checking if the value received from the HMI is 01
if(inp4 == 01)
{
// Send the byte value 0x00 to 8 Digital Output channels starting from the address
0x253
out8(0x240+0x13,0x00);
}
// Checking if the value received from the HMI is 02
else if(inp4 == 02)
{
// Send the byte value 0xaa to 8 Digital Output channels starting from the address
0x253
out8(0x240+0x13,0xaa);
}
// Checking if the value received from the HMI is 03
else if(inp4 == 03)
{
```

```
// Send the byte value 0x55 to 8 Digital Output channels starting from the address
0x253
   out8(0x240+0x13,0x55);
   }
// Checking if the value received from the HMI is 04
   else if(inp4 == 04)
   {
// Send the byte value 0xff to 8 Digital Output channels starting from the address
0x253
   out8(0x240+0x13,0xff);
   }
   }// end of Q
// Receiving character s from the HMI
   else if(ch1 == 's')// 40 O/P Board
   {
   c[0] = 'S';
// Sending 'S' to the HMI as a handshaking variable. Will generate error if failed to
send
   if(write(clntSock,c,1)<0)
   {
   perror("Error Writing to socket of settings page");
   continue;
   }
// Reading the ASCII values from the HMI and stores in the buffer
   if(read(clntSock,buffer,80)<0)
   {
   perror("Read Error");
   continue;
   }
// Converting the ASCII value into the decimal
   inp5 = 10.0*(int)(buffer[0]-'0') + 1.0*(int)(buffer[1]-'0');
// printing the decimal value
   printf("\n inp5 = %d",inp5);
// Checking if the value received from the HMI is 01
   if(inp5 == 01)
   {
```

```
// Send the byte value 0x00 to 8 Digital Output channels starting from the address
0x254
   out8(0x240+0x14,0x00);
   }
// Checking if the value received from the HMI is 02
   else if(inp5 == 02)
   {
// Send the byte value 0xaa to 8 Digital Output channels starting from the address
0x254
   out8(0x240+0x14,0xaa);
   }
// Checking if the value received from the HMI is 03
   else if(inp5 == 03)
   {
// Send the byte value 0x55 to 8 Digital Output channels starting from the address
0x254
   out8(0x240+0x14,0x55);
   }
// Checking if the value received from the HMI is 04
   else if(inp5 == 04)
   {
// Send the byte value 0xff to 8 Digital Output channels starting from the address
0x254
   out8(0x240+0x14,0xff);
   }
   }// end of S
// Receiving the character 'v' from the HMI
   else if(ch1 == 'v')// 24 O/P Board
   {
   c[0] = 'V';
// Sending the character 'V' from the HMI as a means of handshaking. Will generate
error if failed to send
   if(write(clntSock,c,1)<0)
   {
   perror("Error Writing to socket of settings page");
   continue;
```

```
    }
// Reading the ASCII values from the HMI and stores in the buffer
    if(read(clntSock,buffer,80)<0)
    {
    perror("Read Error");
    continue;
    }
// Converting the ASCII values received from the HMI into the decimal
    inp6 = 10.0*(int)(buffer[0]-'0') + 1.0*(int)(buffer[1]-'0');
// printing the decimal value
    printf("\n inp6 = %d",inp6);
// Checking if the value received from the HMI is 01
    if(inp6 == 01)
    {
// Send the byte value 0x00 to 8 Digital Output channels starting from the address
0x300
    out8(0x300,0x00);
    }
// Checking if the value received from the HMI is 02
    else if(inp6 == 02)
    {
// Send the byte value 0xaa to 8 Digital Output channels starting from the address
0x300
    out8(0x300,0xaa);
    }
// Checking if the value received from the HMI is 03
    else if(inp6 == 03)
    {
// Send the byte value 0x55 to 8 Digital Output channels starting from the address
0x300
    out8(0x300,0x55);
    }
// Checking if the value received from the HMI is 04
    else if(inp6 == 04)
    {
```

```
// Send the byte value 0xff to 8 Digital Output channels starting from the address
0x300
   out8(0x300,0xff);
   }
   }// end of V
// Receiving the character 'u' from the HMI
   else if(ch1 == 'u')// 24 O/P Board
   {
   c[0] = 'U';
// Sending the character 'U' to the HMI as a means of handshaking. Will generate
error if failed to send
   if(write(clntSock,c,1)<0)
   {
   perror("Error Writing to socket of settings page");
   continue;
   }
// Reading the ASCII values from the HMI and stores in the buffer
   if(read(clntSock,buffer,80)<0)
   {
   perror("Read Error");
   continue;
   }
// Converting the ASCII value into the decimal
   inp7 = 10.0*(int)(buffer[0]-'0') + 1.0*(int)(buffer[1]-'0');
// Printing the decimal value
   printf("\n inp7 = %d",inp7);
// Checking if the value received from the HMI is 01
   if(inp7 == 01)
   {
// Send the byte value 0x00 to 8 Digital Output channels starting from the address
0x301
   out8(0x300+0x01,0x00);
   }
// Checking if the value received from the HMI is 02
   else if(inp7 == 02)
```

```
{
// Send the byte value 0xaa to 8 Digital Output channels starting from the address 0x301
out8(0x300+0x01,0xaa);
}
// Checking if the value received from the HMI is 03
else if(inp7 == 03)
{
// Send the byte value 0x55 to 8 Digital Output channels starting from the address 0x301
out8(0x300+0x01,0x55);
}
// Checking if the value received from the HMI is 04
else if(inp7 == 04)
{
// Send the byte value 0xff to 8 Digital Output channels starting from the address 0x301
out8(0x300+0x01,0xff);
}
}// end of U
// Receiving the character 't' from the HMI
else if(ch1 == 't')// 24 O/P Board
{
c[0] = 'T';
// Sending the character 'T' to the HMI as a means of handshaking. Will generate error if failed to send
if(write(clntSock,c,1)<0)
{
perror("Error Writing to socket of settings page");
continue;
}
// Receiving the ASCII values from the HMI and stores in the buffer
if(read(clntSock,buffer,80)<0)
{
perror("Read Error");
continue;
```

```
   }
// Converting the ASCII value into the decimal
   inp8 = 10.0*(int)(buffer[0]-'0') + 1.0*(int)(buffer[1]-'0');
// Printing the decimal value
   printf("\n inp8 = %d",inp8);
// Checking if the value received from the HMI is 01
   if(inp8 == 01)
   {
// Send the byte value 0x00 to 8 Digital Output channels starting from the address
0x302
   out8(0x300+0x02,0x00);
   }
// Checking if the value received from the HMI is 02
   else if(inp8 == 02)
   {
// Send the byte value 0xaa to 8 Digital Output channels starting from the address
0x302
   out8(0x300+0x02,0xaa);
   }
// Checking if the value received from the HMI is 03
   else if(inp8 == 03)
   {
// Send the byte value 0x55 to 8 Digital Output channels starting from the address
0x302
   out8(0x300+0x02,0x55);
   }
// Checking if the value received from the HMI is 04
   else if(inp8 == 04)
   {
// Send the byte value 0xff to 8 Digital Output channels starting from the address
0x302
   out8(0x300+0x02,0xff);
   }
   }// end of T
   }// end of else */
   close(clntSock); // close of socket
```

```
    return (NULL);
    } // end of while loop
}
void serial_timer(char ch2)
{
    if(ch2 == 'm')
    {
    printf("\n hello");
    rs_flush(sp);
    // Sprint will write the formatted data into string str
sprintf(str,"m%02x%02x%02x%02x%02x%02x%02x%02x%02x%02x%02x%02x%04d
%04d%04d%04d%04d%04d%04d%04d%04d%04d%04d%04d%04d%04d%04d%04d
%04d%04d%04d%04d%04d%04d%04d%04d%04d%04d%04d%04d%04d%04d%04d
%04d",ii,j,k,l,m,n,o,p,q,r,s,t,(int)(rx_meter1.Vr1*100.0),(int)(rx_meter1.Ir1*100.0),(int)
(rx_meter1.Vy1*100.0),(int)(rx_meter1.Iy1*100.0),(int)(rx_meter1.Vb1*100.0),(int)
(rx_meter1.Ib1*100.0),(int)(rx_meter1.Vbb1*100.0),(int)(rx_meter1.In1*100.0),(int)
(rx_meter1.Vr2*100.0),(int)(rx_meter1.Ir2*100.0),(int)(rx_meter1.Vy2*100.0),(int)
(rx_meter1.Iy2*100.0),(int)(rx_meter1.Vb2*100.0),(int)(rx_meter1.Ib2*100.0),(int)
(rx_meter1.Vbb2*100.0),(int)(rx_meter1.In2*100.0),(int)(rx_meter1.Vr3*100.0),(int)
(rx_meter1.Ir3*100.0),(int)(rx_meter1.Vy3*100.0),(int)(rx_meter1.Iy3*100.0),(int)
(rx_meter1.Vb3*100.0),(int)(rx_meter1.Ib3*100.0),(int)(rx_meter1.Vbb3*100.0),(int)
(rx_meter1.In3*100.0),(int)(rx_meter1.Vr4*100.0),(int)(rx_meter1.Ir4*100.0),(int)
(rx_meter1.Vy4*100.0),(int)(rx_meter1.Iy4*100.0),(int)(rx_meter1.Vb4*100.0),(int)
(rx_meter1.Ib4*100.0),(int)(rx_meter1.Vbb4*100.0),(int)(rx_meter1.In4*100.0));
// Sending the string array str to the HMI
    for(z=0;z<strlen(str);rs_putch(sp,str[z]),z++);
    }
}
void metering()
{
// Mapping the rms values calculated in the measurand function with the structure
variables of metering function.
    rx_meter1.Ir1 =Ir1;rx_meter1.Iy1 =Iy1;rx_meter1.Ib1 =Ib1;rx_meter1.In1 =In1;
rx_meter1.Vr1 =Vr1;rx_meter1.Vy1 =Vy1;rx_meter1.Vb1 =Vb1;rx_meter1.Vbb1
=Vbb1;
    rx_meter1.Ir2 =Ir2;rx_meter1.Iy2 =Iy2;rx_meter1.Ib2 =Ib2;rx_meter1.In2 =In2;
rx_meter1.Vr2 =Vr2;rx_meter1.Vy2 =Vy2;rx_meter1.Vb2 =Vb2;rx_meter1.Vbb2
=Vbb2;
```

```
    rx_meter1.Ir3 =Ir3;rx_meter1.Iy3 =Iy3;rx_meter1.Ib3 =Ib3;rx_meter1.In3 =In3;
rx_meter1.Vr3 =Vr3;rx_meter1.Vy3 =Vy3;rx_meter1.Vb3 =Vb3;rx_meter1.Vbb3
=Vbb3;
    rx_meter1.Ir4 =Ir4;rx_meter1.Iy4 =Iy4;rx_meter1.Ib4 =Ib4;rx_meter1.In4 =In4;
rx_meter1.Vr4 =Vr4;rx_meter1.Vy4 =Vy4;rx_meter1.Vb4 =Vb4;rx_meter1.Vbb4
=Vbb4;
}
// Start of the main function or the program
    int main(int argc, char *argv[])
    {
// Variables declaration
    pthread_t threadID;
    struct ThreadArgs *threadArgs;
    int sockoptval = 1;
    void *Thread_rc;
unsigned long periods[3] = { 1,2,1000};    //500usec,1msec ,500msec
    unsigned long base_period = 500000; //500usec base period
    sp = open("/dev/ser1", O_RDWR ); //Open the serial port 1
// Checking if the port got opened or not
    if (sp == -1)
            {
    perror("open_port: Unable to open /dev/ttyS0 - ");
    return 1;
            }
    else
    {
    //      printf("\n opened the serial Port");
fcntl(sp, F_SETFL, 0);
// Provide control over an open file
    }
// F_SETFL - Set the file status flags
    initport(sp);    // Initialise the port
    rs_flush(sp); // flushing the data if present on the port
    printf("\nHercules Board testing......");
// Making the Ports A,B,C,D and E as the Input Ports by setting 0x1F in the memory
```

```
Location 0x256.
   out8(0x240+0x16,0x1f);
   // Calling the initials function for Board initialization
   initials();
   // Calling the adcinit function for Data acquisition
   adcinit();
   // Calling the dacinit function for converting Digital value into Analog
   dacinit();
/*******************TCP/IP Socket connection initialization star
ts*************************/
   /// Initializing a socket
   if((servSock = socket(AF_INET,SOCK_STREAM,IPPROTO_TCP))<0)
   {
   printf("Failed to create socket");
   }
   // Allowing Immediate Reuse of the Port
   //      setsockopt(servSock,SOL_SOCKET,SO_REUSEADDR,&sockoptval,sizeo
f(int));
   /// Creating a socket
   memset(&echoServAddr,0,sizeof(echoServAddr));
   echoServAddr.sin_family = AF_INET;
   echoServAddr.sin_addr.s_addr = htonl(INADDR_ANY);
   echoServAddr.sin_port = htons(PORT);
   // Binding the socket to the source address
if(bind(servSock,(struct sockaddr *)&echoServAddr,sizeof(echoServAddr))<0)
   {
   printf("Failed to bind\n");
   //sleep(1);
   }
   /// Queuing by the server for having multiple connections from the client
   if(listen(servSock,MAXPENDING <0))
   {
   printf("Failed to listen\n");
   }
// Creation of Threads starts with their base period
```

```
/// Thread of 500 micro secs
periods[0] *= base_period;
   pthread_create( NULL, NULL, Thread1, &periods[0] );
   /// Thread of 1 millisec
   periods[1] *= base_period;
   pthread_create( NULL, NULL, Thread2, &periods[1] );
   /// Thread of 500 milli secs
   periods[2] *= base_period;
   pthread_create( NULL, NULL, Thread3, &periods[2] );
   //// TCP/IP Thread creation of 1 sec
   while(1)
   {
   clntlen = sizeof(echoClntAddr);
   while((clntSock = accept(servSock, (struct sockaddr *)&echoClntAddr,&clntlen))<0)
   {
   printf("Failed to accept client\n");
   }
   printf("\n Accept successful\n");
   pthread_create(&threadID,NULL,ThreadMain,(void *) threadArgs);
   }
   while (1) {};
   // Joining of threads
   pthread_join((pthread_t)Thread1,&Thread_rc);
   pthread_join((pthread_t)Thread2,&Thread_rc);
   pthread_join((pthread_t)Thread3,&Thread_rc);
   // Close of the ports
   rs_close(sp);
   close(clntSock);
   // free up the board
   finally();
return (int)Thread_rc;
   }
```

HMI CODE OF THE WINDOWS FORM APPLICATION DEMO IN VISUAL STUDIO IDE IN C#.NET LANGUAGE

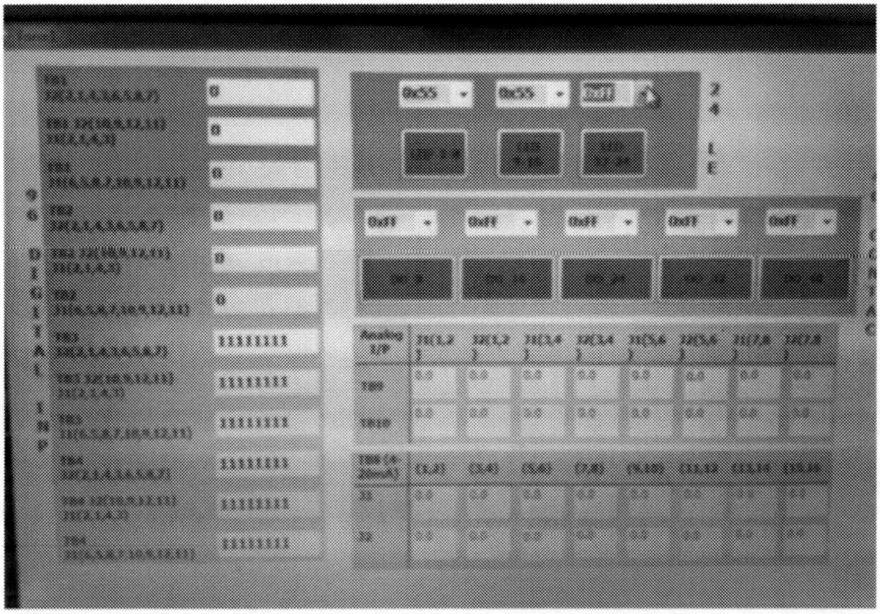

Fig. 11.15 *Screenshot of the HMI*

Source Code

```
using System;
using System.Collections.Generic;
using System.ComponentModel;
using System.Data;
using System.Drawing;
using System.Linq;
using System.Threading;
using System.Timers;
using System.Net;
using System.Net.Sockets;
using System.Text;
using System.Windows.Forms;
// sbctest is the name of the C# Windows Form Program
namespace sbctest
{
// Form1 is the windows Form Application named as a partial class
public partial class Form1 : Form
```

```
{
// IP Address of the Application program in the SBC Board
public string HOSTIP = "192.168.1.81";
// Port Number of the SBC Board
public int PORT = 1234;
// Instance of TCPClient Class
private TcpClient client;
// Instance of Network Stream Class
private NetworkStream stream;
// Counter for the timer count
int count = 0;
// buf is a variable of type string
string buf = "";
// Unparametrized constructor calling the function InitializeComponent
public Form1()
{
// Required method for Designer support - do not modify
/// the contents of this method with the code editor.
InitializeComponent();
}
// Connect function for connecting with the socket
public bool Connect()
{
bool stat = false;
try
{
   Initializes a new instance of the System.Net.Sockets.TcpClient class and
// connects to the specified port on the specified host.
client = new TcpClient(HOSTIP, PORT);
// Gets or sets the amount of time a System.Net.Sockets.TcpClient will wait to receive
data once a read operation is initiated. It is initialized to 1 sec.
client.ReceiveTimeout = 1000;
// Returns the System.Net.Sockets.NetworkStream used to send and receive
data
stream = client.GetStream();
```

```
    // Gets a value that indicates whether the System.Net.Sockets.Network
    Stream supports writing.
stat = stream.CanWrite;
}
catch
{
}
return stat;
}
private void Close1()
{
try
{
    // Close of NetworkStream and TcpClient
stream.Close();
client.Close();
}
catch
{
}
}
public string Read(int BYTES_READ)
{
    // Array for storing the data
byte[] data;
int bytes;
string responsedata = string.Empty;
    // Try Catch is used for catching of the exception if found at run time
    // Tell C# to try some code, and if can't do anything with it then catch the
    // errors
try
{
data = new byte[BYTES_READ];
// Reads data from the System.Net.Sockets.NetworkStream.
bytes = stream.Read(data, 0, data.Length);
```

```
// When overridden in a derived class, decodes a sequence of bytes // from the specified byte array into a string.
responsedata = System.Text.Encoding.ASCII.GetString(data, 0, bytes);
}
catch
{
}
// Return the response data.
return responsedata;
}
public void write(string msg)
{
// Array of message.
byte[] message;
try
{
// Returns the stream for sending and receiving data
stream = client.GetStream();
if (stream.CanWrite)
{
// When overridden in a derived class, encodes all the characters in the specified string into a sequence of bytes.
Message = System.Text.Encoding.ASCII.GetBytes(msg);
// Writes data to the System.Net.Sockets.NetworkStream.
stream.Write(message, 0, message.Length);
}
else
{
MessageBox.Show("Unable to write to stream");
}
}
catch
{
}
}
private void Timer1_Tick(object sender, EventArgs e)
```

```
{
// Variable declaration
string ch = null;
string X_data = "";
string temp = "";
   // Declaring a character 'M' in the variable ch for handhshaking
ch = "M";
double i;
bool stat1 = false;
   // Connecting with the socket and will return true if connected.
stat1 = Connect();
if (stat1 == true)
{
   // Writes data to the System.Net.Sockets.NetworkStream and send it to
the Application side as a start of Handshaking.
write(ch);
}
   // will receive the data bytes from the Application side and store it in the
   X_data variable. The Array of 500 indicates maximum bytes to read
X_data = X_data + Read(500);
   // Storing the X_data in the textBox13
textBox13.Text = X_data;
   // Closing the Socket
Close1();
   // Counting the length of the data
int l = X_data.Length;
label74.Text = l.ToString();
   // Checking if the length of the databytes read is not zero
if (l != 0)
{
   // Splitting the string and reading the first character from the X_data and
   storing in the temp variable
temp = X_data.Substring(0, 1);
}
   // Incrementing the count indicating the timer run count
```

```
label75.Text = (count++).ToString();
    // Checking if the character 'm' is read and the length of the data bytes
    received from the Application side is 153 or not
if ((temp == "m") & (l == 153))
{
// Selects a range of text in the text box from 1 with a length of 2.
textBox13.Select(1,2);
    //Converting the selected text into the 8-bit hexadecimal representation.
    TextBox1.Text=Convert.ToString(Convert.ToInt32(textBox13.SelectedText,16),2);
    // Selects a range of text in the text box from 3 with a length of 2.
textBox13.Select(3, 2);
//Converting the selected text into the 8-bit hexadecimal representation.
    TextBox2.Text = Convert.ToString(Convert.ToInt32(textBox13.SelectedText, 16),
2);
    // Selects a range of text in the text box from 5 with a length of 2.
textBox13.Select(5, 2);
//Converting the selected text into the 8-bit hexadecimal representation.
    TextBox3.Text = Convert.ToString(Convert.ToInt32(textBox13.SelectedText, 16),
2);
    // Selects a range of text in the text box from 7 with a length of 2.
textBox13.Select(7, 2);
    //Converting the selected text into the 8-bit hexadecimal representation.
    TextBox4.Text = Convert.ToString(Convert.ToInt32(textBox13.SelectedText, 16),
2);
    // Selects a range of text in the text box from 9 with a length of 2.
textBox13.Select(9, 2);
    //Converting the selected text into the 8-bit hexadecimal representation.
    TextBox5.Text = Convert.ToString(Convert.ToInt32(textBox13.SelectedText, 16),
2);
    // Selects a range of text in the text box from 11 with a length of 2.
textBox13.Select(11, 2);
    //Converting the selected text into the 8-bit hexadecimal representation.
    TextBox6.Text = Convert.ToString(Convert.ToInt32(textBox13.SelectedText, 16),
2);
    // Selects a range of text in the text box from 13 with a length of 2.
textBox13.Select(13, 2);
```

```
//Converting the selected text into the 8-bit hexadecimal representation.
TextBox7.Text = Convert.ToString(Convert.ToInt32(textBox13.SelectedText, 16), 2);
// Selects a range of text in the text box from 15 with a length of 2.
textBox13.Select(15, 2);
//Converting the selected text into the 8-bit hexadecimal representation.
TextBox8.Text = Convert.ToString(Convert.ToInt32(textBox13.SelectedText, 16), 2);
// Selects a range of text in the text box from 17 with a length of 2.
textBox13.Select(17, 2);
//Converting the selected text into the 8-bit hexadecimal representation.
TextBox9.Text = Convert.ToString(Convert.ToInt32(textBox13.SelectedText, 16), 2);
// Selects a range of text in the text box from 19 with a length of 2.
textBox13.Select(19, 2);
//Converting the selected text into the 8-bit hexadecimal representation.
TextBox10.Text = Convert.ToString(Convert.ToInt32(textBox13.SelectedText, 16), 2);
// Selects a range of text in the text box from 21 with a length of 2.
textBox13.Select(21, 2);
//Converting the selected text into the 8-bit hexadecimal representation.
TextBox11.Text = Convert.ToString(Convert.ToInt32(textBox13.SelectedText, 16), 2);
// Selects a range of text in the text box from 23 with a length of 2.
textBox13.Select(23, 2);
//Converting the selected text into the 8-bit hexadecimal representation.
TextBox12.Text = Convert.ToString(Convert.ToInt32(textBox13.SelectedText, 16), 2);
// Vr1
// Selects a range of text in the text box from 25 with a length of 4.
textBox13.Select(25, 4);
// Converting the string selected into double
i = double.Parse(textBox13.SelectedText) *0.635;
// Formatting the value and storing into Label
Label14.Text = string.Format("{0:0.0}", i);
// Ir1
// Selects a range of text in the text box from 29 with a length of 4.
```

```
textBox13.Select(29, 4);
   // Converting the string selected into double
i = double.Parse(textBox13.SelectedText);
   // Formatting the value and storing into Label
Label15.Text = string.Format("{0:0.0}", i);
// Vy1
   // Selects a range of text in the text box from 33 with a length of 4.
textBox13.Select(33, 4);
   // Converting the string selected into double
i = double.Parse(textBox13.SelectedText) * 0.635;
   // Formatting the value and storing into Label
Label16.Text = string.Format("{0:0.0}", i);
// Iy1
   // Selects a range of text in the text box from 37 with a length of 4.
textBox13.Select(37, 4);
   // Converting the string selected into double
i = double.Parse(textBox13.SelectedText);
   // Formatting the value and storing into Label
Label17.Text = string.Format("{0:0.0}", i);
// Vb1
   // Selects a range of text in the text box from 41 with a length of 4.
textBox13.Select(41, 4);
   // Converting the string selected into double
i = double.Parse(textBox13.SelectedText) * 0.635;
   // Formatting the value and storing into Label
Label18.Text = string.Format("{0:0.0}", i);
// Ib1
   // Selects a range of text in the text box from 45 with a length of 4.
textBox13.Select(45, 4);
// Converting the string selected into double
i = double.Parse(textBox13.SelectedText);
// Formatting the value and storing into Label
Label19.Text = string.Format("{0:0.0}", i);
// Vbb1
   // Selects a range of text in the text box from 49 with a length of 4.
```

```
textBox13.Select(49, 4);
    // Converting the string selected into double
i = double.Parse(textBox13.SelectedText) * 0.635;
    // Formatting the value and storing into Label
Label20.Text = string.Format("{0:0.0}", i);
// In1
    // Selects a range of text in the text box from 53 with a length of 4.
textBox13.Select(53, 4);
    // Converting the string selected into double
i = double.Parse(textBox13.SelectedText);
    // Formatting the value and storing into Label
Label21.Text = string.Format("{0:0.0}", i);
// Vr2
    // Selects a range of text in the text box from 57 with a length of 4.
textBox13.Select(57, 4);
    // Converting the string selected into double
i = double.Parse(textBox13.SelectedText) * 0.635;
    // Formatting the value and storing into Label
Label22.Text = string.Format("{0:0.0}", i);
// Ir2
    // Selects a range of text in the text box from 61 with a length of 4.
textBox13.Select(61, 4);
    // Converting the string selected into double
i = double.Parse(textBox13.SelectedText);
    // Formatting the value and storing into Label
Label23.Text = string.Format("{0:0.0}", i);
// Vy2
    // Selects a range of text in the text box from 65 with a length of 4.
textBox13.Select(65, 4);
    // Converting the string selected into double
i = double.Parse(textBox13.SelectedText) * 0.635;
    // Formatting the value and storing into Label
Label24.Text = string.Format("{0:0.0}", i);
// Iy2
    // Selects a range of text in the text box from 69 with a length of 4.
```

```
textBox13.Select(69, 4);
    // Converting the string selected into double
i = double.Parse(textBox13.SelectedText);
    // Formatting the value and storing into Label
Label25.Text = string.Format("{0:0.0}", i);
// Vb2
    // Selects a range of text in the text box from 73 with a length of 4.
textBox13.Select(73, 4);
    // Converting the string selected into double
i = double.Parse(textBox13.SelectedText) * 0.635;
    // Formatting the value and storing into Label
Label26.Text = string.Format("{0:0.0}", i);
// Ib2
    // Selects a range of text in the text box from 77 with a length of 4.
textBox13.Select(77, 4);
    // Converting the string selected into double
i = double.Parse(textBox13.SelectedText);
Label27.Text = string.Format("{0:0.0}", i);
// Vbb2
    // Selects a range of text in the text box from 81 with a length of 4.
textBox13.Select(81, 4);
    // Converting the string selected into double
i = double.Parse(textBox13.SelectedText) * 0.635;
    // Formatting the value and storing into Label
Label28.Text = string.Format("{0:0.0}", i);
// In2
    // Selects a range of text in the text box from 85 with a length of 4.
textBox13.Select(85, 4);
    // Converting the string selected into double
i = double.Parse(textBox13.SelectedText);
    // Formatting the value and storing into Label
Label29.Text = string.Format("{0:0.0}", i);
// Vr3
    // Selects a range of text in the text box from 89 with a length of 4.
textBox13.Select(89, 4);
```

```
    // Converting the string selected into double
i = double.Parse(textBox13.SelectedText) * 0.635;
    // Formatting the value and storing into Label
Label30.Text = string.Format("{0:0.0}", i);
// Ir3
    // Selects a range of text in the text box from 93 with a length of 4.
textBox13.Select(93, 4);
    // Converting the string selected into double
i = double.Parse(textBox13.SelectedText);
    // Formatting the value and storing into Label
Label31.Text = string.Format("{0:0.0}", i);
// Vy3
    // Selects a range of text in the text box from 97 with a length of 4.
textBox13.Select(97, 4);
    // Converting the string selected into double
i = double.Parse(textBox13.SelectedText) * 0.635;
    // Formatting the value and storing into Label
Label32.Text = string.Format("{0:0.0}", i);
// Iy3
    // Selects a range of text in the text box from 101 with a length of 4.
textBox13.Select(101, 4);
    // Converting the string selected into double
i = double.Parse(textBox13.SelectedText);
    // Formatting the value and storing into Label
Label33.Text = string.Format("{0:0.0}", i);
// Vb3
    // Selects a range of text in the text box from 105 with a length of 4.
textBox13.Select(105, 4);
    // Converting the string selected into double
i = double.Parse(textBox13.SelectedText) * 0.635;
    // Formatting the value and storing into Label
Label34.Text = string.Format("{0:0.0}", i);
// Ib3
    // Selects a range of text in the text box from 109 with a length of 4.
textBox13.Select(109, 4);
```

```
    // Converting the string selected into double
i = double.Parse(textBox13.SelectedText);
    // Formatting the value and storing into Label
Label35.Text = string.Format("{0:0.0}", i);
// Vbb3
    // Selects a range of text in the text box from 113 with a length of 4.
textBox13.Select(113, 4);
    // Converting the string selected into double
i = double.Parse(textBox13.SelectedText) * 0.635;
    // Formatting the value and storing into Label
Label36.Text = string.Format("{0:0.0}", i);
// In3
    // Selects a range of text in the text box from 117 with a length of 4.
textBox13.Select(117, 4);
    // Converting the string selected into double
i = double.Parse(textBox13.SelectedText);
    // Formatting the value and storing into Label
Label37.Text = string.Format("{0:0.0}", i);
// Vr4
    // Selects a range of text in the text box from 121 with a length of 4.
textBox13.Select(121, 4);
    // Converting the string selected into double
i = double.Parse(textBox13.SelectedText) * 0.635;
    // Formatting the value and storing into Label
Label38.Text = string.Format("{0:0.0}", i);
// Ir4
    // Selects a range of text in the text box from 125 with a length of 4.
textBox13.Select(125, 4);
    // Converting the string selected into double
i = double.Parse(textBox13.SelectedText);
    // Formatting the value and storing into Label
Label39.Text = string.Format("{0:0.0}", i);
// Vy4
    // Selects a range of text in the text box from 129 with a length of 4.
textBox13.Select(129, 4);
```

```
    // Converting the string selected into double
i = double.Parse(textBox13.SelectedText) * 0.635;
    // Formatting the value and storing into Label
Label40.Text = string.Format("{0:0.0}", i);
// Iy4
    // Selects a range of text in the text box from 133 with a length of 4.
textBox13.Select(133, 4);
    // Converting the string selected into double
i = double.Parse(textBox13.SelectedText);
    // Formatting the value and storing into Label
Label41.Text = string.Format("{0:0.0}", i);
// Vb4
    // Selects a range of text in the text box from 137 with a length of 4.
textBox13.Select(137, 4);
    // Converting the string selected into double
i = double.Parse(textBox13.SelectedText) * 0.635;
    // Formatting the value and storing into Label
Label42.Text = string.Format("{0:0.0}", i);
// Ib4
    // Selects a range of text in the text box from 141 with a length of 4.
textBox13.Select(141, 4);
    // Converting the string selected into double
i = double.Parse(textBox13.SelectedText);
    // Formatting the value and storing into Label
Label43.Text = string.Format("{0:0.0}", i);
// Vbb4
    // Selects a range of text in the text box from 145 with a length of 4.
textBox13.Select(145, 4);
    // Converting the string selected into double
i = double.Parse(textBox13.SelectedText) * 0.635;
    // Formatting the value and storing into Label
Label44.Text = string.Format("{0:0.0}", i);
// In4
    // Selects a range of text in the text box from 149 with a length of 4.
textBox13.Select(149, 4);
```

```
    // Converting the string selected into double
i = double.Parse(textBox13.SelectedText);
    // Formatting the value and storing into Label
Label45.Text = string.Format("{0:0.0}", i);
}
}
    // On Form Load Event, the timer is enabled
private void Form1_Load(object sender, EventArgs e)
{
Timer1.Enabled = true;
}
    // Whenever the form is activated timer is enabled
private void Form1_Activated(object sender, EventArgs e)
{
Timer1.Enabled = true;
}
    // Whenever the form is deactivated timer is disabled
private void Form1_Deactivate(object sender, EventArgs e)
{
Timer1.Enabled = false;
}
// On Button Click event of DO_8, first the socket is connected, then the
    // character 'p' is sent to the application side and the character 'P'
// is received from the application side as a handshaking variable. If received //
correctly, selecting a hexadecimal value from the combobox which can be either
// 0x00,0xaa,0x55,0xff and sending it to the application side and finally closing //
the Socket and the stream. Depending on the hexadecimal value received in the //
Application side, the respective DO channel bit will be high(+5V) or low(0V).
private void DO_8_Click(object sender, EventArg+5s e)
{
string ch1 = string.Empty;
bool stat1 = false;
string X_data1 = "";
ch1 = "p";
stat1 = Connect();
if (stat1 == true)
```

```
{
write(ch1);
}
MessageBox.Show("sent successfully: p");
X_data1 = X_data1 + Read(1);
if (X_data1 == "P")
{
if (ComboBox1.SelectedItem.ToString() == "0x00")
{
buf = "01";
write(buf);
MessageBox.Show("sent buf: 01 successfully");
}
else if (ComboBox1.SelectedItem.ToString() == "0xAA")
{
buf = "02";
write(buf);
MessageBox.Show("sent buf: 02 successfully");
}
else if (ComboBox1.SelectedItem.ToString() == "0x55")
{
buf = "03";
write(buf);
MessageBox.Show("sent buf: 03 successfully");
}
else if (ComboBox1.SelectedItem.ToString() == "0xFF")
{
buf = "04";
write(buf);
MessageBox.Show("sent buf: 04 successfully");
}
}
Close1();
}
// On Button Click event of DO_16, first the socket is connected, then the
```

```
// character 'o' is sent to the application side and the character 'O'
// is received from the application side as a handshaking variable. If received //
correctly, selecting a hexadecimal value from the combobox which can be either
// 0x00,0xaa,0x55,0xff and sending it to the application side and finally closing //
the Socket and the stream. Depending on the hexadecimal value received in the //
Application side, the respective DO channel bit will be high(+5V) or low(0V).
private void DO_16_Click(object sender, EventArgs e)
{
string ch1 = string.Empty;
bool stat1 = false;
string X_data1 = "";
ch1 = "o";
stat1 = Connect();
if (stat1 == true)
{
write(ch1);
}
MessageBox.Show("sent successfully: o");
X_data1 = X_data1 + Read(1);
if (X_data1 == "O")
{
if (ComboBox2.SelectedItem.ToString() == "0x00")
{
buf = "01";
write(buf);
MessageBox.Show("sent buf: 01 successfully");
}
else if (ComboBox2.SelectedItem.ToString() == "0xAA")
{
buf = "02";
write(buf);
MessageBox.Show("sent buf: 02 successfully");
}
else if (ComboBox2.SelectedItem.ToString() == "0x55")
{
buf = "03";
```

```
write(buf);
MessageBox.Show("sent buf: 03 successfully");
}
else if (ComboBox2.SelectedItem.ToString() == "0xFF")
{
buf = "04";
write(buf);
MessageBox.Show("sent buf: 04 successfully");
}
}
Close1();
}
   // On Button Click event of DO_24, first the socket is connected, then the
   // character 'r' is sent to the application side and the character 'R' is
// is received from the application side as a handshaking variable. If received //
correctly, selecting a hexadecimal value from the combobox which can be either
// 0x00,0xaa,0x55,0xff and sending it to the application side and finally closing //
the Socket and the stream. Depending on the hexadecimal value received in the //
Application side, the respective DO channel bit will be high(+5V) or low(0V).
private void DO_24_Click(object sender, EventArgs e)
{
string ch1 = string.Empty;
bool stat1 = false;
string X_data1 = "";
ch1 = "r";
stat1 = Connect();
if (stat1 == true)
{
write(ch1);
}
MessageBox.Show("sent successfully: r");
X_data1 = X_data1 + Read(1);
if (X_data1 == "R")
{
if (ComboBox3.SelectedItem.ToString() == "0x00")
{
buf = "01";
```

```
write(buf);
MessageBox.Show("sent buf: 01 successfully");
}
else if (ComboBox3.SelectedItem.ToString() == "0xAA")
{
buf = "02";
write(buf);
MessageBox.Show("sent buf: 02 successfully");
}
else if (ComboBox3.SelectedItem.ToString() == "0x55")
{
buf = "03";
write(buf);
MessageBox.Show("sent buf: 03 successfully");
}
else if (ComboBox3.SelectedItem.ToString() == "0xFF")
{
buf = "04";
write(buf);
MessageBox.Show("sent buf: 04 successfully");
}
}
Close1();
}
    // On Button Click event of DO_32, first the socket is connected, then the
    // character 'q' is sent to the application side and the character 'Q' is
// is received from the application side as a handshaking variable. If received //
correctly, selecting a hexadecimal value from the combobox which can be either
// 0x00,0xaa,0x55,0xff and sending it to the application side and finally closing //
the Socket and the stream. Depending on the hexadecimal value received in the //
Application side, the respective DO channel bit will be high(+5V) or low(0V).
private void DO_32_Click(object sender, EventArgs e)
{
string ch1 = string.Empty;
bool stat1 = false;
string X_data1 = "";
ch1 = "q";
```

```
stat1 = Connect();
if (stat1 == true)
{
write(ch1);
}
MessageBox.Show("sent successfully: q");
X_data1 = X_data1 + Read(1);
if (X_data1 == "Q")
{
if (ComboBox4.SelectedItem.ToString() == "0x00")
{
buf = "01";
write(buf);
MessageBox.Show("sent buf: 01 successfully");
}
else if (ComboBox4.SelectedItem.ToString() == "0xAA")
{
buf = "02";
write(buf);
MessageBox.Show("sent buf: 02 successfully");
}
else if (ComboBox4.SelectedItem.ToString() == "0x55")
{
buf = "03";
write(buf);
MessageBox.Show("sent buf: 03 successfully");
}
else if (ComboBox4.SelectedItem.ToString() == "0xFF")
{
buf = "04";
write(buf);
MessageBox.Show("sent buf: 04 successfully");
}
}
Close1();
}
```

```
// On Button Click event of DO_40, first the socket is connected, then the
// character 's' is sent to the application side and the character 'S' is
// is received from the application side as a handshaking variable. If received //
correctly, selecting a hexadecimal value from the combobox which can be either
// 0x00,0xaa,0x55,0xff and sending it to the application side and finally closing //
the Socket and the stream. Depending on the hexadecimal value received in the //
Application side, the respective DO channel bit will be high(+5V) or low(0V).
private void DO_40_Click(object sender, EventArgs e)
{
string ch1 = string.Empty;
bool stat1 = false;
string X_data1 = "";
ch1 = "s";
stat1 = Connect();
if (stat1 == true)
{
write(ch1);
}
MessageBox.Show("sent successfully: s");
X_data1 = X_data1 + Read(1);
if (X_data1 == "S")
{
if (ComboBox5.SelectedItem.ToString() == "0x00")
{
buf = "01";
write(buf);
MessageBox.Show("sent buf: 01 successfully");
}
else if (ComboBox5.SelectedItem.ToString() == "0xAA")
{
buf = "02";
write(buf);
MessageBox.Show("sent buf: 02 successfully");
}
else if (ComboBox5.SelectedItem.ToString() == "0x55")
{
buf = "03";
```

```
write(buf);
MessageBox.Show("sent buf: 03 successfully");
}
else if (ComboBox5.SelectedItem.ToString() == "0xFF")
{
buf = "04";
write(buf);
MessageBox.Show("sent buf: 04 successfully");
}
}
Close1();
}
   // On Button Click event of button6, first the socket is connected, then the
   // character 'v' is sent to the application side and the character 'V' is
// is received from the application side as a handshaking variable. If received //
correctly, selecting a hexadecimal value from the combobox which can be either
// 0x00,0xaa,0x55,0xff and sending it to the application side and finally closing //
the Socket and the stream. Depending on the hexadecimal value received in the //
Application side, the respective DO channel bit will be high(+5V) or low(0V).
// Button 7 is for 1-8 channels of 24 channel adon module used for DO.
private void Button6_Click(object sender, EventArgs e)
{
string ch1 = string.Empty;
bool stat1 = false;
string X_data1 = "";
ch1 = "v";
stat1 = Connect();
if (stat1 == true)
{
write(ch1);
}
MessageBox.Show("sent successfully: v");
X_data1 = X_data1 + Read(1);
if (X_data1 == "V")
{
if (ComboBox6.SelectedItem.ToString() == "0x00")
{
```

```
buf = "01";
write(buf);
MessageBox.Show("sent buf: 01 successfully");
}
else if (ComboBox6.SelectedItem.ToString() == "0xAA")
{
buf = "02";
write(buf);
MessageBox.Show("sent buf: 02 successfully");
}
else if (ComboBox6.SelectedItem.ToString() == "0x55")
{
buf = "03";
write(buf);
MessageBox.Show("sent buf: 03 successfully");
}
else if (ComboBox6.SelectedItem.ToString() == "0xFF")
{
buf = "04";
write(buf);
MessageBox.Show("sent buf: 04 successfully");
}
}
Close1();
}
    // On Button Click event of button7, first the socket is connected, then the
    // character 'u' is sent to the application side and the character 'U' is
// is received from the application side as a handshaking variable. If received //
correctly, selecting a hexadecimal value from the combobox which can be either
// 0x00,0xaa,0x55,0xff and sending it to the application side and finally closing //
the Socket and the stream. Depending on the hexadecimal value received in the //
application side, the respective DO channel bit will be high(+5V) or low(0V).
// Button 7 is for 9-16 channels of 24 channel adon module used for DO.
private void Button7_Click(object sender, EventArgs e)
{
string ch1 = string.Empty;
bool stat1 = false;
```

```
string X_data1 = "";
ch1 = "u";
stat1 = Connect();
if (stat1 == true)
{
write(ch1);
}
MessageBox.Show("sent successfully: u");
X_data1 = X_data1 + Read(1);
if (X_data1 == "U")
{
if (ComboBox7.SelectedItem.ToString() == "0x00")
{
buf = "01";
write(buf);
MessageBox.Show("sent buf: 01 successfully");
}
else if (ComboBox7.SelectedItem.ToString() == "0xAA")
{
buf = "02";
write(buf);
MessageBox.Show("sent buf: 02 successfully");
}
else if (ComboBox7.SelectedItem.ToString() == "0x55")
{
buf = "03";
write(buf);
MessageBox.Show("sent buf: 03 successfully");
}
else if (ComboBox7.SelectedItem.ToString() == "0xFF")
{
buf = "04";
write(buf);
MessageBox.Show("sent buf: 04 successfully");
}
}
```

```
Close1();
}
  // On Button Click event of button8, first the socket is connected, then the
  // character 't' is sent to the application side and the character 'T' is
// is received from the application side as a handshaking variable. If received //
correctly, selecting a hexadecimal value from the combobox which can be either
// 0x00,0xaa,0x55,0xff and sending it to the application side and finally closing //
the socket and the stream. Depending on the hexadecimal value received in the //
application side, the respective DO channel bit will be high(+5V) or low(0V).
// Button 7 is for 17-24 channels of 24 channel adon module used for DO.
private void Button8_Click(object sender, EventArgs e)
{
string ch1 = string.Empty;
bool stat1 = false;
string X_data1 = "";
ch1 = "t";
stat1 = Connect();
if (stat1 == true)
{
write(ch1);
}
MessageBox.Show("sent successfully: t");
X_data1 = X_data1 + Read(1);
if (X_data1 == "T")
{
if (ComboBox8.SelectedItem.ToString() == "0x00")
{
buf = "01";
write(buf);
MessageBox.Show("sent buf: 01 successfully");
}
else if (ComboBox8.SelectedItem.ToString() == "0xAA")
{
buf = "02";
write(buf);
MessageBox.Show("sent buf: 02 successfully");
}
```

```
else if (ComboBox8.SelectedItem.ToString() == "0x55")
{
buf = "03";
write(buf);
MessageBox.Show("sent buf: 03 successfully");
}
else if (ComboBox8.SelectedItem.ToString() == "0xFF")
{
buf = "04";
write(buf);
MessageBox.Show("sent buf: 04 successfully");
                    }
            }
            Close1();
        }
    }
}
```

11.14 EXPLANATION OF THE SOURCE CODE AT THE APPLICATION SIDE

The above source code consists of different files as explained with the source code. Let's start with the file main.c. The main.c file consists of four threads. Thread being a part of operating system is a smallest sequence execution of programmed instructions which can be managed independently by the scheduler. Here, we have used multithreading allowing multiple threads to exist within the context of a single process. There are 4 threads used in the main.c file. Thread1 is of 500 μs. Thread2 is of 1 μs. Thread3 is of 500 μs. The last thread is ThreadMain of 1 sec mainly responsible for communication over TCP/IP and serial port. The communication over serial port is done point to point, i.e., through RS-232 communication media. The communication is done via handshake media. The communication is tested via hyper-terminal device which is opened at a PC and its port is serially wired through Port1 as explained in Figure 11.16. The port settings has been done at the Hyper-terminal device by first selecting the Com Port on the available PC having baudrate as 19200, data bits as 8, parity none, flow control none and stop bit as one. First we will explain about RS-232 communication and then the TCP/IP communication. RS-232 communication has its advantages and disadvantages. The advantages are as follows:

1. The distance for the data acquisition hardware connectivity between the micro-link hardware and computer is up to 50 m.
2. Low cost.

The disadvantages are as follows:

1. RS-232 communication is point to point, i.e., only 1 device can be connected.
2. Cable length is limited.
3. Ground loop noise sensitivity.
4. It is easily damaged from electrical faults.
5. It is slow and runs at 38400 baud rate over short distances as ows to capacitive effects.

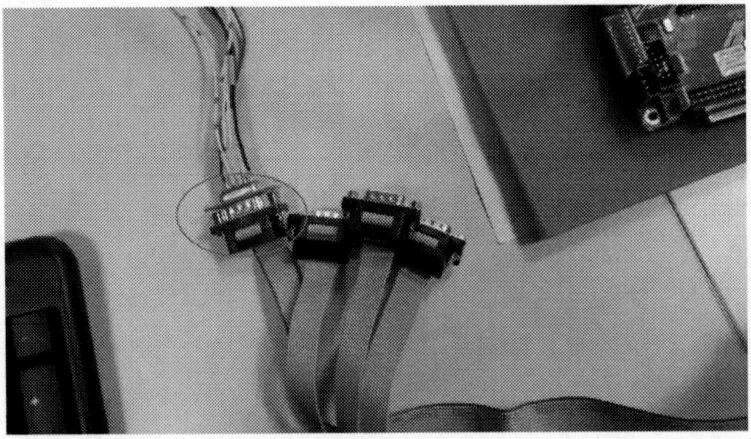

Fig. 11.16 *Serial Port1 encircled in Red of the SBC Board*

There are 4 serial ports in the SBC board. As per the manual of hercules-II board, serial ports 1 and 2 are derived from the south bridge chip. Serial ports 3 and 4 are derived from an Exar 16C2850 UART chip on the ISA bus. Serial ports 3 and 4 also can be BIOS-selected for RS-485 or RS-232. These settings can be overridden by a jumper to select RS422. termination resistors of 120 ohms can be jumper-enabled on these two ports. The SBC board is the hercules-II board of diamond systems. The application program has been written in the QNX OS. For developing QNX applications, an eclipse based IDE has been used, namely, momentics IDE. Whenever a character **'m'** is sent from the hyper-terminal device, application side will first receive and will send the response with the data bytes stored in the **'str'** variable. The **'hello'** string printed in the console window indicates that the character **'m'** has been received from the hyper-terminal as a query. The data bytes sent from the application side and the console window of the momentics side look is shown in Figure 11.17. The one underlined in Figure 11.17 indicates that the string has been sent from the SBC to the hyper-terminal thus indicating the communication between the 2 ports as a means of handshaking. The above example of RS-232 communication will give you a feel of how the data has been exchanged between the 2 serial ports comfortably. The RS-232 communication is a point to point communication and can't be a point to multipoint. The QNX neutrino uses 8250 serial communication manager as devc-ser8250. By default on the BIOS, all the serial ports are configured as RS-232. But can be configured as RS-485 based on the jumper settings of the board. The encircled portion in Figure 11.17 indicates the application program running on the SBC board.

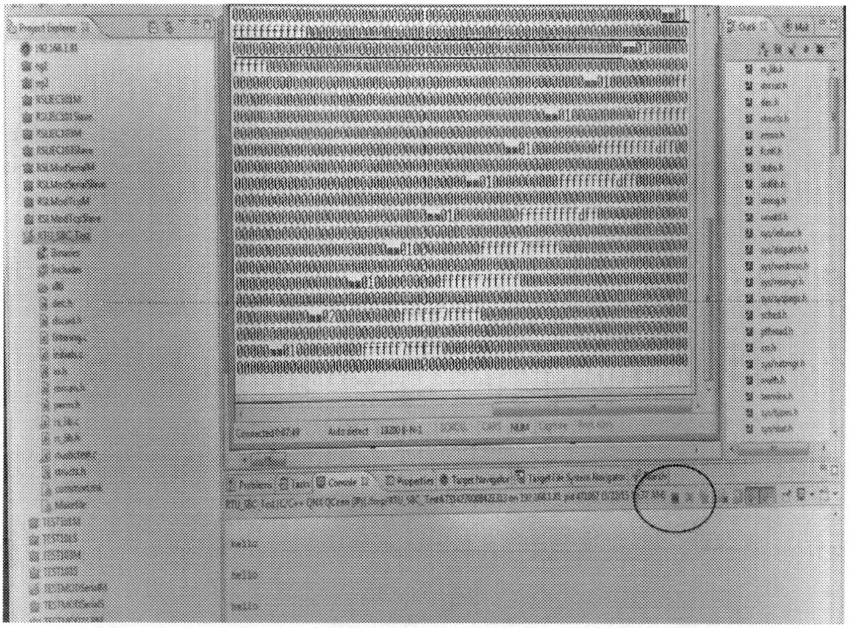

Fig. 11.17 *Hyperterminal and the Board side QNX Momentics Communication using RS-232*

Figure 11.18 shows the top view of the serial communication connection with the SBC board. The one highlighted in the grey circle will show the position of the serial port cable connection.

Fig. 11.18 *Top View of the SBC with the Serial Port Connection encircled in the Red*

11.15 ETHERNET PORT (TCP/IP) COMMUNICATION OVER THE SBC BOARD

Before explaining the TCP/IP communication used in the above program, it is important to know about the TCP/IP communication. TCP/IP is a suite of communication protocol developed by the defense advanced research projects agency (DARPA) of the US department of defense in 1970. It was developed on DARPA's network called ARPANET. It is a short form of transmission control protocol and internet protocol. It can be used as a communication protocol in a private network. Some of the advantages of the TCP/IP over OSI are as follows:

1. It enables Internet working between organizations.
2. It includes a support of a number of routing protocols.
3. It has a scalable client/server architecture.
4. It operates independently of the operating system.
5. It includes support for name and address resolution services like domain name services, dynamic host configuration protocol etc.
6. Ethernet port had good ground noise immunity.

The basic TCP/IP Communication between client and server is shown in Figure 11.19.

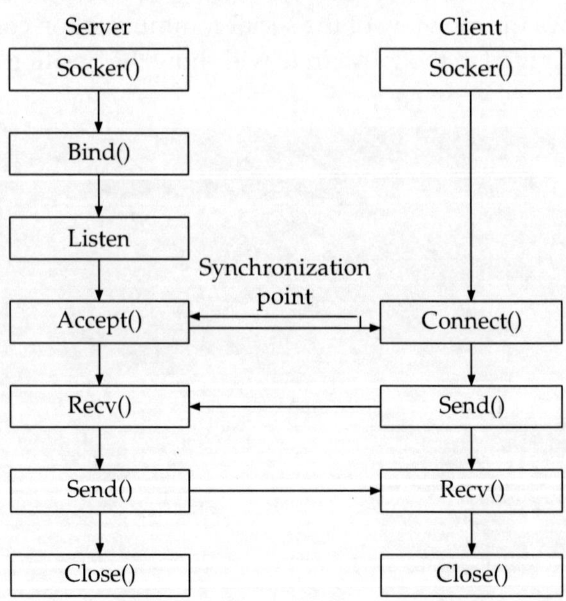

Fig. 11.19 *TCP/IP communication between client and server*

From the above flowchart (Figure 11.19), we can see that the functions like connect, bind, listen, accept and close have been used in the source code explained. The application side of the program is the server where the socket is always on the listening mode and the HMI side is the Master where the Query is sent to the server and the server *n* turns responds to the above query. The purpose of the functions used is as follows:

Primitive	Meaning
Socket	Create a new communication end point.
Bind	Attach a local address to a socket.
Listen	Announce willingness to accept connections.
Accept	Block caller until a connection request arrives.
Connect	Actively attempt to establish a connection.
Send	Send some data over the connection.
Receive	Receive some data over the connection.
Close	Release the connection.

The communication done using TCP/IP is as follows for Different Analog and Digital Channels. There are two external modules inserted into the SBC Board. First is the 96 channel GPIO-MM-21 board. As per the manual, the GPIO-MM-21 is a PC/104 board featuring 96 Digital I/O (DIO) lines. The DIO function is implemented in FPGA cores, emulating quad 82C55A PPI chips. The DIO Pin-Out details are as follows:

I/O Connector Pinout

Digital I/O Header Pinout

Connectors J3 and J4 provide a total of 96 general-purpose DIO interfaces.

They can be pulled-up to +5V or pulled-down to ground using jumpers J8 and J11 respectively.

The following table shows the 82C55A port arrangement. Ports 1 and 2 are on connector J3 and ports 3 and 4 are on connector J4.

J3				J4			
Port 1A7	1	2	Port 2A7	Port 3A7	1	2	Port 4A7
Port 1A6	3	4	Port 2A6	Port 3A6	3	4	Port 4A6
Port 1A5	5	6	Port 2A5	Port 3A5	5	6	Port 4A5
Port 1A4	7	8	Port 2A4	Port 3A4	7	8	Port 4A4
Port 1A3	9	10	Port 2A3	Port 3A3	9	10	Port 4A3
Port 1A2	11	12	Port 2A2	Port 3A2	11	12	Port 4A2
Port 1A1	13	14	Port 2A1	Port 3A1	13	14	Port 4A1
Port 1A0	15	16	Port 2A0	Port 3A0	15	16	Port 4A0
Port 1B7	17	18	Port 2B7	Port 3B7	17	18	Port 4B7
Port 1B6	19	20	Port 2B6	Port 3B6	19	20	Port 4B6
Port 1B5	21	22	Port 2B5	Port 3B5	21	22	Port 4B5

Port 1B4	23	24	Port 2B4	Port 3B4	23	24	Port 4B4
Port 1B3	25	26	Port 2B3	Port 3B3	25	26	Port 4B3
Port 1B2	27	28	Port 2B2	Port 3B2	27	28	Port 4B2
Port 1B1	29	***30***	Port 2B1	Port 3B1	29	30	Port 4B1
Port 1B0	31	32	Port 2B0	Port 3B0	31	32	Port 4B0
Port 1C7	33	34	Port 2C7	Port 3C7	33	34	Port 4C7
Port 1C6	35	36	Port 2C6	Port 3C6	35	36	Port 4C6
Port 1C5	37	38	Port 2C5	Port 3C5	37	38	Port 4C5
Port 1C4	39	40	Port 2C4	Port 3C4	39	40	Port 4C4
Port 1C3	41	42	Port 2C3	Port 3C3	41	42	Port 4C3
Port 1C2	43	44	Port 2C2	Port 3C2	43	44	Port 4C2
Port 1C1	45	46	Port 2C1	Port 3C1	45	46	Port 4C1
Port 1C0	47	48	Port 2C0	Port 3C0	47	48	Port 4C0
+5V	49	50	Ground	+5V	49	50	Ground

NOTE: The connectors on the board are labeled "Port 1" and "Port 2" but do not denote the 82C55A port numbers shown here.

For more information about the board, just visit http://www.diamondsystems.com/ and find GPIO-MM-21 user manual. The 96 channel module GPIO-MM-21 inserted into the board from the top is shown in Figure 19.20. The 8255 base address has been set 0x330 from the following jumper connection.

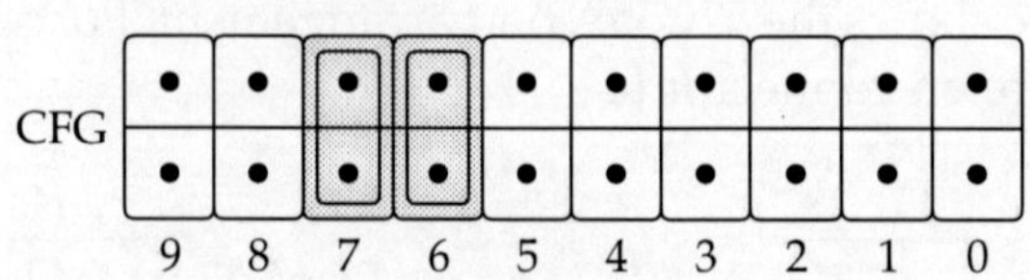

Similarly, another external module MRC-424 –XT which is a 24-channel digital I/O adon module. It is a PC/104-plus I/O module with 2 dual 10/100 ethernet switches and DIO. As per the manual, mercator II is a PC/104-plus I/O module with two dual 10/100 ethernet switches. Each switch occupies a single PCI slot. All four ethernet ports are brought out to a 6-pin header. 24 digital I/O lines are available on a 26-pin header. The board contains both PC/104 (ISA) and PC/104-plus (PCI) stack through connectors. All circuitry is driven by and accessed by the PCI bus, the ISA connector is only used for pass-through. Mercator II requires +5VDC input power and operates over the industrial temperature range of –40 °C to +85 °C. The DIO ports are accessible via the PC/104 address space and the address decoding logic is provided by a CPLD. The DIO ports can be configured at various addresses using 3 jumpers. The board will default at base address of 0x300. The DIO connector details of MRC-424-XT are after Figure 11.21.

Fig. 11.20 *GPIO-MM-21 Top View*

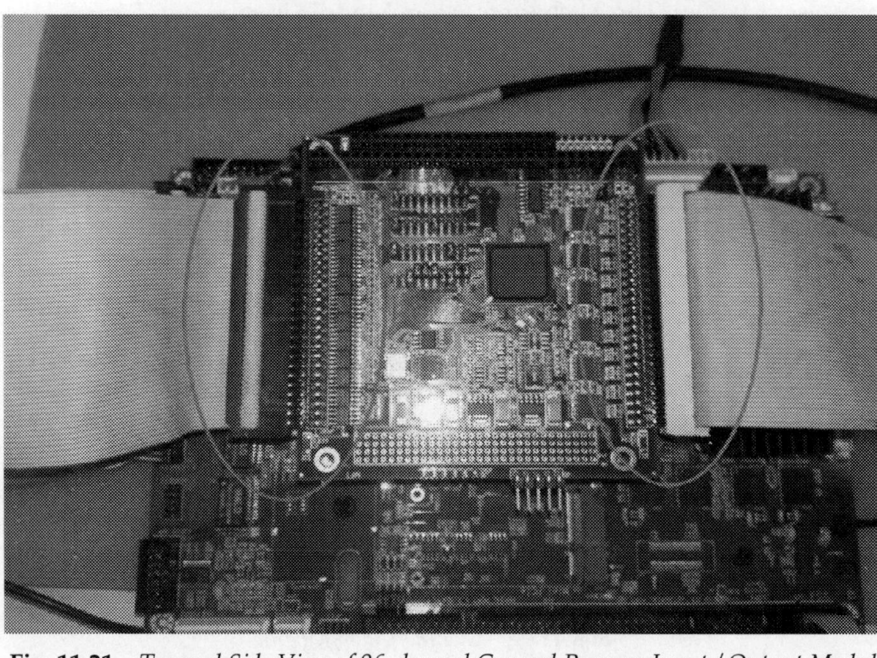

Fig. 11.21 *Top and Side View of 96 channel General Purpose Input / Output Module*

11.15.1 Digital I/O Connector (J12)

The digital I/O is provided on a 26-pin right angle header. The digital I/O connector pin-out is as follows:

C7	13	26	C6
C5	12	25	C4
C3	11	24	C2
C1	10	23	C0
B7	9	22	B6
B5	8	21	B4
B3	7	20	B2
B1	6	19	B0
A7	5	18	A6
A5	4	17	A4
A3	3	16	A2
A1	2	15	A0
+5V	1	14	GND

Connector Type: Right-angle 26-pin (2x13) header
Samtec IPL1-113-01-L-D-RA-K connector

For more information, you can refer to http://www.diamondsystems.com/ and search Mercator 2 User Manual. The 24 channel Adon Module is shown in Figure 11.22.

Fig. 11.22 *Channel Adon Module*

11.16 DIGITAL INPUT CHANNEL TEST USING TCP/IP COMMUNICATION PROTOCOL

As we can see that there are two 50-pin connectors in Figure 11.23. From the DIO Pin details, we can see that each 50-pin connector has 48 channels thus, in total 96 channels. For each pin connector there is a +5V and a ground pin. The channels of each port can be tested by giving one end of a wire to +5V or to ground and another end to the channel number. Depending on the channel selected, we can see high or low at the HMI side. For example, from Figure 11.23 we can see that +5V encircled in grey colour has been connected to port 1B7 encircled at black colour via wire. So we can see from the HMI that, a high bit of value '1' has been monitored in the byte value '10000000' encircled in Figure 11.24. The portion of the code responsible in the application side is in8 (0x331). Another is the ground encircled in grey colour and the port 4B0 encircled in black colour shown in Figure 11.25. We can see from the HMI that:

Fig. 11.23 *+5V Connection at Port 1B7*

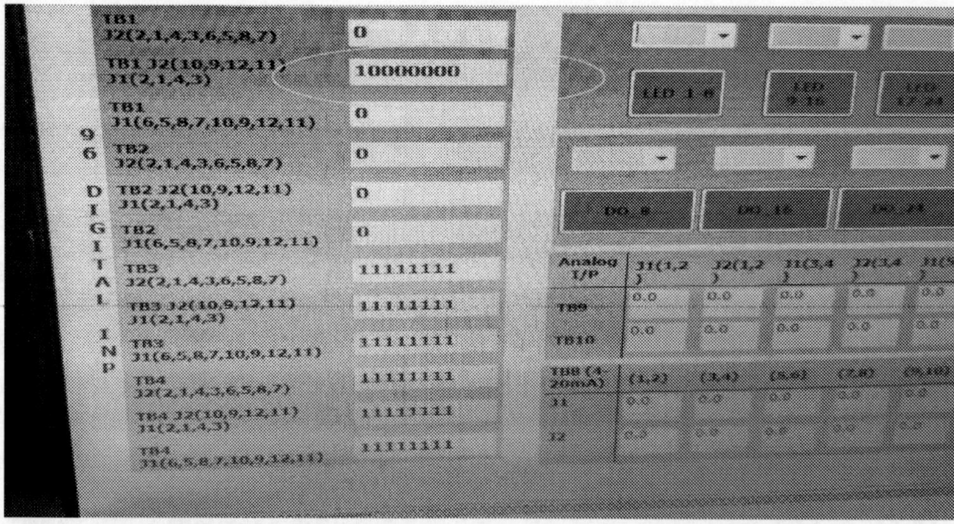

Fig. 11.24 *Byte value 1 due to +5V connection at the Port 1B7*

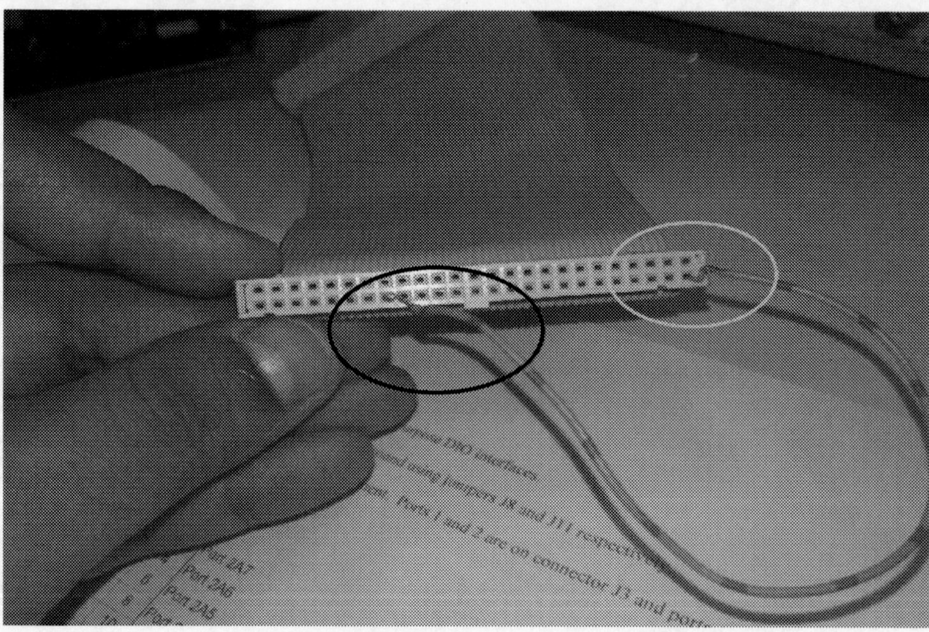

Fig. 11.25 *Ground connected to the channel 4B0*

A low bit value '0' has been monitored in the byte '11111110' highlighted (encircled) as shown in Figure 11.26, thus indicating that a zero has been read in the channel. The portion of the code responsible in the application side is in8 (0x33c).

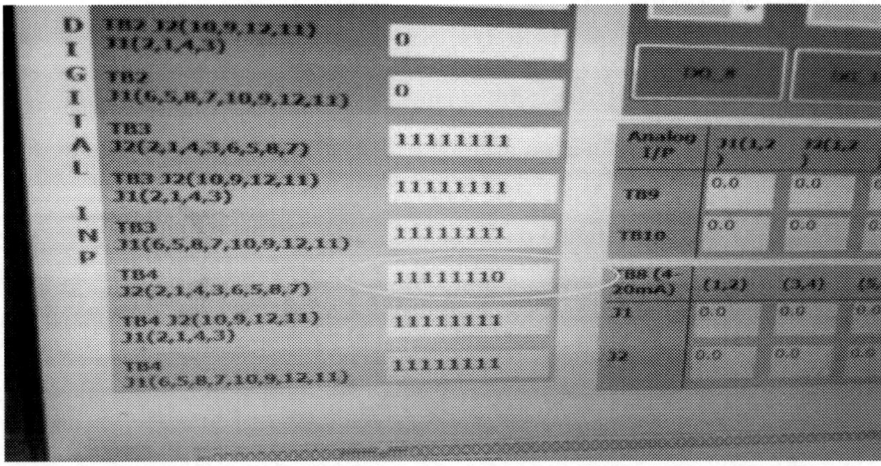

Fig. 11.26 *Byte Value 0 due to Ground Connection at the port 4B0*

At the HMI side, just see the portion of the code where we are splitting the text into 2. Actual portion is the textBox13.Select(19, 2).

11.17 DIGITAL OUTPUT CHANNEL TEST USING TCP/IP COMMUNICATION PROTOCOL

From the hercules-II SBC board, we can see that there is a 50-pin DIO connector. As per the manual, hercules II EBX includes a 50-pin header, J8, for all digital data acquisition I/O. This header is located on the right side of the board. Pin 1 is the lower right pin, as marked on the board. Diamond systems corporation cable no. C-50-18 provides a standard 50-pin connector at each end and mates with this header.

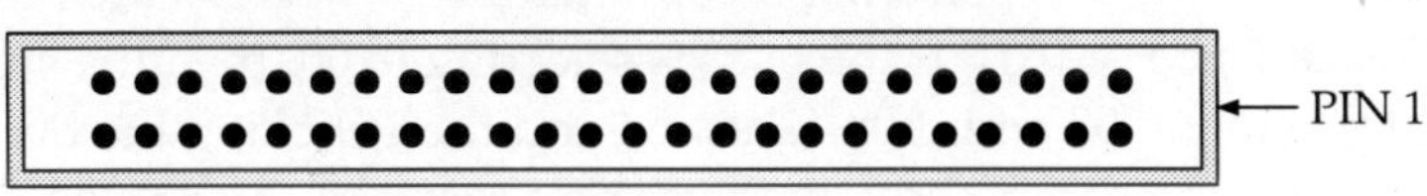

Fig. 11.27 *J8 Connector*

Table 11.5 *Signal description of J8 connector*

Signal	Definition
DIO A7-A0	Digital I/O port A; programmable direction
DIO B7-B0	Digital I/O port B; programmable direction
DIO C7-C0	Digital I/O port C; programmable direction
DIO D7-D0	Digital I/O port D; programmable direction
DIO E7-E0	Digital I/O port E; programmable direction
	Note : E3-E0 may be configured for PWM signals.

PWM3-PWM0	Pulse-Width Modulation Outputs (4 independent channels) Note : E7-E4 may be configured for counter/timer signals.
GATE 1-0	Gate inputs for Counter/Timer 1 and 0
TOUT1	Counter/Timer 1 output
DIOLATCH	Handshaking line used (with ACK signal below) for automated digital data transfers
Ext Trig	External A/D trigger input, External Counter/Timer 1 input
TOUT 0	Counter/Timer 1 output
+5V Out	Connected to switched +5V supply (Output only! Do not connect to external supply)
Digital Ground	Digital ground (0V - reference); used for digital circuitry only
WDO	Watchdog Timer Output NOTE: The watchdog timer circuit may be programmed directly or with Diamond Systems' Universal Driver software.
WDI	Watchdog Timer Input NOTE: The watchdog timer circuit may be programmed directly or with Diamond Systems' Universal Driver software.
ACK	Handshaking line used (with DIOLATCH signal) for automated digital data transfers
FXA, FXB	These lines should be left unconnected

From the HMI, the user is selecting a combination of hexadecimal bytes which may be either 0x00, 0xaa, 0x55 or 0xff. If the user is selecting 0x00, then the string 01 is sent, if 0xaa, then 02, if 0x55 then 03 and if 0xff then 04 and then clicking the button. Depending on the selection from the combo box, the user first receives the handshake variable and then sends his own handshake variable to the HMI and finally sends the string. The string received in the HMI will be in the form of ASCII and then it is subtracted with '-0' so as to convert from the ASCII and finally is converted into decimal. The decimal value condition is checked, and accordingly the hexadecimal byte value is being given "**OUT**" to the port using out8 function. The base address of the SBC board is 0x240 and if the adon module is inserted into the board its base address is 0x300. Since hexadecimal byte value is "outputted" from the port, there are 8 channels for each port and depending on the hexadecimal value received will give a +5V if wired with the ground, i.e., if one end of the wire is connected to the channel and the other end of the wire is connected to the ground. From Figure 11.28, we can see that the user is selecting 0x00 from the combo box, and then on clicking the DO_8 button, the HMI will send the character 'p' to the application side as 'sent successfully p' message being popped out which can be seen in Figure 11.29. The application side will, in turn, send the character 'P' as a handshake variable to the HMI from where the HMI will send the buf value 01 successfully to the application side as seen in Figure 11.30. The SBC side will receive the ASCII value which will be converted into decimal and simultaneously the hexadecimal byte 0x00 corresponding to the Query will be outputted to the Digital Output Port with the base address 0x250. As per the manual, the Digital Output base address is starting from 0x250 (0x240 + 10) for Port A, 0x251 (0x240 + 11) for Port B, 0x252 (0x240 + 12) for Port C, 0x253 (0x240 + 13) for Port D, 0x255 (0x240 + 14) for Port E. In the file "initials.c", we have made confbyte=0x1f; and dscDIOSetCon fig(dscb,&confbyte); ensuring the Port A, Port B, Port C, Port D and Port E to be used as an Output port. See page No. 82 in the Digital I/O Configuration part in the Hercules II Manual. Figure 11.30 depicts the Console Window of the QNX Momentics where inp1 = 1 is getting printed as we are printing the buf value which was received from the HMI. Thus, if the hexadecimal byte 0xAA has been selected from the Combo box, then the inp1 value will be 2 and outputting the byte 0xAA to the DO port with the base address 0x250 (Figures 11.31 and 11.32). If the hexadecimal byte 0x55 has been selected from the Combo box, then the inp1 value will be 3 and outputting the byte 0x55 to the DO port with the base address 0x250 (Figures 11.33 and 11.34). If the hexadecimal byte 0xFF has been selected from the Combo box, then the inp1 value will be 4 and outputting the byte 0xAA to the DO port with the base address 0x250 (Figures 11.35 and 11.36). From Figure 11.37 we have connected one end of the wire to Port B channel 8 and the other end of the wire to the Gnd. From the Combo box hexadecimal byte 0xFF has been selected and we can see that channel 8 of Port B reads +5V. In the multimeter, rms value is shown. Please note that, the multimeter terminals are connected to the two ends which we have not shown in any figure. The figures are as follows.

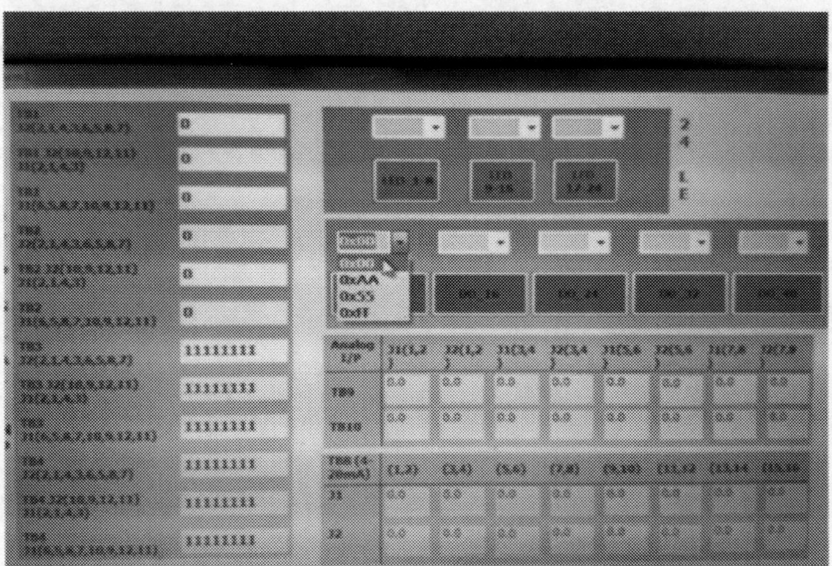

Fig. 11.28 *Selecting the value 0x00 from the Combo box 1*

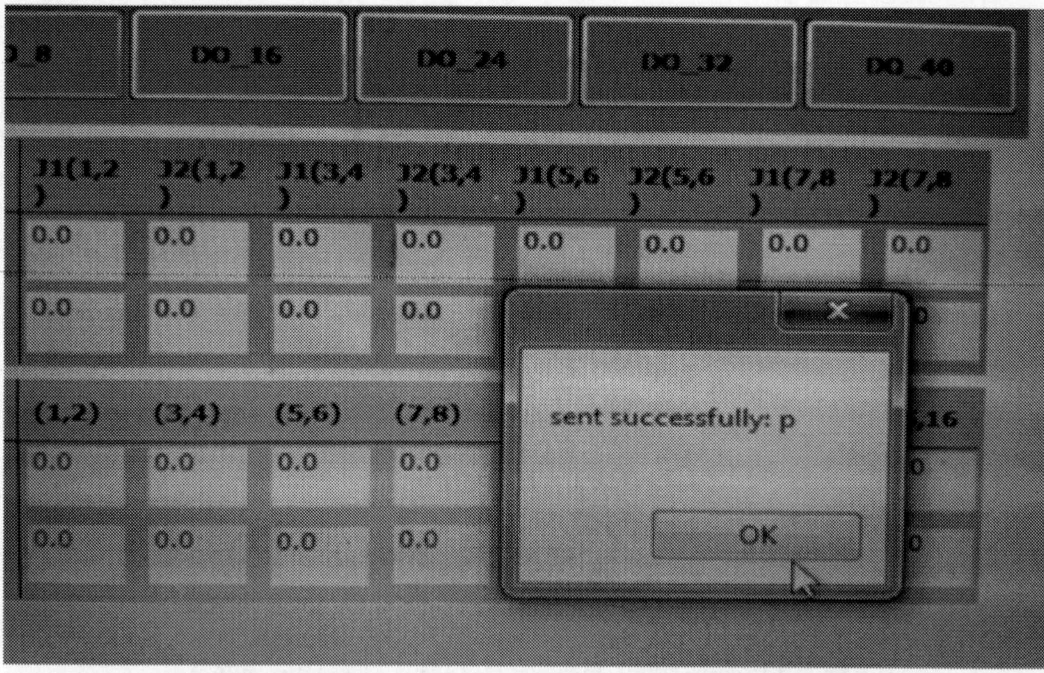

Fig. 11.29 *Sent successfully 'p' when the DO_8 button was clicked*

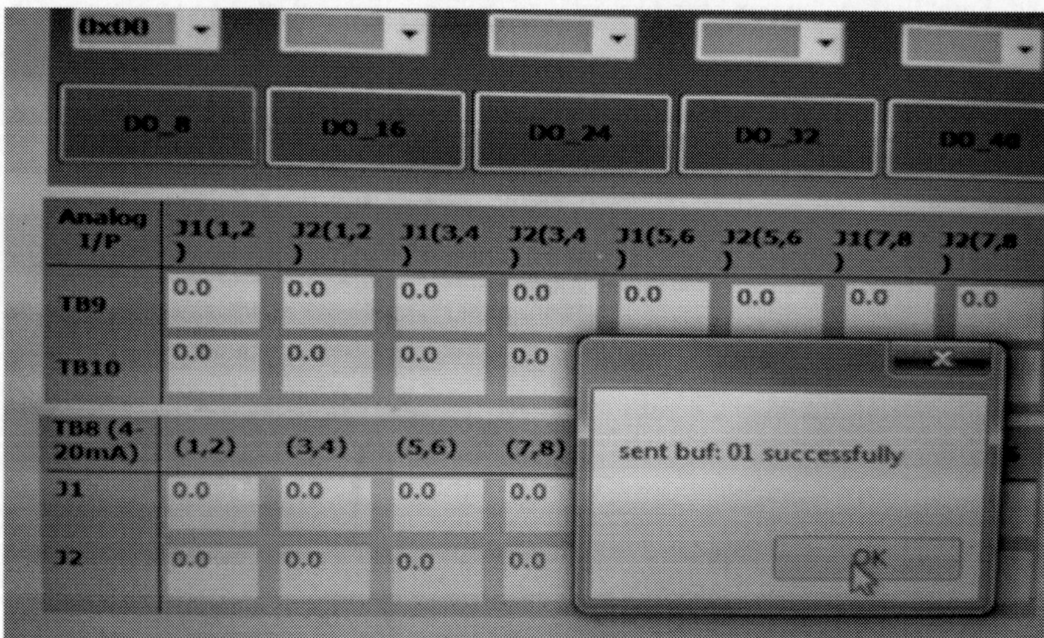

Fig. 11.30 *If 'P' was received as a response from the application side then, sent buf:01*

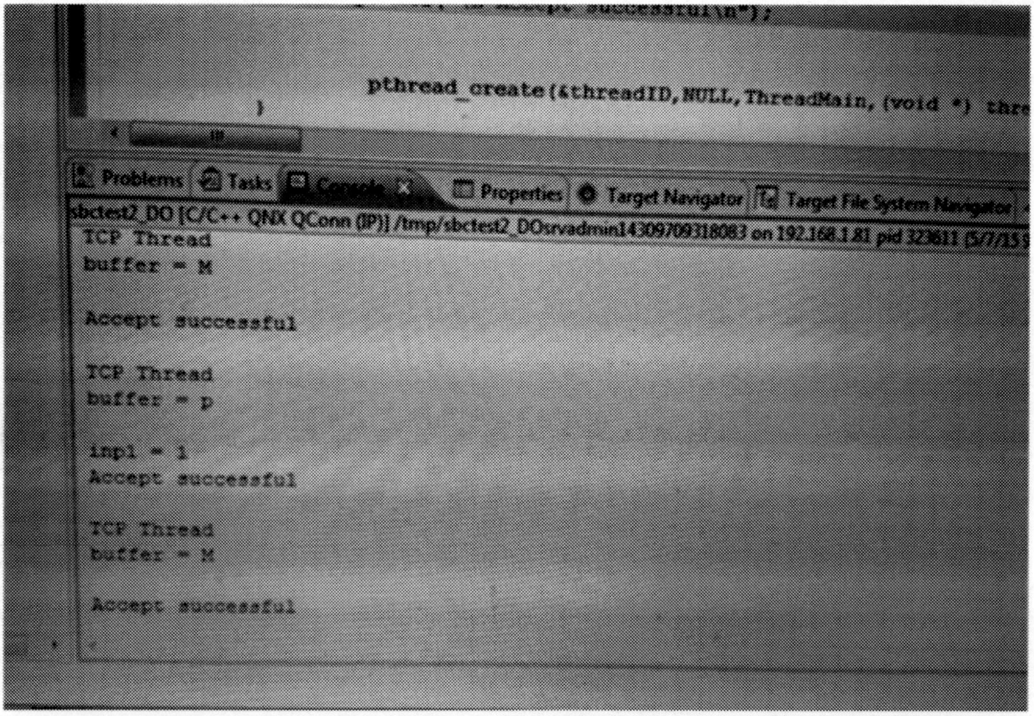

Fig. 11.31 *QNX Console Window displaying inp1 = 1 as the Combo box value was 0x00*

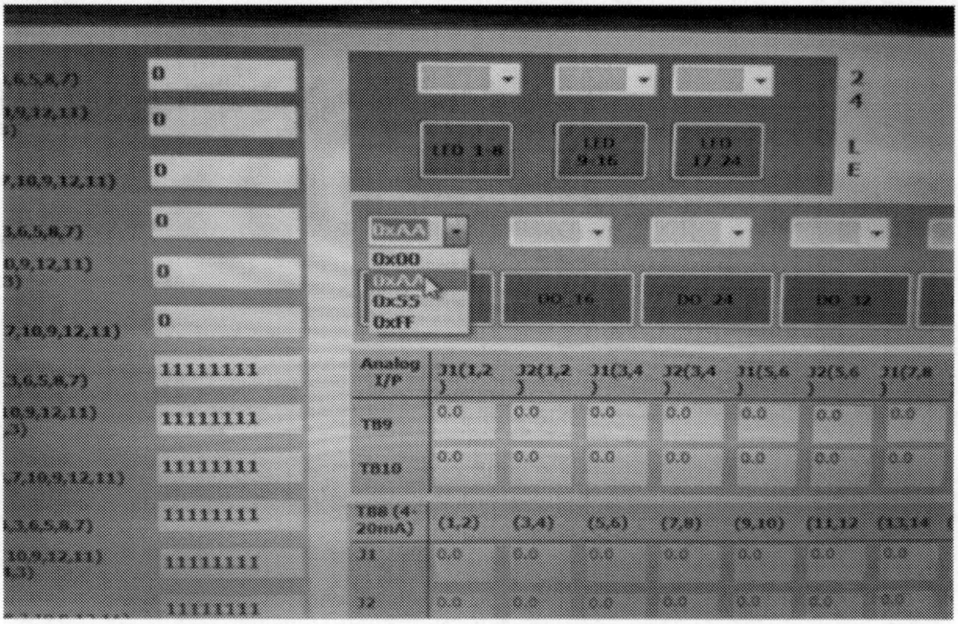

Fig. 11.32 *Combo box value 0xAA selected from the HMI*

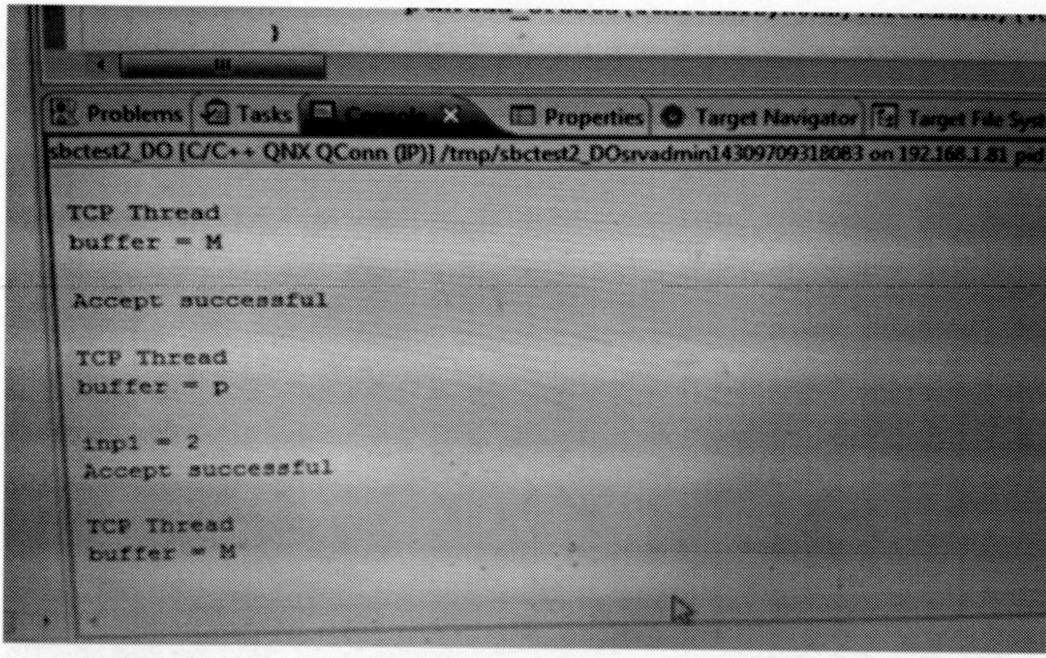

Fig. 11.33 *QNX Console Window displaying inp1 = 2 as 0xAA was selected from the HMI*

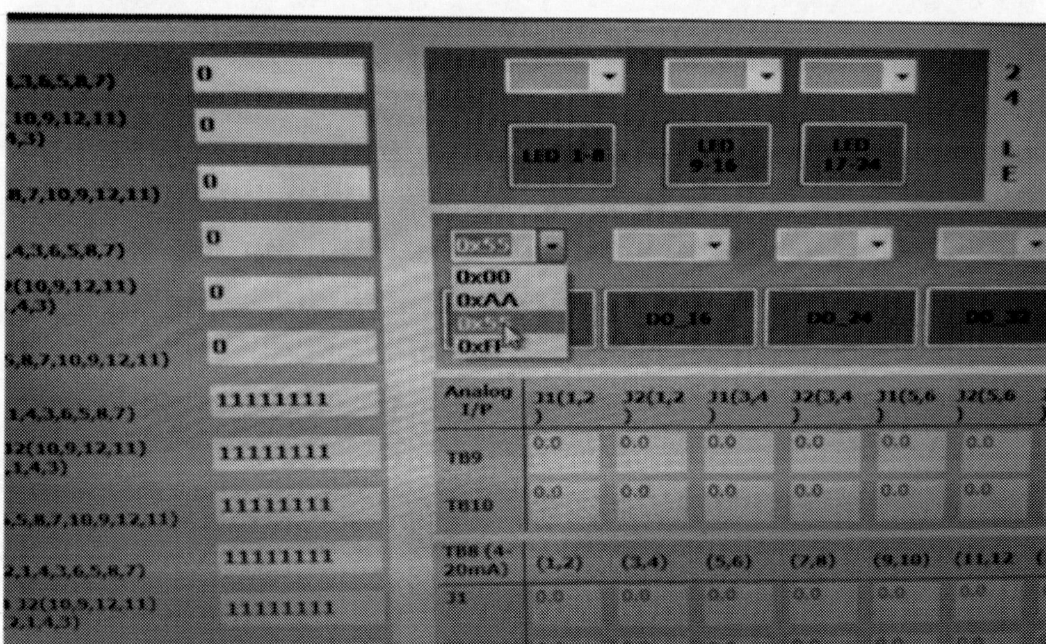

Fig. 11.34 *Combo box value 0x55 selected from the HMI*

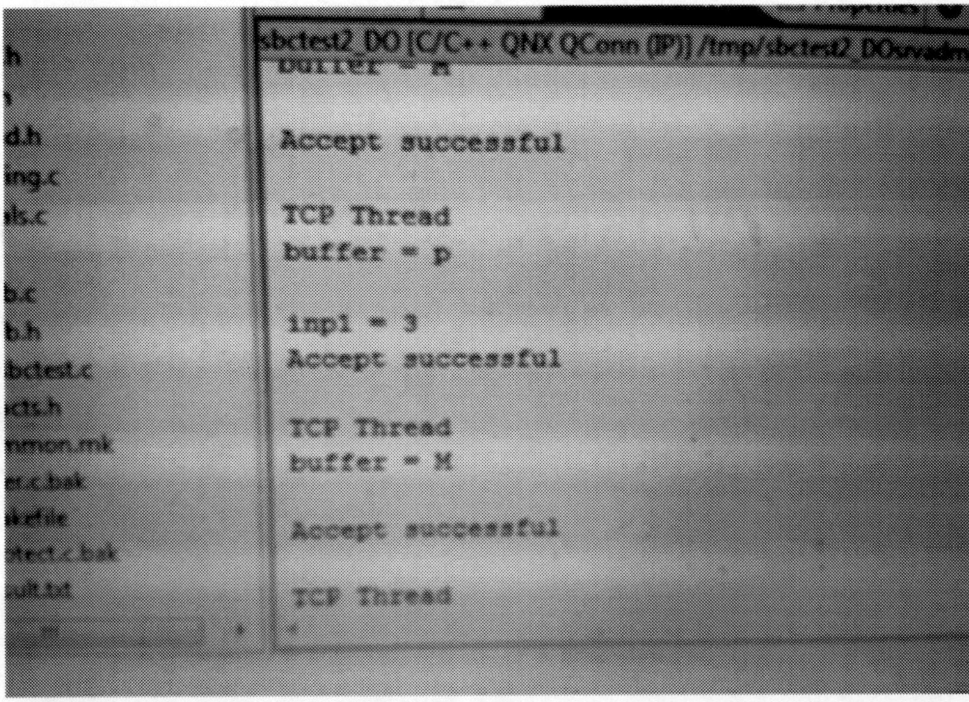

Fig. 11.35 *QNX Console Window displaying inp1 = 3 as 0x55 was selected from the HMI*

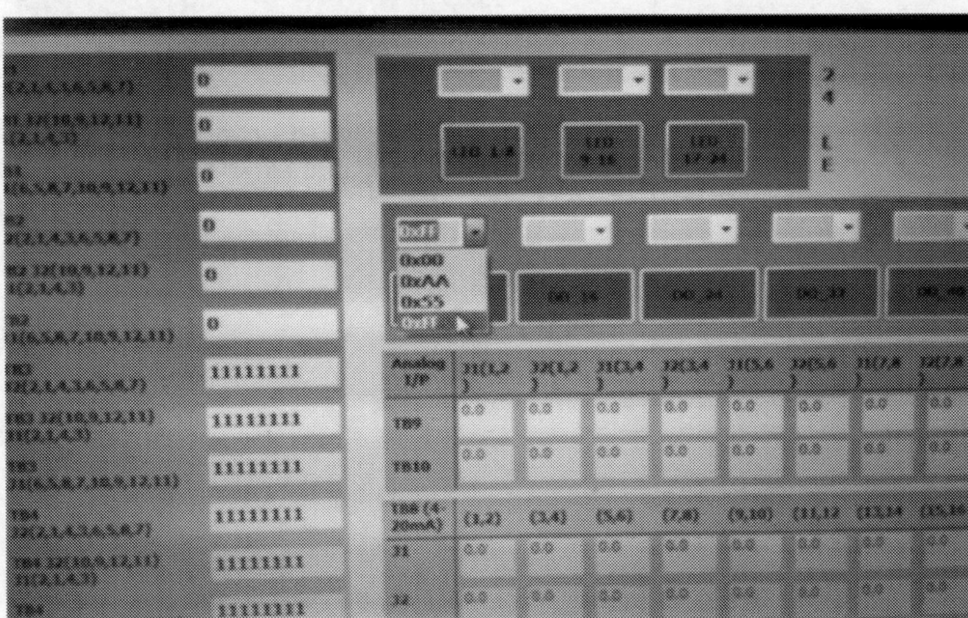

Fig. 11.36 *Combo box value 0xFF selected from the HMI*

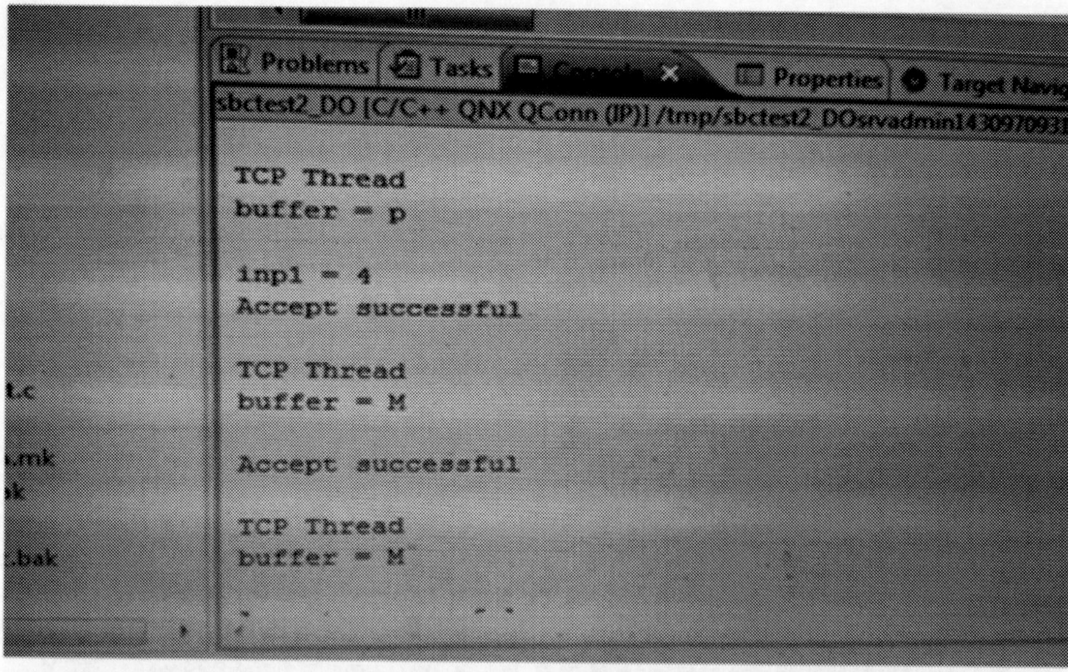

Fig. 11.37 *QNX Console Window displaying inp1 = 4 as 0xFF was selected from the HMI*

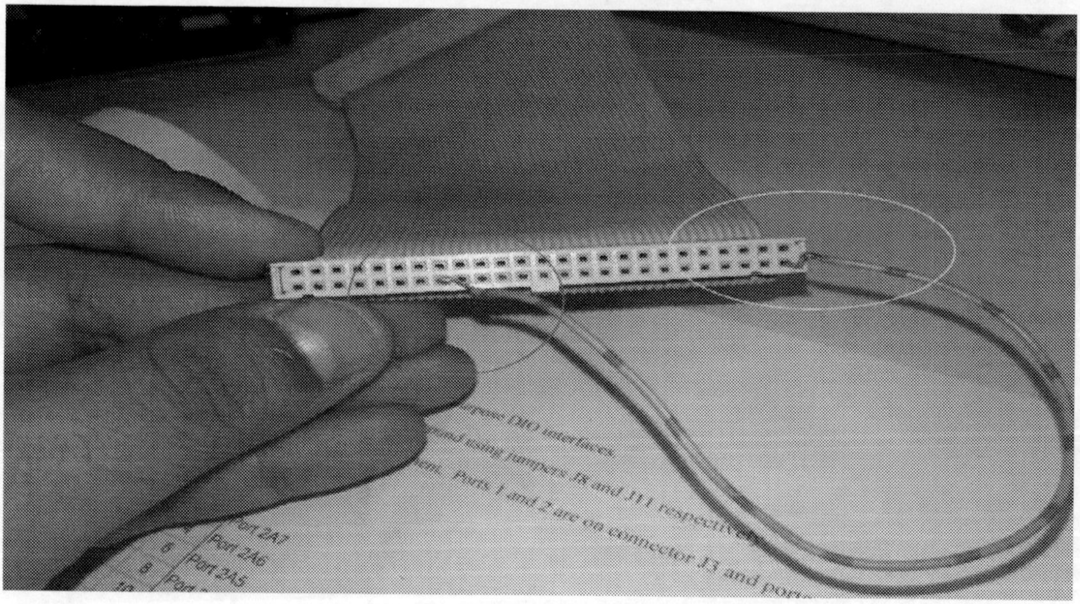

Fig. 11.38 *Connecting the wire to Port B channel 8 with other end ground*

Fig. 11.39 *Displaying the Reading 3.26V when 0xFF was selected from Combo box 2*

Also, the external adon module MRC-424-XT is integrated into the SBC Board having the base address 0x300. You can refer to the manual as per the website link given. The above adon module has been been used for the digital output functionality. Its functionality is same as explained for DO on SBC. Here, the address starts from 0x300 for port A, 0x301 for port B and 0x302 for Port C.

11.17.1 Analog Input and Analog Output Channel Test using TCP/IP Communication Protocol

As per the manual, there are 32 analog input channels from pin 7 to pin 40 with pin No. 39 and pin No. 40 as input ground. There are 4 analog output channels from pin No. 1-6 with Pin No. 5 and 6 as Output ground. For the Analog Input first we have called the adcinit function from the main program in main.c file has been called where the definition of the adcinit has been in initials.c file. Inside the adcinit function we are initializing the range, polarity (bipolar/unipolar), input gain, loading of the calibration values, scan interval, the number of channels to scan, i.e., starting from low channel to high channel. Here, the range has been set for +10V, the polarity is bipolar so ±10V is the range, input gain is 1, not loading the calibration values. The number of channels assigned is 32, i.e., the channel will start from 0 and will scan till 31. Channel No. 1 is Vin 0 and Channel No. 31 is Vin 31.

The A/D section includes a 16-bit A/D converter, 32 input channels and a 2048-sample (4kByte) FIFO. The pin configurations and the details as per the manual are given below.

NOTE: These reference grounds are NOT decoupled from the power grounds – they are indirectly connected to the power supply input (and other on-board ground/0V references). Do not assume that these grounds are floating; do not apply a high voltage input (relative to the power input ground) to these ground signals or to any other board I/O pin.

The data acquisition Analog Connector J9 of the Hercules III board is as follows from the manual.

Data acquisition (analog I/O) connector

Connector J9 is used for analog I/O data acquisition.

Fig. 11.40 *J9 Connector*

Single-ended

Vout 0	1	2	Vout 1
Vout 2	3	4	Vout 3
Output ground	5	6	Output ground
Vin 0	7	8	Vin 16
Vin 1	9	10	Vin 17
Vin 2	11	12	Vin 18
Vin 3	13	14	Vin 19
Vin 3	13	14	Vin 19
Vin 4	15	16	Vin 20
Vin 5	17	18	Vin 21
Vin 6	19	20	Vin 22
Vin 7	21	22	Vin 23
Vin 8	23	24	Vin 24
Vin 9	25	26	Vin 25
Vin 10	27	28	Vin 26
Vin 11	29	30	Vin 27
Vin 12	31	32	Vin 28
Vin 13	33	34	Vin 29
Vin 14	35	36	Vin 30
Vin 15	37	38	Vin 31
Input ground	39	40	Input ground

Differential

Vout 0	1	2	Vout 1
Vout 2	3	4	Vout 3
Output ground	5	6	Output ground
Vin 0 +	7	8	Vin 0 -
Vin 1 +	9	10	Vin 1 -
Vin 2 +	11	12	Vin 2 -
Vin 3 +	13	14	Vin 3 -
Vin 3 +	13	14	Vin 3 -
Vin 4 +	15	16	Vin 4 -
Vin 5 +	17	18	Vin 5 -
Vin 6 +	19	20	Vin 6 -
Vin 7 +	21	22	Vin 7 -
Vin 8 +	23	24	Vin 8 -
Vin 9 +	25	26	Vin 9 -
Vin 10 +	27	28	Vin 10 -
Vin 11 +	29	30	Vin 11 -
Vin 12 +	31	32	Vin 12 -
Vin 13 +	33	34	Vin 13 -
Vin 14 +	35	36	Vin 14 -
Vin 15 +	37	38	Vin 15 -
Input ground	39	40	Input ground

Signal	Definition
Vout3-0	Analog output channels 3 – 0
Output Ground	Analog ground; 0V reference for *Vout3-0*
Vin 31 ~ Vin 0	Analog input channels 31 – 0 in single-ended mode
Vin 15 + ~ Vin 0 +	High side of input channels 15 – 0 in differential mode
Vin 15 - ~ Vin 0 -	Low side of input channels 15 – 0 in differential mode
Input Ground	Analog ground; 0V reference for *Vin31-0*

We are doing DFT computation of all the channels in the sampling function and then calculating the rms value of each channel in the measurand function. Both these functions are defined in the file filtering.c. At every 500 micro-sec thread, DFT computation is done for each channel. At each channel, 40 samples are taken since 1 sinsuoidal wave is of 50Hz and the DFT computation has been done at 500 micro-sec. Thus, the total number of samples stored at each channel is 40. The rms computed value can also be scaled high according to your requirement. The measurand function is running at every 1 milli-sec thread. The resolution of the SBC is of 16-bit. Since bipolar, therefore the digital bit value corresponding to the analog voltage will be from -32768 to + 32767, i.e., –10V corresponds to the bit value –32768 and +10V corresponds to the bit value 32767. The rms value computed of each analog channel has been mapped to the structure variables of metering function which will be later used for the RS-232 and TCP/IP communication. The metering function is running at every 500 micro-sec. In Figure 11.30, we can see that buffer = 'm', i.e., the HMI is sending the character 'm' to the application side as a query and, in turn, the SBC will be sending the data stored in the string to the HMI as a response. The HMI will first split the data string and will check for the handshake character and will count the length of the string data after the handshake variable. If found correct, will split the data into the respective textboxes. The data from count 1 to 24 is the digital input data and the data from count 25 to 153 is the analog input data with 32 channels for each textbox the data has been selected and split into 4. ***Here we have assumed that there are 4 incoming feeders. Each feeder is having 3-phase voltage and 3-phase currents. One is the neutral voltage and the other is neutral current. Thus, each feeder is having 8 inputs. Since there are 4 feeders so in total 32 inputs.***

As explained earlier, there are four analog outputs. Vout0 is the channel 1 and Vout3 is the channel 4. In the main function of main.c file, the dacinit function has been called which is defined in the initials.c file. In the dacinit function, we are initializing the polarity (unipolar/bipolar) and the calibration value needed or not. Then we are calling the dscDAConvert function where the first parameter is the handler to the board, the second parameter is the channel number and the third paramater is the digital output code. Maximum digital output code is 4095 since the D/A section includes 4 12-bit D/A channels. For example, dscDAConvert (dscb, 3, 2048) means the channel number 3 has been converted and the digital output (DO) code is 2048. For DO code 0 means 0V and for DO code +2048 means 2.5V. For DO code 4095 means +5V. You can measure the analog output voltage with the help of multimeter, where the analog voltage will be first measuring 1.76V rms and will start decreasing to 0V.

11.18 CONCLUSION

Thus with the help of SBCs with different external I/O devices integrated into one board, all the required applications can be met. The requirement of the PCB cards which can be either analog I/O card having current transformer (HE055T01)/potential transformer (Magnatech) modules or the isolator cards having 4-20mA modules such as Lp420 or the digital input card with opto-isolator IC's such as MCT6E series or the relay cards with change over relays with the filter card having different stages of filtering can be integrated with the SBC for control, monitoring and protection of the SwitchYard equipment. For teaching the ever-important concepts of computer science to the younger generation, SBCs have served as a cheaply important educational tool. Some mechanical design activity such as enclosure design, cable design and the test procedures establishment, SBC is a tremendous solution for anyone wanting to provide complete control of their product.

A customer can buy whatever input/output is needed for the application from the supplier thus eliminating the need for the circuit design and engineering. As we have discussed different I/O expansions such as ISA/PCI/PCI-Express, etc., most SBCs have used heat sink at the top side for cooling the processor. Since they conduct heat to the surrounding air, the heat sinks are inefficient. The cooling of the processor will get even worse if the SBC is in a sealed box. SBCs are used in a wide variety of applications in the industries. If the customers do not want to design and use their own electronics, they will use SBCs at large. Thus, SBC is not only a solution for the real-world problems but also a boon for solving those problems which require automation, control, monitoring and protection in the real world.

REVIEW QUESTIONS

1. Write a short note on single board computer.
2. Write short notes on cache and system chipset.
3. If the user is having the clock multiplier of 2 with a 2.4 GHz bus, then what is the clock frequency?
4. If the user is having the clock multiplier of 3 with a 2.5 GHz CPU, then what is the processor's bus frequency?
5. Explain the term Ethernet.
6. Explain watchdog timer and expansion interface.
7. Explain in detail BIOS.
8. Write short notes on solid state drive and hard disk drive.

9. Explain power management for a single board computer system.
10. What are the design issues in a single board computer? Explain in details.
11. What are the resources needed to build a small single board computer system?
12. Explain PC – 104 and PC – 104 plus in details.
13. How is the minimum memory requirement like size of RAM and ROM decided for the single board computer system? Explain with suitable example.
14. Does the size of the RAM affect the speed of computing. If yes, explain how.
15. Which one is better solid state drive or hard disk drive and why?
16. Explain what happens when any installed file is opened from the hard disk drive. How can it be processed by a single board computer system?
17. What is the feature that needs to be taken into consideration before choosing a single board computer?
18. Differentiate between RS232 and RS485 communication media; also explain which one of the two is better and why.

Objective Type Questions on Microcontrollers

I

Appendix

1. The basic job of a microcontroller is to
   (*a*) Communicate with its peripheral devices
   (*b*) Control external world
   (*c*) Fetch and execute instructions
   (*d*) None of the above
2. Which feature of the microcontroller 8051 is not available in microprocessor based system?
   (*a*) ADC (*b*) Registers
   (*c*) On-chip program memory (*d*) None of these
3. Microcontroller 8051 was introduced by ___ in ____.
   (*a*) Intel, 1980 (*b*) Fairchild, 1981
   (*c*) Intel, 1978 (*d*) Motorola, 1979
4. The type of program memory available in 87XX series of microcontroller is
   (*a*) EPROM (*b*) OTP
   (*c*) Flash ROM (*d*) None of these
5. The type of program memory available in AT89C51 microcontroller is
   (*a*) EPROM (*b*) OTP
   (*c*) Flash ROM (*d*) None of these
6. The size of on-chip program memory and data memory for XX51 is
   (*a*) 4K bytes, 64 bytes (*b*) 8K bytes, 128 bytes
   (*c*) 4K bytes, 128 bytes (*d*) 16K bytes, 256 bytes
7. One machine cycle of 8051 takes
   (*a*) 6 clock pulses (*b*) 12 clock pulses
   (*c*) 18 clock pulses (*d*) None of these
8. Total number of special function registers in 8051 are
   (*a*) 21 (*b*) 20
   (*c*) 30 (*d*) 31

9. What should be the values of RS1 and RS0 bits of PSW register to select register bank 1?
   (*a*) 0 and 1 (*b*) 0 and 0
   (*c*) 1 and 0 (*d*) 1 and 1
10. What is the reset value of stack pointer of 8051 microcontroller?
   (*a*) 00h (*b*) FFh
   (*c*) 08h (*d*) 07h
11. What is the size of program counter in 8051 microcontroller?
   (*a*) 8-bit (*b*) 16-bit
   (*c*) 32-bit (*d*) None of the above
12. Which of the following is the 16-bit register of 8051 microcontroller?
   (*a*) SP (*b*) R0
   (*c*) DPTR (*d*) Accumulator
13. What is the addressing mode of MOV 30h, A?
   (*a*) Direct (*b*) Indirect
   (*c*) Register (*d*) Immediate
14. What would be done by the following instruction, when executed:
   MOV A, 0D0h?
   (*a*) Load A with data byte D0h (*b*) Copy register B to register A
   (*c*) Copy register PSW to register A (*d*) Copy stack pointer to register A
15. What would happen if the following instruction is executed:
   MOV 0D0h, #18h?
   (*a*) R0 of bank 3 would be copied to PSW register
   (*b*) Register bank 3 would be selected
   (*c*) Contents of the location whose address is in R0 of bank 3 would be copied to PSW
   (*d*) None of the above
16. Which of the following registers in 8051 doesn't have on-chip address?
   (*a*) R0 (*b*) PC
   (*c*) B (*d*) DPL
17. What would be the contents of DPH register after execution of:
   MOV DPTR, #1234h
   INC 83h
   INC 82h?
   (*a*) 13h (*b*) 14h
   (*c*) 12h (*d*) 1234h
18. After system reset MOV A, SP is executed, what are the contents of the accumulator?
   (*a*) 70h (*b*) 07h
   (*c*) 08h (*d*) FFh

19. Which of the following instructions is performed as an indirect RAM to accumulator?
(*a*) MOV A, Rn (*b*) MOV A, @Ri
(*c*) MOV @Ri, A (*d*) MOV Rn, A

20. ACALL instruction allows specifying ______address in the instruction and calling subroutine within ______ program memory block.
(*a*) 2byte, 3K (*b*) 11bit, 2K
(*c*) 9bit, 2K (*d*) 1byte, 3K

21. Which of the following instructions need to be performed to move accumulator to external RAM of 16-bit address?
(*a*) MOV @ DPTR, A (*b*) MOVX @ Ri, A
(*c*) MOV A, @ Ri (*d*) MOVX @ DPTR, A

22. Which of the following instructions performs jump indirect relative to DPTR?
(*a*) JMP A+DPTR (*b*) JMP DPTR
(*c*) JMP @A+DPTR (*d*) SJMP A+DPTR

23. Which of the following instructions is wrong in 8051 microcontroller?
(*a*) INC DPTR (*b*) MOVX @DPTR, A
(*c*) MOVC A, @A+DPTR (*d*) DEC DPTR

24. What instruction performs compare immediate to indirect and jump if not equal?
(*a*) CJNE A, #data, rel (*b*) CJNE Rn, #data, rel
(*c*) CJNE @ Ri, #data, rel (*d*) CJNE A, data, rel

25. What is the result of RR A instruction if the accumulator contains 1000 0000?
(*a*) 0000 0001 (*b*) 0000 0000
(*c*) 0100 0000 (*d*) 0000 0010

26. MOV 45, #4FH
MOV R0, 45H
INC R0
MOV @R0, #30H
What is the location value of R0 and the content at that place?
(*a*) 45H, 4F (*b*) 50H, 30H
(*c*) 30H, 50H (*d*) 50H, 45H

27. Which of the following instructions is used to set bit port directly?
(*a*) SET P1.0 (*b*) MOV P1.0, bit
(*c*) SETB P1.0 (*d*) JB P1.0, bit

28. MOV A, #56H
MOV R1, #50H
MOV 50H, # 45H
XCHD A, @R1

What is the result at A, R1?

(*a*) 56H, 45H (*b*) 45H, 50H
(*c*) 55H, 46H (*d*) 45H, 56H

29. The 8051 has _____ 16 – bit timers / counters.

(*a*) 1 (*b*) 2
(*c*) 3 (*d*) 4

30. The 8051 can handle _______ interrupt sources.

(*a*) 4 (*b*) 5
(*c*) 6 (*d*) 3

31. The following program will receive data from port 1, determine whether bit 2 is high, and then send the number FFh to port 3

```
READ:   MOV A, P1
        ANL A, #02H
        CJNE A, #02H, READ
        MOV P3, #FFH
```

(*a*) True (*b*) False

32. When the 8051 is reset and the EA# line is high, the program counter points to the first program instruction in the

(*a*) External code memory (*b*) Internal code memory
(*c*) Internal data memory (*d*) External data memory

33. An alternate function of port pin P3.4 in the 8051 is

(*a*) Timer 0 (*b*) Timer 1
(*c*) Interrupt 0 (*d*) Interrupt 1

34. Both registers TL0 and TL1 are needed to start timer 0.

(*a*) True (*b*) False

35. The last 96 locations in the internal data memory are reserved for general purpose data storage and stack.

(*a*) True (*b*) False

36. The total external data memory that can be interfaced to 8051 is

(*a*) 32KB (*b*) 64KB
(*c*) 128KB (*d*) 256KB

37. Bit addressable memory locations are

(*a*) 10h – 1Fh (*b*) 20h – 2Fh
(*c*) 30h – 3Fh (*d*) 40h – 4Fh

38. The contents of the accumulator after this operation would be

```
MOV A, #0Bh
ANL A, #2Ch
```

(*a*) 11010111 (*b*) 11011010
(*c*) 00001000 (*d*) 00101000

39. This program code will be executed once

```
START:  MOV A, #01h
        JNZ START
```

(*a*) True (*b*) False

40. This program code will be executed once

```
START:  MOV A, #01h
        DEC A
        JNZ START
```

(*a*) True (*b*) False

41. The I/O ports that are used as address and data for external memory are
(*a*) Port 1 and 2 (*b*) Port 0 and 2
(*c*) Port 1 and 3 (*d*) Port 0 and 1

42. Data transfer from I/O to the external data memory can only be done with the MOV command.
(*a*) True (*b*) False

43. The number of data registers are
(*a*) 8 (*b*) 16
(*c*) 32 (*d*) 64

44. What is the difference between the 8031 and the 8051?
(*a*) 8031 has no interrupt (*b*) 8051 is ROM less
(*c*) 8031 is ROM less (*d*) The 8051 has 64 bytes more memory

45. The I/O port that does not have a dual role is
(*a*) Port 0 (*b*) Port 1
(*c*) Port 2 (*d*) Port 3

46. To interface external program memory for applications, it is necessary to demultiplex the address/data lines of the 8051.
(*a*) True (*b*) False

47. The microcontroller is useful in systems that have non-variable programs for dedicated applications.
(*a*) True (*b*) False

48. The HIGH signal on which pin resets the 8051 is
(*a*) RESET (*b*) PSEN
(*d*) RST (*d*) RSET

49. An alternate function of port pin P3.1 in the 8051 is
(*a*) Serial port input (*b*) Serial port output
(*d*) Memory write strobe (*d*) Memory read strobe

50. The contents of the accumulator after this operation will be
MOV A, #2Bh
ORL A, #00h
(*a*) 1Bh (*b*) 2Bh
(*c*) 3Bh (*d*) 4Bh

51. What would be the contents of DPTR register after execution of INC DPTR instruction? Assume contents of DPTR = 10FFh
(*a*) 1100h (*b*) 00FFh
(*c*) F000h (*d*) 2000h

52. What would be the contents of DPTR register after execution of DEC 82h instruction? Assume contents of DPTR = 10FFh
(*a*) 10F2h (*b*) 10FEh
(*c*) 1200h (*d*) None of the above

53. What would be the contents of DPTR register after execution of DEC 83h instruction? Assume contents of DPTR = 00FF
(*a*) FFFFh (*b*) FE00h
(*c*) F100h (*d*) 0000h

54. What is the limit of LJMP instruction?
(*a*) 2K bytes (*b*) 256 bytes
(*c*) 64K bytes (*d*) 1K bytes

55. What is the limit of AJMP instruction?
(*a*) 2K bytes (*b*) 256 bytes
(*c*) 64K bytes (*d*) 1K bytes

56. How many bits of the jumping address are available in ACALL instruction?
(*a*) 10 bits (*b*) 16 bits
(*c*) 11 bits (*d*) 12 bits

57. The JZ instruction is applicable
(*a*) Only for register R0 to R7 (*b*) Only for special function registers
(*c*) Only for register B (*d*) Only for accumulator

58. Which of the following instructions affects any flag of PSW register?
(*a*) DJNZ (*b*) NOP
(*c*) SJMP (*d*) None of these

59. Which flags are affected by the execution of the instruction POP 0D0h?
(*a*) No flags are affected (*b*) Only carry flag is affected
(*c*) Only AC flag is affected (*d*) All flags are affected

60. What is the maximum number of nesting of subroutines possible in 8051 microcontroller?
(*a*) 64 (*b*) 63
(*c*) 128 (*d*) 27

61. What would happen if the following subroutine is executed?

```
SUBPROG:    PUSH 0E0h
            PUSH 0F0h
            POP 0E0h
            POP 0F0h
            RET
```

(*a*) Data present in registers A and B is interchanged
(*b*) Same contents return in registers A and B
(*c*) No result
(*d*) None of these

62. The ANL instruction is generally used to
(*a*) Set a few bits (*b*) Clear a few bits
(*c*) Complement a few bits (*d*) None of these

63. The ORL instruction is generally used to
(*a*) Set a few bits (*b*) Clear a few bits
(*c*) Complement a few bits (*d*) None of these

64. The XRL instruction is generally used to
(*a*) Set a few bits (*b*) Clear a few bits
(*c*) Complement a few bits (*d*) None of these

65. Which of the following instructions are used for clear accumulator?
(*a*) SUBB A, 0E0h and MOV A, B (*b*) ANL A, 0E0h and XRL A, #0FFh
(*c*) SUBB A, 0E0h and XRL A, 0E0h (*d*) ANL A, #00h and XRL A, #00h

66. Which of the following instructions are used for clear accumulator?
(*a*) SUBB A, A, MOV A, B and ANL A, 0E0h
(*b*) ANL A, A, MOV A, #00h and ORL A, #00h
(*c*) SUBB A, A, XRL A, A and CLR A
(*d*) ANL A, A, CLR A and ORL A, #00h

67. Which is the correct solution for finding two's complement of accumulator?
(*a*) CPL A and INC A (*b*) CPL A and DEC A
(*c*) CPL A and ADD A, #01h (*d*) CPL A and ADD A, #00h

68. Instead of using the instruction XRL A, #0FFh, which of the following instructions may be used to get the same result?
(*a*) CLR A (*b*) CPL A
(*c*) Both (a) and (b) (*d*) None of these

69. Instead of using the instruction ANL A, #00h, which of the following instructions may be used to get the same result?
(*a*) CLR A (*b*) CPL A
(*c*) Both (a) and (b) (*d*) None of these

70. The instructions MOV C, bit and MOV bit, C takes
(*a*) One cycle each (*b*) One and two cycles
(*c*) Two cycles each (*d*) None of these

71. Communication with the external data memory may be achieved only through
(*a*) B register (*b*) Accumulator register
(*c*) PSW register (*d*) R0 register

72. The XCHD instruction exchanges
(*a*) D0 bits only (*b*) Lower nibbles only
(*c*) Higher nibbles only (*d*) D0 and D1 bits

73. After execution of MUL AB instruction, the higher byte of the product would be available in which register?
(*a*) Accumulator (*b*) B
(*c*) R0 (*d*) PSW

74. After execution of DIV AB instruction, the quotient byte is available in which register?
(*a*) Accumulator (*b*) B
(*c*) R0 (*d*) PSW

75. DA A instruction is applicable for
(*a*) BCD numbers (*b*) Binary numbers
(*c*) Hexadecimal numbers (*d*) Decimal numbers

76. The instruction JMP @ A + DPTR is capable of transferring the control within
(*a*) 2K bytes (*b*) 256 bytes
(*c*) 64K bytes (*d*) 1024 bytes

77. The instruction SWAP exchanges a pair of nibbles
(*a*) Between A and an indirectly addressed byte
(*b*) Within the accumulator
(*c*) Between A and directly addressed byte
(*d*) Within register B

78. An interrupt INT1, is enabled by setting
(*a*) IE.7 (*b*) IE.2
(*c*) Both (a) and (b) (*d*) None of these

79. The SFRs of μC 8051 address is of
(*a*) 256 bytes and ranges from 00H to 7FH.
(*b*) 128 bytes and ranges from 80H to FFH.
(*c*) 256 bytes and ranges from 80H to FFH.
(*d*) 128 bytes and ranges from 00H to 7FH.

80. Match the following dual functions of P3 Register.

Bit Addressable Register	Function
(1) P3.0	(i) Timers(T1)
(2) P3.1	(ii) Serial Communication(RXD)
(3) P3.4	(iii) Timers(T0)
(4) P3.5	(iv) Serial Communication(TXD)

(*a*) (1) is performing (iv)
(2) is performing (ii)
(3) is performing (i)
(4) is performing (iii)

(*b*) (1) is performing (ii)
(2) is performing (iv)
(3) is performing (i)
(4) is performing (iii)

(*c*) (1) is performing (iv)
(2) is performing (ii)
(3) is performing (iii)
(4) is performing (i)

(*d*) (1) is performing (ii)
(2) is performing (iv)
(3) is performing (iii)
(4) is performing (i)

81. Match the following dual functions of P3 Register.

Bit Addressable Register	Function
(1) P3.2	(i) External Interrupt (INTO bar)
(2) P3.3	(ii) External Memory Access (WR bar)
(3) P3.6	(iii) External Memory Access (RD bar)
(4) P3.7	(iv) External Interrupt (INT1 bar)

(*a*) (1) is performing (i)
(2) is performing (iv)
(3) is performing (iii)
(4) is performing (ii)

(*b*) (1) is performing (i)
(2) is performing (iv)
(3) is performing (ii)
(4) is performing (iii)

(*c*) (1) is performing (iv)
(2) is performing (i)
(3) is performing (iii)
(4) is performing (ii)

(*d*) (1) is performing (iv)
(2) is performing (i)
(3) is performing (ii)
(4) is performing (iii)

82. After the execution of the above program, the content of the accumulator will be

```
    ORG 0000H
    MOV A, #7FH
    CPL A
    XRL A, #3FH
    RR A
$:  SJMP $
    END
```

(*a*) BFH (*b*) DFH
(*c*) CFH (*d*) EFH

83. After the execution of the above program, the content of the accumulator will be

```
    ORG 0000H
    MOV A, #7FH
    CPL A
    XRL A, #3FH
    RL A
$:  SJMP $
END
```

(*a*) 6FH (*b*) AFH
(*c*) 7FH (*d*) 8FH

84. After the execution of the above program, the content of the accumulator will be

```
    ORG 0000H
    MOV B, #56H
    SETB B.0
    CPL B.1
    MOV A, B
    CLR ACC.0
    ADD A, #A0H
    ADD A, #D0H
$:  SJMP $
END
```

(*a*) A5H (*b*) B5H
(*c*) C5H (*d*) D5H

85. After the execution of the above program, the content of the accumulator will be

```
    ORG 0000H
    MOV B, #56H
    SETB B.0
    CPL B.1
    MOV A, B
    CLR ACC.0
    ADD A, #A0H
    ADD A, #D0H
    RR A
    RR A
    RR A
    CPL A
$:  SJMP $
END
```

(*a*) 17H (*b*) 27H
(*c*) 37H (*d*) 47H

86. After the execution of the above program, the content of the accumulator will be

```
    ORG 0000H
    MOV B, #56H
    SETB B.0
    CPL B.1
    MOV A, B
    CLR ACC.0
    ADD A, #A0H
    ADD A, #D0H
    RRC A
    RRC A
    RRC A
    CPL A
$:  SJMP $
END
```

(*a*) 87H (*b*) 97H
(*c*) A7H (*d*) B7H

87. After the execution of the above program, the content of the accumulator will be

```
    ORG 0000H
    MOV A, #56H
    MOV B, A
    CPL A
    CPL ACC.6
    XRL A,#BFH
    CLR ACC.0
$:  SJMP $
END
```

(*a*) 56H (*b*) 66H
(*c*) 77H (*d*) 86H

88. After the execution of the above program, the content of the accumulator will be

```
    ORG 0000H
    MOV A, #56H
    MOV B, A
    CPL A
    CPL ACC.6
    XRL A, #BFH
    CLR ACC.0
    RL A
    RL A
    RL A
```

```
    CPL A
$:  SJMP $
END
```

(*a*) 1DH (*b*) 2DH
(*c*) 3DH (*d*) 4DH

89. After the execution of the above program, the content of the accumulator will be

```
    ORG 0000H
    MOV A, #56H
    MOV B, A
    CPL A
    CPL ACC.6
    XRL A,#BFH
    SETB C
    RL A
    RL A
    RL A
    CPL A
$:  SJMP $
END
```

(*a*) 1AH (*b*) 2AH
(*c*) 3AH (*d*) 4AH

90. Predict the data byte content inside the Register A after the execution of the program given below:

```
    ORG 0000H
    COUNT EQU 255
    MOV R2, #COUNT
    MOV R1, #00H
    CLR A
L1: ADD A, #03H
    JNC L3
    INC R1
L3: DJNZ R2 , L1
    MOV R3, #45
L2: ADD A, #03H
    DJNZ R3 , L2
$:  SJMP $
END
```

(*a*) 64H (*b*) 74H
(*c*) 84H (*d*) None of the above

91. Predict the data byte content inside the Register A after the execution of the program given below:

```
    ORG 0000H
    COUNT EQU 255
    MOV R2, #COUNT
    MOV R1, #00H
    CLR A
L1: ADD A, #03H
    JNC L3
    INC R1
L3: DJNZ R2 , L1
    MOV R3, #45
L2: SUBB A, #03H
    DJNZ R3, L2
$:  SJMP $
END
```

(*a*) 12H. (*b*) 13H.
(*c*) 14H. (*d*) None of these.

92. In problem 13, indicate the value of Register R1 after the execution of the program.
(*a*) 01H. (*b*) 02H.
(*c*) 03H. (*d*) 04H.

93. In problem 12, indicate the value of Register R1 after the execution of the program.
(*a*) 01H. (*b*) 02H.
(*c*) 03H. (*d*) 04H.

94. Predict the data byte present in the accumulator and the carry flag bit after the execution of the program given below:

```
    ORG 0000H
    MOV R2, #03H
    MOV A,#41H
    CLR C
L1: RRC A
    MOV P2.1, C
    DJNZ R2,L1
$:  SJMP $
END
```

(*a*) 90H and carry flag is set. (*b*) 90H and carry flag is reset.
(*c*) 48H and carry flag is reset. (*d*) 48H and carry flag is set.

95. Predict the data byte present in the accumulator and the carry flag bit after the execution of the below program.

```
    ORG 0000H
    MOV R2, #03H
    MOV A, #BDH
    SETB C
L1: RLC A
    MOV P2.1, C
    DJNZ R2, L1
$:  SJMP $
END
```

(*a*) EEH and carry flag is set. (*b*) EEH and carry flag is reset.
(*c*) FEH and carry flag is reset. (*d*) FEH and carry flag is set.

96. In the program given below estimate the value of the accumulator content and predict which operation is being performed.

```
    ORG 0000H
    MOV A, #31H
    MOV R1, A
    CLR C
    RRC A
    XRL A, R1
$:  SJMP $
END
```

(*a*) 29H AND GREY TO BINARY CONVERTER.
(*b*) 19H AND BINARY TO GREY CONVERTER.
(*c*) 29H AND BINARY TO GREY CONVERTER.
(*d*) 19H AND GREY TO BINARY CONVERTER.

97. In the program given below, predict what the above program is executing?

```
        ORG 0000H
        MOV R1, #20H
        MOV R2, #30H
        MOV R0, #40
        MOV DPTR, #3000H
AGAIN:  MOVX A,@DPTR
        CJNE A, #25H, JUMP
        MOV R2, DPH
        MOV R1, DPL
        SJMP WOW
JUMP:   INC DPTR
        DJNZ R0, AGAIN
WOW:    MOV DPTR, #9000H
```

```
        MOV A, R1
        MOVX @DPTR, A
        INC DPTR
        MOV A, R2
        MOVX @DPTR, A
$:      SJMP $
END
```

(*a*) The above program will store the lower and higher byte address if the required data byte is found.

(*b*) The data bytes stored in memory location 9000H and 9001H are 20H and 30H always.

(*c*) If the required data byte is not found from memory location, then the data bytes stored in memory location 9000H and 9001H are 20H and 30H.

(*d*) Both options (a) and (c) are correct.

98. The ALP given below will exchange a block of data bytes between external memory location starting from 0300H onwards to internal memory location starting from 30H. Complete the missing instructions for correct execution of exchange of the data bytes.

```
        ORG 0000H
        EXT EQU 0300H
        INT EQU 30H
        COUNT EQU 05H
        MOV R2, #COUNT
        MOV DPTR, #EXT
        MOV R0, #INT
LOOP:   --------------
        --------------
        --------------
        INC R0
        INC DPTR
        DJNZ R2, LOOP
$:      SJMP $
        END
```

(*a*) XCH A, @R0, MOVXA,@DPTR, MOVX @DPTR,A.

(*b*) XCH A,@R0, MOVX @DPTR, A, MOVX A,@DPTR.

(*c*) MOVX A,@DPTR , XCH A , @R0 , MOVX @DPTR , A

(*d*) MOVX @DPTR,A , XCH A,@R0 , MOVX A,@DPTR.

99. Predict the data bytes present in the memory locations 40H, 41H and 42H after the execution of the below program. Assume that the data byte present in the memory location 40H is FFH.

```
    ORG 0000H
    NO EQU 40H
    MOV R1, #NO
    MOV A,@R1
    MOV 0F0H, #64H
    DIV AB
    MOV @R1, A
    MOV 0E0H, 0F0H
    MOV 0F0H, #0AH
    DIV AB
    INC R1
    MOV@R1,A
    INC R1
    MOV A, B
    RR A
    RR A
    RR A
    RR A
    MOV @R1, 0E0H
$:  SJMP $
END
```

(*a*) 40H-02H, 41H-50H, 42H-50H.
(*a*) 40H-02H, 41H-05H, 42H-50H.
(*a*) 40H-02H, 41H-05H, 42H-50H.
(*a*) 40H-02H, 41H-05H, 42H-00H.

Common data question for Q. 100 and Q. 101

100. Predict the conversion method after the execution of the below program.

```
    ORG 0000H
    MOV R1, #09H
    MOV A, #86H
    RR A
    XRL A, R1
    MOV B, A
    ANL A, #F0H
    SWAP A
    ORL A, #30H
    MOV @R1, A
```

```
    INC R1
    MOV A, B
    ANL A, #0FH
    CJNE A, #09H, L1
    SJMP L2
L3: ADD A, #07H
    SJMP L2
L1: JC L2
    JNC L3
L2: ADD A, #30H
    MOV @R1, A
$:  SJMP $
END
```

(*a*) ASCII to HEX Conversion. (*b*) HEX to BCD Conversion.
(*c*) Packed BCD to Unpacked BCD. (*d*) None of these.

101. The data bytes stored in memory location 09H and 0AH are
(*a*) 33H and 34H. (*b*) 34H and 33H.
(*c*) 33H and 35H (*d*) None of these.

102. Predict the data bytes stored in memory location 0BH and 0CH after the execution of the program given below:

```
    ORG 0000H
    MOV R1, #0BH
    MOV A, #86H
    RR A
    XRL A, R1
    MOV B, A
    ANL A, #F0H
    SWAP A
    ORL A, #30H
    MOV @R1, A
    INC R1
    MOV A, B
    ANL A, #0FH
    CJNE A, #09H, L1
    SJMP L2
L3: ADD A, #07H
    SJMP L2
L1: JC L2
    JNC L3
```

```
L2: ADD A, #30H
    MOV @R1, A
$:  SJMP $
END
```

The data bytes stored in memory location 16H and 17H are

(*a*) 34H and 39H (*b*) 34H and 38H
(*c*) 34H and 34H (*d*) None of these

Common data question for Q. 103 and Q. 104

103. Predict the conversion method done in the above program. Assume that the data byte present in the DPTR Memory Location is 47H.

```
    ORG 0000H
    MOV DPTR, #2000H
    MOVX A,@DPTR
    MOV R2, A
    ANL A, #0F0H
    RR A
    RR A
    RR A
    RR A
    MOV 0F0H, #10
    MUL AB
    MOV R1, A
    MOV A, R2
    ANL A, #0FH
    ADD A, R1
    INC DPTR
    MOVX @DPTR, A
$:  SJMP $
END
```

(*a*) HEXADECIMAL TO BCD. (*b*) BCD TO BINARY.
(*c*) ASCCI TO BCD. (*d*) BCD TO ASCII.

104. The data byte present in the memory location 2001H after the execution of the above program is

(*a*) 2CH (*b*) 2DH
(*c*) 2EH (*d*) 2FH

Common data question for Q. 105, Q. 106, Q. 107, Q. 108 and Q. 109

105. Identify the type of conversion used in the program given below

```
            ORG 0000H
            NUM EQU 50H
            MOV R1, #NUM
            MOV A, @R1
L1:         CJNE A, #30H, UNMATCH
EXACTLY:    CJNE A, #39H, EXACTLY1
EXACTLY1:   JNC EXACTLY2
            SUBB A, #30H
            MOV @R1, A
            SJMP LAST2
EXACTLY2:   SUBB A, #07H
            SUBB A, #30H
            MOV @R1, A
            SJMP LAST2
UNMATCH:    JC LAST1
            CJNE A, #46H, UNMATCH2
            SJMP EXACTLY
UNMATCH2:   JNC LAST1
            SJMP EXACTLY
LAST1:      MOV A, #0FFH
LAST2:      END
```

(*a*) ASCII TO HEX CONVERSION.
(*b*) HEX TO ASCII CONVERSION.
(*c*) BCD TO ASCII CONVERSION.
(*d*) UNPACKED BCD TO PACKED BCD CONVERSION

106. Assume that the data byte present in the memory location 50H is 38H. Then the data byte present in the accumulator after the execution of the program is
(*a*) 05H (*b*) 06H
(*c*) FFH (*d*) 08H

107. Assume that the data byte present in the memory location 50H is 45H. Then the data byte present in the accumulator after the execution of the program is
(*a*) FFH (*b*) 0DH
(*c*) 0EH (*d*) 0FH

108. Assume that the data byte present in the memory location 50H is 4AH. Then the data byte present in the accumulator after the execution of the program is
(*a*) FFH (*b*) 0DH
(*c*) 0EH (*d*) 0FH

109. Assume that the data byte present in the memory location 50H is 28H. Then the data byte present in the accumulator after the execution of the program is

(*a*) 00H (*b*) FFH
(*c*) 0EH (*d*) 0FH

110. Identify the combinational circuit executed in the program given below:

```
        ORG 0000H
        MOV A, #0FFH
        MOV P2, A
START:  JB P2.3 , L1
        MOV C, P2.0
        MOVP1.0 , C
        SJMP START
L1:     MOV C, P2.1
        MOV P1.0 , C
        SJMP START
        END
```

(*a*) 2:4 DECODER. (*b*) 2:1 MUX.
(*c*) 4:1 MUX. (*d*) 4:2 PRIORITY ENCODER.

111. Identify the combinational circuit executed in the program given below:

```
        ORG 0000H
        MOV A, #0FFH
        MOV P2, A
START:  JB P2.7, L1
        JB P2.6,                L2
        MOV C, P2.0
        MOV P1.0, C
        SJMP START
L1:     JB P2.6, L3
        MOV C, P2.2
        MOV P1.0, C
        SJMP START
L2:     MOV C, P2.1
        MOV P1.0, C
        SJMP START
L3:     MOV C, P2.3
        MOV P1.0, C
        SJMP START
        END
```

(*a*) 2:1 MUX. (*b*) 4:1 MUX.
(*c*) 8:1 MUX. (*d*) 3:8 DECODER.

112. Identify the combinational logic circuit after the execution of the program given below:

```
        ORG 0000H
        MOV A, #0FFH
        MOV P2, A
        MOV P0, A
START:  JB P0.7, L1
        JB P0.6, L2
        JB P0.5, L3
        MOV C, P2.0
        MOV P1.0, C
        SJMP START
L1:     JB P0.6, L4
        JNB P0.6, L6
L2:     JB P0.5, L8
        MOV C, P2.2
        MOV P1.0, C
        SJMP START
L3:     MOV C, P2.1
        MOV P1.0, C
        SJMP START
L4:     JB P0.5, L5
        MOV C, P2.6
        MOV P1.0, C
        SJMP START
L5:     MOV C, P2.7
        MOV P1.0, C
        SJMP START
        L6: JB P0.5, L7
        MOV C, P2.4
        MOV P1.0, C
        SJMP START
L7:     MOV C, P2.5
        MOV P1.0, C
        SJMP START
L8:     MOV C, P2.3
        MOV P1.0, C
        SJMP START
        END
```

(*a*) 2:1 MUX. (*b*) 4:1 MUX.
(*c*) 8:1 MUX. (*d*) 3:8 DECODER.

113. Realize the combinational circuit after the execution of the program given below. Assume the control signals are having such a logic that it is by default enabling the combinational circuit.

```
        ORG 0000H
        MOV A, #0FFH
        MOV P2, A
START:  JB P2.1, L1
        JB P2.0, L2
        SETB P1.0
        SJMP START
L1:     JB P2.0, L3
        SETB P1.2
        SJMP START
L2:     SETB P1.1
        SJMP START
L3:     SETB P1.3
        SJMP START
        END
```

(*a*) 2:4 DECODER.
(*b*) 3:8 DECODER.
(*c*) 4:2 PRIORITY ENCODER.
(*d*) 8:3 PRIORITY ENCODER.

114. Realize the combinational circuit after the execution of the program. Assume the control signals are having such a logic that it is by default enabling the combinational circuit

```
        ORG 0000H
        MOV A, #0FFH
        MOV P2, A
        MOV P3,A
START:  JB P2.2, L0
        JB P2.1, L2
        JB P2.0, L8
        SETB P3.0
        SJMP START
L0:     JB P2.1, L4
        JNB P2.1, L6
L6:     JB P2.0, L7
        SETB P3.4
        SJMP START
```

```
L7:         SETB P3.5
            SJMP START
L4:         JB P2.0, L5
            SETB P3.6
            SJMP START
L5:         SETB P3.7
            SJMP START
L2:         JB P2.0, L3
            SETB P3.2
            SJMP START
L3:         SETB P3.3
            SJMP START
L8:         SETB P3.1
            SJMP START
            END
```

(*a*) 2:4 DECODER. (*b*) 3:8 DECODER.
(*c*) 4:2 PRIORITY ENCODER. (*d*) 8:3 PRIORITY ENCODER.

115. Realize the combinational circuit after the execution of the program. Assume the control signals are having such a logic that it is by default enabling the combinational circuit

```
            ORG 0000H
            MOV A, #0FFH
            MOV P2, A
START:      CLR P1.0
            CLR P1.1
            CLR P1.2
            JNB P2.0, L0
            JNB P2.1, L1
            JNB P2.2, L2
            JNB P2.3, L3
            JNB P2.4, L4
            JNB P2.5, L5
            JNB P2.6, L6
            JNB P2.7, L7
            SJMP START
```

```
L0:     MOV A, #00H
        MOV P1, A
        SJMP START
L1:     MOV A, #01H
        MOV P1, A
        SJMP START
L2:     MOV A, #02H
        MOV P1, A
        SJMP START
L3:     MOV A, #03H
        MOV P1, A
        SJMP START
L4:     MOV A, #04H
        MOV P1, A
        SJMP START
L5:     MOV A, #05H
        MOV P1, A
        SJMP START
L6:     MOV A, #06H
        MOV P1, A
        SJMP START
L7:     MOV A, #07H
        MOV P1, A
        SJMP START
        END
```

(*a*) 2:4 DECODER. (*b*) 3:8 DECODER.
(*c*) 4:2 PRIORITY ENCODER. (*d*) 8:3 PRIORITY ENCODER.

116. The 8051 timers are capable of counting
(*a*) Down (*b*) Up
(*c*) Both up and down (*d*) None of these

117. Timer/counter configuration is selectable by
(*a*) Hardware (*b*) Software
(*c*) Using hardware and software (*d*) None of these

118. How many modes of operation are offered by timer 0?
(*a*) One (*b*) Two
(*c*) Three (*d*) Four

119. To which SFR does the RUN / STOP control bit TR0 for timer 0 belong?
(*a*) PCON (*b*) SCON
(*c*) TCON (*d*) TMOD

120. Pseudo instructions are basically
(*a*) False instructions.
(*b*) Instructions that are ignored by the microprocessor.
(*c*) Assembler directives.
(*d*) Instructions that are treated like comments.

121. If a 1M ×1 DRAM requires 4 ms for a refresh and has 256 rows to be refreshed, no more than __________ of time must pass before another row is refreshed.
(*a*) 64 ms. (*b*) 4 ns.
(*c*) 0.5 ns. (*d*) 15.625μs.

122. A certain SRAM has CS# = 0, WE# = 0 and OE# = 1. In which of the following modes this SRAM is operating?
(*a*) Read (*b*) Write
(*c*) Stand by (*d*) None of the above

123. How many times the following loop is executed?

```
        MOV A, #00H
LOOP:   INC A
        JNZ LOOP
```

(*a*) 64 (*b*) 128
(*c*) 255 (*d*) 256

124. Which of the following is true with respect to EEPROM?
(*a*) Contents can be erased byte wise only.
(*b*) Contents of full memory can be erased together.
(*c*) Contents can be erased using ultraviolet rays.
(*d*) Contents cannot be erased.

125. What will be the contents of register AL after the following has been executed:

```
MOV A, #8Ch
MOV R0, #7Eh
ADD A, R0?
```

(*a*) 0A and carry flag is set (*b*) 0A and carry flag is reset
(*c*) 6A and carry flag is set (*d*) 6A and carry flag is reset

126. SD RAM refers to
(*a*) Synchronous DRAM (*b*) Static DRAM
(*c*) Semi DRAM (*d*) Second DRAM

127. EPROM is generally erased by using
(*a*) Ultraviolet rays (*b*) Infrared rays
(*c*) 12 V electrical pulse (*d*) 24 V electrical pulse

128. The time required to refresh a typical DRAM is
(*a*) 2 – 4 μs (*b*) 2 – 4 ns
(*c*) 2 – 4 ms (*d*) 2 – 4 ps

129. The number of address lines required to address a memory of size 32 K is
(*a*) 15 lines (*b*) 16 lines
(*c*) 18 lines (*d*) 14 lines

130. How many address lines are required to access 1MB memory?
(*a*) 20 (*b*) 21
(*c*) 19 (*d*) 18

131. How many memory ICs of 1K x 4 are required to design 1K x 8?
(*a*) 3 (*b*) 4
(*c*) 2 (*d*) 1

132. To interface 2K x 8 SRAM with microcontroller using the available IC 1K x 4 SRAM, how many 1K x 4 SRAM are required?
(*a*) 4 (*b*) 5
(*c*) 6 (*d*) 7

133. To interface 6K x 8 SRAM with microcontroller using the available IC 1K x 4 SRAM, how many 1K x 4 SRAM are required?
(*a*) 10 (*b*) 11
(*c*) 12 (*d*) 13

134. To interface 16K x 16 EPROM with microcontroller using the available IC 4K x 8 EPROM, how many 4K x 8 EPROMs are required?
(*a*) 4 (*b*) 6
(*c*) 7 (*d*) 8

135. To interface 32K x 8 SRAM with microcontroller using the available IC 8K x 8 SRAM, how many 8K x 8 SRAM are required?
(*a*) 4 (*b*) 5
(*c*) 6 (*d*) 7

136. The address lines and data lines in the memory organization 4K x 4 are
(*a*) 12 address lines and 8 data lines
(*b*) 13 address lines and 8 data lines
(*c*) 12 address lines and 4 data lines
(*d*) 13 address lines and 4 data lines

137. The address lines and data lines in the memory organization 32K x 8 are
(*a*) 14 address lines and 8 data lines
(*b*) 15 address lines and 8 data lines
(*c*) 14 address lines and 4 data lines
(*d*) 15 address lines and 4 data lines

138. The address lines and data lines in the memory organization 1 MB are
(*a*) 19 address lines and 8 data lines
(*b*) 20 address lines and 8 data lines
(*c*) 19 address lines and 4 data lines
(*d*) 20 address lines and 4 data lines

139. The address lines and data lines in the memory organization 8K x 2 is
(*a*) 12 address lines and 8 data lines
(*b*) 13 address lines and 8 data lines
(*c*) 12 address lines and 2 data lines
(*d*) 13 address lines and 2 data lines

140. Identify the memory IC 27C64 is
(*a*) SRAM (*b*) DRAM
(*c*) EPROM (*d*) Flash ROM

141. Identify the memory IC 6164 is
(*a*) SRAM (*b*) DRAM
(*c*) EPROM (*d*) Flash ROM

142. The memory IC 2764 has memory organization of
(*a*) 8k × 8 (*b*) 16k × 8
(*c*) 4k × 8 (*c*) 64k × 8

143. The memory IC 6116 has ______ address lines.
(*a*) 11 (*b*) 12
(*c*) 13 (*d*) 14

144. The memory IC 2114 has ______ address lines.
(*a*) 8 (*b*) 9
(*c*) 10 (*d*) 11

145. To interface 1 MB memory using 16 KB memory, how many 16 KB memory ICs are required?
(*a*) 13 (*b*) 14
(*c*) 15 (*d*) 16

146. If the starting address of 1 KB memory is 0000h, then the final address would be
(*a*) 13FFh (*b*) 0400h
(*c*) 1400h (*d*) 03FFh

147. If the starting address of 8 KB memory is 0000h, then the final address would be
(*a*) 1FFF (*b*) 2FFF
(*c*) 8000h (*c*) 9000h

148. Eight ICs of 8 KB memory are interfaced with the microcontroller based system and the starting address of 1st IC is 0000h, then what is final address of 5th IC?
(*a*) 8FFFh (*b*) 9000h
(*c*) 9FFFh (*d*) A000h

149. Eight ICs of 8KB memory are interfaced with the microcontroller based system and the starting address of 1st IC is 0000h, then what is the initial (starting) and final address of 7th IC?
(*a*) Initial address = 5000h, final address = 6FFFh
(*b*) Initial address = C000h, final address = DFFFh
(*c*) Initial address = E000h, final address = FFFFh
(*d*) Initial address = A000h, final address = AFFFh

150. If the final (end) address of 16 KB memory is BFFFh, then the starting (initial) address would be
(*a*) B000h (*b*) A000h
(*c*) 9000h (*d*) 8000h

151. Shadow addresses are present in the ______ decoding techniques.
(*a*) Full (*b*) Partial
(*c*) Both full and partial (*d*) None of the above

152. Some of the address lines are unused in the _____ decoding techniques.
(*a*) Full (*b*) Partial
(*c*) Both full and partial (*d*) None of the above

153. SRAM is volatile memory.
(*a*) True (*b*) False

154. SRAM is faster than DRAM.
(*a*) True (*b*) False

155. Eight flip flops are required to store 9-bit of information.
(*a*) True (*b*) False

156. For designing of chip, select signal ______ gate is generally used.
(*a*) AND (*b*) NOT
(*c*) EXOR (*d*) NAND

Key Answers

1. (c)	**2.** (c)	**3.** (a)	**4.** (a)	**5.** (c)	**6.** (c)	**7.** (b)	**8.** (a)
9. (a)	**10.** (d)	**11.** (b)	**12.** (c)	**13.** (a)	**14.** (c)	**15.** (b)	**16.** (b)
17. (a)	**18.** (b)	**19.** (b)	**20.** (b)	**21.** (d)	**22.** (c)	**23.** (d)	**24.** (c)
25. (c)	**26.** (b)	**27.** (c)	**28.** (c)	**29.** (b)	**30.** (b)	**31.** (a)	**32.** (b)
33. (a)	**34.** (b)	**35.** (b)	**36.** (b)	**37.** (b)	**38.** (c)	**39.** (b)	**40.** (a)
41. (b)	**42.** (b)	**43.** (c)	**44.** (c)	**45.** (b)	**46.** (a)	**47.** (a)	**48.** (c)
49. (b)	**50.** (b)	**51.** (a)	**52.** (b)	**53.** (a)	**54.** (c)	**55.** (a)	**56.** (c)
57. (d)	**58.** (d)	**59.** (d)	**60.** (b)	**61.** (a)	**62.** (b)	**63.** (a)	**64.** (c)
65. (c)	**66.** (c)	**67.** (c)	**68.** (b)	**69.** (a)	**70.** (b)	**71.** (b)	**72.** (b)
73. (b)	**74.** (a)	**75.** (a)	**76.** (c)	**77.** (b)	**78.** (c)	**79.** (b)	**80.** (d)

81. (b)

82. (b)

Solution:

ORG 0000H	; Initialize the program from the memory location 0000H.
MOV A, #7FH	; Load the accumulator with the data byte 7FH.
CPL A	; Complementing the value of the Accumulator. So, the accumulator data byte becomes 80H.
XRL A, #3FH	; XORing the accumulator data byte 80H with the data ; byte 3FH. The final data byte present in the accumulator after the XORing operation is BFH. 80H: 1000 0000 3FH: 0011 1111 BFH: 1011 1111

RR A: Rotating the value of the Accumulator right by one without carry. The data byte present in the Accumulator after the rotation is DFH.

83. (c)

Solution:

ORG 0000H	; Initialize the program from the memory location 0000H.
MOV A, #7FH	; Load the accumulator with the data byte 7FH.
CPL A	; Complementing the value of the accumulator. So, the accumulator data byte becomes 80H.
XRL A, #3FH	; XORing the accumulator data byte 80H with the data byte 3FH. The final data byte present in the accumulator after the XORing operation is BFH. 80H: 1000 0000 3FH: 0011 1111 BFH: 1011 1111

RR A: Rotating the value of the accumulator left by one without carry. The data byte present in the accumulator after the rotation is 7FH.

84. (c)

Solution:

ORG 0000H	; Initialize the program from the memory location 0000H.
MOV B, #56H	; Loading the data byte of register B with 56H.
SETB B.0	; Setting the D0[th] bit of register B. The data byte present in the register B is 57H.
CPL B.1	; Complementing the D1[th] bit of register B. The data byte present in register B is 55H.
MOV A, B	; Storing the data byte 55H in the accumulator.
CLR ACC.0	; Clearing the D0[th] data bit present in the accumulator. ; The present data byte of the accumulator is 55H.
ADD A, #A0H	; Adding the accumulator data byte 55H with the data byte A0H. The data byte present in the accumulator after addition is F5H. (A): 0101 0101 (B): 1010 0000 (A): 1111 0101
ADD A, #D0H	; Adding the accumulator data byte F5H with the data byte D0H. The data byte present in the accumulator after addition is C5H. The CY flag is set. (A): 1111 0101 (B): 1101 0000 (A): 1100 0101

85. (d)

Solution:

ORG 0000H	; Initialize the program from the memory location 0000H.
MOV B, #56H	; Loading the data byte of register B with 56H.
SETB B.0	; Setting the D0[th] bit of register B. The data byte present in register B is 57H.
CPL B.1	; Complementing the D1[th] bit of register B. The data byte ; present in register B is 55H.
MOV A, B	; Storing the data byte 55H in the accumulator.

```
CLR ACC.0        ; Clearing the D0th data bit present in the accumulator.
                 ; The present data byte of the accumulator is 55H.
ADD A, #A0H      ; Adding the accumulator data byte 55H with the
                   data byte A0H. The data byte present in the
                   accumulator after addition is F5H.
                 (A): 0101 0101
                 (B): 1010 0000
                 (A): 1111 0101
ADD A, #D0h      ; Adding the accumulator data byte F5H with the data
                   byte D0H. The data byte present in the accumulator
                   after addition is C5H. The CY flag is set.
                 (A): 1111 0101
                 (B): 1101 0000
                 (A): 1100 0101
RR A             ; Rotating the value of the accumulator right by one without
                   carry. The result generated after rotation in the
                   accumulator data byte is E2H.
RR A             ; Rotating the value of the accumulator right by one without
                   carry. The result generated after rotation in the
                   accumulator data byte is 71H.
RR A             ; Rotating the value of the accumulator right by one without
                   carry. The result generated after rotation in the
                   accumulator data byte is B8H.
CPL A            ; Complementing the accumulator data byte value. The
                   data byte present in the accumulator after the
                   complementation is 47H.
```

86. (a)

Solution:

```
ORG 0000H        ; Initialize the program from the memory location 0000H.
MOV B, #56H      ; Loading the data byte of register B with 56H.
SETB B.0         ; Setting the D0th bit of register B. The data byte present
                 ; in register B is 57H.
```

```
CPL B.1        ; Complementing the D1th bit of register B. The data byte
                 present in register B is 55H.
MOV A, B       ; Storing the data byte 55H in the accumulator.
CLR ACC.0      ; Clearing the D0th data bit present in the accumulator.
                 The present data byte of the accumulator is 55H.
ADD A, #A0H    ; Adding the accumulator data byte 55H with the data
                 byte A0H. The data byte present in the accumulator
                 after addition is F5H.
               (A): 0101 0101
               (B): 1010 0000
               (A): 1111 0101
ADD A, #D0H    ; Adding the accumulator data byte F5H with the data
                 byte D0H. The data byte present in the accumulator
                 after addition is C5H. The CY flag is set.
               (A): 1111 0101
               (B): 1101 0000
               (A): 1100 0101
RRC A          ; Rotating the value of the accumulator right by one with
                 carry. The result generated after rotation in the
                 accumulator data byte is E2H. The CY flag bit is set.
RRC A          ; Rotating the value of the accumulator right by one without
                 carry. The result generated after rotation in the
                 accumulator data byte is F1H. The CY flag bit is reset.
RRC A          ; Rotating the value of accumulator right by one without
                 carry. The result generated after rotation in the
                 accumulator data byte is 78H. The CY flag bit is set.
CPL A          ; Complementing the accumulator data byte value. The
                 data byte present in the accumulator after the
                 complementation is 87H.
```

87. (a)

Solution:

ORG 0000H	; Initialize the program from the memory location 0000H.
MOV A, #56H	; Loading the data byte 56H into the accumulator.
MOV B, A	; Copying the data byte 56H of register A into the register B.
CPL A	; Complementing the value of the accumulator. The accumulator data byte will be A9H.
CPL ACC.6	; Complementing the D6th bit of the accumulator. The accumulator data byte will be E9H.
XRL A, #BFH	; XORing the data byte of the accumulator E9H with the data byte BFH. The resultant data byte present in the accumulator will be 56H. (A): 1110 1001 (B): 1011 1111 (A): 0101 0110
CLR ACC.0	; Clearing the D0th bit of the accumulator. The resultant data byte present in the accumulator is 56H.

88. (d)

Solution:

ORG 0000H	; Initialize the program from the memory location 0000H.
MOV A, #56h	; Loading the data byte 56H into the accumulator.
MOV B, A	; Copying the data byte 56H of register A into the register B.
CPL A	; Complementing the value of the accumulator. The accumulator data byte will be A9H.
CPL ACC.6	; Complementing the D6th bit of the accumulator. The accumulator data byte will be E9H.
XRL A, #BFH	; XORing the data byte of accumulator E9H with the data byte BFH. The resultant data byte present in the accumulator will be 56H. (A): 1110 1001

(B): 1011 1111

(A): 0101 0110

CLR ACC.0	; Clearing the D0th bit of the accumulator. The resultant data byte present in the accumulator is 56H.
RL A	; Rotating the value of the accumulator left by one without carry. The result generated after rotation in the accumulator data byte is ACH.
RL A	; Rotating the value of the accumulator left by one without carry. The result generated after rotation in the the accumulator data byte is 59H.
RL A	; Rotating the value of accumulator left by one without carry. The result generated after rotation in the accumulator data byte is B2H.
CPL A	; Complementing the data byte value B2H. The resultant data byte after complementing will be 4DH.

89. (d)

Solution:

ORG 0000H	; Initialize the program from the memory location 0000H.
MOV A, #56H	; Loading the data byte 56H into the accumulator.
MOV B, A	; Copying the data byte 56H of register A into the register B.
CPL A	; Complementing the value of the accumulator. The accumulator data byte will be A9H.
CPL ACC.6	; Complementing the D6th bit of the accumulator. The accumulator data byte will be E9H.
XRL A, #BFH	; XORing the data byte of the accumulator E9H with the data byte BFH. The resultant data byte present in the accumulator will be 56H.

(A): 1110 1001

(B): 1011 1111

(A): 0101 0110

SETB C ; Setting the carry flag bit. Hence, the carry flag bit value is 1.

RLC A ; Rotating the value of the accumulator left by one with carry. The result generated after rotation in the accumulator; data byte is ADH. The carry flag bit is reset.

RLC A ; Rotating the value of the accumulator left by one with carry. The result generated after rotation in the accumulator data byte is 5AH. The carry flag bit is set.

RLC A ; Rotating the value of the accumulator left by one with carry. The result generated after rotation in the accumulator data byte is B5H. The carry flag bit is reset.

CPL A ; Complementing the data byte value B2H. The resultant data byte after complementing will be 4AH.

90. (c)

Solution:

Table 1

Decimal Value	Hexadecimal Value	Count Value
15	0F	5
30	1E	10
60	3C	20
90	5A	30
180	B4	60
240	F0	80
255	**FF**	**85**
02	02	86
32	20	96
62	3E	106
182	B6	146
242	F2	166
254	**FE**	**170**
01	01	171
31	1F	181
61	3D	191
181	B5	231
241	F1	251

253	FD	255
00	00	01
30	1E	11
90	5A	31
120	78	41
123	7B	42
126	7E	43
132	**84**	**45**

In the above program, the count value of the number of times the loop has been counted its count value and the data byte value subsequently has been shown in the form of a table. This table will depict the count value and the data byte value. The count value asked in the above program has been highlighted in bold letters and the data byte shown with respect to the count is calculated. So, the flow of the program is as follows:

```
ORG 0000H          ; Initialize the program from the memory location 0000H.
COUNT EQU 255      ; The label Count has been initialized with the count value
                     255.
MOV R2, # COUNT    ; Copying the count value 255 into register R2.
MOV R1, #00H       ; Loading the value 00H into register R1.
LR A               ; Clearing the accumulator value to 00H.
L1: ADD A, #03H    ; Adding the value 03H with the accumulator data byte
                     content.
JNC L3             ; Jump to label L3 if no carry is generated.
INC R1             ; Incrementing the value of R1.
L3: DJNZ R2, L1    ; Decrement the value of R2 by one and jump
                     to label L1 if Zero Flag is reset.
MOV R3, #45        ; Copy the count value 45 to register R3.
L2: ADD A, #03H    ; Add the value 03H with the accumulator data byte
                     content.
DJNZ R3, L2        ; Decrement the value of R3 by one and jump to label L2
                     if Zero Flag is reset.
$: SJMP $          ; Short jump to the label $.
END                ; Terminate the execution of the above program.
```

Thus, after the execution of the above program, the data byte value stored inside the accumulator will be 84H.

91. (d)

Solution:

Table 2:

Decimal Value	Hexadecimal Value	Count Value
15	0F	5
30	1E	10
60	3C	20
90	5A	30
180	B4	60
240	F0	80
255	**FF**	**85**
02	02	86
32	20	96
62	3E	106
182	B6	146
242	F2	166
254	**FE**	**170**
01	01	171
31	1F	181
61	3D	191
181	B5	231
241	F1	251
253	**FD**	**255**
250	00	01
130	82	41
127	7F	42
124	7C	43
121	79	44
118	**76**	**45**

In the above program, the count value of the number of times the loop has been counted and the data byte value subsequently has been shown in the form of a table. This table will depict the count value and the data byte value. The count value asked in the above program has been highlighted in bold letters and the data byte shown with respect to the count is calculated. So, the flow of the program is as follows:

```
ORG 0000H           ; Initialize the program from the memory location 0000H.
COUNT EQU 255       ; The label count has been initialized with the count value
                      255.
MOV R2, # COUNT     ; Copying the count value 255 into register R2.
MOV R1, #00H        ; Loading the value 00H into register R1.
CLR A               ; Clearing the accumulator value to 00H.
L1: ADD A, #03H     ; Adding the value 03H with the accumulator data byte
                      content.
JNC L3              ; Jump to label L3 if no carry is generated.
INC R1              ; Incrementing the value of R1.
L3: DJNZ R2, L1     ; Decrement the value of R2 by one and jump and jump
                      to label L1 if Zero Flag is reset.
MOV R3, #45         ; Copy the count value 45 to register R3.
L2: SUBB A, #03H    ; Subtract the value 03H with the accumulator data byte
                      content.
DJNZ R3, L2         ; Decrement the value of R3 by one and jump to label L2
                      if Zero Flag is reset.
$: SJMP $           ; Short jump to the label $.
END                 ; Terminate the execution of the above program.
```

Thus, after the execution of the above program, the data byte value stored in the accumulator will be 76H.

92. (b)

From Table 2 we can see that at count value 85 and 170, the data byte values are FFH and FEH. At 86th and 171st count value, the data byte values are 02H and 01H and hence the count value of register R1 will increase by one. Hence, the data byte stored in register R1 will be 02H. Hence, the correct answer is (b).

93. (c)

From Table 1 we can see that at count value 85, 170 and 155, the data byte values are FFH, FEH and FDH. At 86th, $171^{st}$and 1st (i.e., the second time loop run) count value, the data byte values are 02H, 01H and 00H and hence the count value of register R1 will increase by one. Hence, the data byte stored in register R1 will be 02H. Hence, the correct answer is (c).

94. (c) **95.** (a) **96.** (c) **97.** (d) **98.** (c) **99.** (c) **100.** (d) **101.** (d)

102. (b) **103.** (b) **104.** (d) **105.** (a) **106.** (d) **107.** (c) **108.** (a) **109.** (b)

110. (b) **111.** (b) **112.** (c) **113.** (a) **114.** (b) **115.** (d) **116.** (b) **117.** (b)

118. (d) **119.** (a) **120.** (c) **121.** (b) **122.** (b) **123.** (d) **124.** (c) **125.** (a)

126. (a) **127.** (a) **128.** (c)* **129.** (a)* **130.** (a) **131.** (c) **132.** (a) **133.** (c)

134. (d) **135.** (a) **136.** (c) **137.** (b) **138.** (b) **139.** (d) **140.** (c) **141.** (a)

142. (a) **143.** (a) **144.** (c) **145.** (d) **146.** (d) **147.** (a) **148.** (c) **149.** (b)

150. (d) **151.** (b) **152.** (b) **153.** (a) **154.** (b) **155.** (b) **156.** (d)

Q.128

Solution:

The capacitor Cs discharges through the internal resistance of the NMOS transistor T1. Typically, Cs = 0.2 pF and the internal resistance Rin = 1010 ohms, so:

Cs x Rin = 0.2 x 10-12 x 1010 x 103 ms = 2 ms

So the typical refresh time interval is 2 ms. Hence, the answer is (c).

Q. 129

Solution:

32K = 32 X 1024 bits = 2^5 X 2^{10} = 2^{15}. Hence, answer is (a).

Appendix II

Summary of Instruction Sets of 8051 Microcontrollers

INSTRUCTION SETS OF MICROCONTROLLER 8051

The following abbreviations are used for the instructions of 8051 microcontroller

#data/#data8	8-bit data
#data16	16-bit data
Addr8/saddr	8-bit address
Addr11	11-bit address
Addr16/laddr	16-bit address
rel	8-bit relative offset address
bit	8-bit bit address (8-bit address of bit addressable RAM or bit addressable SFR address)
direct	8-bit internal RAM/SFR address

SECTION – A

INSTRUCTIONS ARRANGED BY FUNCTIONAL GROUP

A.1 DATA TRANSFER GROUP

S. No.	Opcode	Instruction	Bytes	Cycles	Flags affected
1	74	MOV A, #DATA	2	1	No flags are affected
2	E8	MOV A, R0	1	1	
3	E9	MOV A, R1	1	1	
4	EA	MOV A, R2	1	1	
5	EB	MOV A, R3	1	1	
6	EC	MOV A, R4	1	1	
7	ED	MOV A, R5	1	1	
8	EE	MOV A, R6	1	1	
9	EF	MOV A, R7	1	1	
10	E5	MOV A, DIRECT	2	1	
11	E6	MOV A, @R0	1	1	

12	E7	MOV A, @R1	1	1	
13	78	MOV R0, #DATA	2	1	
14	79	MOV R1, #DATA	2	1	
15	7A	MOV R2, #DATA	2	1	
16	7B	MOV R3, #DATA	2	1	
17	7C	MOV R4, #DATA	2	1	
18	7D	MOV R5, #DATA	2	1	
19	7E	MOV R6, #DATA	2	1	
20	7F	MOV R7, #DATA	2	1	
21	F8	MOV R0, A	1	1	
22	F9	MOV R1, A	1	1	
23	FA	MOV R2, A	1	1	
24	FB	MOV R3, A	1	1	
25	FC	MOV R4, A	1	1	
26	FD	MOV R5, A	1	1	
27	FE	MOV R6, A	1	1	
28	FF	MOV R7, A	1	1	
29	A8	MOV R0, DIRECT	2	2	
30	A9	MOV R1, DIRECT	2	2	
31	AA	MOV R2, DIRECT	2	2	
32	AB	MOV R3, DIRECT	2	2	
33	AC	MOV R4, DIRECT	2	2	
34	AD	MOV R5, DIRECT	2	2	
35	AE	MOV R6, DIRECT	2	2	
36	AF	MOV R7, DIRECT	2	2	
37	F5	MOV DIRECT, A	2	1	
38	75	MOV DIRECT, #DATA	3	2	
39	85	MOV DIRECT, DIRECT	3	2	
40	88	MOV DIRECT, R0	2	2	
41	89	MOV DIRECT, R1	2	2	
42	8A	MOV DIRECT, R2	2	2	
43	8B	MOV DIRECT, R3	2	2	
44	8C	MOV DIRECT, R4	2	2	
45	8D	MOV DIRECT, R5	2	2	
46	8E	MOV DIRECT, R6	2	2	
47	8F	MOV DIRECT, R7	2	2	
48	86	MOV DIRECT, @R0	2	2	

49	87	MOV DIRECT, @R1	2	2	
50	F6	MOV @R0, A	1	1	
51	F7	MOV @R1, A	1	1	
52	76	MOV @R0, #DATA	2	1	
53	77	MOV @R1, #DATA	2	1	
54	A6	MOV @R0, DIRECT	2	2	
55	A7	MOV @R1, DIRECT	2	2	
56	90	MOV DPTR, #DATA16	3	2	
57	E2	MOVX A, @R0	1	2	
58	E3	MOVX A, @R1	1	2	
59	E0	MOVX A, @DPTR	1	2	
60	F2	MOVX @R0, A	1	2	
61	F3	MOVX @R1, A	1	2	
62	F0	MOVX @DPTR, A	1	2	
63	93	MOVC A, @A+DPTR	1	2	
64	83	MOVC A, @A+PC	1	2	
65	C0	PUSH DIRECT	2	2	
66	D0	POP DIRECT	2	2	No flags are affected but for POP 0D0 all flags are modified
67	C8	XCH A, R0	1	1	No flags are affected
68	C9	XCH A, R1	1	1	
69	CA	XCH A, R2	1	1	
70	CB	XCH A, R3	1	1	
71	CC	XCH A, R4	1	1	
72	CD	XCH A, R5	1	1	
73	CE	XCH A, R6	1	1	
74	CF	XCH A, R7	1	1	
75	C5	XCH A, DIRECT	2	1	
76	C6	XCH A, @R0	1	1	
77	C7	XCH A, @R1	1	1	
78	D6	XCHD A, @R0	1	1	
79	D7	XCHD A, @R1	1	1	

A.2 ARITHMETIC INSTRUCTION GROUP

S. No.	Opcode	Instruction	Bytes	Cycles	Flags affected
1	24	ADD A, #DATA	2	1	CY, OV and AC
2	25	ADD A, DIRECT	2	1	CY, OV and AC
3	26	ADD A, @R0	1	1	CY, OV and AC
4	27	ADD A, @R1	1	1	CY, OV and AC
5	28	ADD A, R0	1	1	CY, OV and AC
6	29	ADD A, R1	1	1	CY, OV and AC
7	2A	ADD A, R2	1	1	CY, OV and AC
8	2B	ADD A, R3	1	1	CY, OV and AC
9	2C	ADD A, R4	1	1	CY, OV and AC
10	2D	ADD A, R5	1	1	CY, OV and AC
11	2E	ADD A, R6	1	1	CY, OV and AC
12	2F	ADD A, R7	1	1	CY, OV and AC
13	34	ADDC A, #DATA	2	1	CY, OV and AC
14	35	ADDC A, DIRECT	2	1	CY, OV and AC
15	36	ADDC A, @R0	1	1	CY, OV and AC
16	37	ADDC A, @R1	1	1	CY, OV and AC
17	38	ADDC A, R0	1	1	CY, OV and AC
18	39	ADDC A, R1	1	1	CY, OV and AC
19	3A	ADDC A, R2	1	1	CY, OV and AC
20	3B	ADDC A, R3	1	1	CY, OV and AC
21	3C	ADDC A, R4	1	1	CY, OV and AC
22	3D	ADDC A, R5	1	1	CY, OV and AC
23	3E	ADDC A, R6	1	1	CY, OV and AC
24	3F	ADDC A, R7	1	1	CY, OV and AC
25	94	SUBB A, #DATA	2	1	CY, OV and AC
26	95	SUBB A, DIRECT	2	1	CY, OV and AC
27	96	SUBB A, @R0	1	1	CY, OV and AC
28	97	SUBB A, @R1	1	1	CY, OV and AC
29	98	SUBB A, R0	1	1	CY, OV and AC
30	99	SUBB A, R1	1	1	CY, OV and AC
31	9A	SUBB A, R2	1	1	CY, OV and AC
32	9B	SUBB A, R3	1	1	CY, OV and AC
33	9C	SUBB A, R4	1	1	CY, OV and AC
34	9D	SUBB A, R5	1	1	CY, OV and AC

35	9E	SUBB A, R6	1	1	CY, OV and AC
36	9F	SUBB A, R7	1	1	CY, OV and AC
37	D4	DA A	1	1	CY and AC
38	04	INC A	1	1	No flags are affected
39	05	INC DIRECT	2	1	No flags are affected
40	06	INC @R0	1	1	No flags are affected
41	07	INC @R1	1	1	No flags are affected
42	08	INC R0	1	1	No flags are affected
43	09	INC R1	1	1	No flags are affected
44	0A	INC R2	1	1	No flags are affected
45	0B	INC R3	1	1	No flags are affected
46	0C	INC R4	1	1	No flags are affected
47	0D	INC R5	1	1	No flags are affected
48	0E	INC R6	1	1	No flags are affected
49	0F	INC R7	1	1	No flags are affected
50	A3	INC DPTR	1	2	No flags are affected
51	14	DEC A	1	1	No flags are affected
52	15	DEC DIRECT	2	1	No flags are affected
53	16	DEC @R0	1	1	No flags are affected
54	17	DEC @R1	1	1	No flags are affected
55	18	DEC R0	1	1	No flags are affected
56	19	DEC R1	1	1	No flags are affected
57	1A	DEC R2	1	1	No flags are affected
58	1B	DEC R3	1	1	No flags are affected
59	1C	DEC R4	1	1	No flags are affected
60	1D	DEC R5	1	1	No flags are affected
61	1E	DEC R6	1	1	No flags are affected
62	1F	DEC R7	1	1	No flags are affected
63	A4	MUL AB	1	4	CY and OV
64	84	DIV AB	1	4	CY and OV

A.3 LOGICAL INSTRUCTION GROUP

S. No.	Opcode	Instruction	Bytes	Cycles	Flags affected
1	54	ANL A, #DATA	2	1	No flags are affected
2	55	ANL A, DIRECT	2	1	No flags are affected
3	56	ANL A, @R0	1	1	No flags are affected

4	57	ANL A, @R1	1	1	No flags are affected
5	58	ANL A, R0	1	1	No flags are affected
6	59	ANL A, R1	1	1	No flags are affected
7	5A	ANL A, R2	1	1	No flags are affected
8	5B	ANL A, R3	1	1	No flags are affected
9	5C	ANL A, R4	1	1	No flags are affected
10	5D	ANL A, R5	1	1	No flags are affected
11	5E	ANL A, R6	1	1	No flags are affected
12	5F	ANL A, R7	1	1	No flags are affected
13	52	ANL DIRECT, A	2	1	No flags are affected
14	53	ANL DIRECT, #DATA	3	2	No flags are affected
15	44	ORL A, #DATA	2	1	No flags are affected
16	45	ORL A, DIRECT	2	1	No flags are affected
17	46	ORL A, @R0	1	1	No flags are affected
18	47	ORL A, @R1	1	1	No flags are affected
19	48	ORL A, R0	1	1	No flags are affected
20	49	ORL A, R1	1	1	No flags are affected
21	4A	ORL A, R2	1	1	No flags are affected
22	4B	ORL A, R3	1	1	No flags are affected
23	4C	ORL A, R4	1	1	No flags are affected
24	4D	ORL A, R5	1	1	No flags are affected
25	4E	ORL A, R6	1	1	No flags are affected
26	4F	ORL A, R7	1	1	No flags are affected
27	42	ORL DIRECT, A	2	1	No flags are affected
28	43	ORL DIRECT, #DATA	3	2	No flags are affected
29	64	XRL A, #DATA	2	1	No flags are affected
30	65	XRL A, DIRECT	2	1	No flags are affected
31	66	XRL A, @R0	1	1	No flags are affected
32	67	XRL A, @R1	1	1	No flags are affected
33	68	XRL A, R0	1	1	No flags are affected
34	69	XRL A, R1	1	1	No flags are affected
35	6A	XRL A, R2	1	1	No flags are affected
36	6B	XRL A, R3	1	1	No flags are affected
37	6C	XRL A, R4	1	1	No flags are affected
38	6D	XRL A, R5	1	1	No flags are affected
39	6E	XRL A, R6	1	1	No flags are affected

40	6F	XRL A, R7	1	1	No flags are affected
41	62	XRL DIRECT, A	2	1	No flags are affected
42	63	XRL DIRECT, #DATA	3	2	No flags are affected
43	E4	CLR A	1	1	No flags are affected
44	F4	CPL A	1	1	No flags are affected
45	23	RL A	1	1	No flags are affected
46	33	RLC A	1	1	CY only
47	03	RR A	1	1	No flags are affected
48	13	RRC A	1	1	CY only
49	C4	SWAP A	1	1	No flags are affected

A.4 BOOLEAN INSTRUCTION GROUP

S. No.	Opcode	Instruction	Bytes	Cycles	Flags affected
1	92	MOV BIT, C	2	2	No flags are affected
2	A2	MOV C, BIT	2	1	CY only
3	82	ANL C, BIT	2	2	CY only
4	B0	ANL C, /BIT	2	2	CY only
5	72	ORL C, BIT	2	2	CY only
6	A0	ORL C, /BIT	2	2	CY only
7	C2	CLR BIT	2	1	No flags are affected
8	C3	CLR C	1	1	CY only (CY = 0)
9	D2	SETB BIT	2	1	No flags are affected
10	D3	SETB C	1	1	CY only (CY = 1)
11	B2	CPL BIT	2	1	No flags are affected
12	B3	CPL C	1	1	CY only (CY complemented)

A.5 BRANCHING INSTRUCTION GROUP

S. No.	Opcode	Instruction	Bytes	Cycles	Flags affected
01	11 31 51 71 91 B1 D1 F1	ACALL ADDR11	2	2	No flags are affected

02	01 21 41 61 81 A1 C1 E1	AJMP ADDR11	2	2	No flags are affected
03	12	LCALL ADDR16	3	2	No flags are affected
04	02	LJMP ADDR16	3	2	No flags are affected
05	80	SJMP REL	2	2	No flags are affected
06	73	JMP @A+DPTR	1	2	No flags are affected
07	B4	CJNE A, #DATA, REL	3	2	CY only
08	B5	CJNE A, DIRECT, REL	3	2	CY only
09	B6	CJNE @R0, #DATA, REL	3	2	CY only
10	B7	CJNE @R1, #DATA, REL	3	2	CY only
11	B8	CJNE R0, #DATA, REL	3	2	CY only
12	B9	CJNE R1, #DATA, REL	3	2	CY only
13	BA	CJNE R2, #DATA, REL	3	2	CY only
14	BB	CJNE R3, #DATA, REL	3	2	CY only
15	BC	CJNE R4, #DATA, REL	3	2	CY only
16	BD	CJNE R5, #DATA, REL	3	2	CY only
17	BE	CJNE R6, #DATA, REL	3	2	CY only
18	BF	CJNE R7, #DATA, REL	3	2	CY only
19	D5	DJNZ DIRECT, REL	3	2	No flags are affected
20	D8	DJNZ R0, REL	2	2	No flags are affected
21	D9	DJNZ R1, REL	2	2	No flags are affected
22	DA	DJNZ R2, REL	2	2	No flags are affected
23	DB	DJNZ R3, REL	2	2	No flags are affected
24	DC	DJNZ R4, REL	2	2	No flags are affected
25	DD	DJNZ R5, REL	2	2	No flags are affected
26	DE	DJNZ R6, REL	2	2	No flags are affected
27	DF	DJNZ R7, REL	2	2	No flags are affected
28	60	JZ REL	2	2	No flags are affected
29	70	JNZ REL	2	2	No flags are affected

30	40	JC REL	2	2	No flags are affected
31	50	JNC REL	2	2	No flags are affected
32	20	JB BIT, REL	3	2	No flags are affected
33	30	JNB BIT, REL	3	2	No flags are affected
34	10	JBC BIT, REL	3	2	No flags are affected
35	00	NOP	1	1	No flags are affected
36	22	RET	1	2	No flags are affected
37	32	RETI	1	2	No flags are affected

SECTION – B

INSTRUCTIONS ARRANGED BY ALPHABETICAL ORDER

S. No.	Opcode	Instruction	Bytes	Cycles	Flags affected
1	11 31 51 71 91 B1 D1 F1	ACALL ADDR11	2	2	No flags are affected
2	28	ADD A, R0	1	1	CY, OV and AC
3	29	ADD A, R1	1	1	CY, OV and AC
4	2A	ADD A, R2	1	1	CY, OV and AC
5	2B	ADD A, R3	1	1	CY, OV and AC
6	2C	ADD A, R4	1	1	CY, OV and AC
7	2D	ADD A, R5	1	1	CY, OV and AC
8	2E	ADD A, R6	1	1	CY, OV and AC
9	2F	ADD A, R7	1	1	CY, OV and AC
10	25	ADD A, DIRECT	2	1	CY, OV and AC
11	26	ADD A, @R0	1	1	CY, OV and AC
12	27	ADD A, @R1	1	1	CY, OV and AC
13	24	ADD A, #DATA	2	1	CY, OV and AC
14	38	ADDC A, R0	1	1	CY, OV and AC
15	39	ADDC A, R1	1	1	CY, OV and AC
16	3A	ADDC A, R2	1	1	CY, OV and AC

17	3B	ADDC A, R3	1	1	CY, OV and AC
18	3C	ADDC A, R4	1	1	CY, OV and AC
19	3D	ADDC A, R5	1	1	CY, OV and AC
20	3E	ADDC A, R6	1	1	CY, OV and AC
21	3F	ADDC A, R7	1	1	CY, OV and AC
22	35	ADDC A, DIRECT	2	1	CY, OV and AC
23	36	ADDC A, @R0	1	1	CY, OV and AC
24	37	ADDC A, @R1	1	1	CY, OV and AC
25	34	ADDC A, #DATA	2	1	CY, OV and AC
26	01 21 41 61 81 A1 C1 E1	AJMP ADDR11	2	2	No flags are affected
27	58	ANL A, R0	1	1	No flags are affected
28	59	ANL A, R1	1	1	No flags are affected
29	5A	ANL A, R2	1	1	No flags are affected
30	5B	ANL A, R3	1	1	No flags are affected
31	5C	ANL A, R4	1	1	No flags are affected
32	5D	ANL A, R5	1	1	No flags are affected
33	5E	ANL A, R6	1	1	No flags are affected
34	5F	ANL A, R7	1	1	No flags are affected
35	55	ANL A, DIRECT	2	1	No flags are affected
36	56	ANL A, @R0	1	1	No flags are affected
37	57	ANL A, @R1	1	1	No flags are affected
38	54	ANL A, #DATA	2	1	No flags are affected
39	52	ANL DIRECT, A	2	1	No flags are affected
40	53	ANL DIRECT, #DATA	3	2	No flags are affected
41	82	ANL C, BIT	2	2	CY only
42	B0	ANL C, /BIT	2	2	CY only
43	B4	CJNE A, #DATA, REL	3	2	CY only
44	B5	CJNE A, DIRECT, REL	3	2	CY only
45	B6	CJNE @R0, #DATA, REL	3	2	CY only

46	B7	CJNE @R1, #DATA, REL	3	2	CY only
47	B8	CJNE R0, #DATA, REL	3	2	CY only
48	B9	CJNE R1, #DATA, REL	3	2	CY only
49	BA	CJNE R2, #DATA, REL	3	2	CY only
50	BB	CJNE R3, #DATA, REL	3	2	CY only
51	BC	CJNE R4, #DATA, REL	3	2	CY only
52	BD	CJNE R5, #DATA, REL	3	2	CY only
53	BE	CJNE R6, #DATA, REL	3	2	CY only
54	BF	CJNE R7, #DATA, REL	3	2	CY only
55	E4	CLR A	1	1	No flags are affected
56	C2	CLR BIT	2	1	No flags are affected
57	C3	CLR C	1	1	No flags are affected (CY = 0)
58	F4	CPL A	1	1	No flags are affected
59	B2	CPL BIT	2	1	No flags are affected
60	B3	CPL C	1	1	No flags are affected (CY is complemented)
61	D4	DA A	1	1	CY and AC
62	14	DEC A	1	1	No flags are affected
63	15	DEC DIRECT	2	1	No flags are affected
64	18	DEC R0	1	1	No flags are affected
65	19	DEC R1	1	1	No flags are affected
66	1A	DEC R2	1	1	No flags are affected
67	1B	DEC R3	1	1	No flags are affected
68	1C	DEC R4	1	1	No flags are affected
69	1D	DEC R5	1	1	No flags are affected
70	1E	DEC R6	1	1	No flags are affected
71	1F	DEC R7	1	1	No flags are affected
72	16	DEC @R0	1	1	No flags are affected
73	17	DEC @R1	1	1	No flags are affected
74	84	DIV AB	1	4	CY and OV
75	D5	DJNZ DIRECT, REL	3	2	No flags are affected
76	D8	DJNZ R0, REL	2	2	No flags are affected
77	D9	DJNZ R1, REL	2	2	No flags are affected
78	DA	DJNZ R2, REL	2	2	No flags are affected
79	DB	DJNZ R3, REL	2	2	No flags are affected

80	DC	DJNZ R4, REL	2	2	No flags are affected
81	DD	DJNZ R5, REL	2	2	No flags are affected
82	DE	DJNZ R6, REL	2	2	No flags are affected
83	DF	DJNZ R7, REL	2	2	No flags are affected
84	04	INC A	1	1	No flags are affected
85	05	INC DIRECT	2	1	No flags are affected
86	A3	INC DPTR	1	2	No flags are affected
87	08	INC R0	1	1	No flags are affected
88	09	INC R1	1	1	No flags are affected
89	0A	INC R2	1	1	No flags are affected
90	0B	INC R3	1	1	No flags are affected
91	0C	INC R4	1	1	No flags are affected
92	0D	INC R5	1	1	No flags are affected
93	0E	INC R6	1	1	No flags are affected
94	0F	INC R7	1	1	No flags are affected
95	06	INC @R0	1	1	No flags are affected
96	07	INC @R1	1	1	No flags are affected
97	20	JB BIT, REL	3	2	No flags are affected
98	10	JBC BIT, REL	3	2	No flags are affected
99	40	JC BIT, REL	3	2	No flags are affected
100	73	JMP @A + DPTR	1	2	No flags are affected
101	30	JNB BIT, REL	3	2	No flags are affected
102	50	JNC REL	2	2	No flags are affected
103	70	JNZ REL	2	2	No flags are affected
104	60	JZ REL	2	2	No flags are affected
105	12	LCALL ADDR16	3	2	No flags are affected
106	02	LJMP ADDR16	3	2	No flags are affected
107	E8	MOV A, R0	1	1	No flags are affected
108	E9	MOV A, R1	1	1	No flags are affected
109	EA	MOV A, R2	1	1	No flags are affected
110	EB	MOV A, R3	1	1	No flags are affected
111	EC	MOV A, R4	1	1	No flags are affected
112	ED	MOV A, R5	1	1	No flags are affected
113	EE	MOV A, R6	1	1	No flags are affected
114	EF	MOV A, R7	1	1	No flags are affected
115	E5	MOV A, DIRECT	2	1	No flags are affected

116	E6	MOV A, @R0	1	1	No flags are affected
117	E7	MOV A, @R1	1	1	No flags are affected
118	74	MOV A, #DATA	2	1	No flags are affected
119	F8	MOV R0, A	1	1	No flags are affected
120	F9	MOV R1, A	1	1	No flags are affected
121	FA	MOV R2, A	1	1	No flags are affected
122	FB	MOV R3, A	1	1	No flags are affected
123	FC	MOV R4, A	1	1	No flags are affected
124	FD	MOV R5, A	1	1	No flags are affected
125	FE	MOV R6, A	1	1	No flags are affected
126	FF	MOV R7, A	1	1	No flags are affected
127	A8	MOV R0, DIRECT	2	2	No flags are affected
128	A9	MOV R1, DIRECT	2	2	No flags are affected
129	AA	MOV R2, DIRECT	2	2	No flags are affected
130	AB	MOV R3, DIRECT	2	2	No flags are affected
131	AC	MOV R4, DIRECT	2	2	No flags are affected
132	AD	MOV R5, DIRECT	2	2	No flags are affected
133	AE	MOV R6, DIRECT	2	2	No flags are affected
134	AF	MOV R7, DIRECT	2	2	No flags are affected
135	78	MOV R0, #DATA	2	1	No flags are affected
136	79	MOV R1, #DATA	2	1	No flags are affected
137	7A	MOV R2, #DATA	2	1	No flags are affected
138	7B	MOV R3, #DATA	2	1	No flags are affected
139	7C	MOV R4, #DATA	2	1	No flags are affected
140	7D	MOV R5, #DATA	2	1	No flags are affected
141	7E	MOV R6, #DATA	2	1	No flags are affected
142	7F	MOV R7, #DATA	2	1	No flags are affected
143	F5	MOV DIRECT, A	2	1	No flags are affected
144	88	MOV DIRECT, R0	2	2	No flags are affected
145	89	MOV DIRECT, R1	2	2	No flags are affected
146	8A	MOV DIRECT, R2	2	2	No flags are affected
147	8B	MOV DIRECT, R3	2	2	No flags are affected
148	8C	MOV DIRECT, R4	2	2	No flags are affected
149	8D	MOV DIRECT, R5	2	2	No flags are affected
150	8E	MOV DIRECT, R6	2	2	No flags are affected
151	8F	MOV DIRECT, R7	2	2	No flags are affected

152	85	MOV DIRECT, DIRECT	3	2	No flags are affected
153	86	MOV DIRECT, @R0	2	2	No flags are affected
154	87	MOV DIRECT, @R1	2	2	No flags are affected
155	75	MOV DIRECT, #DATA	3	2	No flags are affected
156	F6	MOV @R0, A	1	1	No flags are affected
157	F7	MOV @R1, A	1	1	No flags are affected
158	A6	MOV @R0, DIRECT	2	2	No flags are affected
159	A7	MOV @R1, DIRECT	2	2	No flags are affected
160	76	MOV @R0, #DATA	2	1	No flags are affected
161	77	MOV @R1, #DATA	2	1	No flags are affected
162	A2	MOV C, BIT	2	1	No flags are affected
163	92	MOV BIT, C	2	2	No flags are affected
164	90	MOV DPTR, #DATA16	3	2	No flags are affected
165	93	MOVC A, @A + DPTR	1	2	No flags are affected
166	83	MOVC A, @A + PC	1	2	No flags are affected
167	E2	MOVX A, @R0	1	2	No flags are affected
168	E3	MOVX A, @R1	1	2	No flags are affected
169	E0	MOVX A, @DPTR	1	2	No flags are affected
170	F2	MOVX @R0, A	1	2	No flags are affected
171	F3	MOVX @R1, A	1	2	No flags are affected
172	F0	MOVX @DPTR, A	1	2	No flags are affected
173	A4	MUL AB	1	4	CY and OV
174	00	NOP	1	1	No flags are affected
175	48	ORL A, R0	1	1	No flags are affected
176	49	ORL A, R1	1	1	No flags are affected
177	4A	ORL A, R2	1	1	No flags are affected
178	4B	ORL A, R3	1	1	No flags are affected
179	4C	ORL A, R4	1	1	No flags are affected
180	4D	ORL A, R5	1	1	No flags are affected
181	4E	ORL A, R6	1	1	No flags are affected
182	4F	ORL A, R7	1	1	No flags are affected
183	45	ORL A, DIRECT	2	1	No flags are affected
184	46	ORL A, @R0	1	1	No flags are affected
185	47	ORL A, @R1	1	1	No flags are affected
186	44	ORL A, #DATA	2	1	No flags are affected
187	42	ORL DIRECT, A	2	1	No flags are affected

188	43	ORL DIRECT, #DATA	3	2	No flags are affected
189	72	ORL C, BIT	2	2	CY only
190	A0	ORL C, /BIT	2	2	CY only
191	D0	POP DIRECT	2	2	No flags are affected but for POP 0D0 all flags are modified
192	C0	PUSH DIRECT	2	2	No flags are affected
193	22	RET	1	2	No flags are affected
194	32	RETI	1	2	No flags are affected
195	23	RL A	1	1	No flags are affected
196	33	RLC A	1	1	CY only
197	03	RR A	1	1	No flags are affected
198	13	RRC A	1	1	CY only
199	D2	SETB BIT	2	1	No flags are affected
200	D3	SETB C	1	1	No flags are affected (CY = 1)
201	80	SJMP REL	2	2	No flags are affected
202	98	SUBB A, R0	1	1	CY, OV and AC
203	99	SUBB A, R1	1	1	CY, OV and AC
204	9A	SUBB A, R2	1	1	CY, OV and AC
205	9B	SUBB A, R3	1	1	CY, OV and AC
206	9C	SUBB A, R4	1	1	CY, OV and AC
207	9D	SUBB A, R5	1	1	CY, OV and AC
208	9E	SUBB A, R6	1	1	CY, OV and AC
209	9F	SUBB A, R7	1	1	CY, OV and AC
210	95	SUBB A, DIRECT	2	1	CY, OV and AC
211	96	SUBB A, @R0	1	1	CY, OV and AC
212	97	SUBB A, @R1	1	1	CY, OV and AC
213	94	SUBB A, #DATA	2	1	CY, OV and AC
214	C4	SWAP A	1	1	No flags are affected
215	C8	XCH A, R0	1	1	No flags are affected
216	C9	XCH A, R1	1	1	No flags are affected
217	CA	XCH A, R2	1	1	No flags are affected
218	CB	XCH A, R3	1	1	No flags are affected
219	CC	XCH A, R4	1	1	No flags are affected
220	CD	XCH A, R5	1	1	No flags are affected

221	CE	XCH A, R6	1	1	No flags are affected
222	CF	XCH A, R7	1	1	No flags are affected
223	C5	XCH A, DIRECT	2	1	No flags are affected
224	C6	XCH A, @R0	1	1	No flags are affected
225	C7	XCH A, @R1	1	1	No flags are affected
226	D6	XCHD A, @R0	1	1	No flags are affected
227	D7	XCHD A, @R1	1	1	No flags are affected
228	68	XRL A, R0	1	1	No flags are affected
229	69	XRL A, R1	1	1	No flags are affected
230	6A	XRL A, R2	1	1	No flags are affected
231	6B	XRL A, R3	1	1	No flags are affected
232	6C	XRL A, R4	1	1	No flags are affected
233	6D	XRL A, R5	1	1	No flags are affected
234	6E	XRL A, R6	1	1	No flags are affected
235	6F	XRL A, R7	1	1	No flags are affected
236	65	XRL A, DIRECT	2	1	No flags are affected
237	66	XRL A, @R0	1	1	No flags are affected
238	67	XRL A, @R1	1	1	No flags are affected
239	64	XRL A, #DATA	2	1	No flags are affected
240	62	XRL DIRECT, A	2	1	No flags are affected
241	63	XRL DIRECT, #DATA	3	2	No flags are affected

SECTION – C

INSTRUCTIONS ARRANGED BY THEIR OPCODES

S. No.	Opcode	Instruction	Bytes	Cycles	Flags affected
1	00	NOP	1	1	No flags are affected
2	01	AJMP ADDR11	2	2	No flags are affected
3	02	LJMP ADDR16	3	2	No flags are affected
4	03	RR A	1	1	No flags are affected
5	04	INC A	1	1	No flags are affected
6	05	INC DIRECT	2	1	No flags are affected
7	06	INC @R0	1	1	No flags are affected
8	07	INC @R1	1	1	No flags are affected

9	08	INC R0	1	1	No flags are affected
10	09	INC R1	1	1	No flags are affected
11	0A	INC R2	1	1	No flags are affected
12	0B	INC R3	1	1	No flags are affected
13	0C	INC R4	1	1	No flags are affected
14	0D	INC R5	1	1	No flags are affected
15	0E	INC R6	1	1	No flags are affected
16	0F	INC R7	1	1	No flags are affected
17	10	JBC BIT, REL	3	2	No flags are affected
18	11	ACALL ADDR11	3	2	No flags are affected
19	12	LCALL ADDR16	3	2	No flags are affected
20	13	RRC A	1	1	CY only
21	14	DEC A	1	1	No flags are affected
22	15	DEC DIRECT	2	1	No flags are affected
23	16	DEC @R0	1	1	No flags are affected
24	17	DEC @R1	1	1	No flags are affected
25	18	DEC R0	1	1	No flags are affected
26	19	DEC R1	1	1	No flags are affected
27	1A	DEC R2	1	1	No flags are affected
28	1B	DEC R3	1	1	No flags are affected
29	1C	DEC R4	1	1	No flags are affected
30	1D	DEC R5	1	1	No flags are affected
31	1E	DEC R6	1	1	No flags are affected
32	1F	DEC R7	1	1	No flags are affected
33	20	JB BIT, REL	3	2	No flags are affected
34	21	AJMP ADDR11	2	2	No flags are affected
35	22	RET	1	2	No flags are affected
36	23	RL A	1	1	No flags are affected
37	24	ADD A, #DATA	2	1	CY, OV and AC
38	25	ADD A, DIRECT	2	1	CY, OV and AC
39	26	ADD A, @R0	1	1	CY, OV and AC
40	27	ADD A, @R1	1	1	CY, OV and AC
41	28	ADD A, R0	1	1	CY, OV and AC
42	29	ADD A, R1	1	1	CY, OV and AC
43	2A	ADD A, R2	1	1	CY, OV and AC
44	2B	ADD A, R3	1	1	CY, OV and AC

45	2C	ADD A, R4	1	1	CY, OV and AC
46	2D	ADD A, R5	1	1	CY, OV and AC
47	2E	ADD A, R6	1	1	CY, OV and AC
48	2F	ADD A, R7	1	1	CY, OV and AC
49	30	JNB BIT, REL	3	2	No flags are affected
50	31	ACALL ADDR11	3	2	No flags are affected
51	32	RETI	1	2	No flags are affected
52	33	RLC A	1	1	CY only
53	34	ADDC A, #DATA	2	1	CY, OV and AC
54	35	ADDC A, DIRECT	2	1	CY, OV and AC
55	36	ADDC A, @R0	1	1	CY, OV and AC
56	37	ADDC A, @R1	1	1	CY, OV and AC
57	38	ADDC A, R0	1	1	CY, OV and AC
58	39	ADDC A, R1	1	1	CY, OV and AC
59	3A	ADDC A, R2	1	1	CY, OV and AC
60	3B	ADDC A, R3	1	1	CY, OV and AC
61	3C	ADDC A, R4	1	1	CY, OV and AC
62	3D	ADDC A, R5	1	1	CY, OV and AC
63	3E	ADDC A, R6	1	1	CY, OV and AC
64	3F	ADDC A, R7	1	1	CY, OV and AC
65	40	JC REL	2	2	No flags are affected
66	41	AJMP ADDR11	2	2	No flags are affected
67	42	ORL DIRECT, A	2	1	No flags are affected
68	43	ORL DIRECT, #DATA	3	2	No flags are affected
69	44	ORL A, #DATA	2	1	No flags are affected
70	45	ORL A, DIRECT	2	1	No flags are affected
71	46	ORL A, @R0	1	1	No flags are affected
72	47	ORL A, @R1	1	1	No flags are affected
73	48	ORL A, R0	1	1	No flags are affected
74	49	ORL A, R1	1	1	No flags are affected
75	4A	ORL A, R2	1	1	No flags are affected
76	4B	ORL A, R3	1	1	No flags are affected
77	4C	ORL A, R4	1	1	No flags are affected
78	4D	ORL A, R5	1	1	No flags are affected
79	4E	ORL A, R6	1	1	No flags are affected
80	4F	ORL A, R7	1	1	No flags are affected

81	50	JNC REL	2	2	No flags are affected
82	51	ACALL ADDR11	2	2	No flags are affected
83	52	ANL DIRECT, A	2	1	No flags are affected
84	53	ANL DIRECT, #DATA	3	2	No flags are affected
85	54	ANL A, #DATA	2	1	No flags are affected
86	55	ANL A, DIRECT	2	1	No flags are affected
87	56	ANL A, @R0	1	1	No flags are affected
88	57	ANL A, @R1	1	1	No flags are affected
89	58	ANL A, R0	1	1	No flags are affected
90	59	ANL A, R1	1	1	No flags are affected
91	5A	ANL A, R2	1	1	No flags are affected
92	5B	ANL A, R3	1	1	No flags are affected
93	5C	ANL A, R4	1	1	No flags are affected
94	5D	ANL A, R5	1	1	No flags are affected
95	5E	ANL A, R6	1	1	No flags are affected
96	5F	ANL A, R7	1	1	No flags are affected
97	60	JZ REL	2	2	No flags are affected
98	61	AJMP ADDR11	2	2	No flags are affected
99	62	XRL DIRECT, A	2	1	No flags are affected
100	63	XRL DIRECT, #DATA	3	2	No flags are affected
101	64	XRL A, #DATA	2	1	No flags are affected
102	65	XRL A, DIRECT	2	1	No flags are affected
103	66	XRL A, @R0	1	1	No flags are affected
104	67	XRL A, @R1	1	1	No flags are affected
105	68	XRL A, R0	1	1	No flags are affected
106	69	XRL A, R1	1	1	No flags are affected
107	6A	XRL A, R2	1	1	No flags are affected
108	6B	XRL A, R3	1	1	No flags are affected
109	6C	XRL A, R4	1	1	No flags are affected
110	6D	XRL A, R5	1	1	No flags are affected
111	6E	XRL A, R6	1	1	No flags are affected
112	6F	XRL A, R7	1	1	No flags are affected
113	70	JNZ REL	2	2	No flags are affected
114	71	ACALL ADDR11	2	2	No flags are affected
115	72	ORL C, BIT	2	2	CY only
116	73	JMP @A+DPTR	1	2	No flags are affected

117	74	MOV A, #DATA	2	1	No flags are affected
118	75	MOV DIRECT, #DATA	3	2	No flags are affected
119	76	MOV @R0, #DATA	2	1	No flags are affected
120	77	MOV @R1, #DATA	2	1	No flags are affected
121	78	MOV R0, #DATA	2	1	No flags are affected
122	79	MOV R1, #DATA	2	1	No flags are affected
123	7A	MOV R2, #DATA	2	1	No flags are affected
124	7B	MOV R3, #DATA	2	1	No flags are affected
125	7C	MOV R4, #DATA	2	1	No flags are affected
126	7D	MOV R5, #DATA	2	1	No flags are affected
127	7E	MOV R6, #DATA	2	1	No flags are affected
128	7F	MOV R7, #DATA	2	1	No flags are affected
129	80	SJMP REL	2	2	No flags are affected
130	81	AJMP ADDR11	2	2	No flags are affected
131	82	ANL C, BIT	2	2	CY only
132	83	MOVC A, @A+PC	1	2	No flags are affected
133	84	DIV AB	1	4	CY and OV
134	85	MOV DIRECT, DIRECT	3	2	No flags are affected
135	86	MOV DIRECT, @R0	2	2	No flags are affected
136	87	MOV DIRECT, @R1	2	2	No flags are affected
137	88	MOV DIRECT, R0	2	2	No flags are affected
138	89	MOV DIRECT, R1	2	2	No flags are affected
139	8A	MOV DIRECT, R2	2	2	No flags are affected
140	8B	MOV DIRECT, R3	2	2	No flags are affected
141	8C	MOV DIRECT, R4	2	2	No flags are affected
142	8D	MOV DIRECT, R5	2	2	No flags are affected
143	8E	MOV DIRECT, R6	2	2	No flags are affected
144	8F	MOV DIRECT, R7	2	2	No flags are affected
145	90	MOV DPTR, #DATA16	3	2	No flags are affected
146	91	ACALL ADDR11	2	2	No flags are affected
147	92	MOV BIT, C	2	2	No flags are affected
148	93	MOVC A, @A+DPTR	1	2	No flags are affected
149	94	SUBB A, #DATA	2	1	CY, OV and AC
150	95	SUBB A, DIRECT	2	1	CY, OV and AC
151	96	SUBB A, @R0	1	1	CY, OV and AC
152	97	SUBB A, @R1	1	1	CY, OV and AC

153	98	SUBB A, R0	1	1	CY, OV and AC
154	99	SUBB A, R1	1	1	CY, OV and AC
155	9A	SUBB A, R2	1	1	CY, OV and AC
156	9B	SUBB A, R3	1	1	CY, OV and AC
157	9C	SUBB A, R4	1	1	CY, OV and AC
158	9D	SUBB A, R5	1	1	CY, OV and AC
159	9E	SUBB A, R6	1	1	CY, OV and AC
160	9F	SUBB A, R7	1	1	CY, OV and AC
161	A0	ORL C, /BIT	2	2	CY only
162	A1	AJMP ADDR11	2	2	No flags are affected
163	A2	MOV C, BIT	2	1	CY only
164	A3	INC DPTR	1	2	No flags are affected
165	A4	MUL AB	1	4	CY and OV
166	A5	UNDEFINED			
167	A6	MOV @R0, DIRECT	2	2	No flags are affected
168	A7	MOV @R1, DIRECT	2	2	No flags are affected
169	A8	MOV R0, DIRECT	2	2	No flags are affected
170	A9	MOV R1, DIRECT	2	2	No flags are affected
171	AA	MOV R2, DIRECT	2	2	No flags are affected
172	AB	MOV R3, DIRECT	2	2	No flags are affected
173	AC	MOV R4, DIRECT	2	2	No flags are affected
174	AD	MOV R5, DIRECT	2	2	No flags are affected
175	AE	MOV R6, DIRECT	2	2	No flags are affected
176	AF	MOV R7, DIRECT	2	2	No flags are affected
177	B0	ANL C, /BIT	2	2	CY only
178	B1	ACALL ADDR11	2	2	No flags are affected
179	B2	CPL BIT	2	1	No flags are affected
180	B3	CPL C	1	1	No flags are affected (CY is complemented)
181	B4	CJNE A, #DATA, REL	3	2	CY only
182	B5	CJNE A, DIRECT, REL	3	2	CY only
183	B6	CJNE @R0, #DATA, REL	3	2	CY only
184	B7	CJNE @R1, #DATA, REL	3	2	CY only
185	B8	CJNE R0, #DATA, REL	3	2	CY only
186	B9	CJNE R1, #DATA, REL	3	2	CY only
187	BA	CJNE R2, #DATA, REL	3	2	CY only

188	BB	CJNE R3, #DATA, REL	3	2	CY only
189	BC	CJNE R4, #DATA, REL	3	2	CY only
190	BD	CJNE R5, #DATA, REL	3	2	CY only
191	BE	CJNE R6, #DATA, REL	3	2	CY only
192	BF	CJNE R7, #DATA, REL	3	2	CY only
193	C0	PUSH DIRECT	2	2	No flags are affected
194	C1	AJMP ADDR11	2	2	No flags are affected
195	C2	CLR BIT	2	1	No flags are affected
196	C3	CLR C	1	1	No flags are affected (CY = 0)
197	C4	SWAP A	1	1	No flags are affected
198	C5	XCH A, DIRECT	2	1	No flags are affected
199	C6	XCH A, @R0	1	1	No flags are affected
200	C7	XCH A, @R1	1	1	No flags are affected
201	C8	XCH A, R0	1	1	No flags are affected
202	C9	XCH A, R1	1	1	No flags are affected
203	CA	XCH A, R2	1	1	No flags are affected
204	CB	XCH A, R3	1	1	No flags are affected
205	CC	XCH A, R4	1	1	No flags are affected
206	CD	XCH A, R5	1	1	No flags are affected
207	CE	XCH A, R6	1	1	No flags are affected
208	CF	XCH A, R7	1	1	No flags are affected
209	D0	POP DIRECT	2	2	No flags are affected (for POP 0D0, all flags are modified)
210	D1	ACALL ADDR11	2	2	No flags are affected
211	D2	SETB BIT	2	1	No flags are affected
212	D3	SETB C	1	1	No flags are affected (CY = 1)
213	D4	DA A	1	1	No flags are affected
214	D5	DJNZ DIRECT, REL	3	2	No flags are affected
215	D6	XCHD A, @R0	1	1	No flags are affected
216	D7	XCHD A, @R1	1	1	No flags are affected
217	D8	DJNZ R0, REL	2	2	No flags are affected
218	D9	DJNZ R1, REL	2	2	No flags are affected
219	DA	DJNZ R2, REL	2	2	No flags are affected
220	DB	DJNZ R3, REL	2	2	No flags are affected

221	DC	DJNZ R4, REL	2	2	No flags are affected
222	DD	DJNZ R5, REL	2	2	No flags are affected
223	DE	DJNZ R6, REL	2	2	No flags are affected
224	DF	DJNZ R7, REL	2	2	No flags are affected
225	E0	MOVX A, @DPTR	1	2	No flags are affected
226	E1	AJMP ADDR11	2	2	No flags are affected
227	E2	MOVX A, @R0	1	2	No flags are affected
228	E3	MOVX A, @R1	1	2	No flags are affected
229	E4	CLR A	1	1	No flags are affected
230	E5	MOV A, DIRECT	2	1	No flags are affected
231	E6	MOV A, @R0	1	1	No flags are affected
232	E7	MOV A, @R1	1	1	No flags are affected
233	E8	MOV A, R0	1	1	No flags are affected
234	E9	MOV A, R1	1	1	No flags are affected
235	EA	MOV A, R2	1	1	No flags are affected
236	EB	MOV A, R3	1	1	No flags are affected
237	EC	MOV A, R4	1	1	No flags are affected
238	ED	MOV A, R5	1	1	No flags are affected
239	EE	MOV A, R6	1	1	No flags are affected
240	EF	MOV A, R7	1	1	No flags are affected
241	F0	MOVX @DPTR, A	1	2	No flags are affected
242	F1	ACALL ADDR11	2	2	No flags are affected
243	F2	MOVX @R0, A	1	2	No flags are affected
244	F3	MOVX @R1, A	1	2	No flags are affected
245	F4	CPL A	1	1	No flags are affected
246	F5	MOV DIRECT, A	2	1	No flags are affected
247	F6	MOV @R0, A	1	1	No flags are affected
248	F7	MOV @R1, A	1	1	No flags are affected
249	F8	MOV R0, A	1	1	No flags are affected
250	F9	MOV R1, A	1	1	No flags are affected
251	FA	MOV R2, A	1	1	No flags are affected
252	FB	MOV R3, A	1	1	No flags are affected
253	FC	MOV R4, A	1	1	No flags are affected
254	FD	MOV R5, A	1	1	No flags are affected
255	FE	MOV R6, A	1	1	No flags are affected
256	FF	MOV R7, A	1	1	No flags are affected

Laboratory Manual of 8051 Microcontroller

III Appendix

EXPERIMENT 1

Exp. Date: 15 Jan 2019

Write an assembly language program to perform the addition of two 8-bit numbers and store the result in register R7 of register bank 1.

Requirement: 8051 IDE Simulation Software

Program:

Assembly Language Program		
Memory Location	Instruction	Comment
2000	MOV A, #20h	*Copy data 20h to the accumulator*
2002	MOV R0, #30h	*Copy data 30h to the register R0*
2004	ADD A, R0	*Addition of the accumulator and register R0*
2005	MOV 30H, A	*Copy accumulator to memory location 30h*
	L1: SJMP L1	*Stop*

Result:

1. **Data used:**

   90h and 75h

2. **Final Result:**

   R6 = 05h

   R7=01h

Submission Date: 16 Jan 2019

Professor Incharge
Prof. Nilesh B. Bahadure

EXPERIMENT 2

EXP. DATE: 15 Jan 2019

Write an assembly language program to add two 16-bit numbers present in register R2 -R1 and at memory location 31h -30h, put the result in the external RAM from 1000h.

Requirement: 8051 IDE Simulation Software

Program:

Assembly Language Program		
Memory Location	**Instruction**	**Comments**
2000	MOV A, R1	Copy the contents of register R1 of register bank 0 to the accumulator
2001	ADD A, 30H	Addition of contents of the accumulator with contents of memory location 30H
2003	MOV DPTR,#1000H	Initialize external memory location 1000h
2006	MOVX @DPTR, A	Copy the result from the accumulator to external RAM location 1000H
2007	MOV A, R2	Copy the contents of register R2 of register bank 0 to the accumulator
2008	ADDC A, 31H	Addition of contents of the accumulator with contents of memory location 31H with carry
200A	INC DPTR	Go to next memory location 1001h
200B	MOVX @DPTR, A	Copy the result from the accumulator to external RAM location 1001H
200C	L1: SJMP L1	End of program

Result:

1. **Data used:**

   R1-R0H = 1234H and 31-30H = 5678H

2. **Final Result:**

   1000H-1001H = 68ACH

   Submission Date: 16 Jan 2019

Professor Incharge

Prof. Nilesh B. Bahadure

EXPERIMENT 3

Exp. Date: 15 Jan 2019

Write an assembly language program to perform x-y, where x and y are 8-bit numbers present in the memory location at 31h and 30h respectively, put result in register R7.

Requirement: 8051 IDE simulation software

Program:

Assembly Language Program		
Memory Location	**Instruction**	**Comments**
2000	CLR C	Clear the contents of carry flag
2001	MOV A, 31H	Copy the contents of memory location 31h to the accumulator
2003	SUBB A, 30H	Subtraction of contents of the accumulator with contents of memory location 30H
2005	MOV R7, A	Copy the contents of the accumulator to the R7 register
2006	L1: SJMP L1	End of program

Result:

1. **Data used:**

   Memory locations : 30H = 1FH and 31H = 54H

2. **Final Result:**

   R7 = 35H

   Submission Date: 16 Jan 2019

Professor Incharge

Prof. Nilesh B. Bahadure

EXPERIMENT 4

Exp. Date: 15 Jan 2019

Write an assembly language program to find one's complement of 8-bit number present in register R7 and put the result in the same register.

Requirement: 8051 IDE simulation software

Program:

Assembly Language Program		
Memory Location	**Instruction**	**Comments**
2000	MOV A,R7	Get the contents of register R7 to the accumulator
2001	CPL A	Find one's complement of the accumulator
2002	MOV R7,A	Store the result in register R7
2003	L1: SJMP L1	End of program

Result:

1. **Data used:**

   R7 = 55h

2. **Final Result:**

   R7 = AAH

   Submission Date: 16 Jan 2019

Professor Incharge

Prof. Nilesh B. Bahadure

EXPERIMENT 5

Exp. Date: 15 Jan 2019

Write an assembly language program to find two's complement of 8-bit number present in register R7 and put the result in the same register.

Requirement: 8051 IDE simulation software

Program:

Assembly Language Program		
Memory Location	**Instruction**	**Comments**
2000	MOV A,R7	Get the contents of register R7 to the accumulator
2001	CPL A	Find one's complement of the accumulator
2002	ADD A, #01H	Increment one's complement by one
2004	MOV R7,A	Store the result in register R7
2005	L1: SJMP L1	End of program

Result:

1. **Data used:**
   R7 = 55h
2. **Final Result:**
   R7 = ABH

   Submission Date: 16 Jan 2019

Professor Incharge
Prof. Nilesh B. Bahadure

EXPERIMENT 6

Exp. Date: 15 Jan 2019

Write an assembly language program to perform multiplication of two 8-bit numbers present in register R2 and R3, put the result in the memory location 31h - 30h.

Requirement: 8051 IDE simulation software

Program:

Assembly Language Program		
Memory Location	**Instruction**	**Comments**
2000	MOV A, R2	Copy the contents of register R2 to the accumulator
2001	MOV B, R3	Copy the contents of register R3 to register B
2003	MUL AB	Multiplication of contents of the accumulator and register B and put the result as higher byte in register B and lower byte in the accumulator
2004	MOV 30H, A	Copy the contents of the accumulator to memory location 30h
2006	MOV 31H, B	Copy the contents of register B to memory location 31h
2009	L1: SJMP L1	End of program

Result:

1. **Data used:**

   Registers: R2 = 12H and R3 = 34H

2. **Final Result:**

   31-30H = 03A8H

   Submission Date: 16 Jan 2019

Professor Incharge
Prof. Nilesh B. Bahadure

EXPERIMENT 7

Exp. Date: 15 Jan 2019

Write an assembly language program to perform $\frac{x}{y}$, where x and y are 8-bit numbers present in the memory location 30h and 31h respectively, put the result in register R7 (Remainder) and R6 (Quotient).

Requirement: 8051 IDE simulation software

Program:

Assembly Language Program		
Memory Location	**Instruction**	**Comments**
2000	MOV A, 30H	Copy the contents of memory location 30H to the accumulator
2002	MOV B, 31H	Copy the contents of memory location 31H to register B
2004	DIV AB	Division of contents of the accumulator and register B and put the result as quotient in the accumulator and reminder in register B
2005	MOV R6, A	Copy the contents of the accumulator to register R6
2006	MOV R7, B	Copy the contents of register B to register R7
2007	L1: SJMP L1	End of program

Result:

1. **Data used:**

   Registers: 30H = 54H and 31H = 1FH

2. **Final Result:**

   R6-R7 = 02H-47H

   Submission Date: 16 Jan 2019

Professor Incharge
Prof. Nilesh B. Bahadure

EXPERIMENT 8

Exp. Date: 15 Jan 2019

Write an assembly language program to find largest byte from the 10 bytes present in the memory location from 30h.

Requirement: 8051 IDE simulation software

Program:

Assembly Language Program		
Memory Location	**Instruction**	**Comments**
2000	MOV R1, #0AH	Initialize the counter
2002	MOV R0, #30H	Initialize the memory pointer
2004	DEC R1	Decrement counter by 1
2005	MOV A, @R0	Load number in the accumulator
2006	MOV B, A	Copy the number to register
2008	UP: INC R0	Increment the memory pointer
2009	MOV A, @R0	Read the next number in the accumulator
200A	CJNE A, B, DOWN	Compare the first two numbers, if not equal then go to DOWN
200D	AJMP NEXT	Else go to NEXT
200F	DOWN: JC NEXT	If number in the accumulator is greater then go to NEXT
2011	MOV B, A	Else move the number in register B
2013	NEXT: DJNZ R1, UP	Decrement the counter by 1, if count not equal to 0, then go to UP
2015	INC R0	Increment the memory pointer
2016	MOV A, B	Increment the memory pointer
2018	MOV 50H, A	Store the result at memory location 50H
201A	L1: SJMP L1	End of program

Result:

1. **Data used:**

   Starting Memory location: 30H = 10H, 3EH, 1FH, AEH, FEH, 12H, 2EH, 65H, 76H, 11H

2. **Final Result:**

   Memory location 50H = FEH

   Submission Date: 16 Jan 2019

Professor Incharge
Prof. Nilesh B. Bahadure

EXPERIMENT 9

Exp. Date: 15 Jan 2019

Write an assembly language program to copy data 10 bytes from source internal memory 30h to destination internal memory 40h.

Requirement: 8051 IDE simulation software

Program:

Assembly Language Program		
Memory Location	**Instruction**	**Comments**
	ORG 0000H	Initialize starting memory location
0000	MOV R0, #30H	Initialize source memory
0002	MOV R1, #40H	Initialize destination memory
0004	MOV R2, #10H	Initialize counter
0006	L1: MOV A, @R0	Copy contents of source memory to the accumulator
0007	MOV @R1, A	Copy accumulator to destination memory
0008	INC R0	Increment source memory
0009	INC R1	Increment destination memory
000A	DJNZ R2, L1	Decrement R2 if not 0 go to L1
000C	L2: SJMP L2	End of program

Data used:

Starting Memory location 30H = 00H, 01H, 02H, 03H, 44H, 55H, 66H, 77H, 88H, 99H

Result

Starting Memory location 40H = 00H, 01H, 02H, 03H, 44H, 55H, 66H, 77H, 88H, 99H

Submission Date: 16 Jan 2019

Professor Incharge
Prof. Nilesh B. Bahadure

EXPERIMENT 10

Exp. Date: 15 Jan 2019

Write an assembly language program to add series 10 bytes of 8-bit numbers present in the memory location from 30h. Put the result in the external RAM memory location 1001h (Carry) – 1000h (Sum).

Requirement: 8051 IDE simulation software

Program:

Assembly Language Program		
Memory Location	**Instruction**	**Comments**
	ORG 0000H	Initialize starting memory location
0000	MOV R2, #10	Initialize counter
0002	MOV R0, #30H	Initialize starting memory location
0004	CLR A	Copy the memory location 30H in register R0
0005	MOV R1, #00H	Copy starting value to the accumulator
0007	MOV DPTR, #1000H	Add 11H to the value stored in the accumulator
000A	L2: ADD A, @R0	Copy the value to register R0
000B	INC R0	Increment R0
000C	JNC L1	Decrement R2 if not 0 go to L1
000E	INC R1	Decrement R1 if not 0 go to L2
000F	L1: DJNZ R2, L2	Decrement R1 if not 0 go to L2
0011	MOVX @DPTR, A	Decrement R1 if not 0 go to L2
0012	INC DPTR	Decrement R1 if not 0 go to L2
0013	MOV A, R1	Decrement R1 if not 0 go to L2
0014	MOVX @DPTR, A	Decrement R1 if not 0 go to L2
0015	L3: SJMP L3	End of program

Data Used:

Starting Memory location 30H = 01H, 02H, 03H, 04H, 05H, 06H, 07H, 08H, 09H, 0AH

Result:

Ext. Memory 1000h = 37h (Sum) and 1001h =00h (Carry)

Submission Date: 16 Jan 2019

Professor Incharge
Prof. Nilesh B. Bahadure

EXPERIMENT 11

Exp. Date: 15 Jan 2019

Write an assembly language program to count even and odd bytes from the 10 bytes present in the memory location from 30h. Put the result of even and odd count in register R6 and R7 respectively.

Requirement: 8051 IDE simulation software

Program:

Assembly Language Program		
Memory Location	**Instruction**	**Comments**
	ORG 0000H	Originate the program at 0000H
0000	MOV R1, #00H	Initialize counter = 0 for odd numbers
0002	MOV R2, #00H	Initialize counter = 0 for even numbers
0004	MOV R3, #0AH	Initialize counter = 10
0006	MOV R0, #30H	Copy memory location 30H to register R3
0008	BACK: MOV A, @R0	Copy the given number to the accumulator
0009	INC R0	Increment memory pointer
000A	ANL A, #01H	Masks all other bits accept LSB
000C	JZ EVEN	If ZF = 0 then go to EVEN
000E	INC R1	Otherwise increment odd number count
000F	AJMP LAST	Jump to LAST
0011	EVEN: INC R1	Increment even number count
0012	LAST: DJNZ R3, BACK	Decrement R2 if not 0 go to BACK
0014	MOV A, R2	Copy even count value to the accumulator
0015	MOV B, R1	Copy odd count value to register B
0016	MOV R6, A	Copy even count value from the accumulator to register R6
0017	MOV R7, B	Copy odd count value from the accumulator to register R7
0018	END	End of program

Result:

1. **Data used:**

   Starting Memory location: 30H = 10H, 3EH, 1FH, AEH, FEH, 12H, 2EH, 65H, 76H, 11H

2. **Final Result:**

   R6 = 7 & R7 = 3

   Submission Date: 16 Jan 2019

Professor Incharge
Prof. Nilesh B. Bahadure

EXPERIMENT 12

Exp. Date: 15 Jan 2019

Write an assembly language program to add only positive numbers from the 10 bytes present in the memory location from 30h and put the result in the register R7-R6.

Requirement: 8051 IDE simulation software

Program:

Assembly Language Program		
Memory Location	**Instruction**	**Comments**
	ORG 0000H	Originate the program at 0000H
0000	MOV R0, #0AH	Initialize counter = 10
0002	MOV R1, #30H	Copy the values from memory location 30H to register R1
0004	BACK: MOV A, @R1	Copy the value from register R1 to the accumulator
0005	INC R1	Increment R1
0006	ANL A, #80H	Masks all other bits accept MSB
0008	JZ POS	If ZF = 0 then go to POS
000A	AJMP LAST	Jump to LAST
000C	POS: ADD A, #00H	Add value of the accumulator with 00H
000D	MOV R6, A	Store the value in register R6
000E	ADDC A, #00H	Add carry with 00H
000F	MOV R7, A	Store higher byte in register R7
0010	LAST: DJNZ R0, BACK	Decrement R0 if not 0 go to BACK
0012	END	End of program

Result:

1. **Data used:**

   Starting Memory location: 30H = 10H, 33H, 1FH, AEH, F1H, 12H, 21H, 65H, 77H, 11H

2. **Final Result:**

   R7-R6 = 00-D0H

   Submission Date: 16 Jan 2019

Professor Incharge
Prof. Nilesh B. Bahadure

EXPERIMENT 13

Exp. Date: 15 Jan 2019

Write an assembly language program to copy data 00h, 01h, 02h, 03h, ... FFh to the external RAM from 2000h.

Requirement: 8051 IDE simulation software

Program:

Assembly Language Program		
Memory Location	**Instruction**	**Comments**
	ORG 0000H	Initialize starting memory location
0000	MOV R2, #0FFH	Initialize counter
0002	CLR A	Clear the accumulator
0003	MOV DPTR, #2000H	Copy the memory location 2000h to DPTR
0006	L1: MOVX @DPTR, A	Copy the accumulator to external RAM
0007	INC A	Increment the accumulator
0008	INC DPTR	Increment DPTR to next memory location
0009	DJNZ R2, L1	Repeat operation
000B	L2: SJMP L2	End of the program

Result:

Starting External RAM Memory location 2000H = 00H, 01H, 02H, 03H,... FFH

Submission Date: 16 Jan 2019

Professor Incharge
Prof. Nilesh B. Bahadure

EXPERIMENT 14

Exp. Date: 15 Jan 2019

Write an assembly language program to copy data 00h, 01h, 02h, 03h, ... 99h (BCD Numbers) to the external RAM from 2000h.

Requirement: 8051 IDE simulation software

Program:

Assembly Language Program		
Memory Location	**Instruction**	**Comments**
	ORG 0000H	Initialize starting memory location
0000	MOV R2, #99H	Initialize counter
0002	CLR A	Clear the accumulator
0003	MOV DPTR, #2000H	Copy the memory location 2000h to DPTR
0006	L1: MOVX @DPTR, A	Copy the accumulator to external RAM
0007	ADD A, #01h	Increment accumulator
0009	DA A	Adjust the accumulator to BCD
000A	INC DPTR	Increment DPTR to next memory location
000B	DJNZ R2, L1	Repeat operation
000D	L2: SJMP L2	End of the program

Result:

Starting external RAM memory location 2000H = 00H, 01H, 02H, 03H, ... 99H

Submission Date: 16 Jan 2019

Professor Incharge

Prof. Nilesh B. Bahadure

EXPERIMENT 15 **Exp. Date:** 15 Jan 2019

Write an assembly language program to copy data 00h, 11h, 22h, 33h, ... EEh, FFh, 00h, 11h, 22h, 33h, ... EEh, FFh in the internal RAM from 30h.

Requirement: 8051 IDE simulation software

Program:

Assembly Language Program		
Memory Location	**Instruction**	**Comments**
	ORG 0002H	Initialize starting memory location
0002	MOV R1, #02H	Initialize counter
0004	MOV R2, #16H	Initialize counter
0006	MOV R0, #30H	Copy the memory location 30H in register R0
0008	L2: MOV A, #00H	Copy starting value to the accumulator
000A	L1: ADD A, #11H	Add 11H to the value stored in the accumulator
000C	MOV @R0, A	Copy the value to register R0
000D	INC R0	Increment R0
000E	DJNZ R2, L1	Decrement R2 if not 0 go to L1
0010	DJNZ R1, L2	Decrement R1 if not 0 go to L2
0012	END	End of program

Result:

Starting memory location 30H = 00H, 11H, 22H, 33H, 44H, 55H, 66H, 77H, 88H, 99H, AAH, BBH, CCH, DDH, EEH, FFH, 00H, 11H, 22H, 33H, 44H, 55H, 66H, 77H, 88H, 99H, AAH, BBH, CCH, DDH, EEH, FFH

Submission Date: 16 Jan 2019

Professor Incharge
Prof. Nilesh B. Bahadure

EXPERIMENT 16

Exp. Date: 15 Jan 2019

Show the interfacing of 8 LED's with 8051 microcontroller and write an assembly language program to blink alternate LED's (Common anode), assume that crystal frequency of 8051 is 11.0592 MHz.

Requirement: 8051 IDE simulation software

Program:

Assembly Language Program		
Memory Location	**Instruction**	**Comments**
	ORG 0000H	
0000	MOV A, #55H	Copy the data 55h to the accumulator
0002	LOOP: CPL A	Complement the contents of the accumulator
0003	MOV P2, A	Copy the contents of the accumulator to port P2
0005	ACALL DELAY	Absolute call the DELAY subroutine program
0007	JMP LOOP	Jumps to the relative address
0010	DELAY: MOV R3, #0FFH	Initialize count value
0012	REP: MOV R2, #0FFH	Initialize the delay time
0014	L1: DJNZ R2,L1	Decrement and jump to L1 if contents of R2 are not equal to 0
0016	DJNZ R3, REP	Decrement and jump to REP if contents of R3 are not equal to 0
0018	RET	Return to the address

Result: LEDs are alternately blinked as shown in Figure 1.

Submission Date: 19 Jan 2019

Professor Incharge

Prof. Nilesh B. Bahadure

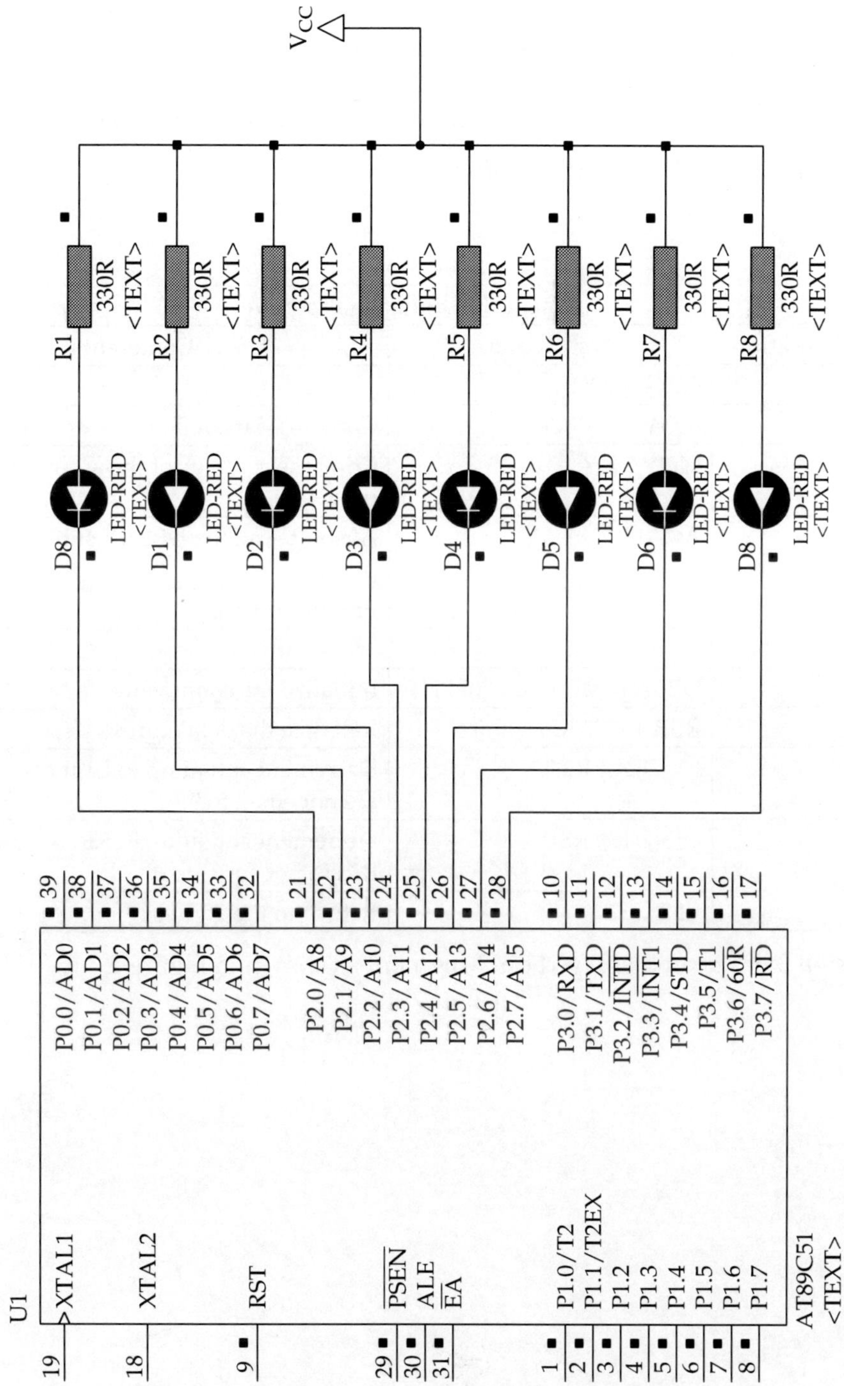

Figure 1: *LED Alternate Blinking*

EXPERIMENT 17

Exp. Date: 17 Jan 2019

Show the interfacing of 7-segment display device (common anode) with 8051 microcontroller and write an assembly language program to display digit 3 on seven segment display device (common anode), assume that crystal frequency of 8051 is 11.0592 MHz.

Requirement: 8051 IDE simulation software

Program:

Assembly Language Program		
Memory Location	**Instruction**	**Comments**
	ORG 0000H	
0000	MOV A, #0B0H	Copy the data 0B0h to the accumulator
0002	MOV P2, A	Copy the contents of the accumulator to port P2
0004	ACALL DELAY	Absolute call the DELAY subroutine program
0006	BACK: SJMP BACK	End of program
0009	DELAY: MOV R3, #0FFH	Initialize the count value
000B	REP: MOV R2, #0FFH	Initialize the delay time
000D	L1: DJNZ R2,L1	Decrement and jump to L1 if contents of R2 are not equal to 0
000F	DJNZ R3, REP	Decrement and jump to REP if contents of R3 are not equal to 0
2011	RET	Return to the address

Result: Digit 3 has been displayed on 7-segment as shown in Figure 2.

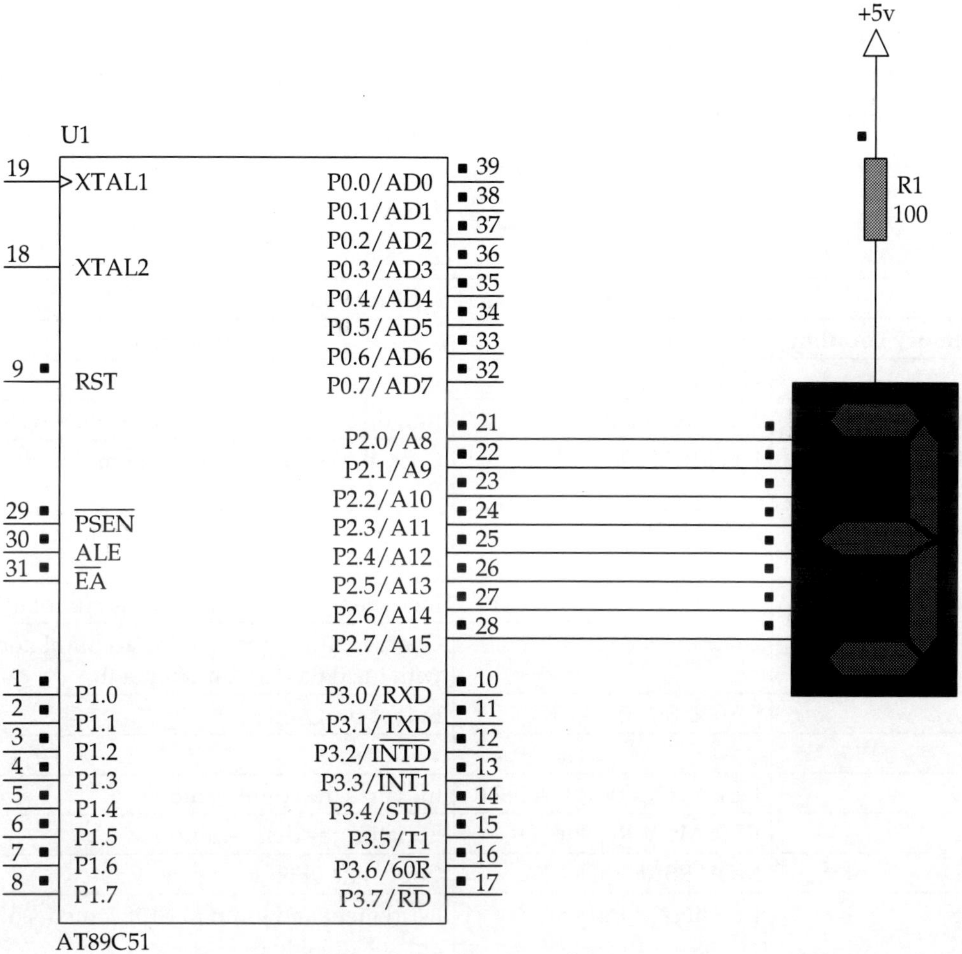

Figure 2: *7-Segment Display 3*

Submission Date: 18 Jan 2019

Professor Incharge

Prof. Nilesh B. Bahadure

EXPERIMENT 18

Exp. Date: 17 Jan 2019

Show the interfacing of 7-segment display device using decoder 7447 and write an assembly language program to display digits 0 to 9 on 7-segment display device, assume that crystal frequency of 8051 is 11.0592 MHz.

Requirement: 8051 IDE simulation software

Program:

Assembly Language Program		
Memory Location	**Instruction**	**Comments**
	ORG 0000H	
0000	MOV A, #00H	Initialize accumulator with starting data
0002	UP: MOV P2, A	Copy the contents of the accumulator to port P2
0004	ACALL DELAY	Absolute call the DELAY subroutine program
0006	INC A	Increment the content of the accumulator
0007	CJNE A, #0AH, UP	Compare the content of the accumulator with the data and jump if not 0
000A	BACK: SJMP BACK	End of program
2009	DELAY: MOV R3, #0FFH	Initialize the count value
200B	REP: MOV R4, #0FFH	Initialize the delay value
200D	MOV R5, #0FFH	Copy data 0FFh to register R5
200F	L1: DJNZ R5,L1	Decrement and jump to L1 if contents of R5 are not equal to 0
2012	L2: DJNZ R4,L2	Decrement and jump to L2 if contents of R4 ar3 not equal to 0
2014	DJNZ R3, REP	Decrement and jump to REP if contents of R3 are not equal to 0
2016	RET	Return to the address

Result: Digits 0 to 9 are being displayed as shown in Figure 3.

Submission Date: 18 Jan 2019

Professor Incharge

Prof. Nilesh B. Bahadure

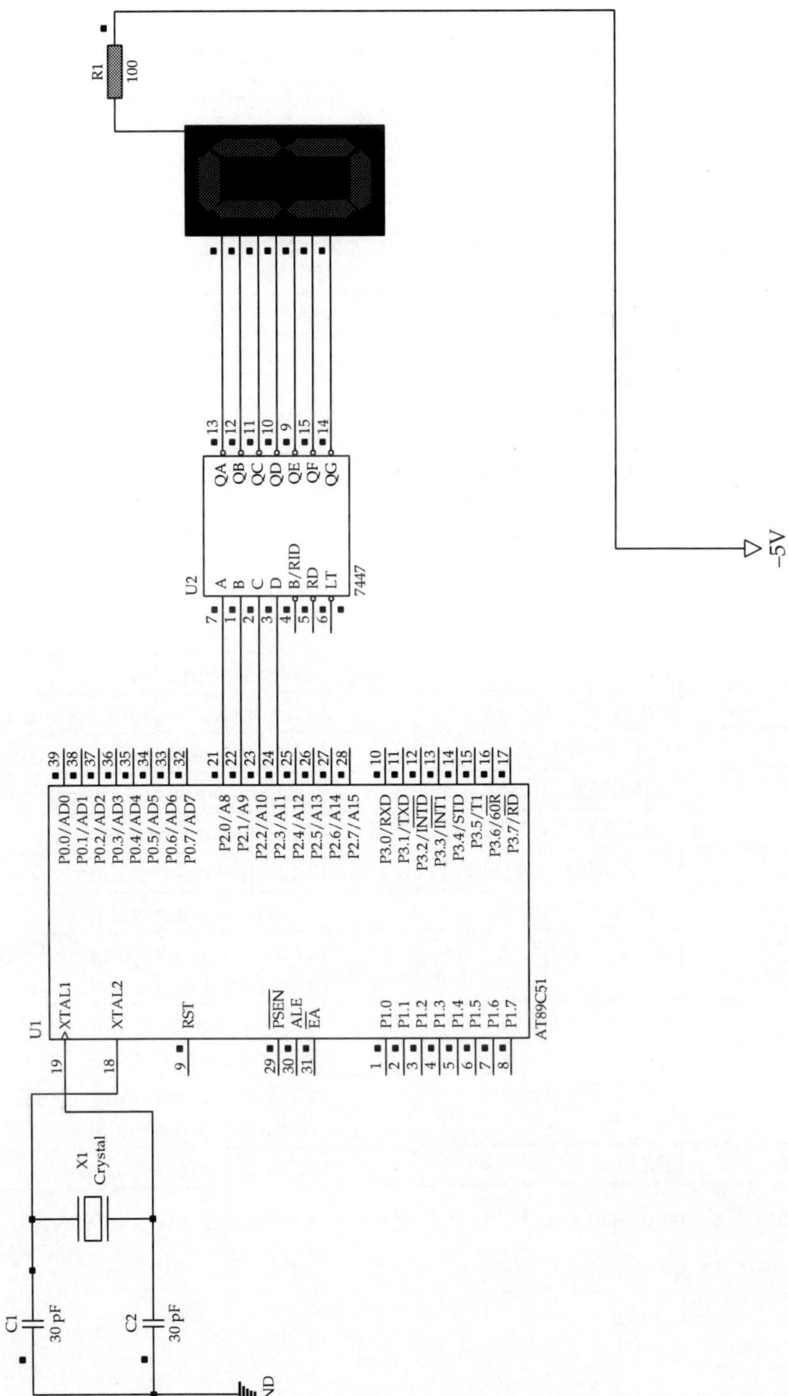

Figure 3: *7-Segment Display 0 TO 9*

EXPERIMENT 19

Exp. Date: 17 Jan 2019

Show the interfacing of two 7-segment LED's (common anode) with 8051 Microcontroller and write an assembly language program to display digit '92' on it.

Requirement: 8051 IDE simulation software

Program:

Assembly Language Program		
Memory Location	**Instruction**	**Comments**
	ORG 0000H	
0000	CLR P3.0	Clear port pin P3.0
0001	CLR P3.1	Clear port pin P3.1
0002	MOV P2, #0FFH	Initializing all pin values high of port P2
0004	LOOP: CLR P3.0	Clears port pin P3.0
0005	SETB P3.1	Set port pin P3.0
0006	MOV P2, #90H	Displays 9 to 1st 7-segment
0008	CALL DEALY	Call subroutine DELAY function
000A	CLR P3.1	Clear port pin P3.1
000B	SETB P3.0	Set port pin P3.0
000C	MOV P2, #0A4H	Displays 2 to 2nd 7-segment
000E	CALL DELAY	Call subroutine DELAY function
0010	AJMP LOOP	Jumps to memory location of LOOP
2009	DELAY: MOV R3, #0FFH	Initialize count value
200B	REP: MOV R4, #0FFH	Initialize delay value
200D	MOV R5, #0FFH	Initialize delay value
200F	L1: DJNZ R5,L1	Decrement and jump to L1 if contents of R5 are not equal to 0
2011	L2: DJNZ R4,L2	Decrement and jump to L2 if contents of R4 are not equal to 0
2013	DJNZ R3, REP	Decrement and jump to REP if contents of R3 are not equal to 0
2015	RET	Return to the address

Result: Digits '92' are displayed in two 7-segments respectively as shown in Figure 4.

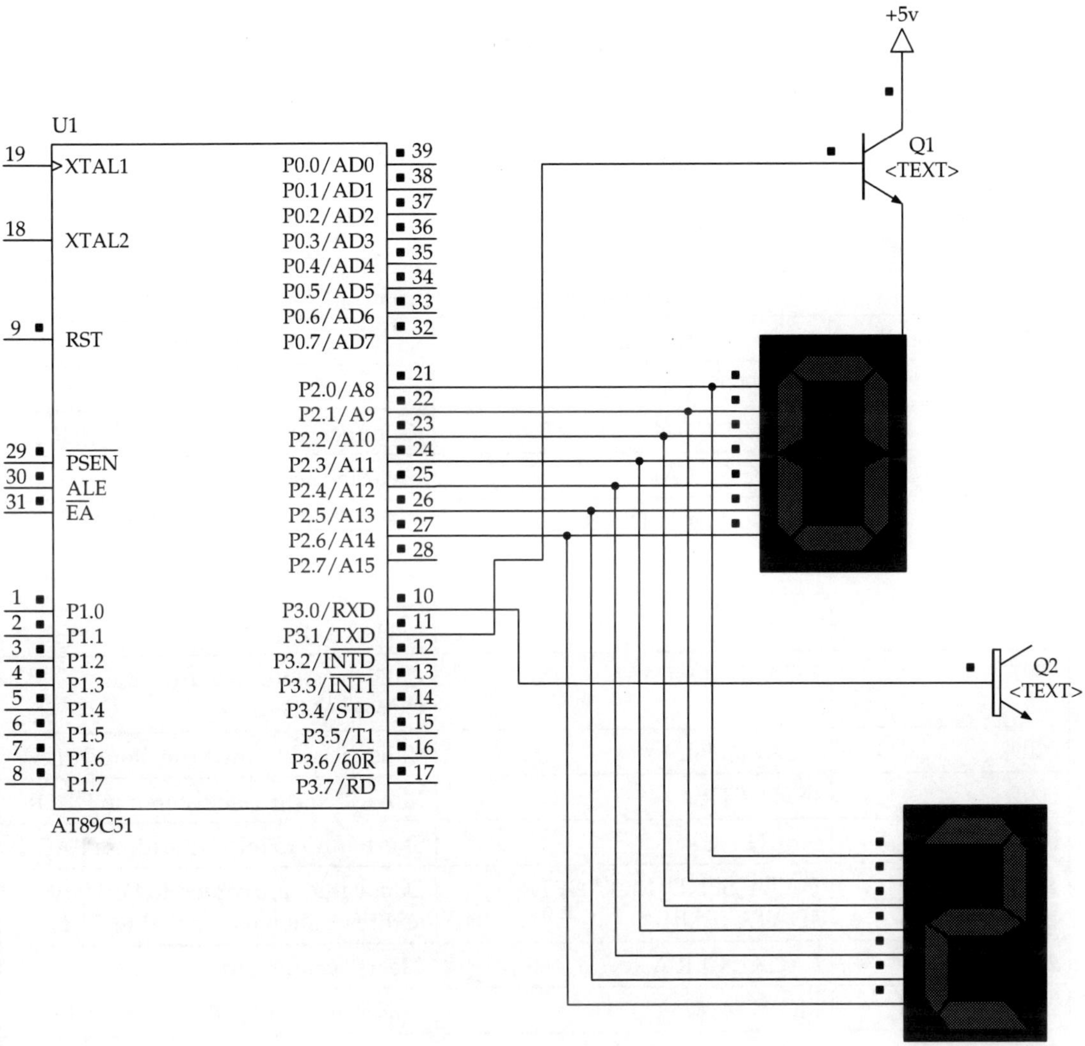

Figure 4: *7-Segment Display*

Submission Date: 18 Jan 2019

Professor Incharge

Prof. Nilesh B. Bahadure

EXPERIMENT - 20 **Exp. Date:** 17 Jan 2019

Show the interfacing of 16 x 2 LCD with 8051 Microcontroller and write an assembly language program to display 'Your Name' on LCD, assume that crystal frequency of 8051 is 11.0592 MHz.

Requirement: 8051 IDE simulation software

Program:

Assembly Language Program		
Memory Location	**Instruction**	**Comments**
	ORG 0000H	
0000	MOV DPTR, #COM_TAB	Command value copied to DPTR
0003	BACK: CLR A	Clear the accumulator
0004	MOVC A, @A+DPTR	Copies the command values to the accumulator
0005	JZ DATA_SEND	Jump when accumulator becomes 0 to DATA_SEND
0007	ACALL COMMAND	Calls the subroutine function COMMAND
0009	ACALL DELAY	Calls the subroutine function DELAY
000B	INC DPTR	Moves to next value stored in DPTR
000C	SJMP BACK	Short jump to relative address BACK
2006	DATA_SEND: MOV DPTR, #DATA_TAB	Data values are copied to DPTR or text to be shown is copied to DPTR
2009	BACK1: CLR A	Clears accumulator
200A	MOVC A, @A+DPTR	Copies the data to the accumulator
200B	JZ STOP	Jump when the accumulator becomes 0, to STOP
200D	ACALL DATA	Calls the subroutine function DATA
200F	ACALL DELAY	Calls the subroutine function DELAY
2011	INC DPTR	Moves to next value stored in DPTR
2012	SJMP BACK1	Short jump to relative address BACK1
2014	STOP: SJMP STOP	End of program
2020	COMMAND: MOV P1,A	Copies the contents of the accumulator to port 1

2022	CLR P2.0	Reset RS pin of LCD
2023	CLR P2.1	Reset RW pin of LCD
2024	SETB P2.2	Sets EN pin of LCD
2025	ACALL DELAY	Calls subroutine function DELAY
2027	CLR P2.2	Reset EN pin of LCD
2028	RET	Return to the address
2029	DATA:MOV P1, A	Copies the contents of the accumulator to port 1
202B	SETB P2.0	Set RS pin of LCD
202C	CLR P2.1	Reset RW pin of LCD
202D	SETB P2.2	Sets EN pin of LCD
202E	ACALL DELAY	Calls subroutine function DELAY
2030	CLR P2.2	Reset EN pin of LCD
2031	RET	Return to the address
2032	DELAY: MOV R3, #0FFH	Initialize count value
2034	REP: MOV R2, #0FFH	Initialize delay value
2036	L1: DJNZ R2,L1	Decrement and jump to L1 if contents of R2 are not equal to 0
2038	DJNZ R3, REP	Decrement and jump to REP if contents of R3 are not equal to 0
203A	RET	Return to the address
2040	COM_TAB: DB 38H, 01H, 06H, 0EH, 80H, 00H	Command values
2047	DATA_TAB: DB 'Deeksha Pathak', 00H	Text to be displayed
2056	END	End of program

Result: Name 'Deeksha Pathak' is displayed in LCD as shown in Figure 5.

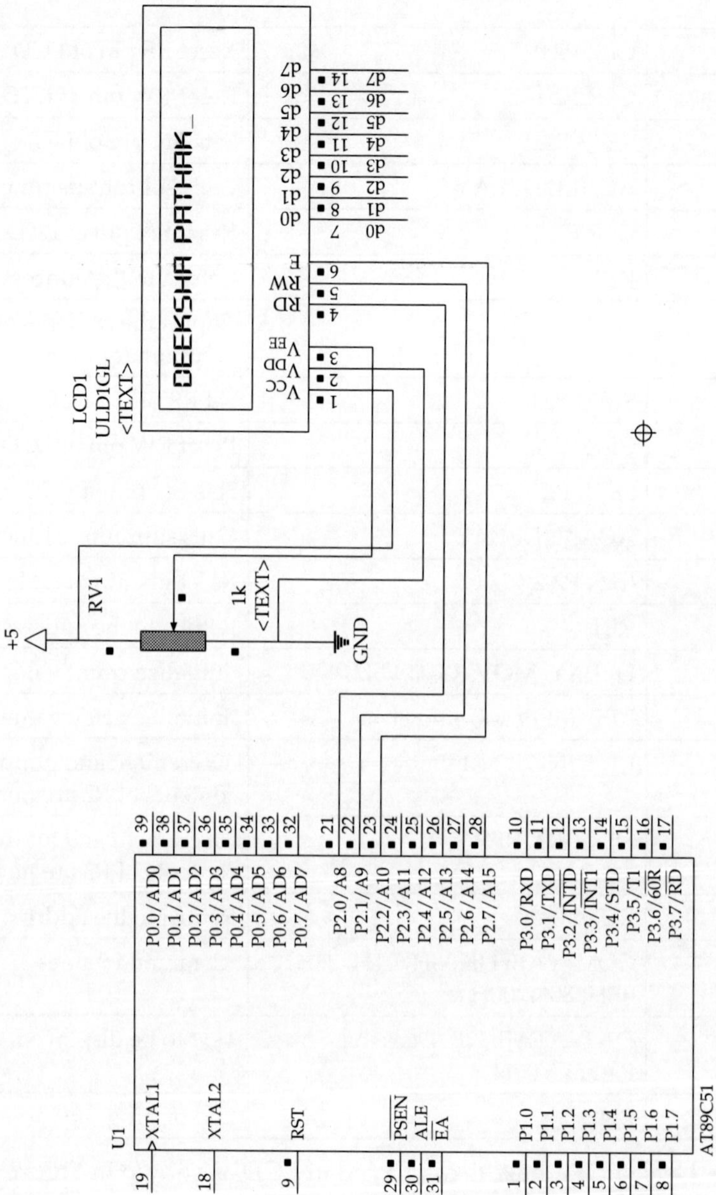

Figure 5: *LCD 16x2*

Submission Date: 18 Jan 2019

Professor Incharge

Prof. Nilesh B. Bahadure

EXPERIMENT 21

Exp. Date: 20 Jan 2019

Show the interfacing of 16 x 2 LCD with 8051 Microcontroller and write an assembly language program to display 'Your initial name' on 1st line of LCD and 'Final name' on the 2nd line of LCD, assume that crystal frequency of 8051 is 11.0592 MHz.

Requirement: 8051 IDE simulation software

Program:

Assembly Language Program		
Memory Location	**Instruction**	**Comments**
	ORG 0000H	
0000	MOV A, #38	Initialize 16x2 LCD
0002	ACALL COMMAND	Calls the subroutine function COMMAND
0004	ACALL DELAY	Calls subroutine function DELAY
0006	MOV A, #0EH	Clears port pin P3.0
0008	ACALL COMMAND	Calls the subroutine function COMMAND
000A	ACALL DELAY	Calls subroutine function DELAY
000C	MOV A, #01H	Clears port pin P3.0
000E	ACALL COMMAND	Calls the subroutine function COMMAND
0010	ACALL DELAY	Calls subroutine function DELAY
0012	MOV A, #06H	Clears port pin P3.0
0014	ACALL COMMAND	Calls the subroutine function COMMAND
0016	ACALL DELAY	Calls subroutine function DELAY
0018	MOV A, #80H	Clears port pin P3.0
001A	ACALL COMMAND	Calls the subroutine function COMMAND
001C	ACALL DELAY	Calls subroutine function DELAY
001E	MOV A, #'D'	Displays 'D'
0020	ACALL DATA	Calls the subroutine function DATA
0022	ACALL DELAY	Calls subroutine function DELAY
0024	MOV A, #'E'	Displays 'E'
0026	ACALL DATA	Calls the subroutine function DATA
0028	ACALL DELAY	Calls subroutine function DELAY
002A	MOV A, #'E'	Displays 'E'
002C	ACALL DATA	Calls the subroutine function DATA
002E	ACALL DELAY	Calls subroutine function DELAY
0030	MOV A, #'K'	Displays 'K'
0032	ACALL DATA	Calls the subroutine function DATA

0034	ACALL DELAY	Calls subroutine function DELAY
0036	MOV A, #'S'	Displays 'S'
0038	ACALL DATA	Calls the subroutine function DATA
003A	ACALL DELAY	Calls subroutine function DELAY
003C	MOV A, #'H'	Displays 'H'
003E	ACALL DATA	Calls the subroutine function DATA
0040	ACALL DELAY	Calls subroutine function DELAY
004A	MOV A, #'A'	Displays 'A'
004C	ACALL DATA	Calls the subroutine function DATA
004E	ACALL DELAY	Calls subroutine function DELAY
0050	MOV A, #' '	Displays ' '
0052	ACALL DATA	Calls the subroutine function DATA
0054	ACALL DELAY	Calls subroutine function DELAY
0056	MOV A, #0COH	Command for 2nd line in LCD
0058	ACALL COMMAND	Calls the subroutine function COMMAND
005A	ACALL DELAY	Calls subroutine function DELAY
005C	MOV A, #'P'	Displays 'P'
005E	ACALL DATA	Calls the subroutine function DATA
0060	ACALL DELAY	Calls subroutine function DELAY
0062	MOV A, #'A'	Displays 'A'
0064	ACALL DATA	Calls the subroutine function DATA
0066	ACALL DELAY	Calls subroutine function DELAY
0068	MOV A, #'T'	Displays 'T'
006A	ACALL DATA	Calls the subroutine function DATA
006C	ACALL DELAY	Calls subroutine function DELAY
006E	MOV A, #'H'	Displays 'H'
0070	ACALL DATA	Calls the subroutine function DATA
0072	ACALL DELAY	Calls subroutine function DELAY
0074	MOV A, #'A'	Displays 'A'
0076	ACALL DATA	Calls the subroutine function DATA
0078	ACALL DELAY	Calls subroutine function DELAY
007A	MOV A, #'K'	Displays 'K'
007C	ACALL DATA	Calls the subroutine function DATA
007E	ACALL DELAY	Calls subroutine function DELAY
0080	AGAIN:SJMP AGAIN	End of program

2000	COMMAND: MOV P1,A	Copies the contents of the accumulator to port 1
2002	CLR P2.0	Reset RS pin of LCD
2003	CLR P2.1	Reset RW pin of LCD
2004	SETB P2.2	Sets EN pin of LCD
2005	ACALL DELAY	Calls subroutine function DELAY
2007	CLR P2.2	Reset EN pin of LCD
2008	RET	Return to the address
2009	DATA:MOV P1, A	Copies the contents of the accumulator to port 1
200B	SETB P2.0	Set RS pin of LCD
200C	CLR P2.1	Reset RW pin of LCD
200D	SETB P2.2	Sets EN pin of LCD
200E	ACALL DELAY	Calls subroutine function DELAY
2010	CLR P2.2	Reset EN pin of LCD
2011	RET	Return to the address
2012	DELAY: MOV R3, #0FFH	Initialize count value
2014	REP: MOV R2, #0FFH	Initialize delay value
2016	L1: DJNZ R2,L1	Decrement and jump to L1 if contents of R2 are not equal to 0
2018	DJNZ R3, REP	Decrement and jump to REP if contents of R3 are not equal to 0
201A	RET	Return to the address
201B	END	End of program

Result: Name 'Deeksha' in 1st line and 'Pathak' in 2nd line is displayed as shown in Figure 6.

Submission Date: 21 Jan 2019

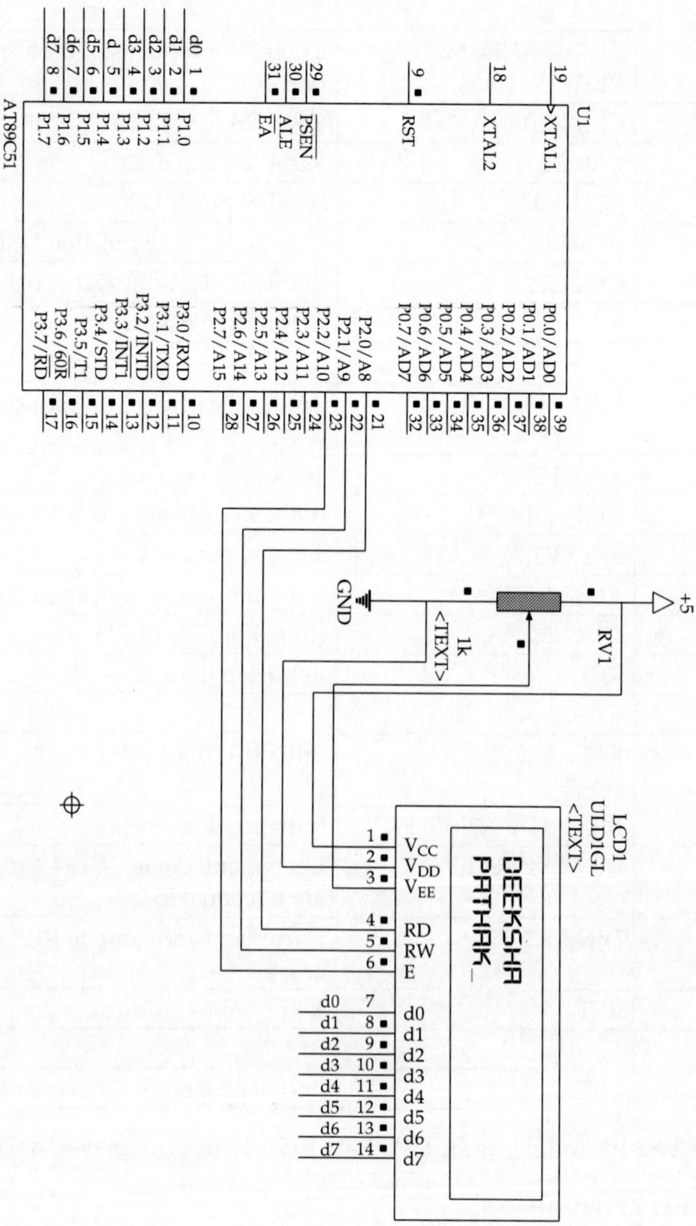

Figure 6: *LCD 16X2*

Professor Incharge

Prof. Nilesh B. Bahadure

Viva Voce Questions

1. What is Microcontroller

Ans. Microcontroller is an electronics device which is fabricated on small piece of silicon. It is used to perform arithmetic, logical and data manipulation operation dictated by software called assembly language program.

2. What is assembler?

Ans. Assembler is a software program used to convert assembly language program to machine language program or binary language program. Compiler and Interpreter are also software programs used to convert high level language program to machine language [inline, color=blue!40].

3. Write short note on Microcontroller 8051.

Ans. Microcontroller 8051 is an integrated circuitry built on a small piece of silicon. Microcontroller 8051 is an 8-bit microcontroller operated on +5V single power supply and on the clock frequency of 11.0592 MHz. It also consists of 16-bit address lines so it can access maximum 64 KB external memory. It consists of on-chip 128 bytes of RAM and 4 KB on-chip ROM. The other features of 8051 include four ports of 8-bit each P0, P1, P2 and P3, serial port, timers/counters and Interrupt sources.

4. What is the size of on-chip RAM and ROM in 8051?

Ans. 8051 has 4K on-chip ROM and 256 bytes of internal RAM.

5. What is OTP?

Ans. OTP means One Time Programmable. Thus when the application is loaded into the device, it can't be changed.

6. Explain differences between Microprocessor and Microcontroller.

Ans. Microprocessor contains ALU, general purpose register, stack pointer, program counter, clock timing circuit and interrupt circuit. It has a bulky and complex system.

Microcontroller contains the circuitry of microprocessor and in addition it has built-in ROM, RAM, input devices, timers and counters. It has a simple system.

7. What is the clock frequency of 8051?

Ans. Clock frequency of 8051 is 11.0592 MHz.

8. Explain the role of B-register.

Ans. The B-register is mainly used for multiplication and division operations along with A register.

9. Explain the difference between 8051 and 8031.

Ans. The only difference between 8051 and 8031 is 8031 doesn't have any on-chip program space.

10. Why microcontroller 8051 is 8-bit microcontroller?

Ans. Micro controller is called 8-bit microcontroller because it is a microcontroller with an 8-bit bus width.

11. Explain stack pointer.

Ans. The stack pointer is a small register in 8051 used to point at the stack. When we push something into the stack memory, the stack pointer increases, and when we retrieve (pop) the contents from stack memory the stack pointer decreases.

12. What are the different addressing modes in microcontroller? Give one example for each.

Ans. The different addressing modes in micro controller are:

- Immediate addressing mode. Ex-MOV A,#23H
- Register addressing mode. Ex-MOV A,R0
- Direct addressing mode. Ex-MOV A,30H
- Register Indirect addressing mode. Ex-MOV A,@R0
- Indexed addressing mode. Ex-MOV A,@A+PC

13. What are different instruction sets?

Ans. The different instruction sets are:

- Data Transfer Group.
- Arithmetic Group.
- Logical group.
- Boolean Instruction Group.
- Branching Instruction Group.

14. What are the various jump instructions in microcontroller?

Ans. The various jump instructions in microcontroller are:

- JC REL.
- JNC REL.
- JB BIT,REL.
- JNB BIT,REL.
- JBC BIT,REL.

15. Explain program counter.

Ans. A Program Counter (PC) is a register in a computer processor that contains the address of the instruction being executed of the current time. As each instruction is fetched, the PC increases its stored value by 1.

16. Explain PUSH instruction.

Ans. In PUSH instruction, SP is first incremented by 1 and then the contents of internal RAM memory location or SFR's is copied in the stack memory pointed by stack pointer register.

17. Explain MUL and DIV instructions.

Ans. MUL-This instruction performs multiplication between register and accumulator. The higher byte of the result is copied to register and lower byte to accumulator.

DIV-This instruction performs division between register and accumulator. The remainder is copied to register and quotient to the accumulator.

18. Explain DJNZ 30h; Label instruction.

Ans. (30H) – 1 ⟶ [30H]

if

Not equal to 00H

then

– 1 –>[30H]

PC + label –> PC

otherwise

PC + 2 –> PC

19. Explain the difference between JMP and CALL.

Ans. A JMP instruction permanently changes the program counter whereas CALL instruction leaves information on the stack so that the original program execution sequence can be resumed.

20. What is Special Function Register?

Ans. Special Function Register is a register within the microprocessor, which controls or monitors various aspects of the microprocessor function depending on the processor architecture.

21. What is stack?

Ans. Stack is a container of objects that is inserted and removed according to the last-in first-out principle.

22. Explain the significance of DPTR register.

Ans. The DPTR is 8051's only user accessible 16-bit register.

23. What is counter and timer? How does timer run in 8051?

Ans. Counter - Counting external pulses (on the corresponding I/O pin), which can be provided by a rotational encoder, an IR barrier sensor, or any device that provides pulses.

Timer - Counting internal clock pulses, which are fixed with time, hence, we can say that it is a very precise timer, whose resolution depends on the frequency of the main CPU clock.

24. Explain TCON special function register.

Ans. TCON register controls the following timer operations:

1. Start and stop timer 0 and timer 1.
2. Provides status of timer/counter overflows.

3. Provides status of external interrupts.
4. Configures external interrupts as either low level triggered or falling edge triggered.

25. Draw SCON and PCON.

Ans. Format of PCON-

D7	D6	D5	D4	D3	D2	D1	D0
SMOD	-	-	-	GF1	GF0	PD	IDL

Format of SCON-

D7	D6	D5	D4	D3	D2	D1	D0
SM0	SM1	SM2	REN	TB8	RB8	TI	RI

26. What do you mean by timer flag interrupt?

Ans. When the timer reaches its maximum limit then overflow flag is automatically set. If the corresponding timer interrupt bit is set in IE register then microcontroller automatically jumps on the vector location.

27. Draw IE and IP registers of 8051.

Ans. Format of IE Register-

D7	D6	D5	D4	D3	D2	D1	D0
EA	-	-	ES	ET1	EX1	ET0	EX0

Format of IP Register-

D7	D6	D5	D4	D3	D2	D1	D0
-	-	PT2	PS	PT1	PX1	PT0	PX0

28. What is PSW?

Ans. The PSW (Program Status Word) register is an 8-bit register. It is also referred as the Flag register; although the PSW register is 8-bit wide only 6-bits of it are used by the 8051. The two unused bits are user definable flags.

29. How does register bank change in 8051?

Ans. In 8051 Register Bank 11 conflicts with the stack memory when the R1=0 and RS0+1. In case of the conflict condition, the stack memory uses the space of scratch pad RAM 60H=7FH. The stack pointer is an 8bit register having default contents as 07H.

30. What do you mean by bit addressable registers?

Ans. Bit addressable registers are registers that its bits can be modified individually. It means if you have register named ACC that is bit addressable, we can change its bits (D0-D7) individually by special instructions SETB and CLR.

In 8051 there are many bit addressable registers such as A, B, SCON, PCON, TCON, P0, P1, P2, P3.

31. Explain the rule of $\overline{EA}$ and $\overline{PSEN}$.

Ans. $\overline{EA}$

$\overline{EA}$ is an I/P pin and must be connected to either V_{CC} or GND, it cannot be left unconnected. When $\overline{EA}$ =0, this pin is activated and external memory (64 KB) will be accessed, when $\overline{EA}$ =1, then internal memory (4KB) will be accessed.

$\overline{PSEN}$

$\overline{PSEN}$ (Program Store Enable) is an O pin and it must be connected to DE pin $\overline{RD}$ of ROM IC. The 8031/51 uses $\overline{PSEN}$ signal, when $\overline{EA}$ pin is connected to ground, the 8031/51 fetches op-code from external ROM by using $\overline{PSEN}$. When E is connected to V_{CC}, these chips do not activate the $\overline{PSEN}$ pin, indicates that the on-chip ROM contains program code.

32. Explain classification of on-chip RAM memory.

Ans. Random Access Memory (RAM) is the best known form of Computer Memory. The Read and Write (R/W) memory of a computer is called RAM. The RAM is a volatile memory, it means information written to it can be accessed as long as power is on. As soon as the power is off, it cannot be accessed. So this means RAM computer memory is essentially empty. RAM holds data and processing instructions temporarily until the CPU needs it.

There are two basic types of RAM :

(*i*) Dynamic RAM

(*ii*) Static RAM

Dynamic RAM

This loses its stored information in a very short time (for milli sec.) even when power supply is on. D-RAMs are cheaper and lower. Similar to a microprocessor chip is an Integrated Circuit (IC) made of millions of transistors and capacitors.

In the most common form of computer memory, Dynamic Memory Cell represents a single bit of data. The capacitor holds the bit of information 0 or 1. The transistor acts as a switch that lets the control circuitry on the memory chip read the capacitor or change its state. A capacitor is like a small bucket that is able to store electrons.

The refresh operation is where dynamic RAM gets its name. Dynamic RAM has to be dynamically refreshed all the time or it forgets what it is holding. The downside of all of this refreshing is that it takes time and slows down the memory.

Static RAM

Static RAM uses a completely different technology. S-RAM retains stored information only as long as the power supply is on. Static RAMs are costlier and consume more power. They have higher speed than D-RAMs. They store information in flip-flop.

In static RAM, a form of flip-flop holds each bit of memory. A flip-flop for a memory cell takes four or six transistors along with some wiring, but never has to be refreshed. This makes static RAM significantly faster than dynamic RAM. However, because it has more parts, a static memory cell takes up a lot more space on a chip than a dynamic memory cell. Therefore, you get less memory per chip, and that makes static RAM a lot more expensive. Static RAM is fast and expensive, and dynamic RAM is less expensive and slower. Static RAM is used to create the CPU to speed sensitive cache, while dynamic RAM forms the larger system RAM space.

33. Explain serial communication of 8051.

Ans. UART are used for serial communication between systems, they can be either half duplex (send or receive but at different interval) or full duplex (sent and receive at same time). Also known as a RS232 connection the microcontroller UART can provide the connection with a PC or another microcontroller based system.

34. Explain power down and idle mode of 8051.

Ans. In idle mode, by setting IDL bit in PCON, it makes the processor in idle state. In this state, clock signal is available to all peripherals but not to CPU; an interrupt can resume the idle state to normal state and CPU starts executing where it left.

In power down mode, clock is gated off to all sections including the peripherals. When put in power down mode, the supply voltage can be reduced to 2V to keep internal RAM intact. To come out of power down mode a hardware reset should be applied. When coming out from power down mode at hardware reset, CPU can execute where it left.

35. Explain how baud rate is doubled in serial communication.

Ans. There are two ways to increase the baud rate-

1. Use a higher frequency crystal.
2. Change a bit in the PCON register.

With a fixed crystal frequency, we can double the baud rate by making SMOD=1.

36. Explain different types of interrupt and exception.

Ans. An exception is a rarely occurring (exceptional!) condition that requires deviation from the program's normal flow. Normally, an exception should not result in total failure, but instead be attended by an exception handler. Exception handling is a built-in construct in many programming languages. Usually, exceptions are handled by unwinding the stack, thus rolling back to a defined state outside the exception's scope, and then invoking a handler routine.

Hardware Exceptions

Certain processors (Power PC for example) define exceptions to indicate that some sort of unusual condition has occurred: System Reset, Invalid Address, some virtual address translation cache miss, etc. These exceptions are also used to implement breakpoints and system calls. In this case, they act almost like interrupts.

OS Exceptions

Some of the hardware exceptions will the handled by the OS. For example, your program accesses invalid memory. This will cause a hardware exception. The OS has a handler for that exception, and odds are that the OS will send a signal to your application (SIGSEGV for example) denoting there is a problem.

If your program has a signal handler installed, the signal handler will run and hopefully deal with the situation. If you don't have a signal handler, the program can be terminated or suspended.

I would consider window's Structured Exception Handlers (SEH) to be such type of exceptions.

Software Exceptions

Some languages like Java, C++ and C# have the concept of software exceptions, where the language provides for the handling of unforeseen or unusual conditions related to the operation of the program. In this case, an exception is raised at some point in the code and some code higher up on the program execution stack would "catch" the exception and execute. This is what try/catch blocks do.

37. How many hardware interrupts are supported by 8051?

Ans. Two hardware interrupts are supported by 8051:

- Level triggered
- Edge triggered

38. Explain vectored and non-vectored interrupts of 8051.

Ans. **Vectored Interrupts:** The interrupts for which the starting address of the interrupts service routine is predefined is called vectored interrupts. Since the starting address of interrupt service routine is predefined, the interrupt response time for vector interrupts is less compared to non-vectored interrupts

Non-Vectored Interrupts: The interrupts for which the starting address of the interrupts service routine is not predefined and it is provided by the external hardware at the time of interrupt is called non-vectored interrupts.

39. What is the vector location of timer interrupt of 8051?

Ans. Vector location for timer interrupt is:

- Timer0 interrupt(TF0) => 000BH
- Timer1 interrupt(TF1) => 001BH

40. Why is only one vector location reserved of serial communication interrupt?

Ans. TI & RI are stored in the same vector location so either transfer or receiving is possible at a time. Hence, only 1 vector location is reserved of serial communication interrupt.

41. How many data lines and address lines are present in the memory IC 8K4?

Ans. Address line = 12
Data line = 4

42. Calculate memory capacity in bits for the memory IC consisting of 12 address lines and 8-bit of data line.

Ans. Memory capacity= 4K8

43. What is embedded system?

Ans. An Embedded System is a system that has software embedded into computer hardware, which makes a system dedicated for an application(s) or specific part of an application or product or part of a larger system.

44. List any five applications of embedded system.

Ans. 1. Consumer Electronics : Camcorders, Cameras.
2. Card Readers : Bar-code, Smart card reader.
3. Telecoms : Cellular Phones, Telephone switches.
4. Healthcare : ECG, EEG machines.
5. Automotive Industry : Anti-lock breaking system(ABS), engine control.

45. What is an assembler?

Ans. An assembler is a type of computer program the interprets software programs written in assembly language into machine language, code and instructions that can be executed by a computer.

46. What is LED? Draw symbol of LED.

Ans. A Light Emitting Diode (LED) is a two-head semiconductor device light source. It is a *p-n* junction diode which emits light when connected to a battery source. Figure 7 shows the symbol of LED.

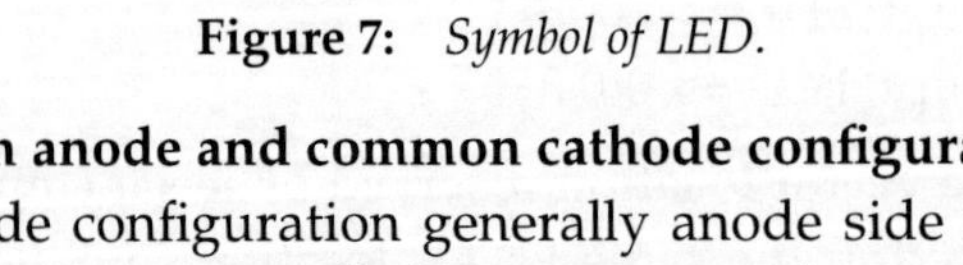

Figure 7: *Symbol of LED.*

47. What is common anode and common cathode configuration of LED?

Ans. In common anode configuration generally anode side is connected to the +5V and the cathode side is controlled by ports pin of microcontroller 8051, whereas in common cathode configuration cathode side ground and the anode side is controlled by port pins of microcontroller 8051.

48. What is 7-segment LED?

Ans. A 7-segment display (SSD), or seven-segment indicator is a form of electronic display device for displaying decimal numerals to the more complex dot matrix displays.

49. List the different types of stepper motors.

Ans. There are three main types of Stepper Motors:

- Permanent magnet stepper (can be subdivided into: 'tin-can' and 'hybrid')
- Hybrid synchronous stepper
- Variable reluctance stepper

50. Write different types of stepper motors based on interfacing type.

Ans. Types of motor based on interfacing:

1. Bipolar stepper motor
2. Unipolar stepper motor
3. Universal stepper motor

51. Write a short note on LCD.

Ans. LCD stands for liquid crystal display. There is actually a liquid compound, liquid crystal, sandwiched between two grids of electrodes. The electrodes can selectively turn on the different cells or pixels in the grid to create images we see.

52. What is the use of BCD to 7-segment LED?

Ans. The BCD to seven segment is used to convert binary coded value i.e. decimal value to the equivalent seven segment code directly.

Additional Information

- Please send your feedback/queries at nbahadure@gmail.com
- For download of source code of Lab Manual in LaTeX format kindly visit https://sites.google.com/site/nileshbahadure/downloads or https://www.ikbooks.com/
- Scan QR code, for download of lab manual in word on your mobile device (see word file Figures.docx for QR code, at the end of file QR code is given)

Index

K

L

M

P

Q

R

S

T

V

W